636
606
598 ✓
597
480
394 *
376 *
363 *
348
339
210
6
21
34 *

Methods in Enzymology

Volume 183
MOLECULAR EVOLUTION: COMPUTER ANALYSIS OF PROTEIN AND NUCLEIC ACID SEQUENCES

METHODS IN ENZYMOLOGY

EDITORS-IN-CHIEF

John N. Abelson   Melvin I. Simon

DIVISION OF BIOLOGY
CALIFORNIA INSTITUTE OF TECHNOLOGY
PASADENA, CALIFORNIA

FOUNDING EDITORS

Sidney P. Colowick and Nathan O. Kaplan

Methods in Enzymology

Volume 183

Molecular Evolution: Computer Analysis of Protein and Nucleic Acid Sequences

EDITED BY

Russell F. Doolittle

CENTER FOR MOLECULAR GENETICS
UNIVERSITY OF CALIFORNIA, SAN DIEGO
LA JOLLA, CALIFORNIA

ACADEMIC PRESS, INC.
Harcourt Brace Jovanovich, Publishers
San Diego   New York   Berkeley   Boston
London   Sydney   Tokyo   Toronto

ACADEMIC PRESS, INC.
San Diego, California 92101

United Kingdom Edition published by
ACADEMIC PRESS LIMITED
24-28 Oval Road, London NW1 7DX

LIBRARY OF CONGRESS CATALOG CARD NUMBER: 54-9110

ISBN 0-12-182084-X (alk. paper)

PRINTED IN THE UNITED STATES OF AMERICA
90 91 92 93 9 8 7 6 5 4 3 2 1

Table of Contents

Section I. Databases

Section II. Searching Databases

Section III. Patterns in Nucleic Acid Sequences

Section VI. Estimating Sequence Divergence

Section VII. Phylogenetic Trees

Contributors to Volume 183

Article numbers are in parentheses following the names of contributors.
Affiliations listed are current.

YVON ABEL (26), *Centre de Recherches Mathématiques, Université de Montréal, Montréal, Quebec H3C 3J7, Canada*

WAYNE F. ANDERSON (27), *Department of Biochemistry, Vanderbilt University, Nashville, Tennessee 37232*

PATRICK ARGOS (21), *European Molecular Biology Laboratory, D-6900 Heidelberg, Federal Republic of Germany*

DAVID J. BACON (27), *Department of Biochemistry, University of Alberta, Edmonton, Alberta T6G 2H7, Canada*

WINONA C. BARKER (3, 20), *Protein Identification Resource, National Biomedical Research Foundation, Georgetown University Medical Center, Washington, D.C. 20007*

GEOFFREY J. BARTON (25), *Laboratory of Molecular Biophysics, University of Oxford, Oxford 0X1 3QU, England*

DAVID BENTON (1), *IntelliGenetics, Incorporated, Mountain View, California 94040*

B. EDWIN BLAISDELL (24), *Department of Mathematics, Stanford University, Stanford, California 94305*

TOM L. BLUNDELL (42), *Laboratory of Molecular Biology, Department of Crystallography, Birkbeck College, University of London, London WC1E 7HX, England*

LYDIE BOUGUELERET (15), *Computer Science Unit, Institut Pasteur, 75724 Paris Cedex 15, France*

VOLKER BRENDEL (24), *Department of Mathematics, Stanford University, Stanford, California 94305*

JOSEPH P. BROWN (7), *Genetic Systems Corporation, Seattle, Washington 98121*

CHRISTIAN BURKS (1), *Theoretical Biology and Biophysics Group, Los Alamos National Laboratory, Los Alamos, New Mexico 87545*

GRAHAM CAMERON (2), *The European Molecular Biology Laboratory Data Library, D-6900 Heidelberg, Federal Republic of Germany*

ROBERT CEDERGREN (19, 26), *Département de Biochimie, Université de Montréal, Montréal, Quebec H3C 3J7, Canada*

JEAN-MICHEL CLAVERIE (15), *Computer Science Unit, Institut Pasteur, 75724 Paris Cedex 15, France*

JOHN F. COLLINS (30), *Biocomputing Research Unit, Department of Molecular Biology, University of Edinburgh, Edinburgh EH9 3JR, Scotland*

ANDREW F. W. COULSON (30), *Biocomputing Research Unit, Department of Molecular Biology, University of Edinburgh, Edinburgh EH9 3JR, Scotland*

JOHN CZELUSNIAK (37), *Department of Anatomy and Cell Biology, Wayne State University School of Medicine, Detroit, Michigan 48201*

RUSSELL F. DOOLITTLE (6, 23, 41), *Center for Molecular Genetics, University of California, San Diego, La Jolla, California 92093*

MANFRED EIGEN (32), *Max-Planck-Institut für Biophysikalische Chemie, D-3400 Göttingen, Federal Republic of Germany*

DAVID EISENBERG (9), *Molecular Biology Institute, University of California, Los Angeles, Los Angeles, California 90024*

DA-FEI FENG (23, 41), *Center for Molecular Genetics, University of California, San Diego, La Jolla, California 92093*

WALTER M. FITCH (38), *Department of Ecology and Evolutionary Biology, University of California, Irvine, Irvine, California 92717*

DANIEL GAUTHERET (19), *Département de Biochimie, Université de Montréal, Montréal, Quebec H3C 3J7, Canada*

DAVID G. GEORGE (3, 20), *Protein Identification Resource, National Biomedical Research Foundation, Georgetown Univer-*

sity Medical Center, Washington, D.C. 20007

TAKASHI GOJOBORI (33), *Department of Evolutionary Genetics, National Institute of Genetics, Mishima 411, Japan*

MORRIS GOODMAN (37), *Department of Anatomy and Cell Biology, Wayne State University School of Medicine, Detroit, Michigan 48201*

MANOLO GOUY (40), *Laboratoire de Biómetrie, Université Lyon I, 69622 Villeurbanne Cedex, France*

MICHAEL GRIBSKOV (9), *BRI-Basic Research Program, National Cancer Institute–Frederick Cancer Research Facility, Frederick, Maryland 21701*

NOMI L. HARRIS (16), *Laboratory of Computer Science, Massachusetts Institute of Technology, Cambridge, Massachusetts 02139*

MASAMI HASEGAWA (34), *The Institute of Statistical Mathematics, Minato-ku, Tokyo 106, Japan*

JOTUN HEIN (39), *Centre de Recherche de Mathematiques Appliques, Universitè de Montrèal, Montrèal, Quebec H3C 3J7, Canada*

STEVEN HENIKOFF (7), *Fred Hutchinson Cancer Research Center, Seattle, Washington 98104*

LOIS T. HUNT (3, 20), *Protein Identification Resource, National Biomedical Research Foundation, Georgetown University Medical Center, Washington, D.C. 20007*

JOHN A. JAEGER (17), *Department of Chemistry, University of Rochester, Rochester, New York 14627*

MARK S. JOHNSON (42), *Laboratory of Molecular Biology, Department of Crystallography, Birkbeck College, University of London, London WC1E 7HX, England*

ROBERT JONES (14), *Thinking Machines Corporation, Cambridge, Massachusetts 02142*

PATRICIA KAHN (2), *The European Molecular Biology Laboratory Data Library,*

D-6900 Heidelberg, Federal Republic of Germany

SAMUEL KARLIN (24), *Department of Mathematics, Stanford University, Stanford, California 94305*

SUZANNE M. KEHOE (37), *Department of Anatomy and Cell Biology, Wayne State University School of Medicine, Detroit, Michigan 48201*

BORIVOJ KEIL (4), *Institut Pasteur, 75015 Paris, France*

MOTOO KIMURA (33), *Department of Population Genetics, National Institute of Genetics, Mishima 411, Japan*

HIROHISA KISHINO (34), *The Institute of Statistical Mathematics, Minato-ku, Tokyo 106, Japan*

CECILIA LANAVE (35), *Centro Studi sui Mitocondri e Metabolismo Energetico, CNR, Bari, 70126 Bari, Italy*

GAD M. LANDAU (31), *Division of Computer Sciences, Polytechnic University, Brooklyn, New York 11201*

CHARLES B. LAWRENCE (8), *Department of Cell Biology, Baylor College of Medicine, Houston, Texas 77030*

WEN-HSIUNG LI (40), *Center for Demographic and Population Genetics, Health Science Center at Houston, University of Texas, Houston, Texas 77225*

ROLAND LÜTHY (9), *Molecular Biology Institute, University of California, Los Angeles, Los Angeles, California 90024*

FRANÇOIS MAJOR (19), *Département d' Informatique et Recherche Opérationnelle, Université de Montréal, Montréal, Quebec H3C 3J7, Canada*

HUGO M. MARTINEZ (18), *Department of Biochemistry and Biophysics, University of California, San Francisco, San Francisco, California 94143*

NANCY D. MONCRIEF (37), *Department of Biology, University of Virginia, Charlottesville, Virginia 22901*

ETSUKO N. MORIYAMA (33), *Department of Evolutionary Genetics, National Institute of Genetics, Mishima 411, Japan*

MITSUO MURATA (22), *Department of Biochemistry, University of Georgia, Athens, Georgia 30602*

RUTH NUSSINOV (31), *Sackler Institute of Molecular Medicine, Tel Aviv University, Tel Aviv 69978, Israel, and Laboratory of Mathematical Biology, National Cancer Institute, National Institutes of Health, Bethesda, Maryland 20892*

WILLIAM R. PEARSON (5), *Department of Biochemistry, University of Virginia, Charlottesville, Virginia 22908*

GRAZIANO PESOLE (35), *Dipartimento di Biochimica e Biologia Molecolare, Università di Bari, 70126 Bari, Italy*

GIULIANO PREPARATA (35), *Dipartimento di Fisica, Università di Milano, 20100 Milano, Italy*

CECILIA SACCONE (35), *Dipartimento di Biochimica e Biologia Molecolare, Università di Bari, 70126 Bari, Italy*

NARUYA SAITOU (36), *Department of Anthropology, Faculty of Science, The University of Tokyo, Hongo, Bunkyo-ku, Tokyo 113, Japan*

ANDREJ ŠALI (42), *Laboratory of Molecular Biology, Department of Crystallography, Birkbeck College, University of London, London WC1E 7HX, England*

DAVID SANKOFF (26), *Centre de Recherches Mathématiques, Université de Montréal, Montréal, Quebec H3C 3J7, Canada*

ISABELLE SAUVAGET (15), *Computer Science Unit, Institut Pasteur, 75724 Paris Cedex 15, France*

PERIANNAN SENAPATHY (16), *Biotechnology Center, University of Wisconsin, Madison, Wisconsin 53706*

MARVIN B. SHAPIRO (16), *Laboratory of Statistical and Mathematical Methodology, Division of Computer Research and Technology, National Institutes of Health, Bethesda, Maryland 20892*

JOHN C. W. SHEPHERD (11), *Cell and Molecular Biology Laboratory, University of Sussex, Brighton BN1 9QG, England*

RODGER STADEN (10, 12), *Laboratory of Molecular Biology, Medical Research Council, Cambridge CB2 2QH, England*

GARY D. STORMO (13), *Department of Molecular, Cellular, and Developmental Biology, University of Colorado, Boulder, Colorado 80309*

WILLIAM R. TAYLOR (29), *Laboratory of Mathematical Biology, National Institute for Medical Research, Medical Research Council, London NW7 1AA, England*

DOUGLAS H. TURNER (17), *Department of Chemistry, University of Rochester, Rochester, New York 14627*

MAUNO VIHINEN (28), *Department of Biochemistry, University of Turku, SF-20500 Turku, Finland*

MARTIN VINGRON (21), *European Molecular Biology Laboratory, D-6900 Heidelberg, Federal Republic of Germany*

UZI VISHKIN (31), *Institute for Advanced Computer Studies, University of Maryland, College Park, Maryland 20742, and Department of Computer Science, School of Mathematical Sciences, Tel Aviv University, Tel Aviv 69978, Israel*

JAMES C. WALLACE (7), *Fred Hutchinson Cancer Research Center, Seattle, Washington 98104*

MICHAEL S. WATERMAN (14), *Departments of Mathematics and Molecular Biology, University of Southern California, Los Angeles, California 90089*

PATRICK L. WILLIAMS (38), *College of Veterinary Medicine, North Carolina State University, Raleigh, North Carolina 27606*

RUTHILD WINKLER-OSWATITSCH (32), *Max-Planck-Institut für Biophysikalische Chemie, D-3400 Göttingen, Federal Republic of Germany*

MICHAEL ZUKER (17), *Division of Biological Sciences, National Research Council of Canada, Ottawa, Ontario K1A OR6, Canada*

Preface

Since its initiation, *Methods in Enzymology* has tried to keep abreast of developments in all areas important to biochemistry. Increasingly, this has meant moving into the realm of molecular biology. The main objective has remained the same, nonetheless: putting ready-to-use methods into the hands of the at-the-bench scientist. This volume is no exception. Still, this may be the first *Methods in Enzymology* volume in which the expression "pH" is never mentioned and the word buffer, if used at all, has nothing to do with aqueous solutions!

The past decade has witnessed nothing less than a flood of nucleic acid and protein sequence data, the management of which would simply be impossible without computers. The chapters in this volume address a variety of areas in which computers are used to manage and manipulate sequences. The manipulations include searching, aligning, and determining the significance of similarities, as well as the construction of phylogenetic trees that show the evolutionary history of related sequences. This relatively new field of sequence-computing has stimulated interest from a variety of more established disciplines as is apparent from the diverse backgrounds of the contributing authors. Thus, investigators in mathematics and statistics, taxonomy and cladistics, and protein and nucleic acid chemistry are all represented, each bringing different perspectives and insights. All the authors were asked, however, to make an effort to write for the biochemist or molecular biologist who wants to analyze, search, or manipulate nucleic acid or protein sequence data. Some have met this charge more aptly than others, and in some instances readers must be prepared to go the extra mile to find genuine satisfaction. Often this will mean mastering new terms and jargon, the troublesome persiflage of all specialized fields. Bootstraps and jackknifes, informative sites and k-tuples, NP-complete and NP-completeness are all words with simple enough meanings, and it would be a mistake for readers to turn away from the methods described merely because of an alien nomenclature or notation. Thus, bootstraps and jackknifes are computer-intensive stastistical procedures for estimating the quality of a data set. Informative site is a term used in strict parsimony analysis for those arrays of characters in a set of sequences that allow a single, simple interpretation. k-Tuples are merely sets of units: duplets, triplets, quartets, and so on. NP-complete is an expression used by computer scientists to indicate a set of problems it may not be possible to solve exactly by dynamic programming. The phylogeny problem may be NP-complete. I must confess that not too long ago I had not the faintest notion of what most of these expressions implied.

It was agreed in advance that the computer programs described in this volume would be made available by the contributing authors, usually only for the cost of materials (tapes or diskettes) and shipping. Readers must understand, nonetheless, that there are often problems in transporting programs from one kind of computer to another and that differences in operating systems often present stumbling blocks. The authors are very interested in the community using their programs, however, and they are prepared to help with any problems that may ensue. Naturally they would like to hear, also, when the programs work well.

The sequence–computer field is a very fast-moving one, and every effort was made to produce this volume as rapidly as possible so the methods would be timely. In this regard, I must thank the authors for adhering to a set of rigid deadlines. Special thanks are due, also, to Karen Anderson, who managed many aspects involving the manuscripts and without whom this volume would likely never have materialized.

RUSSELL F. DOOLITTLE

METHODS IN ENZYMOLOGY

VOLUME XLI. Carbohydrate Metabolism (Part B)
Edited by W. A. WOOD

VOLUME XLII. Carbohydrate Metabolism (Part C)
Edited by W. A. WOOD

VOLUME XLIII. Antibiotics
Edited by JOHN H. HASH

VOLUME XLIV. Immobilized Enzymes
Edited by KLAUS MOSBACH

VOLUME XLV. Proteolytic Enzymes (Part B)
Edited by LASZLO LORAND

VOLUME XLVI. Affinity Labeling
Edited by WILLIAM B. JAKOBY AND MEIR WILCHEK

VOLUME XLVII. Enzyme Structure (Part E)
Edited by C. H. W. HIRS AND SERGE N. TIMASHEFF

VOLUME XLVIII. Enzyme Structure (Part F)
Edited by C. H. W. HIRS AND SERGE N. TIMASHEFF

VOLUME XLIX. Enzyme Structure (Part G)
Edited by C. H. W. HIRS AND SERGE N. TIMASHEFF

VOLUME L. Complex Carbohydrates (Part C)
Edited by VICTOR GINSBURG

VOLUME LI. Purine and Pyrimidine Nucleotide Metabolism
Edited by PATRICIA A. HOFFEE AND MARY ELLEN JONES

VOLUME LII. Biomembranes (Part C: Biological Oxidations)
Edited by SIDNEY FLEISCHER AND LESTER PACKER

VOLUME LIII. Biomembranes (Part D: Biological Oxidations)
Edited by SIDNEY FLEISCHER AND LESTER PACKER

VOLUME LIV. Biomembranes (Part E: Biological Oxidations)
Edited by SIDNEY FLEISCHER AND LESTER PACKER

VOLUME 68. Recombinant DNA
Edited by RAY WU

VOLUME 69. Photosynthesis and Nitrogen Fixation (Part C)
Edited by ANTHONY SAN PIETRO

VOLUME 70. Immunochemical Techniques (Part A)
Edited by HELEN VAN VUNAKIS AND JOHN J. LANGONE

VOLUME 71. Lipids (Part C)
Edited by JOHN M. LOWENSTEIN

VOLUME 72. Lipids (Part D)
Edited by JOHN M. LOWENSTEIN

VOLUME 73. Immunochemical Techniques (Part B)
Edited by JOHN J. LANGONE AND HELEN VAN VUNAKIS

VOLUME 74. Immunochemical Techniques (Part C)
Edited by JOHN J. LANGONE AND HELEN VAN VUNAKIS

VOLUME 75. Cumulative Subject Index Volumes XXXI, XXXII, XXXIV–LX
Edited by EDWARD A. DENNIS AND MARTHA G. DENNIS

VOLUME 76. Hemoglobins
Edited by ERALDO ANTONINI, LUIGI ROSSI-BERNARDI, AND EMILIA CHIANCONE

VOLUME 77. Detoxication and Drug Metabolism
Edited by WILLIAM B. JAKOBY

VOLUME 78. Interferons (Part A)
Edited by SIDNEY PESTKA

VOLUME 79. Interferons (Part B)
Edited by SIDNEY PESTKA

VOLUME 80. Proteolytic Enzymes (Part C)
Edited by LASZLO LORAND

Section I

Databases

[1] GenBank: Current Status and Future Directions

By Christian Burks, Michael J. Cinkosky, Paul Gilna, Jamie
E.-D. Hayden, Yuki Abe, Edwin J. Atencio, Steve Barnhouse,
David Benton, Connie A. Buenafe, Karen E. Cumella, Dan B.
Davison, David B. Emmert, Mary Jo Faulkner, James W. Fickett,
William M. Fischer, Mark Good, Deborah A. Horne, F. Kay
Houghton, Praful M. Kelkar, Tom A. Kelley, Michael Kelly,
Melinda A. King, Bernard J. Langan, Jeffrey T. Lauer, Natalie
Lopez, Conrad Lynch, Janet Lynch, Janet B. Marchi, Thomas G.
Marr, Frances A. Martinez, Mia J. McLeod, Pat A. Medvick,
Santosh K. Mishra, John Moore, Christine A. Munk, Socorro M.
Mondragon, Kevin K. Nasseri, Debra Nelson, Will Nelson, Tan
Nguyen, Gloria Reiss, John Rice, Julie Ryals, Margarita D.
Salazar, Stephen R. Stelts, Brian L. Trujillo, Laurie J.
Tomlinson, Mark G. Weiner, Frank J. Welch, Susan E. Wiig,
Katherine Yudin, and Larry B. Zins

The GenBank[1] database provides a collection of nucleotide sequences as well as relevant bibliographic and biological annotation. We present an updated view of the size and scope of the database, and we also describe recent developments in the strategies, protocols, and software for collecting, maintaining, and distributing the data.

Introduction

GenBank, the genetic sequence data bank, is chartered to provide a computer database of all published (and, increasingly, unpublished) DNA and RNA sequences and related bibliographic and biological information. The project is funded through an NIGMS contract with IntelliGenetics, Inc. (IG) which, in turn, contracts with the DOE acting on behalf of Los Alamos National Laboratory (LANL). The project is funded with cosponsorship from other institutes of the National Institutes of Health, National Library of Medicine (NLM), DRR, USDA, NSF, DOE, and DOD. Data collection and distribution are carried out in collaboration with the EMBL Data Library (EMBL) and the DNA Data Bank of Japan (DDBJ). Inquiries

---

[1] "GenBank" is a registered trademark of the U.S. Department of Health and Human Services.

regarding data distribution and release should be addressed to IG,[2] inquiries regarding data submissions and collection should be addressed to LANL.[3]

Background. Computer databases offer several advantages for collecting, maintaining, analyzing, and distributing data sets: compact storage, low-error rates for data archiving or transmission, more precise specification and organization of data, less effort in varying complex queries, and less effort in reorganizing data. We have previously discussed both the scientific community's move 10 years ago to apply these advantages to nucleotide sequence data and the resulting establishment of the GenBank database. The database was chartered as described above and intended to provide[4,5] (1) an archive, for long-term storage and preservation of sequence data; (2) retrieval of nucleotide sequences, with delineation by sequence-specific, bibliographic, physical, or functional criteria; and (3) a research platform, for examination and analysis of sequences grouped by sequence-specific or annotation-specific criteria.

Over the past several years there have been a number of developments in the database project. GenBank has passed into a second contract period with considerably expanded resources, allowing us to address many of the issues that arose as challenges in the early phase of the project. The nucleotide sequence database collaboration has recently been extended from GenBank and EMBL[6,7] to include DDBJ,[8] which began operations in 1987.

The unanticipated exponential growth in the amount of nucleotide sequence data being published as well as the possibility of another order of magnitude increase in the number of sequences being determined — as emphasized in recent discussions of the possibility of a human genome

[2] Postal service: GenBank; IntelliGenetics, Inc.; 700 El Camino Real East; Mountain View, CA 94040. Electronic mail: "genbank@genbank.ig.com". Telephone: (415) 962-7364.

[3] Postal service: GenBank; T-10, MS K710; Los Alamos National Laboratory; Los Alamos, NM 87545. Electronic mail: "genbank%life@lanl.gov" (general); or "gbsub%life@lanl.gov" (data submissions and questions regarding submissions). Telephone: (505) 665-2177.

[4] C. Burks, J. W. Fickett, W. B. Goad, M. Kaneshisa, F. I. Lewitter, W. P. Rindone, C. D. Swindell, C.-S. Tung, and H. S. Bilofsky, *Comput. Appl. Biosci.* **1,** 225 (1985).

[5] J. W. Fickett and C. Burks, *in* "Mathematical Models for DNA Sequences" (M. S. Waterman, ed.), p. 1. CRC Press, Boca Raton, Florida, 1988.

[6] G. Cameron, *Nucleic Acids Res.* **16,** 1865 (1988).

[7] P. Kahn and G. Cameron, this volume [2].

[8] S. Miyazawa, *in* "Computers and DNA" (G. I. Bell and T. Marr, eds.), p. 47. Addison-Wesley, New York, 1989.

initiative[9,10] and other large-scale sequencing projects—has led us (and others) to focus on the need for alternatives for processing the much larger streams of data coming into the database.[11,12] Similarly, the recognition of a need for improved cross-referencing and, eventually, automatic cross-linking of GenBank and related databases (which figures heavily in recent discussions of a matrix of biological knowledge)[13] has led us to consider alternative data representations that would provide the basis for such cross-linking. Finally, computer science tools (e.g., relational databases)[14] previously in use in other contexts have evolved to the point that it is appropriate to begin exploring their application to GenBank.

Focus of Chapter. We begin by providing an update on the kinds and amount of data present in the database and the objectives (completeness, timeliness, and depth of annotation) of GenBank with respect to the user community. We describe our two primary thrusts at this time: shifting our data collection effort over to a system that relies on direct, computer-readable submissions from authors, and restructuring the database in the context of a relational database management system (RDBMS). An overview of the current media and formats for data distribution is given. We close with a description of the directions in which the database effort is moving as well as a discussion of the challenges facing us now.

Current Span and Size of Database

What Data Are in the Database? The primary datum in the database is nucleotide (DNA and RNA) sequences (i.e., actggcagacagggtcatt . . .). Each contiguous sequence determined by a single research group and submitted (or published) as a single entity is maintained as such. Occasionally, when more than one sequence for the same gene in the same organism is available, and the sequences overlap, a merged view of the sequences is presented in the distributed versions of the database.

[9] B. M. Alberts, D. Botstein, S. Brenner, C. Cantor, R. F. Doolittle, L. Hood, V. A. McKusick, D. Nathans, M. V. Olson, S. Orkin, L. E. Rosenberg, F. H. Ruddle, S. Tilghman, J. Tooze, and J. D. Watson, "Mapping and Sequencing the Human Genome." National Academy Press, Washington, D.C., 1988.

[10] U.S. Congress, Office of Technology Assessment, "Mapping Our Genes—The Genome Projects: How Big, How Fast?" U.S. Govt. Printing Office, Washington, D.C., 1988.

[11] L. Phillipson, *Nature (London)* **332,** 676 (1988).

[12] G. I. Bell and T. Marr, eds., "Computers and DNA." Addison-Wesley, New York, 1989.

[13] H. J. Morowitz and T. F. Smith, "Report of the Matrix of Biological Knowledge Workshop." Santa Fe Institute, Santa Fe, New Mexico, 1987.

[14] C. J. Date, "An Introduction to Database Systems," 4th Ed. Addison-Wesley, Reading, Massachusetts, 1986.

The database provides the bibliographic context of the sequence. This is most often represented by a specific journal citation, although GenBank accepts and cites submissions of original data published in books, theses, and other sources. Increasingly, unpublished nucleotide sequences are appearing in the database,[15] although most often directly tied to a publication that describes the sequence without presenting it explicitly. Unpublished submissions are cited as such.

The physical context of a sequence, i.e., the organism, chromosome, map position, etc., that describes its origin *in vivo* is included when known and made available to GenBank. Official nomenclature lists and map assignments are beginning to be made available to us in a form that allows GenBank to dynamically maintain proper, uniform values for these data items. The functional context of a sequence (or parts of a sequence) is also annotated in the database. A good example is protein-coding regions, which are annotated to support automatic extraction[16,17] of the coding regions from the sequences in which they are embedded (we use such a tool for checking the integrity of coding regions in the database). Most of this information is presented in the GenBank FEATURES table. Finally, a limited number of data items providing the administrative context of a sequence are provided. These indicate when data were last revised, the degree of review the data have received from GenBank staff, etc.

A sample GenBank entry in the current magnetic tape flat file distribution format is presented in Fig. 1. A more detailed presentation of the various data items has been given previously[5]; up-to-date descriptions and planned changes can be found in the release notes accompanying GenBank releases.[2]

How Many Data Are in the Database? The current release (Release 61.0) of the database contains about 3.5×10^7 nucleotides (based on over 4.0×10^7 nucleotides, with the reduction being due to the merging described above) in about 2.9×10^4 entries. The exponential growth rate of nucleotide sequence data over the past 10 years has been characterized elsewhere[18] (see also Fig. 3). The database includes 150 entries containing 10^4 or more contiguous nucleotides; 5 of these entries contain over 10^5 contiguous nucleotides. There are about 1200 organisms (counting individual animal, plant, and microorganism species as well as viruses) repre-

[15] C. Burks, *in* "Biomolecular Data: A Resource in Transition" (R. R. Colwell, ed.), p. 327. Oxford University Press, Oxford, 1989.

[16] J. W. Fickett, *Trends Biochem. Sci.* **11**, 190 (1986).

[17] J. W. Fickett, *Trends Biochem. Sci.* **11**, 382 (1986).

[18] C. Burks, *in* "Biomolecular Data: A Resource in Transition" (R. R. Colwell, ed.), p. 17. Oxford University Press, Oxford, 1989.

```
LOCUS         HUMCRYGA1      465 bp ds-DNA              PRI       15-MAR-1989
DEFINITION    Human gamma-A-crystallin gene (gamma-G5), exons 1 and 2.
ACCESSION     M17315
KEYWORDS      crystallin; gamma-crystallin.
SEGMENT       1 of 2
SOURCE        Human fetal liver DNA, clone lambda-16G3.
  ORGANISM    Homo sapiens
              Eukaryota; Metazoa; Chordata; Vertebrata; Tetrapoda; Mammalia;
              Eutheria; Primates; Anthropoidea; Hominoidea; Hominidae.
REFERENCE     1  (bases 1 to 465)
  AUTHORS     Meakin,S.O., Du,R.P., Tsui,L.-C. and Breitman,M.L.
  TITLE       Gamma-crystallins of the human eye lens: Expression analysis of
              five members of the gene family
  JOURNAL     Mol. Cell. Biol. 7, 2671-2679 (1987)
  STANDARD    full staff_entry
COMMENT
FEATURES         from   to/span      description
     pept         109        117     gamma-A-crystallin, exon 1 /nomgen="CRYG5"
                                     /map="2q33-q35"
                  219   +    461     gamma-A-crystallin, exon 2
     IVS          118        218     CRY-g-A intron A
     IVS          462   >    465     CRY-g-A intron B
     refnumbr       2          2     numbered 1 in [1]
BASE COUNT        91 a      137 c       111 g      126 t
ORIGIN        Chromosome 2q33-35.
        1 aggtcccttt tgtgttgttt ttgccaacac agcagcttcc ctgctatata taccagttgc
       61 ccctttgtcc ctatcatact agatgctaat caccctctgt caacaaccat ggggaaggtg
      121 agcctgtgga ggtgctgtgc catgtctatt gggggtctgt ggtgtgtggg gatgttcctt
      181 ccagctgact gtctactgtc accttatttc tacctcagat caccttctac gaggaccgag
      241 actttcaggg tcgctgctac aattgcatca gtgactgccc caacctgcgg gtctacttca
      301 gccgctgcaa ctccatccga gtagacagcg gctgctggat gctctatgag cgtcccaatt
      361 accagggcca ccagtacttc ctgcgccgag gcaagtaccc cgactatcag cactggatgg
      421 gcctcagcga ctcggtccaa tcctgccgta taattcctca tgtga
```

FIG. 1. Sample GenBank entry. The entry is in the standard line-type format used in the magnetic tape distribution of GenBank.

sented in the database. Of the sequences for which feature annotation is available, roughly one-half represent protein-coding regions. However, many other functional categories are also well represented: for example, there are over 400 Alu repeats annotated in the database and over 1600 tRNA gene sequences.

There has been a greatly enhanced interest in "complete" genomes over the past few years. There are enough *Escherichia coli* DNA sequence data in GenBank to account for roughly 28% of the *E. coli* genome (though the apparent percentage of the genome covered would be smaller if potentially overlapping segments were merged). On the other hand, there are only enough *Homo sapiens* DNA sequence data in GenBank to account for, at most, 0.2% of the human genome. If one includes viruses, plasmids, and organellar chromosomes, there are about 150 "complete" genomes represented in the database, but the highly parasitic nature of these latter entities would require that one include a significant amount of host genetic material in the complete specification of the life cycle.

Update on Objectives

From the point of view of the scientific community, there are three important criteria for measuring the usefulness of GenBank: completeness (are all the known sequences present in the database?); timeliness (are the sequence data getting into the database rapidly?); and amount of annotation provided (is the context provided on which one wants to base one's query?).

Completeness. GenBank strives to include every known nucleotide sequence in the database. The standard way of accomplishing this has been to scan physically new journal issues and identify articles containing original reports of new sequence data. Of course, for a number of reasons, a small number of appropriate articles are occasionally missed using this approach, and thus the data they contain do not end up in the database. More significantly, missed data result from the fact that physical scanning (which is currently divided among GenBank, EMBL, and DDBJ) is limited to the less than 100 journals that include large numbers of articles containing sequence data (in fact, a handful of journals publish most of the data GenBank collects).[15] The physical scanning method therefore misses many articles appearing in other journals that may report only a few new sequences each year; however, the total summed over many such journals adds up to many articles.

We have established several new initiatives since 1987 to overcome these problems. First (as discussed below), many journals have adopted publication protocols either requiring or requesting direct data submission prior to publication; this greatly reduces the number of articles missed by us during physical scanning. Second, in collaboration with the NLM, we have begun a system whereby the low-sequence-frequency journals can be scanned by use of new Medline MeSH headings that have been developed to point specifically to data appropriate for GenBank. This enables us to identify appropriate articles in journals that are not physically scanned. Finally, we have mounted a major effort[19] to identify and enter all articles that were published in the past and still not in GenBank; this is being achieved by means of scanning keyword terms in abstracting services (like the mechanism described above for scanning the new MeSH terms in Medline). Anyone aware of previously published articles that should be in GenBank but that appear to be missing should submit them for data entry.[3]

Timeliness. For many users of the database, it is crucial that new data arrive in distributed versions of the database as soon as possible after they

---

[19] D. Benton and C. Burks, *BRL Focus,* **11**(4), 69 (1989).

are published or otherwise become publically available. Several years ago, when it became clear to us that the amount of data being published was beginning to greatly outstrip our ability to enter sequence data and the associated annotation in a timely fashion, we adopted several lines of action to counter the growing lag time between appearance of a sequence in the literature and appearance in GenBank.

We began an effort to greatly increase the percentage of direct, structured, and computer-readable submissions (see section on direct submissions) of both sequence data and annotation. This had and has the effect of greatly reducing the in-house lag time for data to appear in our quarterly update queue. We have also encouraged authors to submit their data to GenBank prior to publication. In this way, the minimal time lag imposed on processing and distribution of new data by noncontinuous updates and even highly automated processing of submissions is hidden from the user's view. We have rebuilt the data flow protocols for logging in and initial processing of new articles and/or submissions. For example, entries originally were held from the quarterly update queue until completely annotated; because the extraction of annotation from articles (and subsequent verification of the quality of that annotation) can be very time consuming, entries for which sequence and citation were already available might wait many months before becoming visible in a release. To counter this, we introduced several levels of annotation in legal database entries (see below), allowing an entry to appear in a release initially at even the minimally defined level of annotation. Finally, we have most recently taken steps to make new additions to the update queue available by electronic means (through EMBL or GenBank), thereby eliminating the lag associated with the off-line quarterly update cycle.

These efforts have greatly reduced the time lag for data to appear in the database, as indicated in Fig. 2. In the early 1980s data were taking on average over 1 year to appear initially in a database release; today, the lag time has dropped to a few months and will continue to drop, especially as direct submission prior to publication becomes the norm for the publication process.

Depth of Annotation. The utility of a nucleotide sequence is minimal when the sequence is viewed without its context. We consequently devote considerable effort to providing the contexts discussed in the previous section. Our goal in annotating a given sequence is to provide as much meaning as possible through these various contextual fields; this goal must often be traded off against the constraints associated with the other two objectives discussed above. We constantly evaluate annotation with regard to its usefulness to the scientific community and in light of the amount of effort required to abstract it.

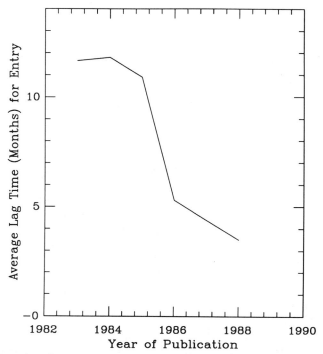

FIG. 2. Lag time for data collected by GenBank to enter the database. The lag time corresponds to the delay between appearance of sequence data in a published journal article and the first appearance of those data in a GenBank release. Note that figures are based on only articles appearing in journals scanned directly by GenBank.

As we have grown more familiar with the data structures associated with nucleotide sequence context, and in the face of the ongoing explosion of data, we have (1) more precisely delineated and atomized the data items we provide when annotating and (2) defined several levels of annotation that provide useful guidelines for seeking a balance between depth of annotation and resources available for providing such annotation in a timely way. This has led to three categories of entry, classified roughly by the amount of annotation present: (1) unannotated entries, with sequence, citation, and a few other data items; (2) simple entries, which include (1) and also organism name, delineation of transcription and translation, and a few other data items; and (3) full entries, with all data items included if the information is available.

In gauging the needs of the scientific community we determined that the annotation of the many outweighed the needs of the few. That is, it is more important to keep up with the minimal annotation of all entries than it is to have a smaller number of entries annotated in great detail. Accordingly, and as discussed above, we introduced "unannotated" entries into the database several years ago to decrease the time lag for at least sequence and citation to appear in the database; while this change accomplished the desired decrease in lag time, it also resulted in data residing in the database for a considerable period with little associated annotation. Thus, after a brief period of time, the unannotated division of the database had become one of our largest (see Fig. 3), and the time lag for annotation to be developed was considerable. The increase of direct submissions and increased resources since 1987 have allowed us to begin to address this problem. As can be seen in Fig. 3, the relative and even the absolute sizes of the unannotated division of the database have stopped growing (and, in fact, begun shrinking) for the first time since we introduced unannotated entries.

We also constantly monitor the utility of the syntax and scope of existing annotations. This has led, for example, to the development of a new FEATURES table for the distributed line-type versions of the database. Our goal was both to provide a more flexible framework in which to express the biological context of nucleotide sequences and to allow greater ease in extracting data. The new FEATURES table design,[20,21] was developed in collaboration with EMBL and DDBJ; we plan that the GenBank FEATURES table will reflect this revised view within the next year.

In addition to encouraging the scientific community to increase its role in making direct submissions of new data to the database, we are also encouraging (and developing the tools and protocols to support) scientists to take a more direct hand in the maintenance of data in the database, with respect to both depth and currency of annotation. We are calling on experts in various biological fields to bring their expertise to bear on the data in GenBank and to act as curators of entries in their fields (see also the discussion of curators below).

Direct Submission of Data

We determined several years ago that the exponentially increasing rate of sequencing and our model for data collection (i.e., in-house abstraction and entry of data) were incompatible. We revised our model so that the primary role of GenBank in data collection would be the design and provision of tools to the scientific community that would allow them to

[20] GenBank staff, The DDBJ/EMBL/GenBank Feature Table: Definition, unpublished GenBank document, 1989.
[21] GenBank staff, The Features Table Annotation Standards Guide, unpublished GenBank document, 1989.

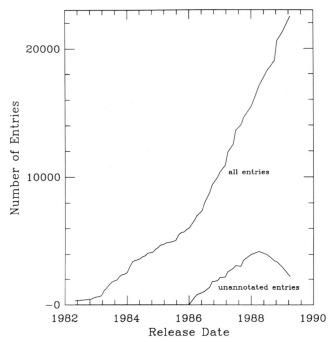

FIG. 3. Number of unannotated entries in GenBank. The total number of entries and the number of unannotated entries are plotted as a function of database release date. The rate of increase of the number of nucleotides in the database is steeper than that for the total number of entries because the average sequence length per entry has also been increasing. Note the initial, and subsequent lack of, correlation between the number of unannotated entries (this figure) and the drop in entry lag time (Fig. 2).

submit their new data directly to GenBank in computer-readable and automatically processable form. Over the past two years, we have set about conforming GenBank's operation to this model.

There are several key elements on which the success of this changeover depends. The community must come to see direct submission of their data to GenBank as a requisite part of the publication process. Uniform criteria for what constitutes a valid submission must be made available to the community. The processing of submissions should be as fully automatic as possible. Finally, GenBank must be able to process and acknowledge receipt of submissions both quickly and reliably. The following is an account of progress on these several requirements, including the efforts of those entities in the community in the best position to effect this progress.

In-House Efforts. Several years ago, we developed, in collaboration with EMBL and DDBJ, a common Data Submission Form, which is available in both printed and computer-readable form. This provides uniform, though limited, criteria to the community for what constitutes a valid submission. It is limited in that, as a simple hardcopy form to be filled in by hand or as an on-line form to be edited by hand, the responses to such a form cannot be checked for internal consistency or consistency with Gen-Bank standards until they are processed in-house after submission. However, GenBank is now releasing[2] "data entry at source" software, Authorin, that will greatly extend the efficiency of direct submission by exporting many of these validity/integrity checks to the sites where the initial data entry is being done. This menu-driven, window-based package (which is initially being provided for use on MS-DOS-based systems) prompts the correspondent for the data items that constitute a data submission to GenBank and performs the syntactic and vocabulary checks associated with these data items.

The relative amount of submitted data being processed automatically (or semiautomatically) has increased considerably, owing in large part to the more frequent use of computer-readable media by the community when submitting data to GenBank, and also because of the development of in-house tools that can transcribe these transmissions into standard Gen-Bank format. However, and as indicated above, the ability to converge on completely automatic processing of submissions is currently limited by the less-than-rigorous definition (in the Submission Form) of the syntax and content of data items in a submission. The database restructuring effort (see following section) has been directed toward this problem, which will be solved in large part by the use of a standard transaction protocol for transmitting data to and from the database. For example, Authorin will provide output that conforms to this standard in preparation for sending a direct submission to GenBank.

Finally, we have addressed the need for rapid checking and acknowledgment of direct submissions. In-house procedures have been developed for initial processing of submissions so that an acknowledgment of receipt of submitted data, in the form of an accession number, is issued after the completeness and internal consistency of a submission are checked, but in any case within 7 days of receipt of a direct submission (within 48 h if received by electronic mail). The appearance of an accession number in a published article in those journals requiring direct submission thus indicates that the sequence data in the article have been submitted to GenBank by the author.

Cooperation with Journals. The scientific community has made great strides in taking responsibility for both the timeliness and quality of data entry into the database. The support provided by the editors and editorial staffs of the journals (beginning with the *Journal of Biological Chemistry* several years ago) publishing these sequences has been critical to this step forward. The majority of journals scanned by GenBank now provide as part of their instructions to authors at least some encouragement for authors to submit their nucleotide sequence data directly to GenBank. (EMBL and DDBJ have worked out analogous arrangements with the journals that they scan.) Journals have also published articles and editorials on the mechanisms[22] and desirability (see, for example, the recent editorial[23] in *Proceedings of the National Academy of Sciences of the United States of America*) of direct submission. Documentation of the details for initiating and supporting these protocols is available on request.[3]

An increasing number of journals seeking to strengthen their support of direct submission are now implementing (or considering implementing) a requirement for submission as an integral component of the manuscript review and publication process. The long-term benefits of such a policy will be manifest in the enhanced timeliness of availability of the sequence data to the scientific public (as restructuring of the database leads to a more continuously updated GenBank, sequence information will be available at the time of publication). Additionally, this shift of editorial policy gives journals greater flexibility in deciding how to deal with the ever-increasing amount of nucleotide sequence data being included with submitted manuscripts: electing to omit part or all of these data from explicit presentation in a published figure is less controversial when the same data can be assumed to be available through GenBank.

Role of Large-Scale Sequencing Projects. Work is already under way to sequence orders-of-magnitude larger amounts of DNA in a number of different organisms. We are in the process of establishing links with these sequencing efforts with the goal of facilitating direct flow of data into the database. We actively encourage those who are planning such projects to contact GenBank at as early a stage in the planning process as possible.

Role of Software Developers. The majority of scientists who view the information in the GenBank database do so with the aid of commercial or public domain software packages that provide them with the tools to make complicated queries or extensive searches for homology. For this reason, the developers and maintainers of these packages are in an excellent position to facilitate direct submission by providing "submit nucleotide sequence data" modules in their packages; it will be especially useful if the output from these packages conforms to GenBank's transaction protocol and provide the user with aid in transmitting transactions through electronic mail or via on-line access (e.g., using Kermit) to GenBank com-

---

[22] C. Burks and L. J. Tomlinson, *Proc. Natl. Acad. Sci. U.S.A.* **86,** 408 (1989).
[23] I. Dawid, *Proc. Natl. Acad. Sci. U.S.A.* **86,** 407 (1989).

puters. This topic was one of several discussed at the GenBank Software Developers' Workshop in October 1988. In addition to including representatives from most of the major software developers, the meeting was attended by a number of groups involved in the development of large-scale sequencing stations. GenBank will continue to host these workshops, certainly to keep developers informed of project advances, but also to solicit their feedback and aid, especially in assisting in direct submission through software integrated in their sequence analysis packages.

Current Numbers for Direct Submission. Our and the community's initial efforts have resulted in a great jump in the percentage of direct submissions. During the calender year 1988, approximately 50% of the sequences entered into GenBank were received as direct submissions from the scientific community; this fraction represents a considerable improvement over previous years. Some 70% of the sequences in these submissions were in computer-readable form. The obvious goal in the future is to bring these numbers much closer to 100%.

Restructuring the Database

Our experience maintaining the GenBank database led us several years ago to the conclusion that the database needed a fundamental restructuring if it were both to keep up with the ever-increasing flow of data and to remain genuinely useful to the molecular biology community. Much of our effort since 1987 has been directed toward redesign of the database as well as implementation of the software needed to support the new structure. At this writing, we are completing the first round of software development and data translation and are testing the new database along side the flat file version in preparation for the final conversion.

Relational Schema. At the heart of the new design (see Fig. 4) is a relational schema describing the various tables in which the data will be stored. The design of the schema is critical to the success of the conversion because it is the schema that determines both what the database can store and what questions can be asked of the database. The design of the schema for GenBank has been particularly challenging because the community's understanding of molecular biology has been evolving rapidly in recent years and will continue to do so in the future. This has meant not only considerable effort in keeping the schema current, but also in designing it to allow for future changes and to minimize the impact of these changes on the overall schema.

Transaction Protocol. Another essential component in the redesign effort has been the development of a transaction protocol for database

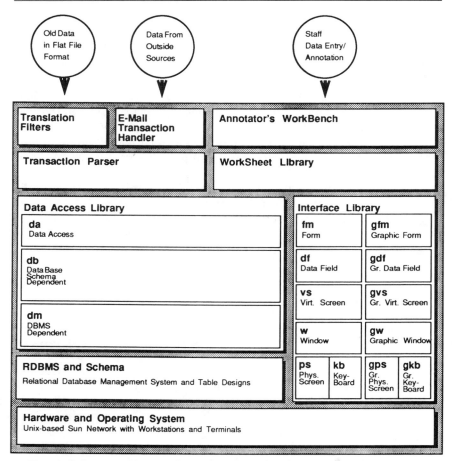

Fig. 4. Software architecture for the restructured GenBank database. The top-down dependency of the software and database design[24] is illustrated. GenBank data have historically been maintained in flat-file, line-type format[5]; these data will be restructured using translation filters and deposited in the RDBMS. Once we have made the transition to the new design, data revisions and entry of new data will be mediated either interactively through the Annotator's Workbench or by submission of ASCII file transactions through the Transaction Parser.

updates. The protocol describes the syntax of transaction files that can be submitted to the database and processed automatically, eliminating the bottleneck associated with human intervention. In addition to supporting the direct submission of data by researchers, we hope that the transaction protocol will become the primary format for the exchange of data between

GenBank and its data bank collaborators. This will dramatically improve both the quality and efficiency of these collaborations.

Conversion of Flat Files to DBMS Version. In order to convert the data from the current, flat-file format to the relational database management system (DBMS) model, it has been necessary to write a number of programs to convert the current entries into transactions that can then be submitted to the new database. This code has had to tackle such problems as unmerging entries containing more than one sequence. These filters will continue to be used even after the initial conversion until we have put the Annotator's Workbench (see below) on-line.

Data Access Library. For reasons such as DBMS independence and greater internal control over the data, we have designed a library of functions that fits between our application programs and the DBMS. This library is called the Data Access Library. The functions in the Data Access Library provide what is basically an object-oriented interface to the programmer so that applications do not need to understand the internal complexities of the schema. This library, as well, allows for consistent processing of transactions.

Transaction Parser. The Transaction Parser is a program which reads transactions from ASCII files and hands them over to the Data Access Library for processing. We soon will be able to accept transactions through electronic mail from other databases and from researchers directly, which will be automatically processed by the Transaction Parser with results (e.g., whether the transaction was legal and/or completed) being returned to the originator without requiring human intervention.

Annotator's Workbench. At the core of all of our on-line tools for locally updating the database is the Interface Library. This is a terminal-independent, operating system-independent library of routines for building window-based interfaces. The Annotator's Workbench, built on the Interface Library, will be the primary interface to the database for both annotators and curators. This Workbench will allow data to be retrieved and manipulated, with the database being updated immediately. A more detailed description of the restructuring strategy, design, and associated software will appear elsewhere.[24]

Access to GenBank Data

As the needs of the user community and available computer technology have evolved, so have the media for distributing the GenBank database to

[24] M. J. Cinkosky, D. Nelson, J. W. Fickett, T. G. Marr, and C. Burks, Los Alamos National Laboratory, in preparation (1989).

the user community. As new media for data distribution are developed, and new and/or more sophisticated categories of queries recognized, the presentation of the data to the user community has been modified and often expanded in scope. The following are the current options for accessing the database; future developments are discussed in the following section. For more details on media specification, text formats, etc., one should consult the current GenBank release notes or contact IG[2] directly.

Magnetic Tape. The data are currently available as (1) industry-standard 9-track unlabeled ASCII, written in fixed-length records of 80 characters; (2) VAX/VMS Backup saveset (on 9-track tapes or TK-50 cartridges), or (3) compressed Unix tar archives (on 9-track tapes or Sun $\frac{1}{4}$-inch QIC 24 format tape cartridges). In all three cases, uploaded data appear in standard GenBank line-type format.[5]

Floppy Diskettes. The data are available on (1) $5\frac{1}{4}$-inch 360-kbyte, double-sided, double-density disks, which can be read by IBM or IBM-compatible computers running PC-DOS or MS-DOS (versions ≥ 2.00); (2) $5\frac{1}{4}$-inch, 1.2-Mbyte, double-sided, high-density disks, which can be read on the same systems as the double-density disks; or (3) $3\frac{1}{2}$-inch, 800-kbyte, Macintosh format microdisks, which can be read by a Macintosh computer. The uploaded data appear in a modified version of the standard GenBank line-type format.

Dial-up On-line Service. Ten-minute on-line sessions are currently offered at no charge, and they are available through the Telenet public data network. Entries can be identified and extracted by accession number, author name, entry name, keyword phrase, or organism name. Extracted entries can be downloaded to your local computer using the Kermit file transfer protocol.

Hardcopy. Though several years ago we maintained an annual, published version of the database in book form, the size of the database has made this impractical. The last such version,[25] undertaken in collaboration with EMBL, appeared in 1987; at this time, there are no plans to publish another. The format in these compendiums was a modification of standard GenBank line-type format.

Secondary Sources. The GenBank project was not chartered to develop or distribute software for manipulation and analysis of data by the user

[25] E. J. Atencio, H. S. Bilofsky, J. Bossinger, C. Burks, G. N. Cameron, M. J. Cinkosky, C. E. England, V. I. Esekogwu, J. W. Fickett, B. T. Foley, W. B. Goad, G. H. Hamm, D. J. Hazeldine, P. Kahn, L. Kay, F. I. Lewitter, N. Lopez, K. A. MacInnes, M. J. McLeod, D. L. Melone, G. Myers, D. Nelson, J. L. Nial, J. K. Norman, E. D. Rasmussen, A. A. Revels, W. P. Rindone, C. R. Schermer, M. T. Smith, G. Stoesser, C. D. Swindell, B. L. Trujillo, and C.-S. Tung. "Nucleotide Sequences 1986/1987: A Compilation from the GenBank and EMBL Data Libraries." Academic Press, Orlando, Florida, 1987.

community; however, such software can be acquired[26-29] from a number of independent distributors and vendors. In many, if not most, cases, these software packages also include a redistributed (and often reformatted) version of GenBank data as part of the package one acquires. The user opting for the advantages in handling data that these software packages provide should carefully investigate (1) how often and how rapidly the distributor of the package provides updates to the GenBank database and (2) the extent to which the package takes advantage of the syntax and content of the annotation in the distributed database.

Future Directions

Expected Growth of Sequence Data. We have presented elsewhere[18,30] a detailed discussion of the past and future growth rates and the implications of the future growth rate for data bank operations. In brief, we expect that by 1995 the GenBank database will contain 10 to 50 times as much nucleotide sequence data as in the current database, and we expect that the annual rate of appearance of new sequence data will have increased 5 to 70 times the current rate. We believe that our plans for the database adequately anticipate these numbers (and, in fact, much larger numbers in the more distant future), and we are moving as rapidly as possible toward their implementation.

Data Access. The development of distribution of the GenBank database on floppy diskettes several years ago addressed the needs of PC-based users (especially those on stand-alone systems) to gain facile access to the data. The data-compacting technology available on floppies has not kept abreast of the growth rate of the GenBank database (leading to the database releases being spread over very large numbers of floppies), so we began last year to explore the feasibility of CD-ROM-based distribution. With input from the community of software developers supporting access to GenBank, we have begun development of a CD-ROM format that will be initially released in the Spring of 1990.[2] Knowing that we are in the process of converting to an RDBMS-based maintenance scheme for the data, we have also begun discussing the possibility of distributing the GenBank data

[26] F. I. Lewitter and W. P. Rindone, this series. Vol. 151, p. 582.
[27] C. Rawlings, "Software Directory for Molecular Biologists." MacMillan, London, 1986.
[28] R. Beynon, J. Modelevsky, R. Roberts, and D. Soll, eds., *Nucleic. Acids Res.* **16,** 1655 (1988).
[29] R. Beynon, J. Modelevsky, R. Roberts, and D. Soll, eds., *Comput. Appl. Biosci.* **4,** 1 (1988).
[30] C. Burks, *in* "Biotechnology and the Human Genome: Innovations and Impact" (A. D. Woodhead and B. J. Barnhart, eds.), p. 51. Plenum, New York, 1988.

in a form that would allow users at remote sites to upload data directly into their own RDBMSs.

Cross-Links and Collaborations. There has been a dramatic increase in the number of databases containing information pertaining to molecular biology over the recent years.[31,32] A large number of these resources contain information relevant to but not necessarily explicitly represented in GenBank entries. As the number of molecular biological data banks increases, so does the community's desire to migrate from one data set to another, automatically, where possible, through natural cross-links. GenBank has[33,34] and will continue to place pointers in entries that will allow scientists and, ultimately, software to follow cross-references to as many other databases as possible. Note, however, that most of the current protocols supporting cross-referencing are labor-intensive and inexact; there is an outstanding need to develop dynamic, automatic updating schemes for maintaining links.

GenBank currently inserts the gene names approved by the Human Gene Mapping Library[35] (HGML) and the associated map locations in all relevant human entries.[34] An important and current goal of the ongoing collaboration with HGML has been the development of mechanisms for accurate exchange of update information. This is currently being carried out at a manual level, but protocols and data structures whereby changes made at HGML are propagated automatically to the relevant entries in GenBank are currently in the design phase. Once implemented, such a scheme will allow GenBank to perform similar informational exchanges with other databases. Among those planned are the *E. coli* stock center and mapping databases,[36] the ATCC probe database[37] (a three-way link including HGML), and the protein databases (e.g., the Protein Identification

[31] C. Burks, J. R. Lawton, and G. I. Bell, *Science* **241**, 888 (1988).
[32] J. R. Lawton, F. A. Martinez, and C. Burks, *Nucl. Acids Res.* **17**, 5885 (1989).
[33] B. T. Foley, D. Nelson, M. T. Smith, and C. Burks, *Trends Genet.* **2**, 233 (1986).
[34] C. R. Schermer, M. T. Smith, D. Nelson, L. Miller, M. J. McLeod, F. I. Lewitter, J. E.-D. Hayden, B. T. Foley, J. W. Fickett, V. I. Esekogwu, I. H. Cohen, and C. Burks, *Cytogenet. Cell Genet* **46**, 688 (1987).
[35] I. H. Cohen, D. A. Consiglio, U. Francke, K. K. Kidd, R. L. Miller, A. J. Pakstis, C. W. Partridge, S. R. Perretta, F. H. Ruddle, R. K. Track, T. B. Shows, and P. J. McAlpine, *in* "Genetic Maps 1987: A Compilation of Linkage and Restriction Maps of Genetically Studied Organisms" (S. J. O'Brien, ed.), Vol. 4, p. 518. Cold Spring Harbor Laboratory, Cold Spring Harbor, New York, 1987.
[36] B. J. Bachman, *in* "Genetic Maps 1987: A Compilation of Linkage and Restriction Maps of Genetically Studied Organisms" (S. J. O'Brien, ed.), Vol. 4, p. 178. Cold Spring Harbor Laboratory, Cold Spring Harbor, New York, 1987.
[37] Database for the Repository of Human Probes and Libraries; American Type Culture Collection; 12301 Parklawn Drive; Rockville, MD 20852.

Resource[38]). Plans are already under way to provide new and to strengthen and expand existing cross-links to the NLM bibliographic databases,[39] the Human Retrovirus and AIDS database,[40] and the Human Genome Information Resource database.[41]

Direct Submissions. Though there is still much effort to be completed, the move to expand the role of journals in supporting direct submissions is well on track. On the other hand, there is a current need to examine the mechanism for interaction with the large-scale nucleotide sequencing efforts that are now getting off the ground. As such projects are being described in planning discussions and funding proposals, they often raise questions about timing and frequency of data transmission to the data banks, proprietary concerns, and data representation in the database.[42] We look forward to addressing these issues collaboratively with those planning such projects.

Curator Program. The GenBank annotation staff can extract much information from individual articles in the literature, and presumably the authors of these articles are in an even better position to provide such information (e.g., using Authorin). However, much of the annotation (e.g., gene names, product names, delineation of sequence functionally) transcends individual articles, if for no other reason than that the community's understanding of sequence data and vocabulary for depicting this understanding is constantly evolving. We have recognized the need to look beyond GenBank's internal staff resources for help in maintaining and improving the quality of thoroughness of annotation in this context.

We have initiated an effort to provide experts, scientists who have extensive knowledge covering features of nucleotide sequences within a specific domain of molecular biology, with the tools to impart that knowledge to the database. Using interfaces designed to work in the foreground of the relational database structure and equipment allowing remote use of such tools, curators will access the database from their laboratories and elaborate upon the existing annotation. Once the relational DBMS structure is fully installed, and over the next 3 years, about 15 curators will be brought on-line. It is likely that some curators will be associated with the

[38] K. E. Sidman, D. G. George, W. C. Barker, and L. T. Hunt, *Nucleic Acids Res.* **16,** 1869 (1988).
[39] R. Haynes, K. McKibbon, D. Fitzgerald, G. H. Guyatt, C. J. Walker, and D. Sackett, *Ann. Intern. Med* **105,** 810 (1986).
[40] D. Barnes, *Science* **235,** 634 (1987).
[41] Human Genome Information Resource; T-10, MS K710: Los Alamos National Laboratory; Los Alamos, NM 87545.
[42] C. Burks, *in* "Computers and DNA" (G. I. Bell and T. Marr, eds.), p. 35. Addison-Wesley, New York, 1989.

large-scale sequencing projects mentioned above, where it is hoped that their presence will greatly improve the speed and quality of sequence data coming from these projects.

One True Database. One of the most-often expressed frustrations in the user community has been the lack of identity among the comprehensive nucleotide sequence data banks: GenBank, DDBJ, and EMBL. These data banks have traditionally used different distribution formats, have occasionally defined (and filled) their data items differently, and have always, in any particular release, represented data sets from the literature that do not completely overlap. We have made some progress toward addressing these issues over the past few years, as discussed in the previous sections.

On the other hand, we are far short of a satisfactory state. The current restructuring effort (which is also being undertaken by DDBJ and EMBL) is intended to provide a much better platform for addressing the problems associated with maintaining identical data sets at multiple sites. Problems that have yet to be fully addressed include communications bandwidth and reliability, common data definitions and integrity checks, and parallel updating (and tracking of updating) of multiple copies of the same data set. Once the databases have been restructured in the current effort, we will be turning our full attention to these challenges.

Acknowledgments

Much of the technical effort for GenBank has been undertaken in and benefited from close collaboration with staffs of the Human Genome Information Resource, the EMBL Data Library, the DNA Data Bank of Japan, the Human Gene Mapping Library, the National Library of Medicine Lister Hill Center, and the Protein Identification Resource; we are also grateful to the staffs of these efforts for many useful discussions. We are indebted to past members of the GenBank staff for their many contributions to the project, as we are to past and present members of the advisory groups to the National Institutes of Health who have given thoughtful insights to the needs of the user community. This work was funded by a contract (NO1-GM-7-2110) with the NIH. The authors at LANL were also supported under the auspices of the U.S. Department of Energy.

[2] EMBL Data Library

By PATRICIA KAHN and GRAHAM CAMERON

Background

The EMBL Data Library was established in 1980 to collect, organize, and distribute a database of nucleotide sequences and related descriptive information extracted from publications in scientific journals. Since 1982 this work has been done in collaboration with GenBank (Los Alamos, NM, and Mountain View, CA), and recently the DNA Database of Japan (Mishima) joined the collaboration. Each of the three groups collects a portion of the total reported sequence data and exchanges it with the others on a regular basis. Since 1987, the Data Library has begun to provide additional data sets useful to molecular biologists.

Databases

EMBL Nucleotide Sequence Database

The Nucleotide Sequence Database was the original motivation for the establishment of the group, and it continues to be the main endeavor of the Data Library. The data are presently distributed as "flat" text files where each entry comprises a single contiguous sequence and accompanying descriptive information (annotation). Different line types, each with their own two-letter code, are used to make up an entry. A sample entry is shown in Fig. 1. Each entry is uniquely identified within a release by its name (DMSHAKE3 in Fig. 1) and across releases by its accession number list (X07133 and Y00847 in Fig. 1). References to database entries should always cite the primary (first) accession number.

Release 18 of the database (February 1989) contained 27.2×10^6 bases, approximately 40 times more than release 1 (June 1982). As illustrated in Fig. 2, this increase has not been linear; rather, it reflects a rapidly accelerating rate of growth, a trend which will undoubtedly continue. Clearly, the resources of the databases cannot increase at the same rate, and, therefore, coping with the data flow demands that we revise and streamline our data processing procedures. As part of an effort to achieve this we have just finished a restructuring of our database, including installation into the ORACLE relational database management system. This should result in a more efficient service to users in the short term and will enable us to offer

METHODS IN ENZYMOLOGY, VOL. 183

```
ID   DMSHAKE3     standard; RNA; 1176 BP.
XX
AC   X07133; Y00847;
XX
DT   21-APR-1988  (publication)
DT   20-APR-1988  (ref. correction)
DT   08-APR-1988  (annotation)
XX
DE   D. melanogaster larval Shaker mRNA for put. K(+)-channel
XX
KW   Shaker gene; potassium channel protein.
XX
OS   Drosophila melanogaster (fruit fly, drosophile, Fruchtfliege)
OC   Eukaryota; Metazoa; Arthropoda; Insecta; Diptera.
XX
CC   *source: strain=Canton S; library= 1. and  2. larval instar gt10;
CC   *source: clone=ShD
CC   **map: chromosome=X/16F; map position=57.7
XX
XX
RN   [1]  (bases 1-1176; enum. 1 to 1176)
RA   Pongs O.;
RT   ;
RL   Submitted (15-MAR-1988) on tape to the EMBL data library by:
RL   Pongs O., Lehrstuhl fuer Biochemie, Ruhr Universitaet Bochum,
RL   Univeritaetsstr. 150, 4630 Bochum-Querenburg, West Germany.
XX
RN   [2]  (bases 1-1176; enum. -72 to 1108, no zero)
RA   Pongs O., Kecskemethy N., Mueller R., Krah-Jentgens I.,
RA   Baumann A., Kiltz H.H., Canal I., LLamazares S., Ferrus A.;
RT   "Shaker encodes a family of putative potassium channel proteins in
RT   the nervous system of Drosophila";
RL   EMBO J. 7:1087-1096(1988).
XX
FH   Key          From     To        Description
FH
FT   CDS            73     984       K(+)-channel
FT   SITE          619     675       transmembrane segment S1
FT   SITE          772     837       transmembrane segment S2
FT   SITE          871     933       transmembrane segment S3
XX
SQ   Sequence  1176 BP;    381 A;   238 C;    254 G;    303 U;
     aaucccagug gagaucaaag ugcacgugca gagagagaaa aaguggagua ggucgaugaa
     uggauugccu guauggcaca caucacgacg acgcacggca gcuuaagcca agcgacaagg
     ucuuugccca aauugagcag ucaagacgaa gaaggggggg cuggucaugg cuuuggguggc
     ggaccgcaac acuuugaacc cauuccucac gaucaugauu ucugcgaaag agucguuaua
     aauguaagcg gauuaagguu ugagacacaa cuacguacgu uaaaucaauu cccggacacg
     cugcuugggg auccagcucg gagauuacgg uacuuugacc cgcuuagaaa ugaauauuuu
     uuugaccgua gucgaccgag cuucgaugcg auuuuauacu auuaucagag ugguggccga
     cuacggagac cggucaaugu cccuuuagac guauuuuagug aagaaauaaa auuuuaugaa
     uuaggugauc aagcaauuaa uaaauucaga gaggaugaag gcuuuauuaa agaggaagaa
     agaccauuac cggauaaaga gaaacagaga aaagucuggc ugcucuucga guauccagua
     aguucgcaag ccgccagagu uguagccaua auuaguguau uuguuauauu gcuaucaauu
     guuuauauuu gucuagaaac auuacccgaa uuuaagcauu acaaggugu caauacaaca
     acaaauggca caaaaaucga ggaagacgag gugccugaca ucacagaucc uuucuuccuu
     auagaaacgu uauguauacu uugguuuuaca uuugaacuaa cugucaggu ccucgcaugu
     ccgaacaaau uaaaauuucug caggggauguc augaauguua ucgacauaau cgccaucauu
     ccguacuuua uaacacuagc gacugucgu gccgaagagg aggauacgu aaaucuucca
     aaagcgccag cguauggagu uucuguuugc ccaaugccaa cuacaccaaa
     aauaucaaua uuaaucaccc acacacgacc acacacacac acaugcauau acaguugaac
     cgauuaaagu aauaaaaugu gccaucauca uuuaugaguu gccugacaug auaaagaagg
     gguugccugc ccgccccgaa accaaaaaaa aaaaaa
//
```

FIG. 1. Sample entry from the EMBL Nucleotide Sequence Database.

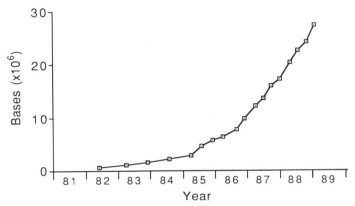

Fig. 2. Growth of the EMBL Nucleotide Sequence Database.

an on-line query system in the longer term. In the future the relational tables will also be available to interested users.

Although the EMBL and GenBank databases regularly exchange data, the problem has existed that the content of the two databases does not completely overlap. Unfortunately for the user, this has meant that a complete collection of nucleotide sequences was obtainable only by using both the EMBL and GenBank databases. The problem of incomplete overlap arises because the two groups initially set up their own (different) procedures for processing data and because the development of procedures to exchange all data (including corrections and updates) has proceeded slowly. Difficulties have arisen, for instance, in situations where one group enters a new sequence and merges it with an existing sequence which they believe is contiguous with the new data. When these merged data arrive at the other database, the procedures for determining whether they are new or are already in the database will be confounded, since they are in fact a mixture. Until recently, in an attempt to avoid duplication of data, we have taken a conservative approach: if any of the data in an entry were already present, the entire entry would not be added to the database. However, at this point in time the majority of researchers seem more concerned about missing data than about duplicates, and we have therefore decided to change our approach. As of Release 19 (May 1989) any entry whose status with respect to the database is unclear will be included in a separate, "ancillary" database; together, the main and ancillary databases should constitute a complete collection of data from the latest EMBL and GenBank releases. This represents a temporary solution; a major effort is underway by both groups to resolve this problem in a more elegant manner.

SWISS-PROT Protein Sequence Database

The SWISS-PROT database, maintained collaboratively by the EMBL Data Library and Amos Bairoch (University of Geneva), is a collection of amino acid sequences from the Protein Identification Resource collection (PIR, Washington, D.C.) along with translations of coding sequences in our Nucleotide Sequence Database. Its releases are coordinated with those of the Nucleotide Sequence Database such that they include translations of the newest data. SWISS-PROT is essentially identical in format to the Nucleotide Sequence Database and therefore the two collections can easily be used together. ORACLE-based data management systems presently under development aim to integrate the two collections more fully.

SWISS-PROT entries contain pointers to related data in the EMBL Nucleotide Sequence Database, the Protein Data Bank of three-dimensional structures (maintained at the Brookhaven National Laboratory, Upton, NY), and the PIR. These cross-references will soon be listed in the corresponding nucleotide sequence entries.

Eukaryotic Promoter Database (EPD)

In 1988 we began to distribute a database of eukaryotic promoters, prepared by Philipp Bucher (presently at Stanford University, Stanford, CA). This database contains detailed annotation of eukaryotic transcription start sites present in the Nucleotide Sequence Database and documented in the research literature. The database itself contains no sequences, but rather references to the sequences. Its releases are coordinated with the designed to be used with the Nucleotide Sequence Database.

The EPD is important not only for the scientific information it contains but also as a prototype for the future. Rapid growth in both the volume and complexity of sequence data makes it increasingly impractical for the data banks to maintain a pool of expertise capable of annotating all sequences; indeed, it has been argued that such annotation is interpretive work of a kind inappropriate to the major sequence databases. The EPD, as a database maintained remote from, but coordinated with, a centralized sequence collection, provides a model whereby the detailed biological annotation can be carried out at remote sites where the appropriate expertise is present.

Restriction Enzyme Data

The database of restriction enzymes provided by Dr. R. Roberts (Cold Spring Harbor Laboratory, Cold Spring Harbor, NY) is distributed with all releases of the nucleotide sequence data.

Data Acquisition

Since 1987 the staff of the nucleotide sequence databases have devoted a great amount of effort to developing systems which encourage researchers to submit their newly determined sequences and related data directly to the databases, preferably in computer-readable form. In doing so, we are attempting to replace our traditional method of data acquisition, in which database staff identify sequence-containing articles in the scientific literature, enter the sequence data into our computer, and then annotate these sequences by reading through the corresponding articles to locate the relevant information.

Direct submission is important for many reasons: (1) Abstracting the information from the research literature is labor intensive. (2) Entries can appear in the database much sooner if we get the information from the authors early in the publication process. (3) Machine-readable submissions reduce the chance of introducing errors. (4) Authors can bring far more expertise to bear on annotating their own data than the database staff can. (5) Interpretive annotation is more a research than a database activity. (6) The scale of present-day sequencing is reaching the point where journals are finding it inappropriate to print the actual sequences. If no mechanism to ensure their deposition in the databases is in place, research papers will be published whose underlying data are unavailable to the research community. (7) With an increasing number of journals publishing sequence-related papers without printing the sequences, scanning the literature to locate sequence data will no longer be feasible.

Progress with direct submissions has been good, with almost 60% of the new data coming as submissions in the latter half of 1988. Since 1989 two journals *(Nucleic Acids Research* and *Plant Molecular Biology)* have begun not only to request that authors submit sequence data to the database but to require it as a condition for publication. While these developments are encouraging, the accelerating rate at which sequence data are generated ensures that the work of entering and annotating the remaining (nonsubmitted) data will present an enormous, and eventually unmanageable, workload. The goal of ensuring more direct submissions therefore remains a high priority.

How to Submit Data to EMBL Data Library

Researchers who intend to submit data to any of the sequence databases should get a copy of a Sequence Data Submission Form, which solicits all the information needed for a nucleotide or protein sequence entry and provides instructions on how to submit the data. The form exists both in a paper version and as a computer-readable text file which can be completed using a text editor. Many molecular biology journals distribute the paper version to authors reporting sequence data, and a few journals publish it periodically.[1,2] The computer-readable version of the form is distributed with all releases of the EMBL and GenBank databases and can also be obtained via computer network (using the EMBL File Server, see below). Alternatively, either of these versions can be obtained by contacting the Data Library in any of the ways listed in the Appendix. A data submission should include the sequence data in computer-readable form (computer network mail, magnetic tape, or MS-DOS or Macintosh floppy diskette) and a completed data submission form for each submitted sequence. Data can be sent to the Data Library via computer network, telefax, or normal post (see Appendix).

Upon receiving the data, Data Library staff check whether the submission is complete. If so, it is processed within a few days, and the submittor is then informed what accession number(s) has been assigned to the data. This accession number serves as a reference that permanently identifies these data in the database, and we recommend that it be cited when referring to the data in publications. If the submission is incomplete, the submittor is notified what additional information is needed. The submission form also asks submittors whether the data should be made available to the public as soon as we have finished the corresponding entry or be withheld until publication.

Data Distribution

Magnetic Tape Subscriptions

The main way of distributing copies of the entire database continues to be by mailing magnetic tapes. Much of this distribution is done by the Data Library, and some is done by secondary distributors, such as groups which supply the data along with sequence analysis software. The four releases of nucleotide sequence data in 1988 were supplied to an estimated total user community of approximately 10,000 scientists.

The sequence databases produced by the Data Library are available either singly or as a yearly subscription. Users who subscribe to the EMBL Nucleotide Sequence Database releases for 1 year receive one release every 3 months along with the corresponding release of the EPD and the Restriction Enzyme database (see above). The four yearly SWISS-PROT releases can also be obtained by subscription. Subscribers are charged an annual fee

---

[1] P. Kahn and D. Hazledine, *Nucleic Acids Res.* **16**, i (1988).
[2] P. Kahn, D. Hazledine, and G. Cameron, *Plant Mol. Biol.* **11**, 541 (1988).

which covers our distribution costs (see Table I). Users can also order the latest release of the nucleotide or SWISS-PROT collections at any time. Further information about subscriptions can be obtained by contacting us at the address given in the Appendix.

CD-ROM

CD-ROM is attractive as a medium on which to distribute the sequence databases because it represents an inexpensive way to store large quantities data and because the devices required to read it are within financial reach of the typical personal computer user. In early 1989 the Data Library produced a prototype disk which includes prototype CD-ROM retrieval software developed in collaboration with Philips Du Pont Optical and Circle Information Systems. The design of a common GenBank/EMBL/ PIR CD-ROM format should be finalized soon, and thereafter disks will also contain the data in this format.

Network Access

The rapid pace of research in molecular biology has generated a requirement for better and more rapid access to the databases than that provided by quarterly releases. As part of an attempt to meet this need, EMBL set up in early 1988 a file server which enables researchers worldwide to retrieve entries from the major databases available at EMBL via computer network. Data are available over the file server as soon as Data Library staff have completed the preliminary entry, while the indices are updated daily. This is particularly attractive in combination with the data submission arrangements we have with *Nucleic Acids Research* and *Plant Molecular Biology,* since it enables readers of these journals to access the

TABLE I
DISTRIBUTION CHARGES

Subscriber category	Fee per release[b]		Fee per year[b]	
	Tape	CD-ROM	Tape	CD-ROM
Academic users				
EMBL member states[a]	DM 50	DM 100	DM 200	DM 400
Nonmember states	DM 100	DM 200	DM 400	DM 800
Industrial/commercial				
users	DM 250	DM 500	DM 1000	DM 2000

[a] The member states of EMBL (as of May 1989) are Austria, Denmark, Federal Republic of Germany, Finland, France, Greece, Israel, Italy, The Netherlands, Norway, Spain, Sweden, Switzerland, and the United Kingdom.

[b] Charges given in West German marks.

sequence data in computer-readable form as soon as the issue containing them appears. The file server can be used by anyone with access to the BITNET/EARN network or to any other network which has a gateway into BITNET/EARN (e.g., JANET in the United Kingdom or ARPANET in the United States). It is provided free of charge, though users may have to meet some or all of the communication costs, depending on the accounting system of their local computer service.

The following brief instructions should enable interested users to get started. Use of the facility is very simple and involves sending file server commands as standard electronic mail to the address NETSERV@EMBL.BITNET. Each line of the mail message should consist of a single file server command and nothing else. The most important file server command, to get users started, is HELP. If the file server receives this command, it will return a help file to the sender, explaining in some detail how to use the facility. In order to send electronic mail to a BITNET/EARN address, users must find out which command they have to use on their own machine and how they should format the address NETSERV@EMBL.BITNET. Users who do not already know how to do this should contact their local computer service or, if all else fails, contact the Data Library and we will do our best to help. Below are some examples which illustrate how to send commands to the file server using a VAX/VMS system that is a BITNET/EARN node running JNET software.

To send a HELP command to the file server, use the operating system command MAIL as follows:

    $ MAIL ⟨filename⟩ "JNET%" "NETSERV@EMBL"""

where ⟨filename⟩ is the name of a file containing file server commands. To request help information the file should contain the following command:

    HELP

To request a copy of the data submission form, the file should contain the following GET command:

    GET DATALIB:DATASUB.TXT

To request a copy of a nucleotide sequence entry with the accession number X00596, the file should contain the following GET command:

    NUC:X00596

EMBNet

Another way in which EMBL is attempting to increase the availability and usefulness of the various databases is by establishing a European molecular biology network (EMBNet) consisting of EMBL connected via DECNet to a series of national centers in western European countries. [DECNet was chosen because DEC (Digital Equipment Corporation) has made a firm commitment to migrate to the ISO-OSI standard for networking as soon as the standard is finalized.] The long-term plan is for EMBL to provide the national centers with copies of the latest sequence databases and other relevant data collections and with retrieval software. The national centers will then make these tools, along with analytical software, available to researchers within their countries and will offer training and support in their use.

In 1988 a trial phase of the EMBNet project was initiated with the following four centers: SERC Laboratory (Daresbury, UK), CITI2 (Paris, France), COAS/CAMM Center (Nijmegen, The Netherlands), and Hoffman-La Roche (Basel, Switzerland). Progress so far includes establishment of network connections with the centers and implementation of software which updates their copy of the database on a daily basis. It is planned to add several additional nodes to the system in 1989, after which a gradual expansion will lead to the inclusion of at least one node in each of the EMBL member countries.

Appendix: How to Contact EMBL Data Library

Computer network: datasubs@embl.earn (for data submissions); datalib@embl.earn (for questions requiring a personal response)
Postal address: Data Submissions, EMBL Data Library, Postfach 10.2209, 6900 Heidelberg, Federal Republic of Germany
Telephone: +49-6221-387-258
Telefax: +49-6221-387-306
Telex: 461613 (embl d)

[3] Protein Sequence Database

By WINONA C. BARKER, DAVID G. GEORGE, and LOIS T. HUNT

Introduction

The Protein Sequence Database has been maintained by researchers at the National Biomedical Research Foundation (NBRF) since the early 1960s. The database was originally compiled by the late Margaret O. Dayhoff as a collection of sequences for the study of evolutionary relationships between proteins, and it continues to be maintained by a scientific staff as a research database. The database has become truly international with the recent establishment of PIR-International, an association of protein sequence data collection centers including NBRF, the Martinsried Institute for Protein Sequences (MIPS), and the International Protein Information Database in Japan (JIPID). All three centers are working cooperatively to produce a single protein sequence database.

Currently the NBRF effort is supported as part of the Protein Identification Resource (PIR) project funded by the NIH Division of Research Resources, the National Library of Medicine, and the National Institute of General Medical Sciences. The main purpose of this resource is to aid the research community in the identification and interpretation of protein sequence information. The PIR consists of an integrated computer system composed of a number of protein and nucleic acid sequence databases and software designed for the identification and analysis of protein sequences. The technology of the PIR is available to the scientific community in several forms: on-line access to the PIR computer system, magnetic tapes of databases and software, and a confidential sequence identification service.[1] While there are many sources of software for sequence analysis and other computer centers serving molecular biology, the unique contributions of the PIR are maintenance of a comprehensive, carefully edited, and widely disseminated database of protein sequences and associated information and development of software specifically tailored to the requirements of this database. Moreover, the resource is operated by a group of scientists with many years of experience in the application and interpretation of sequence comparison methods. Thus, the PIR is a center providing not only the necessary facilities for protein identification but also practical guidance in the interpretation of the results of these methods.

What Kind of Database?

There are three types of sequence databases that can be useful to biological researchers (Lennart Philipson at an EMBL/NIH Workshop, "Future Databases for Molecular Biology," February 25–27, 1987). The first is a "raw sequence" database, containing all available sequence data but compiled with little effort at standardization or other quality control measures. An example is the PSeqIP database,[2] which is assembled by merging a number of publicly and privately available protein sequence collections and automatically eliminating all but one of any entries that

---

[1] K. E. Sidman, D. G. George, W. C. Barker, and L. T. Hunt, *Nucleic Acids Res.* **16,** 1869 (1988).

[2] J.-M. Claverie and L. Bricault, *Proteins: Struct. Funct. Genet.* **1,** 60 (1986).

contain duplicate or nearly duplicate sequences. Another example would be a database consisting of translations in all reading frames of all the nucleotide sequences in the GenBank Genetic Sequence Database. Because current sequence database searching methods are very sensitive, considerable error can be tolerated in the raw sequence data, and related sequences will still be recognized. When a researcher is primarily interested in establishing the identity and function of a newly determined sequence, the availability of as many data as possible may be more critical than the accuracy of the data. After a possible relationship is found, the accuracy of the sequences identified can be checked and conflicts resolved.

The second type is a scientific abstract database, in which strict quality control is exercised so that the information in an entry will correspond exactly to that reported in one published article or data submission. Such a database may be highly redundant because data from different sources often overlap. However, the data entries can be indexed, the nomenclature of molecules and organisms can be standardized, and the data entries can be ordered to facilitate the location of similar data. The "new data" file of the NBRF Protein Sequence Database and many of the entries in Gen-Bank are examples of this type of data. An abstract database provides the research community with a comprehensive collection of sequences and associated annotations in a form that can be examined and manipulated by computer. The abstract database can also serve as an archive for the permanent storage of information that may not be published elsewhere; the existence of such an archive is essential to ensure that critical information is not lost. However, because the entire sequence of a molecule is not always represented in a single report, the entries in such a database may not correspond to biologically functional units; the complete sequence may not be explicitly represented even when it is known. The inherent redundancy in this type of database makes database searching time longer and meaningful statistical analyses of the data impossible; moreover, significant relationships are more likely to be missed. Many database searching methods select a predetermined number of top-scoring segments. If the database redundancy is high, the scores saved may correspond to a small number of overrepresented proteins (or nucleic acids). A partial solution to this problem is to increase the number of top-scoring segments saved, but the investigator will be required to examine large amounts of output and, as a result, may still miss the important relationships.

The third type of database is the reference data compendium that normally includes only one "value" for each item of data or perhaps the range of values covered by sources considered reliable. In such a collection, the information is examined, analyzed, and reviewed by a scientific staff, and emphasis is placed on reporting scientific data rather than on accurately reproducing the results of individual research reports.[3] Traditionally, the Protein Sequence Database of the PIR has been maintained as a reference data compendium rather than as an abstract database. The policy has been to minimize the inclusion of redundant information by combining into a single entry information concerning multiple determinations of the sequence of the same molecule. This policy has made the PIR Protein Sequence Database the least redundant of the comprehensive databases.[2]

Uses of Database

Although sequence searching is perhaps the most well known use of the Protein Sequence Database, other types of major uses involve the text information or the alignment of related sequences. The text portion of the Protein Sequence Database provides information useful for evaluating the results of database searches. In addition to literature citations and information concerned with the experimental determination of the sequences, the text may include alternative nomenclature, domain structure, functional sites in the sequence, genetic information, and general information about the source, function, tertiary and quaternary structures, and physicochemical properties of the molecule. This information may be used to locate particular types of proteins in the database or to compare other features when two proteins are found to have similar sequences.

The intercomparison of related sequences is extremely useful and powerful. From an alignment of related sequences one can infer information about evolution, secondary and tertiary structure, function, active site or functional site localization, and chemical mechanisms. The use of alignments of related proteins has improved secondary[4,5] and tertiary structure predictions.[6] Despite much work devoted to the development of predictive methods, the most effective method of inferring biological function based on primary sequence information is by comparisons with protein homologs and/or analogs. If the crystal structure is known for any protein from a group of closely related proteins, it is possible to make a best fit of the related sequences to the known three-dimensional structure.[6] Very good success is obtained for proteins up to 50% different,[6,7] which corresponds to a family of proteins in the database. Different sites along a protein chain

[3] D. R. Lide, Jr., *Science* **212**, 1343 (1981).

[4] M. J. Zvelebil, G. J. Barton, W. R. Taylor, and M. J. E. Sternberg, *J. Mol. Biol.* **195**, 957 (1987).

[5] W. R. Taylor, *in* "Nucleic Acid and Protein Sequence Analysis, A Practical Approach" (M. J. Bishop and C. J. Rawlings, eds.), p. 285. IRL Press, Oxford and Washington, 1987.

[6] C. Chothia and A. M. Lesk, *EMBO J.* **5**, 823 (1986).

[7] W. R. Taylor, *Protein Eng.* **2**, 77 (1988).

show considerable variability in their ability to accommodate replacement amino acids without significantly affecting the overall function of the protein. Highly conserved regions (those resistant to change) can generally be identified as important for biological activity and/or structural integrity; loop regions connecting regular secondary structures tend to be less highly conserved and often correspond to regions of insertion/deletion events in an alignment.[4]

These observations have provided the basis for defining characteristic sequence patterns that may prove to be of biological significance.[4] Alignments are also being employed in attempts to develop more sensitive sequence comparison methods.[7] In one approach, a profile characterizing the alignment is deduced and the test sequence is compared with the profile in order to examine the relationship of the test protein with the entire group of sequences.[8]

It follows that a sequence database that is organized according to biological principles and from which related information can readily be extracted will be an invaluable information resource, having the essential characteristics recently called for in second generation databases.[9] Scientists using a raw or abstract database may need to spend considerable time verifying, comparing, and combining related data from a number of reports in order to extract the information necessary to formulate or support a hypothesis. Moreover, the existence of a highly verified, nonredundant data collection will stimulate further analysis of the data and support the types of statistical analyses that require a complete or representative collection. Such information can be compiled much more efficiently and cost-effectively in a centralized environment.

It has always been the goal of the PIR to compile a well-verified database of high quality that is organized in accordance with the major research uses of the data. The scientific review and analysis necessary to organize such a database has been labor intensive and expensive. Recently the rate of accumulation of protein sequences has exceeded the capacity of our research group. Increasing the database staff in proportion to the increase in data was not feasible because of funding limitations and other practical considerations. Consequently, the PIR is currently making major changes in data structures and modes of operation, as are the other comprehensive sequence database centers. As these changes are being implemented gradually, it is necessary to describe both the Protein Sequence Database as it has been and the blueprint for its future structure.

[8] M. Gribskov, M. Homyak, J. Edenfield, and D. Eisenberg, *CABIOS* **4,** 61 (1988).
[9] C. O. Pabo, *Nature (London)* **327,** 467 (1987).

Early History

The NBRF Protein Sequence Database was started in the early 1960s by Margaret Dayhoff, a pioneer in the use of computers to store, analyze, and compare sequence data. The fundamental philosophy on which the database was designed and has been operated includes many of the concepts that are being proposed for future sequence databases. From the beginning the data have been organized according to functional and evolutionary relationships. The discovery of these relationships and the development of computer tools to detect and display them have been major research thrusts. The researchers at NBRF have used the database for many investigations in the field of protein evolution, but probably the most well known of their early publications is the series of volumes of the *Atlas of Protein Sequence and Structure,* edited by M. O. Dayhoff and published by the NBRF. In addition to all of the known protein (and nucleic acid) sequences, these volumes contained alignments of sequences, tables giving the amount of difference between sequences, and evolutionary trees for many groups of related proteins. Further, they included chapters on aspects of molecular evolution and methods of sequence analysis as well as many useful tables of related data such as hemoglobin mutants, enzyme active sites, rates of mutation acceptance in proteins, average amino acid percentage compositions for families of proteins, and mutation probabilities for amino acids.

By 1972, the staff found it necessary to rely on computer methods to establish the identity of newly sequenced peptides and to search for previously unsuspected protein relationships. To allow others access to this useful body of information, the NBRF made the sequences available on magnetic tape. In the late 1970s, the Protein Sequence Database entered a period of accelerated growth caused by the advent of methods to sequence nucleic acids quickly and accurately. With the continued improvement of these methods and the worldwide interest in completely sequencing the human and other genomes, this rate of growth can be expected to continue to increase.

In the early 1980s, the database was restructured into a versatile, fully integrated software–data system.[10] The Protein and Nucleic Acid Sequence Query programs (PSQ and NAQ) were initially designed to aid NBRF scientists in the maintenance of the databases; the programs provided rapid access to the database information and, with the addition of several sequence manipulation routines, soon became extremely powerful

[10] B. C. Orcutt, D. G. George, and M. O. Dayhoff, *Annu. Rev. Biophys. Bioeng.* **12,** 419 (1983).

research tools. Because of the general utility and power of these programs, they were made available to the scientific community. They now constitute the main components of the PIR software system. Dissemination of the database is now primarily via magnetic media, and the information is accessed mainly through computer systems involving specialized retrieval and analysis software. Although the computerized database contains much information in addition to the sequence data and citations, cessation of publication of the database in book form essentially removed the alignments and derived evolutionary trees from the publicly distributed database.

Superfamily Concept

The origin of the database as a set for the study of protein evolution has had a profound effect on the way the data are organized and presented. Integral to the organization is the concept of a protein superfamily.[10,11] A protein superfamily comprises a group of proteins whose amino acid sequences can be shown to be evolutionarily related. In principle, an evolutionary tree can be constructed for each protein superfamily; inclusion of a protein in a specific superfamily implies that the protein is homologous (evolved from a common ancestral sequence) with the other proteins in that superfamily. In general, sequence comparison methods alone cannot prove that distantly related proteins are homologous but can only indicate that the similarity between their sequences is unusual. Functional relationships must be demonstrated experimentally, and homology must be inferred in conjunction with other biological reasoning. Thus, although similarity scores obtained from sequence comparisons[12] are used extensively in defining the superfamily organization, there is no single criterion on which the classification is based.

After every database update, each entry is assigned a set of five numbers, the first of which represents the superfamily. The family, subfamily, and entry numbers subdivide a superfamily into groups of proteins that are less than 50%, 20%, and 5% different, respectively. The subentry number allows each entry in the database to be assigned a unique set of numbers. It is not possible to adhere strictly to these criteria: two sequences may be less than 50% different from a third but more than 50% different from each other; moreover, when only partial sequences are available for some members of a superfamily, their classifications represent educated guesses. The

---

[11] M. O. Dayhoff, P. J. McLaughlin, W. C. Barker, and L. T. Hunt, *Naturwissenschaften* **62,** 154 (1975).

[12] M. O. Dayhoff, W. C. Barker, and L. T. Hunt, this series, Vol. 91, p. 524.

superfamily organization is helpful in the interpretation of database searches inasmuch as, if a new sequence is truly homologous with one already in the database, it should show unusual sequence similarity to the corresponding regions of other members of the superfamily as well. Furthermore, it is possible to choose a representative subset of the database for searching by removing all sequences that are, for example, less than 20% different from another sequence.

This superfamily classification was designed long before it became apparent that many large proteins are composed of domains and that these distinct regions may have different evolutionary origins and histories. Not only have genes duplicated to give rise to new proteins, but they have fused, gained and lost exons, changed reading frames, and incorporated DNA from other parts of the genome, from other genomes within the organism, or even from other organisms. Moreover, the policy of the database is to store explicitly the primary transcripts when known; these may be posttranslationally processed to yield different protein sequences. They often contain several physiologically active proteins or peptides. The viral polyproteins represent a special case of this problem; often a large portion of the genome is translated into a single polypeptide chain, which is subsequently cleaved to yield a number of functionally distinct proteins. Up to now these situations have been handled on an ad hoc basis: some sequences have been divided to form two entries (immunoglobulin V and C regions; viral *gag* and *pol* polyproteins); others have been grouped on the basis of one type of domain that they share (trypsin-related serine proteases); and some sequences are in different superfamilies although they contain homologous domains (epidermal growth factor precursor and the low density lipoprotein receptor).

Fully Annotated Database Entry

Figure 1 shows a data entry as it appears on the ASCII card-image tape distributed by PIR. The recently introduced ASCII format adheres closely to the CODATA standardized format for data exchange, which was developed collaboratively[13] and is recommended for the communication of protein sequence data. It has several advantages over the previous card-image format: (1) As all data items are labeled with specific identifiers or subidentifiers, it is much easier to write computer programs to access and manipulate the information. (2) It is more "human" readable. (3) It includes information previously available only to VAX/VMS users. (4) It is more readily convertible to other formats for use by programs and retrieval

[13] D. G. George, H. W. Mewes, and H. Kihara, *Protein Sequences Data Anal.* **1,** 27 (1987).

```
ENTRY           OKBOG          #Type Protein
TITLE           cGMP-dependent protein kinase - Bovine #EC-number
                2.7.1.37

DATE            17-May-1985 #Sequence 17-May-1985 #Text 28-May-1986
PLACEMENT       160.0      1.0      1.0      1.0      1.0
SOURCE          Bos primigenius taurus #Common-name cattle
ACCESSION       A00619

REFERENCE       (Sequences of residues 1-17, 89-374, and 407-670)
    #Authors    Takio K., Wade R.D., Smith S.B., Krebs E.G., Walsh
                K.A., Titani K.
    #Journal    Biochemistry (1984) 23:4207-4218

REFERENCE       (Sequence of residues 13-104)
    #Authors    Takio K., Smith S.B., Walsh K.A., Krebs E.G., Titani
                K.
    #Journal    J. Biol. Chem. (1983) 258:5531-5536

REFERENCE       (Sequence of residues 373-409)
    #Authors    Hashimoto E., Takio K., Krebs E.G.
    #Journal    J. Biol. Chem. (1982) 257:727-733

COMMENT         This protein, isolated from lung, is a dimer of
                identical chains.
SUPERFAMILY     102-340 #Name cAMP-dependent protein kinase
                regulatory chain\
                475-599 #Name kinase-related transforming protein
KEYWORDS        acetylation\ phosphoprotein\ cGMP\ serine-specific
                protein kinase

FEATURE
    1                           #Modified-site acetylated amino end\
    42                          #Disulfide-bonds interchain\
    58                          #Binding-site phosphate\
    1-101                       #Domain dimerization <DIM>\
    102-219                     #Domain cGMP-binding 1 <GB1>\
    220-340                     #Domain cGMP-binding 2 <GB2>\
    341-474                     #Domain ATP-binding <APB>\
    475-599                     #Domain catalytic <CAT>
COMMENT         These boundaries are approximate.

SUMMARY         #Molecular-weight 76287  #Length 670  #Checksum 5503
SEQUENCE
                    5       10      15      20      25      30
      1 S E L E E D F A K I L M L K E E R I K E L E K R L S E K E E
     31 E I Q E L K R K L H K C Q S V L P V P S T H I G P R T T R A
     61 Q G I S A E P Q T Y R S F H D L R Q A F R K F T K S E R S K
     91 D L I K E A I L D N D F M K N L E L S Q I Ĵ E I V D C M Y P
    121 V E Y G K D S C I I K E G D V G S L V Y V M E D G K V E V T
      .
      .
      .
```

FIG. 1. Entry from the Protein Sequence Database (CODATA format).

systems based on these formats; for example, the entry shown in Fig. 1 is currently displayed by the PSQ program as shown in Fig. 2. (5) It is extensible by the addition of new identifiers and subidentifiers. In the CODATA standard format the entries explicitly include some information that is stored in auxiliary files in the PIR computer system.

```
OKBOG
cGMP-dependent protein kinase (EC 2.7.1.37) - Bovine

Species: Bos primigenius taurus (cattle)

Accession: A00619

Takio, K., Wade, R.D., Smith, S.B., Krebs, E.G., Walsh, K.A.,
    and Titani, K., Biochemistry 23, 4207-4218, 1984
    (Sequences of residues 1-17, 89-374, and 407-670)

Takio, K., Smith, S.B., Walsh, K.A., Krebs, E.G., and Titani,
    K., J. Biol. Chem. 258, 5531-5536, 1983 (Sequence of
    residues 13-104)

Hashimoto, E., Takio, K., and Krebs, E.G., J. Biol. Chem.
    257, 727-733, 1982 (Sequence of residues 373-409)

This protein, isolated from lung, is a dimer of identical
    chains.

Superfamily: 102-340/cAMP-dependent protein kinase regulatory
    chain; 475-599/kinase-related transforming protein

Keywords: acetylation; phosphoprotein; cGMP; serine-specific
    protein kinase

Residues            Feature
1                   Modified site: acetylated amino end
42                  Disulfide bonds: interchain
58                  Binding site: phosphate
1-101               Domain: dimerization <DIM>
102-219             Domain: cGMP-binding 1 <GB1>
220-340             Domain: cGMP-binding 2 <GB2>
341-474             Domain: ATP-binding <APB>
475-599             Domain: catalytic <CAT>
    These boundaries are approximate.

Mol. wt. unmod. chain = 76,287      Number of residues = 670

            5        10        15        20        25        30
    1 S E L E E D F A K I L M L K E E R I K E L E K R L S E K E E
   31 E I Q E L K R K L H K C Q S V L P V P S T H I G P R T T R A
   61 Q G I S A E P Q T Y R S F H D L R Q A F R K F T K S E R S K
   91 D L I K E A I L D N D F M K N L E L S Q I Q E I V D C M Y P
  121 V E Y G K D S C I I K E G D V G S L V Y V M E D G K V E V T
    .
    .
    .
```

Fig. 2. Alternative display of the entry of Fig. 1, using the PSQ program.

A sequence entry consists of an entry identification code, an entry title containing the name of the protein and its biological source, a block of text consisting of citations and summarizing the experimental details of the sequence determinations, a block of descriptive information concerning the properties of the molecule, and the protein sequence. The amino acid composition and molecular weight are calculated from the sequence. The identification code is a unique code of four to six alphanumeric characters used by the computer to identify each entry. The identification code for the entry in Figs. 1 and 2 is OKBOG.

To facilitate computer searching, key identifiers (descriptors) are used to label explicitly specific types of information. Subidentifiers (preceded by the # symbol) are allowed in the CODATA format, and some of the identifiers are not identical with the descriptors currently displayed in the PIR system (see Fig. 2). Many entries contain protein names designated by an ALTERNATE-NAME (Alternate names) identifier and/or an IN-CLUDES (Contains) identifier, which list other commonly used names for the same protein or other activities or functions that are included in the protein. For example, *Neurospora* anthranilate synthase component II contains three domains, each with a different enzymatic activity, and mammalian corticotropin–lipotropin precursor contains the peptides melanotropin, β-endorphin, and Met-enkephalin. Any computer system that maintains a file of entry titles for rapid searching should include these additional names. The ACCESSION identifier lists permanent accession numbers assigned to the data in the entry.

The SOURCE (Species) identifier gives the scientific and common names of the organism. Several subidentifiers are used for GENETIC information such as map position, gene name, and translational start codon (if other than ATG). The #Introns subidentifier designates the positions where introns occur in the coding region for the protein sequence (#Introns 25/2 indicates that the intron occurs after the second nucleotide of the codon for amino acid number 25). The superfamily to which the protein belongs is shown following the SUPERFAMILY identifier and #Name subidentifier; for some proteins, additional superfamily names will be indicated corresponding to discrete domains of the protein. A KEY-WORDS identifier is used to provide keyword indexing terms that refer to various attributes of the protein, such as modifications, time and place of expression, and function. These keywords specifically refer to properties of the molecule and not of its biological source. For example, the keyword photosynthesis is used only for proteins that function in the photosynthetic pathway. The PLACEMENT data item contains the currently assigned superfamily numbers for each entry. These numbers are used only to denote the hierarchical relationships among the sequences; their exact values change at each release.

Regions or sites of the protein sequence that are of biological interest are listed in a feature table, which includes the name of each feature and a specification of the residues it comprises. Special subidentifiers are used to denote specific types of features. The #Protein and #Peptide subidentifiers indicate the residues that correspond to the mature protein or peptides derived from the protein. The #Domain subidentifier indicates the residues that make up discrete, evolutionarily distinct, functional domains within a protein. Examples of other subidentifiers are #Region, #Duplication, #Active-site, #Binding-site, #Modified-site, and #Disulfide-bonds. The characters enclosed by angled brackets (⟨ and ⟩) can be used as extensions to the entry identification code to allow direct specification of regions of the sequence.

Minimal Entry

Beginning in 1985, the Protein Sequence Database included a separate "new data" file, a small file of sequence entries under preparation for the main database. This file of minimally reviewed entries remained small for 2 years, but in 1987 it more than quadrupled in size to over 1800 sequences, about 27% of the total. In 1988, the PIR staff severely cut back on review, annotation, and organizational activities in order to concentrate on reducing the backlog of published sequences not yet in the database. Although over 4000 sequences were added to the database in 1988, 80% of these are minimally annotated and not organized by homology.

It is now obvious that the new data file is not going to disappear but has in fact become a file of minimally annotated entries. These entries have been compared with other sequences in the database and with the translations of the corresponding nucleotide sequences (when these are available) and have been examined by computer to detect certain syntax errors. When the name in the database is different from that used by the author, the alternate names are added to the entry. The scientific name of the organism is also added to allow for sorting entries according to taxonomic classification; these names are checked against standardized lists. The title of the article from which the sequence was taken is also provided as it contains additional information. For posttranslationally processed sequences, a one-line feature table is added to show the extent of the mature form. After each update, the entries in this file are sorted taxonomically and, within each species, alphabetically according to protein name. Additional information in these entries, when present, has not been subjected to critical staff review and should not be relied on.

The staff time necessary to prepare a fully reviewed and annotated entry and place it in the database according to degree of similarity to other sequences is about 4 times that needed for preparation of a minimal entry; all of the additional effort requires highly trained scientists. The temporary switch in emphasis from fully annotated and organized to minimal entries will clearly result in the database containing more sequences with less information about each sequence and fewer correlations between sequences. Users will be additionally inconvenienced by the increased amount of redundancy as each determination of the same sequence will be a separate entry. The international collaboration described below will increase the number of highly trained scientists available to annotate and organize the database. Moreover, new database structures will reduce the effort required to produce an annotated and organized database. These factors should allow a return to the traditional standard in which database entries are carefully reviewed, nonredundant, and organized based on biological principles.

Protein Sequence Database System

The Protein Sequence Database, as supplied for VAX/VMS systems, fits the classic definition of a database as an integrated software–data system. The software is designed to allow rapid and efficient access to the data and is as much a part of the database as are the data. In such a system, knowledge concerning the data is encoded in the software, and there is no clear-cut distinction between software and data.

The Protein Sequence Query (PSQ) retrieval[10] program has relational aspects and limited hierarchical tree-searching capabilities. Much of the information in the text and feature tables in the Protein Sequence Database is compiled as a set of index files that can be accessed using a relational scheme. The index-searching routines of PSQ allow the results of operations on the indexes to be compared and combined in a variety of ways. Data derived from the primary sequences, such as sequence length, molecular weight, and composition, are also compiled and are stored in relational form. The superfamily numbers and the taxonomic classification scheme are hierarchical structures, and limited hierarchical tree-searching capabilities have been built into PSQ to utilize this information. The SCAN command of PSQ allows rapid retrieval of sequence entries based on exact matching of peptides 3 to 30 residues in length. This command achieves its speed through access to a special Tripeptide Catalog File. The PSQ program, all of the files necessary to support its retrieval, sequence manipulation, and display capabilities, as well as programs to create databases accessible by PSQ are supplied with the Protein Sequence Database for VAX/VMS systems.

Some of the information that is stored in auxiliary files in the PSQ system is incorporated into the entries on the CODATA version of the database, e.g., length, molecular weight, and dates that entries were added or revised. Index files are supplied for accession numbers, authors, and species for both the main and new data files, and for features, superfamily names, and keywords in the annotated file. The hierarchical taxonomic scheme and a list of the titles of all sequences in each superfamily are also included.

Future Enhancements to PIR Protein Sequence Database

Since the 1960s, in conjunction with evolutionary and other studies, researchers at NBRF have maintained alignments of homologous groups of proteins. Because of the importance of alignments to the study of protein sequences, the organization of the Protein Sequence Database is being redesigned so that its fundamental structure will include alignments of related sequences. The current superfamily organization will be replaced by one based on alignments, i.e., each alignment will define a set of related sequences or subsequences. The proposed structure of the alignment files allows groups of sequences within alignments to be hierarchically clustered in a manner reminiscent of the superfamily, family, and subfamily structuring of the current database. The hierarchical clustering may include as many levels as desired, and, unlike the current superfamily clustering, there will be no rigid rules for delineating subgroups. The clustering can be tailored specifically to the properties of each individual group of sequences.

The shift of the organizational structure to alignments also provides a mechanism for generalizing the superfamily concept to handle multidomain sequences and to represent relationships other than evolutionary. The current criteria assume that sequences in the same superfamily are related along their entire lengths. This is problematic for sequences composed of several domains, each of which may have evolved separately. In the context of the current sequential organization, this problem must be resolved by the unsatisfactory strategy of choosing one type of domain on which to base the superfamily assignment. Under the proposed scheme, although the entire multidomain protein will be shown in a regular entry, each domain can also be represented separately on an alignment of related domains.

This new organization provides a flexible environment for depicting the interrelationships among sequences and subsequences. The basis for an alignment (homology, function, structure, etc.) will be explicitly indicated, thus allowing alignments to accommodate relationships other than evolutionary. Sequences may be represented on as many alignments as appropriate, and alternative alignments of the same group of sequences are allowable. For example, an alignment based on superposition of three-dimensional structures as revealed by X-ray crystallography, one based on replacement frequencies of amino acids, and one based on minimal mutational distance may differ in some details. The set of alignments will not be restricted to those developed at the PIR; contributions from the scientific community will be encouraged.

The shift of emphasis toward sequence alignments as opposed to individual sequences also provides a mechanism for alleviating the burden of keeping pace with an extremely rapid influx of new sequence data. The most scientifically demanding tasks in the processing of data for entry into the Protein Sequence Database are annotating the entries and elucidating the relationships of the new sequence to other entries in the database. Under the present scheme, to annotate fully a group of proteins requires the repetition of virtually the same information in every entry in the group. Most of this information can instead be included as annotations to the alignment.

Individual sequence entries will still be the primary unit of sequence information; however, the ancillary information in the sequence entries will be kept to a minimum. As before, the set of sequences will be sequentially organized by protein function, but the detailed superfamily organization will no longer be imposed; therefore, a full investigation of sequence interrelationships will not be required to assign a position in the database to a new sequence. By minimizing the amount of scientific judgment necessary for maintaining the sequence portion of the database, new sequence information can be incorporated at a much faster rate.

Described above are only those aspects of the new database structures that will be most obvious to the users. These will be augmented by other new features such as linkages between the various database components (e.g., sequences and alignments), comprehensive term indexes, and semantic networks including a detailed taxonomic hierarchy. The database retrieval systems, the software for data entry and verification, and the database-updating programs will be enhanced to accommodate and make use of the new features.

Database as International Collaboration

With the development of protein microanalytical techniques and automated DNA sequencing, the rate of determination of sequences of proteins and protein coding regions will continue to accelerate. Whatever the level of funding for the human genome sequencing project, the very existence of technology capable of attempting a project on such a scale virtually assures an acceleration in sequence data several orders of magnitude greater than the rate of the past few years.

These factors made it critical to establish a framework for international cooperation in the coordination of existing and emerging protein sequence data banks.[14] The PIR has joined with recently established centers in Japan (the International Protein Information Database in Japan, JIPID, headed by Dr. Akira Tsugita at the Science University of Tokyo) and in Germany (the Martinsried Institute for Protein Sequences, MIPS, headed by Dr. Werner Mewes at the Max Planck Institute) to establish a single, truly international Protein Sequence Database. Data collection is being shared; most North American information sources are the responsibility of the PIR, most European sources are the responsibility of MIPS, and those in Asia and Oceania, as well as specialty publications in plant science, biotechnology and peptides, and selected other journals, are the responsibility of JIPID. JIPID has also established relationships with groups in Australia, India, China, Korea, and Taiwan, which serve as local collection sites. For example, scientists in Beijing and Shanghai are preparing entries for protein sequences reported in Chinese journals. These are sent in computer-readable form to JIPID and forwarded by electronic mail to the PIR, where they are added to the database.

Currently the processing of a submitted entry involves nearly as much staff effort as one that was entered in-house: vocabulary and format must be made consistent with database conventions, the new data may be combined with overlapping data, and the sequence is tested for homology to other sequences and placed with its closest relatives in the database. However, submitted entries can be added to the new data file with much less effort. Scientists at MIPS and JIPID will also concentrate on the organization and annotation of selected groups of proteins in cooperation with researchers with expertise in fields related to the study of these molecules. MIPS and JIPID have taken over distribution of the Protein Sequence Database (as well as the PIR newsletter) in Europe and Asia, respectively. PIR distributes the VecBase database,[15] developed at MIPS. Both JIPID and MIPS are developing other related databases that may also be distributed by PIR. JIPID has sponsored an international project to establish a protein NMR database;[16] its format is based on the CODATA standardized format for data exchange.

The three groups are currently examining ways to correlate the Protein Sequence Database with other related databases such as the Protein Data Bank for X-ray crystallographic data[17] and the nucleotide sequence data-

[14] B. Keil, *Protein Sequences Data Anal.* **1**, 123 (1987).
[15] F. Pfeiffer and W. A. Gilbert, *Protein Sequences Data Anal.* **1**, 269 (1988).
[16] E. L. Ulrich, J. L. Markley, and Y. Kyogoku, *Protein Sequences Data Anal.* **2**, 23 (1989).
[17] F. C. Bernstein, T. F. Koetzle, G. J. B. Williams, E. F. Meyer, Jr., M. D. Brice, J. R. Rodgers, O. Kennard, T. Shimanouchi, and M. Tasumi, *J. Mol. Biol.* **112**, 535 (1977).

bases. Special attention is being focused on intercommunication between, and compatible organization with, these related databases. A forum for exchange with other groups has been provided by the CODATA Task Group on the Coordination of Protein Sequence Data Banks,[14,18] recently reorganized as the Task Group on Biological Macromolecules and expanded to include representatives from the major nucleic acid databases. Both the establishment of the international collaboration for the Protein Sequence Database and the standardized format for the exchange of protein sequence data (Fig. 1) were direct results of participation in the CODATA Task Group. The annual meeting of this group provides an opportunity for formal and informal meetings between managers of the protein and nucleic acid database centers. These interactions will continue to foster exchange of information to reduce redundant activity, promote standardization in nomenclature, and encourage more compatibility in data representation.

Each of the protein database centers is collaborating with one of the centers for nucleotide sequence data collection: PIR with GenBank, MIPS with EMBL (European Molecular Biology Laboratory, Heidelberg), and JIPID with DDBJ (DNA Data Bank of Japan, Mishima). The protein and nucleic acid database centers have agreed to use a common form for data submission and to coordinate their relationships with journals. GenBank forwards electronically submitted data to PIR. Software for the extraction of protein coding regions from the GenBank database was developed at Los Alamos National Laboratory;[19] after every update, GenBank supplies the PIR with the coding regions and their translations. These are used at PIR to implement the "Lospro" data file, which is consulted by PIR staff for cross-checking preliminary PIR entries with the data in GenBank. This procedure reveals errors in both databases as well as discrepancies in the original reports. EMBL forwards its newly prepared entries to MIPS, which can then immediately begin preparing corresponding protein database entries. These data are checked, and the reported coding regions are verified; discrepancies are reported to EMBL and corrected in the nucleic acid collection also.

Steps toward Common Nomenclature

A small step in assembling the Matrix of Biological Knowledge[20] will be to correlate the knowledge contained in the databases of protein and

[18] B. Keil, this volume [4].
[19] J. W. Fickett, *Trends Biochem. Sci.* **11**, 190 (1986).
[20] H. J. Morowitz and T. Smith, Report of the Matrix of Biological Knowledge, Santa Fe Institute, July 13–August 14, 1987.

nucleotide sequences. To this end, researchers at PIR have developed a sequence information retrieval system that operates simultaneously on several protein and nucleotide sequence databases. With this system, one can rapidly search all of the resident databases for text terms such as protein names, keywords, author names, and organism names. Thus, one can locate corresponding information from several sources as long as its representation is similar in the various databases. Unfortunately, each of these databases has developed its own system of nomenclature and syntax, distinct from each other as well as from other biomedical information resources such as MEDLINE. The PIR is actively participating in developmental efforts toward the establishment of a common nomenclature among the biomedical information resources on an international level.

The National Library of Medicine also promotes greater coordination between the macromolecular sequence databases. They have compared the journal abbreviations used by PIR, GenBank, EMBL, and MEDLINE with those formed according to the rules of the American National Standards Institute, using the word lists of the International Standards Organization. The cooperating databases will change their nonstandard abbreviations. In addition, PIR staff are comparing the keyword and taxonomic terms of the Protein Sequence Database with MEDLINE MeSH terms. This study will complement similar comparisons between GenBank keywords and MeSH terms[21] and provide information for the eventual development of a semantic network structure for molecular biology data.

Expanded Role of Scientific Community

In order to keep pace with the rapid growth in sequence information, much of the data collection and review that in the past was done by database staff must become the responsibility of the scientific community. Scientists will be asked to prepare draft entries for their own data and submit these in computer-readable form to the database. This should speed data entry, improve the accuracy of sequences and interpretive comments, and ensure a more complete and up-to-date database. Scientists must also be responsible for communicating revisions of their data. Mechanisms need to be established for the systematic review of related information to ensure that nomenclature and annotations for particular groups of sequences are consistent and up to date.

Although efficient software for data entry and verification and new methods for data representation, storage, analysis, and retrieval will dramatically increase the effectiveness of the scientists who review and organize the database, the scientific effort will remain an essential aspect of the project. It is doubtful that even the combined staffs of PIR, JIPID, and

MIPS will be sufficient to integrate all of the data, although we will do so in selected areas. The assistance of many other researchers will be required if the Protein Sequence Database is to remain an example of a second generation database,[9] containing information that can readily be incorporated into a future Matrix of Biological Knowledge.[20]

How to Obtain PIR Databases, Software, and Newsletters

For information on currently available database releases, or other services, or for a copy of the PIR newsletter, contact the PIR Technical Services Coordinator, National Biomedical Research Foundation, 3900 Reservoir Road NW, Washington, D.C. 20007; telephone +1 202 687 2121; FAX +1 202 687 1662; electronic mail PIRMAIL@GUNBRF.BITNET. In Europe, contact MIPS: Martinsrieder Institute fuer Proteinsequenzen, Max-Planck-Institut fuer Biochemie, D-8033 Martinsried bei Muenchen, FRG; telephone +49 89 8578 2656; FAX +49 89 8578 3777; electronic mail MEWES@MIPS.BITNET. In Asia or Australia, please contact JIPID: International Protein Information Database in Japan, Science University of Tokyo, 2641 Yamazaki, Noda 278, Japan; telephone +81 471 241501; FAX +81 471 221544; electronic mail TSU-GITA@JPNSUT31.BITNET or, on DIALCOM, 42:CDT0079. The NBRF PIR Protein Sequence Database has been incorporated into or used as the primary source for other protein sequence databases and is also distributed by many other vendors in conjunction with software packages. The PIR is not responsible for the versions of the database supplied by these secondary sources. Although users may find these software–data packages convenient, they should be aware that the database supplied may not be the latest release and may not include all of the information available in the original.

Acknowledgments

The Protein Identification Resource is supported by National Institutes of Health Grant RR01821.

$

[21] E. J. Bicknell, R. Rada, S. Davidson, and R. Stander, *Nucleic Acids Res.* **16,** 1667 (1988).

[4] Cooperation between Databases and Scientific Community

By BORIVOJ KEIL

Projects to sequence entire genome (human, yeast, bacteria, rice) have been discussed at great length, and concrete steps have been taken to elaborate them in a realistic way. In spite of extensive discussions concerning technical, financial, scientific, and ethical problems, we have to realize that, sooner or later, these projects will be put on track and that we must anticipate such a development. Simultaneously and independently, a still-growing volume of sequence data on proteins and nucleic acids is being created as the result of current research projects.

Retrieval Capacity

With the present state of the art, an individual scientist can determine and treat about 10 kilobases of nucleic acid sequence per year (microsequencing). Productivity is changing rapidly, and the adoption of new automated techniques and improvement in the organization of work will raise the output without difficulty to about 100–200 kilobases per year per scientist (macrosequencing). To launch the human genome project, it is conceivable to organize high technology centers producing about 6,000 kilobases per year per person (megasequencing). The volume of data already incorporated in the set of all data banks is about 20,000 kilobases; presently these data banks can acquire up to 10,000 kilobases per year. Figure 1 compares the state of the art with the challenge.

The minimum set allowing for the life and reproduction of the simplest cellular organisms is estimated at about 1000 kilobases. This is represented on the left-hand side of Fig. 1 by a corresponding area. A complete nucleotide set of a typical microorganism like *Escherichia coli,* is 4.700 kilobases (4.7 megabytes). If printed, this would be equivalent to a book of about 1500 pages, but *Escherichia coli* is only a unicellular prokaryote, without cellular differentiation. A simple eukaryote like yeast contains 15 megabytes, and the genome of a simple plant like cress has a length of 70 Mbyte. In comparison, the human genome is represented by the frame of the entire Fig. 1. If printed in form of a text, the human genome would take about 1,000,000 pages.

As to the actual state of the art, the existing databases cannot absorb the volume of data presently produced every day, even before genome projects are started. It is evident that the most serious problem, i.e., the explosion of

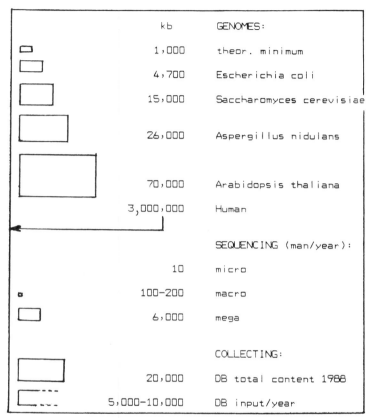

FIG. 1. Comparison of data volumes in nucleotide sequences. The framed areas are proportional to the volume in kilobytes.

data created by the genome projects, cannot be mastered merely by improving the efficiency of the existing structures and by better cooperation. Nevertheless, a solution cannot be found without a consensus among database creators and managers, who are the most competent to make decisions about future needs and interests.

Evolutionary Trends

The nucleotide sequence databases will undoubtedly grow in the near future at higher speeds than other bases dealing with biological macromolecules (amino acid sequences and protein structures, carbohydrate sequences, enzyme specificity and kinetic data, biological activity of peptides and proteins, NMR data, specialized collections). Nevertheless, these other databases will also continue to develop, because they store and interpret data which cannot be deduced from a nucleotide sequence database. The corresponding data are determined by independent measurements, and they reveal different aspects of the problem.

The three most important data banks dealing with protein and nucleic acid sequences, NBRF, EMBL, and GenBank, are treated in detail elsewhere in this volume. The U.S. and European data banks on nucleotide sequences (GenBank and EMBL) have developed a close partnership, joined recently by the corresponding data bank in Japan (DDBJ). Close cooperation also exists among protein databases.

In 1985, and with active help of the CODATA Task Group, three major protein databases [the Protein Information Resource (PIR) in the United States, the Martinsried Institute Protein Sequence database (MIPS) in West Germany, and the International Protein Information Database in Japan (JIPID)] agreed to share the responsibilities of data collection by allocating geographical catchment areas and pooling their combined data files. In 1987, these databases made a trilateral agreement to further integrate and form an international database called the Protein Information Resource (PIR)-International. This meant that the databases would in effect act as parts of a larger body. This international effort is supported at the governmental level with funds allocated for transfer of PIR-International staff between the databases to cooperate in various projects such as software and database development. One result of this collaboration is publication of the journal *Protein Sequences and Data Analysis* (Springer Verlag), which contains recent entries into the database and families of related sequences.

There is much interest in the scientific community in having access to databases that maintain repositories of useful data on proteins and nucleic acids, other than purely sequence information. The classic database for enzymes, which exists both as a monograph and on tape, is *Enzyme Nomenclature*[1] (an updated edition is now in preparation). Moreover, four specialized databases or collections were founded recently. BRENDA collects systematic data on enzymes, as well as on their specificity and kinetic properties, whereas DBEMP concentrates on their metabolic pathways. RED deals with restriction enzymes; LYSIS treats systematic data on endopeptidases and their bond specificity.

JIPID, in collaboration with the other PIR-International database

---

[1] "Enzyme Nomenclature 1984." Recommendations of the Nomenclature Committee of the International Union of Biochemistry on the Nomenclature and Classification of Enzyme-Catalysed Reactions. Academic Press, Orlando, San Diego, New York, London, Toronto, Montreal, Sydney, and Tokyo, 1984.

members and research institutes, is developing several new databases: the Biological Activity Database (JIPIDB), the Variant Database (JIPIDA), and the NMR database (JIPIDM). The Biological Activity Database contains data on the biological and chemical activity of proteins. Presently under development are data files for enzymes, toxins, and electron-carrier proteins. Eventually, all entries in the sequence databases will have corresponding entries in the other databases. The Variant Database contains information about artificially created mutant molecules. These data may be compared with the JIPIDB entries which act as the wild-type molecules. The NMR database contains both a literature database and NMR data repository containing coordinate data. It is being developed in collaboration with the University of Wisconsin and the Protein Research Institute (Japan).

Unlike the case for protein and nucleic acid data banks, data on carbohydrate sequences and structures have been without a bank until recently. Supported by the U.S. Department of Energy and by the large community of carbohydrate chemists and biochemists, the database CSD is now in full development.

This short survey traces only some more recent trends in development. Table I presents a list of databases covering miscellaneous fields of molecular biology. More complete information can be found either in the last release of the data bank LIMB[2] or in the CODATA Directory of Protein Sequence and Nucleic Acid Sequence Data Sources.[3]

The information in Table I, as well as the above-mentioned development of the CSD database, seems to answer the question of whether the evolution of databases on natural polymers will follow the way of a few centralized projects or of cooperation within networks of numerous centers. As long as the sponsorship is not centralized, the data bank system will have a tendency to spread. In countries with strong scientific communities there is a growing tendency to create and support new local and regional data collections to complete existing resources, to retrieve data independently, to resorb smaller local collections, and to meet particular needs. Even if we would prefer centralization for the sake of efficiency, independent systems may already be on the horizon.

Task for International Scientific Community

As it is improbable that a single national institution will be able to take care of the international network of biological databases in the future, the

---

[2] See Ref. (8) in Table I.
[3] "Directory of Protein Sequence and Nucleic Acid Data Sources," *CODATA Bull.* **65**, Pergamon, New York, 1987.

TABLE I

DATABASES FOR BIOLOGICAL MACROMOLECULES AND THEIR COMPONENTS

Database	Name	Type[a]	Data	Volume (kbyte)	Address[b]	Ref.[c]
AANSPII	Amino Acid and Nucleotide Sequences of Proteins of Immunological Interest (Kabat DB)	3,4	Immunoglobulins	20,000	a	
AMINODB	Amino Acid Database	9	Molecular properties of amino acids	2	b	(1)
BMRB	BioMagResBank, NMR Data Respository	6	Protein NMR data		c	(2,3)
BRD	Berlin RNA Data Bank	2	rRNA sequences	1,390	d	
BRENDA	Braunschweig Enzyme Database	7	Enzymes, kinetics		e	
CCD	Cambridge Structural Database	10	Small molecule atomic coordinates		f	
*CSD	Carbohydrate Structure Database	5	Carbohydrate sequences		g	(4)
CSRS	Compilation of Small RNA Sequences	2	Small RNA nucleotide sequences		h	
*DBEMP	Data Bank on Enzymes and Metabolic Pathways	7	Enzymes, kinetics, pathways	6,000	i	
DDBJ	DNA Data Bank of Japan	3	Nucleotide sequences		j	
*EMBL	EMBL Data Library	3	Nucleotide sequences	20,000	k	(5)
*GENBANK	GenBank Genetic Sequence Data Bank	3	Nucleotide sequences	47,000	l	(6)
HIVSSA	HIV Sequence and Sequence Analysis Database	4	AIDS virus amino acid and nucleotide sequences	1,000	m	(7)
IHIC	International Hemoglobin Information Center	4	Sequences of natural Hb variants		n	
*IIPIDA	Artificial Variant Database	7	Artificial protein variants		o	
*IIPIDB	Biological Database	7	Protein properties		o	
*IIPIDM	NMR Database of Polymers	6	Protein NMR data		o	
*IIPIDN	Natural Variant Database	4	Natural protein variants		o	
*IIPIDS	Sequence Database	3,4	Amino acid and nucleotide sequences		o	
*LIMB	Listing of Molecular Biology Databases	1	Information on access to molecular		p	(8)

biology databases

*LYSIS	Proteolysis Specificity Database	7	Cleavage sites of endopeptidases	4,000	q	(9)
*MIPS	Martinsried Institute for Protein Sequences	4	Amino acid sequences	1,000	r	
NEWAT	NEWAT	4	Amino acid sequences	1,000	s	(10–12)
*PDB	Protein Data Bank	6	Crystallographic structure data for biological macromolecules	55,000	t	(13,14)
*PIR	NBRF Protein Identification Resource	4	Amino acid sequences	6,500	u	(15)
*PRF-SEQDB	Protein Research Foundation Sequence Database	8	Amino acid sequences	5,350	v	(16)
*PSEQIP	PseqIP	4	Nonredundant amino acid sequence collection	6,880	w	(17)
QTDGPD	Quest 2D Gel Protein Database	7	Protein identification from 2D gel electrophoresis		x	(18)
RED	Restriction Enzyme Database	7	Restriction endonucleases	10,000	y	(19)
SIGPEP	Signal Peptide Database	4	Preprotein signal peptide sequence	200	z	(20)
SRSRSC	Small Ribosomal Subunit RNA Sequence	2	Small RNA sequences	292	aa	(21)
*SWISS-PROT	SWISS-PROT Protein Sequence Data Bank	3,4	Amino acid sequences	5,000	bb	
TRNAC	tRNA Compilation	2	tRNA gene nucleotide sequence and alignment	1,000	cc	(22,23)
*VECBASE (VECTOR)	Cloning Vector Sequence Database	3	Cloning vector nucleotide sequence	1,500	dd,ee	(24)

* Represented in the CODATA Task Group on Biological Macromolecules as member or observer.

[a] Type of database: 1, general information on databases; 2, RNA sequences; 3, DNA sequences; 4, amino acid sequences; 5, carbohydrate sequences; 6, macromolecular structure; 7, macromolecules, specific activity, other characteristics; 8, peptides; 9, amino acids, characteristics; 10, crystallography, small molecules.

[b] Key to addresses: (a) E. Kabat, National Institutes of Health, Bldg. 8, Room 126, Bethesda, MD 20892; (b) C. Froemmel, Institut fr Biochemie, Hessische Str. 3, DDR-1040 Berlin GDR; (c) J. L. Markley, National Magnetic Resonance Facility, Department of Bochemistry, College of Agriculture and Life Sciences, University of Wisconsin—Madison, 420 Henry Mall, Madison, WI 53706; (d) V. Erdmann, Institut für Biochemie,

FB Chemie, Freie University of Berlin, Otto-Hahn-Bau, Thielallee 63, D-1000 Berlin 33 (Dahlem), FRG; (e) D. Schomburg, GBF, Mascheroder Weg 1, D-3300 Braunschweig, FRG; (f) O. Kennard, Cambridge Crystallographic Data Centre, University Chemistry Lab, Lensfield Road, Cambridge CB2 1EW, U.K.; (g) S. Doubet, Complex Carbohydrate Research Centr, Russell Lab, P.O. Box 5677, Athens, GA 30613; (h) R. Reddy, Department of Pharmacology, Baylor College of Medicine, Houston, TX 77030; (i) E. E. Selkov, Institute of Biological Physics, USSR Academy of Sciences, 142292 Pushchino, Moscow region, URSS; (j) S. Miyazawa, National Institute of Genetics, Mishima 411, Japan; (k) G. Cameron, EMBL Data Library, Postfach 10.2209, D-6900 Heidelberg, FRG; (l) D. Benton, IntelliGenetics Inc., 700 El Camino Real East, Mountain View, CA 94040; (m) K. MacInnes,T-10, MS K710, Los Alamos National Lab., Los Alamos, NM 87545; (n) R. N. Wrightstone, International Hemoglobin Information Center, Medical College of Georgia, Augusta, GA 30912-1100; (o) A. Tsugita, JIPID, Life Science Institute, Science University of Tokyo, Yamazaki Noda 278, Japan; (p) C. Burks, Theoretical Biology and Biophysics Group, T-10, MS K710, Los Alamos National Lab., Los Alamos, NM 87545; (q) B. Keil, Institute Pasteur, 28 Rue du Dr. Roux, 75015 Paris, France; (r) H. W. Mewes, Max-Planck Institut für Biochemie, Am Klopferspitz 18, D-8033 Martinsried, FRG; (s) R. F. Doolittle, Chemistry Department, D006, University of California, San Diego, La Jolla, CA 92093; (t) F. C. Bernstein, Chemistry Department, Brookhaven National Lab., Upton, NY 11973; (u) K. E. Sidman, National Biomedical Research Foundation, 3900 Reservoir Road NW, Washington, DC 20007; (v) Y. Seto, Protein Research Foundation, 4-1-2 Ina, Minoh-Shi, Osaka 562, Japan; (w) I. Sauvaget, Institute Pasteur, 25 Rue du Dr. Roux, 75724 Paris Cedex 15, France; (x) C. Chang, Cold Spring Harbor Lab., P.O. Box 100, Cold Spring Harbor, NY 11724; (y) R. Roberts, Cold Spring Harbor Lab., P.O. Box 100, Cold Spring Harbor, NY 11724; (z) G. von Heijne, Theoretical Physics, Royal Institute of Technology, S-10044 Stockholm, Sweden; (aa) R. DeWachter, Dept. Biochemie, Univ. Antwerpen (UIA), Universiteitsplein 1, B-2610 Antwerpen, Belgium; (bb) A. Bairoch, Dept. de Biochimie Medicale, CMU, 1 Rue Michel Servet, 1211 Geneve 4, Switzerland; (cc) M. Sprinzl, Department of Biochemistry, University of Bayreuth, Universitatsstrasse 30, D-8580 Bayreuth, FRG; (dd) W. Gilbert, Whitehead Institute, Room 211, Nine Cambridge Center, Cambridge, MA 02142; (ee) F. Pfeiffer, Max-Planck Institut für Biochemie, Am Klopferspitz 18, D-8033 Martinsried, FRG.

[c] Key to references: (1) E. L. Ulrich, J. L. Markley, and Y. Kyogoku, *Protein Seq. Data Anal.* **2**, 23 (1989); (2) J. Wolters and V.A. Erdmann, *Nucleic Acids Res.* **141**, r1 (1986); (3) V. A. Erdmann and J. Wolters, *Protein Seq. Data Anal.* **1**, 127 (1987); (4) R. Reddy, *Nucleic Acids Res.* **14**, r61 (1986); (5) G. Hamm and G. Cameron, *Nucleic Acids Res.* **14**, 5, (1986); (6) H. S. Bilofsky, C. Burks, J. W. Fickett, W. B. Goad, F. I. Lewitter, W. P. Rindone, C. D. Swindel, and C.-S. Tung, *Nucleic Acids Res.* **14**, 1 (1986); (7) D. Barnes, *Science* **235**, 634 (1987); (8) G. van Heijne, in "Sequence Analysis in Molecular Biology," p. 153. Academic Press, New York, 1987; (9) B. Keil, *Protein Seq. Data Anal.* **1**, 13, (1987); (10) R. F. Doolittle, *Science* **214**, 149 (1981); (11) R. F. Doolittle, in "Protein Engineering," p. 15. Academic Press, New York, 1986; (12) R. F. Doolittle, in "Of URFS and ORFS," p. 4. Univ. Science Books, Mill Valley, California, 1987; (13) F. Bernstein, T. F. Koetzle, G. J. B. Williams, E. F. Myers, Jr., M. D. Brice, J. R. Rodgers, O. Kennard, T. Shimanouchi, and M. Tasumi, *J. Mol. Biol.* **112**, 535 (1977); (14) E. Abola, F. Bernstein, and T. Koetzle, in "The Role of Data in Scientific Progress" (P. S. Glaeser, ed.), p. 139. Elsevier/North Holland, Amsterdam, 1985; (15) D. G. George, W. C. Barker, and L. T. Hunt, *Nucleic Acids Res.* **14** 11 (1986); (16) Y. Seto, in "Computational Molecular Biology" (A. M. Lesk, ed.), p. 28. Oxford Univ. Press, Oxford, New York, Tokyo, 1988; (17) J. M. Claverie and L. Bricault, *Proteins* **1**, 60 (11986); (18) B. J. Proujan, *Res. Resour. Rep.*, Feb. 13 (1988); (19) R. Roberts, *Nucleic Acids Res.* **13**, r165 (1985); (20) G. von Heijne, *Protein Seq. Data Anal.* **1**, 41 (1987); (21) E. Huysmans and R. DeWachter, *Nucleic Acids Res.* **14**, r73 (1986); (22) M. Sprinzl, J. Moll, F. Meissner, and T. Hartmann, *Nucleic Acids Res.* **13**, r1 (1985); (23) M. Sprinzl, T. Vorderwuelbecke, and T. Hartmann, *Nucleic Acids Res.* **13**, r51 (1985); (24) F. Pfeiffer and W. A. Gilbert, *Protein Seq. Data Anal.* **1**, 269 (1988).

problem of cooperation is international. What are the ways to improve contacts between data banks, to foster their rational development, and to establish their place as a major fundamental source of information for the scientist? What structure can solve the problems of compatibility, terminology, communicability, user education, permanent information, and contact with neighboring scientific disciplines?

The database creators and managers are bound primarily to develop their own projects in their local national scientific environment, and they have only limited means to develop activities external to their goal. The funding agencies are naturally interested in the return on their investment. They are anxious that their support will not be wasted by duplicity and that the action will further larger long-term goals. A supranational, intergovernmental action for financing and coordination of biological databases does not exist; the national agencies wisely leave the initiative in the tasks mentioned above to professional scientific bodies, either on national levels (academies, councils) or those organized in international scientific unions.

What help can we expect from the International Unions? They are organized according to scientific disciplines and are headed by the ICSU (International Commission of Scientific Unions), their representative to intergovernmental agencies. Data banks specializing in the structure and function of biological macromolecules are of great interest to several Unions (Pure and Applied Chemistry, Pure and Applied Biophysics, Biochemistry, Biological Sciences, Microbiological Societies, Physiological Sciences, etc.).

For projects which in the past needed cooperative effort of different disciplines, the ICSU created twelve specialized committees (COSPAR for coordination of space research, SCOR on oceanic research, etc.), and CODATA for data for science and technology. Since 1966, CODATA has promoted and encouraged on a worldwide basis the production and distribution of collections of reliable numerical data of importance in science and technology. Initially concerned primarily with physics and chemistry, it has broadened its scope to include data from the bio- and geosciences as well.

In 1984, CODATA established a Task Group "for coordination of protein sequence data banks."[4] From the very beginning it was clear that the primary role of the group was not to "coordinate" well-developed and highly competent teams of specialists, but to help to generate approaches and proposals for unresolved problems. Specific achievements of this task group included (1) encouraging the complementary relations between computer-based protein sequence data collectors, stimulating them to de-

[4] B. Keil, *Protein Sequences Data Anal.* **1**, 123 (1987)

velop and agree on a standardized format recommended for exchange of protein sequence data.[5] This effort resulted in a trilateral agreement among the three data banks in the United States, Japan, and the Federal Republic of Germany to create a single integrated protein sequence data bank (PIR-International). (2) Publication of the *CODATA Directory of Protein Sequence and Nucleic Acid Sequence Data Sources,*[3] and of the book *Computational Molecular Biology,*[6] introducing sources and methods of sequence data analysis to researchers; (3) fostering the establishment of a new journal, *Protein Sequences and Data Analysis;*[7] (4) encouraging the development of new computer data collections, for example, the NMR Data Repository, the Artificial Variant Database, and the Proteolysis Database LYSIS; (5) planning for new modes of data collection and distribution; and (6) presenting lectures at conferences and symposia to apprise users of the state of the field.

The Task Group has elaborated its methodology and programs for promotion and cooperation of protein data banks on the international level which did not interfere with any other actions in this domain. However, as protein data banks represent only a subgroup of databases in molecular biology, it became evident that cooperative efforts can be efficient only if protein, nucleic acid, and carbohydrate databases and their users are in permanent contact.

CODATA Task Group on Biological Macromolecules

Cooperation among protein sequence data banks is now well established, as is cooperation among nucleic acid data banks. It became appropriate to foster communication between these sets of data banks and with other data banks in related fields. Since 1987 the Task Group has focused on proteins, and representatives of the major nucleic acid data banks have nevertheless participated in its activities; the process of transformation to deal with methodological problems common to all databases in molecular biology was smooth.

The newly constituted Task Group on Biological Macromolecules[8]

---

[5] D. G. George, H. W. Mewes, and H. Hikara, *Protein Sequences Data Anal.* **1**, 27 (1987).

[6] A. M. Lesk, ed., "Computational Molecular Biology." Oxford Univ. Press, Oxford, 1988.

[7] A. Tsugita, Executive Editorial Chairman, *Protein Sequences and Data Analysis.* Springer, Berlin and New York, 1987.

[8] Task Group on Biological Macromolecules, 1989. Chairman: B. Keil (France); Secretary: A. Tsugita (Japan); Members: W. C. Barker (United States), C. Burks (United States), G. Cameron (EMBL), A. Henschen (FRG), M. Kotani (Japan), A. M. Lesk (U.K.), H. W. Mewes (FRG), S. Miyazaki (Japan); Consulting and Corresponding Members: K. Bock (Denmark), J. M. Claverie (France), D. George (United States), P. Kahn (EMBL), T. Koetzle (United States), A. S. Kolaskar (India), K. Loening (United States), T. Marr (United States), F. Pfeiffer (FRG), J. Rodgers (Canada), E. E. Selkov (USSR), R. Simpson (Australia).

started work in February 1989. It continues the tradition of an informal and nonvoting round-table group of database creators, noncommercial distributors, and users. Its main goal is to become a neutral host for cooperation among the information resources and to prepare common courses of action with other task groups and databases in biology. Some of these issues have already been addressed by individual data banks,' but satisfactory solutions will require a general forum for discussion of problems common to all computer data banks and their users. The Task Group will try to address in particular the problems discussed below.

Enhancement of Communication between Data Banks and Laboratories

Contact between data banks and laboratories collecting the data is a two-way process. In most cases, researchers publish their data in various journals. After a production delay, the sequences are retrieved by data banks provided, however, that the particular journal is included in their routine scans. Following an additional processing time, data banks then make the annotated data available to the laboratories, usually by way of an intermediate server or distributor.

The work of collecting from journals and of annotating the exponentially growing volume of primary sequence data has become overwhelming for database staff. The delay between the time a sequence has been determined in a laboratory and the time it is available to the research community is considerable. Moreover, sequences are generally not published in journals unless they solve a biological problem directly. The only way out of this dilemma is to shift a large part of the collecting and annotating work to the authors. A system is being set up by the major data banks whereby the scientists will prepare the sequences in a standardized, computer-readable form and submit them to the designated databases simultaneously or prior to the submission of the manuscript to a journal.

As the content and method of access to the data banks differ from the traditional documentation (journals, monographs), the worldwide community of potential users still has not worked out a routine access to the information, even to the most important data collections. Other issues to be addressed include the following: Can one data bank satisfy everyone? What additional collections or banks should be created or developed for public circulation? What is the accessibility and policy of local servers or distributors? What is the legal status of intellectual property as concerns determination, publication, and collection of sequences?

Problems Common to all Databases

Within the Task Group, there are discussions among database representatives on cross-referencing, indexing, adopting common standards, and coordinating data input. The diversity in hardware, software, and underlying technology presents a problem. The Task Group recognizes that organizers of already existing data banks would hardly be ready to replace their different routine formats by a common one; on the other hand, an exchange format could be of help for communication between them. Another important problem which is also being discussed is the system of quality control, not only consistency checks by computer but also scientific review by database staff and external curators.

The domain of terminology, nomenclature, vocabularies, and abbreviations must also be addressed by the Task Group. The source of the macromolecular sequence must be described using standardized vocabularies. There are many organizations and authorities in biology that develop nomenclatures; however, reliable pathways to the body of information are not organized. The nomenclature of many protein groups is not yet established, and many recommended names are not easily entered into a computer because of subscripts, greek letters, etc. A data bank editor often has to make a choice between several names, spellings, or abbreviations proposed by different laboratories for the same protein or organism. One of the roles of the Task Group is to identify the major problems, to elaborate an inventory of redundancy, and to seek solutions by working in close contact with the Nomenclature Commissions of the International Unions and, in particular, with NC-IUB of IUPAC-IUB and NCBJ.

By working with specialists in data treatment in molecular biology and a worldwide audience of users, the Task Group may be able to provide a leadership role in a field which heretofore has had goals and objectives too large to be solved by a single scientific union, a single country, or a single decision-making authority.

Section II

Searching Databases

[5] Rapid and Sensitive Sequence Comparison with FASTP and FASTA

By WILLIAM R. PEARSON

Introduction

Rapid computer algorithms for comparing DNA and protein sequences have dramatically decreased the amount of time required to compare an unidentified sequence to a DNA or protein sequence database. These methods have come at a time when the protein and DNA sequence libraries are growing almost 50% per year, owing to more efficient cloning techniques and more productive sequencing procedures. Sequence searches have led to the discovery of many new families of proteins, including the tyrosine kinase oncogene family, the steroid receptor and v-*erbA* oncogene family, the growth factor receptor family, the G-protein-coupled receptor family, and transcription factors containing a zinc-finger motif. Often a database search provides the first insight into the mechanism of action of a newly sequenced protein.

In 1985, David Lipman and I described the FASTP program for searching protein sequence libraries.[1] FASTP combines a rapid technique for focusing on those regions in a pair of sequences that share a high density of identities with a scoring procedure that uses the PAM250 scoring matrix[2] (Fig. 3) for high sensitivity. FASTP decreased the computer time required for a protein database search from about 6 h on a VAX11/780 to about 10 min on an IBM-PC. Shortly thereafter, I modified FASTP for DNA sequence comparison and distributed the FASTN program. More recently, we described an improved version of FASTP called FASTA, which combines the functions of the FASTP and FASTN programs and provides a more sensitive sequence comparison algorithm.[3] In addition, the FASTA package (Table I) includes programs for comparing a protein sequence to a DNA sequence database (TFASTA), for identifying local sequence similarities or duplications in sequences (LFASTA, PLFASTA), and for evaluating the statistical significance of a similarity score (RDF2). The FASTA programs can be tailored to specific comparison problems by

---

[1] D. J. Lipman and W. R. Pearson, *Science* **227**, 1435 (1985).

[2] M. Dayhoff, R. M. Schwartz, and B. C. Orcutt, *in* "Atlas of Protein Sequence and Structure" (M. Dayhoff, ed.), Vol. 5, Suppl. 3, p. 345. National Biomedical Research Foundation. Silver Spring, Maryland, 1978.

[3] W. R. Pearson and D. J. Lipman, *Proc. Natl. Acad. Sci. U.S.A.* **85**, 2444 (1988).

TABLE I
FASTA SEQUENCE COMPARISON PROGRAMS

Program	Description
FASTA	Compares a protein sequence to another protein sequence or to a protein database, or a DNA sequence to another DNA sequence or to a DNA library.
TFASTA	Compares a protein sequence to a DNA sequence or DNA sequence library. The DNA sequence is translated in all six reading frames, and the protein query sequence is compared to each of the six derived protein sequences. The DNA sequence is translated from one end to the other; no attempt is made to edit out intervening sequences. Termination codons are translated into unknown ("X") amino acids. The eukaryotic nuclear genetic code is used for all translations.
LFASTA	Compares two sequences to identify regions of sequence similarity. While FASTA and TFASTA report a single alignment between two sequences, LFASTA will report several sequence alignments if there are several similar regions. LFASTA can identify similarities arising from internal repeats or similar regions that cannot be aligned by FASTA because of gaps (Fig. 6). LFASTA reports actual sequence alignments and similarity scores.
PLFASTA	PLFASTA is identical to LFASTA, but it presents a dot-matrix-like plot of the similar regions, rather than the actual alignments. PLFASTA plots are shown in Figs. 6 and 7.
RDF2	Evaluates the significance of pairwise similarity scores using a Monte Carlo analysis. Similarity scores for the two sequences are calculated, and then the second sequence is shuffled 50 to 200 times and compared with the first sequence. RDF2 can use one of two shuffling strategies. One strategy simply keeps the amino acid composition of the entire shuffled sequence identical to the unshuffled sequence. The second local shuffle destroys the order but preserves the composition of small segments (10–25 residues) of the shuffled sequence.

changing the similarity scoring matrix and gap penalties. Thus, the same program can be used to compare protein sequences with the PAM250 matrix or a matrix based on the genetic code and to compare DNA sequences. In addition, FASTA provides several output options that can be used to highlight similarities and differences in aligned sequences.

In this chapter, I show an example of a simple FASTA library search, describe the FASTA algorithm, and then discuss in detail a more problematic search, namely, one for members of the G-protein-coupled receptor family. Additional information about how to customize the scoring parameters and output from the FASTA programs is included in the appendices.

Much of this chapter focuses on the evaluation of distant alignments that have ambiguous similarity scores. In any database search, there is always a library sequence with the best score, regardless of whether that sequence shares common ancestry or some other significant similarity with the query sequence. In sequence comparison, there is a trade-off between sensitivity — the ability to identify distantly related sequences — and selectivity — the avoidance of false positives (unrelated sequences with high similarity scores). The perfect sequence comparison method would be both sensitive and selective; it would rank all the members of a protein family that share a common ancestor above all the sequences that are similar but nonhomologous. No such program exists, because proteins evolve at very different and sometimes very rapid rates, so that a sequence may contain only a trace of its evolutionary ancestry. In many cases, the problem is to differentiate between high scoring sequences that share common ancestry or significant similarity with the query sequence and sequences with high scores that are due to local sequence composition and random chance. The FASTA program is more sensitive than FASTP, so more distantly related sequences can be identified with FASTA. Nevertheless, methods that increase sensitivity decrease selectivity, and additional care is required when interpreting the results of a FASTA search.

Throughout this chapter, the emphasis is on protein sequence comparison. FASTA can compare either DNA or protein sequences, but protein sequence comparison is far more useful, because distant sequence relationships can best be identified at the protein sequence level. While DNA sequences that encode structural RNAs have been successfully used to examine ancient evolutionary relationships, DNA sequences from repeated sequences, intervening sequences, upstream regions, or untranslated regions of messages rarely allow the demonstration of common ancestry for sequences that diverged more than 100 to 200 million years ago. In contrast, common ancestry can frequently be demonstrated for protein sequences that diverged 1 to 2 billion years ago. Protein sequence comparisons are more useful both because of the degeneracy of the genetic code (a change in a DNA sequence may not change the encoded protein) and also because of the biochemical information in the amino acid itself (arginines are very similar to lysines, but glycines and isoleucines are very different). If there is ever a question about the relationship between two DNA sequences which encode proteins, the comparison should always be done with the derived amino acid sequences.

Using FASTA Programs

Although the FASTA programs provide a number of options for customizing searches, most of the time only three entries are required: the name of the file containing the query sequence, the name of the file containing the library or a second sequence, and the value of the *ktup* parameter. FASTA and TFASTA compare the first (query) sequence to all the sequences in the second file (there need only be one), reporting the one best similarity score and alignment for each pairwise comparison. All of the FASTA programs calculate a "local" similarity score, i.e., the best region of similarity is found between the two sequences being compared. The score of the local region is not affected by poorly aligned portions of the sequences outside the best region. Thus, programs in the FASTA package can be used to find conserved or shuffled protein domains, such as the epidermal growth factor (EGF) precursorlike domains in the low density lipoprotein (LDL) receptor[4] (Fig. 7).

FASTA and TFASTA report only the similarity score for the one best pairwise alignment between two sequences. In the case of proteins with repeated domains, there may be several alignments with high similarity scores that are of biological interest. Multiple regions of similarity can be displayed as alignments or as a dot-matrix-style plot by LFASTA and PLFASTA, respectively. LFASTA and PLFASTA use a slight modification of the FASTA algorithm that focuses more tightly on local regions of similarity, so that regions of strong similarity do not overshadow neighboring regions with lower similarity scores.

RDF2 is designed to evaluate the statistical significance of a pairwise similarity score. The program calculates similarity scores for the best pairwise alignment, using the FASTA algorithm, then randomly shuffles the second sequence and calculates pairwise scores for the query sequence and each of the shuffled sequences. By examining the distribution of similarity scores obtained with randomly shuffled sequences with the same length and amino acid composition, one can evaluate the likelihood that the similarity scores for the unshuffled sequence are due to unusual sequence composition or other random fluctuations.

Figure 1 shows an example of a search of the National Biomedical Research Foundation Protein Identification Resource (PIR) library, using a murine glutathione transferase as the query sequence. After the program is started, it asks for three entries: the query sequence file name (gst87.aa in Fig. 1), selection of the sequence library (p for the NBRF protein database), and the *ktup* parameter (2 in this case by default). In this example, the user is prompted for the file names and *ktup* parameter, but each of these entries can be specified on the command line. For example, the command

    fasta gst87.aa p 1

would do the same search with *ktup* = 1.

After the search is finished, the program asks for a file name for the results (gt875.k2 in Fig. 1). If a file name is specified, the histogram, the list of top scoring library sequences, and the sequence alignments are written

---

[4] T. C. Sudhof, J. L. Goldstein, M. S. Brown, and D. W. Russell, *Science* **228**, 815 (1985).

to the file as well as to the display terminal. If the results are written to a file, the sequence alignments are not shown on the terminal. The program then asks for the number of sequences to be displayed initially. In this example, there were 2634 library sequences with initial similarity scores greater than 28, and as many as 2634 scores could be displayed.[5] The meanings of the terms initial similarity score *(initn), init1* score, and optimized score are discussed in the next section.

The example in Fig. 1 is a simple one; the 20 top scoring sequences have similarity scores that indicate common ancestry (homology), and the rest of the sequences are unrelated. Mammalian glutathione transferases can be grouped into three classes; members of the same class share 80–95% amino acid sequence identity, while interclass alignments show 25–30% sequence identity.[6] Members of all three glutathione transferase classes are found in the list of top scoring sequences. Many searches of the protein sequence library, particularly with soluble proteins, are similar to this one, i.e., related library sequences have similarity scores that are well separated from the main distribution of similarity scores for unrelated sequences. When this occurs, one can be confident that a homologous sequence has been found. Alternatively, as was the case when this search was done in 1985, there may be no sequences with similarity scores greater than 60 because there are no related sequences in the library. Unfortunately, there are sometimes related sequences with similarity scores of 60 or less, and unrelated sequences with scores greater than 100.

FASTA Implementation

The programs are written in the C programming language and run on IBM-PC microcomputers under the DOS operating system, on the Macintosh, on computers running the System V, BSD, and Xenix versions of the Unix operating system, and on VAX computers running the VMS or Unix operating systems. VMS versions of the program are designed to be used in conjunction with the National Biomedical Research Foundation Protein Identification Resource PSQ and NAQ programs and with the University of Wisconsin Genetics Computer Group program package. The program code is very portable; exactly the same source code compiles on all the

---

[5] This search was done on a Sun 3/260 workstation. On an IBM-PC, no more than 2000 scores would be saved. The older FASTP program would save the first 1000–2000 scores greater than the CUTOFF value and then fail to save additional high scoring sequences. The FASTA program saves the first 6000 (2000 on an IBM-PC) scores greater than the CUTOFF value, and, if additional high scoring sequences are found, it saves the top 75% of the sequences and adjusts the CUTOFF value upward. Thus, FASTA always saves the top scoring library sequences; in some cases FASTP does not.

[6] B. Mannervik and U. H. Danielson, *CRC Crit. Rev. Biochem.* **23**, 283 (1988).

```
Visage 2000 % fasta
fasta 1.3 [Feb, 1989] searches a sequence data bank

Please cite:
W.R. Pearson & D.J. Lipman PNAS (1988) 85:2444-2448

test sequence file name: gst87.aa

Choose sequence library:

    G: GENBANK Translated Protein Database
    P: NBRF Protein Database
    A: NBRF Protein Database + Genbank
    S: Swiss-Prot Release 8

Enter letter or filename: p
ktup? (1 to 2) [2] __
>GT8.7 transl. of pa875.con, 19 to 675 : 217 aa    sequence description
vs NBRF Protein Database library                   Library description
searching aabank.19 library                        Library file name
 2802054 aa in 10526 sequences                     Search is finished

      initn   init1
<  2     4     4:==
   4     0     0:
   6     6     6:===
   8    16    16:========
  10    42    42:=====================
 -----//-----
  40   140   124:=================================================
  42   115    72:------------------------------------++++++++++++++
  44    86    42:--------------------+++++++++++++++++++++++
  46    63    29:---------------+++++++++++++++++
  48    59    15:--------+++++++++++++++++++++
  50    36    10:-----++++++++++++++
  52    21     9:-----++++++
  54    13     2:-++++++           '-' indicates the number of init1 scores
  56     9     5:---++
  58    16     2:-+++++++          '+' indicates the number of initn scores
  60     7     0:++++
  62     2     0:+
  64     7     0:++++
  66     6     0:+++
  68     1     0:+
  70     0     0:
  72     0     0:
  74     3     0:++
  76     0     0:
  78     0     0:
  80     0     1:-
> 80    20    19:==========
2802054 residues in 10526 sequences
 mean initn score:  25.2 (7.08)
 mean init1 score:  24.9 (6.38)
 2634 scores better than 28 saved, ktup: 2, fact: 8  scan time:  0:01:14
 Enter filename for results : gt875.k2
 How many scores would you like to see? [20] 25
```

FIG. 1. Sample FASTA library search. A transcript of a protein sequence library search with the FASTA program is shown. Entries typed by the user are underlined. A large portion of the histogram of library similarity scores has been removed; complete histograms are shown in Fig. 4. The alignments are written to a file in this example. The selection of the library to be searched from a list of libraries is made possible by defining the FASTLIBS environment symbol.

```
The best scores are:                                        initn init1   opt
A25510 - Glutathione S-transferase (EC 2.5.1.18), Yb c 1133  1133  1133
A24085 - Glutathione S-transferase (EC 2.5.1.18), Yb1  1127  1127  1127
B26187 - Glutathione S-transferase (EC 2.5.1.18), Yb-2 1018  1018  1038
A26307 - Glutathione S-transferase (EC 2.5.1.18), Yb2  1018  1018  1038
XURTG4 - Glutathione S-transferase (EC 2.5.1.18), chai  999   999  1019
A29036 - Glutathione S-transferase (EC 2.5.1.18) Yb3 -  988   988  1010
A26484 - Glutathione S-transferase (EC 2.5.1.18) - Flu  503   416   545
XURTGP - Glutathione S-transferase P (EC 2.5.1.18) - R  198   109   289
B20831 - Glutathione S-transferase (EC 2.5.1.18) minor  189   189   200
A20831 - Glutathione S-transferase (EC 2.5.1.18) major  174   174   180
H24735 - Glutathione transferase (EC 2.5.1.18), MIII -  137   137   137
A28562 - Glutathione S-transferase (EC 2.5.1.18), clas  134   134   144
E24735 - Glutathione transferase (EC 2.5.1.18), class   130   130   130
B22457 - Glutathione S-transferase (EC 2.5.1.18) mu -   130   130   130
J24735 - Glutathione transferase (EC 2.5.1.18), GT-9.3  126   126   126
G24735 - Glutathione transferase (EC 2.5.1.18), 4-4 -   125   125   126
I24735 - Glutathione transferase (EC 2.5.1.18), GT-8.7  116   116   120
K24735 - Glutathione transferase (EC 2.5.1.18) - Bovin  107   107   107
F24735 - Glutathione transferase (EC 2.5.1.18), 3-3 -   104   104   114
A26598 - 28K antigen precursor - Fluke (Schistosoma ma   87    79   131
A29944 - Chaoptin precursor - Fruit fly                  74    51    82
A29352 - SST2 protein - Yeast (Saccharomyces cerevisia   73    52    54
QQECO3 - Hypothetical protein F-300 - Escherichia coli   73    45    59
A29949 - Glycogen phosphorylase (EC 2.4.1.1), brain -    68    52    70
A25026 - Chloramphenicol acetyltransferase (EC 2.3.1.2   66    43    47
  More scores? [0] __
  Display alignments also? y
  number of alignments [20]? 20
Library scan:  0:01:14  total CPU time:  0:01:20
```

FIG. 1.

machines except the Macintosh. Versions of the program for the IBM-PC or Macintosh can search the library with query sequences up to 2,000 residues in length and can search library sequences of any length. (For the IBM-PC, this limit is set so that the programs can run using the faster small memory model, which limits data to 64 kilobytes). On other machines, the query sequence can be up to 10,000 residues, but this value can be increased by editing and recompiling the programs. There is no limit to the length of the library sequence on large or small machines. If the library sequence is too long ($> 10,000$ residues for small machines, $> 50,000$ for large), it is scanned in overlapping pieces. The complete source code is available for all versions of the program from William R. Pearson.

The searching programs FASTA and TFASTA can search libraries in a variety of different formats, including: (1) FASTP/DM or query sequence format; (2) GenBank magnetic tape format; (3) the Protein Identification Resource CODATA format; (4) EMBL and SWISS-PROT format; (5) IntelliGenetics sequence file format; and (6) Compressed GenBank format for floppy disk distribution (Appendix 1). Although the programs can be run immediately after they are copied onto a computer, they are easier to use if an additional file is installed. This file, which is referred to by the UNIX, DOS, or VMS environment symbol FASTLIBS, allows FASTA and TFASTA to list the libraries that are available to be searched. FASTA can also search a library made up of several files containing data, as is often the case for libraries that are distributed on floppy diskettes. To indicate that a library file contains a list of file names, rather than actual sequence data, the library file name is preceded by the symbol @.

The behavior of all the FASTA programs can be modified by specifying a different scoring matrix file or CUTOFF value, and the alignments displayed by FASTA, TFASTA, and LFASTA can be modified by specifying options on the command line or with environment symbols. A complete list of input and output options is described in Appendix 3.

FASTA Algorithm

FASTA uses four steps to calculate three scores that characterize sequence similarity; these steps are outlined in Table II. The first step uses a rapid technique for finding identities shared between two sequences; the method is similar to an earlier technique described by Wilbur and Lipman.[7] FASTP and FASTA achieve much of their speed and selectivity in this first step by using a lookup table[8] to locate all identities or groups of identities between two DNA or amino acid sequences during the first step of the comparison.[9] The *ktup* parameter determines how many consecutive identities are required in a match. A *ktup* value of 2 is frequently used for protein sequence comparison, which means that the program examines only those portions of the two sequences being compared that have at least two adjacent identical residues in both sequences. More sensitive searches can be done using *ktup* = 1. For DNA sequence comparisons, the *ktup* parameter can range from 1 to 6; values between 4 and 6 are recommended. When the query sequence is a short oligonucleotide or oligopep-

---

[7] W. J. Wilbur and D. J. Lipman, *Proc. Natl. Acad. Sci. U.S.A.* **80**, 726 (1983).

[8] A lookup table is a rapid method for finding the position of a residue in a sequence. One way to find the "A" in the sequence "NDAPL" is to compare the "A" to each residue in the sequence. A faster way (if many residues are to be checked), is to make a table of all possible residues (23 for proteins) so that the computer representation for the residue (e.g., "A" is 1, "R" is 2, "N" is 3) is the same as its position in the table. A value is then placed in the table that indicates whether the residue is present in the sequence and, if it is, where it is present. For this example, the table has the value 1 at position 3 in the table ("N" is the third amino acid), 2 at position 4, 3 at position 1, 4 at 15, 5 at 11, and the remaining 18 positions are 0. The presence and position of the "A" in the sequence can then be determined in a single step by looking it up at position 1 in the table.

[9] J. P. Dumas and J. Ninio, *Nucleic Acids Res.* **10**, 197 (1982).

TABLE II
CHARACTERIZATION OF SEQUENCE SIMILARITY BY FASTA

Step 1	Identify regions shared by the two sequences with the highest density of identities ($ktup = 1$) or pairs of identities ($ktup = 2$).
Step 2	Rescan the ten regions with the highest density of identities using the PAM250 matrix shown in Fig. 3. Trim the ends of the region to include only those residues contributing to the highest score. Each region is a partial alignment without gaps.
Step 3	(FASTA only) If there are several initial regions with scores greater than the CUTOFF value, check to see whether the trimmed initial regions can be joined to form an approximate alignment with gaps. Calculate a similarity score that is the sum of the joined initial regions minus a penalty (usually 20) for each gap. This initial similarity score (*intin*) is used to rank the library sequences. The score of the single best initial region found in Step 2 is reported (*init1*); it is the same as the initial similarity score calculated by FASTP.
Step 4	Construct a NWS optimal alignment of the query sequence and the library sequence, considering only those residues that lie in a band 32 residues wide centered on the best initial region found in Step 2. FASTA and FASTP both report this score as the optimized (*opt*) score.

tide, $ktup = 1$ should be used. A sequence comparison using the traditional Needleman–Wunsch–Sellers (NWS) algorithm[10,11] requires a number of residue comparisons proportional to the product of the lengths of the sequences being compared, for example, 33,434 for a comparison of hemoglobin β chain (146 amino acids) with trypsin (229 amino acids). (Methods that compare all the fixed length segments of one sequence with another require a similar amount of time.) FASTP and FASTA require only 94 comparisons (1/355th as many) to examine β-globin and trypsin with $ktup = 2$; this number increases to 1921 with $ktup = 1$.[12]

In conjunction with the lookup table, we use the "diagonal" method to find all regions of similarity between the two sequences, counting $ktup$ matches and penalizing for intervening mismatches.[1,7] This method identified regions of a diagonal that have the highest density of $ktup$ matches. The term diagonal refers to the diagonal line that is seen on a dot matrix plot[13] when a sequence is compared with itself, and it denotes an alignment between two sequences without gaps (see Figs. 6 and 7). For example, the alignment of residue 8 in sequence one (on the x axis) and 13 in sequence

[10] S. Needleman and C. Wunsch, *J. Mol. Biol.* **48**, 444 (1970).
[11] P. Sellers, *SIAM J. Appl. Math.* **26**, 787 (1974).
[12] The ratio of comparisons with $ktup = 2$ to $ktup = 1$ is 20, exactly the value predicted by the number of different amino acids. Additional processing is required for each $ktup = 2$ match, however, and in practice the time required for protein searches with $ktup = 1$ is only about 5 times that with $ktup = 2$.
[13] J. Maizel and R. Lenk, *Proc. Natl. Acad. Sci. U.S.A.* **78**, 7665 (1981).

two (y axis) specified a diagonal; the alignment of residue 16 (sequence one) and 21 (sequence two) falls on the same diagonal, as do 27 and 32, and so on.

As shown in Fig. 2, FASTP uses a simple formula to identify portions of a diagonal with a high density of identities, referred to as a local region of

For each *ktup* group of residues in the library sequence {

    A. For each position in the query sequence with the identical residues {

           1. Calculate the diagonal based on the current position in the library and query sequences; (diagonal = library_position - query_position)

           2. Check to see if an initial region has been found in this diagonal;

           3. If no region has been found on this diagonal before, save the current query_position as the start of a region; the score for the initial region is 16 (for *ktup* =2 in FASTP).

           4. Otherwise {

                  a. If there already was an region on this diagonal, calculate the distance from the end of the previous initial region to the current *ktup* match.

                  b. If that distance is greater than the score of the previous region, save the previous region and start a new region.

                  c. Otherwise {

                       i. Save the score of the previous region;

                      ii. Extend the previous region to the current *ktup* match;

                      iii. Decrease the score of the previous region by the distance the region was extended and increase it by 16 for the new *ktup* match.

                      }

                }

        }

}

Each time the program saves a region, it determines whether the score of the region is better than the lowest score of the ten best regions that have already been saved. If new score is better, it replaces the lowest scoring saved region with the new region, and finds new lowest scoring region in the updated list of regions.

FIG. 2. FASTP/FASTA scanning algorithm. The logic used to scan sequences in the library and identify regions of similarity is shown. The values shown for increasing the score of a region at a *ktup* match are used in FASTP. In FASTA, the values used to increment the score of a region at a *ktup* match are based on the PAM250 matrix. Thus, matching a Leu-Leu in the query sequence with a Leu-Leu in the library sequence would increase the region score by 24 instead of 16, while matching an Ala-Ser with an Ala-Ser would increase the region score by 8 instead of 16. As a result, the score of the initial region includes the PAM250 scores of all the *ktup* identities, with a constant penalty (-1) for each residue that separates *ktup* identities on the same diagonal. For DNA sequences, a constant value is used, which is equal to the square of the *ktup* value.

similarity, or simply a region. As a result of this formula, a group of $ktup = 2$ amino acid matches separated by fewer than 16 residues would be combined into a region, but $ktup = 2$ matches separated by more than 30 residues from the previous region would start a new region. FASTA uses a formula for scoring $ktup$ matches that incorporates the actual PAM250 values for the aligned residues. Thus, groups of identities with high similarity scores contribute more to the local diagonal score than do identities with low similarity scores. This more sensitive formula is used for protein sequence comparisons; the constant value for $ktup$ matches is used for DNA sequence comparisons. The earlier NUCALN programs[7] use a constant value for $ktup$ matches, but these programs save the one best local region in each diagonal. FASTA saves the ten best local regions, regardless of whether they are on the same or different diagonals.

After the ten best local regions are found in the first step, they are rescored using a scoring matrix that allows runs of identities shorter than $ktup$ residues and conservative replacements to contribute to the similarity score. For protein sequences, this score is usually calculated using the PAM250 matrix[2] (Fig. 3), although scoring matrices based on the minimum number of base changes required for a specific replacement, on identities alone, or on an alternative measure of similarity, can also be used with FASTA. The PAM250 scoring matrix was derived from the analysis of the amino acid replacements occurring among related proteins, and it specifies a range of positive scores for replacements that commonly occur among related proteins and negative scores for unlikely replacements. FASTA can also be used for DNA sequence comparisons, and matrices can be constructed that allow separate penalties for transitions and transversions. For each of the best diagonal regions rescanned with the scoring matrix, a subregion with the maximal score is identified.

The FASTP program uses the single best scoring initial region to characterize pair-wise similarity; the initial scores are used to rank the library sequences. The FASTP initial score is also calculated by FASTA, and it is referred to as the *init1* score. FASTA goes one step further during a library search; it checks to see whether several initial regions can be joined together in a single alignment to increase the initial score. Thus, FASTA improves on the sensitivity of FASTP by allowing multiple high scoring initial regions to be joined. Given the locations of the initial regions, their respective scores, and a "joining" penalty (analogous to a gap penalty), FASTA calculates an optimal alignment of initial regions as a combination of compatible regions with maximal score. This optimal alignment of initial regions can be rapidly calculated using a dynamic programming algorithm similar to that described for NUCALN;[7] FASTA uses the resulting score, referred to as the *initn* score, to rank the library sequences.

	Ala	Arg	Asn	Asp	Cys	Gln	Glu	Gly	His	Ile	Leu	Lys	Met	Phe	Pro	Ser	Thr	Trp	Tyr	Val	Asx	Glx	???
A Ala	2																						
R Arg	-2	6																					
N Asn	0	0	2																				
D Asp	0	-1	2	4																			
C Cys	-2	-4	-4	-5	12																		
Q Gln	0	1	1	2	-5	4																	
E Glu	0	-1	1	3	-5	2	4																
G Gly	1	-3	0	1	-3	-1	0	5															
H His	-1	2	2	1	-3	3	1	-2	6														
I Ile	-1	-2	-2	-2	-2	-2	-2	-3	-2	5													
L Leu	-2	-3	-3	-4	-6	-2	-3	-4	-2	2	6												
K Lys	-1	3	1	0	-5	1	0	-2	0	-2	-3	5											
M Met	-1	0	-2	-3	-5	-1	-2	-3	-2	2	4	0	6										
F Phe	-4	-4	-4	-6	-4	-5	-5	-5	-2	1	2	-5	0	9									
P Pro	1	0	-1	-1	-3	0	-1	-1	0	-2	-3	-1	-2	-5	6								
S Ser	1	0	1	0	0	-1	0	1	-1	-1	-3	0	-2	-3	1	2							
T Thr	1	-1	0	0	-2	-1	0	0	-1	0	-2	0	-1	-3	0	1	3						
W Trp	-6	2	-4	-7	-8	-5	-7	-7	-3	-5	-2	-3	-4	0	-6	-2	-5	17					
Y Tyr	-3	-4	-2	-4	0	-4	-4	-5	0	-1	-1	-4	-2	7	-5	-3	-3	0	10				
V Val	0	-2	-2	-2	-2	-2	-2	-1	-2	4	2	-2	2	-1	-1	-1	0	-6	-2	4			
B Asx	0	-1	2	3	-4	1	3	0	1	-2	-3	1	-2	-5	-1	0	0	-5	-3	-2	2		
Z Glx	0	0	1	3	-5	3	3	0	2	-2	-3	0	-2	-5	0	0	-1	-6	-4	-2	2	3	
X ???	0	0	0	0	0	0	0	0	0	0	0	0	0	0	0	0	0	0	0	0	0	0	0

FIG. 3. PAM250 scoring matrix. The amino acid replacement matrix developed by Dayhoff *et al.* [M. Dayhoff, R. M. Schwartz, and B. C. Orcutt, *in* "Atlas of Protein Sequence and Structure" (M. Dayhoff, ed.), Vol. 5, Suppl. 3, p. 345. National Biomedical Research Foundation, Silver Spring, Maryland, 1978] and used by FASTP and FASTA is shown. The numbers indicate the score for a match between the amino acids in the same row and column. Tryptophan and cysteine receive the highest scores when paired with themselves in an alignment (17 and 12, respectively); the tryptophan/cysteine substitution receives the lowest score (−8). The PAM250 matrix scores the valine/isoleucine substitution with the same value (4) as the valine/valine identity.

This third "joining" step in the computation of the initial score increases the sensitivity of the search method because it allows for insertions and deletions as well as conservative replacements. The modification does, however, decrease selectivity, as can be seen in Fig. 4. We limit the degradation of selectivity by including in the optimization step only those initial regions whose scores are above an empirically determined threshold.[14]

After a complete search of the library, FASTA plots the initial scores of each library sequence in a histogram, calculates the mean similarity score for the query sequence against each sequence in the library, and determines the standard deviation of the distribution of initial scores (Fig. 4). The initial scores are used to rank the library sequences, and, in the fourth and final step of the comparison, the highest scoring library sequences are aligned using a modification of the standard NWS optimization method[10,15] (Tables III and IV). The optimization employs the same scoring matrix used in determining the initial regions; the resulting optimized alignments are calculated for further analysis of potential relationships, and the optimized similarity score is reported. With the FASTP program, optimization frequently improved the similarity scores of related sequences by factors of two or three; these improvements can be seen by comparing the *init1* score with the optimized score in Tables III and IV. Because FASTA calculates an initial similarity score based on an optimization of initial regions during the library search, the initial score is much closer to the optimized score for many sequences. In fact, unlike FASTP, the FASTA method may yield initial scores that are higher than the corresponding optimized scores.

Searching with **FASTA**: G-Protein-Coupled Receptors

In the late 1980s, molecular cloning of rhodopsin, the β-adrenergic receptor, and the acetylcholine receptor demonstrated the existence of a large superfamily of membrane-bound receptors that interact with guanine nucleotide regulatory proteins[16] (G-protein-coupled receptors). A common structure with seven conserved transmembrane regions connected by variable length cytoplasmic and extracellular loops has been predicted for these

---

[14] FASTA joins an initial region only if its similarity score is greater than the CUTOFF value, a value that is approximately one standard deviation above the average score expected from unrelated sequences in the library. For a 200-residue query sequence and $ktup = 2$, this value is 28.

[15] T. Smith and M. S. Waterman, *J. Mol. Biol.* **147**, 195 (1981).

[16] R. J. Lefkowitz and M. G. Caron, *J. Biol. Chem.* **263**, 4993 (1988).

A. *ktup* =2

```
        initn   init1
<   2     1     1:=
    4     0     0:
    6     1     1:=
    8     2     2:=
   10     5     5:===
   12    28    28:==============
   14    40    40:====================
   16   122   122:==================================================
   18   202   202:==================================================
   20   415   415:==================================================
   22   836   836:==================================================
   24  1079  1079:==================================================
   26   884   884:==================================================
   28  1067  1067:==================================================
   30  1021  1097:==================================================
   32   435   519:==================================================
   34   365   460:==================================================
   36   212   301:==================================================
   38   183   203:==================================================
   40   190   168:==================================================
   42   126   103:==================================================
   44   104    47:--------------------++++++++++++++++++++++++
   46    74    50:------------------------+++++++++++
   48    85    39:------------------+++++++++++++++++++++++
   50    68    13:-------+++++++++++++++++++++++++++
   52    36    12:------+++++++++++
   54    38     2:-++++++++++++++++++
   56    15     2:-+++++++
   58    12     1:-+++++
   60    15     2:-+++++++
   62     8     5:---+
   64    10     1:-++++
   66     6     0:+++
   68     3     1:-+
   70     2     0:+
   72     6     1:-++
   74     5     0:+++
   76     2     0:+
   78     1     2:=
   80     1     0:+
   82     1     0:+
   84     0     0:
   86     1     0:+
   88     1     0:+
   90     0     0:
   92     0     0:
   94     1     0:+
   96     0     0:
   98     0     0:
  100     0     0:
>100    15    13:-------+
2224465 residues in  7724 sequences,   scan time:  0:03:07
 mean initn score:  27.8 (7.36), mean init1 score:  27.2 (6.14)
```

FIG. 4A. See legend on p. 78.

B. *ktup* =1

```
        initn    init1
<   2      0      0:
    4      0      0:
    6      2      2:=
    8      0      0:
   10      1      1:=
   12      3      3:==
   14      7      7:====
   16      9      9:=====
   18     28     28:==============
   20     44     44:======================
   22     92     92:==============================================================
   24    191    191:=====================================================================
   26    337    337:=====================================================================
   28    513    513:=====================================================================
   30    785    785:=====================================================================
   32    773    773:=====================================================================
   34    854    865:=====================================================================
   36    712    820:=====================================================================
   38    620    806:=====================================================================
   40    449    644:=====================================================================
   42    385    591:=====================================================================
   44    245    387:=====================================================================
   46    142    264:=====================================================================
   48    135    198:=====================================================================
   50    144    110:=====================================================================
   52    121     64:----------------------------++++++++++++++++++
   54    179     66:----------------------------++++++++++++++++++
   56    148     30:---------------+++++++++++++++++++++++++++++++++++++
   58    140     22:-----------++++++++++++++++++++++++++++++++++++++++++++
   60    118     19:---------++++++++++++++++++++++++++++++++++++++++++++++
   62     94     12:------+++++++++++++++++++++++++++++++++++++++++++++++++
   64     69      9:-----++++++++++++++++++++++++++++++++
   66     57      4:--++++++++++++++++++++++++++++
   68     47      3:--++++++++++++++++++++++
   70     42      4:--+++++++++++++++++++
   72     50      1:-+++++++++++++++++++++++++
   74     28      0:++++++++++++++
   76     26      1:-+++++++++++++
   78     26      0:+++++++++++++
   80     15      1:-+++++++
   82     17      0:+++++++++
   84     11      1:-+++++
   86     11      0:++++++
   88      6      0:+++
   90      6      1:-++
   92      3      0:++
   94      6      0:+++
   96      1      0:+
   98      3      0:++
  100      2      0:+
>100     27     16:-------++++++
2224465 residues in  7724 sequences, scan time:   0:17:08
  mean initn score:   36.1 (8.91), mean init1 score:  34.6 (6.62)
```

FIG. 4B. See legend on p. 78.

proteins by analogy with bacteriorhodopsin. Figures 4 and 5 and Tables III and IV show the results of a search of the SWISS-PROT protein sequence database using the β_2-adrenergic receptor as the query sequence. Examples are shown with both $ktup = 2$ and $ktup = 1$ to highlight the trade-offs between sensitivity and selectivity encountered during library searches.

Figure 4 shows the distribution of initial similarity scores calculated by FASTA for searches of the protein database with $ktup = 2$ and $ktup = 1$. Two distributions of scores are plotted, the distribution of the FASTA initial similarity score used to rank the library sequences *(initn),* and the distribution of the older FASTP initial score *(init1).* For example, in Fig. 4A there were 15 library sequences with *initn* scores of 59 or 60, but there were only 2 library sequences with *init1* scores in this interval. The values in the *initn* and *init1* columns are identical in the intervals including scores of 28 and below, because these scores fall below a threshold value (29 for this query sequence with $ktup = 2$). This example begins to show the effect of increasing the sensitivity of the search. With $ktup = 2$, there are 63 library sequences with *initn* scores greater than 60, while there are only 23 library sequences with *init1* scores greater than 60. The FASTA joining has moved scores out of the intervals between 29 and 38 into the intervals 39 and above.

The top ranking library sequences from the search with $ktup = 2$ are shown in Table III. In this example, all 17 of the top scoring sequences are G-protein-coupled receptors; only 2 of the 20 sequences with the highest similarity scores do not belong to this receptor family. Nevertheless, there are additional G-protein-coupled receptors that are not found in the top 40 sequences. One *Drosophila* opsin is ranked 98th, and a second (OPS3$DROME) is not found in the top 200 sequences with $ktup = 2$.

FIG. 4. Identification of sequences related to the β-adrenergic receptor. The β-adrenergic receptor (PIR entry QRHYB2, SWISS-PROT entry B2AR$MESAU, 418 amino acids) was used to search the Swiss-Prot protein sequence database (Release 8, August 1988), using the PAM250 scoring matrix. A total of 2,224,465 residues in 7,724 sequences were compared in 3 min ($ktup = 2$) or 17 min ($ktup = 1$) on a Sun 3/50 workstation. (A) Distribution of initial scores with $ktup = 2$. Three numbers are shown to the left of the histogram: the score reported in the histogram interval (< 2, $3-4$, . . . , $99-100$, > 100); the number of library sequences that obtained an initial similarity score in the histogram interval *(initn)*; and the number of sequences in the library with a best single initial region similarity score in the histogram interval *(init1).* The *init1* value is identical to the initial score reported by the FASTP program. When there is a difference between the number of library sequences reported in columns two and three, the column two *(initn)* values are graphed with a $+$ and the column three *(init1)* values with a $-$. The mean of the distribution of *initn* similarity scores was 27.8, with a standard deviation of 7.4. The mean best single *(init1)* similarity score was 27.2 ± 6.1. (B) Distribution of scores with $ktup = 1$. The mean initial score was 36.1 ± 8.9, and the mean *init1* score was 34.6 ± 6.6.

TABLE III
RECEPTOR SEQUENCES RELATED TO β-ADRENERGIC RECEPTOR, $ktup = 2$

SWISS-PROT entry	Definition	Score		
		initn	*init1*	Optimized
B2AR$MESAU	β_2-Adrenergic Receptor	2177	2177	2177
B2AR$HUMAN	β_2-Adrenergic Receptor	1919	1781	1917
B1AR$MELGA	β_1-Adrenergic Receptor	1140	798	1155
B1AR$HUMAN	β_1-Adrenergic Receptor	1088	768	794
ACM3$RAT	Muscarinic acetylcholine receptor M3	429	302	390
ACM1$RAT	Muscarinic acetylcholine receptor M1	411	260	353
ACM1$PIG	Muscarinic acetylcholine receptor M1	409	260	353
ACM2$PIG	Muscarinic acetylcholine receptor M2	330	259	356
ACM2$HUMAN	Muscarinic acetylcholine receptor M2	330	259	356
10 ACM4$HUMAN	Muscarinic acetylcholine receptor M4	221	183	336
ACM4$RAT	Muscarinic acetylcholine receptor M4	217	180	334
OPSR$HUMAN	Red-sensitive opsin (red cone)	128	114	186
OPSG$HUMAN	Green-sensitive opsin (green cone)	126	117	187
OPSD$BOVIN	Rhodopsin	108	67	185
OPS2$DROME	Opsin RH2 (Ocellar opsin)	101	77	256
OPSD$HUMAN	Rhodopsin	93	71	189
OPS4$DROME	Opsin RH4 (inner R7 photorecep-tor cells)	88	77	211
SYI$ECOLI	Isoleucyl-tRNA synthetase (isoleucine–RNA ligase)	86	43	47
CIN2$RAT	Sodium channel protein II, brain	81	61	61
20 OPSB$HUMAN	Blue-sensitive opsin (blue cone)	79	50	152
NUO5$DROYA	NADH ubiquinone oxidoreductase [NADH dehydrogenase (ubiquinone)]	77	47	58
A2MG$RAT	α_2-Macroglobulin precursor	76	37	39
VE1$HPV8	Probable E1 protein	76	52	56
OPSD$SHEEP	Rhodopsin (fragments)	74	42	62
CIN3$RAT	Sodium channel protein III, brain	74	61	61
POLG$POL3L	Genome polyprotein	74	45	45
Y590$TRYBR	Hypothetical protein C-590	74	51	69
UDP2$RAT	UDPGlucuronosyltransferase precursor	73	44	44
CP32$RAT	Cytochrome *P*-450IIIa2 (*P*-450PCN2)	71	51	57
30 UMUA$ECOLI	MucA protein (gene name: *mucA*)	71	58	60
98 OPS1$DROME	Opsin RH1 (outer R1–R6 photoreceptor cells)	63	47	232

TABLE IV

RECEPTOR SEQUENCES RELATED TO β-ADRENERGIC RECEPTOR, $ktup = 1$

SWISS-PROT entry	Definition	Score		
		initn	*init1*	Optimized
B2AR$MESAU	β_2-Adrenergic Receptor	2177	2177	2177
B2AR$HUMAN	β_2-Adrenergic Receptor	1925	1784	1917
B1AR$MELGA	β_1-Adrenergic Receptor	1155	813	1155
B1AR$HUMAN	β_1-Adrenergic Receptor	1135	794	794
ACM3$RAT	Muscarinic acetylcholine receptor M3	483	302	390
ACM2$PIG	Muscarinic acetylcholine receptor M2	435	250	356
ACM2$HUMAN	Muscarinic acetylcholine receptor M2	435	250	356
ACM1$RAT	Muscarinic acetylcholine receptor M1	427	260	353
ACM1$PIG	Muscarinic acetylcholine receptor M1	425	260	353
ACM4$HUMAN	Muscarinic acetylcholine receptor M4	424	247	336
ACM4$RAT	Muscarinic acetylcholine receptor M4	420	244	334
OPS1$DROME	Opsin RH1 (outer R1–R6 photoreceptor cells)	190	147	232
OPS2$DROME	Opsin RH2 (ocellar opsin) (gene name: *RH2*)	186	147	256
OPS4$DROME	Opsin RH4 (inner R7 photoreceptor cells)	166	83	211
OPSG$HUMAN	Green-sensitive opsin (green cone)	153	133	187
OPSD$BOVIN	Rhodopsin	146	102	185
CIN2$RAT	Sodium channel protein II, brain	136	61	61
OPSD$SHEEP	Rhodopsin (fragments)	131	64	109
OPSD$HUMAN	Rhodopsin	131	90	189
OPSR$HUMAN	Red-sensitive opsin (red cone)	130	130	186
RCEM$RHOVI	Reaction center protein M chain	123	69	78
CIN1$RAT	Sodium channel protein I, brain	122	61	61
CIN3$RAT	Sodium channel protein III, brain	120	61	61
NUO4$DROYA	NADH ubiquinone oxidoreductase	112	52	92
ARG2$CANUT	Arginine metabolism regulation protein II	105	56	56
POLG$COXB4	Genome polyprotein	103	48	48
CINA$ELEEL	Sodium chanel protein	102	60	74
HIP1$YEAST	Histidine permease (gene name: *HIP1*)	100	44	47
GER2$BACSU	Spore germination protein II (GERA)	99	54	54

TABLE IV *Continued*

	SWISS-PROT entry	Definition	Score		
			initn	*init1*	Optimized
30	CO8B$HUMAN	Complement C8 β chain precursor	98	53	59
	MELB$ECOLI	Melibiose carrier protein	97	47	50
	BGAL$KLEPN	β-Galactosidase	97	51	51
	CYB$DROYA	Cytochrome *b*	95	43	49
	VGV7$ROTAS	Glycoprotein VP7	94	53	53
	COX1$TRYBR	Cytochrome-*c* oxidase polypeptide I	93	55	62
	RCEM$RHOSH	Reaction center protein M chain	93	66	76
	VME1$MHVJH	E1 glycoprotein (matrix glycoprotein)	93	47	47
	Y590$TRYBR	Hypothetical protein C-590	93	65	69
	CP32$RAT	Cytochrome *P*-450IIIa2 (*P*-450PCN2)	93	56	57
40	YCO1$PARDE	Cox locus hypothetical protein 1	92	53	68
45	OPS3$DROME	Opsin RH3 (inner R7 photoreceptor cells)	90	62	142
60	OPSB$HUMAN	Blue-sensitive opsin (blue cone)	86	69	152

Table III also shows the effect of the "joining" step in the calculation of the initial similarity score used for ranking. All the top 20 library sequences have *initn* scores that are greater than the *init1* score, and for several muscarinic acetylcholine receptors the initial FASTA score is higher than the optimized score. For the muscarinic receptor sequences there are two high scoring initial regions that require a gap longer than 32 residues for alignment (Fig. 6), and the lower scoring region is not included in the optimal score.

One can best identify members of the G-protein-coupled receptor family by examining all three of the scores reported by FASTA. The 2 unrelated sequences ranked in the top 20 with *ktup* = 2 [isoleucyl-tRNA synthetase (isoleucine–tRNA ligase) and the sodium channel protein II] have high *initn* scores, but the *init1* scores do not increase when gaps are allowed in the calculation of the optimized score. Likewise, the 98th ranked *Drosophila* opsin is striking because the *init1* score increases almost 5-fold when gaps are introduced. While the *initn* score, which is used to rank the library sequences, provides a very sensitive measure of protein sequence similarity, the relationship between the *init1* score and the optimized score provides a more selective perspective.

The effect of increasing the sensitivity of the search can be seen in Fig.

OPS3$DROME   Opsin RH3 (inner R7 photoreceptor cells)
19.8% identity in 243 aa overlap

```
                                 10        20        30        40        50
b2adren                   MGPPGNDSDFLLTTNGSHVPDHDVT-EERDEAWVVGMAILMSVIVLAIVFGNVLVI
                                      :.:.: .: :. :.    ...X    ..: ..:: :::
OPS3$D    MESGNVSSSLFGNVSTALRPEARLSAETRLLGWNVPPEELRHIPEHWLTYPEPPESMNYLLGTLYIFFTLMSMLGNGLVI
               10        20        30        40        50        60        70        80

             60        70        80        90       100       110       120       130
b2adren   TAIAKFERLQTVTNYFITSLACADLVMGLAVVPFGASHILMKMWNFGNFWCEFWTSIDVLCVTASIETLCVIAVDRYIAI
           ... ..:.: .: ....:: :..: .. ..:^V . . ... ...: :.... :.... : ..:. .:  :: ::. .:
OPS3$D    WVFSAAKSLRTPSNILVINLAFCDFMM-MVKTPIFIYNSFHQGYALGHLGCQIFGIIGSYTGIAAGATNAFIAYDRFNVI
               90       100       110       120       130       140       150

            140       150       160       170       180       190       200       210
b2adren   TSPFKYQSLLTKNKARMVILMVWIVSGLTSFLPIQMHWYRATHQKAIDCYHKETCCDFFTNQAYAIASSIVSFYVPLVVM
           :.:.  .. .:..::  ...   ..         : :  ...   . : :... ..    . :  .: ...
OPS3$D    TRPM--EGKMTHGKAIAMIIFIYMYATPWVVACYTETWGRFVPEGYLTSCTFDYLTDNFDTRLFVACIFFFSFVCPTTMI
          160       170       180       190       200       210       220       230

            220       230       240       250       260       270       280       290
b2adren   VFVYSRVFQVAKRQLQKIDKSEGRFHSPNLGQVEQDGRSGHGLRRSSKFCLKEHKALKTLGIIMGTFTLCWLPFFIVNIV
          .. ::..  . .  . ...  ..  .:     ....... ..:
OPS3$D    TYYYSQIVGHVFSHEKALRDQAKKMNVESLRSNVDKNKETAEIRIAKAAITICFLFFCSWTPYGVMSLIGAFGDKTLLTP
          240       250       260       270       280       290       300       310

            300       310       320       330       340       350       360       370
b2adre    HVIQDNLIPKEVYILLNWLGYVNSAFNPLIYCRSPDFRIAFQELLCLRRSSSKAYGNGYSSNSNGKTDYMGEASGCQLGQ

OPS3$D    GATMIPACACKMVACIDPFVYAISHPRYRMELQKRCPWLALNEKAPESSAVASTSTTQEPQQTTAA
          320       330       340       350       360       370       380

            380       390       400       410
b2adren   EKESERLCEDPPGTESFVNCQGTVPSLSLDSQGRNCSTNDSPL
```

FIG. 5. Alignment of the β_2-adrenergic receptor and *Drosophila* opsin. The FASTA alignment ($ktup = 1$) of the β_2-adrenergic receptor (labeled b2adren) and a very distantly related opsin (OPS3$DROME, labeled OPS3$D), ranked 45th in Table IV, is shown. The initial score for this comparison is 90, the score of the best single initial region is 62, and the score of the aligned amino acids in the optimized region denoted by ":" and "." is 142. Aligned amino acid identities are denoted by ":"; substitutions with PAM250 scores of zero or greater are denoted by ".".

4B and Table IV, which show the results of a search with $ktup = 1$. In the search with $ktup = 2$, there were only 19 library sequences with *initn* scores greater than 80. With $ktup = 1$, that number increases to 93. Of those 93 sequences, only 21 belong to the G-protein-coupled receptor family (Table IV). However, the *Drosophila* opsin sequence OPS1$DROME, which was ranked 98th with $ktup = 2$, is now ranked 12th, and *Drosophila* opsin sequence OPS3$DROME, which was not found in the top 200 seqeunces with $ktup = 2$, is now ranked 45th. (With $ktup = 2$, the OPS3$DROME similarity scores were: $initn = 45$, $init1 = 34$, $opt = 34$.) Nevertheless, the human *mas* oncogene, which is also a member of the G-protein-coupled receptor family, is ranked 203th with $ktup = 1$. Thus, increasing the sensitivity of the search, both by joining initial regions and by looking for

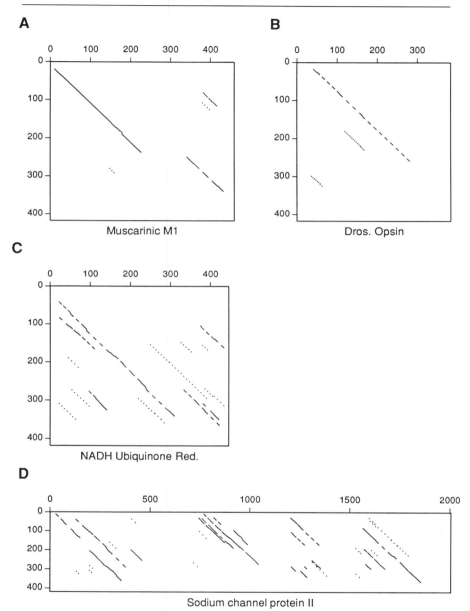

FIG. 6. Local sequence similarities between the β_2-adrenergic receptor and related and
unrelated membrane proteins. The β_2-adrenergic receptor was compared with two related
G-protein-coupled receptors with the PLFASTA program: (A) the M1 subtype of the muscarinic acetylcholine receptor (SWISS-PROT entry ACM1$PIG) and (B) the *Drosophila* opsin

similarities with *ktup* = 1, raises the scores of distantly related sequences but also dramatically increases the scores of unrelated sequences.

Figure 5 shows the alignment of a distantly related G-protein-coupled receptor, the *Drosophila* opsin found in the inner R7 photoreceptor cells (OPS3$DROME), with the β_2-adrenergic receptor. The optimization procedure uses the standard PAM250 matrix to score identities and substitutions, and it penalizes -12 for the first residue in a gap and -4 for each additional residue in a gap. These gap penalties can be changed by modifying the scoring matrix file (Appendix 2). The amino-terminal boundary of the *init1* initial region is denoted by an "X" at residue 39 in b2adren (Fig. 5). The carboxy-terminal boundary of this region is denoted by a "^" at residue 89 in b2adren and a "v" at residue 114 in OPS3$D. These two residues were aligned without gaps when the region was identified in the initial FASTA comparison, but the introduction of the gap after residue 117 in OPS3$D caused the two phenylalanines to be offset from one another. The best initial region found with *ktup* = 1 includes transmembrane regions I and II of the β_2-adrenergic receptor and *Drosophila* opsin; optimization extends the alignment to include transmembrane regions III, IV, and V. The alignment of regions II and III reveals that the *Drosophila* opsin has several residues that are conserved in almost all members of the G-protein-coupled receptor superfamily in the appropriate relative positions (Leu[75], Asp[79], Arg[131]).

In this example, FASTA is able to align only five of the seven membrane-spanning regions in the two receptors. However, FASTA can align all seven of the transmembrane regions of the β_2-adrenergic receptor with another *Drosophila* opsin, OPS4$DROME, which shares 72% identity with OPS3$DROME. FASTA will align all seven transmembrane regions of the β_2-adrenergic receptor with OPS3$DROME if the penalty for each residue in a gap is decreased from -4 to -2.

(Swiss-Prot entry OPS3$DROME) shown in Fig. 5. Also shown are similarities to (C) NADH ubiquinone oxidoreductase and (D) a high-ranking but unrelated rat sodium channel protein II (SWISS-PROT entry CIN2$RAT). The vertical scale refers to the β_2-adrenergic receptor sequence in each case. Lines are drawn to indicate the residues aligned in the local similar regions found by the LFASTA program. The sequences were compared with *ktup* = 1. (A) The aligned region denoted by the solid line from residue pair 11, 20 (ACM1$PIG, b2adren) to 227, 238 has an initial score of 260 and an optimized score of 353. The region denoted by the offset diagonal broken line in the lower right corner (341, 250 to 340, 433) has an initial score of 139 and an optimized score of 173. The other off-diagonal regions have initial scores ranging from 38 to 56 and optimized scores ranging from 39 to 63. (B) The region denoted by the dashed line near the central diagonal (11, 20 to 227, 238) has an initial score of 62, and an optimized score of 140. (D) Eleven similar regions are shown, with initial scores ranging from 35 to 61, and optimized scores ranging from 35 to 90. The region from 130, 39 to 378, 290 has the highest optimized score, 90, and an initial score of 39.

The partial alignment (Fig. 5) includes two additional values that can be used to evaluate the significance of the sequence similarity: the percent identity and the length of the aligned region. In general, the length of the aligned region is a better indicator of significance than the actual percent identity. It is not uncommon to find short regions of sequences (15–40 amino acids) that share 30–50% identity in unrelated proteins. However, sequences that share more than 20–25% identity over their entire length almost always share a common ancestor, and it is possible to show convincingly that sequences which share as little as 15% identity over their entire length are homologous.

Evaluating Sequence Similarities: RDF2 and LFASTA

Examination of the 40 top scoring sequences in Table IV suggests that the β-adrenergic receptor shares strong similarity, and almost certainly common evolutionary ancestry, with the muscarinic acetylcholine receptors, rhodopsin, and the opsins.[17] In addition, many of the highly ranked sequences that do not belong to the G-protein-coupled receptor family are membrane proteins, which presumably share sequence similarity because of the requirement to form structures that span membranes. Often, these high ranking but unrelated library sequences have high *initn* scores but much lower *init1* scores, and the *init1* scores increase very little, if at all, when the band around the best initial region is optimized. However, if the G-protein-coupled receptors were not in the library, one might propose that sodium channel protein II or NADH ubiquinone oxidoreductase [NADH dehydrogenase (ubiquinone)] share common ancestry with the β_2-adrenergic receptor. The sodium channel protein is tempting because it has such a high initial score, but one must be cautious because the *init1* score does not increase with optimization. The NADH ubiquinone reductase is even more tantalizing because it has a high *initn* score, and its *init1* score almost doubles with optimization.

The RDF and RDF2 programs test whether high similarity scores are likely to reflect sequence similarity that is due to common ancestry or simply a locally biased amino acid composition. These programs compare two sequences, calculating initial and optimized scores, and then shuffle the second sequence a specified number of times (100 to 200 shuffles are recommended), again calculating the initial and optimized scores. The earlier RDF program shuffled the second sequence by moving each of its residues to a random position in the shuffled sequence, thus preserving the length and amino acid composition of the sequence, and calculated *init1*

[17] R. A. F. Dixon, B. K. Kobilka, D. J. Strader, J. L. Benovic, H. G. Dohlman, T. Frielle, M. A. Bolanowski, C. D. Bennett, E. Rands, R. E. Diehl, R. A. Mumford, E. E. Stater, I. S. Sigal, M. G. Caron, R. J. Lefkowitz, and C. D. Strader, *Nature (London)* **321**, 75 (1986).

and optimized scores using the FASTP procedure. With this shuffling procedure, however, local regions of sequence bias are spread throughout the shuffled sequence. The RDF2 program provides a more stringent shuffling procedure in which the second sequence is shuffled in short blocks (usually 10 residues). With RDF2, residues 1–10 of the original sequence become residues 1–10 of the second sequence, but in a different order, and so on for 21–30, 31–40, etc. A local shuffling procedure is particularly appropriate for examining sequences where local sequence bias is expected, as in the membrane-spanning helices of membrane proteins. RDF2 then calculates three similarity scores for each shuffled sequence using the FASTA procedure.

RDF2 was used to evaluate the similarities between the β_2-adrenergic receptor and either members of the G-protein-coupled receptor family or other unrelated but high scoring membrane proteins (Table V). RDF2 provides several perspectives from which one can evaluate the statistical significance of a similarity score. One perspective is the z value, which is calculated by subtracting the mean score of the randomly shuffled sequences from the score of the unshuffled sequence and then dividing by the standard deviation of the distribution of shuffled scores. RDF2 calculates z values for each of the three similarity scores calculated by FASTA, but the z value for the optimized score is the most informative (Table V). For the example of the β_2-adrenergic receptor versus *Drosophila* opsin OPS3$DROME, the optimized score is 142, the mean optimized score for the shuffled sequences is 50.3, and the standard deviation of the distribution of optimized shuffled sequence scores is 10.4. Therefore, the unshuffled score is 8.8 standard deviations above the mean, or has a z value of 8.8. In an earlier paper,[1] we suggested that one should be skeptical of conclusions based on sequence similarity scores with z values less than 3, and more confident when the z values are greater than 6. The z values in Table V tend to support this guideline: the lowest scoring member of the G-protein-coupled receptor family has a z value of 4.3, and the highest scoring unrelated sequence has a z value of 2.9.

Unfortunately, the z values determined in a Monte Carlo analysis become less informative as the distribution of similarity scores diverges from a normal distribution. Since FASTA searches have a more asymmetric distribution of scores than FASTP ones, an alternative perspective focuses on the highest scores of the shuffled sequences (the maximum column in Table V). For example, after 200 shuffles, the highest optimized similarity score between the β_2-adrenergic receptor and a shuffled copy of the NADH ubiquinone reductase was 100; 3 of the 200 shuffled sequences obtained optimized similarity scores greater than the value of 92 found for the unshuffled sequence. In contrast, the highest optimized similarity score

TABLE V
STATISTICAL SIGNIFICANCE OF SIMILARITY SCORES[a]

Shuffled Sequence	initn score				Optimized score			
	Unshuffled	Average	Maximum	z value	Unshuffled	Average	Maximum	z value
Drosophila opsin OPS4$DROME	166	51.5 ± 14.9	102	7.7	211	48.8 ± 9.9	96	16.3
Drosophila opsin OPS3$DROME	90	52.3 ± 13.8	94	2.7	142	50.3 ± 10.4	93	8.8
Human opsin (blue-sensitive)	86	56.3 ± 16.0	113	1.8	153	50.7 ± 9.5	90	10.7
Human *mas* oncogene	75	55.0 ± 15.0	101	1.3	101	53.1 ± 10.9	122	4.3
NADH ubiquinone oxidoreductase NUO4$DROYA	112	67.5 ± 18.8	127	2.4	92	59.6 ± 11.0	100	2.9
Rat sodium channel CIN2$RAT	136	70.5 ± 18.0	118	3.66	61	54.0 ± 9.8	93	0.7
Eel sodium channel CINA$ELEEL	102	78.3 ± 18.5	122	1.29	74	59.7 ± 12.3	112	1.2

[a] RDF2 was used to compare the hamster β_2-adrenergic receptor with the sequences listed using *ktup* = 1. Each sequence was shuffled 200 times, using a local shuffle with a window of 10 residues.

for the shuffled *Drosophila* opsin sequence was 93, substantially lower than the optimized score of 142 for the unshuffled sequence. Thus, the *Drosophila* opsin sequence appears to be related to the β_2-adrenergic receptor, and the NADH ubiquinone reductase unrelated, by both the criteria of z values and the highest shuffled scores.

Another way to evaluate sequence similarity is to compare the sequences with LFASTA and PLFASTA. While FASTA reports a similarity score based on the alignment of multiple initial regions that do not overlap one another and displays the one best alignment between the two sequences, LFASTA displays all the local regions of similarity shared between two sequences. LFASTA displays the actual local alignments and similarity scores, and PLFASTA plots the alignments in a form similar to a dot matrix plot (Figs. 6 and 7). Plots of local similarities between the β_2-adrenergic receptor and four related and unrelated sequences are shown in Fig. 6. The PLFASTA comparison of the β_2-adrenergic receptor with the muscarinic receptor shows why the initial similarity score for this comparison is higher than the optimized score; there are two regions with strong similarity that are shared between these two sequences, but they are separated by an insert of 70 amino acids in the muscarinic acetylcholine receptor sequence. These two regions of similarity are joined in the calculation of the *initn* initial similarity score by FASTA. Both of the protein sequences that are related to the β_2-adrenergic receptor show a strong region of similarity along a single major diagonal, with a few weak off-diagonal regions. In contrast, the NADH reductase and the sodium channel have a large number of regions with modest similarity scores. In the case of the sodium channel protein, four distinct regions of similarity are found. These regions correspond to the four clusters of six transmembrane segments that have been predicted.[18] Although none of the local regions of similarity in the unrelated sequences has a very high score, there are so many of them that they can be joined together. This large number of similar regions accounts for the high initial similarity scores calculated for these sequences as well as the much lower *init1* and optimized similarity scores.

LFASTA, PLFASTA, and Local Similarity

LFASTA and PLFASTA can also be used to examine repeated structures in proteins, or to map exons in genomic clones using cDNA sequences (see Fig. 2 in Ref. 3). LFASTA used the same first two steps for finding initial regions that FASTA uses. Instead of saving ten initial re-

[18] W. A. Catterall, *Science* **242**, 50 (1988).

A

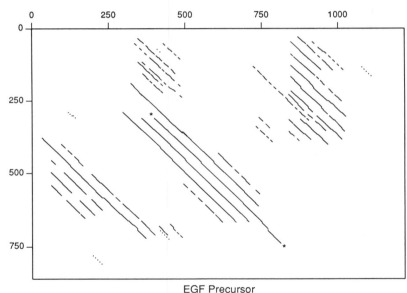

B

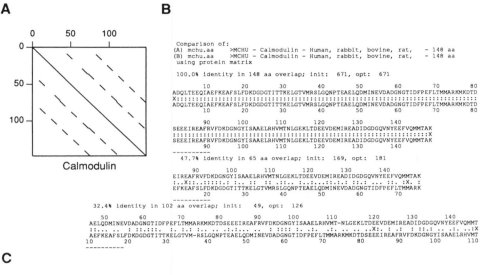

FIG. 7. Examination of local sequence similarities with LFASTA and PLFASTA. (A) The PLFASTA program was used to compare the human calmodulin sequence (PIR entry MCHU) with itself (*ktup* = 2). The ordinate and abscissa indicate the position in the calmodulin sequence. (B) Sample output from the LFASTA program comparing calmodulin to itself. Only two of the four off-diagonal alignments are shown, since the other two are symmetrically identical. As in Fig. 5, "X" denotes the boundaries of the initial region found by

gions, however, LFASTA saves all diagonal regions with similarity scores greater than a threshold. In addition, instead of focusing on a single region, LFASTA computes a local alignment for each initial region. The ends of the optimized local region of similarity are determined by scanning backward, then forward, in the sequences. Starting at the end of the initial region, a Smith and Waterman type of optimization[15] is performed in a band that is centered around the initial region. The optimization continues past the beginning of the initial regions until all possible alignment scores have gone to zero. The location where the maximal local similarity score occurred during the backward scan is saved and used as the starting position of a second optimization that proceeds in the forward direction. The forward optimization proceeds in the same way until all possible alignment scores have gone to zero; then the position where the new maximal local similarity score occurred is saved. An optimal path starting at that maximum is then generated by a standard traceback procedure.[10] In some cases, because of the dynamic boundaries of the optimization, several initial regions may be included in a single optimization. A check is therefore made to prevent the output of identical alignments.

Figure 7 shows two examples of how LFASTA and PLFASTA can be used to examine internal duplications and repeated domains in protein sequences. Figure 7A,B displays duplications within the human calmodulin sequence. Calmodulin is a member of a superfamily of calcium-binding proteins that share an E-F hand structure; different members of this family contain from two to six E-F hand calcium-binding domains. Calmodulin contains four E-F hand domains, which can be numbered 1 through 4. The similarity scores and percent identities determined by LFASTA (Fig. 7B) suggest that domains 1 and 2 diverged from 3 and 4 more recently (they share 48% identity) than 1 diverged from 2 or 3 from 4.

Figure 7C shows the complex pattern of exon-shuffling that apparently took place during the evolution of the LDL receptor and the EGF precursor.[4] The four solid diagonal lines that align residues 250–750 of the LDL receptor with residues 300–750 of the EGF precursor represent alternative alignments of an EGF precursor domain that is present three times in the

LFASTA. (C) The PLFASTA program was used to compare the human LDL receptor (PIR entry QRHULD) and the mouse EGF precursor (PIR entry EGMSMG) with $ktup = 1$. The position in the LDL receptor is shown on the ordinate; the position in the EGF precursor is shown on the abscissa. The highest scoring local alignment, indicated with an asterisk, has an initial score of 274 and an optimized score of 654. The other solid diagonal lines indicate regions with optimized similarity scores ranging from 252 to 350, the long-dashed diagonal lines show regions with scores ranging from 110 to 144, the short-dashed lines show regions with scores between 50 and 100, and the dotted lines indicate regions with scores less than 50.

LDL receptor and four times in the EGF receptor. The diagonal indicated by the asterisk would be the only alignment reported by FASTA.

TFASTA and DNA Library Searches

Included in the FASTA package of programs is TFASTA, a program for comparing a protein sequence to a DNA sequence by translating the DNA sequence in all six reading frames. The value of doing sequence searches with protein rather than DNA sequences cannot be overemphasized. A search of the mammalian portion of GenBank (Release 58) with the mRNA sequence encoding the hamster β_2-adrenergic receptor (GenBank locus HAMARBR, 2015 nucleotides), which required 35 min on a Sun 3/260 computer, did not detect any G-protein-coupled receptors except other β-adrenergic and muscarinic acetylcholine receptors. In contrast, a search of the same database with TFASTA and $ktup = 2$ took 17 min and revealed the mammalian G-protein-coupled receptors shown in Table III. Since TFASTA compares a protein to a DNA sequence in all six reading frames, it can also be used to check for frameshifts in cDNA sequences when other homologs for the protein coded by the cDNA are known.

Short Sequences

Although FASTA was originally designed to search protein sequence libraries for homologous sequences, it can also be used to search for oligonucleotides and oligopeptides. Nevertheless, a few adjustments to the normal search parameters may be required for satisfactory results. When searching with short query sequences, the *ktup* parameter should almost always be set to 1. In addition, care must be taken to make certain that the CUTOFF score is not set too high by default. The CUTOFF score is a value that is used to decide whether a particular library sequence should be saved for later display and optimization. The CUTOFF score is calculated by a predetermined formula based on the length of the sequence and the *ktup* value. If the sequence is so short that the CUTOFF score is less than three times the length of the sequence, the program warns that the CUTOFF value may be too high and prompts the user for a new value. Unfortunately, this value may be inappropriate for DNA sequences. In these cases, the program may not prompt even though the CUTOFF value is too high, with the result that even though all the library is scanned, no sequences are saved. To prevent this from happening, the CUTOFF value can be set to a very low number such as 5 by using the symbol CUTOFF. In addition, the default values in the DNA scoring matrix may be too conservative for oligonucleotide sequences ($+4$ for an identical match, -3 for a mismatch,

+2 for a match to an ambiguous base, −12 for the first residue in a gap, −4 for additional residues).

Searching with Different SMATRIX

It is possible to use other scoring matrices with the FASTA programs, and files for alternative matrices are included with the program package. To use an alternative matrix, one can either define the environment symbol SMATRIX to the name of the file that contains the alternative matrix or enter the alternative SMATRIX file name on the command line preceded by "−s." By setting SMATRIX to a different matrix file at the beginning of a work session, one can use the same scoring parameters for FASTA, RDF2, or LFASTA. However, setting SMATRIX can cause FASTA to treat a DNA sequence as a protein sequence, or vice versa, so one can also change the SMATRIX by including its file name on the command line using the "−s" option. For example,

FASTA −s codaa.mat

will cause FASTA to use a matrix based on the genetic code. Our experience has been that different scoring matrices can change the sensitivity of a FASTA search (most matrices are less sensitive than the PAM250 matrix), but that the relative ranking of the library scores remains about the same. Alternative scoring matrices and modification of the gap penalties have more effect on the precise sequence alignments than on relative similarity scores.

Considerable flexibility has been built into FASTA, so that virtually every aspect of the search process can be modified. For example, line 6 in the scoring matrix file (Appendix 2) specifies the relationship between the residues of a sequence and the number used in the lookup table calculation. Thus, during the lookup calculation, it is possible to cause several different residues to have the same value and appear to be the same. This mapping of residues to lookup values is used for DNA searches so that ambiguous residues such as R (purine, A or G) or Y (pyrimidine, C or T) can match sequences with no ambiguity codes. In this case, an R in the query or library sequence would match an A or R (but not a G), and a Y would match a C or Y. While this mapping method is not perfect, since a match between an R and G is not found during the first step of the search (the R–G match would be scored correctly in the second rescanning step), it allows ambiguous codes to be recognized part of the time even when they are rare in the library.

This mapping for the lookup table can also be used to change the way FASTA looks for initial regions within protein sequences. For example, one might classify amino acids into six groups: acidic, basic, small side chain, large hydrophobic side chain, aromatic, and cysteine. Each of the amino acids could then be placed in a group and FASTA would search for initial regions with high densities of "identities," but any member of the same group would be considered an "identity."

A similar strategy can be used to scan a set of consensus DNA binding sites. FASTA used the IUPAC–IUB code for ambiguous nucleotides (Appendix 4), which allows all 15 possible nucleotides and ambiguities to be specified. Thus, one could make a library of known transcription factor binding sites and compare binding sites for a newly characterized factor by comparing the new binding site to the library. Some judgment would have to be used in encoding of the ambiguous residues, and FASTA cannot distinguish between an A–R match at one position and an A–R match at a different position (although a modified version of FASTA has been developed that does allow position-specific scoring). Searches with short sequences against libraries with large numbers of ambiguous residues should always be done with $ktup = 1$.

Output Options

FASTA also includes a variety of other input and output options, which are listed in Appendix 3. Several output options control how much of the sequence alignment is shown, how identities and substitutions are highlighted, and the number of residues displayed on each line. However, one must remember that FASTA highlights only identities and substitutions in the aligned region of the two sequences, and this region may not contain some residues at the ends of the sequences. In addition, the "−Q" (quiet) option can be used to allow the FASTA program to run without requesting any additional input. This option allows one to do FASTA searches in the background on some computers, or to do several FASTA searches, one after another, on an IBM-PC.

Summary

The FASTA program can search the NBRF protein sequence library (2.5 million residues) in less than 20 min on an IBM-PC microcomputer and unambiguously detect proteins that shared a common ancestor billions of years in the past. FASTA is both fast and selective because it initially considers only amino acid identities. Its sensitivity is increased not only by using the PAM250 matrix to score and rescore regions with large numbers of identities but also by joining initial regions. The results of searches with FASTA compare favorably with results using NWS-based programs that are 100 times slower. FASTA is slightly less sensitive but considerably more selective. It is not clear that NWS-based programs would be more successful in finding distantly related members of the

G-protein-coupled receptor family. The joining step by FASTA to calculate the *initn* score is especially useful for sequences that share regions of sequence similarity that are separated by variable-length loops.

FASTP and FASTA were designed to identify protein sequences that have descended from a common ancestor, and they have proved very useful for this task. In many cases, a FASTA sequence search will result in a list of high scoring library sequences that are homologous to the query sequence, or the search will result in a list of sequences with similarity scores that cannot be distinguished from the bulk of the library. In either case, the question of whether there are sequences in the library that are clearly related to the query sequence has been answered unambiguously. Unfortunately, the results often will not be so clear-cut, and careful analysis of similarity scores, statistical significance, the actual aligned residues, and the biological context are required. In the course of analyzing the G-protein-coupled receptor family, several proteins were found that, because of a high *initn* score and a low *init1* score that increased almost 2-fold with optimization, appeared to be members of this family which were not previously recognized. RDF2 analysis showed borderline z values, and only a careful examination of the sequence alignments that focused on the conserved residues provided convincing evidence that the high scores were fortuitous. As sequence comparison methods become more powerful by becoming more sensitive, they become more likely to mislead, and even greater care is required.

Appendix 1. FASTA File Formats

FASTA and TFASTA can search library files in the following formats on non-VAX/VMS systems:

(1)   FASTA/DM (query sequence), library type 0

```
>SEQID1 - title line
either protein sequence or DNA sequence
>SEQID2 - comment line
AGTHKPRY...
```

(2)   GENBANK tape format, library type 1

```
LOCUS       HUMHBB  ....
DEFINITION  ....
ORIGIN      .....
         1  ACGT....
```

The GENBANK DNA sequence library is available on tapes in this format from:

```
GenBank
c/o IntelliGenetics, Inc.
700 El Camino Real East
Mountain View, CA 94040
```

(3)   PIR Codata format (library type 2)

```
ENTRY               CCHU        #Type Protein
TITLE               Cytochrome c - Human
SEQUENCE
                    5 ...
          1 A F T G H I E W ...
```

The NBRF/PIR protein sequence library is available in this format from:

Protein Identification Resource
National Biomedical Research Foundation
Georgetown University
3900 Reservoir Rd., N. W.
Washington, D.C. 20007

(4)   EMBL/SWISS-PROT Format, library type 3

```
ID   16K$TRVPS       STANDARD;      PRT;   141 AA.
DE   16 KD PROTEIN.
SQ   SEQUENCE   141 AA;  16297 MW;  93420 CN;
     DVYNCCGRSH LEKCRKRVEA RNREIWKQIR RIQAESSSAT RKKSHNSKNS KKKFKEDREF
```

DNA and protein sequence libraries in this format are distributed by:

EMBL Data Library
European Molecular Biology Laboratory
Postfach 10 2209
D-6900 Heidelberg
Federal Republic of Germany

(5)   IntelliGenetics format, library type 4

```
;comment
;comment
SEQID
ABCDEFG...
```

(6)   GenBank compressed floppy disk format, library type 9. Files in this format containing the Gen-Bank DNA sequence library are distributed on IBM-PC and Macintosh floppy disks by IntelliGenetics.

On VAX/VMS systems, the FASTA programs can read sequences in the NBRF/PIR VAX/VMS file format, and the University of Wisconsin Genetics Computer Group format.

Appendix 2. SMATRIX file

An sample SMATRIX file for DNA sequences is shown. The line numbers are referred to in the text below.

```
;D standard DNA scoring matrix                              1
 1 45 80 5 6 80 4                                           2
 -12 -4                                                     3
 *  @  0  1  2                                              4
 A  C  G  T  R  Y  M  W  S  K  D  H  V  B  N               5
 0  1  2  3  0  1  0  0  1  2  0  0  0  1  0               6
 4                                                          7
-3  4
-3 -3  4
-3 -3 -3  4
 2 -1  2 -1  2
-1  2 -1  2 -2  2
 2  2 -1 -1  0  0  2
 2 -1 -1  2  0  0  0  2
-1  2  2 -1  0  0  0  0  2
-2 -2  1  1  0  0  0  0  0  2
 1 -2  1  1  1  0  0  1  0  1  1
 1  1 -2  1  0  1  1  1  0  0  0  1
 1  1  1 -2  1  0  1  0  1  0  0  0  1
-2  1  1  1  0  1  0  0  1  1  0  0  0  1
 0  0  0  0  0  0  0  0  0  0  0  0  0  0  0
```

1.   ;P or ;D, this comment, if present, is used to determine whether sequences should be labeled as amino acids (aa) or nucleotides (nt).

2.   Scoring parameters:

       FACT BESTOFF BESTSCALE BKFACT BKTUP BESTMAX HISTSIZ

       KFACT is used in the "diagonal method" search for the best initial regions, for proteins, KFACT = 4, for DNA, KFACT = 1 (used only if PAMFACT=0).

       BESTOFF, BESTSCALE, BKFACT, BKTUP and BESTMAX are used to calculate the cutoff score. The bestcut parameter is calculated from parameters 2 - 6. If N0 is the length of the query sequence:
       BESTCUT = BESTOFF + N0/BESTSCALE + BKFACT*(BKTUP-KTUP)
       if (BESTCUT>BESTMAX) BESTCUT=BESTMAX
       HISTSIZ is the size of the histogram interval.

3.   Deletion penalties. The first value is the penalty for the first residue in a gap, the second value is the penalty charged to each subsequent residue in a gap.

4.   End of sequence characters. These are not required, since FASTA knows how to find the beginning of a library sequence, but they can be used if sequences have additional comments after the end. If not used, the line must be left blank.

5.   The alphabet. The program automatically converts upper to lower case and vice-versa.

6.   The lookup table values for each letter in the alphabet. This allows several characters to be hashed to the same value, e.g. a DNA sequence alphabet with A = adenosine, R = purine, N = any base, would have each of these characters treated as 0. The lowest hash value should be 0.

7ff. The lower triangle of the symmetric scoring matrix. There should be exactly as many lines as there are characters in the alphabet, and the last line should have n-1 entries. The program does not check for the length of each line, so it is easy to use an incorrect matrix by having fewer entries in the scoring matrix than in the alphabet.

Appendix 3. FASTA Options

Scoring parameters, output line lengths, and other features of FASTA can be modified either by setting an environment symbol or on the command line. For example, to have alignments be displayed with 80 residues per line, one can either set an environment variable:

```
set LINELEN=80 (PC-DOS)   or  FASTA -1 80
```

Command Line Option	Environment Symbol	Function
-a	SHOWALL=1	Normally the optimized region of an alignment is shown in context, but the complete sequence may not be shown if the optimized region does not extend near the end of the sequence. With this option, complete sequences are always shown.
-c #	CUTOFF=#	The CUTOFF value (#) is the threshold for saving a sequence in a list of sequences to be sorted and optimally aligned after the search. This value is also used as the threshold for the optimal alignment of initial regions in the second step of FASTA.
-f	PAMFACT=1	Use the newer FASTA variable SMATRIX score for a *ktup* match. Default for protein comparisons.
-k	PAMFACT=0	Use a constant (FASTP) score in scan for a *ktup* match. Default for DNA.
-l file	FASTLIBS	File name for the location of library menu file.
-m #	MARKX = #	(0, 1, 2) MARKX modifies the way aligned residues are highlighted. MARKX=0 (default) MARKX=1 MARKX=2 MWRTCGPPYT MWRTCGPPYT MWRTCGPPYT ::..:: ::: xx X ..KS..Y... MWKSCGYPYT MWKSCGYPYT
-o #		The number of scores and alignments to be reported by default. (Used in conjunction with -Q).
-p #	GAPPEN	The gap threshold for joining two initial regions in the calculation of the *initn* score. Normally set to the CUTOFF value.

-Q		Quiet mode - FASTA does not prompt for any input. The default number of scores and alignments are displayed on the terminal or written to the standard output file. This option is used for running in batch mode, or in the background.
-s *file*	SMATRIX	The scoring matrix is read from *file*.
-w #	LINLEN=#	Number of residues per line for sequence alignments. This value must be less than 200.
-3		(TFASTA only) translate only three forward frames.

Not all of these options are appropriate for all of the programs. The options above are used by FASTA and TFASTA. RDF2 uses -c, -f, -k, and -s.

Appendix 4. Codes for Ambiguous Nucleotides*

Code	Nucleotide
A	Adenine
C	Cytosine
G	Guanine
T	Thymine
U	Uracil, treated as T
R	Purine, A or G
Y	Pyrimidine, C or T
M	A or C
W	A or T
S	C or G
K	G or T
D	A, G, or T (not C)
H	A, C, or T (not G)
V	A, C, or G (not T)
B	C, G, or T (not A)
N	A, G, C, or T

* FASTA also recognizes X.

[6] Searching through Sequence Databases

By RUSSELL F. DOOLITTLE

Introduction

Certainly, among the first things an investigator must do upon determining a new sequence is to compare it with all available sequences to see
if it resembles something already known.[1] The search need not wait until
the sequence is known in every detail; indeed, the sequence determination
itself often can be guided by comparison with homologous sequences. At
the same time, there is at present an almost excessive zeal in the community for finding "homologies," to the extent that some researchers would
rather find the second member of some new family than the first. Worse,
some investigators are misled by marginal resemblances that are likely not
due to common ancestry. In either event, the results of a sequence search
usually require that judgments be made about the significance of what has
or has not been found. The primary aim of this chapter is to provide a few
simple guidelines and hints about how to make these judgments. When it
comes to low-level similarity, caution is always warranted.

Another topic touched on is how to interpret homologies when they are
found. Even though the discovery of an unexpected homology is often
exceedingly useful in providing insights about the function of a protein, a
still greater significance lies in reconstructing the history of present-day
systems. The construction of sequence-based phylogenies is another area
laden with pitfalls, however, and as I shall try to show, the beginner must
tread lightly. Thus, although the computer is a wonderful helpmate for the
sequence searcher and comparer, biochemists and molecular biologists
must guard against the blind acceptance of any algorithmic output; given
the choice, think like a biologist and not a statistician.

Tools

In general, there are three variables involved in any sequence search:
the nature of the sequence being searched, the size and quality of the

---

[1] R. F. Doolittle, "Of URFs and ORFs: A Primer on How to Analyze Derived Amino Acid
Sequences." University Science Books, Mill Valley, California, 1987.

sequence collection available for comparison, and the capability of the computer program and the computer that will be used. The query sequence may be nucleic acid or protein, it may be short or long, known precisely or perhaps not so precisely. Similarly, the database may be composed of either nucleic acid or protein sequences. Moreover, the collection may be small or large, redundant or representative, up-to-date or not up-to-date. The searching program may be fast or slow, sensitive or insensitive, capable of identifying reasonably short runs of similarity or not. It remains for the individual investigator to assess his particular situation as the first step in determining the significance of any similar sequences that may be retrieved.

As discussed elsewhere in this volume, the databases are growing at an alarming rate. At the same time, searching programs are becoming faster and, in some cases, more sensitive. Nevertheless, some of the most remarkable sequence matches have been made with rather limited data banks and relatively simple search routines. Of course, the impact of finding certain kinds of matches is lessening. It is no longer such a surprise to find protein sequences falling into well-defined groups. What was remarkable several years ago may be routine today. Thus, it behooves us to ask just what is being sought and what can be expected realistically.

In this regard, it is not uncommon to hear laments about "how far behind" the sequence banks are, the implication being that current research is thus handicapped. In my view, such complaints are ill-founded. Quite apart from the fact that the major sequence banks have done a remarkable job in a time of limited resources, the parimutuel nature of the balance between how many sequences have been determined and how many matches have been made seems to me to blunt this criticism effectively.

Nucleic Acid versus Protein Sequence Data Banks

Currently the two biggest sequence data banks, GenBank and the EMBL Data Library, are mostly collections of DNA sequences. Nonetheless, most of the interesting matches that have been made in the last decade have involved protein sequences. Why is this so, and how does it bear on the maintenance of sequence data banks? Certainly, the overwhelming majority of new "protein sequences" are being determined on the basis of DNA sequences, so it is altogether appropriate, and, indeed, desirable, to bank them as the DNA sequences. The DNA sequence provides essential information, in the resource sense, for molecular geneticists and others; there is an enormous amount of molecular biology not directly concerned with the gene product per se. Nonetheless, there are good reasons for searching expeditions to begin at the protein sequence level.

Some investigators are under the erroneous impression that there is more to be gained by searching the actual DNA sequence rather than the amino acid sequence derived from it. That view is greatly mistaken, and resemblances will be missed if it is adopted. The reason is, of course, that there are only four bases but 20 amino acids. As such, the so-called signal-to-noise ratio is improved greatly when the DNA sequence is translated; the "wrong-frame" information is set aside and third-base degeneracies consolidated. As a general rule, then, searchers dealing with potential gene products should translate their DNA sequences into the protein equivalents. Obviously, this implies that the sequence bank to be searched should be in the form of protein sequences also.

Several of the earlier chapters in this volume speak to the point. Thus, although GenBank maintains only nucleic acid sequences on-line, they do provide the Protein Identification Resource with translated versions of all open reading frames. The EMBL Data Library has a similar agreement with SWISS-PROT.

Short versus Long Query Sequences

Generally speaking, the longer a sequence is, the easier it will be to establish a level of confidence about its relationship to other sequences (Fig. 1). Practically speaking, however, the researcher should search whatever is available, as the following example should make clear.

Some time ago, during the course of characterization of a melanocyte tumor cell antigen by workers at the Fred Hutchinson Cancer Research Center, a small amount of amino-terminal sequence was determined. Although fewer than 13 residues were available for searching, a scan of the modest-sized protein sequence bank denoted NEWAT[2] retrieved a single sequence: human transferrin. Seven of 12 matchable residues were identical; was this significant? The answer to the question came not from statistics, but from experiment. Because transferrin is an iron-binding protein, tests were undertaken to see if the tumor antigen could bind iron also; it did, and to the same extent as transferrin.[3] Since that time, the cDNA sequence for the tumor antigen has been determined and the overall protein sequences found to be about 40% identical with serum transferrins.[4] The point is that the initial search with a very short sequence set the

[2] R. F. Doolittle, *Science* **214,** 149 (1981).

[3] J. P. Brown, R. M. Hewick, I. Hellström, K. E. Hellström, R. F. Doolittle, and W. J. Dreyer, *Nature (London)* **296,** 171 (1982).

[4] T. M. Rose, G. D. Plowman, D. B. Teplow, W. J. Dreyer, K. E. Hellström, and J. P. Brown, *Proc. Natl. Acad. Sci. U.S.A.* **83,** 1261 (1986).

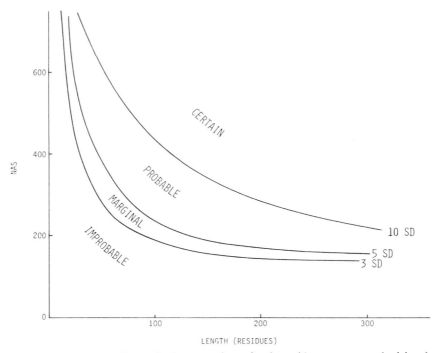

Fig. 1. Guide to significance in the comparison of amino acid sequences, emphasizing the importance of sequence length. NAS is a "normalized alignment score," in which the number of identities between two sequences is multiplied by 10 (20 for cysteines) and the number of gaps is multiplied by 25. After subtraction of the latter from the former, the score is normalized by dividing by the average length of the two sequences and multiplying by 100. (From Ref. 1.)

investigation on a proper course which would not have been possible by other means.

There are many other examples where searches of partial sequences have allowed great leaps forward, including matching of a partial sequence of platelet-derived growth factor (PDGF) with the v-*sis* oncogene[5,6] and identification of the epidermal growth factor (EGF) receptor with the v-*erb* oncogene.[7] On a less happy note, more than a few studies have been curtailed when a preliminary search of a sequence revealed it to be a

[5] R. F. Doolittle, M. W. Hunkapiller, L. E. Hood, S. G. Devare, K. C. Robbins, S. A. Aaronson, and H. G. Antoniades, *Science* **221,** 275 (1983).
[6] M. D. Waterfield, G. T. Scrace, N. Whittle, P. Stroobant, A. Johnsson, A. Wasteson, B. Westermark, C.-H. Heldin, J. S. Huang, and T. F. Deuel, *Nature (London)* **304,** 35 (1983).
[7] J. Downward, Y. Yarden, E. Mayes, G. Scrace, N. Totty, P. Stockwell, A. Ullrich, J. Schlessinger, and M. D. Waterfield, *Nature (London)* **307,** 521 (1984).

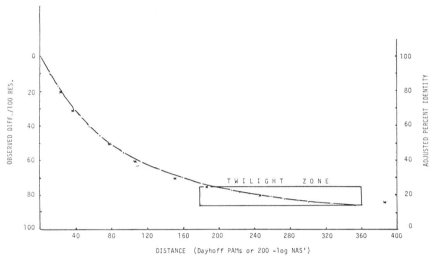

FIG. 2. Evolutionary distance versus sequence dissimilarity. Two randomly diverging sequences change in a negatively exponential fashion. After the insertion of gaps to align two random amino acid sequences, it can be expected that they will be 10–20% identical. (From Ref. 1.)

common contaminant such as albumin from the serum or concanavalin A from the lectin column used in purification. As a rule, then, the experimentalist should search early and often.

Significance of Resemblances

Often, one is faced with the problem of judging the similarity of a pair of amino acid sequences. Alliteratively put, the question is one of chance, convergence, or common ancestry. As a rule of thumb, if two sequences of 100 residues or more in length are more than 25% identical, common ancestry will almost certainly be the case. It is when two sequences are less than 25% identical that one must be cautious about inferring homology (Fig. 2). On the other hand, it is never possible to show that two sequences are not related by common ancestry. They may just have been changed so much by repeated mutation that common heritage is no longer apparent, at least to the point where any statistical confidence can be expressed.

Sometimes, however, it is possible to make a case by considering several sequences at a time rather than just pairs.[8] As an example, consider the ever-expanding immunoglobulin-type family. As is well known, vertebrate antibodies are composed of various combinations of variable and

[8] R. A. Jue, N. Woodbury, and R. F. Doolittle, *J. Mol. Evol.* **15**, 129 (1980).

constant regions, the prototypic units of which are about 100 residues long and each containing two cysteine residues.[9,10] Early on it was noticed that the histocompatibility-involved protein β_2-microglobulin[11] conformed to this general structural arrangement, and later, a host of other immune response-related proteins were found to fall into the same general family.[12] The really surprising finding was that the family includes certain receptor proteins like the PDGF receptor, not ordinarily thought to be an agent involved in the self–nonself recognition that underlies immunology.[13,14] The sequence resemblances in these instances are very low level, often amounting to as few as 10 or 12 residues per 100, but they are consistent enough across a large number of comparisons that the statistical likelihood is compelling.[13]

Worrying about Convergence

Sequence resemblance owing to convergence, that is, selection for a similar order of amino acids in order to build a specific three-dimensional structure, is rare, mostly because there are so many different combinations of amino acids that can be assembled into equivalent structures. Still, one can imagine instances where similar sequences might be observed in proteins not as a result of common ancestry: in proteins rich in amphipathic helices, which are known for a rhythmic occurrence of polar and nonpolar residues, as one example, or in integral membrane proteins that have a preponderance of nonpolar residues, for another. Again, the case for or against common ancestry will usually be made on nonstatistical grounds. In this regard, common function is more persuasive than standard deviations. As an example, consider the remarkable case of the G-protein (transducer)-linked receptor proteins.

The story begins with the visual pigment protein rhodopsin, the first full sequence of which was reported for cattle.[15] Not too long after, a

[9] S. J. Singer and R. F. Doolittle, *Science* **153**, 13 (1966).

[10] R. L. Hill, R. Delaney, R. E. Fellows, and H. E. Lebovitz, *Proc. Natl. Acad. Sci. U.S.A.* **56**, 1762 (1966).

[11] P. A. Peterson, B. A. Cunningham, I. Berggard, and G. M. Edelman, *Proc. Natl. Acad. Sci. U.S.A.* **69**, 1697 (1972).

[12] A. F. Williams, *Immunol. Today* **8**, 298 (1987).

[13] H. Hayashida, K. Kuma, and T. Miyata, *Proc. Jpn. Acad. (Ser. B)* **64**, 113 (1988).

[14] A. F. Williams and A. N. Barclay, *Annu. Rev. Immunol.* **6**, 381 (1988).

[15] Y. A. Ovchinnikov, N. G. Abdulaev, M. Y. Feigina, I. D. Artamanov, A. S. Zoletarev, M. B. Kostina, A. S. Bogachuk, A. I. Miroshinikov, V. I. Martinov, and A. B. Kudelin, *Bio. org. Khim.* **8**, 1011 (1982).

rhodopsin sequence was obtained from the fruit fly,[16,17] and not long after that came the publication of three human cone pigment sequences.[18] Then a sequence was determined for the β_2-adrenergic receptor, a binder of adrenalin; a computer search revealed that it was an astonishing 25% identical with bovine rhodopsin.[19] Because these sequences were of the order of 350 residues in length, there was no doubt that they were homologous. That both proteins transmitted extracellular signals via cytoplasmic G proteins (transducins) seemed to cement the case beyond a doubt. Since that finding, over a dozen more receptor sequences have been assigned to the same family.[20-26]

The percent identity of the most distant pairs in this group is well below 25%, however, and the question arises whether convergence might lie at the heart of some of the similarities. As it happens, all of these proteins appear to have seven segments rich in hydrophobic amino acids, as revealed by hydropathy plots.[27] A general model has been proposed whereby the amino-terminal segment is exposed extracellularly and, given seven traverses of the cytoplasmic membrane, the carboxy-terminal segment ends up intracellular.

For the sake of argument, let us suppose that there is a strong selective advantage for a receptor protein to cross the cytoplasmic membrane seven times in order to form a stable seven-pillared structure. Could not the restriction of having to use only the nonpolar subset of amino acids, in segments long enough to cross the lipid bilayer (a minimum of 20 residues

[16] J. E. O'Tousa, W. Baehr, R. L. Martin, J. Hirsh, W. L. Pak, and M. L. Applebury, *Cell* **40**, 839 (1985).

[17] C. S. Zuker, A. F. Cowman, and G. M. Rubin, *Cell* **40**, 851 (1985).

[18] J. Nathans, D. Thomas, and D. S. Hogness, *Science* **232**, 193 (1986).

[19] R. A. F. Dixon, B. K. Kobilka, D. J. Strader, J. L. Benovic, H. G. Dohlman, T. Frielle, M. A. Bolanowski, C. D. Bennett, E. Rands, R. E. Diehl, R. A. Mumford, E. E. Slater, I. S. Segal, M. E. Caron, R. J. Lefkowitz, and C. D. Strader, *Nature (London)* **321**, 75 (1986).

[20] T. Kubo, K. Fukuda, A. Mikami, A. Maeda, H. Takahashi, M. Mishina, T. Haga, K. Haga, A. Ichiyama, K. Kangawa, M. Kojima, H. Matsuo, T. Hirose, and S. Numa, *Nature (London)* **323**, 411 (1986).

[21] Y. Masu, K. Nakayama, H. Tamaki, Y. Harada, M. Kuno, and S. Nakanishi, *Nature (London)* **329**, 836 (1987).

[22] B. K. Kobilka, H. Matsui, T. S. Kobilka, T. L. Yang-Feng, U. Francke, M. G. Caron, R. J. Lefkowitz, and J. W. Regan, *Science* **238**, 650 (1987).

[23] D. Julius, A. MacDermott, R. Axel, and T. M. Jessell, *Science* **241**, 558 (1988).

[24] J. R. Bunzow, H. H. M. Van Tol, D. K. Grandy, P. Albert, J. Salon, M. Christie, C. A. Machida, K. A. Neve, and O. Civelli, *Nature (London)* **336**, 783 (1988).

[25] L. Marsh and I. Herskowitz, *Proc. Natl. Acad. Sci. U.S.A.* **85**, 3855 (1988).

[26] P. S. Klein, T. J. Sun, C. L. Saxe III, A. R. Kimmel, R. L. Johnson, and P. N. Devreotes, *Science* **241**, 1467 (1988).

[27] J. Kyte and R. F. Doolittle, *J. Mol. Biol.* **157**, 105 (1982).

at 1.5 Å per residue of helix to stretch across 30 Å of lipid bilayer), be responsible for the similar amino acid sequences? Indeed, the resemblances between some of these sequences are limited to the predicated membrane-spanning segments.[25] Were it not for the fact that all these diverse receptors interact with similar transducer proteins, an argument for convergence might be made. Thus, in the case of the most distant members of the group, the major evidence for divergence is not statistical; it is physiological.

Shuffled Exons

If all extant proteins were descended from a conventional, average-sized parental type consisting of about 350 amino acids, the reconstruction of lineages would be a straightforward problem. As it happens, however, in many instances various parts of gene products appear to have been exchanged. Often, the shuffled segments involve exons, and these in turn frequently correspond to structural domains in the protein.[28] Obviously, "exon shuffling" can confound attempts to reconstruct the history of events.

A number of such genetically mobile units have been identified in eukaryotic proteins. These include the EGF (epidermal growth factor) domain, several classes of units found in fibronectin, the "kringles" characteristic of certain blood-clotting proteins, and several others.[29] Typically, these involve sequences of 40–80 amino acids that seem to have the wherewithal to fold into compact units; intrasegment disulfide bonds are common.[30]

The EGF motif, for example, has been found in dozens of animal proteins.[31] The segments are only 40–45 residues long, however, and the question arises, how confident are we that all these units share common ancestry? What they have in common, for the most part, is an array of six cysteines. Cysteines that participate in disulfide bonds are known to be among the slowest changing residues in a protein, since the replacement of either of the linked single residues leaves an unpaired partner. As a result, such cysteines weigh more heavily in decisions about ancestry. The bulk of the present data favors common ancestry for all the identified EGF types, as opposed to convergence or chance.

Cases of segment resemblance have been found that do not involve

---

[28] W. A. Gilbert, *Science* **228**, 823 (1985).
[29] L. Patthy, *Cell* **41**, 652 (1985).
[30] R. F. Doolittle, *Trends Biochem. Sci.* **10**, 233 (1985).
[31] R. F. Doolittle, *in* "Prediction of Protein Structure and the Principles of Protein Conformation" (G. Fasman, ed.), p. 599. Plenum, New York, 1989.

cysteines, of course. For example, a carboxy-terminal skein of 66 residues of the oncogene known as v-*jun* was found to resemble the carboxy-terminal segment of the yeast regulatory protein GCN4.[32] It was known that this region of the yeast protein binds DNA, and this provided an important clue to understanding how v-*jun* and its normal cellular counterpart c-*jun* function. The underlying message is that one must be alert to regions of similarity even when they occur embedded in an overall background of dissimilarity.

Reverse Searching

Up to this point, we have been dealing with the situation in which an unknown sequence is "searched" against a large database. There is another circumstance in which the sequence is examined to see if it contains particular diagnostic or consensus sequences. Consensus sequence searching itself has two overlapping but distinctly different applications. In the one case, an investigator examines the newly determined sequence to see if it contains subsequences typical of some previously characterized families. This might include a scan for active sites or binding motifs (Table I), or longer prototypic sequences, including those of EGF domains, "zinc fingers," [33] or immunoglobulins. If one is dealing with raw nucleic acid sequence data, and especially if noncoding regions are suspected, a search for highly conserved but short regulatory elements[34] or less conserved but longer "high-repeat" motifs like the primate Alu sequences[35] is in order.

We call this reverse searching in that extracts from the database are being used to screen the unknown rather than the other way around. The distinction is not trivial. Thresholds can be set much lower when a screen is being conducted against a single new sequence, and things will be found that would be lost in the noise in a forward direction search. As a simple example, it would be very cumbersome, it not impossible, to detect potential asparagine-linked carbohydrate attachment sites in a conventional search of a new protein sequence versus a protein sequence data bank, but it is obviously easy to identify any such sites merely by scanning the sequence directly for the presence of the consensus NXS or NXT sequence (Table I). Similarly, once any consensus is established, say, for two cysteines separated by 30–40 other noncysteine residues, it is a simple matter to examine a new sequence for its presence.

---

[32] P. K. Vogt, T. J. Bos, and R. F. Doolittle, *Proc. Natl. Acad. Sci. U.S.A.* **84**, 3316 (1987).
[33] J. M. Berg, *Science* **232**, 485 (1986).
[34] T. Wirth, L. M. Staudt, and D. Baltimore, *Nature (London)* **329**, 174 (1987).
[35] P. L. Deininger and C. W. Schmid, *J. Mol. Biol.* **127**, 437 (1979).

TABLE I
SIMPLE SEQUENCES THAT TYPIFY CERTAIN
ACTIVE SITES AND BINDING SITES

Protein Type	Sequence
Serine protease[a]	GDSG
Zinc-metallopeptidase[b]	HEXXH
Acid protease[c]	$D_S^T G$
Asn-linked carbohydrate[d]	NX_T^S
Cell adhesion[e]	RGD
Nucleotide binding[f]	$_A^G XXXGK_T^S$
Cytochrome c[g]	CXXCH

[a] H. Neurath and G. H. Dixon, *Fed Proc. Fed. Am. Soc. Exp. Biol.* **16,** 791 (1957).

[b] C. V. Jongeneel, J. Bouriver, and A. Bairoch, *FEBS Lett.* **242,** 211 (1989).

[c] J. Sodek and T. Hofmann, *Can. J. Biochem.* **48,** 1014 (1970).

[d] L. T. Hunt and M. O. Dayhoff, *Biochem. Biophys. Res. Commun.* **39,** 757 (1970).

[e] E. Ruoslahti and M. D. Pierschbacher, *Science* **238,** 491 (1987).

[f] J. E. Walker, M. Saraste, M. J. Runswick, and N. J. Gay, *EMBO J.* **1,** 945 (1982).

[g] R. E. Dickerson, *J. Biol.* **57,** 1 (1971).

The other mode of consensus sequence searching arises when one has identified a particular motif and is interested in finding other potential members in the data bank. In this case, the consensus sequence is run through the entire collection in a search for candidates.

Reconstructing History

At the beginning of this chapter I remarked that the most interesting aspect of sequence comparison is the reconstruction of events leading to present-day proteins and organisms. As new proteins evolved, new environments were conquered and novel adaptations enacted. In searching for a good example, I settled on a subset of the transducin-linked receptor proteins discussed in an earlier section. The sequences of a number of visual pigment proteins are available,[15-18] and a phylogenetic tree based on these sequences ought to provide insights into the evolution of color vision.

Accordingly, two distantly related rhodopsins (bovine and fruit fly) and three human cone pigments were subjected to three different tree-building regimens.[36-38] Other members of the transducin-linked receptor family were used as "outliers" in order to root the trees. All three trees, obtained by the completely different methods, had the same topology and indicated that color vision in primates has evolved incrementally as the result of widely spaced gene duplications on different lineages. The first of these, which ought to have occurred in some early vertebrate, led to the elaboration of a red–green type diverging from the rhodopsin prototype (denoted by a in Fig. 3). The second key duplication also involved the parental rhodopsin gene and led to a blue type opsin (b in Fig. 3). Assuming an approximately constant rate of change, this duplication ought to have occurred well along in vertebrate evolution, but probably before the divergence of reptiles and mammals. Finally, there has been a very recent duplication on the primate lineage that gave rise to the red–green splitting (c in Fig. 3).

By coincidence, Yokoyama and Yokoyama[39] recently analyzed these same visual pigments, but using the genomic DNA sequences. They constructed a topology by yet a fourth method and generated a quite different branching order. In their tree, the duplication denoted b in Fig. 3 does not involve the parental rhodopsin gene but involves instead the color-sensitive daughter descendant of the first duplication (a).

How can we tell which is the correct interpretation? Once again, in my view statistics will not make the case. Rather, opsin sequences must be examined from a variety of vertebrates. If the depiction in Fig. 3 is correct, some lower vertebrates, perhaps fish or amphibia, will be found to have a red–green-sensitive pigment in addition to rhodopsin, but more recently diverged vertebrates (reptiles and birds) will have in addition a blue-sensitive pigment. Moreover, when the phylogenies of these sequences are determined they will mirror Fig. 3. If the alternative tree of Yokoyama and Yokoyama[39] is correct, then the color-sensitive pigment found in lower vertebrates could well be either red–green or blue sensitive. In either case, the additional sequences should reveal the real history of events. The point I would make is that sequence-based topologies can often be verified by examining a judiciously chosen set of organisms which have diverged before and after the duplication events of interest.

[36] D.-F. Feng and R. F. Doolittle, *J. Mol. Evol.* **25**, 351 (1987).
[37] J. Hein, this volume, [39].
[38] R. F. Doolittle and D.-F. Feng, this volume, [41].
[39] S. Yokoyama and R. Yokoyama, *Mol. Biol. Evol.* **66**, 186 (1989).

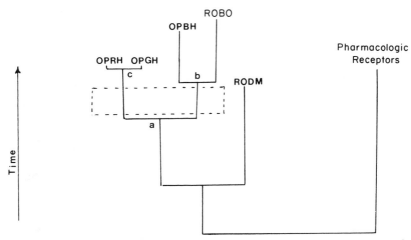

Fig. 3. Phylogeny of visual pigment proteins[15-18] and their relationship to pharmacologic receptors of the β_2-adrenergic receptor type. RODM, Rhodopsin from *Drosophila melanogaster;* ROBO, rhodopsin, bovine; OPRH, opsin red, human; OPGH, opsin green, human; OPBH, opsin blue, human. The arrow marked Time indicates the direction of evolution; the vertical segment lengths are proportional to amounts of evolutionary change (horizontal lengths are arbitrary). The depiction shows that color vision must have evolved incrementally. The prediction is that there are extant creatures which diverged from other vertebrates in the time zone set off by dashed lines before the duplications leading to blue and green (red) pigments and which possess only a red (or green) pigment and rhodopsin. A significantly different version of events has been described by others.[39]

Summary Comment

The vast majority of extant proteins are the result of a continuous series of genetic duplications and subsequent modifications. As a result, redundancy is a built-in characteristic of protein sequences, and we should not be surprised that so many new sequences resemble already known sequences. All the same, some sequences may have changed so much that their histories are blurred beyond recognition, and investigators must be cautious in interpreting marginal similarities. Furthermore, exon shuffling can confuse situations to the point where the simplest interpretations may be misleading. Still, searching a database with a new sequence is always an exciting proposition. As the data accumulate and the searching becomes more systematic, it should be possible not only to reconstruct the history of most existing proteins but also to relate their appearance to the adaptations that underlie all organismal existence.

[7] Finding Protein Similarities with Nucleotide Sequence Databases

By STEVEN HENIKOFF, JAMES C. WALLACE, and JOSEPH P. BROWN

Introduction

The relative ease with which DNA can be cloned and sequenced has led to rapid expansion of the nucleotide sequence databases. As a consequence, the amino acid sequences of the large majority of new proteins have been deduced from nucleotide sequences rather than determined directly. Since these sequences are initially deposited in the DNA databases, the protein databases have become mostly secondary sources of information. This trend promises to increase as larger scale nucleotide sequencing projects get under way. Therefore, it is worthwhile to search nucleotide sequence databases for protein similarities, since these databases are more complete and up-to-date. As illustrated in this chapter, it is advantageous to search these databases for amino acid rather than nucleotide sequence similarities. Therefore, we have adapted amino acid sequence searching procedures to detect similarities within nucleotide sequence databases.[1] This involves translating the DNA database sequence in each possible reading frame into protein prior to alignment with the protein sequence used as query. Not only is this simple procedure very sensitive, but it also allows access to sequence data that are not known or suspected to encode proteins, and hence would not appear in protein databases.

Searches using translated database sequences can be more sensitive than nucleotide sequence searches of the same data because (1) the genetic code is degenerate and (2) regularities in proteins can be used as selective criteria. The codon degeneracy problem is best illustrated in the case of serine: Since serine can be encoded in six ways (UCN, AGU, or AGC), a perfect match between serines in two aligned proteins is very likely to be mismatched at the level of nucleotide sequence. In fact, many of these nucleotide mismatches are at all three positions, such as UCA (Ser) versus AGU (Ser). Codon degeneracy is even more of a problem where organisms show different biases in the choice of codons for any particular amino acid. For example, in mammals, CAG (Gln) is preferred over CAA (Gln) about

---

[1] S. Henikoff and J. C. Wallace, *Nucleic Acids Res.* **16**, 6191 (1988).

75% of the time, whereas in yeast this preference is reversed.[2] Therefore, the large majority of glutamine matches in homologous proteins between yeast and vertebrates are mismatched at the third position. Translating a database sequence prior to comparison overcomes this difficulty. Translation also allows one to take into account amino acid preferences and conservative substitutions, such as the lod (log odds) matrix scoring system.[3]

There are potential drawbacks to the translation of nucleotide database sequences for searching. Some of these are due to the greater number of comparisons required to search a translated nucleotide sequence database when compared to a protein sequence database. Since all six possible reading frames must be searched, there are at least 6-fold as many comparisons necessary in order to search a coding sequence when compared to the derived amino acid sequence. Furthermore, the nucleotide sequence databases include large amounts of sequence not encoding protein. Therefore, the number of computations involved are typically more than an order of magnitude greater for translated database searching than for protein database searching, with a corresponding increase in computer time. In addition, the greater amount of data requires more disk storage space. Nevertheless, we have found that even relatively modest personal computers are adequate for translated searches. By 4-fold compression of the nucleotide databases and translation "on-the-fly," disk storage requirements are not excessive [about 8 megabytes (MB) for EMBL 17 or GenBank 58). Typically, personal computers are turned off at night, so that even the most exhaustive search can be carried out with minimal cost and inconvenience by using an otherwise idle machine. The most intensive of the searches that we describe here requires about 8 hr on an IBM-compatible personal computer with an Intel 80286 processor or about 4 hr on a "386" machine. The rapid improvement in technology, the availability of useful general purpose software, and the reduction in hardware prices have put these machines in many laboratories where sequence data are being collected and analyzed. Although nucleotide sequence databases have increased in size over the past several years, the increase in speed and storage space of general purpose personal computers has been even greater.

Another problem for translated searches is the increased level of chance similarities detected when more data are analyzed. This background level of spurious matches reduces sensitivity, making distant relationships po-

[2] T. Maruyama, T. Gojobori, S. Aota, and T. Ikemura, *Nucleic Acids Res.* **14**, r151 (1986).
[3] M. O. Dayhoff, R. M. Schwartz, and B. C. Orcutt, *in* "Atlas of Protein Sequence and Structure" (M. O. Dayhoff, ed.), Vol. 5, suppl. 3, p. 345. National Biomedical Research Foundation, Silver Spring, Maryland, 1978.

tentially difficult to detect. Generally, sensitivity can be increased by extension of an alignment following detection.[4,5] However, extension is not always appropriate for increasing sensitivity of translated searches. Therefore, we have relied on an alternative means of increasing sensitivity: identifying a short conserved sequence motif and using it for sensitive detection. In this way, one does not rely on extension of a similarity for detection, although it can be useful for confirmation. We illustrate these methods, individually and in combination, using examples taken from our recent analyses of bacterial activator protein families.

The programs NUTSS (Nucleotide Translation Similarity Searcher) and PATMAT (PATtern MATrix builder) which carry out the procedures described in this chapter are available on a 360K floppy diskette from the authors on request. These programs are also part of a comprehensive package, GENEPRO, available from Riverside Scientific Enterprises (18332 57th Avenue N.E., Seattle, WA 98155). A 100% IBM-PC compatible computer with a hard disk and 640K of memory is recommended. A version for the Apple Macintosh II is planned. The searching programs utilize uncompressed databases available from GenBank or highly compressed GenBank, EMBL, and NBRF-PIR databases available from Riverside Scientific.

Methods and Applications

Method 1: Standard Similarity Search Using a Single Amino Acid Sequence as Query

Standard similarity searches are performed by fetching an individual nucleotide sequence, translating each reading frame into protein, comparing that reading frame with the query, repeating the comparison for the next reading frame, and then repeating the entire operation for the next nucleotide sequence.

The comparison strategy[6] is to align a fixed length of sequence (a window) from the query with the same length from a translated database sequence and calculate a lod (log-odds) score, which measures the likelihood that two aligned amino acids are functionally equivalent.[3] The window is then aligned with the next stretch of translated sequence and a lod score calculated. For greater speed, only alignments in which one or more dipeptides match between the query and the translated database sequences

---

[4] R. F. Doolittle, *Science* **214,** 149 (1981).

[5] D. J. Lipman and W. R. Pearson, *Science* **227,** 1435 (1985).

[6] W. J. Wilbur and D. J. Lipman, *Proc. Natl. Acad. Sci. U.S.A.* **80,** 726 (1983).

are considered. To reduce the likelihood of missing optimal matches, all possible alignments including one or more dipeptide matches are made. Gaps are not allowed. Generally, a window of 30 amino acids is effective.[1] In some cases, however, larger windows are useful for more sensitive detection.[7] Use of a fixed searching window rather than a flexible one[5] allows the typical investigator to evaluate more readily the meaning of a high score, as no hidden decisions have been made by the program. The top scoring sequences are then visually examined for extension of each detected match using standard computer alignment[8] and dot matrix[9] methods. Visual inspection at this point takes advantage of the investigator's judgment in deciding whether an alignment reflects common ancestry or a chance similarity. Extension is only one criterion used in this decision; other information, such as function, species, intron–exon structure, and DNA sequence context, can be considered at the same time.

Application 1: Detection of LysR Family by Standard Similarity Searching. An example of the standard searching approach is the previous identification of members of a large family of bacterial activator proteins related to LysR.[7] Using the *Salmonella typhimurium* MetR protein as query of GenBank 52 and EMBL 14, and a window of 90, 14 different bacterial sequences were detected with higher lod scores and number of matches than the next best matches, which appeared to be chance similarities (Table I). Excluding proteins with similar known functions in different species, this group consisted of five proteins (LysR, NodD, IlvY, CysB, and AmpR) which, like MetR, were known to activate other genes, one protein of unknown function, and two partial open reading frames which were not known to encode proteins. Considering that each of the searches examined the equivalent of about 100,000 translated sequences the size of MetR (276 amino acids), evidence for similarity was very strong. Several criteria were used to demonstrate that all of these predicted proteins are actually related and not the result of chance similarity to MetR. Dot matrix analysis indicated that multiple regions of similarity could be detected between pairs of the proteins. Regions that were similar in one pair tended to be similar in the other pairs. Alignments extended essentially from end to end, suggesting a similar overall fold. The full sequence of one of the incomplete predicted sequences (*Escherichia coli* LeuO) was determined, extending the alignment to its carboxy terminus. As LysR is thought to have a helix–turn–helix DNA-binding domain between residues 21 and 40, each aligned protein was evaluated for the likelihood of containing this

[7] S. Henikoff, G. W. Haughn, J. M. Calvo, and J. C. Wallace, *Proc. Natl. Acad. Sci. U.S.A.* **85**, 6602 (1988).

[8] S. Needleman and C. Wunsch, *J. Mol. Biol.* **48**, 443 (1970).

[9] J. Maizel and R. Lenk, *Proc. Natl. Acad. Sci. U.S.A.* **78**, 7665 (1981).

TABLE I

BEST DATABASE MATCHES USING *S. typhimurium* MetR ACTIVATOR PROTEIN AS QUERY[a]

Lod score[b]	Matches	Protein	Database[c]
92	24	*Escherichia coli* LysR activator protein	P, G
91	28	*Enterobacter cloacae* AmpR activator protein	E
91	24	*Alcaligenes eutrophus* TfdO ORF	E
90	31	*S. typhimurium* ORF upstream of *leu* operon (LeuO)[d]	G, E
90	27	*E. coli* IlvY activator protein	G, E
88	31	*E. coli* ORF upstream of the *leu* operon (LeuO)[d]	G, E
87	22	*E. coli* CysB activator protein	G, E
86	23	*E. coli* ORF downstream of the *ant* operon (AntO)	E
85	23	*Rhizobium meliloti* NodD activator protein	P, G, E
85	23	*Rhizobium leguminosarum* NodD activator protein	G, E
85	22	*Rhizobium trifolii* NodD activator protein	G, E
85	21	*Bradyrhizobium* sp. NodD activator protein	E
85	20	*Rhizobium* sp. NodD1 activator protein	E
85	20	*S. typhmurium* CysB activator protein	G, E
85	18	Mouse ubiquitin mRNA, inverted sequence	G
84	21	HSV-a glycoprotein b mRNA, inverted	G, E
84	18	Rabbit β-myosin heavy chain mRNA, inverted	G, E
84	18	Bovine protein C mRNA, inverted	E

[a] Window = 90.

[b] For computations, +8 has been added to each PAM250 matrix value (Ref. 3) to eliminate negative values, and the mean log-odds (lod) score of each alignment has been multiplied by 10 and rounded off to the nearest integer.

[c] P, NBRF-PIR 14; G, GenBank 54; E, EMBL 14.

[d] Using *R. meliloti* NodD protein as query.

motif using the parameters of Dodd and Egan.[10] In nearly every case, the best predicted region of the protein aligned precisely with residues 21–40 of LysR.[7]

This detection of nine members of a single family using DNA database searching for protein similarities demonstrates several features of the procedure: (1) Access to a more complete database. Only two of the proteins, LysR and *Rhizobium meliloti* NodD, were present in the NBRF-PIR database (Table I). These were the only members of the family previously known to be related. (2) Detection of unrecognized open reading frames. In the case of TfdO and LeuO, the published sequences were for genes adjacent to the ones detected. Since this family is usually characterized by divergent transcription, the amino-terminal portion of a family member might fall within the presumed regulatory region of a known gene. (3)

[10] I. B. Dodd and J. B. Egan, *J. Mol. Biol.* **194**, 557 (1987).

Confirmation of detected matches using other criteria. In this case, the use of a predictive scheme for a DNA-binding motif[10] was particularly valuable. (4) Sensitivity of the procedure. The overall level of similarity of the proteins of the various family members to MetR ranges from 15 to 24%. To achieve sufficient sensitivity for a standard translated database search, it was necessary to use a searching window of 90. Gaps could confound a search that uses such a large window. The procedure described next is an alternative means of achieving higher sensitivity while using a small searching window.

Method 2: Matrix Searching

Gribskov *et al.*[11] have demonstrated that greater sensitivity can be obtained in a search using a position-specific scoring matrix derived from related sequences (profile analysis) rather than using a single query. The matrix consists of values for each amino acid at each position reflecting the frequency with which that residue appears at that position among the aligned sequences. As demonstrated below, this procedure can be effective even when there are only two sequences contributing to the matrix. Furthermore, it appears to be advantageous to use a relatively short region to generate the scoring matrix so that the remaining information can be used to confirm candidate matches and so that variable gaps need not be considered in the search. One advantage of searching for a short pattern is that no computational shortcuts are necessary, since every stretch of translated database sequence the size of the pattern can be searched in about the same time that a standard search can be run for a typical query. Another advantage is the conceptual simplicity of the procedure: The precise pattern is decided on beforehand by the investigator who does not need to use any scoring matrix other than the one he constructs.

A matrix-building program uses aligned amino acid sequences (e.g., Fig. 1a) to construct a scoring matrix (e.g., Fig. 1b). The individual matrix entries can be weighted to compensate for some of the nonrandomness of database sequences. For protein database searches, one can divide each matrix entry by the average frequency with which that amino acid is found in proteins in order to penalize the more common residues which occur more frequently by chance. For DNA database searches, one can divide each matrix entry by the frequency with which a codon for that amino acid appears in the genetic code in order to penalize chance occurrences for database sequences translated in all frames. For the examples described below, weighting by codon frequency is found to improve performance. After this optional weighting, each position is normalized so that the sum

[11] M. Gribskov, A. D. McLachlan, and D. Eisenberg, *Proc. Natl. Acad. Sci. U.S.A.* **84**, 4355 (1987).

a

Protein	1st residue	Amino acid sequence
MetR (*S. typhimurium*)	22	AAAVLHQTQSALSHQFSDLEQRLGFRLFVR
LysR (*E. coli*)	24	AAHLLHTSQPTVSRELARFEKVIGLKLFER
TfdO (*A. eutrophus*)	21	AARRLHISQPPVTRQIHALEQHLGVLLFER
LeuO (*E. coli*)	22	AAHVLGMSQPAVSNAVARLKVMFNDELFVR
LeuO (*S. typhimurium*)	22	AAHTLGMSQPAVSNAVARLVVMFNDVLFVR
NodD (*R. leguminosarum*)	26	AARSINLSQPAMSAAISRLRDYFRDDLFIM
NodD1 (*R. meliloti*)	29	AARRINLSQPAMSAAIARLRTYFGDELFSM
NodD (*R. trifolii*)	26	AARSINLSQPAMSAAIGRLRAYFNDELFLM
NodD (*Bradyrhizobia*)	25	AARKINLSQPAMSAAIARLRSYFRDELFTM
NodD2 (*R. meliloti*)	26	AARRVKLSQPAMSAAIARLRTYFGDELFSM
CysB (*E. coli*)	22	TAEGLYTSQPGISKQVRMLEDELGIQIFSR
CysB (*S. typhimurium*)	22	TAEGLYTSQPGISKQVRMLEDELGIQIFAR
AmpR (*E. cloacae*)	26	AAIELNVTHSAISQHVKTLEQHLNCQLFVR
IlvY (*E. coli*)	21	SARAMHVSPSTLSRQIQRLEEDLGQPLFVR
AntO (*E. coli*)	26	AAEALYLTPQTITGQIRALEDALQAKLFKR

b

Amino acid

Pos.	A	C	D	E	F	G	H	I	K	L	M	N	P	Q	R	S	T	V	W	Y	*
1	81	0	0	0	0	0	0	0	0	0	0	0	0	0	0	8	12	0	0	0	0
2	100	0	0	0	0	0	0	0	0	0	0	0	0	0	0	0	0	0	0	0	0
3	8	0	0	32	0	0	32	11	0	0	0	0	0	0	16	0	0	0	0	0	0
4	22	0	0	22	0	11	0	0	4	7	0	0	0	0	10	3	5	16	0	0	0
5	0	0	0	0	0	0	0	11	0	47	40	0	0	0	0	0	0	2	0	0	0
6	0	0	0	0	0	0	6	47	0	2	0	0	21	0	0	0	0	0	0	24	0
7	0	0	0	0	0	0	0	11	0	11	32	0	0	16	0	0	16	16	0	0	0
8	0	0	0	0	0	0	0	0	0	0	0	0	0	0	0	57	43	0	0	0	0
9	0	0	0	0	0	0	12	0	0	0	0	0	12	75	0	0	0	0	0	0	0
10	0	0	0	0	0	0	0	0	0	0	0	0	56	22	0	22	0	0	0	0	0
11	44	0	0	0	0	11	0	0	0	0	0	0	11	0	0	33	0	0	0	0	0
12	0	0	0	0	0	0	0	32	0	11	32	0	0	0	0	0	0	24	0	0	0
13	0	0	0	0	0	0	0	0	0	0	0	0	0	0	0	70	30	0	0	0	0
14	8	0	0	0	0	8	17	0	17	0	0	17	0	17	17	0	0	0	0	0	0
15	12	0	0	12	0	0	12	0	0	0	0	0	0	62	0	0	0	0	0	0	0
16	0	0	0	0	18	0	0	48	0	6	0	0	0	0	0	0	0	27	0	0	0
17	24	0	0	0	0	2	18	0	18	0	0	0	0	18	12	7	0	0	0	0	0
18	17	0	17	0	0	0	0	0	0	0	34	0	0	0	23	0	9	0	0	0	0
19	0	0	0	0	27	0	0	0	0	73	0	0	0	0	0	0	0	0	0	0	0
20	0	0	0	87	0	0	0	0	0	6	0	0	0	0	4	0	0	3	0	0	0
21	1	0	27	12	0	0	0	0	12	0	0	0	0	37	0	0	2	6	0	0	0
22	6	0	12	12	0	0	24	0	0	0	24	0	0	0	4	0	0	6	0	12	0
23	0	0	0	0	43	0	0	14	0	43	0	0	0	0	0	0	0	0	0	0	0
24	0	0	0	0	0	45	0	0	0	0	0	36	0	17	2	0	0	0	0	0	0
25	7	14	29	0	14	0	0	10	0	5	0	0	0	14	0	0	0	7	0	0	0
26	0	0	3	19	0	0	0	0	29	5	0	0	7	29	5	0	0	4	0	0	0
27	0	0	0	0	0	0	0	20	0	80	0	0	0	0	0	0	0	0	0	0	0
28	0	0	0	0	100	0	0	0	0	0	0	0	0	0	0	0	0	0	0	0	0
29	4	0	0	34	0	0	0	2	17	1	0	0	0	0	0	5	2	34	0	0	0
30	0	0	0	0	0	0	0	0	0	0	43	0	0	0	57	0	0	0	0	0	0

FIG. 1. Alignment (a) and scoring matrix (b) of LysR family members within a highly conserved region. The names used for putative proteins, such as LeuO, AntO, and TfdO, are adopted for heuristic purposes and do not connote any functional relationship to the *leu, ant,* or *tfdA* operons.

of scores is 100. Stop codons can be penalized with negative values if desired. In addition to weighting of residues, optional grouping of closely related sequences is allowed. Grouping allows the contributions of multiple closely related sequences to be averaged prior to the calculation of a matrix entry. In this way, multiply represented subfamilies contribute only a single value to a scoring matrix that is otherwise composed of distantly related sequences.

The resulting matrix is then used to score each stretch of translated sequence equal to the size of the window by assigning the value found in the matrix for each residue at each position. The sum of values for the whole stretch is the pattern score. Pattern scores can be thought of as equivalent to the number of matches in a simple alignment multiplied by 100, except that fractional matches are allowed. The searching program notes the single best score for each reading frame of a database entry and reports the 200 best pattern scores in a search, along with the 200 aligned sequences. Extension and inspection of the top scoring matches is carried out as for standard similarity searches. Confirmed new members can be added to the list of aligned sequences and the scoring matrix recalculated. In this way, starting with a crude scoring matrix derived from only two proteins, a more refined matrix can be constructed, allowing for an even more sensitive search. The following three applications illustrate the practical use of these searching tools and reveal several previously unknown relationships among bacterial activator proteins.

Application 2: Expansion of LysR Family by Matrix Searching. It was apparent from visual inspection of the multiply aligned LysR family members[7] that the most highly conserved ungapped region corresponded to MetR residues 22 through 51, which includes part of the predicted helix–turn–helix DNA-binding motif. Figure 1a shows this aligned stretch of 30 amino acids, where true homologs, representing proteins of similar function in different species, are grouped. A scoring matrix weighted according to codon frequency was generated (Fig. 1b). Consider, for example, the first matrix entry, 81 for alanine at position 1, present in 7 of 9 groups. This entry is calculated by first dividing the fraction of groups that have alanine at position 1 by the fractional occurrence of alanine codons in the genetic code ($7/9 \div 4/64 = 12.44$). Similar calculations are carried out for serine and threonine, each of which occurs in one group in position 1. For serine, the corresponding value is 1.19 ($1/9 \div 6/64$) and for threonine, 1.78 ($1/9 \div 4/64$). Matrix entries are calculated by normalizing these three values so that their sum is 100. For example, the matrix entry for alanine is $12.44 \div (12.44 + 1.19 + 1.78) \times 100 = 81$.

It is worth noting that an alanine match is given a relatively low score in the lod scoring matrix of Dayhoff *et al,*[3] in part because alanine is not usually a conserved residue. However, in the case of this particular region of 30 amino acids, alanines in the first and second positions are among the most highly conserved residues in the protein. Using a matrix that is customized to take this peculiarity into account helps to increase the sensitivity of the search. Also, we do not allow "conservative replacements," but rather assign values of zero for all residues that do not occur at a position in any of the known family members: What is a conservative replacement in most situations might not be in the particular position being searched. For example, the highest scoring substitution in the lod scoring matrix is between phenylalanine and tyrosine.[3] Nevertheless, their chemical properties are quite different; for example, substitution of a phenylalanine for the active site tyrosine of a tyrosine kinase will inactivate the enzyme. Our procedure focuses on the most highly conserved regions of a protein where such special residues are often found. Furthermore, a scoring procedure that makes fewer assumptions allows easier interpretation of the results.

The searching program scored all possible segments of 30 amino acids derived from GenBank 58 and EMBL 17 translated in all six reading frames using the scoring matrix shown in Fig. 1b. Since there are about 20 million bases in each database, each search involves about 40 million comparisons (20 million bases per frame ÷ 3 bases per amino acid × 6 frames = 40 million amino acid positions that can be aligned with position 1 in the scoring matrix). The output (Fig. 2) shows the top 28 scores along with the aligned stretch of 30 amino acids for each corresponding database sequence. As expected, the sequences used to construct the matrix were among the top scores, ranging from 1319 for *Alcaligenes eutrophus* TfdO down to 1093 for *Rhizobium leguminosarum* NodD. The highest score for what is clearly a chance occurrence is 750 for an open reading frame derived from the opposite strand of the *Caenorhabditis elegans* myosin heavy chain gene. Of the 11 new sequences scoring above 750, 3 are NodD sequences from other sources, and 1 is an incomplete sequence encoding *S. typhimurium* IlvY. The latter has a relatively low score (820) because the coding sequence for only 22 amino acids was present in the database entry. This illustrates how the searching algorithm will align translated sequence with the scoring matrix out to the end of a database entry, allowing for detection of sequences that do not include the entire match.

The remaining 7 matches represent new members of the LysR family. One match is to a predicted sequence of 77 amino acids upstream of and oppositely oriented to the *Pseudomonas putida clcABD* operon: Residues 21–50 score 1314. This corresponds to the amino terminus of the ClcR activator protein (A. Chakrabarty, personal communication). The second match is to an ORF of 144 amino acids upstream of and oppositely

EMBL ID	Sequence entry	Frame	Score	Window alignment
AETFDA	*A. eutrophus tfdA* operon	-3	1319	AARRLHISQPPVTRQIHALEQHLGVLLFER
M16964	*P. putida clc* operon	-2	1314*	AARRLHISQPPITRQIQALEQDLGVVLFER
STMETR	*S. typhimurium metR* gene	3	1280	AAAVLHQTQSALSHQFSDLEQRLGFRLFVR
ECLEUP	*E. coli leu* operon	-1	1250	AAHVLGMSQPAVSNAVARLKVMFNDELFVR
ECGALLYS	*E. coli lysR* gene	1	1223	AAHLLHTSQPTVSRELARFEKVIGLKLFER
STLEUP	*S. typhimurium leu* operon	-1	1221	AAHTLGMSQPAVSNAVARLVVMFNDVLFVR
ECAMPR	*E. cloacae ampR* gene	1	1218	AAIELNVTHSAISQHVKTLEQHLNCQLFVR
ECCYSB	*E. coli cysB* gene	1	1203	TAEGLYTSQPGISKQVRMLEDEDLGIQIFSR
STCYSB	*S. typhimurium cysB* gene	3	1202	TAEGLYTSQPGISKQVRMLEDELGIQIFAR
EAALDS	*E. aerogenes aldc* operon	-1	1188*	AAKALGISQPPLSQQIKRLEEEVGTPLFRR
ECILVYC	*E. coli ilvY* gene	-2	1186	SARAMHVSPSTLSRQIQRLEEDLGQPLFVR
ECRPSTB	*E. coli rpsT-ant* region	-2	1179	AAEALYLTPQTITGQIRALEDALQAKLFKR
M21093	*P. aeruginosa trpI* gene	2	1172*	AAEELHVTHGAVSRQVRLLEEDLGVALFGR
RMNOD	*R. meliloti nodD* gene	-2	1154	AARRINLSQPAMSAAIARLRTYFGDELFSM
M18972	*R. japonicum nodD1* gene	2	1128	AARSINLSQPAMSAAIARLRTYFGDDLFTM
RMNODD2	*R. meliloti nodD2* gene	2	1126	AARRVKLSQPAMSAAIARLRTYFGDELFSM
RTNODG	*R. trifolii nodD* gene	2	1111	AARSINLSQPAMSAAIGRLRAYFNDELFLM
BSNOD1	*Bradyrhizobium nodD* gene	-1	1100	AARKINLSQPAMSAAIARLRSYFRDELFTM
RLNOD	*R. leguminosarum nodD* gene	1	1093	AARSINLSQPAMSAAISRLRDYFRDDLFIM
M18971	*R. japonicum nodD2* gene	2	1077	AARSINLSQPAMSAAITRLRTYFRDELFTM
ECOTDC	*E. coli tdc* operon	1	1072*	AAKELGLTQPAVSKIINDIEDYFGVELVVR
AETFDCD	*A. eutrophus tfdCD* operon	-1	1038*	AAQRMHISQPPLTRQIQALERDIGAKL
RSNODD1	*Rhizobium sp. nodD1* gene	1	984	ASRRINLSQPAMSAAITRLRTYFRDELFTM
PFASPA	*P. fluorescens aspA* gene	-1	944*	AAERRFVTQPAFSRRIRSLEAALGLTLVN
M19460	*P. putida catBC* gene	-1	856*	AAELLHIAQPPLSRQISQLE
STILVYCR	*S.typhimurium ilvC* upstream	-3	820	SARAMHVSPSTLSRQIQRLEED
CEMYUNC	*C.elegans* major MHC gene	2	750	TAFFLLKSQVLMSTMQERSEEGLGLRLSGR
HSRGM	Human 28S rRNA gene	-1	731	AAPGSGSSVRHMSRAPRGGDSALGSSLFTR

FIG. 2. Results of a pattern search using the scoring matrix of Fig. 1b. Asterisks indicate new LysR family members.

oriented to the *Enterobacter aerogenes aldc* operon: Residues 21–50 score 1188. No information is available concerning its function. The third match is to the *Pseudomonas aeruginosa* TrpI activator protein which is known to regulate the *trpBA* operon.[12] Residues 26–45 score 1172. The fourth match is to a protein of unknown function in the *E. coli tdc* operon. Residues 27–46 score 1072. Extension of the alignment beyond the searching window further indicates that this is a LysR family member. The fifth match is to an incomplete ORF of 46 amino acids upstream of and oppositely oriented to the *A. eutrophus tfdCDEF* operon: Residues 21–46 (26 amino acids) score 1038. This ORF shows 65% identity to the first 46 amino acids of the ORF upstream of and oppositely oriented to the *A. eutrophus tfdA* operon, a previously detected member of the LysR family.[7] The relationships between the putative ORFs upstream of TrpI, *clcABD*, *tfdA*, and *tfdCDEF* and various known members of the LysR family have been noted by others.[12,13] The sixth match is to an incomplete ORF of 51 amino acids upstream of and oppositely oriented to the *Pseudomonas fluorescens aspA* operon: Residues 23–51 (29 amino acids) score 944. No information is available concerning its function. The seventh match is to

---

[12] M. Chang, A. Hadero, and I. P. Crawford, *J. Bacteriol.* **171,** 172 (1989).
[13] E. J. Perkins, G. W. Bolton, M. P. Gordon, and P. F. Lurquin, *Nucleic Acids Res.* **16,** 7200 (1988).

an ORF of 39 amino acids upstream of and oppositely oriented to the *P. putida catBC* operon: Residues 20–39 (20 amino acids) score 856. This corresponds to the amino terminus of the CatR activator protein (A. Chakrabarty, personal communication).

One way to test the assertion of common ancestry for the new sequences is to use each ORF as query in a standard similarity search. When this was done, each of the 7 new members detected several other members of the LysR family as best matches using a window of 30 (data not shown). Furthermore, each new sequence shows characteristics found in several known LysR family members: (1) the location of this region of similarity about 20 amino acids from the likely amino terminus, (2) a high score using the helix–turn–helix DNA-binding predictive scheme[10] in the region that aligns with the other proteins (data not shown), and (3) divergent transcription from a promoter region for a known gram-negative bacterial operon in 6 of 7 cases. These features further support the assertion of common ancestry. Since only 20 predicted amino acids were necessary in one case to detect a distant relationship, it is clear that the matrix searching strategy is very sensitive. In addition, the absence of high scoring chance occurrences demonstrates that the strategy also is extremely selective.

Application 3: Successive Searching Allows Expansion of AraC Family. The previous example demonstrated the ability to detect short sequence similarities by a single round of matrix searching. The following application demonstrates the use of successive standard similarity and matrix searches for detecting, confirming, and expanding a family. The procedure is summarized in Fig. 3 (S. Henikoff, unpublished results).

Step 1. The *E. coli* AraC protein regulates the *araBAD* operon encoding structural components for arabinose metabolism. A standard similarity search of GenBank 58 was carried out using *E. coli* AraC as query. The resulting matches included known AraC proteins from three other bacteria and a striking similarity (31% identity) to an incomplete ORF (110 amino acids) upstream of the *Streptomyces lividans xp55* gene (Xp55O). This previously unrecognized ORF (T. Eckhardt, personal communication) would correspond to the carboxy-terminal portion of a presumptive member of the AraC family.

Step 2. Inspection of the alignment revealed a segment of 43 amino acids near the carboxy terminus that is most similar between the AraCs and Xp55O (data not shown). A scoring matrix was constructed and used to search GenBank 58 and EMBL 17. The six highest scoring matches were *E. coli* RhaR (1679), MelR (1573), transposon Tn*10* TetD protein (1375), M5 protein (1364), RhaS (1354) and PhoO, an incomplete and apparently unrecognized ORF upstream of and divergent from the *E. coli phoM* operon (1354). The best spurious match was 1216.

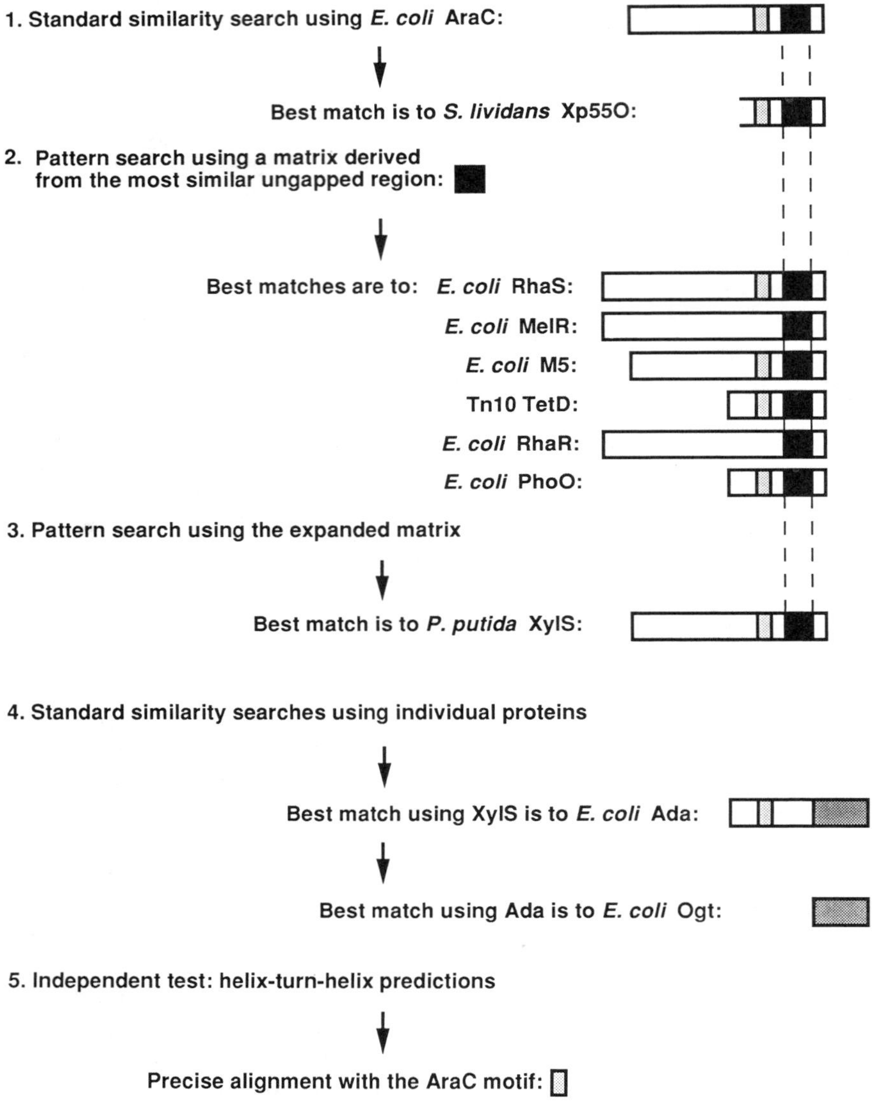

1. Standard similarity search using *E. coli* AraC:

Best match is to *S. lividans* Xp550:

2. Pattern search using a matrix derived
   from the most similar ungapped region:

Best matches are to: *E. coli* RhaS:

E. coli MelR:

E. coli M5:

Tn10 TetD:

E. coli RhaR:

E. coli PhoO:

3. Pattern search using the expanded matrix

Best match is to *P. putida* XylS:

4. Standard similarity searches using individual proteins

Best match using XylS is to *E. coli* Ada:

Best match using Ada is to *E. coli* Ogt:

5. Independent test: helix-turn-helix predictions

Precise alignment with the AraC motif:

FIG. 3. Expansion of the AraC family. Solid boxes indicate alignment with the 43-amino acid window used in the matrix searches. Predicted helix–turn–helix regions are shaded.

The overlapping genes encoding the rhamnose regulatory proteins RhaS and RhaR regulate and are oppositely oriented to the *rhaBAD* operon encoding structural components for rhamnose metabolism.[14] The similarity of RhaR and RhaS to one another and to AraC has been noted previously.[14] MelR resembles AraC in that it is a sugar-sensitive regulatory protein encoded divergently from a gene that it regulates.[15] This relationship appears not to have been noted previously. The other three predicted proteins also were not previously reported to be members of a family and are of unknown function.[16–18]

Step 3. A new scoring matrix including the six new sequences was constructed and used to search GenBank 58 and EMBL 17. The highest scoring match (958 versus 907 or less for likely spurious matches) was *P. putida* XylS, a known bacterial activator protein.[19]

Step 4. Standard similarity searches were carried out for each of the new sequences. Using XylS as query, the best match was to the *E. coli* Ada protein, a known regulatory protein that includes a carboxy-terminal O^6-methylguanine methyltransferase domain.[20,21] Using Ada as query, an excellent match was to an unrecognized reading frame upstream of the *E. coli* *nirR* gene. In this case, the similarity was confined to the methyltransferase portion of Ada. This sequence coincides with that for the *E. coli* *ogt* gene[22] except for seven differences, including two frameshifts which are probably errors in the database entry.

Step 5. An independent test of the hypothesized relationships among these proteins was carried out. AraC is a helix–turn–helix DNA-binding protein.[10] The predicted 20-amino acid DNA-binding motif begins 38 amino acids upstream of the segment used to construct the scoring matrix. The helix–turn–helix predictive scheme[10] was carried out for each of the proteins. For RhaS, M5, TetD, PhoO, XylS, and Ada, the highest scoring stretch of 20 amino acids aligns precisely with the predicted helix–turn–helix region of AraC (data not shown). Since the segment detected in the matrix searches does not overlap the predicted helix–turn–helix region,

[14] J. F. Tobin and R. F. Schlief, *J. Mol. Biol.* **196**, 789 (1987).

[15] C. Webster, K. Kempsell, I. Booth, and S. Busby, *Gene* **59**, 253 (1987).

[16] G. Braus, M. Argast, and C. F. Beck, *J. Bacteriol.* **160**, 504 (1984).

[17] E. H. Kemp, N. P. Minton, and N. H. Mann, *Nucleic Acids Res.* **15**, 3924 (1987).

[18] M. Amemura, K. Makino, H. Shinagawa, and A. Nakata, *J. Bacteriol.* **168**, 294 (1986).

[19] S. Inouye, A. Nakazawa, and T. Nakazawa, *Gene* **44**, 235 (1986).

[20] Y. Nakabeppu, H. Kondo, S. Kawabata, S. Iwanaga, and M. Sekiguchi, *J. Biol. Chem.* **260**, 7281 (1985).

[21] B. Demple, B. Sedgwick, P. Robins, N. Totty, M. D. Waterfield, and T. Lindahl, *Proc. Natl. Acad. Sci. U.S.A.* **82**, 2688 (1985).

[22] P. M. Potter, M. C. Wilkinson, J. Fitton, F. J. Carr, J. Brennand, D. P. Cooper, and G. P. Margison, *Nucleic Acids Res.* **15**, 9177 (1987).

this is an independent test of similarity. Therefore, these diverse proteins are likely to resemble AraC in structure and function. Table II summarizes known features of the nine AraC family members.

Application 4: Multiple Searches to Detect Complex Relationships among Families. The following application further demonstrates the ability to detect short sequence similarities, where failure to extend an alignment is due to the limited nature of the homologous region or to frameshifts in database entries. In addition, this application shows how multiple searches can be used to detect complex relationships.

The LuxR protein controls genes for bioluminescence in *Vibrio fischeri.*[23,24] Using the predicted amino acid sequence of LuxR as query in a standard similarity search of GenBank 58 and EMBL 17 with a window of 30, we detected a striking similarity to a 28-kDa protein of suspected regulatory function upstream of the *E. coli* sequence encoding UvrC.[25] Alignment of the two amino acid sequences essentially end to end (Fig. 4a) shows that they are identical at 24% of aligned residues with the insertion of four small gaps. Visual inspection of this alignment indicates that the region of greatest ungapped similarity begins at residue 184 in LuxR and 182 in UvrC-28K and extends for 43 amino acids. A scoring matrix derived from this aligned stretch was used to search GenBank 58 and EMBL 17. After LuxR, five bacterial sequences scored the highest (Fig. 5a). The highest scoring sequence of these five encodes GerE, a *Bacillus subtilis* protein involved in regulation of spore formation. The 74-amino acid GerE protein aligns with the carboxy-terminal one-third of LuxR and UvrC-28K.

The next highest scoring sequence encodes *E. coli* UhpA, a known regulatory protein. Only the carboxy-terminal one-third of UhpA aligns. The next highest scoring sequence is derived from a region of unknown function downstream of the *Pseudomonas aeruginosa trpAB* gene (TrpO). Examination of the sequence just upstream in a different reading frame reveals a striking similarity to *E. coli* UhpA. Deletion of 19 bases leads to the alignment of TrpO with UhpA (Fig. 4b). The two sequences show 26% identity of aligned residues with only four gaps. Subsequent reexamination of the nucleotide sequence data obtained for this region suggests an intact 205-amino acid ORF (I. Crawford, personal communication).

The next highest scoring sequence is the *R. meliloti* FixJ protein, an activator protein that was previously shown to be related to UhpA (30%

[23] J. Engebrecht and M. Silverman, *Nucleic Acids Res.* **15**, 10455 (1987).

[24] J. H. Devine, C. Countryman, and T. O. Baldwin, *Biochemistry* **27**, 837 (1988).

[25] S. Sharma, T. F. Stark, W. G. Beattie, and R. E. Moses, *Nucleic Acids Res.* **14**, 2301 (1986).

TABLE II
FEATURES OF PREDICTED BACTERIAL PROTEINS RELATED TO AraC

Predicted protein	Aligned helix–turn–helix	Regulatory protein	Divergent transcription	Function
AraC (various species)	Yes	Yes	Yes	Regulates arabinose metabolism
Xp55O (*Streptomyces lividans*)	Yes	?	?	Unknown (unrecognized ORF)
RhaS (*E. coli*)	Yes	Yes	Yes	Regulates rhamnose metabolism
RhaR (*E. coli*)	No	Yes	Yes	Regulates rhamnose metabolism
MelR (*E. coli*)	No	Yes	Yes	Regulates melibiose metabolism
M5 (*E. coli*)	Yes	?	?	Possibly regulates polysaccharide biosynthesis
TetD (Tn*10*)	Yes	?	Yes	Unknown
PhoO (*E. coli*)	Yes	?	Yes	Unknown (unrecognized ORF)
XylS (*P. putida*)	Yes	Yes	Yes	Activates xylene-degradative functions
Ada (*E. coli*)	Yes	Yes	?	Activates response to mutagenic alkylation

```
a

LuxR  MKNINADDTYRIINKIKACRAYDINQCLSDMTKMVHCEYYLTLAIIYPHSMVKSDISILDNYPKKWRQYYDDANLIKYDPIVDYSNSNHSPINWNIFE      98
         : : :: :   :   : :  :  :        .     :    .  : : :. . :       :  .  :  :.: ::     .
28K   MQDKDFFSWRRTMLLRFQRMETAEEVYHEIELQAQQLEYDY-SLCVRHPVPFTRPKVAFYTNYPEAWVSYYQAKNFLAIDPVLNPENFSQGHLMWN---      96
                                                                     ++++++++++++
LuxR  NNAVNKKSPNVIKEAKTSGLITGFSFPIHTANNG-FGMLSFAHSEKDNYIDSLFLHACMNIPLIVP-SLVDNYRKINIANNKSNNDLTKREKECLAWACE   196
      .   .   .  :.. .:  :  :.  :   :::   ::.::.  .::: : :.   . :  :: . .:  ::.::  : : :::    ::::: : : :
28K   -DDLFSEAQPLWEAARAHGLRRGVHSVFNAAQTGALGFLSFSRGSRRE-IPILSDELQLKMQLLVRESLMALMRLNDEIVMTFPEMNFSKREKEILRWTAE   194
      ++++++++++++++++++++++++++++++++
LuxR  GKSSWDISKILGCSERTVTFHLTNAQMKLNTTNRCQSISKAILTGAIDCPYFKN   250
      :::: .: :: :.:  .  :. :  :   .  .  .:: ::.  .
28K   GKTSAEIAMILSISENTVNFHQKNMQKKINAPNKTQVACYAAATGLI         241

b

TrpO  MSKVLIVDDHPAIRLAVRLLFERD-GFTMSREADNGAEALQVARKKSPDLAILDIGIPKIDGLEVIARLKSLKLDTKVLVLTRQNRSQFARRLQAGPWA-SS   100
      ::::::::::  :   .  :: :       :  :::::: : .   ::: :.: : :  :.: ::  .   :  . .:::: :: .::.       :   .
UhpA  MITVALIDDHLIVRSGFAQLLGLEPDLQVVAEFGSGREALAGLPGRGVQVCICDISMPDISGLELLSQLP--KGMATIMLSVHDSPALVEQALNAGARGFLS   100
                                                          ++++++++++++++++++++++++++++++++++++++++++++++
                      - 19 bp
TrpO  AKGKPLRAAARRQGVLAG↑GALRSINQQSRDNEARMLESLSDREMTVLQYLANGNTNKAIAQQLFLSEKTVSTYKSRIMLKLNAHSLAGLIDFARRHELT...   198
      .::: ::  :.   .::   : :  . :.       :.::       :. :         :.  :..  ::. : :   ::: .:   : .::::::
UhpA  KRCSPDELIAAVHTVATG GCYLIPDIAIKLASGRQ-DPLTKRERQVAEKLAQGMAVKEIAAELGLSPKTVHVHRANLMEKLGVSNDVELARRMFDGW      196
```

Fig. 4. (a) Alignment between *V. fischeri* LuxR and *E. coli* UvrC-28K (28K). Colons denote identity, and single dots denote conservative replacements. Pluses indicate the location of the 43-amino acid scoring window. (b) Alignment between translated sequence downstream of *Pseudomonas aeruginosa trpAB* (TrpO) and *E. coli* UhpA. The indicated insertion is necessary to bring the former sequence into frame.

a) **Scoring matrix derived from LuxR and UvrC-28K**

EMBL ID	Sequence entry	Frame	Score	Window alignment
ECUVRC	E. coli uvrC operon	2	3404	SKREKEILRWTAEGKTSAEIAMILSISENTVNFHQKNMQKKIN
M19039	V. fischeri luxR operon	-3	3196	TKREKECLAWACEGKSSWDISKILGCSKRTVTFHLTNAQMKLN
M17642	B. subtilis gerE gene	2	1723	TKREREVFELLVQDKTTKEIASELFISEKTVRNHISNAMQKLG
M17102	E. coli uhpA gene	1	1496	TKRERQVAEKLAQGMAVKEIAAELGLSPKTVHVHRANLMEKLG
PATRPAB	P. aeruginosa trpAB operon	3	1426	SDREMTVLQYLANGNTNKAIAQQLFLSEKTVSTYKSRIMLKLN
JO3174	R. meliloti fixLJ operon	3	1365	SERERQVLSAVVAGLPNKSIAYDLDISPRTVEVHRANVMAKMK
ECUVRC	E. coli uvrC operon	1	1350	SERELQIMLMITKGQKVNEISEQLNLSPKTVNSYRYRMFSKLN
DMTHB1	D. melanogaster HB1	2	1243	QGKRMLILKLREGKTYKDIQKTLKCSAKMVSNAIKYKWKPEN
KARCSA	K. aerogenes rcsA gene	2	1220	SKTESNMLQMWMAGHGTSQISTQMNIKAKTVSSHKGNIKKKIQ
MMHOMMH3	Mouse MH-3 homeo box gene	2	1202	LELEKEFHFNRYLTRRRIEIAHTLCLSERQVKIWFQNRRMKWK
ECMALT	E. coli malT gene	3	1193	TQREWQVLGLIYSGYSNEQIAGELEVAATTIKTHIRNLYQKLG

b) **Scoring matrix derived from LuxR, UvrC-28K, GerE, UhpA, TrpO, FixJ and UvrC-23K**

EMBL ID	Sequence entry	Frame	Score	Window alignment
ECUVRC	E. coli uvrC operon	2	2657	NFSKREKEILRWTAEGKTSAEIAMILSISENTVNFHQKNMQKKINA
M17642	B. subtilis gerE gene	2	2459	SLTKREREVFELLVQDKTTKEIASELFISEKTVRNHISNAMQKLGV
M19039	V. fischeri luxR operon	-3	2397	DLTKREKECLAWACEGKSSWDISKILGCSKRTVTFHLTNAQMKLNT
M17102	E. coli uhpA gene	1	2007	PLTKRERQVAEKLAQGMAVKEIAAELGLSPKTVHVHRANLMEKLGV
PATRPAB	P. aeruginosa trpAB operon	3	1824	SLSDREMTVLQYLANGNTNKAIAQQLFLSEKTVSTYKSRIMLKLNA
JO3174	R. meliloti fixLJ operon	2	1712	TLSERERQVLSAVVAGLPNKSIAYDLDISPRTVEVHRANVMAKMKA
ECUVRC	E. coli uvrC operon	1	1511	SLSERELQIMLMITKGQKVNEISEQLNLSPKTVNSYRYRMFSKLNI
ECMALT	E. coli malT gene	3	1351	PLTQREWQVLGLIYSGYSNEQIAGELEVAATTIKTHIRNLYQKLGV
KARCSA	K. aerogenes rcsA gene	2	1234	SLSKTESNMLQMWMAGHGTSQISTQMNIKAKTVSSHKGNIKKKIQT
MMHOMMH3	Mouse MH-3 homeo box gene	2	1160	QVLELEKEFHFNRYLTRRRIEIAHTLCLSERQVKIWFQNRRMKWKK
DMTHB1	D. melanogaster HB1	2	1113	YSQGKRMLILKLREGKTYKDIQKTLKCSAKMVSNAIKYKWKPENR

FIG. 5. Results of successive pattern searches for similarities to LuxR.

identity), aligning essentially from end to end.[26] The next highest scoring sequence is a 23K protein of unknown function just downstream from UvrC-28K and upstream of UvrC.[25] Alignment of this sequence with UhpA, FixJ, and TrpO shows, respectively, 25, 22, and 25% identical aligned residues (data not shown). Therefore, these four predicted proteins are members of a single family. However, they do not align from end to end with the LuxR – UvrC-28K pair. Rather, the carboxy-terminal portions of all six proteins are similar and align with the short GerE protein, but the amino termini differ between the two groups.

Two known bacterial activator proteins, *Klebsiella aerogenes* RcsA and *E. coli* MalT, score among the best spurious matches in this search. When the scoring matrix is expanded to include GerE, UhpA, TrpO, FixJ, and UvrC-23K and the window extended by 3 amino acids, RcsA scores 74 points above the best spurious matches (Fig. 5b). Likewise, MalT improves to 191 points above the best spurious matches. The similarity is confined to the carboxy terminus of MalT, a 901-amino acid protein. Alignment of MalT with GerE is particularly striking, with 33% identical residues (data not shown). In addition, standard similarity searches using both MalT and GerE detect another likely homolog, a partial reading frame downstream of and oppositely oriented to the *B. subtilis* gerA operon (GerO). This 58-amino acid incomplete sequence shares 26 identical residues with GerE and 20 with MalT (data not shown). Therefore, a total of ten diverse proteins, including LuxR and five other known activator proteins, are each identified as having a similar carboxy-terminal domain (Fig. 6).

Previous work has shown that the amino termini of UhpA and FixJ are related to a family of activator proteins that includes *E. coli* OmpR, SfrA, and PhoB, *Agrobacterium tumefaciens* VirG, *S. typhimurium* CheY and CheB, *B. subtilis* Spo0F and Spo0A, *R. leguminosarum* DctD, and *Klebsiella pneumoniae* NtrC.[26] These similarities are seen for the amino-terminal portions of UhpA and FixJ. Therefore, the proteins that align from end to end with UhpA appear to be composites different from LuxR: the amino-terminal portion of each is related to the family of activator proteins that includes Spo0F while the carboxy terminal portion is related to the family that includes GerE (Fig. 6).

Visual examination of the multiply aligned regulatory proteins related to OmpR and Spo0F[26] reveals that the most strongly conserved ungapped region aligns with residues 77 – 120 of OmpR. A scoring matrix of this aligned stretch was used to search GenBank 58 and EMBL 17. The results are shown in Fig. 7. Scores for members of the family used to construct the

[26] M. David, M.-L. Daveran, J. Batut, A. Dedieu, O. Domergue, J. Ghai, C. Hertig, P. Boistard, and D. Kahn, *Cell* **54,** 671 (1988).

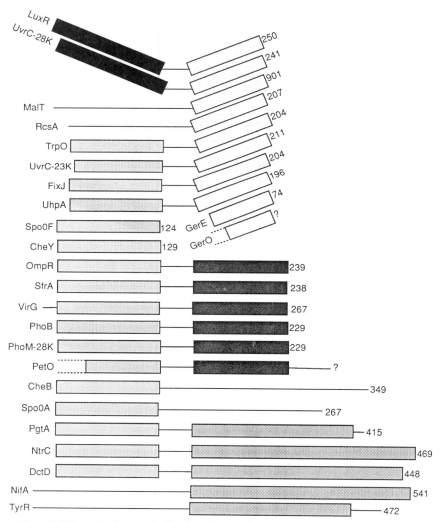

FIG. 6. Schematic diagram indicating the complex relationship among several bacterial regulatory proteins and homologs of unknown function. Different regions of similarity among aligned segments are indicated by different shadings. Numbers indicate protein size in amino acids.

EMBL ID	Sequence entry	Frame	Score	Window alignment
ECPHOB	E. coli phoB gene	1	1903	VVMLTARGEEEDRVRGLETGADDYITKPFSPKELVARIKAVMRR
ECOMPB	E. coli ompR operon	3	1891	IIMVTAKGEEVDRIVGLEIGADDYIPKPFNPRELLARIRAVLRR
ECDYE	E. coli dye gene (SfrA)	2	1664	LMFLTGRDNEVDKILGLEIGADDYITKPFNPRELTIRARNLLSR
M16775	E. coli phoP gene	3	1662	ILMLTAKDEEFDKVLGLELGADDYMTKPFSPREVNARVKAILRR
ATVIR	A.tumefaciens virG gene	3	1607	IIISGDRLETDKVVALELGASDFIAKPFSIREFLARIRVALRV
ECGLN	E. coli glnALG operon (NtrC)	2	1578	VIIMTAHSDLDAAVSAYQQGAFDYLPKPFDIDEAVALVERAISH
KPNTRC	K. pneumoniae ntrC gene	1	1578	VIIMTAHSDLDAAVSAYQQGAFDYLPKPFDIDEAVALVDRAISH
ATVIRBCG	A.tumefaciens virG gene	1	1565	IIISGARLEADKVIALELGATDFIAKPFGTREFLARIRVALRV
J03174	R. meliloti fixLJ gene	2	1542	SIVITGHGDVPMAVEAMKAGAVDFIEKPFEDTVIIEAIERASEH
ECPHOM	E. coli phoM operon	1	1536	VLFLTARSEEVDRLLGLEIGADDVAKPFSPREVCARVRTLLRR
BSSPO0AA	B.subtilis spo0A gene	3	1518	VIMLTAFGQEDVTKKAVDLGASYFILKPFDMENLVGHIRQVSGN
ECCHE3	E. coli cheRBYZ operon	1	1373	VLMVTAEAKKENIIAAQAGASGYVVKPFTPATLEEKLNKIFEK
BSSPO0F	B. subtilis spo0F gene	1	1355	VIIMTAYGELDMIQESKELGALTHFAKPFDIDEIRDAVKKYLPL
ECCHEY	E. coli cheY gene	1	1351	VLMVTAEAKKENIIAAQAGASGYVVKPFTAATLEEKLNKIFEK
M15810	R. meliloti ntrC gene	3	1239	VLVMSAQNTFMTAIKASEKGAYDYLPKPFDLTELIGIIGRALAE
M17102	E. coli uhpA gene	1	1213	TIMLSVHDSPALVEQALNAGARGFLSKRCSPDELIAAVHTVATG
STPGTA	S. typhimurium pgtA gene	3	1159*	ILLITGHGDVPMAVDAVKKGAWDFLQKPSIRAKLLILIEDALRQ
M14227	B. paraponiae ntrBC homologue	2	1133	VIVMSAQNTFMTAIRPSERGAYEYLPKPFDLKELITIVGRALAE
RCNIFR12	R. capsulatus nifR genes	2	1119	VIVISAQNTIMTAIQAAEADAYDYLPKPFDLPDLMKRAARALEL
ECUVRC	E. coli uvrC operon	1	1108*	IIMLTVHTENPLPAKVMQAGAAGYLSKGAAPQEVVSAIRSVYSG
RCFBC	R.capsulata fbc (petABC) operon	2	1036*	DPAFDRPRRDPRAIEGLEAGADDYLPKPFEPNFCCASMRSCGA
ECCHE3	E. coli cheRBYZ operon	2	1028	MVSSLTGKGSEVTLRPLELGAIDFVTKPQLGIREGMLAYNEMIA
SV5PVA	Simian virus 5 V and P mRNA	2	903	VRKVPPQKDLTGLKITLEQLAKDCISKPKMREYLLKINQASSE
BSSPO2A	B.subtilis spoIIA gene	3	889	SIHETVYENDGDPITLLDQIADNSEEKWFDKIALKEAISDLEER
RCPETG	R. capsulata petABC operon	3	886	DPASDRARRNPASDRGARGRGDDYLPKPFEPNELLLRINAILRR

. . .

| RCPETG | R. capsulata petABC operon | 1 | 829 | ILLLTARGETPRAIEGLEAGAMTTCPNRSSRTNFCCASTRSCGG |

Fig. 7. Results of a search using a scoring matrix derived from 11 proteins related to OmpR.[26] Asterisks indicate new members of the family.

matrix ranged from 1903 for PhoB down to 1028 for CheB. In addition, other proteins that were previously shown to be members of the family scored high: PhoP, a *B. subtilis* homolog of PhoB, a 28K protein encoded at the *E. coli phoM* operon, and both the *B. parasponiae* and *Rhodopseudomonas capsulata* homolog of NtrC. The UvrC-23K protein discussed above also scored high. The *S. typhimurium* PgtA protein which regulates phosphoglycerate transport scored 1223. PgtA aligns with NtrC in both its amino- and carboxy-terminal domains.

The next highest score is for a database sequence upstream of the *R. capsulata petABC* operon. In addition, two other matches to this same operon sequenced in a different laboratory are among the highest scoring spurious matches. These latter two matches are shifted in frame with respect to one another by one base. All three matches are derived from the same region of the *petABC* operon. Since the two database sequences represent independent determinations of nearly identical DNA sequences, it is possible to deduce an approximation of the correct sequence assuming that they encode a homolog of this protein family. Figure 8 shows that, when these adjustments are made, a coding sequence (PetO) can be identified that is clearly homologous to OmpR, extending from the beginning of each database entry to the carboxy terminus. Several different sets of frameshifts are necessary in each of the two sequences. That these are

```
OmpR    MQENYKNLVVDDDMRLRALLERYLTEQGFQVRSVANAEQMDRLLTRESFHLMVLDLMLPGEDGLSICRRLRSQ
              :                 :        .    :::.     : : ::::.: :::::::. : ::.
RCFBC                     ...AEVPDPQRLSGHRRARRGPHARRLLSGLEFNLIVLDVMMPGEDGLSLTRDLRTK
RCPETC                                        ...EFNLIVLDVMMPGEDGLSLTRDLRTK


              +++++++++++++++++++++++++++++++++++++++++++++
OmpR    SNPMPIIMVTAKGEEVDRIVGLEIGADDYIPKPFNPRELLARIRAVLRRQANELPGAPSQEEAVIAFGKFKLNL
        ::.   .::.:: .:: ::: :::::.::: ::.::: :: :.:::    . ...:          .: :
RCFBC   -MTTPILLLTARGETRER(+1 bp)      (+1 bp)ELLLRINAILRRVPEAVTAGPKYLSLG(-1 bp)
                        IEGLEAGADDYLPKPFEPK                              -----RALDL
RCPETC  -----ILLLTARGETPRAIEGLEAG(+1 bp)     (+1 bp)RRVPEAVTAGAKYLSLG----PLRYDL
                        GDDYLPKPFEPNELLLRINAIL


OmpR    GTREMFREDEPMPLTSGEFAVLKALVSHPREPLSRDKLMNLARGREYSA-----MERSIDVQISRLRRMVEEDP
         :    :  :: : :.  . :       :  . ::       :        .: .:::::.:::: .::::
RCFBC   DRGELSQGDQPVRLTATEAAALMRSLRADAGE      MRSWAGTRCRQRRGCPGDRAVDVQITRLRRKIEPDP
                        (+1 bp)VIGRTEL(-1 bp)
RCPETC  DRGELSQGDQPVRLTPTEAAALMRIVRASAGEVIGRTELGRDRSASAERR-----GDRAVDVQITRLRRK(-1 bp)
                                                                              IEPDR


OmpR    AHPRYIQTVNGLGYVFVPDGSKA*
         :::::.  :.   :
RCFBC   REPRYLQSCADLATCLHPIEACKRLPQRQKSANRAEPRLFASPACECMLDCGFPGR*
RCPETC  RNRATCRRCADLATCLHPIEACKRLPQRQKSANRAEPRLFASPSCECMLDCGFPGR*
```

FIG. 8. Alignment between *E. coli* OmpR and the two translated database entries (RCFBC and RCPETC) of the region upstream of the *Rhodopseudomonas capsulata petABC* operon. The several frameshifts indicated are necessary to obtain a consistent coding sequence aligning with OmpR. Pluses indicate the region detected in the pattern search.

sequencing errors is the only plausible interpretation in this case, since none of the frameshifts coincide between the two sets of data. In contrast, within extensive regions known to code for the *petABC* proteins, the two sets of data differ only with respect to nucleotide substitution.

It should be noted that in both *petABC* entries, the sequence from only one DNA strand was obtained for the region in question; in fact, the high error rate of these data has already been acknowledged by the investigators.[27] In both cases, two frameshifts occurred within the searching window (Fig. 8). Clearly, even very uncertain sequence data can be of great value when present in nucleotide sequence databases. Perhaps investigators should be encouraged to deposit uncertain sequence (duly noted) into the databases. In summary, Fig. 6 shows the complex relationships among these and other bacterial regulatory proteins determined in this and in previous studies.[1,26,28,29]

Summary

In this chapter we describe strategies for the searching of translated nucleotide sequence databases. By applying standard searching techniques developed for protein databases,[6] we have found that previously unrecognized homologies can be detected. In addition, we have shown that extremely high sensitivity can be obtained using the scoring matrix strategy[11] for short regions of similarity. The latter approach is particularly effective for detecting homologs found at the ends of sequences and within data of poor quality. These individual methods are demonstrated for the LysR family of bacterial activator proteins. Successive applications of these methods allow for sensitive detection of complex relationships, as demonstrated for the AraC family and for the complex LuxR–OmpR–NtrC families of bacterial activator proteins. Although our examples are drawn from bacterial sequences, these methods are likewise effective for higher eukaryotic genomic sequences, where protein-coding sequences are usually interrupted by introns. This should be particularly important in the future, since much of the expected increase in nucleotide sequence databases is likely to come from eukaryotic genomic sequencing projects.

[27] E. Davidson and F. Daldal, *J. Mol. Biol.* **195**, 13 (1987).
[28] S. C. Winans, P. R. Ebert, S. E. Stachel, M. P. Gordon, and E. W. Nester, *Proc. Natl. Acad. Sci. U.S.A.* **83**, 8278 (1986).
[29] B. T. Nixon, C. W. Ronson, and F. M. Ausubel, *Proc. Natl. Acad. Sci. U.S.A.* **83**, 7850 (1986).

[8] Use of Homology Domains in Sequence Similarity Detection

By Charles B. Lawrence

Introduction

This chapter describes an approach to identifying sequence similarities that complements other standard methods. It is in the class of methods that finds local optimal alignments. The main strength of this approach is that it can identify the boundaries of homologous regions between two sequences with great precision.

The recent availability of gene and protein sequence data banks in machine-readable form has stimulated considerable interest in the comparison of sequences and methods for the detection of homology (common ancestry) between different sequences. The most widely used and accepted method for sequence comparison is dynamic programming, first described for biological sequences in 1970 by Needleman and Wunsch.[1] Since then, many variations of the original algorithm have been suggested, as well as algorithms based on methods other than dynamic programming.[2,3] The model that is used by dynamic programming methods to represent compared sequences is two strings of characters that differ from one another by mismatches or by the insertion or deletion of other characters. This is an appropriate model for gene and protein sequences because they change during evolution by single-base mutation and by the insertion or deletion of genetic material. This allows the problem of identifying homology to be viewed as a string-edit problem. Dynamic programming methods find the minimum distance or maximum similarity between two sequences, given values for the cost of a mismatch and insertion or deletion (indel). Each comparison results in a score that represents the distance between the two sequences and is used as a measure of their relatedness. The success of the method can be attributed to its mathematical rigor, the generation of an easily understood score, and the fact that it is successful at identifying distant evolutionary relationships.

We have used a slightly different model for the evolution of a sequence, resulting in an alternate strategy for characterizing sequence simi-

---

[1] S. B. Needleman and C. D. Wunsch, *J. Mol. Biol.* **48**, 444 (1970).

[2] M. S. Waterman, *Bull. Math. Biol.* **46**, 473 (1974).

[3] D. Sankoff and J. B. Kruskal, *"Time Warps, String Edits, and Macromolecules: The Theory and Practice of Sequence Comparison."* Addison-Wesley, Reading, Massachusetts, 1983.

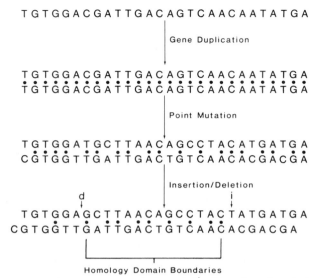

FIG. 1. Definition of homology domain boundaries. Homology domain boundaries are defined by the position of insertion/deletion events in homologous sequences. [Reprinted from *CABIOS* **4**, pp 25–33 (1988).]

larity that is also successful at identifying distant evolutionary relationships. In this model, indels define the boundaries of what we have termed *homology domains*[4] as illustrated in Fig. 1. At the time of gene duplication the two resulting sequences are identical and can be aligned such that each nucleotide is opposite the homologous nucleotide in the other sequence. The sequences then drift by single-base mutations and indel events to form the boundaries of the homology domain. When the two sequences are aligned such that homologous positions within the homology domain are opposed, the frequency of matches within the domain is greater than that expected by chance, while outside of the domain the frequency of matches will equal that expected by chance, given the base composition of the sequences. When sequences are modeled in this manner, the problem of identifying homology becomes the accurate localization of the boundaries of all authentic homology domains.

Altschul and Erickson[5] have described a method that we have found to be successful for the identification of the boundaries of homology domains as defined above. In this chapter, we demonstrate the sensitivity of this method and provide several examples of its use in analyzing possible sequence homology.

[4] C. B. Lawrence and D. A. Goldman, *CABIOS* **4**, 25 (1988).
[5] S. F. Altschul and B. W. Erickson, *Bull. Math. Biol.* **48**, 617 (1986).

Similarity Functions

The method for identifying homology domains takes advantage of two innovations described by Altschul and Erickson.[5,6] The first is the use of a nonlinear similarity function to evaluate the relative significance of sequence subalignments having no gaps. A useful similarity function for identifying homology domains yields a value related to the probability of finding a given subalignment by chance. In this chapter we use a modification of the similarity function:

$$s_2(l,c) = (Tl - c - 0.5)/l^{1/2} \qquad (1)$$

where c is the cost of a subalignment of length l and T is the expected value of the cost at any position in the subalignment [Eq. (1) from Ref. 6)]. The modification to Eq. (1) that we employ is

$$s_3(l,c) = (Tl - c - 0.5)/lv^{1/2} \qquad (2)$$

where v is the cost variance. The value of s_3 is the number of standard deviations that the cost of a subalignment is below the expected cost. We have also found that a normalized similarity measure (see Ref. 6) is useful for reporting the relative significance of subalignments.

The second innovation is the use of an algorithm (referred to as the DD algorithm)[5] to find all locally optimal subalignments of two sequences using a reasonable similarity function to assess their relative significance. In our implementation, we employ a variation of the DD algorithm[7] using the similarity function described above.

Cost Matrices

The cost c of a subalignment is the sum of the cost of each pair of sequence elements in the subalignment. The simplest cost matrix assigns a cost of 0 for identical sequence elements and a cost of 1 for nonidentical elements. Other cost matrices are possible. For example, we have used[4] a cost matrix for protein sequence comparison based on the log-odds matrix of Dayhoff et al.[8,9] that scales costs to integer values between 0 and 6 and a cost matrix based on the genetic distance between amino acids. In the latter example, either the metric distance matrix described by Erickson and

[6] S. F. Altschul and B. W. Erickson, *Bull. Math. Biol.* **48**, 633 (1986).

[7] C. B. Lawrence, D. A. Goldman, and R. T. Hood, *Bull. Math. Biol.* **48**, 569 (1986).

[8] M. O. Dayhoff, R. M. Schwartz, and B. C. Orcutt, in *"Atlas of Protein Sequence and Structure" (M. O. Dayhoff, ed.), Vol. 5, Suppl. 3*, p. 345. National Biomedical Research Foundation, Washington, D.C., 1978

[9] S. F. Altschul, Ph.D. thesis, Massachusetts Institute of Technology, Cambridge, Massachusetts, 1987

Sellers[10] or a nonmetric variation in which mutation costs involving a silent mutation are assigned a cost of 3 can be used. The expected cost t of a pair of sequence elements at any position in a subalignment is calculated by the formula

$$t = \sum_{i,j} f(i) f(j) c(i,j) \qquad (3)$$

where $f(i)$ and $f(j)$ are the frequencies of sequence elements (nucleotides or amino acids) i and j and $c(i,j)$ is the cost of replacing i with j in a sequence. The variance v is given by

$$v = \left[\sum_{i,j} f(i) f(j) c(i,j)^2 \right] - t^2 \qquad (4)$$

For protein sequences, the amino acid frequencies reported by Dayhoff are used and for nucleotide sequences a frequency of 0.25 for each nucleotide is assumed.[8]

Detection of Homology Domains

Homology domains are identified by using the DD algorithm to find locally optimal subalignments whose similarity score exceeds a specified threshold score or "acceptance threshold." The acceptance threshold is the minimum similarity score that a human is willing to accept as indicating potential evolutionary ancestry. We have determined acceptance thresholds empirically by evaluating the similarity scores that result from a search of a set of unrelated sequences with a randomized sequence. For protein sequences, one member each of 397 different superfamilies was compared with a random 500-residue sequence of average amino acid composition[11] using three different cost matrices; for nucleotide sequences one member each of the mRNA sequence for 93 different superfamilies was compared with a random 500-base nucleotide sequence. The mean and standard deviation of the best similarity scores found in each set of pairwise comparisons were calculated to determine the threshold. The acceptance threshold is set to eliminate most scores expected to occur by chance when comparing a large number of sequences and to include most scores resulting from

[10] B. W. Erickson and P. H. Sellers, *in* *"Time Warps, String Edits, and Macromolecules: The Theory and Practice of Sequence Comparison"* (D. Sankoff and J. B. Kruskal, eds.), p. 55. Addison-Wesley, Reading, Massachusetts, 1983.

[11] M. O. Dayhoff, L. T. Hunt, and S. Hurst-Calderone, *in* *"Atlas of Protein Sequence and Structure"* (M. O. Dayhoff, ed.), *Vol. 5. Suppl. 3*, p. 363. National Biomedical Research Foundation, Washington, D.C., 1978

true homology. In practice, we use a similarity score approximately 4.5 standard deviations above the mean of the best scores resulting from the comparison of the superfamily sequences with a randomized sequence. Other investigators have discussed alternative methods of evaluating the significance of sequence similarity scores.[9,12,13] These methods could be applied to evaluating homology domain scores instead of the approach we have adopted.

A key component of the method is the use of the nonlinear similarity score to evaluate the relative significance of potential homology domains. Because the score is a function of both the length of the region and its percent identity (as reflected in the calculated cost), the similarity score of a long region that has a relatively low percent identity may be equivalent to a short region with a higher percent identity.

To test the ability of the DD algorithm to find the boundaries of homology domains, we have constructed pairs of sequences having simulated homology domains of different length and identical sequence embedded in regions of nonhomology. One member of the pair was then partially shuffled to simulate the introduction of point mutations while maintaining base composition. The DD algorithm using the similarity score $s_3(l,c)$ and the $0-1$ cost matrix was employed to find mutated homology domains whose scores were equal to or greater than the acceptance threshold. We found that the left and right domain boundaries are identified to within a few bases of the initial boundaries until the percent identity in the domain falls below about 60% for nucleotide sequences or about 50% for protein sequences. As the percent identity decreases, the boundaries are found with less precision because of local heterogeneities in the distribution of the simulated point mutations. That is, when the local frequency of identical matches in the region of the domain adjacent to a boundary approaches that expected by chance, that portion of the original domain is excluded from the domain identified by the DD algorithm.

Figure 2 illustrates the detection of the boundaries of a 100-amino acid residue homology domain after introducing point mutations to produce approximately 70% (Fig. 2A,B) or 32% (Fig. 2C,D) sequence identity within the selected domains. At both 70 and 32% identity the boundaries are accurately found. However, at 32% identity local heterogeneity in the distribution of mutations results in the identification of boundaries of a significant fraction of domains at a distance from the original boundaries. This is not an unreasonable property of the method as it indicates that portions of a domain which have been mutated to a level of sequence

[12] T. F. Smith, M. S. Waterman, and C. Burks. *Nucleic Acids Res.* **13,** 645 (1985).
[13] R. Arratia, L. Gordon, and M. S. Waterman, *Ann. Stat.* **14,** 971 (1986).

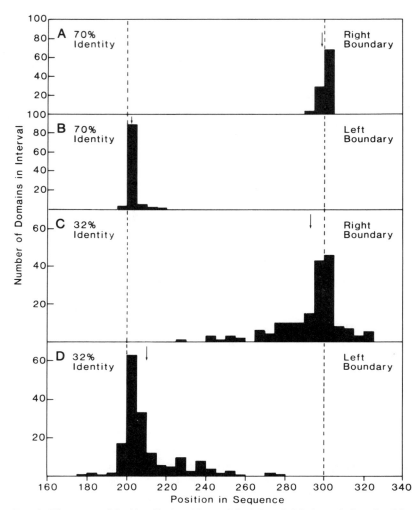

FIG. 2. Histogram of the identified positions of the left and right boundaries of a 100-residue homology domain mutated to 70%. (A, B) and 32% (C, D) identity. The original domain was mutated 100 (A, B) or 200 (C, D) times, and the resulting boundaries were identified using the DD algorithm and the s_3 similarity score. Bars indicate the number of boundaries found in a 5-residue interval. Dashed lines indicate the positions of the left and right boundaries of the original domain. Arrows indicate the positions of the mean of the position of all boundaries. [Reprinted from *CABIOS* **4**, 25–33 (1988).]

TABLE I
COMPARISON OF HUMAN AND CHICK β-Globin Genes

Similarity score	Percent identity	Length of region	Domain boundaries			
			Human[a]		Chick[b]	
			Left[c]	Right[c]	Left[c]	Right[c]
18.1	77.0	230	62459	62688	647	876
12.9	71.4	147	63520	63666	1668	1814
12.0	73.4	115	62222	62336	448	562
6.1	62.3	53	62006	62058	1369	1421
5.8	49.5	109	62299	62407	1105	1213
5.7	54.8	73	62539	62611	1735	1807
5.7	48.3	118	62215	62332	1805	1922
5.7	69.7	33	63615	63647	522	554
5.7	100.0	12	62501	62512	767	778

[a] GenBank Accession #J00179, Release 44.0.
[b] GenBank Accession #J00858, Release 44.0.
[c] Position in GenBank sequence.

identity indistinguishable from that expected by chance will not be included within the detected domain boundaries.

Reporting and Display of Homology Domains

The boundaries of homology domains define ungapped, aligned regions of two sequences. These regions are conveniently reported as a list such as that in Table I from a comparison of the human and chicken β-globin genes. Only the first three domains in the list have a score that exceeds the acceptance threshold. These three domains define the boundaries of the three globin exons to within a few bases of the exon borders (Table II).

A convenient way to display regions of similarity between sequences is the "graph matrix" plot.[14] Homology domains can also be displayed in this manner by plotting the positions of sequence identities that occur only in homology domains whose score exceeds a specified threshold. In Fig. 3A the homology domains listed in Table I are plotted; and in Fig. 3B only the three significant domains are displayed. By plotting graphic matrices in this way, only regions of significant similarity are displayed without the background observed using other methods.

[14] J. Maizel and R. Lenk, *Proc. Natl. Acad. Sci. U.S.A.* **78,** 7665 (1981).

TABLE II
COMPARISON OF HOMOLOGY DOMAIN AND EXON BOUNDARIES
FOR HUMAN β-Globin Gene

Region	Left boundary	Right boundary	Length
Homology domain (score = 12.0)[a]	62222	62336	114
Exon I[b]	62239	62330	91
Difference[c]	17	6	23
Homology domain (score = 18.1)[a]	62459	62688	229
Exon II[b]	62461	62683	222
Difference[c]	2	5	7
Homology domain (score = 12.9)[a]	63520	63666	146
Exon III[b]	63534	63662	128
Difference[c]	14	4	18

[a] Position of homology domain boundaries.
[b] Position of exon boundaries.
[c] Difference in bases between homology domain and exon.

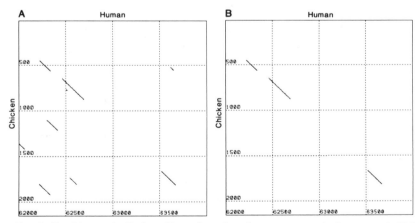

FIG. 3. Graphic matrix display of homology domains common to the human and chicken β-globin genes listed in Table I. (A) All domains with a similarity score of at least 5.7. (B) Domains with a similarity score no lower than the acceptance threshold (6.28). [Reprinted from *CABIOS* **4**, 25–33 (1988).]

Detection of Modules in Genes and Proteins

One theme that has emerged from the analysis of gene and protein sequences is that many genes have been assembled from modules shared by other genes. Structural domains of glycolytic enzymes have boundaries defined by the intron–exon structure of their genes, reflecting their origin in separate ancestral genetic components.[15] The low density lipoprotein (LDL) receptor is a mosaic of modules that appear in proteins having different functions.[16] These modular genetic components can be thought of as homology domains as we have defined here.

To illustrate the detection of homology domains that represent genetic modules we have compared the human LDL receptor sequence to human complement factor C9 and mouse epidermal growth factor (EGF) precursor using the Dayhoff cost matrix. Table III summarizes the homology domains found whose similarity score exceeds the acceptance threshold, and a graphic matrix representation of the homology domains is presented in Fig. 4. The only significant domains found in the comparison between the LDL receptor and complement factor C9 are the seven 30- to 37-residue domains previously noted to be homologous to C9[16] (Table IIIA and Fig. 4A). Several significant domains are observed in the comparison of LDL receptor to the EGF precursor (Table IIIB and Fig. 4B). When the threshold for display by graphic matrix analysis is raised to a similarity score of 8.0 or greater, the major homology domains identified by Südhof et al.[17,16] are evident (Fig. 4C). Inspection of Fig. 4B reveals a number of homology domains having significant similarity scores that lie on diagonals approximately 40 residues apart. Such a pattern is the hallmark of repeated sequence elements whose length is the relative offset of the diagonals (in this case about 40 residues). Repeats like this are best visualized by graphic matrix analysis of a sequence with itself. Figure 4D, which shows a comparison of human EGF precursor with itself using the Dayhoff cost matrix, reveals a repeated sequence in regions between residues 54–205 and 493–698. These correspond approximately to the homologous regions designated x and y by Doolittle and co-workers.[18]

Using Aggregate Similarity Scores for Sequence Comparison

When comparing two sequences it is useful to generate a score that takes into account all significant homology domains found, in order to

[15] W. Gilbert, M. Marchionni, and G. McKnight, *Cell* **46**, 151 (1986).
[16] T. C. Südhof, J. L. Goldstein, M. S. Brown, and D. W. Russell, *Science* **228**, 815 (1985).
[17] T. C. Südhof, J. L. Goldstein, M. S. Brown, and D. W. Russell, *Science* **228**, 893 (1985).
[18] R. F. Doolittle, D. F. Feng, and M. S. Johnson, *Nature (London)* **307**, 558 (1984).

TABLE III
COMPARISON OF LDL RECEPTOR WITH COMPLEMENT C9 AND EGF PRECURSOR

Similarity score	Percent identity	Length of region	Domain boundaries[a]			
			LDL receptor[b]		C9[c]	
			Left	Right	Left	Right
A. LDL versus C9						
7.9	42.9	35	201	235	105	139
7.7	45.2	31	30	60	104	134
7.6	38.9	36	110	145	102	137
7.2	40.5	37	238	274	103	139
6.9	38.7	31	78	108	109	139
6.5	43.3	30	150	179	103	132
6.0	43.8	32	279	310	103	134
					EGF precursor[d]	
					Left	Right
B. LDL versus EGF						
12.1	33.3	138	466	603	553	690
9.3	24.2	120	542	661	542	661
8.9	30.5	59	372	430	462	520
8.1	38.9	36	314	349	362	397
8.0	34.4	61	601	661	686	746
7.8	16.7	203	457	659	501	703
7.4	58.8	17	368	384	895	911
7.1	33.3	42	358	399	367	408
7.1	31.1	61	314	374	403	463
6.8	43.8	32	314	345	747	778
6.4	60.0	10	572	581	92	101
6.1	31.3	16	3	18	2	17
5.9	27.1	48	467	514	116	163
5.9	30.0	60	471	530	162	221

[a] Position in data bank sequence entry.
[b] Protein Identification Resource Entry QRHULD, Release 10.0.
[c] Protein Identification Resource Entry C9HU, Release 10.0.
[d] Translated from GenBank Accession No. J00380, Release 44.0.

reflect the total amount of similarity. We use a simple method to obtain an aggregate score from several homology domains.

1. Sort all domains having a similarity score equal to or greater than the acceptance threshold.

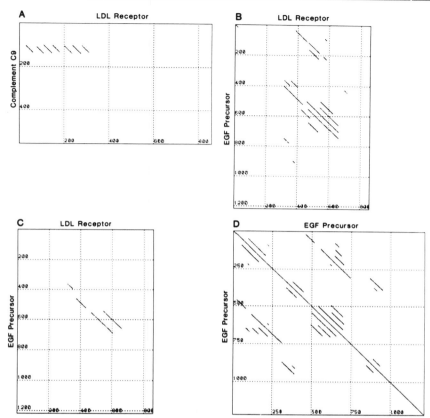

Fig. 4. Graphic matrix display of homology domains. (A) LDL receptor versus complement C9, score ≥ 6.0. (B) LDL receptor versus EGF precursor, score ≥ 5.9. (C) LDL receptor versus EGF precursor, score ≥ 8.1. (D) EGF precursor versus EGF precursor, score ≤ acceptance threshold (5.75). Comparisons were done using the Dayhoff cost matrix. [Reprinted from *CABIOS* **4**, 25–33 (1988).]

2. Mark all domains that lie on diagonals within a certain distance of the diagonal containing the domain having the highest score.
3. Obtain the total cost and length of all marked domains and calculate an aggregate similarity score using Eq. (2).

This is equivalent to considering all aligned elements to be a part of a single homology domain without penalties for gaps. Aggregate scores are useful in searches of the sequence data banks when it is important that similarity scores reflect the total similarity between two sequences.

TABLE IV

SIGNIFICANT SCORES FOUND IN SEARCH OF PROTEIN
DATA BANK USING *E. coli* RNaseH AS A PROBE

Aggregate similarity score	Similarity score of best individual homology domain	Compared sequence	PIR entry[a]	Jumble comparison[b]
11.3	7.9	Ribulose biphosphate carboxylase	RKYCL	5.9
11.2	7.9	HIV *pol* protein	GNVWH3	9.2
10.6	7.7	MLV *pol* protein	GNMV1M	9.8
9.6	6.7	Aspartate aminotransferase	XNPGDM	2.3

[a] Protein Identification Resource, Release 10.0.

[b] Jumble comparison was performed as described by Barker and Dayhoff[20] (scores are given in number of standard deviations).

Table IV lists some of the significant aggregate scores found in a search of the protein sequence data bank using *Escherichia coli* RNaseH as a query. RNaseH has previously been noted to be related to retrovirus *pol* proteins.[19] Comparison of RNaseH with the human immunodeficiency virus I (HIV) and murine leukemia virus (MLV) *pol* proteins gave significant scores as expected. The high scores observed with ribulose biphosphate carboxylase (*Synechococcus sp.)* and pig aspartate aminotransferase were surprising and indicated possible homology with RNaseH. To test more rigorously the significance of the similarity, the regions of interest were subjected to a type of Monte Carlo simulation suggested by Barker and Dayhoff[20] in which one sequence is compared to several jumbled (to randomize) versions of the other sequence. To evaluate the significance, the number of standard deviations that the similarity score of the original comparison differs from the mean of the comparison with the randomized sequences is determined. A dynamic programming method[21] was used to generate similarity scores for this analysis. The comparison of RNaseH with the *pol* proteins yielded scores of 9.2 and 9.8, indicating that they share a common ancestor. The comparison with aspartate aminotransferase resulted in a score of 2.3, a score not high enough to prove common ancestry. However, the score of 5.9 obtained in the comparison with

[19] M. S. Johnson, M. A. McClure, D. Feng, J. Gray, and R. F. Doolittle, *Proc. Natl. Acad. Sci. U.S.A.* **83,** 7648 (1986).

[20] W. C. Barker and M. O. Dayhoff, *in "Atlas of Protein Sequence and Structure"* (M. O. Dayhoff, ed.), p. 101. National Biomedical Research Foundation, Washington, D.C., 1972.

[21] S. F. Altschul and B. W. Erickson, *Bull. Math. Biol.* **48,** 603 (1986).

ribulose biphosphate carboxylase is high enough to propose that the protein shares a common ancestor with RNaseH. Whether or not the possible evolutionary relationship noted above is confirmed in subsequent studies, these results illustrate that scores obtained from aggregated homology domains can be used to detect possible distant relationships between sequences.

Searches of Sequence Data Banks

The identification of homology domains can be used effectively to search sequence data banks for sequences having potential homology to a probe sequence. However, a complete search using the DD algorithm is too inefficient to make such searches practical. We have implemented a search strategy that combines the preprocessing strategy of Wilbur and Lipman[22] and Lipman and Pearson[23,24] with the DD algorithm to provide efficient and sensitive searches of the sequence data banks.[7]

Program Availability

Programs based on this method are available as a part of a package of software tools for sequence analysis. The programs are written in "C" for UNIX systems. Additional information can be obtained by writing the author at the Molecular Biology Information Resource, Department of Cell Biology, Baylor College of Medicine, One Baylor Plaza, Houston, TX 77030, or by electronic mail to chas@mbir.bcm.tmc.edu.

Summary

We have found the detection of homology domains using a nonlinear similarity score and the DD algorithm to be a useful approach for identifying similarity between sequences and evaluating potential homology. There are several reasons for the success of the method. (i) Homology domains are identified by a rigorous method that guarantees they will be locally optimal. (ii) The relative significance of different homology domains can be directly compared using the nonlinear similarity score. (iii) Different cost matrices can be used in the calculation of the similarity

[22] W. J. Wilbur and D. J. Lipman, *Proc. Natl. Acad. Sci. U.S.A.* **80,** 726 (1983).
[23] D. J. Lipman and W. R. Pearson, *Science* **227,** 1435 (1985).
[24] W. R. Pearson and D. J. Lipman, *Proc. Natl. Acad. Sci. U.S.A.* **85,** 2444 (1988).

score. (iv) Relatively long sequences can be compared in a single pass as storage requirements are proportional to the shorter of the two sequences being compared. (v) The method has proved to be very sensitive in practice. (vi) The boundaries of authentic regions of homology are accurately identified. (vii) The information required to define a homology domain (its location, size, similarity score, etc.) can be stored in a compact data structure, facilitating the sharing of homology domain data among different software tools. (viii) The method can be applied to similarity searches of the nucleotide and protein sequence data banks. These properties make the identification of homology domains for studying sequence similarity a useful companion to other accepted methods, such as dynamic programming based analyses.

[9] Profile Analysis

By Michael Gribskov, Roland Lüthy, and David Eisenberg

Introduction

Because of the continuing growth in the number of available protein sequences, where we once had only the sequence of a single protein, we frequently have the sequences of a family of related proteins. Just as individual sequences have been examined for similarity (and inferred homology), in an effort to understand the structure and function of the protein, similar questions can and are being asked about families of proteins. The profile method[1,2,3] provides a convenient way to represent information about groups or families of sequences as well as a means to ask questions about the definition of protein families, the relationships between distantly related proteins, and the presence of sequence or structural motifs in proteins.

When aligned, the sequences of a homologous family of proteins provide information that is not present in any sequence alone. There is information concerning the conservation of residues, which residues are

---

[1] M. Gribskov, A. D. McLachlan, and D. Eisenberg, *Proc. Natl. Acad. Sci. U.S.A.* **84,** 4355 (1987).

[2] M. Gribskov, M. Homyak, J. Edenfield, and D. Eisenberg, *CABIOS* **4,** 61 (1988).

[3] M. Gribskov and D. Eisenberg, *in* "Techniques in Protein Chemistry" (T. E. Hugli, ed.), p. 108. Academic Press, San Diego, California, 1989.

conserved and which are variable, and what kinds of changes are allowed at each position. This information is intimately related to the structure and function of the protein. Residues that comprise the structural core of the protein, or the active site, are tightly constrained, while residues on the surface of the protein are more free to change during evolution. The observed positions of insertions and deletions in the sequences provide similar structural information. This information is incorporated in the profile, and used to improve the detection of sequence patterns that represent structural motifs.

It is clear that certain three-dimensional structural motifs are shared by many proteins. These structural motifs have patterns in their amino acid sequences that permit them to be recognized from the sequences alone. Profile analysis allows us to create probes sensitive to structural patterns. When a sequence is found to be significantly similar to a profile that is specific for a three-dimensional structure, i.e., detected by a profile generated from the structurally aligned sequences characteristic of the structure, one can immediately pose hypotheses about the structure of the newly sequenced protein.

Another use of profile analysis is based on profiles generated from sequences aligned using sequence information alone. In this case, the profile can be considered to be specific for the protein family or superfamily as defined by sequence criteria. Significant similarity of a sequence to this kind of profile suggests that the protein is homologous to the family or superfamily represented by the profile. For both kinds of profile, because the comparison is made to a probe representing the overall sequence characteristics of a group of sequences, it is more sensitive to the overall similarity of the protein and the probe and less dependent on finding a single sequence with relatively high similarity.

Methods

The profile analysis package now consists of the five programs shown in Table I. The programs are written in FORTRAN 77, using a limited number of the language enhancements available on VAX[4] computers and the procedure library of the Genetics Computer Group (GCG).[5] Versions of the programs in ANSI FORTRAN 77 are in preparation. The programs and source code are available on request.

Rather than a single monolithic program, the profile analysis package is divided into a suite of programs, each with a clearly defined function. The

[4] Digital Equipment Corp., Maynard, Massachusetts.
[5] J. Devereux, P. Haeberli, and O. Smithies, *Nucleic Acids Res.* **12**, 387 (1984).

TABLE I
Programs Used in Profile Analysis

Program	Function
PROFILEMAKE	Calculate a profile from a group of aligned sequences.
PROFILEGAP	Align a sequence with a profile using dynamic programming algorithm
PROFILESEARCH	Compare a profile to each sequence in the PSQ database using dynamic programming algorithm
PROFILENORMAL	Normalize the results of **PROFILESEARCH** to remove systematic effects of sequence length
PROFILESCAN	Compare a sequence to each profile in the library of validated profiles and test for significant similarity

relationships between the programs are shown in Fig. 1. The profile is the common data structure used by the package to store the information derived from a group of aligned sequences. The programs are used to calculate a profile from aligned sequences (**PROFILEMAKE**),[1,3] align a sequence and profile (**PROFILEGAP**[3]), search a database for sequences similar to a profile (**PROFILESEARCH**,[1,3] **PROFILENORMAL**[3]), or, when significant thresholds are known, determine if a sequence is significantly similar to a known motif (**PROFILESCAN**[2,3]).

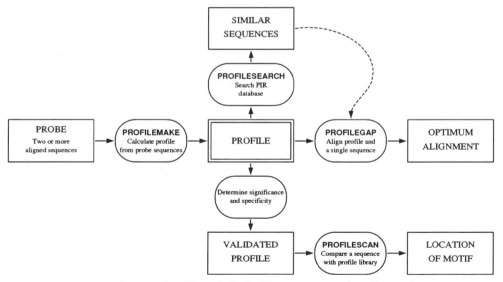

Fig. 1. Flow diagram of profile analysis. Ovals represent operations. Rectangles represent data or results of operations. Sans serif type shows the action of the programs in the profile package.

Calculating the Profile

The profile is a position-specific scoring table (Fig. 2) generated from a group of aligned sequences and a comparison table giving scores for the comparison of each pair of amino acid residues. We use a comparison table derived from the MDM78 table.[6,7] Each row in the profile corresponds to one position in the aligned sequences. The columns of the profile contain the scores for the alignment of each possible amino acid residue with each position in the aligned sequences.

PROFILEMAKE calculates the profile by a simple method. For a group of N aligned sequences of length L

$$
\begin{matrix}
a_{1,1} & a_{1,2} & a_{1,3} & \cdots & a_{1,L} \\
a_{2,1} & a_{2,2} & a_{2,3} & \cdots & a_{2,L} \\
a_{3,1} & a_{3,2} & a_{3,3} & \cdots & a_{3,L} \\
& & \cdot & & \\
& & \cdot & & \\
& & \cdot & & \\
a_{N,1} & a_{N,2} & a_{N,3} & \cdots & a_{N,L}
\end{matrix}
$$

where $a_{i,j}$ is the amino acid residue in sequence i at position j. The profile has a column for each character that occurs in the aligned sequences; for proteins there are 20 columns. The scores in the profile at row r and column c are given by

$$
\mathrm{Profile}(r,c) = \sum_{d=1}^{20} W_d \mathrm{Comp}(\mathrm{residue}_d, \mathrm{residue}_c) \tag{1}
$$

where **Comp** is the value for the comparison of two amino acid residues: each possible residue ($\mathrm{residue}_d$) and the amino acid residue represented by the column of the profile ($\mathrm{residue}_c$). A weight (W_d) is applied that depends on the number of occurrences of each kind of residue at each position, and the sum of the weighted number of sequences. We have used both linear weighting [Eq. (2)],

$$
W_d = \frac{\sum_{i=1}^{N} w_i \delta_d}{\sum_{i=1}^{N} w_i} \qquad \delta_d = \begin{cases} 1 \text{ if } a_{i,r} = \mathrm{residue}_d \\ 0 \text{ if } a_{i,r} = \mathrm{residue}_d \end{cases} \tag{2}
$$

---

[6] M. O. Dayhoff and R. M. Schwartz, *in* "Atlas of Protein Sequence and Structure" (M. O. Dayhoff, ed.), Vol. 5, Suppl. 3, p. 353. National Biomedical Research Foundation, Washington, D.C., 1979.

[7] M. Gribskov and R. R. Burgess, *Nucleic Acids Res.* **14,** 6745 (1986).

Profile

Position	Consensus	A	C	D	E	F	G	H	I	K	L	M	N	P	Q	R	S	T	V	W	Y	Gap Opening	Gap Extension
1	T	15	-3	12	13	-8	11	4	10	12	2	6	11	9	7	6	14	32	12	-22	-8	25	25
2	E	16	-8	22	24	-14	22	8	4	8	-1	3	15	8	16	3	15	13	9	-30	-12	25	25
3	Q	11	-16	23	27	-16	27	22	0	15	-1	2	17	8	35	14	18	7	0	-19	-11	25	25
4	R	12	-16	13	13	-15	5	9	6	31	0	11	16	8	15	21	12	15	5	-8	-12	100	100
5	P	17	-6	13	13	-12	11	8	4	8	3	10	16	8	15	12	17	12	13	-21	-15	100	100
6	Y	-5	20	-23	-19	57	-21	10	20	-20	29	10	-3	-28	-20	-18	-5	-5	25	38	58	100	100
7	V	15	-3	-48	-49	-10	8	-7	20	-51	-73	14	-5	24	6	-27	12	16	30	-27	-8	100	100
8	C	32	142	25	22	-24	19	7	29	-23	-8	-53	19	-6	-53	16	78	28	2	-129	101	100	100
9	D	16	-13	11	11	-10	15	21	3	11	2	-3	21	9	20	15	20	15	2	-26	-18	100	100
10	H	6	-7	40	37	-4	2	14	-7	12	-8	-3	13	6	6	16	11	16	2	-3	-1	24	24
11	Q	27	-15	40	30	-26	25	4	-7	11	-16	-9	24	12	13	-6	16	22	2	-42	-15	24	24
12	S	32	-3	43	45	-31	62	-4	-7	12	-8	2	26	14	24	-6	29	8	10	-53	-28	24	24
13	C	9	-1	-1	5	0	0	-7	9	6	-16	-3	6	4	17	-27	12	8	10	-9	-1	24	24
14	D	32	142	-48	-49	-10	19	-4	29	5	-73	-53	-25	5	-53	3	78	28	30	-129	101	100	28
15	K	19	-13	40	31	-25	31	-4	-1	11	-10	-5	27	9	21	3	18	10	4	-38	-14	28	28
16	S	-15	3	9	8	-13	0	7	2	35	-7	13	14	8	16	30	11	12	16	-1	-10	100	100
17	F	-19	1	9	9	-13	12	3	12	16	-7	13	13	13	20	20	27	33	3	-5	-10	100	100
18	T	-19	1	-46	-34	83	-33	4	41	-35	62	27	-19	-36	-36	-25	-10	22	16	63	81	100	100
19	T	13	1	11	10	-9	10	7	10	11	-2	-5	15	12	8	8	19	13	12	-17	-9	100	100
20	K	13	-20	9	9	-3	3	9	9	11	-1	10	21	8	18	7	15	11	3	-14	-6	100	100
21	S	10	-7	7	7	-20	12	10	1	42	4	13	11	11	7	25	14	6	3	-10	-15	100	100
22	N	5	-8	7	6	-1	5	6	0	9	-4	3	23	4	9	11	15	11	-1	-6	2	25	25
23	T	8	-3	5	5	-3	2	16	6	14	9	8	8	3	5	5	9	6	7	-9	3	25	25
24	L	7	-6	5	3	4	5	5	8	5	4	5	6	3	5	11	7	6	8	-1	0	25	25
25	S	6	-1	3	4	3	2	7	8	3	5	9	6	2	4	4	7	6	6	-5	1	25	25
26	L	-1	-58	-32	-17	77	-28	-10	59	-17	107	92	-19	-16	-4	-19	-20	-3	61	23	18	25	25
27	K	-7	-18	15	16	-22	14	14	-1	39	-6	-1	18	9	21	33	13	14	-3	-18	-18	100	100
28	R	0	-10	12	11	-16	3	18	-1	24	-8	-5	15	12	17	33	15	8	0	0	-13	100	100
29	H	0	-12	28	28	-7	-7	104	-13	11	-8	-18	37	16	50	35	-4	-2	-11	10	17	30	30
30	I	17	-3	17	18	6	-1	12	12	2	10	-8	5	1	3	3	7	11	10	-6	5	100	100
31	S	9	-7	17	18	-11	13	5	4	11	1	2	18	11	21	15	22	11	10	-15	-10	100	100
32	K	9	-7	12	9	-8	5	12	8	19	2	10	11	6	9	-1	15	11	10	-10	-5	100	100
33	V	17	-7	17	18	6	1	-7	23	1	10	-13	2	1	9	3	8	20	23	-12	10	30	30
34	H	-4	40	17	18	-11	13	152	4	11	2	10	18	6	21	15	22	11	10	-15	-10	100	100
35	·	9	-7	17	9	-8	5	12	8	19	2	10	15	6	9	-1	22	11	10	-10	-5	100	100
36	K	9	-7	-4	-2	2	13	-7	23	1	-17	-13	2	1	9	33	10	20	23	-12	10	100	100
37	V	-4	-17	40	39	10	-13	152	-27	15	16	-27	53	21	73	50	-9	-7	-26	25	25	100	100
38	H	11	19	13	14	14	12	21	2	25	16	-8	7	8	7	16	13	15	11	2	9	100	100

Probe

Position	247 - 276	216 - 246	189 - 214	160 - 188	130 - 159	68 - 98	38 - 67	8 - 37

FIG. 2. Profile of the *Xenopus laevis* transcription factor TFIIIA zinc finger. The eight repeats of the zinc finger sequence that form the probe are shown descending vertically at the left, labeled with the positions where they occur in the complete sequence. Insertions made to align the sequences are shown as periods. The profile calculated by **PROFILEMAKE** is shown in the box. The rows correspond to the positions in the aligned sequences, and the columns contain the score for each possible amino acid residue when aligned at that position. The position-specific gap penalties are given in the two right-hand columns. The consensus sequence is shown immediately to the left of the box, and represents the highest scoring column at each row in the profile. In other words, the consensus residue is the amino acid that would receive the highest score when aligned with that position in the aligned probe sequences.

and logarithmic weighting [Eq. (3)].

$$W_d = \frac{\ln\left[1 - \sum_{i=1}^{N} w_i \delta_d \middle/ \left(1 + \sum_{i=1}^{N} w_i\right)\right]}{\ln\left[1 \middle/ \left(1 + \sum_{i=1}^{N} w_i\right)\right]} \tag{3}$$

Using either kind of weighting, $W_d = 0$ when a residue does not occur at a given position, r, and $W_d = 1$ when all residues at a position are identical. The weight on each sequence (w_i) is usually 1.0, but may be larger when it is desired to emphasize one of the aligned sequences more than the others. This is common when the probe sequences contain several closely related sequences and other sequences that are more distantly related. In this case the weights on the more distantly related sequences should be increased to equal the sum of the weights of the closely related sequences. Logarithmic weighting results in a profile that is more specific for the probe sequences used to generate it; we normally use logarithmic weighting.

This method of calculating a profile can be used for any property of amino acids that can be expressed as a similarity between amino acid pairs. More complicated schemes for calculating profiles are possible. For instance, the predicted secondary structure of the proteins, or the inferred ancestral sequence, could be used in the calculation. Part of the rationale for dividing the profile package into individual programs was to allow for flexibility in calculating the profile.

The two right-hand columns of the profile specify position-specific multipliers for the gap-opening (p_{open}) and gap-extension (p_{extend}) penalties used in the alignments (see below). The values for these multipliers are calculated based on L_{gap}, the maximum length of a gap crossing each position. The maximum length of the gap is the size of the gap starting at the first gap position (leftmost as aligned sequences are normally shown, topmost as shown in Fig. 2) in any sequence, and ending at the last gap position (normally rightmost, bottommost in Fig. 2) in any sequence. For instance in Fig. 2, a gap with a $L_{gap} = 4$ begins with the period at position 10 of the 189–214 repeat and ends with the periods at position 13 in six of the eight sequences. The penalty multipliers are calculated as

$$\text{multiplier} = m = G_{max}/(1.0 + G_{inc}L_{gap}) \tag{4}$$

where G_{max} is the maximum multiplier for a position for which there are insertions or deletions in the probe sequences, and G_{inc} determines the rate at which the multipliers decrease with increasing gap length. The default values for G_{max} and G_{inc} are 33.3 and 0.1, respectively, but are user adjustable. The multipliers are always 100 for positions at which no insertions or deletions are observed. Multipliers for both the gap-opening and gap-extension penalties are provided, but **PROFILEMAKE** currently uses the same value for both.

The last row of the profile is shown with consensus residue (marked with an asterisk). This row contains the total numbers of the residues present in the probe sequences, which can be used to predict the score for alignment with unrelated sequences with a given composition. It is important, when comparing a profile to the database, to consider the extent to which the score for comparison of a profile and a sequence is due merely to similarity in composition. The scores calculated by **PROFILESEARCH** are adjusted for this effect.

Aligning Sequences and Database Searching

PROFILEGAP and **PROFILESEARCH** use the dynamic programming algorithm to calculate a score for the similarity of the profile and a sequence. **PROFILEGAP** also displays the alignment of the sequence and the profile. The implementation of the dynamic programming algorithm follows Smith and Waterman.[8] However, any version of the dynamic programming algorithm is easily modified for use with the profile system. Letting $S_{i,j}$ be the score for the alignment of the sequence and profile such that position i of the sequence is aligned with row j of the profile. The score $S_{i,j}$ for the alignment is defined recursively as

$$S_{i,j} = \text{Profile}(j,\text{column}_{a_i}) + \max \begin{bmatrix} S_{i-1,j-1}, \\ \max_{2 \leq k \leq j-1}(S_{i-1,j-k} - w_k), \\ \max_{2 \leq l \leq i-1}(S_{i-l,j-1} - w_l) \end{bmatrix} \qquad (5)$$

where the penalties w_k and w_l are given by

$$w_k = m_{\text{open}}(j-k) \cdot p_{\text{open}} + \sum_{j'=j-k}^{j-2} m_{\text{extend}}(j') \cdot p_{\text{extend}} \qquad (6)$$

$$w_l = m_{\text{open}}(j-1) \cdot p_{\text{open}} + (l-1) \cdot m_{\text{extend}}(j-1) \cdot p_{\text{extend}} \qquad (7)$$

with m_{open} and m_{extend} being the position-specific penalty multipliers in the profile, and p_{open} and p_{extend} the gap-opening and gap-extension penalties entered when the alignment is performed.

Normalization for Length Effects

We have found empirically that, when there are no sequences related to the profile in the database, the distribution of comparison scores can be modeled by $S = Ce^{AL+B}$ where S is the score for comparison of a sequence

[8] T. F. Smith and M. S. Waterman, *Adv. Appl. Math.* **2**, 482 (1981).

of length L to a profile and A, B, and C are constants. The program **PROFILENORMAL** uses a nonlinear curve-fitting algorithm to fit the results of a database search to this equation and estimate the constants A, B, and C. The normalized results are generally easier to interpret, especially when comparing the scores for sequences of different lengths.

Detection of Defined Motifs

PROFILESCAN uses a dynamic programming algorithm modified to allow the production of all distinct alignments greater than a threshold score. During construction of the score matrix, a list of all scores greater than the threshold and their locations is saved. After completion of the score calculation, the list is sorted, and a traceback performed beginning at each location in the list, in decreasing score order. As the traceback is carried out, the point at which the score falls below the threshold is stored in a second list; any other traceback that would contain this position is aborted and no alignment produced. This ensures that the alignments will be distinct, i.e., that they will not contain segments with identical alignments of the sequence and profile unless the alignments diverge before the threshold is reached.

Profile Library

We are assembling a library of profiles (Table II), each specific for a different structural or sequence motif. Each profile included in the library represents a structural or sequence motif that has been validated. The process of validation is described in detail elsewhere,[2,3] but it can be briefly described as the determination of levels of significance based on searches of the PSQ[9] database and confirmation that the profile is specific for the desired motif.

Results

Profiles calculated using **PROFILEMAKE** have the expected properties for sequence patterns characteristic of structural motifs or protein families. Rows of the profile that correspond to conserved positions in the aligned probe sequences (e.g., the conserved cysteines in Fig. 2) have large positive scores for the conserved residues, smaller scores for residues similar to the conserved residues, and large negative scores for residues that are

---

[9] Protein Identification Resource, Protein Sequence Database Release 18.0, PSQ—New Release 35.0, National Biomedical Research Foundation, Georgetown University Medical Center, Washington, D.C. 20007.

TABLE II
Profile Library

Profile	Specificity
Globin.prf[a]	Globin fold[b]
Ig-constant.prf[a]	Immunoglobulin constant regions[c]
Ig-variable.prf[a]	Immunoglobulin variable regions[c]
ZincI.prf[a]	Type I zinc finger[d]
ZincII.prf[a]	Type II zinc finger[e]
Ca-EFHand.prf[a]	Calcium-binding EF-hand[f]
Homeo.prf[g]	Homeotic proteins[g]
Cyclic.prf[g]	cAMP and cGMP binding structure[h]
Nuc.prf[g]	Dinucleotide binding fold[i]
HTH.prf[g]	Helix – turn – helix DNA binding structure[j]
Kinase.prf	Protein kinases, catalytic domain[k]
P450.prf	Cytochrome P-450-related proteins[l]
Chaperonin.prf	Chaperonins[m]
Perforin.prf	Perforin, complement C8 and C9[n]

[a] M. Gribskov, M. Homyak, J. Edenfield, and D. Eisenberg, *CABIOS* **4,** 61 (1988).

[b] A. M. Lesk and C. Chothia, *J. Mol. Biol.* **136,** 225 (1980).

[c] E. A. Kabat, T. T. Wu, M. Reid-Miller, H. M. Perry, and K. S. Gottesman, "Sequences of Proteins of Immunological Interest," 4th Ed. U.S. Department of Health and Human Services, Washington, D.C., 1987.

[d] J. Miller, A. D. McLachlan, and A. Klug, *EMBO J.* **4,** 1609 (1985).

[e] M. Johnston and J. Dover, *Proc. Natl. Acad. Sci. U.S.A.* **84,** 2401 (1987).

[f] R. H. Kretsinger, *Cold Spring Harbor Symp. Quant. Biol.* **52,** 499 (1987).

[g] M. Gribskov and D. Eisenberg, *in* "Techniques in Protein Chemistry" (T. E. Hugli, ed.), p. 108. Academic Press, San Diego, California, 1989.

[h] I. T. Weber, T. A. Steitz, J. Bubis, and S. S. Taylor, *Biochemistry* **26,** 343 (1987).

[i] J. J. Birktoft and L. J. Banaszak, *Peptide Protein Rev.* **4,** 1 (1984).

[j] C. O. Pabo and R. T. Sauer, *Annu. Rev. Biochem.* **53,** 293 (1984).

[k] S. K. Hanks, A. M. Quinn, and T. Hunter, *Science* **241,** 42 (1988).

[l] V. F. Kalb and J. C. Loper, *Proc. Natl. Acad. Sci. U.S.A.* **85,** 7221 (1988).

[m] S. M. Hemmingsen, C. Woolford, S. M. van der Vies, K. Tilly, D. T. Dennis, C. P. Georgopoulos, R. W. Hendrix, and R. J. Ellis, *Nature (London)* **333,** 330 (1988).

[n] M. G. Lichtenheld, K. J. Olsen, P. Lu, D. M. Lowrey, A. Hameed, H. Hengartner, and E. R. Podack, *Nature (London)* **335,** 448 (1988).

dissimilar to the conserved residues. Rows corresponding to positions that show little or no conservation in the aligned probe sequences (e.g., first row, Fig. 2), have relatively small scores for all residues, indicating that alignment of any residue at this position will make little difference in the overall alignment score. Rows of the profile that correspond to positions of insertions or deletions in the aligned probe sequences have smaller values for the gap-opening and gap-extension penalty multipliers (Fig. 2, two right-hand columns), indicating that the penalty for starting an insertion or deletion at, or extending an insertion or deletion past, these positions is lower. When a sequence is aligned with the profile, this lower penalty tends to cause insertions or deletions to be placed in positions where insertions or deletions have been previously observed to occur in the probe sequences.

Resolving Power: Profile, Single Sequence, and Consensus Sequence

In Fig. 3 we compare the abilities of a profile, a heavy chain variable region sequence, and a variable region consensus sequence, to distinguish immunoglobulin variable region sequences. The profile was calculated based on 20 κ, λ, and heavy chain variable region sequences from human and mouse.[10] The calculation of this profile was previously discussed in greater detail.[2] For the purpose of this experiment, sequences were defined to contain the variable region motif if they were listed as members of the "immunoglobulin variable region" superfamily in PSQ. Of the 8072 sequences longer than 50 residues in the PSQ and New databases, 383 are members of this superfamily and 7689 are unrelated.

As shown in Fig. 3a the profile clearly resolves all of the immunoglobulin and T-cell receptor α, β, and γ chain, variable region sequences (complete sequences $Z \geq 8.89$) from unrelated sequences. T-Cell receptor surface glycoprotein CD8 sequences ($Z = 4.33 - 5.47$) are clearly distinguished, and even the more distantly related T-cell surface glycoprotein CD4 ($Z = 2.79 - 3.52$) and Thy-1 membrane glycoprotein sequences ($Z = 1.73 - 2.95$) fall on the upper edge of scores for unrelated proteins ($Z = -5.34 - 5.14$).

Any single sequence, although still efficient at detecting the most similar sequences, is less effective than a profile at distinguishing more distantly related sequences (Fig. 3b). The sequence of an immunoglobulin heavy chain, PSQ entry H3HUTL, was compared to all sequences in the database to test its ability to distinguish variable region sequences (Fig. 3b). It effectively distinguishes immunoglobulin heavy chain sequences ($Z \geq 13.37$), but many light chain sequences ($Z = 0.95 - 9.98$) are poorly resolved from unrelated sequences ($Z = -5.82 - 4.33$). The CD8 ($Z = 1.44 - 1.52$), CD4 ($Z = 0.23 - 0.95$), and Thy-1 ($Z = -0.90 - 0.39$) sequences are not distinguished from unrelated sequences. We have obtained similar results with other single immunoglobulin variable region sequences (not shown).

[10] E. A. Kabat, T. T. Wu, M. Reid-Miller, H. M. Perry, and K. S. Gottesman, *"Sequences of Proteins of Immunological Interest,"* 4th Ed. U.S. Department of Health and Human Services, Washington, D.C., 1987.

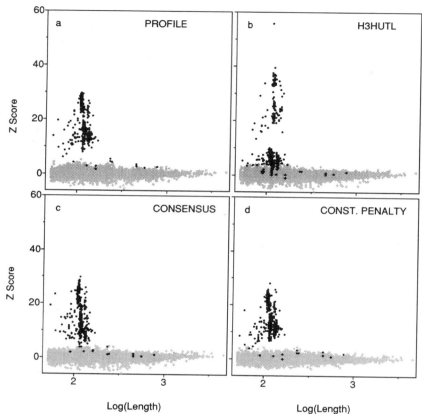

FIG. 3. Selectivity of a profile, compared to simpler alternatives. Each panel shows the scores from comparison of the specified profile or sequence to each sequence in the PSQ and New databases. The scores have been normalized by **PROFILENORMAL** and are shown on a Z scale, i.e., they have been scaled such that unrelated sequences have a mean of 0.00 and a standard deviation of 1.0. The Z scores are plotted against the log of the length of the database sequence. Sequences related to the immunoglobulin variable region motif are shown as black circles; unrelated sequences are lightly shaded. (a) Profile derived from 20 human and mouse κ, λ, and heavy chain variable regions. (b) Heavy chain variable region, PSQ entry H3HUTL. (c) Consensus sequence derived from 20 sequences in (a). (d) Profile (a) with insertion/deletion penalties set to a constant value at all positions.

A characteristic of profile analysis is that the score for aligning a residue at a given position varies depending on the observed conservation of residues at that position. This may be contrasted with the approach of deriving a single consensus sequence for a family and using it, as a kind of family-specific probe, in standard alignment algorithms. To test the significance of this difference we compared the consensus sequence, calculated by **PROFILEMAKE** for the profile in Fig. 3a, to the database as if it was a single sequence. The consensus sequence (Fig. 3c) distinguishes immuno-globulin and T-cell receptor sequences ($Z \geqslant 4.96$) more clearly than a single sequence, but more poorly than the profile. Furthermore, it is rather ineffective at distinguishing the CD8 ($Z = 1.11 - 3.93$), CD4 ($Z = 0.43 - 1.37$), and Thy-1 sequences ($Z = 2.14 - 2.56$).

Effects of Position-Specific Scoring versus Position Specific Gap Penalties

To determine if the improved resolution achieved by the profile can be attributed more to the position-specific variation in scoring or to the position-specific variation in insertion/deletion penalties, a profile was made with the gap-opening and gap-extension penalty multipliers held constant at all positions (Fig. 3d). For comparison to the database, the gap-opening and gap-extension multipliers were set equal to the average value of the multipliers in the original profile ($m = 56$). This constant penalty profile distinguishes the immunoglobulin and T-cell receptor about as well as the original profile ($Z \geqslant 7.60$), but it is not as efficient as the profile in distinguishing the CD8 ($Z = 2.28 - 3.34$), CD4 ($Z = 2.79 - 3.52$), or Thy-1 ($Z = 0.03 - 1.80$) sequences from unrelated sequences ($Z = -3.50 - 4.41$).

Discussion

Profile analysis achieves increased sensitivity by applying additional information extracted from multiple aligned sequences to sequence align-ment problems. The increased sensitivity is due to position-specific varia-tion in the comparison scores as well as to position-specific variation in the insertion/deletion penalties. The variation in insertion/deletion penalties appears to be particularly important in improving the detection of the most distantly related sequences. This seems reasonable since these sequences are the most likely to have large insertions or deletions with respect to the other sequences.

Motifs

As the library of validated profiles grows, the **PROFILESCAN** pro-gram will provide an increasingly powerful method of checking a new sequence for the presence of sequence and structural motifs. The current library (Table II) contains 14 profiles specific for motifs ranging from the supersecondary structure level (helix – turn – helix motif, calcium-binding EF-hand) to whole protein patterns (globins, cytochrome P-450 proteins).

Because the library of validated profiles is small, a new sequence can be rapidly checked (in seconds) for the presence of all defined motifs. Because thresholds of significance have been determined for each profile during the validation process, it is simple to determine if the new sequence is significantly related to a known sequence pattern.

Superfamilies

Profile analysis is well adapted to asking specific questions about the relationship of a sequence to a family. For instance, if a database search suggested that a new sequence was related to trypsin, but the score was not high enough to be significant, and examination of other high but not significant scores showed several other serine proteases, we can use profile analysis to test if there is significant similarity between the new sequence and the serine proteases. To do this we first make a profile from several serine proteases, trying to include the least similar sequences that are available. Next, this profile is compared to all sequences in the database; this will both tell us the distribution of scores for unrelated sequences and allow us to confirm that all serine proteases are distinguished by the profile. Finally, the new sequence is compared to the profile and the significance of the similarity tested by comparison to the distribution of scores from the database search.

Database Searching

The choice of a method for identifying sequences distantly related to a new sequence is constrained by the available computational resources. In the process of testing and validating profiles, we frequently compare a profile to the entire protein sequence database. In our case, this is reasonable because we have access to fast computers, and because we need the information for comparative purposes. However, because **PROFILE-SEARCH** performs a complete dynamic programming alignment of the profile and every sequence in the database, it may not be the best choice for routine database searching. Complete dynamic programming alignment is time consuming in comparison to fast searching programs such as FASTA.[11] Although increases in the speed of available computers could ameliorate this problem, gains in speed tend to be offset by the growth of the sequence database.

---

[11] W. R. Pearson and D. J. Lipman, *Proc. Natl. Acad. Sci. U.S.A.* **85**, 2444 (1988).

The Future

The goal of the profile method is to incorporate information from related sequences into sequence analysis. As the sequence database grows, it will become more the rule than the exception that a new sequence will be a member of an identified family of homologous proteins. We think that profile analysis provides an increasingly powerful way of organizing and using the information from homologous families to make evolutionary, structural, and functional conclusions.

Acknowledgments

Research was sponsored by the National Cancer Institute, DHHS, under Contract N01-CO-74101 with Bionetics Research, Inc., and by National Science Foundation Grant CHE-85-09657 to D.E. The contents of this publication do not necessarily reflect the views or policies of the U.S. Department of Health and Human Services, nor does mention of trade names, commercial products, or organizations imply endorsement by the U.S. government.

Section III

Patterns in Nucleic Acid Sequences

[10] Finding Protein Coding Regions in Genomic Sequences

By RODGER STADEN

Introduction

Those engaged in determining the sequence of even quite modest lengths of genomic DNA are aware of the problems of identifying which segments code for proteins. During the project to map and sequence the human genome, the interpretation of the finished sequences is likely to be a major difficulty. Even if cDNA libraries are generated in addition to the genomic library, it will be hard to ensure that they contain the mRNAs for proteins involved in development and for all the different tissue-specific genes. It is important to push the interpretation of the genomic sequences as far as possible using computer methods.

This chapter outlines some of the methods currently available to aid in the location of protein coding regions. We concentrate here on the ideas behind the methods, because their application is simply a matter of using the appropriate program, whereas the interpretation of their results requires an understanding of the underlying principles. First, it should be realized that there are three separate questions which have to be answered when looking for protein coding regions: which regions are coding, which strand or orientation of the sequence, and which reading frame?

Currently there are two types of information that can be used for finding protein coding regions. The first is to look for the special, so-called signal sequences, such as splice junctions and promoters, that surround coding regions. This is often called gene search by signal. The second is to examine long sections of the DNA to see if they look more like coding sequence than noncoding sequence. These latter methods are often described as gene search by content and are the subject of this chapter.

We examine those properties of protein coding regions that can be used to distinguish them from noncoding sequences. Of these, the simplest is that the coding frame, on the coding strand, will contain no stop codons. Such regions are called open reading frames. There are three stop codons, so that in a random sequence of even base composition we expect a mean open reading frame length of 21 codons. The problem is to distinguish those open reading frames that exist by chance from those that code for proteins.

Constraints on Protein Coding Regions

The function of any particular coding region is to encode, with some level of accuracy and efficiency, using the genetic code and a given pool of tRNAs, a specific sequence of amino acids. The same region of DNA will have other functions. It may contain sequences involved in the positioning of histones and related to DNA bending; the mRNA it produces may have to fold in a certain way, and the sequence of amino acids it encodes may contain secondary structures such as α helices and β sheets. In addition, some genomes have a very marked bias in their base composition, and this will influence the choice of codons.

The observed order of bases depends on the differing strengths of these constraints, but each constraint will tend to produce regularities in the sequence. The assumption made in the methods described below is that the overriding requirement of the DNA sequence is to encode a particular sequence of amino acids. In general, the methods ignore other constraints, but they may increase or diminish the nonrandomness of the sequence and should not be forgotten by those employing the techniques. All the methods examine the sequence as nonoverlapping three-letter words or codons, and rely on the unequal use of codons in the coding sequence to produce the departures from randomness that are sought. Noncoding sequences are considered, when examined in the same ways, to be closer to random.

Statistical Properties of Protein Coding Regions

We examine how coding for a protein influences the amino acid, codon, and base compositions in the coding region, and how these compositions can be measured and used to distinguish coding from noncoding sequences. The three factors that contribute to the unequal usage of codons in a coding sequence are (1) the unequal use of amino acids (some amino acids are used more frequently than others: for example, on average leucine is used many more times than tryptophan), (2) the unequal numbers of codons for different amino acids, for example, leucine has six codons and tryptophan only one, (3) in any gene, for any given amino acid, the codons will generally not be equally used (this is termed codon preference, although note that sometimes the term is employed instead of codon usage, which in fact is the combined outcome of amino acid composition and codon preference).

If reading frame 1 encodes a protein it will influence the following factors: (1) the amino acid composition in both the coding frame and the other two frames (frames 2 and 3); (2) the codon composition in all three

TABLE I
AMINO ACID COMPOSITIONS FOR FRAMES 1, 2, AND 3[a]

	A	C	D	E	F	G	H	I	K	L
Frame 1	83	17	53	62	39	72	22	52	57	90
Frame 2	48	27	14	23	27	50	23	50	49	101
Frame 3	55	37	35	37	29	87	34	35	34	60

	M	N	P	Q	R	S	T	V	W	Y	*
Frame 1	24	44	51	40	57	69	58	66	13	32	0
Frame 2	25	31	60	36	108	99	76	48	24	25	59
Frame 3	7	32	53	36	129	89	51	46	18	34	65

[a] Assuming an average amino acid composition in frame 1 and no codon preference.
* Stop codon.

frames; (3) the frequency with which each of the four bases occupies each of the three positions within codons. This last parameter (3) is called the positional base frequency and below we examine it in two ways. The first way is to measure, for each base, the difference in its abundance in each position. For example we examine the relative abundance of base T in positions 1, 2, and 3 of codons. The greater these relative abundances differ from one another the more likely it is that the sequence is coding for a protein in the sampled region. The gene detection method that measures these relative abundances is known as the uneven positional base frequencies method. The second way is to look for preferences for certain bases to occupy particular positions in codons, that is, to compare the observed positional base frequencies with an expected distribution. We call the gene detection method using this measurement the positional base preferences method.

To examine the effects produced by unequal use of amino acids, independently of those due to codon preference or base composition, we chose a particular amino acid composition. The composition chosen is that calculated to be an average for all proteins.[1] (Note that in the original work[2] a different average composition was used). The new average composition expressed as amino acids per 1000 is shown in Table I. There are differences of up to a factor of 5 (leucine/tryptophan) in this composition. We can produce a codon table (Table II) which contains no codon preference but which corresponds to this average amino acid composition. As-

[1] P. McCaldon and P. Argos, *Proteins* **4**, 99 (1988).
[2] R. Staden, *Nucleic Acids Res.* **12**, 551 (1984).

TABLE II

CODON COMPOSITION FOR FRAMES 1, 2, AND 3[a]

	Frame				Frame				Frame				Frame		
	1	2	3		1	2	3		1	2	3		1	2	3
F TTT	20	12	17	S TCT	12	12	15	Y TAT	16	11	19	C TGT	9	12	17
F TTC	20	14	12	S TCC	12	14	13	Y TAC	16	13	15	C TGC	9	15	21
L TTA	15	19	8	S TCA	12	19	16	* TAA	0	18	25	* TGA	0	20	29
L TTG	15	23	8	S TCG	12	23	10	* TAG	0	21	11	W TGG	13	24	18
L CTT	15	11	17	P CCT	13	11	15	H CAT	11	11	19	R CGT	10	11	17
L CTC	15	13	12	P CCC	13	13	13	H CAC	11	13	15	R CGC	10	13	21
L CTA	15	16	8	P CCA	13	16	16	Q CAA	20	16	25	R CGA	10	16	29
L CTG	15	20	8	P CCG	13	20	10	Q CAG	20	20	11	R CGG	10	20	18
I ATT	17	13	17	T ACT	15	13	14	N AAT	22	14	18	S AGT	12	14	16
I ATC	17	16	11	T ACC	15	16	12	N AAC	22	17	14	S AGC	12	17	20
I ATA	17	21	8	T ACA	15	21	15	K AAA	29	22	24	R AGA	10	22	28
M ATG	24	25	7	T ACG	15	25	9	K AAG	29	27	10	R AGG	10	27	17
V GTT	17	8	18	A GCT	21	8	16	D GAT	27	7	20	G GGT	18	9	17
V GTC	17	10	12	A GCC	21	10	13	D GAC	27	8	15	G GGC	18	11	22
V GTA	17	13	8	A GCA	21	13	16	E GAA	31	10	26	G GGA	18	14	30
V GTG	17	16	8	A GCG	21	16	10	E GAG	31	12	11	G GGG	18	17	19

[a] Assuming an average amino acid composition in frame 1 and no codon preference.
* Stop codon.

TABLE III
BASE COMPOSITIONS FOR FRAMES 1, 2, AND 3[a]

Frame	T	C	A	G
1	17.68	21.08	27.67	33.57
2	27.07	23.78	30.97	18.18
3	25.06	25.06	23.96	25.92
Mean	23.27	23.30	27.53	25.89

[a] Assuming an average amino acid composition in frame 1 and no codon preference.

suming a random distribution of the codons in Table II we can calculate codon tables, amino acid compositions, and base compositions for reading frames 2 and 3. The calculated compositions for amino acids in frames 2 and 3 are also shown in Table I.

It is interesting to study such tables. There are marked differences in the amino acid compositions for each of the three reading frames. For example, the frequency of basic and acidic residues in frames 2 and 3 is quite different from that in frame 1. These values are reflected in the calculated codon table (Table II) along with some more extreme differences for particular codons. Note that some authors[3,4] have suggested that high observed codon frequecies in frames 2 and 3 and the preference for G in the first position of codons in genes imply that they have some role in mRNA translation, but, as we have shown here, these nonrandom features can arise from encoding a protein of average amino acid composition in frame 1. Also we see that there are more stop codons than expected in frames 2 and 3.

The base compositions are shown in Table III, which gives the percentage of occurrences of each of the four bases in each of the three positions in the codons. Notice the preferences for G, and to a lesser extent A, and the low use of T in the first position and the preference for A and the low use of G in position 2. Expressing these values in another way, note the marked difference between the overall frequency for each particular base and its positional frequencies. In summary, we see that there are marked compositional variations in all three reading frames for amino acids, codons, and bases when a gene codes for a protein sequence of average amino acid composition, even when there is no codon preference.

[3] E. N. Trifonov, *J. Mol. Biol.* **194,** 643 (1987).
[4] A. K. Konopka and G. W. Smythers, *CABIOS* **3,** 193 (1987).

TABLE IV

Codon Compositions for *rpoC2* Gene of *Marchantia*
Chloroplast

F TTT	77	S TCT	38	Y TAT	51	C TGT	9
F TTC	6	S TCC	6	Y TAC	3	C TGC	4
L TTA	98	S TCA	24	* TAA	1	* TGA	0
L TTG	8	S TCG	1	* TAG	0	W TGG	16
L CTT	32	P CCT	16	H CAT	27	R CGT	7
L CTC	0	P CCC	1	H CAC	4	R CGC	0
L CTA	5	P CCA	21	Q CAA	63	R CGA	14
L CTG	1	P CCG	2	Q CAG	2	R CGG	2
I ATT	84	T ACT	40	N AAT	113	S AGT	21
I ATC	5	T ACC	4	N AAC	12	S AGC	2
I ATA	58	T ACA	35	K AAA	154	R AGA	12
M ATG	18	T ACG	4	K AAG	2	R AGG	1
V GTT	29	A GCT	15	D GAT	33	G GGT	21
V GTC	3	A GCC	4	D GAC	3	G GGC	7
V GTA	26	A GCA	19	E GAA	68	G GGA	37
V GTG	4	A GCG	3	E GAG	6	G GGG	5

* Stop codon.

Effect of Codon Preference and Base Composition Bias

In coding sequences, codon preference will cause more extreme variations in the parameters that we have identified. In previous accounts[5] we have given examples to demonstrate this, so here we concentrate on the influence of base composition. Generally, we would expect base composition biases to influence mostly the third positions of codons where they do not change the amino acid. However, although we are not aware of any relevant analysis, it is also likely that such a DNA bias might result in a significant constraint on the use of amino acids. Below we show a codon usage table for the *rpoC2* gene of the liverwort *Marchantia* chloroplast[6] (Table IV).

The genome has a very biased base composition (composition A, 35.5; T, 35.7; C, 14.3; G, 14.5), and we can see from Table IV how this overall bias influences the choice of codons: there is a very strong preference for codons with A or T in third positions. In Table V we show the positional

[5] R. Staden and A. D. McLachlan, *Nucleic Acids Res.* **10,** 141 (1982).
[6] K. Ohyama, H. Fukuzawa, K. Kochi, H. Shirai, T. Sano, S. Sano, K. Umesono, Y. Shiki, M. Takeuchi, Z. Chang, S. Aota, H. Inokuchi, and H. Ozeki, *Nature (London)* **322,** 571 (1986).

TABLE V
BASE COMPOSITION FOR *rpoC2* GENE OF
Marchantia CHLOROPLAST

Frame	T	C	A	G
1	24.66	14.20	40.74	20.40
2	32.73	16.80	39.08	11.39
3	44.20	4.61	45.78	5.41
Mean	33.86	11.87	41.86	12.40

base composition for this gene. Note that the differences between the use of each base in each position are much more marked than those for a protein of average base and amino acid compositions, but that, apart from those for adenine, the rank order of usage is unchanged for the first two positions of codons. So, for example, in Table III the ratio of usage for T, position 1/position 2, is 17.68/27.07 whereas in Table V it is 24.66/32.73 and for C we have 21.08/23.78 and 14.20/16.80. However, in this extreme case of base composition bias the most obvious influence is on the third position values, with about 90% of third positions being occupied by A or T, whereas the overall base composition of the coding region is 75% A and T.

Overview of Analytical Methods

We have seen that there are compositional effects which are due to a sequence coding for a protein, and now we examine methods of measuring these effects in order to locate protein genes. We can measure codon compositions and positional base frequencies. First we explain some general characteristics of the sequence analysis methods and the presentation of results.

A window is a length of sequence, say n bases or n codons, over which we perform a calculation. This gives us the value of our function for this section of the sequence, and then we move the window by one unit of length (one base or one codon) and repeat the calculation. By so doing we get a detailed analysis of each section of the sequence because each window overlaps the previous one by its length minus one unit. The length of window chosen depends on the resolution of the method being employed.

All results are presented graphically. This allows us to overlay the plots resulting from different and independent analytical methods and which can hence be seen to reinforce or contradict one another. The sequence is represented by the x axis, and the analytical results are plotted in the y direction. Some methods give only a single value for each position along the sequence whereas others give a separate value for each of the three reading frames. For the latter we plot the three probabilities in separate boxes arranged above one another (for example, see Fig. 3). For these we also indicate which of the three reading frames is the highest scoring at each position along the sequence. This is done by plotting a single dot at the mid-height of the box that contains the highest scoring plot. Then if one frame is the highest scoring for many consecutive positions, the points will produce a solid line at the midpoint of its box. We also mark the positions of stop codons. These are represented as short vertical lines and are positioned so that they bisect the midpoint of each box. Start codons are marked at the base of the box for each reading frame.

Below we describe three basic methods in their order of increasing assumptions and with corresponding improvements in resolution. First we look simply at the randomness of the sequence when examined in terms of nonoverlapping three-letter words. The method is similar to that of Fickett.[7]

Uneven Positional Base Frequencies Method[2]

Above we demonstrated that coding sequences show unequal use of the four bases in the three positions of codons. We expect that this inequality of use will be more pronounced in coding than in noncoding regions. If we know the levels of inequality in these two classes of sequence (i.e., the coding class and the noncoding class) we can give the probability that any section of a sequence lies in either of the two classes, i.e., the probability of being coding or noncoding.

For each of the four bases i we calculate N_{ij}, the number of times it occurs in each of the three positions j of a codon when summed over a window length of the sequence. For every window position we then calculate an expected value E_{ij} for each base in each position of a codon, assuming an even distribution. We use the formula $E_{ij} = (N_{i1} + N_{i2} + N_{i3})/3$ [which is simply the (base composition)/3]. Now we measure the divergence D from even positional usage with the formula $D = \Sigma |E_{ij} - N_{ij}|$, i.e., we measure the absolute differences between the observed and expected positional base frequencies. Notice that this measure D is independent of the base composition of the window. The operation performed by the program is to calculate D for each position in which we place the window, and the window is slid along the whole length of the sequence one codon at a time.

The calculation described above was performed on all the sequences in

---

[7] J. W. Fickett, *Nucleic Acids Res.* **10**, 5303 (1982).

the 1984 version of the EMBL sequence library. Using a window of 67 codons we calculated the D values for all of the coding and then for all of the noncoding regions and plotted histograms for each. The two distributions obtained are shown in Fig. 1. The frequency with which coding sequences obtained scores is plotted as a solid line, and the noncoding values are plotted with a dotted line.

Although there is some overlap between the two distributions, there is sufficient difference to be useful for distinguishing coding from noncoding sequences. When the method is used to analyze new sequences the values from these two distributions are used to scale the resulting plots. The lowest value on the scale was exceeded by all coding regions but was not reached by 14% of windows in noncoding sequence; the top value was not reached by any window in noncoding sequence, but was exceeded by 16% of coding windows. A bar is drawn across the plot at a height corresponding to the value that was exceeded by 76% of windows in coding sequences but that was not reached by 76% of windows in noncoding sequence.

To illustrate the application of the method we have chosen a section of

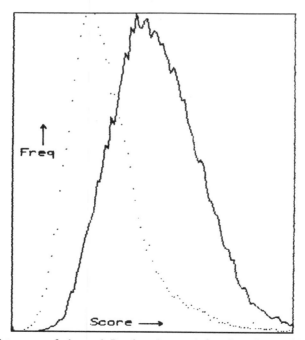

Fig. 1. Histograms of observed D values (see text) for all coding (solid line) and all noncoding (dotted line) sequences in the 1984 EMBL nucleotide library.

the liverwort *Marchantia* chloroplast genome (bases 100,000 to 121,024) that encodes at its 5' end a group of proteins and at the 3' end a set of tRNAs and rRNAs. All these genes are on the complementary strand. The results are shown in Fig. 2. For publication a window of 151 codons has been used to give a clear picture of such a long section of sequence, but normally 67 codons would be employed and gives the same interpretation. In Fig. 2 the positions of the protein genes are labeled P and the RNA genes R, and a scale marking every 100th base is shown below. These simple maps of the sequence features are drawn automatically by the program using the sequence libraries feature tables. It can be seen how the analysis distinguishes between the two regions and clearly indicates the protein genes. It will be noticed, however, that there are two large gaps between genes in the protein map. The left-hand gap coincides with a prediction of not coding, but the right-hand gap is predicted to be coding, and we believe this to be an omission in the sequence library feature tables rather than an incorrect prediction by the program.

We summarize the uneven positional base frequencies method. We have shown that the method can distinguish coding from noncoding sequences, and although it cannot give information about reading frame or even which strand of the DNA is coding (in the example shown all genes were on the complementary strand), it makes very few assumptions and hence is a very useful and unbiased indicator of whether a sequence is coding.

Positional Base Preferences Method[2]

We demonstrated in the introductory sections that a DNA sequence which encoded a protein of average amino acid composition would exhibit bias in its use of the four bases in the first two positions of codons. We also

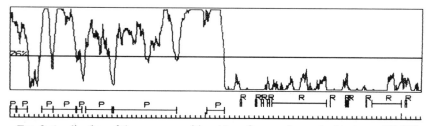

FIG. 2. Application of the uneven positional base frequencies method to bases 100,000 to 121,024 of the liverwort *Marchantia* chloroplast genome. The horizontal scale marks every 100th base, and the bars above indicate the extent of known protein coding segments (P) and known RNA genes (R).

noted that some genomes have marked overall base composition biases and expected these preferences to be mostly confined to the third positions of codons. An example of a coding sequence from a genome with such a bias was shown. Here we concentrate initially on the preferences produced in the first two positions of codons by an average amino acid composition.

We have shown previously[2] that the bias produced by such a composition is sufficiently typical, and adequately strong, to serve as the basis for a method of predicting reading frames. We show in this chapter that, when present, third position preferences can also be diagnostic. We call this the positional base preferences method. In outline we take the positional base frequencies shown in Table III and use them as an expected set. We examine windows along the sequence to see how close their observed positional base frequencies are to the expected set, and plot their level of correlation.

As above, let the expected positional base frequency in a coding sequence be E_{ij}, where i is the base A, C, G, or T and j is the position 1, 2, or 3, in codons, but here the E_{ij} are the values contained in Table III. Let the observed positional base composition in frame f ($f = 1$, 2, or 3) be O_{if}. Then for any codon in frame f we can calculate a correlation score $E_{ij}O_{ig}$, where g is a rotated value dependent on f. If we choose a window of length L codons we can calculate a value for each reading frame f, for each position along the sequence, by summing over the window length to give

$$C_f = \sum E_{ij}O_{ig} \quad \text{(sum within } L\text{)}$$

To estimate the expected scores in each of the three frames we can employ the expected positional base frequencies in the following way. If we divide each of the expected values E_{ij} by 100 we can use them as estimates of the probability of each score occurring. If we multiply the probabilities by the expected values we find the expected average values in each reading frame. Let $P_{ij} = E_{ij}/100$ and use only the first and second positions of codons. Then summing for $i = 1$, 4 the expected score is

Frame 1: $\sum P_{i1}E_{i1} + \sum P_{i2}E_{i2}$

Frame 2: $\sum P_{i2}E_{i1} + \sum P_{i3}E_{i2}$

Frame 3: $\sum P_{i3}E_{i1} + \sum P_{i1}E_{i2}$

These calculations give average expected values per codon on the coding strand of 52.37, 49.35, and 49.52, and on the complementary strand 50.46, 49.56, and 48.90. So on the coding strand there is a 5.4% difference per codon between the coding frame and the next highest scoring. On the noncoding strand, the frame in register with the coding frame scores most highly and is 3.6% lower than that on the coding strand. Notice that we cannot take into account the influence of codon preference or base composition bias when calculating these expected values.

The purpose of this method is to decide which reading frame is coding. Therefore, as a test, we applied it to all the *Escherichia coli* protein genes in the 1984 version of EMBL nucleotide library and counted the number of times it correctly chose the coding frame. We did not therefore compare coding with noncoding values, but only the ability of the methods to choose the correct reading frame. A window of 67 codons was slid, one codon at a time, along the whole length of each gene, so that each window was entirely within the coding sequence. This gave a total of 22,831 window positions, and the calculation identified the correct frame for 91% of these. Note that the *E. coli* genes were chosen only to avoid the difficulty of using the feature tables of the sequence libraries to determine the correct reading frame for spliced genes. We assume, for the purposes of this method, that these genes are representative of all genes from genomes which do not have a very biased base composition, and that the method, as just described, is therefore applicable to sequences from all such organisms.

It is convenient to display the plots on a scale where the top corresponds to the maximum expected value plus 50% of the range, and the bottom is equivalent to the minimum expected value minus 50% of the range (although see below about users choosing scales). The program will plot either absolute values (i.e., the actual values C_f obtained for each window position) or relative values given by $C_f/\Sigma C_f$. Relative values give smoother plots and tend to emphasize the differences between the three frames more clearly than plots of absolute values. Above we have described the method as first published (apart from the use of a new average amino acid composition), but a number of improvements have been made and these are outlined below.

First, we have added the option to take advantage of any base composition bias that might be present in the sequence being analyzed. This is done by allowing the user to specify any region of the sequence to use as a "standard" to define an expected set of positional base preferences. The program will calculate the positional base frequencies over the whole of the standard and the user can employ them in one of two ways. Either the values for all three positions can be used instead of those from the average amino acid composition, or, alternatively, the third positions alone can be substituted for the (effectively neutral) third position values of the average amino acid composition, hence producing a hybrid of the two. The latter is the usual choice, although if, for example, it were suspected that a gene encoding a very hydrophobic protein was being sought, then it would be worth using a similar gene as a standard for all three positions. The second positions in codons of genes encoding hydrophobic proteins have a higher

T content than average. It is also sensible to employ all three positions from an internal standard if a strong genomic base compositional bias does significantly influence the use of amino acids. However a known outcome of such a base composition is strong preferences in the third positions of codons, and this alone can give quite marked improvements in resolution. In addition to the above, the user may also specify weights between 0.0 and 1.0 to apply to each of the three positions in codons in order to increase or diminish their contributions. Finally, the scaling of the plots is now under the user's control.

To demonstrate the method, Fig. 3 shows an analysis of base positions 10,001 to 20,000 of the liverwort *Marchantia* chloroplast DNA, which includes the eight coding regions shown in the map. The plot is divided into three horizontal boxes: one for each of the three reading frames. Within each box is plotted the relative probability of coding $C_f/\Sigma\ C_f$ and the in-frame stop codons. The analysis shown employed the third position values from an internal standard taken over the long open reading frame of *rpoC2* (bases 11811–15971), combined with the values for the first two positions of an average amino acid composition. That is, we employed the values for the first two positions of codons shown in Table III and the third positions from Table V. A window of 67 codons was used, and the scaling was chosen to emphasize the coding regions. Notice that the strong predictions of coding frame coincide with the positions of the genes shown by the

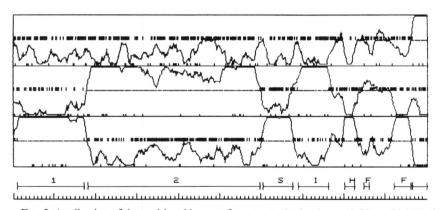

FIG. 3. Application of the positional base preferences method to bases 10,001 to 20,000 of the liverwort *Marchantia* chloroplast genome. The horizontal scale marks every 100th base, and the bars above indicate the extent of known protein coding segments. The three boxes above contain plots of the probability that each of the three reading frames is coding for a protein. The short vertical lines that bisect the mid-height of each box mark the positions of the stop codons in the corresponding reading frames.

map. In noncoding regions there are no strong peaks, but there is a small peak coincident with an open reading frame of 75 codons starting at position 17,693.

In summary, the positional base preferences method makes few assumptions but cannot give an absolute probability of coding. However, it has been shown to predict the correct reading frame in 91% of positions using only the first two positions of codons and would be expected to do better given a strong bias in genomic base composition such as that used here.

Codon Usage Method[5]

The basic assumption of the codon usage method is that all the genes within the sequence we are analyzing have similar codon usage and that this usage is sufficiently strong to be used as a means of discrimination. Previous articles[5] have given references to justify this basic assumption, and below we give the detailed assumptions that allow us to derive the necessary equations for calculation. These assumptions are as follows: (1) We have a given DNA sequence which comes from a collection of genes with certain common features. (2) The given sequence is coding throughout its length in only one reading frame, but we do not know which frame this is. (3) The collection of genes has a characteristic pattern of codon usage of the following type: (a) In the actual reading frame the frequency of codon abc has a definite value $fabc$. (Note this implies both a typical amino acid composition for the proteins coded, and a certain set of codon preferences for any given amino acid.) (b) The codons in the coding frame (and hence the amino acids in the protein) occur in random order. Note that this assumption applies only to the reading frame. In the other two frames (assumption 3a) may apply, but in general there will be some tendency for consecutive triplets to be correlated. (4) Out of all the triplets in the gene collection a fraction Q_1 are read in frame 1, Q_2 in frame 2, and Q_3 in frame 3.

If we choose a sequence randomly from the collection we can translate it in three frames. Suppose we select the following sequence:

$$a_1 b_1 c_1 a_2 b_2 c_2 a_3 b_3 c_3 \ . \ . \ . \ a_n b_n c_n a_{n+1} b_{n+1} c_{n+1}$$

Then, from assumptions 3a and 4, the probability of selecting it in each of the three frames if they were coding is

Frame 1: $p_1 = Q_1 fa_1 b_1 c_1 fa_2 b_2 c_2 \ . \ . \ . \ fa_n b_n c_n$

Frame 2: $p_2 = Q_2 fb_1 c_1 a_2 fb_2 c_2 a_3 \ . \ . \ . \ fb_n c_n a_{n+1}$

Frame 3: $p_3 = Q_3 fc_1 a_2 b_2 fc_2 a_3 b_3 \ . \ . \ . \ fc_n a_{n+1} b_{n+1}$

We want to know the probability that selection of a particular sequence was "caused" by its being a coding sequence, and this is found by using Bayes' formula, which gives the probability that the outcome of an experiment was due to a particular one of the assumed caused of the outcome. The probabilities that the selected sequence is in fact coding in each of the three frames are

$$P_1 = p_1/(p_1 + p_2 + p_3)$$
$$P_2 = p_2/(p_1 + p_2 + p_3)$$
$$P_3 = p_3/(p_1 + p_2 + p_3)$$

It is convenient to work with logarithms of the frequencies and so we define $Fabc = \log fabc$

In practice, when we analyze a sequence we slide a window of length L triplets along it, moving the window by one triplet at a time. For each of the positions of the window we calculate the sum of the F values over the length of the window for each of the three reading frames:

Frame 1: $H_1 = \sum Fa_ib_ic_i$

Frame 2: $H_2 = \sum Fb_ic_ia_{i+1}$

Frame 3: $H_3 = \sum Fc_ia_{i+1}b_{i+1}$

The probability that any window is in fact coding in each of the three frames is then calculated using, for example, for frame 1:

$$P_1 = Q_1 e^{H_1}/(Q_1 e^{H_1} + Q_2 e^{H_2} + Q_3 e^{H_2})$$

Before applying the method to any sequence we need to select a suitable set of expected codon frequencies. They can be calculated from related sequences extracted from the nucleotide libraries or, often, from another part of the sequence being analyzed for which the interpretation is already clear. These frequencies are the combined outcome of amino acid composition and codon preference, and we have previously demonstrated[2] that amino acid composition alone can be sufficient to make correct choices of reading frames. Following from this, a previously unpublished improvement in the method is described below.

After the table of codon usage has been defined, the user is now given the choice of (1) leaving the table unchanged (the original method); (2) removing from the table the component due to amino acid composition (this leaves behind only the codon preferences, so that the sum of frequencies for each amino acid is the same, and was also possible in the original method); (3) modifying the component due to amino acid composition so that it becomes equal to that for a protein of average amino acid composition. The third option results in a table of codon usage that would be produced by a protein having average amino acid composition, and exactly the codon preferences observed in the given standard. In general, the latter is the best choice because the table of usage cannot contain an extreme amino acid bias, but does retain a useful predictive component from the amino acid frequencies. The second option means removing helpful information but may be useful in special circumstances. The first option is particularly beneficial if the standard is chosen to have an amino acid component similar to that of the proteins whose genes are being sought. However, in the most favorable circumstances, i.e., when the codon preferences are very strong, the amino acid component is less important than that from codon preferences.

The program performs the calculation in two stages. First, the standard frequencies have to be chosen and operated on to produce the values used in the probability calculation. Next, the sequence is scanned, and the probability for each position of the window calculated and plotted.

We have to deal with two classes of special codons. These are the stop codons and those codons that have zero frequency in the standard. Often the standard will contain a few codons with zero frequency, and we set these to have a frequency of 1/(number of codons in the standard). To make stop codons neutral we set their frequency to that of the mean value of the standard. This means that, when scanning a sequence, the probabilities based on codon usage and the positions of stop codons can be used as independent criteria in assessing the likelihood that a particular region is coding for a protein.

To demonstrate the application of the codon usage method we selected bases 1 to 10,000 of the liverwort *Marchantia* chloroplast genome, which encodes a number of proteins. As a standard for codon usage we took bases 11,811 to 15,971, the *rpoC2* gene (see Table IV), and used a window of 31 codons. The layout of the plots shown in Fig. 4 is identical to that for the positional base preferences method, and a simple bar map of the protein genes is shown below the results. Along the left edge of each plot is a scale of $\log[P/(1 - P)]$ so that, for example, 4 points up the scale from the midpoint corresponds to 99.99% probability. There is good agreement between the coding map and the positions of the peaks. There is, however, an extra peak above an open reading frame lying between positions 5088 and 5258, which also appears if the positional base preferences method is applied to the same region. Notice that the window is much shorter than for the other methods and yet the resolution is much greater. For some problems we have successfully used windows as short as 11 codons.

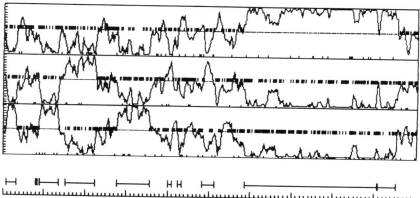

FIG. 4. Application of the codon usage method to bases 1 to 10,000 of the liverwort *Marchantia* chloroplast genome. The horizontal scale marks every 100th base, and the bars above indicate the extent of known protein coding segments. The three boxes above contain plots of the probability that each of the three reading frames is coding for a protein. The short vertical lines that bisect the mid-height of each box mark the positions of the stop codons in the corresponding reading frames. The vertical scale within each box is of $\log[P/(1 - P)]$ so that, for example, 4 points up the scale from the mid-height corresponds to 99.99% probability.

Discussion

We have outlined the constraints imposed on a sequence when coding for a protein, and the ways in which these constraints lead to measurable differences between coding and noncoding sequences and between reading frames. Three methods, all of which require some assumptions, are described. The codon usage method assumes that there is sufficient similarity between the codon usage in related genes for one gene to be used as a standard to find another. Use of the effect that a typical amino acid composition and no codon preference has on the positional base frequencies in coding regions is less dependent on a particular expected codon composition. If an overall genome base composition bias is present, we can take advantage of it, but again make the assumption that it will affect all genes in similar ways. The method that looks for uneven positional base frequencies requires fewest assumptions but will not indicate reading frame or strand.

The methods should be used in conjunction with one another: they all have their uses and limitations. In general we tend to work from the fewest to the highest number of assumptions; that is, we use the uneven positional base frequencies method to get an idea about the regions most likely to be coding. Then the positional base preferences method is used to confirm these results and to show the most likely reading frames. From these two analyses (usually produced in a single diagram with the plots arranged one above the other) we choose standards for the codon usage method. This method is then employed to help define the ends of coding regions and to find small exons that would not show up when the other methods were used. Note that application of such methods is also useful during a sequencing project, as they can indicate possible errors, particularly insertions or deletions.

There is scope for improvement. The methods can be made more quantitative than at present, when the annotations of the sequence libraries are changed to allow correct automatic translation of coding regions. However, the general principles of the existing methods are unlikely to change, and, as they stand, they have already been very useful.

In the introduction we noted that gene search by content, the subject of this chapter, was only one component of the techniques available to search for genes. Apart from marking the positions of stop codons on the plots shown above, we have ignored the extra information derived from searches for signals, such as mRNA splice junctions, that could also have been employed. Methods for describing and searching for such signals are given elsewhere in this volume,[8] and the results of these searches can be readily superimposed on the plots shown above. All the methods described here are contained in the single program ANALYSEQ, which is available from the author. Further details about the requirements for running the program are given elsewhere.[8]

[8] R. Staden, this volume, [12].

[11] Ancient Patterns in Nucleic Acid Sequences

By JOHN C. W. SHEPHERD

Introduction

DNA sequences not only provide an ever increasing wealth of information with regard to present biological function but also represent a very extensive source of knowledge about the past, paralleling the fossil record and even predating it by tracing remnants of patterns existing at the very beginnings of life. Here a brief summary is given of some of the evidence and reasoning leading to the conclusion that remnants of a primeval coding system still exist in present-day DNA sequences from all types of living organisms. Most of the genes now sequenced are still largely read in their original reading frame, but exceptions are found where the ancient messages have been displaced or are otherwise too heavily mutated. A simple computer program is then described which looks for remnants of these primeval messages. Not only is it useful as a quick method of analysis of newly determined sequences by predicting the likely reading frame and, in some cases, the extent of existing genes, but it can be a guide to the nature of the genes and their past history.

Primeval Coding Pattern

When the first longer DNA sequences were published, a detailed study of the positional correlations between purines (R) and pyrimidines (Y) indicated the presence of mutated sections of older message with the codon form RNY (N is R or Y).[1] This pattern was seen especially strongly in existing genes, sometimes with short stretches of displaced pattern within a particular gene. The pattern was also well maintained in degenerate codons, which was one strong argument against the effect being due to a reflection of the amino acid proportion requirements of the proteins.

Searching for these old messages revealed that they are usually in the reading frames of present-day genes, and this meant that the search method could be adapted to show the positions and reading frames of existing genes.[2] Correct predictions in the first longer DNA sequence determined, that of the virus ϕX174, was one of the first encouraging examples of the method's success. As more DNA sequences became available from very varied origins, it became evident that these RNY relics were detectable almost everywhere in coding regions and sometimes outside these present genes.

A comparison was then made with the conclusions independently reached from a study of the central problem of how the synthesis of polypeptides, directed by a nucleic acid template, could have first started on primeval earth. From consideration of a primitive tRNA and the sequence regularities in the anticodon loops of present tRNAs, it had been concluded that the primeval message was of the form $(RRY)_n$, with $(RNY)_n$ as a less likely alternative.[3] Such a message is a purine–pyrimidine

---

[1] J. C. W. Shepherd, *J. Mol. Evol.* **17**, 94 (1981).

[2] J. C. W. Shepherd, *Proc. Natl. Acad. Sci. U.S.A.* **78**, 1596 (1981).

[3] F. H. C. Crick, S. Brenner, A. Klug, and G. Pieczenik, *Origins Life* **7**, 389 (1976).

version of a "comma-less" message (i.e., not needing a start signal to indicate in which frame it should be read), a concept developed much earlier.[4] The problem was further considered, and the most probable primeval message was thought to be $(RNY)_n$ with symmetrical forward and reverse strands, and the purines and pyrimidines more evenly matched.[5] Additional evidence has also been found for remnants of an $(RNY)_n$ message (to be termed RNY message) in an ancestral tRNA, derived from a large number of present tRNA sequences by a phylogenetic analysis.[6,7] Thus, the early RNY nucleic acid strands, thought to have been RNA, could have acted as both message and primitive adaptor. Later work confirmed this conclusion and also found evidence for the former use of RNY patterns in ribosomal 5 S RNAs.[8,9] Other reasons also account for the suitability of RNY coding in early self-replicating systems; for example, on the basis of the present RNY coding assignments, amino acids likely to have been abundant on the primeval earth would be used.

As many more DNA sequences were determined, the primeval RNY analysis was further developed and extended,[10-12] and so far there has been no reason to doubt these original conclusions. (Viral and phage genomic RNA sequences may be similarly analyzed, but a higher rate of mutation away from RNY would generally be expected.) Surprising to many is the fact that the remnants of such messages should survive so long, but this can be easily checked by noting the low rate of $Y \rightleftarrows R$ mutations in examples of genes which diverged long ago. Assuming this rate of mutation was the same in such genes before and after their point of divergence, the last use of this comma-less code can be dated to very approximately 3000 million years ago.[10] Also, by simulating the mutation away from original RNY messages, it can be shown how the present proportions of codon types in existing genes and of the corresponding amino acids in many present proteins could have been derived.[10,12] Particularly striking are the effects in the degenerate codons of some genes of ancient origin which are heavily transcribed. In such genes there has been a strong constraint to continue the use of a set of plentiful tRNAs, and, consequently, the RNY codon

[4] F. H. C. Crick, J. S. Griffith, and L. E. Orgel, *Proc. Natl. Acad. Sci. U.S.A.* **43**, 416 (1957).

[5] M. Eigen and P. Schuster, *Naturwissenschaften* **65**, 341 (1978).

[6] M. Eigen and R. Winkler-Oswatitsch, *Naturwissenschaften* **68**, 217 (1981).

[7] M. Eigen and R. Winkler-Oswatitsch, *Naturwissenschaften* **68**, 282 (1981).

[8] M. Eigen, B. Lindemann, R. Winkler-Oswatitsch, and C. H. Clarke, *Proc. Natl. Acad. Sci. U.S.A.* **82**, 2437 (1985).

[9] R. Winkler-Oswatitsch, A. Dress, and M. Eigen, *Chem. Scr.* **26B**, 59 (1986).

[10] J. C. W. Shepherd, *Cold Spring Harbor Symp. Quant. Biol.* **47**, 1099 (1982).

[11] J. C. W. Shepherd, *Trends Biochem. Sci. (Pers. Ed.)* **9**, 8 (1984).

[12] J. C. W. Shepherd, *Chem. Scr.* **26B**, 75 (1986).

form is generally well preserved. In all or almost all degenerate codons, moreover, mutations away from RNY are restricted to an absolute minimum. The likely reasons for this phenomenon are explained elsewhere.[12]

Method of Analysis

The simplest possible method to look for remnants of the primeval purine–pyrimidine coding pattern is described. A test window length of DNA is taken and moved along the sequence in steps. In the test length the $Y \rightleftharpoons R$ mutations away from RNY in the first and third positions are counted in each reading frame, and the RNY relics are considered to be in the frame with the least such mutations. The existence of a reasonably conserved RNY message is indicated when the frame determined stops changing randomly and the mutations recorded are significantly beyond the limits to be expected from a random sequence.

A computer program to calculate these mutations is easily written for whatever computer may be available. A simple version (called SEARCH) written in Fortran is given in the Appendix and is largely self-explanatory. The READSEQ subroutine needs sequence files written in NBRF format, and reads them into the character array A, the size of which may be altered to suit local requirements. Such NBRF formatted files can be easily obtained from the GenBank or EMBL sequence banks by using the NBRF-PIR software written for a VAX computer (obtainable from the National Biomedical Research Foundation of Washington, DC). Failing this, any sequence can be fed into a computer in a suitable form by using the editor. The first two lines should contain the sequence code and description, respectively. Then the sequence follows, written in lines not longer than 132 characters and with a star (★) to indicate the end of the sequence. The bases can be grouped in tens by using spaces, but no numbering is allowed. If any other format of sequence file is used, such as the STADEN format from Cambridge or the GCG format from Wisconsin, the READSEQ subroutine can easily be adapted to read these.

The RNY subroutine looks in each window for the frame with the least mutations MINM away from an RNY message and records in the printout the extent of the test window, the mutations in each frame, the midpoint MID of the window, and this reading frame (as shown in the Appendix). If the mutations away from RNY are equal in two or more frames, a star is printed at the end of the line, and a decision on this part of the sequence may be possible by varying the window length L or the step size S.

With a random sequence, one would expect a binomial distribution of mutation values centered about their mean MEANM, which would be equal to half of MXMUTE, the maximum number of mutations possible in each window. As a measure of whether there is a significant indication of an RNY message in any window, '+' signals (up to six) are printed alongside the frame, corresponding to the number of standard deviations by which MINM is less than MEANM. When using this program, some trials are necessary to determine a suitable window length for the analysis of a particular sequence. Often a length L of 120 moved along in steps S of 15 or 30 is convenient. With much mutated messages, a longer length may be necessary, but with well-preserved relics more detail can be obtained by taking L to be 90 or 60, or using smaller steps S.

Plots

A better overall picture of the results of the analysis may be obtained by plotting the mutation values and the frame determined against the position of the window's midpoint. To do this, plotting software suitable for the plotter or laser printer locally available will be needed.

Four such plots are shown in Figs. 1–4. These have been selected arbitrarily from the large number of cases in which the RNY ancient pattern search still yields relatively accurate information about the present genes. The $Y \rightleftharpoons R$ mutations away from RNY in each of the three reading frames, as obtained from the SEARCH program, are expressed as percentages of the maximum of such mutations and plotted at the tops of Figs. 1–4 against the midpoint b of the test window. The reading frame f nearest to RNY is plotted against b at the bottom of Figs. 1–4 and enables one to compare the frame of the best preserved RNY message with the actual reading frame, as known from the sequence. For example, in Fig. 1, the gene[13] for the rat neuronal growth protein GAP-43 is analyzed. This is a gene likely to be plentifully transcribed when neurites are developing or regenerating. The coding sequence extends from an ATG start signal at base 70 (reading frame 1) to a stop signal at base 748, as shown by a horizontal arrowed line alongside the predicted frame. In this case, the continuous low values of mutations away from RNY in this frame give a good indication of the extent of the gene. As soon as the coding sequence ends, the 3' end of the DNA is characterized in all frames by mutation values almost within the standard deviations from a random sequence. Figure 2 shows similar correct predictions obtained for the chloroplast *psbA* gene of barley,[14] coding for a chlorophyll or quinone binding protein. The rate of mutation is likely to have been low in this gene, since it is highly homologous to *psbA* genes in other plants.

[13] L. R. Karns, S.-C. Ng, J. A. Freeman, and M. C. Fishman, *Science* **236**, 597 (1987).
[14] J. Mundy and N.-H. Chua, *EMBO J.* **7**, 2279 (1988).

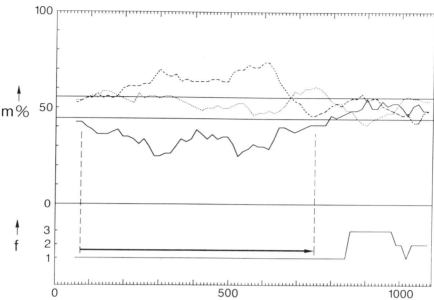

Fig. 1. Plot made from the SEARCH program for detecting remnants of RNY messages in the gene for the rat neuronal growth-related protein GAP-43. A test window length L (120 bases) has been moved forward in steps S (15 bases). The transversion counts ($m\%$) away from an RNY message in the length L are plotted at the top against b, the midpoint of L. Frame 1 values, solid line; frame 2 values, dashed line; frame 3 values, dotted line. The extent of the gene (bases 70–747) is indicated by vertical dashed lines, and the horizontal arrow appears alongside the graph of the predicted frame f, showing that the gene is correctly predicted to be in frame 1.

Figures 3 and 4 illustrate how well in some cases the reading frame and positions of coding sequences in exons can be detected. The rice gene[15] for the water-stress inducible protein RAB21 is analyzed in Fig. 3. The first exon starts at base 1507 and the coding sequence at base 1599 (frame 3). This coding sequence and that in the second exon both stand out as regions of regions of low mutation away from RNY, compared with the 5′ sequence, where the mutations again correspond to those expected for a random sequence. Correct predictions for the glucoamylase gene[16] of the filamentous fungus *Aspergillus niger* are similarly made in Fig. 4. For each of the five exons the correct reading frame of the coding sequence is predicted, apart from one region in the fourth exon and for a short stretch

[15] S. K. Boyer and J. E. Mullet, *Nucleic Acids Res.* **16**, 8184 (1988).
[16] E. Boel, M. T. Hansen, I. Hjort, I. Hoegh, and N. P. Fiil, *EMBO J.* **3**, 1581 (1984).

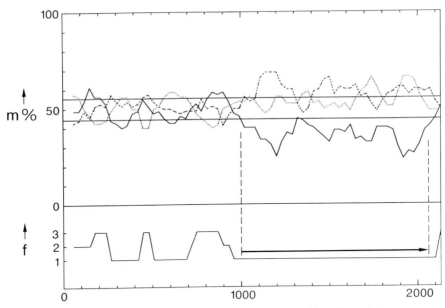

FIG. 2. SEARCH plot ($L = 120$, $S = 30$) for the barley chloroplast *psbA* gene.

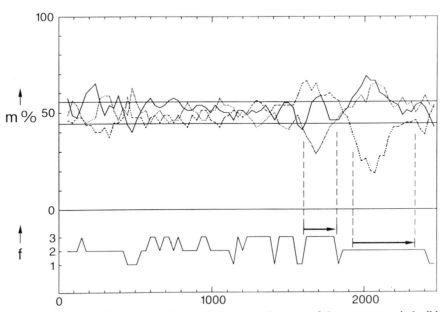

FIG. 3. SEARCH plot ($L = 120$, $S = 30$) for the rice gene of the water-stress inducible protein RAB21. It is coded in the first exon by bases 1599–1808 (frame 3) and in the second exon by bases 1892–2433 (frame 2, 1892 is the first base in codon).

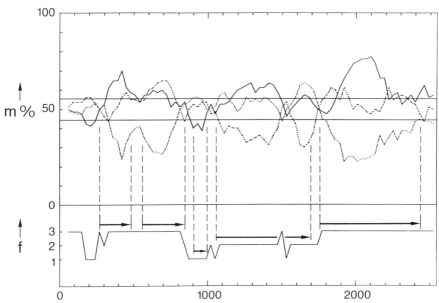

FIG. 4. SEARCH plot ($L = 120$, $S = 30$) for the *Aspergillus niger* glucoamylase (glucan 1,4-α-glucosidase) gene. The coding region base numbers are as follows: 270–483 (frame 3 in exon 1); 559–845 (frame 3 in exon 2, 559 is the second base in codon); 901–997 (frame 1 in exon 3, 901 is the first base in codon); 1059–1696 (frame 2 in exon 4, 1059 is the second base in codon); 1755–2438 (frame 3 in exon 5, 1755 is the first base in codon).

out of frame at the beginning of the gene, which may give an indication of an RNY message being added from a different origin to code for a signal peptide. Such a frameshift is also seen elsewhere, for example, in yeast mitochondrial cytochrome-*c* oxidase.[12] In general, introns appear as if any RNY messages there have been almost mutated to randomness, although sometimes the exons appear to be parts of longer regions of low mutation extending into the introns.[10] In mitochondria, some genes are known to be coded in introns of other genes and can be detected there by this method.

Printout from Search Program

An example of the output given by the above SEARCH program is shown in the Appendix. The region near the start of the coding region in the first exon of the rice RAB21 gene is investigated by moving a test length of 120 bases forward in steps of 3 bases. A change of frame from 1 to 3 is seen when the midpoint of the window is at base 1599, in agreement with the position of the ATG start signal.

As a regular procedure, it is usually advisable to survey the whole gene first in larger steps. Only in relatively few cases can the limits of the coding areas then be resolved to within a few bases by taking smaller steps. Another example is in the virus ϕX174, where an overlapping gene is similarly well defined.[2]

Discussion

In interpreting results of the analysis, it is important always to remember that the method is looking for relics of the past. Its usefulness in determining present reading frames will clearly depend on the type of nucleic acid sequence being examined. Based on the given hypothesis, one would not normally expect to see a clear long primeval message in genes that are known to have had a high mutation rate or in genes that have acquired their present use in relatively recent times, for example, in homeotic developmental control genes for multicellular organisms, which may have evolved within the last 1000 million years. Puzzling features and exceptions to the general rules will undoubtedly be seen in any complete survey. The evolution of many genes is likely to have been enormously complicated, with many examples of rearrangements through such processes as recombination, transposition, or exon reshuffling. Once a more efficient translational mechanism had enabled the comma-less coding system to be abandoned, a change of frame in part or all of the coding sequence was possible. For example, a one-base deletion would result in reading codons of the type NYR, which code more hydrophobic residues than RNY and thus immediately give a selective advantage, if hydrophobicity is required in this part of the protein. Cases where this appears to have happened are known.[10] Another localized evolutionary pressure might be for a high GC content needed for mRNA structure or other reasons. Interesting, also, are sequences where the RNY message extends beyond the limits of the present gene, and stop signals, which seem to have a more recent origin, are seen in this extension.

If these and many other considerations are kept in mind, the RNY search can give not only many indications of past history but also a surprisingly good guide to present reading frames. Taking over 2000 sequences randomly from the EMBL bank and including all types of organisms, the correct reading frame is predicted in 90% of these, if the whole length of the gene is taken as the window length. When considering well-conserved genes of ancient origin and especially those constitutively or periodically heavily expressed, including some such genes for structural proteins, detailed information about the reading frame and the extent of the gene is quickly obtained. The method also gives surprisingly good predictions for a variety of other genes, some of which are not heavily transcribed and yet the RNY is well preserved for other reasons. The presence of an open reading frame with limits coinciding with a strong RNY message can lead to the discovery of additional transcripts in a newly determined DNA sequence (of which both strands should be checked).

The method can be applied to nucleic acid sequences from all types of living organisms, whether prokaryotes, eukaryotes, viruses, or phages, and even to chloroplast and mitochondrial DNA. It needs no prior knowledge of codon frequency in the organism and merely uses purine–pyrimidine information. The program gives information about a newly determined sequence within a few minutes. This often includes not only some important indication about present function but also some evolutionary information. This search for remnants of the long distant past has been termed bioarchaeology,[17] and, stimulated by the first indications obtained from this simple program, the bioarchaeologist may continue the search and analysis by other methods, in order to attempt to unravel the evolutionary processes involved in linking the past with the present.[10,12]

```
         APPENDIX

C        SEARCH is a program to search for remnants of RNY messages
C        in DNA or RNA sequences.
C        The subroutine READSEQ requires the sequence to be given in
C        NBRF format.
C
         CHARACTER*1 A(300000)
         CALL READSEQ(A,LEN)
         CALL RNY(A,LEN)
         STOP
         END
C*****************************************************************************
         SUBROUTINE RNY(A,LEN)
C
C        This RNY subroutine searches for RNY messages in a DNA or RNA
C        sequence of length LEN contained in the character array A.
C        A sequence window of length L is moved in steps of S bases along the
C        sequence. At each position, the mutations (transversions) away from
C        an RNY message are counted in each reading frame of this window
C        and placed in the array MUTE(3).
C        The frame IFR with the least mutations away from the RNY message
C        is then found. The IFR and MUTE values are taken to correspond to the
C        midpoint of the window in the analysis (or to use for a plot).
C
         CHARACTER*1 A(300000),STAR,TEMP,FLAG(6)
         INTEGER S,MUTE(3),FIRST,LAST
         REAL VALUE(6)
C
211      WRITE(6,201)
201      FORMAT('$WINDOW LENGTH FOR SEARCH (MULTIPLE OF 3)=  ')
         READ(5,300,ERR=211)L
300      FORMAT(I10)
212      WRITE(6,202)
202      FORMAT('$STEP USED WHEN MOVING THE WINDOW (MULTIPLE OF 3)= ')
         READ(5,300,ERR=212)S
```

[17] A. Klug, *Cold Spring Harbor Symp. Quant. Biol.* **47**, 1215 (1982).

```
C          If not starting at base 1, start at multiple of 3 plus 1 if the
C          usual frame numbers are desired.
213        WRITE(6,214)
214        FORMAT('$FIRST BASE IN SEQUENCE FOR ANALYSIS (MULTIPLE OF 3 PLUS
     &     1) ')
           READ(5,300,ERR=213)FIRST
215        WRITE(6,216)
216        FORMAT('$LAST BASE IN SEQUENCE FOR ANALYSIS ')
           READ(5,300,ERR=215)LAST
C
C          Convert sequence into R and Y form. Also count NOTHER, the number
C          of bases other than C,A,G,T or U in the sequence.
           NOTHER=0
           DO 101 I=FIRST,LAST
           TEMP=A(I)
           IF((A(I).EQ.'T').OR.(A(I).EQ.'C'))A(I)='Y'
           IF((A(I).EQ.'A').OR.(A(I).EQ.'G'))A(I)='R'
           IF(A(I).EQ.'U')A(I)='Y'
           IF(A(I).EQ.TEMP)NOTHER=NOTHER+1
101        CONTINUE
C
C          Print a warning if necessary.
           IF(NOTHER.GT.0)WRITE(6,107)NOTHER
107        FORMAT(/1X,'WARNING: THERE ARE',I5,1X,'BASES OTHER THAN C,A,G,T
     &     OR U IN THE LENGTH ANALYSED'/)
C
C          Move window in steps S along the sequence and find the MUTE array
C          values.
           DO 102 I=FIRST,LEN,S
C
C          Stop if window extends beyond the end of the sequence.
           IF((I+L-1).GT.LAST)RETURN
C
C          Take each frame in turn.
           DO 103 J=1,3
C
           MUTE(J)=0
C          Take each triplet in turn.
           DO 104 K=1,L,3
C
C          NB1 and NB3 are the sequence positions of the first and third
C          positions of this triplet.
           NB1=I+J+K-2
           NB3=I+J+K
C
C          Count the transversions in these positions.
           IF(A(NB1).NE.'R')MUTE(J)=MUTE(J)+1
           IF(A(NB3).NE.'Y')MUTE(J)=MUTE(J)+1
C
104        CONTINUE
103        CONTINUE
C
C          Find minimum mutations (transversions) away from RNY message.
           MINM=MIN(MUTE(1),MUTE(2),MUTE(3))
C
C          Calculate SIGMA the standard deviation to be expected about the
C          mean, if a random sequence is analysed in the same way.
           MXMUTE=2*L/3
           SIGMA=SQRT(FLOAT(MXMUTE)*0.25)
C
C          Set FLAG(J) equal to '+' if MINM is more than J standard deviations
C          lower than mean mutation value MEANM for a random sequence.
           MEANM=L/3
           DO 109 J=1,6
           VALUE(J)=FLOAT(MEANM)-FLOAT(J)*SIGMA
           FLAG(J)=' '
           IF(FLOAT(MINM).LT.VALUE(J))FLAG(J)='+'
109        CONTINUE
```

```
C
C           Find frame with minimum MUTE value.
            IFLAG=0
            DO 106 IM=1,3
            IF(MINM.EQ.MUTE(IM)) THEN
                    IFR=IM
                    IFLAG=IFLAG+1
                    END IF
106         CONTINUE
C
C           Determine the warning STAR if 2 or 3 frames have the same
C           minimum MUTE value.
            STAR=' '
            IF(IFLAG.GT.1)STAR='*'
C           Calculate the midpoint of the window.
            MID=I+L/2-1
C           In the printout, the number of + signs after the frame IFR
C           give the number of standard deviations lower than the mean.
            WRITE(6,110)I,I+L-1,(MUTE(IP),IP=1,3),MID,IFR,
     &      (FLAG(IP),IP=1,6),STAR
110         FORMAT(1X,'TEST',I6,1X,'TO',1X,I6,2X,'MUTES=',3I5,2X,
     &      'MID=',I6,2X,'FRAME=',I2,1X,6A1,1X,A1)
102         CONTINUE
C
            RETURN
            END
C*******************************************************************

            SUBROUTINE READSEQ(A,LEN)
C
C           A sequence reader subroutine for an NBRF formatted sequence,
C           which reads the sequence into the array A and its length into LEN.
            CHARACTER*1 A(300000),BUF(132)
            CHARACTER*50 NAME
            CHARACTER*132 EMPTY
            EQUIVALENCE (EMPTY,BUF)
C
            WRITE(6,101)
101         FORMAT('$FILENAME FOR SEQUENCE IN NBRF FORMAT ')
            READ(6,102)NAME
102         FORMAT(A)
C
            OPEN(UNIT=2,FILE=NAME,STATUS='OLD',READONLY)
C
C           Set counter LEN for bases to zero.
            LEN=0
C           Set line count LINE to zero.
            LINE=0
C
C           Empty the buffer and start here to read each line.
51          EMPTY=' '
            LINE=LINE+1
            READ(2,103,ERR=111,END=999)(BUF(I),I=1,132)
103         FORMAT(132A1)
            IF(LINE.EQ.1.OR.LINE.EQ.2) THEN
                    WRITE(6,104)(BUF(I),I=1,132)
104                 FORMAT(1X,70A1)
                    GO TO 51
                    END IF
C
111         DO 105 I=1,132
            IF(BUF(I).EQ.' ')GO TO 105
C
C           * indicates end of sequence in NBRF format.
            IF(BUF(I).EQ.'*')GO TO 999
C
            LEN=LEN+1
            A(LEN)=BUF(I)
105         CONTINUE
            GO TO 51
999         WRITE(6,998)LEN
```

```
998        FORMAT(1X,'LENGTH OF SEQUENCE=',I6)
           RETURN
           END
C*********************************************************************

           TRIAL RUN OF SEARCH PROGRAMME


           RUN SEARCH
           FILENAME FOR SEQUENCE IN NBRF FORMAT
           OSRAB21.NBN
           >DL;OSRAB21

           Rice rab21 gene for water-stress inducible protein RAB21

           LENGTH OF SEQUENCE=  2537
           WINDOW LENGTH FOR SEARCH (MULTIPLE OF 3)=
           120
           STEP USED WHEN MOVING THE WINDOW (MULTIPLE OF 3)=
           3
           FIRST BASE IN SEQUENCE FOR ANALYSIS (MULTIPLE OF 3 PLUS  1)
           1510
           LAST BASE IN SEQUENCE FOR ANALYSIS
      1692
      TEST 1510 TO  1629  MUTES=  36  46  37  MID=  1569  FRAME= 1
      TEST 1513 TO  1632  MUTES=  37  47  37  MID=  1572  FRAME= 3
      TEST 1516 TO  1635  MUTES=  35  48  37  MID=  1575  FRAME= 1 +
      TEST 1519 TO  1638  MUTES=  34  49  36  MID=  1578  FRAME= 1 +
      TEST 1522 TO  1641  MUTES=  34  50  35  MID=  1581  FRAME= 1 +
      TEST 1525 TO  1644  MUTES=  34  52  36  MID=  1584  FRAME= 1 +
      TEST 1528 TO  1647  MUTES=  32  51  36  MID=  1587  FRAME= 1 +
      TEST 1531 TO  1650  MUTES=  33  52  34  MID=  1590  FRAME= 1 +
      TEST 1534 TO  1653  MUTES=  33  54  34  MID=  1593  FRAME= 1 +
      TEST 1537 TO  1656  MUTES=  31  53  36  MID=  1596  FRAME= 1 ++
      TEST 1540 TO  1659  MUTES=  33  51  35  MID=  1599  FRAME= 1 +
      TEST 1543 TO  1662  MUTES=  35  51  33  MID=  1602  FRAME= 3 +
      TEST 1546 TO  1665  MUTES=  36  52  31  MID=  1605  FRAME= 3 ++
      TEST 1549 TO  1668  MUTES=  36  53  30  MID=  1608  FRAME= 3 ++
      TEST 1552 TO  1671  MUTES=  36  53  29  MID=  1611  FRAME= 3 ++
      TEST 1555 TO  1674  MUTES=  38  54  29  MID=  1614  FRAME= 3 ++
      TEST 1558 TO  1677  MUTES=  38  54  29  MID=  1617  FRAME= 3 ++
      TEST 1561 TO  1680  MUTES=  37  53  30  MID=  1620  FRAME= 3 ++
      TEST 1564 TO  1683  MUTES=  38  53  31  MID=  1623  FRAME= 3 ++
      TEST 1567 TO  1686  MUTES=  38  51  30  MID=  1626  FRAME= 3 ++
      TEST 1570 TO  1689  MUTES=  39  51  29  MID=  1629  FRAME= 3 ++
      TEST 1573 TO  1692  MUTES=  40  52  29  MID=  1632  FRAME= 3 ++
```

Acknowledgments

Thanks are due to the European Molecular Biology Laboratory, Heidelberg, for making available their DNA data bank, and to the National Biomedical Research Foundation, Washington, D.C., for their PIR software, to R. Doelz for arranging the use of these facilities on our VAX 8830 computer, and to T. Bickle for useful discussions. Acknowledgment is also made of the support given by the Roche Research Foundation.

[12] Searching for Patterns in Protein and Nucleic Acid Sequences

By RODGER STADEN

Introduction

There are many different reasons why we may wish to define and search for patterns of motifs in protein or nucleic acid sequences, and also numerous starting points. We might start from a known or presumed three-dimensional (3D) structure, perhaps of some particular fold or binding site, write down a one-dimensional representation using what are believed to be the critical residues, and search to see if similar patterns exist in other sequences. The aim would be to characterize the 3D structure and use it to propose the function of new proteins whose crystal structure has not been determined. Alternatively we might start from an alignment of similar sequences (none of which has a solved crystal structure), and proceed to write down a representation of these aligned sequences in a form that enables a search for other similar sequences. Similarly, DNA sequences may contain a loosely defined pattern of bases within the control regions of related genes, and we may wish to know whether this pattern is present in equivalent positions in other genes. Finding such matches would be further evidence in favor of the pattern being functional. Whenever any pattern is established as defining a region of interest, it is useful to be able to store it in some standard way so that it can easily be used to search other sequences.

There is a rapidly growing number of well-established patterns, especially in nucleic acid sequences, and some readers may wish only to know how to search for these. This chapter, however, describes a set of programs[1] that not only perform searches for known patterns but which also enable users to define their own patterns. The patterns can be defined in many different ways, in both nucleic acid and protein sequences, and the search programs operate on individual sequences as well as whole libraries of sequences.

When families of sequences are aligned, gaps are introduced to bring similar segments into register. The final alignment will be free of gaps in all sequences over some sections, while other regions will contain gaps of varying sizes. A pattern that represented the whole sequence family would

[1] R. Staden, *CABIOS* **4**, 53 (1988).

METHODS IN ENZYMOLOGY, VOL. 183

need to describe the similar sections that were free of gaps, and allow for the variation in spacings between them. So patterns are characterized by segments of fixed length interspersed by gaps of variable length. The similar segments need to be defined in ways that depend on their degree of similarity and on the type of sequence. In addition, there are other similarities between sequences where the individual residues are not conserved but instead some pattern of interaction is common, for example, the base pairing in tRNAs or rRNAs or the ability to form a bend in a protein. These considerations lead us to the following definitions.

A MOTIF is a continuous segment of sequence; it can be defined in many ways. A PATTERN is a higher order of structure. It comprises a list of motifs, states how each is described, and defines the allowed ranges of spacing between them. In addition it specifies, using the logical operators AND, OR, and NOT, whether each motif must be present, is an alternative to another, or must be absent. To cater for both the wide variety of motif types and the differing approaches of users, we currently employ the methods shown in Fig. 1 for defining motifs.[1,2] The way a motif is defined is called its CLASS. In Fig. 1 classes 1 to 4, 8, and 9 apply to protein sequences, and classes 1 to 8 to nucleic acids.

In previous publications we have shown how the programs can be used to define and locate a eukaryotic promoter consisting of a CCAAT box, a TATA box, and a "capsite" sequence, as well as some unusual nematode mitochondrial tRNAs and globin protein sequences. Here we give further examples but mostly concentrate on how the programs are used.

Range Setting and Logical Operators

When motifs are combined into a pattern, their ranges of relative separations are defined. Presently a pattern can contain up to 50 different motifs, but let us suppose we had four motifs A, B, C, and D combined by the AND operator. Motif A can occur anywhere in the sequence, but the allowed positions for B must be defined relative to the position of motif A. However, the positions for motif C can be defined relative to A or B, and those for D relative to A, B, or C. That is, positions are defined relative to any motif further up the list.

As motifs are included into a pattern they are assigned consecutive numbers, beginning with 1. Ranges are defined using three values: a number, a start distance, and a length. For example, a range definition of 5, 20, 15 means that the range starts 20 residues to the right of motif 5, and

[2] M. O. Dayhoff, R. M. Schwartz, and B. C. Orcutt, in "Atlas of Protein Sequence and Structure" (M. O. Dayhoff, ed.), Vol. 5, Suppl. 3, p. 345. National Biomedical Research Foundation, Washington, D.C., 1978.

CLASS   DESCRIPTION

1   Exact match to a short defined sequence. The IUB symbols can be used for DNA sequences.

2   Percentage match to a defined short sequence. In nucleic acids, the IUB symbols can be used.

3   Match to a defined sequence, using a score matrix and a cutoff score. The DNA matrix[1] gives scores to IUB symbols depending on the amount of overlap between them. MDM78[2] is used for proteins.

4   Match to a weight matrix (see below) with a cutoff score. Scores can be added or multiplied (by adding their logarithms).

5   As class 4 but on the complementary strand.

6   Inverted repeat or stem-loop. Fixed stem length, range of loop sizes, and cutoff score using A-T, G-C=2; G-T=1.

7   Exact match to short sequence but with a defined step size.

8   Direct repeat. Fixed repeat length, range of gap sizes, cutoff score. For protein sequences scores are calculated using MDM78, but an identity matrix is used for nucleic acids.

9   Membership of a set. A list of sets of allowed amino acids for each position in the motif. When they are first entered into the computer they can be typed in with the sets separated by commas. For example IVL,,,DEKR,FYWILVM defines a motif of length 5 residue positions in which one of I,V, or L must be found in the first position, then anything in the next two; D,E,K,or R in the fourth position and F,Y,W,I,L,V, or M in the fifth. Alternatively the sets can be written in the form of a weight matrix, which is how the programs will store them. This class only applies to protein sequences because for nucleic acids "membership of a set" can be achieved using IUB symbols.

FIG. 1. Current list of available motif classes.

extends for a further 15 residues. If the start distance was -20 it would signify 20 residues to the left of motif 5.

As motifs are included into a pattern their logical operator AND, OR, or NOT is specified. Notice that these operate on the whole motif and are separate from the logical operators that are applied implicitly within a motif. For example, a DNA motif might be a consensus sequence that includes the symbol Y meaning T or C, but the motif may be included into the pattern using AND, OR, or NOT.

Motif and Pattern Probabilities

Searches for individual motifs often give a high background of false matches, but when they are combined together with limitations on separations, the number of false positives can be greatly reduced. In specifying a pattern we can often include quite poorly defined elements that are nevertheless important in a particular context. Also, by giving low cutoff scores for each motif we can allow good matches to some motifs, to compensate for poor ones with others. That is, it is the level of match to the pattern, not to the individual motifs, which is important. To achieve this we need low scores for individual motifs but a sufficiently high overall score. If a pattern were formed from motifs that were all of the same class we would be able to write down a suitable overall cutoff score, because all the motif scores would be defined in the same way, and on the same scale. But, if the pattern contains motifs defined using different classes, their scores cannot be combined because they are, in general, defined on different scales.

To overcome this problem, and allow sensible overall cutoffs to be applied, we use probability to normalize the motif scores. When the user defines a motif together with its cutoff score, the program immediately calculates the probability of achieving at least this score. When the whole pattern has been defined, the user can then specify an overall probability cutoff. During a search, the program calculates the probability of achieving the found level of score for each potential match to the pattern, and only if the probability is lower (i.e., the match is less likely) than the given cutoff will it be declared a match. Before the user supplies the probability, the program will give an estimate of the expected number of matches. The probabilities are calculated using probability-generating functions. The calculations are slightly different for each of the nine motif classes, but below we outline the functions used for class 4, the weight matrix.

All equations for motifs assume that the sequences are random and sufficiently long that the probability of finding any base or amino acid is simply equal to its frequency in the particular sequence being analyzed. All equations for patterns assume that the motifs in a pattern are independent and that the probability of finding a motif at position y is independent of the probability of finding it at $y - 1$.

We define the following symbols: character set size, k; sequence composition as frequencies, f_i, where i equals $1, k$; weight matrix length, J; weight matrix weights, w_{ij}, $0 \leq w_{ij} \leq T$. The probability generating function for column j of the weight matrix is

$$Gj(x) = \sum f_i x^{w_{ij}}$$

The probability-generating function for the whole matrix is

$$F(x) = \prod \sum f_i x^{w_{ij}} \qquad (i = 1, k; j = 1, J)$$

and the probability of obtaining score S_N for the weight matrix is given by the coefficient of x^N in this equation.

If a pattern consists of m motifs with ranges r_m and probabilities p_m, then the probability for the pattern P is given by $P = \prod p_m$. If the motifs are all combined using the AND operator, the expected number of matches is $E = \prod p_m r_m$. If the OR operator is used, $E = \sum p_m r_m$, and for the NOT operator $E = (1 - p_m)^{r_m}$. In practice all three operators will be employed together, and so these last three relationships are used as appropriate.

Defining Individual Motifs

Table I shows an alignment of 21 sequences thought to contain the helix–turn–helix motif,[3] and Table II gives the frequencies for each amino acid at each position. How would the methods described here be used to define and search for further occurrences in other sequences? First, there are no gaps in the alignment so our pattern can consist of a single motif, but which motif classes are suitable?

The consensus sequence is QRELAKKLGVSQSTVSRWVN. We could use it as class 1 but would be very unlikely to find any matches, and none of those in the list match. We could rewrite the consensus to include only the most highly conserved residues, say A---G, where - matches any residue type, but clearly it would still be unsatisfactory because we would get too many false matches. As a class 2 motif we could ask for, say, a score of 15, hence allowing any 5 of the 20 residues to mismatch, and might be more successful. In several positions in the consensus, however, other residues are equally as frequent as the ones chosen, or several similar residues are almost as common. Also, we do not know how representative of the helix–turn–helix motif this small collection of sequences is. So the first sensible motif class to use for this problem would be class 3, where we employ a score matrix in conjunction with the consensus sequence, which hence allows for residues similar to those in the consensus to match (and also those which are dissimilar to diminish the score effectively). When we apply this method to search a sequence the consensus sequence is effectively slid along it, one position at a time, in the normal way, but instead of

---

[3] C. O. Pabo and R. T. Sauer, *Annu. Rev. Biochem.* **53**, 293 (1984).

TABLE I
ALIGNMENT OF SEQUENCES CONTAINING THE HELIX–TURN–HELIX MOTIF

	Sequence Description
QESVADKMGMGQSGVGALFN	LAMBDA.REP
QTKTAKDLGVYQSAINKAIH	LAMBDA.CRO
QAALGKMVGVSNVAISQWQR	P22.REP
QRAVAKALGISDAAVSQWKE	P22.CRO
QAELAQKVGTTQQSIEQLEN	434.REP
QTELATKAGVKQQSIQLIEA	434.CRO
RQEIGQIVGCSRETVGRILK	CAP
RGDIGNYLGLTVETISRLLG	Fnr
LYDVAEYAGVSYQTVSRVVN	LAC.R
IKDVARLAGVSVATVSRVIN	GAL.R
TEKTAEAVGVDKSQISRWKR	LAMBDA.CII
QRKVADALGINESQISRWKG	P22.CI
KEEVAKKCGITPLQVRVWCN	MAT.ALPHA
TRKLAQKLGVEQPTLYWHVK	TETR.TN10
TRRLAERLGVQQPALYWHFK	TETR.pSC1
QRELKNELGAGIATITRGSN	TRP.REP
RQQLAIIFGIGVSTLYRYFP	H-INVERSN
ATEIAHQLSIARSTVYKILE	TN3.RESOL
ASHISKTMNIARSTVYKVIN	GD.RESOLV
IASVAQHVCLSPSRLSHLFR	ARA.C
RAEIAQRLGFRSPNAAEEHL	LEX.R

asking if the residues it aligns with are identical, we look up their similarity score in the MDM78 score matrix. If the scores summed over the whole length of the consensus exceed some predetermined cutoff, we announce a match. In choosing a cutoff score we would need to select one that was exceeded by all the sequences in Table I. For this kind of problem, though, class 4, the weight matrix, is generally the most suitable.

A weight matrix is simply a table of scores (or weights) for finding each residue type in each position of the motif. The scores represent the relative importance of each residue type at each position, and, unlike MDM78, a weight matrix allows each residue type to have different scores at every position in the motif. For the example given here the simplest set of scores we could use would be those in Table II. To work out the score for any piece of sequence of this length (20 amino acids) we take the 20 relevant residue scores and either add them or multiply them. As an example, if we found the consensus sequence, and if we were adding, we would get $8 + 5 + 7 + 7 + 16 + 5 + 9 + 18 + 8 + 6 + 6 + 8 + 9 + 8 + 8 + 8 + 5 + 4 + 7 = 152$. Obviously, this is the highest possible score for this motif using these weights, and again we would need to choose a cutoff score

TABLE II
Amino Acid Frequencies for Each of 20 Residue Positions of Aligned Helix–Turn–Helix Sequences[a]

Amino acid	Frequency at position																			
	1	2	3	4	5	6	7	8	9	10	11	12	13	14	15	16	17	18	19	20
C	0	0	0	0	0	0	0	1	1	1	0	0	0	0	0	0	0	0	1	0
S	0	1	2	0	1	0	0	0	1	0	6	1	8	2	0	8	0	0	1	0
T	3	3	0	2	0	1	1	0	0	1	3	0	0	9	0	1	0	0	0	0
P	0	0	0	0	0	0	0	0	0	0	0	2	3	0	0	0	0	0	0	1
A	2	4	2	0	16	0	3	3	0	1	2	0	3	4	1	1	1	1	0	1
G	0	1	0	0	3	0	0	0	18	0	3	0	0	1	0	2	0	1	0	2
N	0	0	0	0	0	2	0	0	1	0	1	1	0	1	0	1	0	0	0	7
D	0	0	3	0	0	2	1	0	0	0	1	1	0	0	0	0	0	0	0	0
E	0	3	7	0	0	3	1	0	0	0	1	1	2	0	0	1	1	1	2	2
Q	8	2	1	0	0	5	1	0	0	0	1	6	3	3	0	1	3	0	1	0
B	0	0	0	0	0	0	0	0	0	0	0	0	0	0	0	0	0	0	0	0
Z	0	0	0	0	0	0	0	0	0	0	0	0	0	0	0	0	0	0	0	0
H	0	0	1	0	0	1	1	0	0	0	0	0	0	0	0	0	1	2	1	1
R	4	5	1	0	0	1	2	0	0	0	1	3	0	1	0	1	8	0	0	3
K	1	1	4	0	1	5	5	0	0	0	1	1	0	0	0	0	3	0	3	3
M	0	0	0	0	0	0	1	2	0	1	0	0	0	0	0	0	0	0	0	0
I	2	0	0	5	0	1	2	0	0	6	0	1	0	0	8	0	0	3	3	0
L	1	0	0	7	0	0	1	9	0	2	0	0	1	0	4	0	1	4	3	1
V	0	0	0	7	0	0	0	5	0	8	0	3	1	0	8	0	1	3	2	0
F	0	0	0	0	0	0	0	1	0	1	0	0	0	0	0	0	0	0	4	0
Y	0	1	0	0	0	0	2	0	0	0	1	1	0	0	0	5	0	1	0	0
W	0	0	0	0	0	0	0	0	0	0	0	0	0	0	0	0	2	5	0	0

[a] As shown in Table I.

that was low enough to accommodate those achieved by the sequences in Table I.

Suppose we had a file containing the alignment of helix–turn–helix sequences shown in Table I and wanted to use it to define a pattern that could be applied to other sequences. For a class 4 motif we need to calculate suitable weights and a cutoff score and to write a file that defines the motif for a search program. A program called WEIGHTS is used for this purpose, and the input data are a sequence alignment exactly as shown in Table I. A copy of the dialogue produced when using WEIGHTS for this problem is shown in Fig. 2, and we now explain the steps. Note that lines containing questions to which the user must respond are preceded by a question mark.

Once started, WEIGHTS needs to know if we want to process nucleic acid or protein sequences, and then asks for the name of the file containing the aligned sequences (here HTH.SEQ). It then offers the opportunity to

```
WEIGHTS   V2.0   AUTHOR: RODGER STADEN
MANAGES WEIGHT MATRICES FOR DNA AND PROTEIN MOTIFS.
SETS OF ALIGNED SEQUENCES CAN BE USED TO CREATE NEW
MATRICES, OR FOR RESCALING EXISTING MATRICES.

  ? FOR PROTEIN SEQUENCES TYPE 1   1
  ? NAME OF ALIGNED SEQUENCES FILE = HTH.SEQ
  ? TO USE AN EXISTING WEIGHT MATRIX TYPE 1
  ? TO ADD WEIGHTS TYPE 1
  ? TO APPLY A MASK TYPE 1

APPLYING WEIGHTS TO INPUT SEQUENCES

      1 QESVADKMGMGQSGVGALFN LAMBDA.REP -44.27
      2 QTKTAKDLGVYQSAINKAIH LAMBDA.CRO -44.45
      3 QAALGKMVGVSNVAISQWQR P22.REP     -44.62
      4 QRAVAKALGISDAAVSQWKE P22.CRO     -40.93
      5 QAELAQKVGTTQQSIEQLEN 434.REP     -42.31
      6 QTELATKAGVKQQSIQLIEA 434.CRO     -45.52
      7 RQEIGQIVGCSRETVGRILK CAP         -44.33
      8 RGDIGNYLGLTVETISRLLG Fnr         -44.73
      9 LYDVAEYAGVSYQTVSRVVN LAC.R       -42.82
     10 IKDVARLAGVSVATVSRVIN GAL.R       -42.53
     11 TEKTAEAVGVDKSQISRWKR LAMBDA.CII  -42.85
     12 QRKVADALGINESQISRWKG P22.CI      -40.97
     13 KEEVAKKCGITPLQVRVWCN MAT.ALPHA   -46.05
     14 TRKLAQKLGVEQPTLYWHVK TETR.TN10   -41.86
     15 TRRLAERLGVQQPALYWHFK TETR.pSC1   -44.06
     16 QRELKNELGAGIATITRGSN TRP.REP     -46.10
     17 RQQLAIIFGIGVSTLYRYFP H-INVERSN   -45.44
     18 ATEIAHQLSIARSTVYKILE TN3.RESOL   -45.25
     19 ASHISKTMNIARSTVYKVIN GD.RESOLV   -48.59
     20 IASVAQHVCLSPSRLSHLFR ARA.C       -47.17
     21 RAEIAQRLGFRSPNAAEEHL LEX.R       -50.62

TOP AND BOTTOM OBSERVED SCORES, MEAN AND STANDARD DEVIATION
     -40.940      -50.625      -44.551         2.345
  MEAN + AND - 1*SD =   -42.20540       -46.89582
  MEAN + AND - 2*SD =   -39.86020       -49.24102
  MEAN + AND - 3*SD =   -37.51498       -51.58624
THE MEAN + AND - 3*SD WILL BE USED AS A CUT-OFF
YOU CAN CHANGE IT BY TYPING 1 NOW
YOU CAN GIVE YOUR MOTIF A TITLE OF < 60 CHARACTERS
  ? TITLE=HTH 13-01-1989
  ? FILE NAME FOR NEW WEIGHT MATRIX =HTH.WTS
```

FIG. 2. Dialogue produced when using the program WEIGHTS to process a set of aligned sequences containing the helix–turn–helix motif.

read in an existing weight matrix, which in this case we decline, and the program reads the sequence alignment in Table I. It immediately calculates the residue frequencies as shown in Table II. We can add residue frequencies or add their logarithms, the latter (as here) being the usual choice. We can also apply a mask to the alignment, which enables us to only use certain column positions to define the motif. If we decide that certain positions were quite arbitrary, but which because of the probable unrepresentative nature of the small subset of sequences we are using, could cause false matches to be found, we could mask them out. The mask is defined by the user typing a string of hyphens (-) and Xs, so that -X--XXX- means exclude positions 1, 3, 4, and 8 but use all others. Here we elect to use all positions. The program then works out the weights and applies them to each of the input sequences, hence producing a set of expected scores for sequences known to contain the motif. It also calculates the mean and standard deviation of these scores. Note that such a display also enables us to see how close each of the sequences is to the average, and possibly to remove any whose scores are exceptionally low. After these scores are displayed the user is given the opportunity to select a cutoff score for use when the motif is applied to search other sequences. In this example we have declined, and the program will automatically use the mean score minus 3 standard deviations. The user supplies a title and a file name, and the program then writes a file that contains all the information a search program would need for defining this motif as class 4.

When we then use the search program we are asked which motif class we wish to use. If we select class 4 the program will ask for the name of the file containing the appropriate weight matrix (here HTH.WTS), and this supplies all the necessary information. A cutoff of -50.7, i.e., just low enough to find all the original sequences, has a probability of 0.22×10^{-5} and finds 68 matches in version 8 of the SWISS-PROT[4] protein sequence library. A cutoff of the mean score minus 3 standard deviations finds 86 matches.

The final motif class suitable for this problem is 9, membership of a set. Here we give a set of allowed residue types for each position in the motif and a minimum score, which is the minimum number of residue positions within a motif that we require to be matched by a sequence. Ignoring some of the less well-conserved positions we might use QR,,NDEQHRK, ILV,AG,NDEQRKH,,AMLV,G,MILVF,STAG,NDEQRK,S,STA,ILV, STAG,HRKNDE. This means Q or R in the first position, anything in the next, N, D, E, Q, H, R, or K in the next, and so on. Notice that in some

[4] SWISS-PROT, Bairoch A, Dept. de Biochimie Medicale, Centre Medical Universitaire, 1211 Geneva 4, Switzerland.

positions we have omitted some residues types that appear in Table I, and that in others we have included some that do not appear in Table I but which look compatible, such as N in the third position. We might try this list with a score of 11, hence allowing 4 mismatches. Note that the search programs can read membership of a set motifs in two formats: the user can type them as above, with the allowed residues and commas, or a file identical to that for a weight matrix can be used, with only the cutoff score being changed. So Table II could also be used directly to define this motif as class 9. Used as it stands, it would allow any nonzero residue type at each position to count as a match, but it could be edited to count only those we have just listed.

If we use Table II as input, hence allowing all nonzero residue types at each position to count as a match, together with a cutoff score of 20, i.e., permitting no mismatches but capable of finding all the sequences in Table I, the program calculates a probability of 0.382×10^{-7} and finds 26 matches in SWISS-PROT. For a cutoff score of 19, i.e., permitting one mismatch, the program calculates a probability of 0.115×10^{-5} and finds 37 matches in SWISS-PROT. In the light of searches of the library, we might refine our definition of helix–turn–helix, edit the weight matrix accordingly, using a word processor, then use WEIGHTS again to apply it to the original set of aligned sequences to obtain a new set of expected scores, and repeat the search.

Defining a Pattern Containing Several Motifs

The purpose of this section is to illustrate the dialogue when a user defines a quite complicated pattern made up of several motifs. The pattern chosen is for tRNA genes and consists of the four conserved stem loops and several of the conserved bases. The actual parts of the tRNA structure used, and their corresponding nine motif numbers, are shown in Fig. 3. The stems selected are shown by lines drawn parallel to their axes, and the conserved bases used are in uppercase letters. All others are shown as lowercase n. For example, motif 1 is the aminoacyl stem, motif 6 is the pair of conserved bases YT in the anticodon loop, and motif 7 is the position in the anticodon where T does not occur (hence the IUB symbol V, meaning not T). The dialogue with the user is shown in Fig. 4.

The procedure starts with the program offering the eight different motif classes available for nucleic acid sequences. Again questions to the user are preceded by a question mark. The user has selected class 6, a stem loop for the first motif, and the program describes the scoring system, and requests the stem length. The user defines the aminoacyl stem and loop, which has 7 base pairs in the stem, and, ignoring the possibility of an intron, the loop

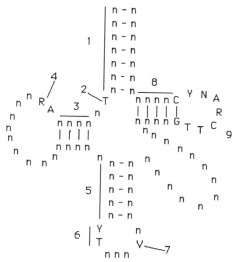

FIG. 3. Diagram representing the tRNA cloverleaf. The positions of the nine motifs chosen to define the structure as a search pattern are marked.

can vary from 56 to 80 bases. A minimum score of 12 is asked for, and the program responds with the probability of finding such a score (0.166×10^{-2}), before again offering the eight motif classes for the user to select from.

This time the user chooses class 1, and the program asks to which motif the relative positions of this second motif should be defined. The user responds with motif 1, and the program states that motif 1 is a stem loop, and, as such, positions can be defined relative to either its 5' or 3' ends. Here the 5' end is chosen, with the range starting 7 bases away and having a length of 1. The program informs the user that the string buffer has space for 210,000 more characters (it is empty at this stage) and asks for the string of base types to be entered followed by an @ symbol as a terminator. The user types T@, and the program gives the probability as 0.25. Note that this dialogue is from the version of the program that searches libraries of sequences, and it assumes an even base composition; however, the programs that search individual sequences calculate probabilities which depend on the particular composition. To save space, the next section of the dialogue dealing with the definition of motifs 3 to 8 has been omitted, and we continue with the definition of the ninth motif. Here the user has chosen class 2, goes through similar dialogue as above, and then types the string GTTCRANYC@ and asks for a score of 8, hence allowing one mismatch. The program calculates the probability to be 0.128×10^{-2}.

```
CLASSES OF MOTIF ARE:
 0 = NO MORE MOTIFS
 1 = EXACT MATCH
 2 = PERCENTAGE MATCH
 3 = CUT-OFF SCORE AND SCORE MATRIX
 4 = CUT-OFF SCORE AND WEIGHT MATRIX
 5 = COMPLEMENT OF WEIGHT MATRIX
 6 = INVERTED REPEAT, OR STEM-LOOP
 7 = EXACT MATCH, DEFINED STEP SIZE
 8 = DIRECT REPEAT.
     PRECEDE THE CLASS NUMBER WITH A,O OR N TO SPECIFY
     THE LOGICAL OPERATOR.  FOR EXAMPLE: A4, O4 OR N4.
? OPERATOR AND CLASS FOR MOTIF   1 = 6
A CLASS 6 MOTIF IS AN INVERTED REPEAT
THE 5 PRIME START POSITIONS HAVE ALREADY BEEN DEFINED
NOW WE NEED THE RANGE OF LOOP SIZES, THE REPEAT OR STEM
LENGTH AND CUTOFF SCORE  (G-C = 2, A-T = 2, G-T = 1)
? STEM LENGTH = 7
? MINIMUM LOOP SIZE = 56
? MAXIMUM LOOP SIZE = 80
? CUTOFF SCORE =12
PROBABILITY OF SCORE     12.0000 = 0.166E-02

CLASSES OF MOTIF ARE:
 0 = NO MORE MOTIFS
 1 = EXACT MATCH
 2 = PERCENTAGE MATCH
 3 = CUT-OFF SCORE AND SCORE MATRIX
 4 = CUT-OFF SCORE AND WEIGHT MATRIX
 5 = COMPLEMENT OF WEIGHT MATRIX
 6 = INVERTED REPEAT, OR STEM-LOOP
 7 = EXACT MATCH, DEFINED STEP SIZE
 8 = DIRECT REPEAT.
     PRECEDE THE CLASS NUMBER WITH A,O OR N TO SPECIFY
     THE LOGICAL OPERATOR.  FOR EXAMPLE: A4, O4 OR N4.
? OPERATOR AND CLASS FOR MOTIF   2 = 1
RANGES ARE DEFINED RELATIVE TO OTHER MOTIFS
AND BY STARTS AND DISTANCES.
? NUMBER OF REFERENCE MOTIF = 1
MOTIF NUMBER        1 IS A BASEPAIRED STEM
TO DEFINE POSITIONS RELATIVE TO ITS 3 PRIME
END TYPE 3, ELSE TYPE CARRIAGE RETURN ONLY
? TYPE 3 OR 0
? START DISTANCE FROM 5 PRIME END =7
? RANGE (INCLUDING LENGTH OF MOTIF) = 1
MAXIMUM STRING LENGTH IS 210000
TYPE IT NOW. FINISH WITH AN @
T@

PROBABILITY OF SCORE     1.0000 = 0.250E+00
  .
  .
  .   Missing dialogue here, skip to the 9th motif
  .
```

FIG. 4. Dialogue produced when using a program to define a tRNA cloverleaf as a search pattern. Only the dialogue for motifs one, two, and nine is shown.

```
CLASSES OF MOTIF ARE:
 0 = NO MORE MOTIFS
 1 = EXACT MATCH
 2 = PERCENTAGE MATCH
 3 = CUT-OFF SCORE AND SCORE MATRIX
 4 = CUT-OFF SCORE AND WEIGHT MATRIX
 5 = COMPLEMENT OF WEIGHT MATRIX
 6 = INVERTED REPEAT, OR STEM-LOOP
 7 = EXACT MATCH, DEFINED STEP SIZE
 8 = DIRECT REPEAT.
     PRECEDE THE CLASS NUMBER WITH A,O OR N TO SPECIFY
     THE LOGICAL OPERATOR.  FOR EXAMPLE: A4, O4 OR N4.
? OPERATOR AND CLASS FOR MOTIF   9 = 2
RANGES ARE DEFINED RELATIVE TO OTHER MOTIFS
AND BY STARTS AND DISTANCES.
? NUMBER OF REFERENCE MOTIF = 8
MOTIF NUMBER      8 IS A BASEPAIRED STEM
TO DEFINE POSITIONS RELATIVE TO ITS 3 PRIME
END TYPE 3, ELSE TYPE CARRIAGE RETURN ONLY
? TYPE 3 OR 0
? START DISTANCE FROM 5 PRIME END =4
? RANGE (INCLUDING LENGTH OF MOTIF) = 9
A CLASS 2 MOTIF IS DEFINED BY A SHORT SEQUENCE
(WHICH MAY INCLUDE IUB SYMBOLS)
AND A CUTOFF SCORE DEFINED AS A MINIMUM NUMBER OF MATCHES
MAXIMUM STRING LENGTH IS 209994
TYPE IT NOW. FINISH WITH AN @
GTTCRANYC@

? CUTOFF SCORE =8
PROBABILITY OF SCORE     8.0000 = 0.128E-02
```

FIG. 4.

To terminate the definition of the pattern the user selects a motif of class 0, and the program responds by writing a description of the whole pattern (shown in Fig. 5) so that it can be checked by the user. After this the probability for the whole pattern (0.113×10^{-9}) is displayed along with the expected number of matches. The user can apply an overall probability cutoff if required.

Dialogue such as this enables the user to define accurately complex patterns without needing to worry about any rigid format: the program asks for the information it needs, in the order it wants it. However, it would be tedious to have to go through such a dialogue every time a pattern is used or, indeed, if only a small change to an existing pattern were required. For these reasons the programs have the facility to store patterns as simple annotated disk files. The programs can read patterns from disk files or from the keyboard as shown above. Figure 6 shows the file for the tRNA pattern just defined. If necessary, users can employ a word processor to edit these files and change the pattern. Notice that when our pattern was

```
THIS PATTERN IS DESCRIBED BY THE FOLLOWING MOTIFS


MOTIF    1 IS OF CLASS       6
WHICH IS A STEM-LOOP STRUCTURE WITH STEM LENGTH    7 AND SCORE     11.
THE LOOP CAN HAVE SIZES       56 TO       80

MOTIF    2 IS OF CLASS       1
WHICH IS AN EXACT MATCH TO
T
AND THE 5 PRIME END CAN START BETWEEN DISTANCES       7 AND  7     OF THE    5 PRIME END OF
MOTIF    1
IT IS ANDED WITH THE PREVIOUS MOTIF

MOTIF    3 IS OF CLASS       6
WHICH IS A STEM-LOOP STRUCTURE WITH STEM LENGTH    4 AND SCORE      5.
THE LOOP CAN HAVE SIZES        4 TO       13
AND THE 5 PRIME STEM CAN START BETWEEN DISTANCES      9 AND 9     OF THE    5 PRIME END OF
MOTIF    1
IT IS ANDED WITH THE PREVIOUS MOTIF

MOTIF    4 IS OF CLASS       1
WHICH IS AN EXACT MATCH TO
AR
AND THE 5 PRIME END CAN START BETWEEN DISTANCES      13 AND 13    OF THE    5 PRIME END OF
MOTIF    1
IT IS ANDED WITH THE PREVIOUS MOTIF

MOTIF    5 IS OF CLASS       6
WHICH IS A STEM-LOOP STRUCTURE WITH STEM LENGTH    5 AND SCORE     8.
THE LOOP CAN HAVE SIZES        7 TO        7
AND THE 5 PRIME STEM CAN START BETWEEN DISTANCES      2 AND 2     OF THE    3 PRIME END OF
MOTIF    3
IT IS ANDED WITH THE PREVIOUS MOTIF

MOTIF    6 IS OF CLASS       1
WHICH IS AN EXACT MATCH TO
YT
AND THE 5 PRIME END CAN START BETWEEN DISTANCES      5 AND  5    OF THE    5 PRIME END OF
MOTIF    5
IT IS ANDED WITH THE PREVIOUS MOTIF

MOTIF    7 IS OF CLASS       1
WHICH IS AN EXACT MATCH TO
V
AND THE 5 PRIME END CAN START BETWEEN DISTANCES      10 AND 10 OF THE    5 PRIME END OF
MOTIF    5
IT IS ANDED WITH THE PREVIOUS MOTIF

MOTIF    8 IS OF CLASS       6
WHICH IS A STEM-LOOP STRUCTURE WITH STEM LENGTH    5 AND SCORE     8.
THE LOOP CAN HAVE SIZES        6 TO        7
AND THE 5 PRIME STEM CAN START BETWEEN DISTANCES      4 AND 18    OF THE    3 PRIME END
OF MOTIF    5
IT IS ANDED WITH THE PREVIOUS MOTIF

MOTIF    9 IS OF CLASS       2
WHICH IS A MATCH OF SCORE      8. TO
GTTCRANYC
AND THE 5 PRIME END CAN START BETWEEN DISTANCES      4 AND  4    OF THE    5 PRIME END OF
MOTIF    8
IT IS ANDED WITH THE PREVIOUS MOTIF
```

FIG. 5. Program description of the tRNA search pattern.

A6	CLASS
7	LENGTH
56	MINIMUM LOOP
80	MAXIMUM LOOP
11.00000	CUTOFF
A1	CLASS
1	RELATIVE MOTIF
5	RELATIVE END
7	RANGE START
1	RANGE LENGTH
T	
@ END OF STRING	
A6	CLASS
1	RELATIVE MOTIF
5	RELATIVE END
9	RANGE START
1	RANGE LENGTH
4	LENGTH
4	MINIMUM LOOP
13	MAXIMUM LOOP
5.00000	CUTOFF
A1	CLASS
1	RELATIVE MOTIF
5	RELATIVE END
13	RANGE START
2	RANGE LENGTH
AR	
@ END OF STRING	
A6	CLASS
3	RELATIVE MOTIF
3	RELATIVE END
2	RANGE START
1	RANGE LENGTH
5	LENGTH
7	MINIMUM LOOP
7	MAXIMUM LOOP
8.00000	CUTOFF
A1	CLASS
5	RELATIVE MOTIF
5	RELATIVE END
5	RANGE START
2	RANGE LENGTH
YT	
@ END OF STRING	
A1	CLASS
5	RELATIVE MOTIF
5	RELATIVE END
10	RANGE START
1	RANGE LENGTH
V	
@ END OF STRING	
A6	CLASS
5	RELATIVE MOTIF
3	RELATIVE END
4	RANGE START
15	RANGE LENGTH
5	LENGTH
6	MINIMUM LOOP
7	MAXIMUM LOOP
8.00000	CUTOFF
A2	CLASS
8	RELATIVE MOTIF
5	RELATIVE END
4	RANGE START
9	RANGE LENGTH
GTTCRANYC	
@ END OF STRING	
8.00000	CUTOFF

FIG. 6. tRNA pattern as stored in a file.

defined at the keyboard we always took the default logical operator AND (i.e., no operator was specified, and so AND was assumed), but in the pattern file the operator is defined explicitly by the letter A, which precedes the motif class.

Everything described thus far is common to searches of both individual sequences and libraries, but below we outline the differences between the two types of programs. There are four programs: one for single nucleic acid sequences, one for libraries of nucleic acid sequences, and an equivalent pair for protein sequences. First we deal with the programs that operate on single sequences.

Searching Individual Sequences

The program that operates on single nucleic acid sequences (ANALY-SEQ) currently contains 66 options besides the one for pattern searching. Similarly, the equivalent protein program (ANALYSEP) currently contains 28 options. The programs produce graphical output for many analyses, the idea being to allow users to superimpose the results of different analyses into a single diagram. Here, however, we are concerned only with pattern searching.

The programs can read sequences stored in all the major formats: EMBL, GenBank, and PIR, and also simple text files. For convenience we convert all libraries to PIR format. We run the programs on VAX computers, but they are easily moved to other machines. Currently we are using a Macintosh terminal with a terminal emulator,[1] but many other terminals are also capable of handling both the text and graphics output.[1]

The choices for output for the pattern searches are (1) motif by motif, (2) inclusive, and (3) graphics. For any pattern match, output 1 simply lists out the sections of sequence that match each constituent motif. Base-paired stems and repeats are shown with their two component halves aligned. Output 2 lists out all the sequence between, and including, the two outermost motifs. Output 3 produces graphical results on the user's terminal screen. The x axis of the screen represents the sequence, and a vertical line will be drawn for each match to the pattern. The x coordinate of this line corresponds to the position in the sequence that the leftmost motif in the pattern matches, and the line height represents the degree of match. The positioning of these plots on the screen is entirely under the user's control. If the patterns being searched for were nucleic acid control sequences such as mRNA splice sites or promoters, then the user would position the plots so that they could be helpful when used in conjunction with the graphical output from the gene search by content methods de-

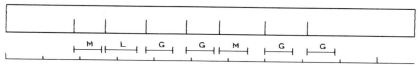

Fig. 7. Results from applying the tRNA pattern to the *E. coli* operon *supb-e*. The operon contains seven tRNA genes, the positions of which are marked by the horizontal bars. The horizontal scale is marked at every 100th base, and the search results are shown by vertical lines in the box. Each vertical line marks the left end position of a match to the tRNA pattern.

scribed elsewhere in this volume.[5] A simple example of the graphical output is given in Fig. 7, which shows the result of applying the tRNA pattern described above to the *Escherichia coli* operon *supb-e*[6] which is 1100 bases long and encodes seven tRNAs. The horizontal scale is marked at every 100th base, and the map indicates the positions of the seven tRNA genes: two for methionine, one leucine, and four glutamine. The box above contains the output from the search and shows that the tRNA pattern exactly locates the seven genes and finds no false matches.

Library Searching

The programs for searching libraries are called ANALYSEQL and ANALYSEPL, and they require the libraries to be stored in PIR format, in which each sequence is identified by an entry name. If required, users can restrict a library search in one of two ways, using a list of names contained in a file. They can either search only those sequences whose entry names are in the list file, or, alternatively, all but those sequences whose names are in the list are scanned. Files containing entry names can be conveniently created using the PSQ or NAQ programs[7] or from files containing the results of previous searches.

For library searches the following forms of output are available: (1) motif by motif (as for single sequences), (2) inclusive (as for single sequences), (3) scores only, (4) padded sections, (5) complete padded sequences. All give the library entry name and a one-line description of the entry. Outputs 1 and 2 have been dealt with, 3 is obvious, and 4 and 5 are described below.

As stated in the introduction, patterns are characterized by sections of

[5] R. Staden, this volume [10].
[6] N. Nakajima, H. Ozeki, and Y. Shimura, *Cell* **23,** 239 (1981).
[7] B. C. Orcutt, D. G. George, J. A. Fredrickson, and M. O. Dayhoff, *Nucleic Acids Res.* **10,** 157 (1982).

fixed length separated by gaps of varying sizes. When we align families of sequences we introduce just enough gaps to get them into register over sections of similarity. The purpose of outputs 4 and 5 is to produce alignments of matching sequences from searches of libraries. This is achieved by the programs writing out the matching sequences with suffi-cient gaps introduced into the sections between the motifs. When the sequences written out in this way are lined up one above the other, all the motif matches will be in register, as the sections between equivalent motifs will be of the same length. Some of these sections will be completely filled with the equivalent segment from the original sequence, but others will be padded out with gaps so as to align with them. The difference between options 4 and 5 is simply that 4 writes only the sections between the outmost motifs, whereas 5 writes complete sequences.

The program actually creates a new file for each match and writes the padded sequence into the file. The file is given the same name as the entry name of the sequences in the library. To bring all the sequences into register we use a program called MAKECON which automatically copies them all into a special database. The databases and their associated soft-ware were originally designed for use during shotgun DNA sequencing projects[8] but have since been modified to permit operations on protein sequences. They can handle up to 1000 sequences, each of up to 4096 characters in length. Once they are put into such a database, it is easy to display, edit, and manipulate the sequences in many ways using all the tools developed for handling shotgun sequencing data.

Discussion

We have described a set of very versatile programs. They give users the ability to effectively "program" their own patterns and then search for them. In the past we have written specific individual routines to search for mRNA splice junctions,[9] E. coli promoters,[9] tRNA genes,[10] etc., but the programs described here easily permit users to define equally complex patterns for themselves. Libraries of patterns can be created, stored in files, and applied to any sequence.

Because of the search strategy employed the programs are very efficient. All searches commence with the first motif in a pattern and only proceed to look for the next motif when a match to the first is found. The search for

---

[8] R. Staden, in "Nucleic Acid and Protein Sequence Analysis, a Practical Approach" (M. J. Bishop and C. J. Rawlings, eds.), p. 173. IRL Press, Oxford, 1987.
[9] R. Staden, Nucleic Acids Res. 12, 521 (1984).
[10] R. Staden, Nucleic Acids Res. 8, 817 (1980).

the second motif is restricted to its permitted range relative to the first motif. Similarly, the third motif is searched for only if the second matches, and again only in its permitted range. In this way, for all motifs, the minimum amount of the sequence is searched. Obviously, when possible, speed can be gained by arranging the order of motifs in a pattern so that those with the lowest probability of matching are searched for first. The programs are available from the author.

Acknowledgments

I thank Bill Turnell for useful discussions and comments, and Dick Harrison for critical reading of the manuscript.

[13] Consensus Patterns in DNA

By GARY D. STORMO

Introduction

A major challenge in molecular biology is understanding regulation of gene expression. A common mechanism of regulation involves a protein binding to a region of DNA to affect transcription of the adjacent gene. Understanding any particular regulatory system requires knowing both the regulatory proteins and the sites of their actions. This chapter describes computer-aided methods useful for the identification and analysis of regulatory sites. The goal of these methods is to extract from a set of known binding sites a pattern which describes the sites and serves to distinguish them from other regions of the genome that are not bound by the protein. A typical procedure is to determine a consensus sequence for the DNA-binding protein, which provides a representative binding site that should be similar to any other binding site. This method has some inherent limitations that are alleviated by a more general approach involving matrix representations of the binding site pattern. Our methods for determining appropriate matrix representations, and some justifications for them, are described.

The problem of analyzing binding site patterns has two fundamental components. The first problem is determining an alignment of the binding sites so that patterns in the sequences correspond to interactions between the DNA and protein. This can be accomplished through biochemical techniques but is most often accomplished, or at least completed, by the comparison of several binding sites and the identification of partially conserved sequences. This problem is an example of the multiple alignment problem described in other chapters and is not dealt with here except for a brief description of a recent method that takes advantage of constraints seen in typical protein binding site patterns. The second fundamental problem is distillation of the information in the set of aligned sequences into a pattern which represents the target specificity of the protein and which can be used to predict the location of new binding sites. This is the problem that is primarily addressed in this chapter. It is generally assumed that the alignment of the binding sites is given and that the goal is to extract the pattern required for a sequence to be included in the set of binding site sequences. A related, more ambitious goal is to determine a pattern that reliably predicts the relative quantitative binding activity of different binding sites.

Sequence Patterns of Regulatory Sites

A common approach to representing the specificity of a DNA-binding protein is to determine a consensus sequence for the protein. This is usually a single sequence, perhaps containing symbols for multiple bases (such as R, which means A or G).[1] Typically, the consensus sequence is derived from the aligned binding sites by choosing the most common base at each position that is significantly nonrandom, where the determination of what constitutes significant is often variable and arbitrary. A more rigorous method is to identify "words," short subsequences, that are common to the set of binding sites, allowing for some mismatches.[2,3] This method is particularly useful if the alignment of the binding sites is only approximately known.

If regulatory sites were like restriction enzyme cleavage sites, a consensus sequence representation would be wholly adequate. For example, the restriction enzyme *Eco*RI acts at sites containing the sequence GAATTC. It cuts only at such sites, and all such sites are susceptible to cleavage by the enzyme (unless the bases are modified, which effectively changes the sequence). In contrast, regulatory proteins typically bind to many sequences other than their consensus sequences. The consensus sequence merely serves as an average binding site, similar to each of the known binding sites but perhaps identical to none of them. The number of mismatches from the consensus that are permitted while still retaining

---

[1] G. D. Stormo, *Annu. Rev. Biophys. Biophys. Chem.* **17,** 241 (1988).
[2] D. J. Galas, M. S. Waterman, and M. Eggert, *J. Mol. Biol.* **186,** 117 (1985).
[3] G. Mengeritsky and T. F. Smith, *Comput. Appl. Biosci.* **3,** 323 (1987).

binding activity and whether all mismatches are equivalent to each other are issues that are usually left unanswered by consensus sequences.

Promoters from *Escherichia coli* provide a well-studied example of a regulatory site. Based on the first six promoter sequences to be determined, a consensus sequence of TATRATG was proposed to occur about 10 bases from the transcription start point.[4] This "−10 box" was joined by a "−35 box" of TTGACA when several more promoter sequences were obtained.[5] More recent and extensive compilations, including a recent survey of nearly 300 promoters,[6] have mostly verified the early proposals. The current consensus sequence is TTGACA-17N-TATAAT, where 17N is the consensus spacing between the −35 and −10 consensus sequences. Unfortunately for our ability to predict promoter sites, neither the spacing nor any of the consensus bases is absolutely conserved. In fact, none of the nearly 300 promoters in the survey is an exact match to the consensus sequence. Different positions within the consensus sequence are conserved to varying degrees. For example, the −10 bases TAxxxT each occur in 80–90% of all promoters, and nearly every promoter has at least two of those three bases, while the other −10 consensus bases occur in only 49–59% of the promoters. A similar variation in the amount of conservation occurs in the −35 region. This makes it very difficult to define a consensus sequence with which to identify new promoters reliably.

Matrix Patterns

Searching with Matrices

Figure 1 shows how a matrix is used to evaluate each position in a sequence to locate the regulatory sites. A matrix clearly supersedes the capabilities of a consensus sequence, since the consensus can be considered a special case of a matrix. For example, assigning a value of 1 to each element of the matrix corresponding to the consensus base at each position and 0 to all the other elements creates a consensus matrix. The evaluation of each position of a sequence is then equal to the number of matches to the consensus sequence. The advantage of the matrix is that each change from a consensus base can be assigned a different penalty. This is consistent with known binding sites where changes at different positions can have drastically different effects on the binding. At some positions there may be very little cost to a substitution for the consensus base, while at other

[4] D. Pribnow, *Proc. Natl. Acad. Sci. U.S.A.* **72**, 784 (1975).
[5] M. Rosenberg and D. Court, *Annu. Rev. Genet.* **13**, 319 (1979).
[6] C. B. Harley and R. P. Reynolds, *Nucleic Acids Res.* **15**, 2343 (1987).

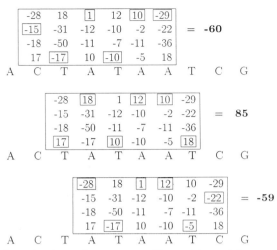

FIG. 1. Matrix evaluations of a sequence. The matrix contains an element for each possible base at six positions of a binding site. The matrix rows are in the order A, C, G, T from top to bottom. For each alignment of the matrix above the sequence a score is calculated as the sum of the matrix elements corresponding to the sequence, which are boxed. The sequence TATAAT scores the highest with this matrix.

positions the consensus base may be essential for activity of the site. In the example of Fig. 1, a T at position 6 is substantially better than any other base, while at position 5 a change from the consensus base A is not so severe. A matrix can easily incorporate this type of information.

The matrix of Fig. 1 assumes the contributions from each position of the binding site are independent and additive. This assumption is certainly going to be an approximation for most proteins, although it may be a reasonably good approximation for many. However, matrices may be constructed that do not assume independence of the bases in a sequence if that is necessary to provide a good representation for the specificity of some protein. For example, if the adjacent bases interact with the protein in a nonadditive way, the matrix could be made with 16 rows instead of 4, one for each dinucleotide, to accommodate those interactions. Some form of matrix must be an accurate representation of the interaction between the protein and the DNA. Consider the worst case example in which there is no additivity between parts of the binding site. Then the only accurate representation of the protein's specificity is a list containing the binding energy of the protein to each possible binding site. Such a list is itself a one-column matrix (or vector). We assume that, in general, protein specificity will be well represented by fairly simple matrices.

Proteins that have two or more recognition regions separated by variable spacing, such as *E. coli* promoters with their -10 and -35 boxes, also can be well represented by matrices. In these cases the binding site is determined by multiple matrices, one for each important region, with a combining function that specifies the effects of spacing between the regions. For *E. coli* promoters a spacing of 17 would be optimal, and different spacings would be penalized by various amounts. It should be clear that matrices represent a particular type of sequence pattern, or motif, and can be combined with other types to make very general search descriptions.[7]

Choosing Appropriate Matrices

Given that matrices can be good representations of protein binding sites, the important issue is how to determine an appropriate matrix for any particular protein. Our method was originally derived from information theory considerations[8] but also can be derived in other ways. Alternative justifications based on likelihood statistics and thermodynamics are described below. Figure 2 shows the three important matrices for describing the specificity of the protein, using *E. coli* -10 sequences as the example. Figure 2A simply contains the count of each base at each position in the aligned sequences.[6] These numbers we call $n_{b,i}$, where b and i refer to the base and the position of the element, respectively. Figure 2B contains the frequency of the bases at each position, $f_{b,i} = n_{b,i}/N$, where N is the total number of sites (242 in this example). The elements in Fig. 2C are derived from Fig. 2B as $\log(f_{b,i}/p_b)$, where p_b represents the genomic frequencies of each base. This normalizes the frequencies to the *a priori* frequencies of obtaining each base. In *E. coli* each base occurs with approximately equal frequency ($p_b \approx 0.25$ for all b). We arbitrarily take logarithms to the base 2 to measure information in *bits*. The dot product of the second and third matrices gives the information content of the binding sites, shown for each position independently in Fig. 2D.[8] Note that the matrix in Fig. 1 is equivalent to that in Fig. 2C, all the elements having been multiplied by 10 and rounded to provide efficient searching with integers. If the same procedure is used to generate a matrix for the -35 region and an analogous procedure to determine the spacing penalties, these can be used to search for unknown promoter sites. This is essentially equivalent to the method of promoter homology index.[6,9] The next two subsections give short justifications for the use of these particular matrices.

[7] R. Staden, *Comput. Appl. Biosci.* **4**, 53, (1988).
[8] T. D. Schneider, G. D. Stormo, L. Gold, and A. Ehrenfeucht, *J. Mol. Biol.* **188**, 415 (1986).
[9] R. Staden, *Nucleic Acids Res.* **12**, 505 (1984).

A

A	9	214	63	142	118	8
C	22	7	26	31	52	13
G	18	2	29	38	29	5
T	193	19	124	31	43	216

B

A	0.04	0.88	0.26	0.59	0.49	0.03
C	0.09	0.03	0.11	0.13	0.22	0.05
G	0.07	0.01	0.12	0.16	0.12	0.02
T	0.80	0.08	0.51	0.13	0.18	0.89

C

A	-2.76	1.82	0.06	1.23	0.96	-2.92
C	-1.46	-3.11	-1.22	-1.00	-0.22	-2.21
G	-1.76	-5.00	-1.06	-0.67	-1.06	-3.58
T	1.67	-1.66	1.04	-1.00	-0.49	1.84

D

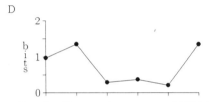

FIG. 2. Matrices from *E. coli* -10 promoter sequences.[6] (A) Number of occurrences of each base at each position of the 242 nonmutant promoters. (B) Fraction of each base at each position, from the above data. (C) Logarithms (base 2) of those fractions divided by 0.25, the genomic frequency of each base in *E. coli*. (D) Information content[8] of each position, which is calculated as the dot product of the vectors in each column of the matrices in B and C.

Thermodynamics. If the only data available are the collection of known binding sites, we expect that matrices derived as that in Fig. 2C will generally provide the best representation of the binding site specificity. This expectation is based on the following thermodynamic analysis of an *in vitro* binding experiment. Imagine there is a protein, E, that binds to mononucleotides, $b \in \{A,C,G,T\}$. The diagram for this reaction is

$$E + b \rightleftharpoons Eb \tag{1}$$

The equilibrium binding constant of the protein for each base, $K(b)$, is defined as

$$K(b) = \frac{[Eb]}{[E][b]} \tag{2}$$

where the brackets indicate concentrations of the reactants they enclose. The relative binding constants to the different bases is independent of the free protein concentration, $[E]$. For example, the ratio of the binding constants for A and C is

$$K(A)/K(C) = \frac{[EA]}{[A]} \Big/ \frac{[EC]}{[C]} \tag{3}$$

The differences in the free energies of binding to the different bases is proportional to the logarithms of those ratios, such that

$$\Delta G^0(A) - \Delta G^0(C) \propto \ln\frac{[EA]}{[A]} - \ln\frac{[EC]}{[C]} \tag{4}$$

The $[Eb]$ and $[b]$ terms are proportional to the fraction of each base bound to the protein and free in solution, respectively. These are analogous to the f_b and p_b terms in the derivation of Fig. 2C. Therefore, the elements of the specificity matrix, $\log f_{b,i}/p_b$, can be thought of as being proportional to the differences in binding free energies to each base at each position in the binding site. Of course this rationalization is completely true only if the interactions at the positions of the site are independent and additive, and if the collection of known sites is a random sample of all possible binding sites, taken from a genome which contains a random sample of all possible sequences. While these conditions are not going to be entirely satisfied, in the absence of other data these assumptions introduce the least bias into the analysis and should lead to the best representation of the specificity of the protein under study.

Likelihood. As in any statistical analysis, the significance of the results is dependent on the sample size. Likelihood statistics is a convenient way to measure the significance of a proposed matrix representation for a protein's specificity because it involves the same calculations as those used in determining the matrices.[10] If there is some population of bases with individual proportions q_b, then the probability of selecting N bases at random with n_b of each different base is

$$P(n_b|q_b) = \frac{N!}{\Pi_b\, n_b!} \Pi\, q_b^{n_b} \tag{5}$$

The likelihood ratio is defined as the relative probabilities of obtaining a particular data set under two different models of the population, i.e., for two different assumptions of q_b. If one model is that the data are just a sample of bases chosen randomly from the genome, then p_b are the appro-

[10] A. W. F. Edwards, "Likelihood," Cambridge University Press, Cambridge, England, 1972.

priate probabilities. Another model is that the bases have been selected by a "recognizer" which chooses bases according to the probabilities f_b. Then the log likelihood ratio of the data being obtained from these two different population models is

$$\log \frac{P(n_b|f_b)}{P(n_b|p_b)} = \sum_b n_b \log \frac{f_b}{p_b} \qquad (6)$$

This is equal to the dot product of the count and specificity matrices, such as those in Fig. 2A,C for the *E. coli* promoter example. The relative probability of obtaining the data from any models of the population probabilities can be calculated by substituting for f_b and p_b in Eq. (6). The maximum likelihood estimate for the population probabilities is that of f_b.

Information Content. The three matrices we calculated for a set of binding sites, as exemplified in Fig. 2A–C for *E. coli* promoters, each contribute to the analysis of the sites. The specificity matrix, as in Fig. 2C, is an estimate of the change in binding energy for any change in the sequence. Searching based on this matrix, as in Fig. 1, should evaluate potential binding sites by their relative affinities for the protein. The dot product of the count and specificity matrices, as in Fig. 2A,C, gives a statistic about the probability of obtaining the observed data by chance, given the genomic proportions of the bases. This is also N times the information content of the sites, obtained from the dot product of the frequency and specificity matrices as in Fig. 2B–D. The information content of the sites is independent of the sample size, although the reliability of the measure is not. It is a measure of the discrimination between different sites by the protein. The maximum discrimination at any position in a binding site corresponds to only one base being allowed at that position, which would have an information content of 2 *bits.* (More than that is possible in genomes which are not equiprobable for each base, but the specificity numbers can be normalized to the case of equally likely bases.[11]) In a number of studied examples, the information content for a binding site pattern is closely correlated with the frequency of binding sites in the genome.[8,11]

Quantitative Matrices

Regulatory sites vary considerably in their binding activities. For example, different *E. coli* promoters can have at least 1000-fold differences in

[11] T. D. Schneider and G. D. Stormo, *Nucleic Acids Res.* **17,** 659 (1989).

their transcription initiation rates.[12] One of the advantages of matrix representations of specificity is that a quantitative evaluation of any sequence is possible, as in Fig. 1. It would be especially useful if the evaluation of any sequence were correlated to its activity. Mulligan *et al.* first attempted to find a matrix that would correlate site scores with promoter activities.[13] Using a statistical approach similar to that described above, they derived a matrix that gave correlations of 83% between the evaluations and the measured *in vitro* promoter activities. Berg and von Hippel used a statistical mechanical derivation, entirely analogous to the thermodynamic one described above, to predict promoter activities and achieved an 84% correlation with the measurements.[14] The fact that these methods do this well in predicting activities based solely on the frequencies of base occurrences in a collection of known sites indicates that the assumptions of additivity and random samples are probably not grossly wrong. In fact, it is now known that *in vivo* promoter activities may not be well correlated with these matrix evaluations.[15] This result is not too surprising since promoter activity requires more than a simple binding reaction. It also includes catalytic steps and the release of the binding site during the elongation phase of transcription activity. Regulatory sites that do involve only a binding reaction might be expected to be predicted even better; although many fewer examples exist than for promoter sequences, predicting quantitative binding activities by these methods can work quite well.[16,17]

A matrix can be derived that is a best fit to quantitative data for a binding site, provided enough data are available.[1,18] This requires that one know a large number of sequences and the binding activity of each one. The matrix elements are unknowns that are to be determined such that the evaluation of each sequence is a best fit to the measured activities. Multiple regression is a common means of obtaining parameters of this type. One might also systematically vary the bases in a binding site and determine the change in binding energy for each substitution. When this is done for λ operators, it is found that Cro protein can be represented very well by an additive matrix.[19]

[12] W. R. McClure, *Annu. Rev. Biochem.* **54**, 171 (1985).
[13] M. E. Mulligan, D. K. Hawley, R. Entriken, and W. R. McClure, *Nucleic Acids Res.* **12**, 789 (1984).
[14] O. G. Berg and P. H. von Hippel, *J. Mol. Biol.* **193**, 723 (1987).
[15] D. Graña, T. Gardella, and M. M. Susskind, *Genetics* **120**, 319 (1988).
[16] O. G. Berg and P. H. von Hippel, *J. Mol. Biol.* **200**, 209 (1988).
[17] O. G. Berg, *Nucleic Acids Res.* **16**, 5089 (1988).
[18] G. D. Stromo, T. D. Schneider, and L. Gold, *Nucleic Acids Res.* **14**, 6661 (1986).
[19] Y. Takeda, A. Sarai, and V. M. Rivera, *Proc. Natl. Acad. Sci. U.S.A.* **86**, 439 (1989).

Multiple Alignments

The general problem of finding the optimal alignment of multiple sequences, in which gaps may be inserted in the sequences and the extent of the alignment is unknown in advance, is very difficult. However, in searching for alignment of binding sites one can take advantage of typical constraints in their patterns to make the problem more tractable. The most useful constraint is that binding sites can be represented by ungapped matrices. While the complete binding pattern for a protein may include gaps, as in the variable spacing between the -10 and -35 patterns of *E. coli* promoters, the separate binding elements described by a matrix for each region are of fixed size. The other useful constraint is that the direct interactions represented by the matrices tend to be short. Owing to the helical nature of DNA, if the protein interacts with more than 10 consecutive bases it must wrap around the DNA, an unlikely occurrence. Therefore, highly informative pattern matrices usually come in blocks of fewer than 10 positions. Prokaryotic operator sequences are most often about 20 bases in length, but are usually composed of two or more high information regions with relatively random sequence specificity in between.[8]

If one is given a set of DNA fragment sequences, each not more than a few hundred base pairs in length, and it is known that each fragment contains the binding site for a particular protein, one can use the constraints typical of binding site patterns to identify the sites.[20] The rationale for our method is based on two principles. The first is that a matrix provides the best representation for the binding pattern of the protein. That is, if one could compare all possible alignments of the fragments, the best candidate for the binding sites would be the one with the highest information content over a region the size of a binding domain. It is usually not feasible to compare all possible alignments, because if there are N sequences of length L the total number of possible alignments is approximately L^N, but information content would be an appropriate method of comparison if one could. The second principle reduces the number of alignments that are compared, hopefully without eliminating the correct alignment. The idea is that, given any two of the fragments, the optimal alignment between them may not be at the binding sites, but the optimal alignment of the binding site in the first sequence is likely to be with the binding site on the second sequence. Since we do not know the binding site on the first sequence, and any of the sites within it are equally likely, we save a matrix for each possible site with its best match on the second sequence. We continue this procedure, saving a progeny of each original

[20] G. D. Stormo and G. W. Hartzell III, *Proc. Natl. Acad. Sci. U.S.A.* **86,** 1183 (1989).

possible binding site as a matrix. When all of the sequences have been added to the analysis, the one with the highest information content is the best guess for the pattern of the binding site. The method works well on typical prokaryotic binding sites, and it is robust enough to work even with some erroneous data included.[20,21]

Summary

Matrices can provide realistic representations of protein/DNA specificity. In many cases simple mononucleotide-based matrices are adequate representations, but more complex matrices may be needed for other cases. Unlike simple consensus sequences, matrices allow for different penalties to be assessed for different changes to a binding site, a property that is essential for accurate description of a binding site pattern. When only a collection of binding site sequences is known, the best representation for the pattern is an information content formulation, based on both thermodynamic and statistical considerations. Quantitative data on relative binding affinities may be used to determine matrices that provide a best fit to the data. Matrix representations also provide an efficient method of aligning multiple sequences to identify binding site patterns that they have in common.

Acknowledgments

I thank Calvin Harley for sending me the *E. coli* promoter sequences on a floppy disk and George Hartzell for helping with the figures and Gerald Hertz for critical comments on the manuscript. Supported by National Institutes of Health Grant GM28755.

[21] G. Z. Hertz, G. W. Hartzell III, and G. D. Stormo, *Comput. Appl. Biosci.,* in press.

[14] Consensus Methods for DNA and Protein Sequence Alignment

By Michael S. Waterman and Robert Jones

Introduction

The increasing body of nucleic acid sequence data has created interest among many scientists in computational approaches to macromolecular sequence analysis. Several international databases have been created in order to store the data in a useful format, both for archival and analysis purposes.[1] Both DNA and protein sequences databases are maintained. The value of simply having easy access to all membrane protein sequences, for example, is not to be underestimated. The quantity of data has naturally led to the development of computer approaches to sequence analysis.[2] The purpose of this chapter is to present some of the tools that we have created in order to analyze multiple sequences in a rigorous, efficient, and systematic way.

Much computer analysis of molecular sequences is directed toward discovery of biologically significant patterns. These patterns include homologous genes, RNA secondary structure, tRNA or structural RNAs, palindromes in DNA sequences, regulatory patterns in promoter regions, and protein structural patterns. Once the patterns have been located they can often be tested by experiment, as in the case of promoter elements. Evolutionary relationships, however, cannot be directly tested, and increasing emphasis is being attached to the discovery and interpretation of sequence evolution.

Sequence alignment is a popular approach to pattern analysis.[2] Computer alignments are often based on an explicit optimization function, rewarding matches and penalizing mismatches, insertions, and deletions. Sequence alignment often gives useful information about evolutionary or functional relationships between sequences. Our approach is based on what we refer to as consensus analysis.[2-4]

Consensus sequence analysis is usually performed by visual inspection of the sequences and by experiment. Of course, a protein binding site can only be verified by experiment, and analysis by "eye" can be biased. Thus, it is useful to have computer methods that can find consensus patterns best fitting explicitly stated criteria. Some algorithms have been developed along these lines,[2-4] and they are described here, along with some biological examples. Our earlier methods applied only to DNA; here we also describe recent extensions to protein sequences.

In 1970 Needleman and Wunsch[5] published an approach sequence comparison (alignment) using a dynamic programming algorithm. Their algorithm find maximum similarity between two sequences, where matches score positive weight and mismatches, insertions, and deletions

[1] C. Burks, J. W. Fickett, W. B. Goad, M. Kanehisa, F. I. Lewitter, W. P. Rindone. C. D. Swindell, C.-S. Tung, and H. S. Bilofsky, *CABIOS* **1**, 225 (1985).

[2] M. S. Waterman, ed., "Mathematical Methods for DNA Sequences." CRC Press, Boca Raton, Florida, 1988.

[3] M. S. Waterman, D. Galas, and R. Arratia, *Bull. Math. Biol.* **46**, 515 (1984).

[4] D. J. Galas, M. Eggert, and M. S. Waterman, *J. Mol. Biol.* **186**, 117 (1985).

[5] S. B. Needleman and C. Wunsch, *J. Mol. Biol.* **48**, 444 (1970).

score nonpositive weight. Mathematicians began to attempt to define a distance between sequences and so to construct a metric space. Sellers[6] obtained these results for single insertions and deletions, and later workers extended the work to multiple insertions and deletions.[7] While dynamic programming methods are very widespread in sequence analysis, there are severe restrictions in computation time with the extension of the dynamic programming methods to allow more than two sequences. A great many biological problems do involve more than two sequences. The consensus methods we have developed avoid the computational difficulties of dynamic programming by using a very different approach to sequence analysis.

The basis of the consensus method is an algorithm to find consensus words, with the degree of matching and alignment specified by the user of the program. We give the specifications of this method in the next section for DNA and protein sequences, along with examples. In this setting the consensus method finds patterns or words that are conserved in an unusual number of sequences. Then the basic method is extended, both for DNA and protein sequences, to an algorithm for sequence alignment. To illustrate the behavior of the algorithms, we have chosen two sequence sets, one DNA and the other protein. The DNA sequences are 19 promoters from the genome of vaccinia virus.[8] The protein sequence set is 16 proteins related to the *Escherichia coli ntrC* gene product.[9] We use these sequences to illustrate the use and power of the programs and the effect of varying certain parameters. We do not attempt to interpret the consensus pattern found in any biological context, but we invite anyone interested in these specific sequences and patterns to contact us for more detailed information.

Consensus Patterns

Now we give a general description of the consensus word algorithm. To begin, take a set of R sequences of length N

$$
\begin{array}{cccc}
a_{1,1} & a_{1,2} & \cdots & a_{1,N} \\
a_{2,1} & a_{2,2} & \cdots & a_{2,N} \\
& & \vdots & \\
a_{R,1} & a_{R,2} & \cdots & a_{R,N}
\end{array}
$$

[6] P. Sellers, *SIAM J. Appl. Math.* **26,** 787 (1974).

[7] M. S. Waterman, T. F. Smith, and W. A. Beyer, *Adv. Math.* **20,** 367 (1976).

[8] M. Mars and G. Beaud, *J. Mol. Biol.* **198,** 619 (1987).

[9] B. T. Nixon, C. W. Ronson, and F. M. Ausubel, *Proc. Natl. Acad. Sci. U.S.A.* **83,** 7850 (1986).

These sequences can be taken to be initially aligned on some biologically or statistically determined feature. The true alignment is, of course, unknown except approximately. Now we give an algorithm for locating consensus words of a given size. By way of comparison, the usual methods of sequence analysis align on single letters, that is words of length 1.

Of course, a concept basic to our algorithm is that of consensus word. The definition has been given in earlier work[2-4] and will be briefly reviewed here. First, take a fixed word size k and a word w of length k; there are 4^k such words in DNA and 20^k in proteins. Next, define the window width W which gives the width of sequence in which a consensus word can be found and thus defines the amount of shifting allowed in matching consensus words. The sequences starting at column $j + 1$ with window width W appear as

$$
\begin{matrix}
a_{1,j+1} & a_{1,j+2} & \cdots & a_{1,j+W} \\
a_{2,j+1} & a_{2,j+2} & \cdots & a_{2,j+W} \\
& & \vdots & \\
a_{R,j+1} & a_{R,j+2} & \cdots & a_{R,j+W}
\end{matrix}
$$

To begin, we search the first sequence of the window for matches to our word w. An exact match to w is called a $d = 0$ neighbor while a 1-letter mismatch from w is called a $d = 1$ neighbor, and so on. For protein sequences, for example, it is desirable to distinguish the many types of $d = 1$ mismatches by different weightings based on amino acid similarity. It is possible to include insertions and deletions in this list of neighbors. We may decide, e.g., to limit the amount of mismatch to $d = 0, 1, 2$ and not find w in a portion of sequence unless it is within this neighborhood. Let $q_{w,d}$ equal the number of lines that the best occurrence of w is as a dth neighbor. Each of these occurrences receives weight λ_d. The score of word w in this window is

$$
s_{j+1,j+W}(w) = \sum_d \lambda_d q_{w,d}
$$

A best scoring word is word w^* satisfying

$$
s_{j+1,j+W}(w^*) = \max_w s_{j+1,j+W}(w)
$$

DNA Consensus Patterns

In the case of DNA all $d = 1$, $d = 2$, . . . mismatches are considered identical in weight. While more complex weighting schemes are easy to incorporate, we have found that it is adequate to score a word by the fraction of letters matching the consensus word. Thus, for a d-letter mismatch to a k-letter consensus word, $\lambda_d = 1 - d/k$.

To perform the computations, each word in the window is read and its neighborhood calculated. The possible words in a neighborhood are found by a simple combinatoric scheme; since all $d = 2$ mismatches receive the same weight it is only necessary to enumerate all $k(k - 1)/2$ of these mismatches. When the portion of each sequence in the window has been examined, the best score each of the 4^k possible consensus words is retained.

Our consensus method for DNA sequences has been implemented in a program called RTIDE written in C and using the SunView window system. Figure 1 shows a typical screen display from this program. The aligned sequences are displayed at the bottom, the displayed consensus word and score of the word are shown above, along with parameters for the run. At top is a plot of the consensus word score against window position for the alignment. Peaks in this plot indicate regions of conservation that

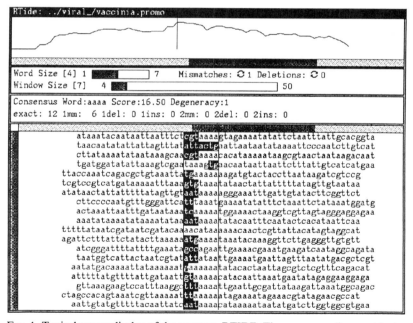

FIG. 1. Typical screen display of the program RTIDE. The sequence alignment is shown in the bottom window. At top is a plot of maximum score against window position in which peaks represent conserved regions.

may be of biological interest; these patterns can then be examined in detail and sequences realigned on the consensus words in any particular window. Through cycles of realignment and analysis it is possible to identify and refine conserved sequence patterns. The nature of the method involves variation of several parameter settings during an analysis session. It is possible to adjust window width, word size, and degree of matching required as well as alignment of the sequences. To facilitate this interaction all features of the program are controlled through the mouse.

The effect of varying the word size on the consensus score is shown in Fig. 2. No mismatches are allowed in these runs, and the window size is increased along with word size to keep the number of words per window constant. There are 16 words of size 2, and the score is the maximum scoring word over all sequences. Since there are so few words, all window positions have a high score and no features are clearly resolved. At word size 3 the graph shows a few features, with the highest score indicated by the dashed line. At word size 4 the central peak is resolved from the background. As word size increases further this peak becomes smaller and

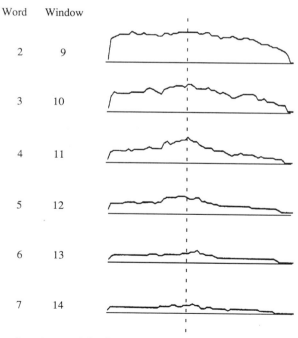

FIG. 2. Effect of varying word size from 2 to 7 for DNA sequence alignment. The window size is varied to maintain a fixed number of words per window. No mismatches are allowed.

merges into the background of nonconserved sequence. The word size at which a peak is most evident is a clear indication of the size of the feature that is conserved; in the present case that word size is 4 nucleotides.

In many cases the initial alignment in which sequences are supplied will not be optimal for a given conserved sequence feature. Varying the window size is a way to accommodate poorly aligned sequence sets. Figure 3 shows the effect of variation of window size from 4 to 20 for the DNA data set. Using a window equal to word size permits no misalignment and for our sequences gives a very low graph of scores. Widening the window brings more instances of a consensus word into a window. This is demonstrated in Fig. 3 by the appearance of a peak in the graphs, most clearly resolved at window sizes 10 and 12. Extending the window further may not bring any new instances of the consensus word, but it does increase the number of window positions that achieve a high score. This is shown by the plateau in the graphs for window sizes 16 and 20.

The effect of varying the neighborhood of words that can contribute to the score of a consensus word is shown in Fig. 4 for our DNA data set. We fix a window size of 12 and a word size of 6 and vary the number of mismatches permitted from 0 to 2, with no insertions or deletions. Requiring exact matches (0 mismatches) with the consensus word results in low scores with no distinct features. Permitting a single mismatch in general causes scores to increase, but the central conserved region emerges as a

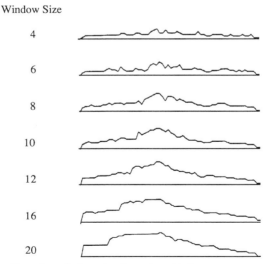

FIG. 3. Effect of varying window size from 4 to 20 for DNA sequence alignment. The word size is 4, and no mismatches are allowed.

Mismatches

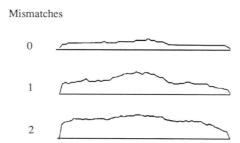

FIG. 4. Effect of varying the number of mismatches from 0 to 2 for DNA sequence alignment. The word size is 6, and the window size is 12.

distinct feature. When the neighborhood is increased to 2 mismatches, however, the background scores are almost equivalent to that of the conserved feature.

A very useful feature of our program is that once a consensus word has been identified, the sequences can be realigned on that word and the new alignment reevaluated for additional conserved features. Figure 5 shows the refinement of a consensus word using this technique. The graph of the scores after cycle 1 shows a maximum score of 14.25 at the position marked by the solid line. The sequences were realigned on the words that

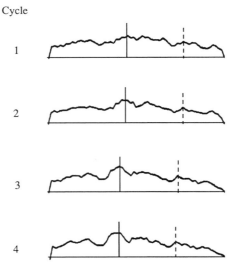

FIG. 5. Effect of realigning the sequences on a consensus word for DNA sequence alignment. The word size is 4, window size is 6, and 1 mismatch is allowed.

	A	C	D	E	F	G	H	I	K	L	M	N	P	Q	R	S	T	V	W	Y
A	8																			
C	6	8																		
D	3	3	8																	
E	1	2	6	8																
F	4	5	1	2	8															
G	7	6	2	1	4	8														
H	1	2	2	3	4	1	8													
I	4	5	1	2	5	4	2	8												
K	2	3	3	4	3	2	6	3	8											
L	4	5	1	2	5	4	2	7	3	8										
M	5	6	2	3	6	5	3	6	4	6	8									
N	4	5	5	4	3	4	2	3	3	3	4	8								
P	4	5	3	2	3	4	0	3	1	3	4	5	8							
Q	3	4	4	5	4	3	3	4	4	4	5	6	4	8						
R	1	2	4	5	2	1	5	2	6	2	3	4	2	5	8					
S	5	4	4	3	2	5	1	2	2	2	3	6	4	5	3	8				
T	5	6	4	3	4	5	3	4	4	4	5	6	4	5	3	5	8			
V	5	6	2	1	4	5	1	6	2	6	5	4	4	3	1	3	5	8		
W	3	4	2	3	6	3	5	4	4	4	5	4	2	5	3	3	5	3	8	
Y	3	4	2	3	6	3	5	4	4	4	5	4	2	5	3	3	5	3	7	8

A C D E F G H I K L M N P Q R S T V W Y

FIG. 6. Similarity matrix used in weighting sequence mismatches, based on the representation of amino acid similarity of Taylor.[10]

contributed to that score and the program run again. In the second cycle the marked peak is more clearly resolved and its score has risen to 15.25. An additional change is that the minor peak marked by the dashed line has become more evident. At the end of the third cycle the main peak has become more distinct and the score is 16.75, resulting from the realignment bringing more related words into the conserved window. The final cycle of scoring does not increase the maximum score and indicates the end of the process.

Protein Consensus Patterns

In the case of protein sequences various mismatches are weighted according to a matrix (Fig. 6), which is derived from Taylor.[10] The mismatches for each letter of a word are arranged according to weight, the nearest or smallest being first. Then in a systematic fashion we allocate mismatches until the limit or cutoff is reached. Then that letter is reduced to identity and the next letter is increased. The algorithm is similar to that of counting with the branch and bound feature we have described.

[10] W. R. Taylor, *J. Theor. Biol.* **119,** 205 (1986).

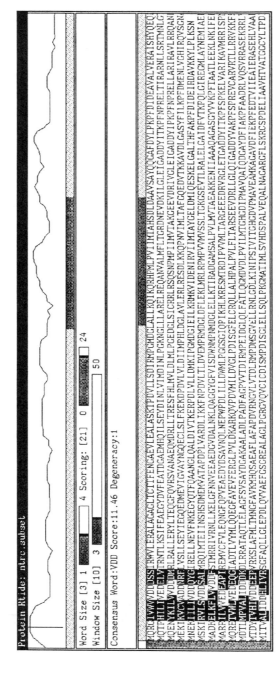

FIG. 7. Typical screen display from the program PRTIDE.

Cycle

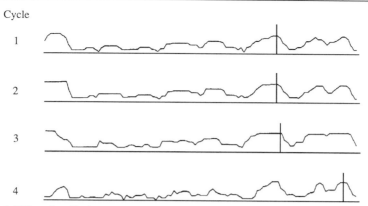

FIG. 8. Effect of realigning sequences on a consensus word over four cycles for the protein sequence data set. The word size is 3, window size is initially 10, and the neighborhood is set to 21, where 24 implies an exact match.

In the protein version of the consensus program, PRTIDE, the only substantial modification is in the definition of neighborhood. Whereas in the DNA version we specify the number of mismatches, insertions or deletions allowed, in the case of proteins we specify a similarity score that a word must attain when scored against a candidate consensus word in order to be included in the neighborhood. Figure 7 shows a typical display of this output.

In most practical aspects the DNA and protein programs are very similar. The larger alphabet of amino acids relative to nucleotides precludes use of words with more than 4 amino acids. Figure 8 shows an example of the protein version in which the sequences are realigned on a consensus word, with the result of resolving other conserved words. At top is shown the graph of scores for the protein data set in which the sequences are aligned at their left ends. The largest peak is found close to this end; aligning on the consensus word and rescoring creates the second plot in which three peaks near the right end of the plot have become more clearly resolved. Aligning on the largest of these and rescoring cause the formation of plateaus. This indicates too large a window size, and reducing the window size from 10 to 5 results in the final plot in which the three peaks are now well resolved.

Consensus Alignment

The idea of the algorithm builds on the previous section.[11] We align on consensus words, attempting to maximize the sum of the scores of the

[11] M. S. Waterman and M. Eggert, *Nucleic Acids Res.* **14,** 9095 (1986).

words. Before the practical algorithms are presented, a more general concept of alignment on words is presented.

We define a partial order on words. The words w_1 and w_2 satisfy $w_1 < w_2$ if the occurrences of w_1 in sequence i are to the left of the occurrences of w_2 in sequence i (and do not intersect) for $i = 1$ to R. It is not necessary for w_1 or w_2 to have occurrences in all sequences. Implicit in the definition is a window width W and neighborhood specification. The goal of an optimal alignment is to find words w_i which satisfy

$$\max\left\{\sum_{i \geq 1} s(w_i) : w_1 < w_2 < \ldots\right\}$$

(It is frequently desirable to require $s(w_i) \geq c$ for all i, where c is some cutoff value.) It is not possible to accomplish this goal in reasonable time, but it is possible to come quite close. We now define two practical algorithms.

Next, $w_1|w_2$ means that consensus words w_1 and w_2 can be found in nonoverlapping windows, each word satisfying as usual the window width and neighborhood constraints. The modified optimization problem is to satisfy

$$T = \max\left\{\sum_{i \geq 1} s(w_i) : w_1|w_2| \ldots\right\}$$

There is a straightforward recursion to find T. Let T_i be the maximum sum for the sequences from base 1 to base i:

$$
\begin{matrix}
a_{1,1} & \cdots & a_{1,i} \\
a_{2,1} & \cdots & a_{2,i} \\
& \vdots & \\
a_{R,1} & \cdots & a_{R,i}
\end{matrix}
$$

Then T_i satisfies

$$T_i = \max\{T_j + s_{j+1,i} : i - W + 1 \leq j \leq i - k\}$$

and $T_{-W} = T_{-W+1} = \cdots = T_0 = T_1 = \cdots = T_{k-1} = 0$. Also, $s_{x,y} = 0$ if $y - x + 1 < k$. This algorithm runs in time approximately proportional to NW^2RB where B is the neighborhood size. Here the factor WRB accounts for the consensus word algorithm with a window width W. (This is an overestimate since the actual windows vary from k to W in width.)

If much shifting is necessary to match the sequences, T is an underestimate and misses some of the relevant matching. To overcome this problem, the definition of T is modified to

$$S_i = \max\{S_j + \hat{s}_{j+1,i} : i - W + 1 \leq j \leq i - k\}$$

where $\hat{s}_{j+1,i}$ is the largest scoring consensus word in the window from $j + 1$ to i such that all occurrences of the consensus word are to the right of the consensus words for S_j. This algorithm is not guaranteed to be equal to the global maximum, but it is much more useful than T. Alignment for each case, DNA and protein, proceeds as just described with the modifications given next.

DNA Alignment

We have written a program, RALIGN, to align multiple DNA or RNA sequences.[11] The sequences are supplied in some initial alignment, which usually consists of the sequences being left justified. As in all our consensus methods, the parameters for window size, word size, and neighborhood are set. We require a consensus word to have a score at least equal to one-half the number of sequences before it can be used for alignment. This requirement is to eliminate the "junk" regions in alignments that are common with programs that optimize a total score.

To illustrate RALIGN for nucleic acids we take a set of 15 tRNA sequences from *Escherichia coli*. These sequences are difficult to align as their relationship is largely determined by conserved helices (secondary and tertiary structure), not the primary sequence itself. Analyses that fold tRNAs by minimum free energy are not too successful, usually folding about 50% of the tRNAs into the correct cloverleaf shape. An analysis based on consensus helices is successful for many structural RNAs, and a study of tRNA by consensus folding appears in Ref. 2. The only universal primary sequence patterns in tRNAs are the CCA at the acceptor arm and the GTTC in the TψC stem and loop. Our analyses (Fig. 9) find these invariant patterns along with other conserved words. Figure 9a has window size 7, word size 3, and up to 1 mismatch (total score equal to 179), while Fig. 9b has window size 8, word size 4, and up to 2 mismatches (total score equal to 127). The window is 7 in the first case and 8 in the second to allow shifts of 4 in each analysis. Notice that CCA is located at the 3' end of all the sequences; in the case of Fig. 9 CCA is generally found overlapping the pattern CACC. In Fig. 9a,b CCA is not found by the program in all the sequences. In several cases an earlier, equally strong pattern is chosen in accord with the algorithm. Of course in Fig. 9b the object is to optimize a 4-letter consensus word, and this weakens the contribution of the 3-letter pattern CCA. GTTC is always located in Fig. 9b between consensus words GGTT and CGAA. In 14 of 15 sequences in Fig. 9a GTTC is located

a

```
       ...gct....  tag....  ...cag.  .tgg.  .tag....  ...agc.     ...ctt...        ...cgg...      .tgg...   ...ggt.  .tcg...  ...atc....       ...ctc....     ...cca
ggg    gcta        tagct    cagc     tgg    gag       agcgcctg    ctttgcacg        caggagg        tctgc     ggt      tcg      atc              cgcatagct      ccaccca
g      gcgcgt      taa      caaag    cggt   tatgt     agcgg       attgcaaatc       cgtc           tagtcc    ggt      tcga     ctcccgaaccgcct                  cca
gga    gcgg        tagtt    cag      tcggttagaat      acctg       cctgtacg         cagggg         tcgcg     ggt      tcga     gtccc            gtccgtt        ccgcca
gtcccct             tcgtc    tag      aggcccaggac      accgcc      ctttcacgg        cggtaacagg     ggt       tcga     atc      ccctgggggacg                    cca
ggtggcta            tagct    cagg     ttgg   tag       agccctgg    attgtgattc       cagttg         tcgtg     ggt      tcga     atcccattag                      ccacccca
ag     gcttg       tagct    cagg     tgg    tag       agcgcacc    cctgata          ggggtgaggtcgt  ggt       tcaa     gtcca    ctcagg           cctaccca
g      ggtcgt      tagct    cagt     tgg    tag       agcagttga   ctttaatcaat      tgg            tcgca     ggt      tcga     atc              ctgcacgac      ccacca
g      gctacg      tagct    cagt     tggt   tag       agcacatca   ctcataatga       tgggg          tcaca     ggt      tcga     atccc            gtcgtag        ccacca
t      cctctg      tagtt    cag      tcgg   tag       aacggcgga   cgtttaat         ccgtatg        tcact     ggt      tcga     gtcca            gtcagaggag     cca
tgg    ggta        tcgc     caag     cgg    taaggc    accgg       attctgattc       cggcat         tccga     ggt      tcga     atc              ctcgtacccag    cca
gcatccg             tagct    cagc     tgg    tag       agtactcgg   ctgcaaccgag      cgg            tcga      ggt      tcga     atc              ctccgatgca     cca
g      ggtgata     tagct    cagc     tgg    tag       agcgcacc    cttggtaagggtg    agg            tcggc     agt      tcg      atctgc           gtatcagca      cca
g      ggtgat      tagct    cagt     gag              agcacctcc   cttacaagg        agggg          tcggt     ggt      tcg      atccc            gtcatcac       ccacca
gcgtccg             tagct    cagt     tggt   tag       agcaccac    cttgactgg        tgggg          tccgt     ggt      tcga     gtcca            ctcggac        gcacca
aggggcg             tagtt    caat     tgg    tag       aacaccgt    ctccaaaac        cgggtg         ttg       ggagt    tcga     gtct             ctccgcccctg    cca
```

b

```
        .....tagc.....  tcag....  .tggt.  .agag.  ...cacc.              ...ggtc.    ...ggtt.  ...cgaa.  ...tccc....      ...cacc.
gggccta  tagc            tcagc     tggg     agag    cgccgcttgcacgcagga    ggtctgc     ggt       cgat      ccgcatagctc      cacc
gcgcgt   taac            aaag      cgttatg          tagcggatgcaaatccctct  agtcc       ggtt      cgac      tccggaacgcgc     ctcca
gtcccct  tcgt            ctag      aggc     ccagga  tacctgcctgtcacgcagg   ggtaacagg   ggtt      cgag      tccggtccgttc     cgcca
ggtggcatago tagc        tcagt     tggg     agag    ccctgattgtgattccagt   tgtcgtg     ggt       cgaa      tccattagc        cacccca
aggcttg  tagc            tcagg     tggtt    agagcg  caccctgataaggtga      ggtcgt      ggt       caag      tccactcaggcc     tacca
gggtcgt  tagc            tcagt     tggt     agag    cagttgactttaatcaatt   ggtcgca     ggt       cgaa      tcctcgacgacc     cacca
ggctacg  tagc            tcagt     tggtt    agag    cacatcactcataatgatg   ggtcaca     ggt       cgaa      tccgtcgtagc      cacca
tcctctg  tagt            tcagt     cggt     agaa    cggcggactgttaatccgta  tgtcact     ggt       cgag      tccagtcagag      gagcca
tgggg    tatcg           ccaag     cggt     aagg    cacggattctgattccggc   attccga     ggt       cgaa      tcctcgtaccc      cagcca
gcatccg  tagc            tcagc     tggt     agag    tactcggctgcgaaccgac   ggtcgga     ggt       cgaa      tcctccgatg       cacca
gctgata  tagc            tcagt     tggt     agagcg  caccctggtaagggtga     agtt        agtt      cgaatc    tgcctatcac       cacca
ggtggat  tagc            tcagt     tggg     agag    cacctccttacaaggagg    ggtcgc      ggtt      cgat      ccgtcatcacc      cacca
ggtcccg  tagc            tcagt     tggtt    agag    caccacctgacatggtggg   ggtcgt      ggt       cgag      tccactcggacg     cacca
aggggcg  tagt            caat      tggt     agaa    cacggtcccaaaccg       ggtgttgag   agtt      cgag      tctctcgc         cctgcca
```

FIG. 9. Multiple sequence alignment of 15 tRNA sequences by the program RALIGN. (a) The word size is 3 with up to 1 mismatch in a window of 7. (b) The word size is 4 with up to 2 mismatches in a window of 8.

between consensus words GGT and TCG. Note the failure in the last sequence where the alignment is

Consensus pattern: . . . . ggt . . . tcg
Sequence 16:   ggagt   tcg

This can be accounted for by the "greedy" nature of our algorithm. To allow more chances the consensus matching, whenever there are ties in scoring the algorithm chooses the 5' or leftmost pattern. Therefore, gga is aligned rather than agt, the biologically correct alignment.

Protein Alignment

Figure 10 shows an example of the program PRALIGN applied to nine protein sequences. The sequences represent the amino-terminal 70 residues from a number of regulatory proteins related to the *E. coli ntrC* gene product. This region is fairly well conserved among the proteins and is believed to be responsible for interaction with proteins related to the *E. coli* gene product.

Figure 10 uses a word size of 3 and a window of 6. The neighborhood is limited to a similarity score of 18 or more. Since a perfect match receives a score of 8, an exact matching 3-letter word has score $24 = 3 \times 8$. As with Fig. 9, the top line of the alignment shows the conserved words identified at each position, and the instances of those words are shown in upper case in the sequence alignment below. With the neighborhood as specified, the conserved words are essentially those that contain conservative replacements from the consensus words. The majority of the amino acids in the sequences have been identified as part of conserved words and have been brought into alignment. The sequences are sufficiently dissimilar, however, that a number of short segments are not part of any conserved word.

Conclusion

The programs are available from Waterman. The programs RTIDE and PRTIDE are both written in C and run on a SUN using the SunView windows system. The programs RALIGN and PRALIGN are written in C and do not require the special graphics interface.

Several generalizations of these ideas are possible. One of the most obvious is to apply the methods to single rather than multiple sequences. Our programs for single sequences find the maximal nonoverlapping repeat pattern; as for the programs described in this chapter there are two programs, one for DNA and one for proteins. Elsewhere we have reported methods to find consensus palindromes in DNA, both for multiple and

```
mqr.ktl.lvd..dnc...irq...lve..cln.qeg...fqv..qav...ena...ecl....lnk.....pdv...lll...dim...mpt....
MQRgivw VVDd DSS   IRW   VLEr ALAgagltcttf   ENGa EVLeal ASKt       PDV   LLS  DIR  MPG
MQTpHIL IVEdelv  TRNtlksi    FEAeg YDVf EAT  DGAemhqi  LSEydin           LVIm DIN  LPG
MQEnyki LVVd DDMr LRA  LLEr YLT  EQG  FQV  RSVana   EQMdrl LTResfh        LMVl DLM  LPGedg
MEKiKVC VAD  DNRelvsl   LSEy   IEGqedMEV IGVa YNGq  ECLs  LFKekd   PDV   LVL  DII  MPH
MNE KIIilVD DQYg IREl  LNE  VFN  KEG  YQTf QAAngl   QALdi VTKer    PDL   VLL  DMK  IPG
MSKiRVL SVD  DSAl MRQi MTE  IINshsdm    EMVatap  DPLvardlIKkfn PDV  LTLdve         MPR
mad KELkflvv DDFstmrr   IVRnllkelgf    NNV  EEAe DGVda LNKlqaggyg  FVIsdwnm        PNMdglel
MAR RIL VVE  DEAp IREmvcf   VLE  QNG  FQPveaedy DSAvnq  LNEpw      PDL   ILL  DWM  LPG
MQR ETVwLVEd EQGiadt   LVY  MLQ  QEG  FAVevf  ERGlpvldkarkqv      PDV   MIL  DVG  LPD
```

```
cng...lql...lrr....lkn...nip..vmm....ltv....hge...dei..glq..iga..dfa...spf.cpk...eic..akv.kgl
MDG  LAL  LKQ  IKQrhp MLP  VII   MTA  HSDldaavs  AYQ  QGAf DYLp  KPFdid  EAV  ALVeRAIs
KNG  LLLare    LREqa  NVA  LMF   LTGrdnev  DKIl  GLE  IGAd DYIt  KPF  NPR  ELTirARNllsr
     LSI  CRR  LRSqsn PMP  IIM   VTA  KGEev DRIv  GLE  IGAd DYIp  KPF  NPR  ELL  ARI RAVl
LDGlav    LER  LREsdlkkqpnv    IMLtafgq  EDVtkkav       DLGa SYFilkpfdmenlvghir
MDG  IEI  LKR  MKVide NIR  VII   MTA  YGEl  DMIqeske  LGAlt HFA  KPFdid  EIRdavk KYLplksn
MDGldf    LEKlmrlr    PMP  VVMvssltgkg    SEV  TLRal ELGaidf  VTKp QLGir EGMlayn
     LKTlirad    GAMsa LPVlmvTAEakkeni    IAAaqaga  SGY  VVKpfta  ATLeekl
GSG  IQPikhLKResmtr    DIP  VVM  LTA  RGEee DRVr  GLE  TGAd DYIt  KPF  SPK  ELV  ARI KAVmr
ISG  FEL  CRQlla LHP   ALP  VLF  LTA  RSEev DRLl  GLQ  IGAd DYVa  KPF  SPR  EVC  ARV RTLlr
```

FIG. 10. Multiple sequence alignment of nine protein sequences using the program PRALIGN. The word size is 3 in a window of 6 positions. The neighborhood is limited to words with a similarity score of at least 18, where a score of 24 represents an exact match.

single sequences.[2] In addition, these methods can also be applied to consensus secondary structure. Finally, although we have not done so, it is possible to include other ideas of consensus such as gap length to improve the alignments.

Acknowledgments

This research was supported by grants from the System Development Foundation, the National Science Foundation, and the National Institutes of Health.

[15] k-Tuple Frequency Analysis: From Intron/Exon Discrimination to T-Cell Epitope Mapping

By Jean-Michel Claverie, Isabelle Sauvaget, and Lydie Bougueleret

Introduction

To determine the function of a gene or a protein from mere inspection of its sequence is one of the ultimate goals of research in sequence analysis. In principle, the solution of the protein folding problem, among others, stands as a prerequisite to this accomplishment. Meanwhile, shortcuts have been found, and several methods, mostly based on the recognition of conserved features in genes or proteins with similar functions, can be used to give at least a partial answer to this general problem. For instance, a straightforward way to assign a function to a given sequence is to run exhaustive homology searches on available data banks. As a rule of thumb, an overall 30% amino acid identity (over several hundred residues) between two protein sequences is taken as strongly suggestive of a similar three-dimensional fold and, thus, function.[1,2]

Alternatively, the simultaneous consideration of a set of divergent sequences of known similar function is used to establish characteristic consensus patterns which can be searched for in unassigned candidate sequences. These patterns can be defined on a small number of positions, be highly degenerate, and yet retain a good discriminative power. Such functional signatures usually correspond to key residues in the active (or binding) site of proteins or key nucleotides to be recognized within genes

---

[1] C. Chothia and A. M. Lesk, *EMBO J.* **5,** 823 (1986).

[2] R. F. Doolittle, *in* "Of URFs and ORFs." University Science Books, Mill Valley, CA, 1986.

(regulatory sequences, enhancers, etc.). Good examples of such short, degenerate, but discriminative signals are found in protein kinases,[3] helicases,[4] nucleotide-binding proteins,[5] and for nucleotide sequences in homeoboxes,[6] heat-shock, or hormone-specific enhancers.[7]

However, there are many classes of functions for which neither a sufficient overall homology nor a discriminant functional signature can be found. The signature might exist but be so degenerate as to disappear among the mutational noise; or it may have multiple forms owing to a variety of underlying mechanisms. Finally, the signature can be absent if no local constraint is associated with the function studied. For instance, the classic consensus GT/AG rule[8] and the occurrence of sequence patterns implied by the lariat model[9,10] are, in general, not sufficient to correctly predict the intron–exon boundaries and splicing schemes of a number of genes.[10]

For such problems, we have developed an alternative approach which does not depend on the *a priori* recognition of a single or a few specific, highly discriminant, patterns. Instead, the methods presented in this chapter take advantage of the frequencies of occurrence of all subsequences of length k (k-tuples) as computed from the sequence of interest.[11] Thus, cases where the correlation between function and sequences is only recognizable as a slight statistical bias on the distribution of a large number of k-tuples (and not only by the specific occurrence of a single pattern) become accessible to discriminant analysis. These methods, first developed in the context of intron–exon discrimination[12] have been successfully applied to other subjects like the mapping of T-cell epitopes[13] in protein sequences, as

---

[3] A. Bairoch and J.-M. Claverie, *Nature (London)* **331**, 22 (1988).

[4] T. C. Hogman, *Nature (London)* **333**, 22 (1988).

[5] R. H. Lathrop, T. A. Webster, and T. F. Smith, *Commun. ACM* **30**, 909 (1987).

[6] W. J. Gehring, *Science* **236**, 1245 (1987).

[7] J.-M. Claverie and I. Sauvaget, *CABIOS* **1**, 95 (1985).

[8] R. Breatnach, C. Benoist, K. O'Hare, F. Gannon, and P. Chambon, *Proc. Natl. Acad. Sci. U.S.A.* **75**, 4853 (1978).

[9] M. Konarska, P. Grabowski, R. Padgett, and P. Sharp, *Nature (London)* **313**, 552 (1985).

[10] L. Bougueleret and J.-M. Claverie, *Eur. J. Biochem.* **160**, 279 (1986).

[11] Throughout this article, we insist on using "k-tuple" instead of "word (of length k) because the definition of the latter suggests a potential significance (here biological), which is not at all implied by our analysis. On the contrary, we want to point out that an efficient discriminant analysis is still possible in absence of any recognizable isolated functional signature. However, any subsequence strongly associated with the function under scrutiny will naturally emerge (because of its anomalous distribution) from such an analysis.

[12] J.-M. Claverie and L. Bougueleret, *Nucleic Acids Res.* **14**, 179 (1986).

[13] J.-M. Claverie, P. Kourilsky, P. Langlade-Demoyen, A. Chalufour-Prochnicka, G. Dadaglio, F. Tekaia, F. Plata, and L. Bougueleret, *Eur. J. Immunol.* **18**, 1547 (1988).

well as for the determination of coding regions and reading frames within eukaryotic genes. In this chapter, we first describe the general principle of *k*-tuple frequency analysis methodology and then review some of its applications by order of increasing complexity.

General Presentation of Method

The sequence under scrutiny is studied in connection with one or several reference sets. These reference sets (usually from data banks) are made up of sequences sharing a common property (being an intron, a human protein, etc.) pertinent to a specific analysis (intron–exon discrimination, T-cell immunogenicity). Instead of being regarded as a simple string of individual characters, both the test and reference sequences are considered as being formed of overlapping subsequences of length *k* (*k*-tuples). For instance, the 11-nucleotide sequence ATGCTCAGGAT can be read as a suite of 10 2-tuples: AT,TG,GC,CT,TC,CA,AG,GG,GA,AT, as well as a suite of 9 3-tuples: ATG,TGC,GCT,CTC,TCA,CAG, AGG,GGA,GAT. The alphabets constituting the nucleotide or amino acid sequences are fixed, and the number of distinct *k*-tuples which can be formed are 4^k and 20^k, respectively. For instance, there are 8000 amino acid 3-tuples and 4096 nucleotide 6-tuples.

For a given value of *k*, a table of the frequencies of occurrence of all possible *k*-tuples is computed from the relevant reference set(s) and stored away. Such catalogs of *k*-tuple usage associated with a given function or property can then be used (individually or in combination) for the immediate analysis of any test sequence. This is done as a classic profile analysis: each successive position in the sequence is associated with a numerical figure corresponding to the current *k*-tuple in the relevant table(s). The resulting distribution is a histogram, often smoothed over by the use of a sliding averaging window of variable span.

The mathematical expression of the method is thus as follows. Given a sequence of character $S = \{s_i\}_{i=1,L}$ of length L, we transform it into a sequence of *k*-tuple $S_k = \{s_{k,i}\}_{i=1,L-k+1}$ of length $L - k + 1$. We then consult a *k*-tuple frequency table F containing the statistics of occurrence on all possible *k*-tuples and form the numerical suite:

$$f = \{f_i = F(s_{k,i})\}_{i=1,L-k+1}$$

When a discriminant analysis is performed using two different tables F_1 and F_2 (e.g., intron and exon reference sets), each f_i is computed from an arithmetic combination of $F_1(s_{k,i})$ and $F_2(s_{k,i})$. The distribution f is finally smoothed over to the profile P using an averaging window of $2w + 1$ (odd) consecutive points:

$$P = \left\{ P_i = \sum_{j=i-w}^{j=i+w} f_j \right\}_{i=w+1, L-k-w+1}$$

The optimal choice for the value of k, length of the elementary motif, and $2w + 1$, span of the averaging window, is empirical and depends on the biological context of the sequence analysis. We think of the two parameters k and w as adjusting knobs used to get the profile to resonate well (i.e., become informative) for a given biological property. Nevertheless, the finite size of the protein and nucleotide data banks imposes a limitation on the largest k-tuple which is meaningful to use: the theoretical number of all different k-tuples (20^k and 4^k, respectively) must approximately match the size of the reference set. The present limits are 5-tuples (pentapeptides) for the analysis of proteins and 12-tuples (dodecanucleotides) for the nucleic acids.

As pointed out by Dumas and Ninio,[14] the k-tuple representation of sequences is efficiently handled by a simple coding scheme. First, each individual character of the sequence alphabet is given a collating rank r, for instance A = 1, C = 2, G = 3, and T = 4. Each k-tuple is then represented by a vector $\mathbf{r} = \{r_1, \cdots, r_i, \cdots, r_k\}$. Then each vector is uniquely associated with an integer value $C(\mathbf{r})$ computed as

$$C(\mathbf{r}) = \sum_{i=1,k} (r_i - 1) N^{i-1} + 1$$

where N, the size of the alphabet, is, respectively, 4 or 20 for nucleotide or amino acid sequences. This numerical code is then used to rapidly build the k-tuple frequency table from the sets of reference sequences as well as to consult them while analyzing a test sequence. In the case of nucleotide sequences, this explicit coding can be replaced by an equivalent bit-wise manipulation allowing an even faster access to the k-tuple usage catalogs.

A set of FORTRAN 77 programs has been designed to perform all the tasks involved in the general methodology described here. This includes the computation of k-tuple frequency catalogs from data bank subsets or private sequence collections, software tools for the consultation, editing, and manipulation of these catalogs as well as for the manipulation of k-tuple coded sequences, and interactive programs for the computation and display of the sequence frequency profiles. This set of programs makes use of the data bank implementation of the SASIP sequence analysis package.[15,16] It is available on written request.

[14] J. P. Dumas and J. Ninio, *Nucleic Acids Res.* **10**, 197 (1982).
[15] J.-M. Claverie, *Nucleic Acids Res.* **12**, 397 (1984).
[16] J.-M. Claverie, I. Sauvaget, and L. Bougueleret, *Biochimie* **67**, 437 (1985).

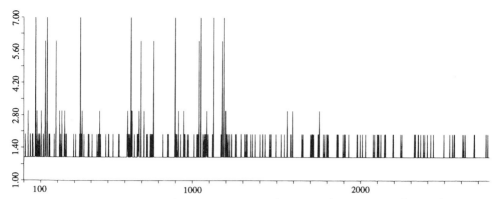

FIG. 1. Repeat density of the 4-tuples composing the bovine thyroglobulin protein sequence. The number of occurrences of each 4-tuple is indicated on the Y axis, and the position in the sequence is marked on the X axis.

Applications Involving Single *k*-Tuple Frequency Catalog

Analysis of Sequence against Itself

At the very first level, the *k*-tuple reference catalog can be constituted from the test sequence itself. The method then provides clear graphical information about the repeated versus unique regions of the molecule. Such a display is especially useful to visualize at once the overall organization of very large proteins (1000 or more amino acids) and genes (several kilobases). The repeat profile can suggest features of primary importance for guiding the detailed study of the sequence, such as the existence of different domains and the evolutionary relationships. As an illustration, Fig. 1 presents the profile obtained with the bovine thyroglobulin amino acid sequence. This profile was computed as follows. The sequence was analyzed as overlapping 4-tuples. For each 4-tuple, the frequency of occurrence (here from 1 up to 7) was recorded in the reference catalog. At each location in the sequence the frequency of occurrence of the current 4-tuple was then plotted. This profile clearly suggests two different parts in the protein: the fragment comprising the first 1200 amino acids is rich in highly repeated motifs, in sharp contrast with the remainder of the molecule. This feature would suggest that a gene fusion occurred around location 1200 at some point during the evolution of thyroglobulin. This hypothesis is further supported by the organization of the gene, which after that location exhibits a high content of very large introns (up to 64 kb).[17]

---

[17] F. Baas, G. B. van Ommen, H. Bikker, A. C. Arnberg, and J. J. M. de Vijlder, *Nucleic Acids Res.* **14,** 5171 (1986).

Use of k-Tuple Frequency Catalogs Computed from Large Reference Sets: Information Profiles

The profiles computed from any type of k-tuple frequency catalog can always be used as a graphical tool to represent very large sequences (several kilobases) in a way allowing one to pick up at a glance some characteristic features. In most applications, the catalogs of k-tuple frequencies are built from large sets of sequences extracted from the available data banks. For instance, one can use all available nucleotide sequences from a given species. In such a case, the frequencies computed for each k-tuple become a good estimate of their true relative abundance within the genome, i.e., their *a priori* probability of occurrence. Accordingly, the frequency profile will now give an image of the local information content at each position of the test sequence: highly informative (significant) regions of the molecules should correspond to low frequency (low probability) values, while peaks in the profile will denote relatively noncharacteristic motifs. This again, can provide useful guidance in the initial steps of sequence analysis. However, the present content of the data banks is neither complete nor random. For instance, the proportion of promoter and coding regions in nucleic acid data banks is certainly much higher than in a regular genome, and this bias is to be kept in mind when interpreting the profile accidents. Yet, the k-tuple frequency analysis methodology is among the rare ones whose results should naturally improve with the foreseen explosion of data bank volume.

Information Profiles of Nucleic Acid Sequences: CpG Islands

We now present a specific example of the use of information profiles in the analysis of a group of polymorphic nucleic acid sequences. Figure 2 shows a 2-tuple (dinucleotide) frequency analysis of the first 2500 nucleotides of the HLA-A3 human major histocompatibility (MHC) gene. The reference set used to compute the frequencies of the 16 dinucleotides was constituted of all human sequences in GenBank[18] (Release 57.00). In this case, the computed frequency profile was not smoothed over ($w = 0$). The profile clearly points out a region (200–1250) exhibiting a high density of lowest frequency dinucleotides: CpG. More precisely, a simple statistical analysis shows that CpGs are strongly avoided outside this precise region while exactly present in the amount expected from the G and C content in the anomalous region.[12] Such a feature is common to all known MHC class I genes,[12] and the location of the CpG islands always coincides with the first three exons of the proteins, which are highly polymorphic.[19] A similar

---

[18] GenBank is a registered trademark of the U.S. Department of Health and Human Services.
[19] P. Kourilsky, *Biochimie* **65**, 85 (1983).

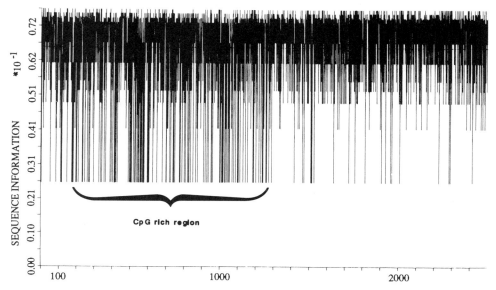

FIG. 2. Dinucleotide frequency histogram of the HLA-A3 gene (for legibility only the first 2500 bases are shown). The Y axis shows the frequency of each 2-tuple in the reference set (here, human sequences from GenBank 57). The CpG-rich region is easily identified. Other such graphs can be found in Ref. 12.

anomalous region, mapping within a known gene conversion region, is found in the α-globin genes. This finding suggested that gene conversion might preferentially occur at highly methylated locations, perhaps through the repair of methylcytosine $\rightarrow$ thymine mutations and, thus, generate polymorphism.[12] Alternatively, DNA polymerases might simply make more errors when replicating rare sequences.

Information Profiles of Protein Sequences: Mapping T-Cell Epitopes

The notion of probability of a given motif particularly makes sense in immunology, where the immune response is naturally directed against macromolecules (mostly proteins) bearing foreign determinants. Because any individual maintains a state of tolerance to its own constituents, antigens can be defined as containing motifs which never (or rarely) occur in the host. Thus, the epitopes, i.e., within the antigen the precise molecular motif subject to immune recognition, are low-probability structures in the host. This notion is not operational at the sequence level for B-cell epitopes (the target of antibody recognition), however, since most of them are conformational determinants made up of noncontiguous residues, as well as oligosaccharides, etc. Recent advances in cellular immunology (see Kourilsky and Claverie[20] and references therein) have demonstrated that T cells recognize protein antigens after their processing into small peptides presented at the cell surface by MHC proteins. Thus, T-cell epitopes (the nominal structure recognized by the T-cell antigen receptor analog to the antibody) are small peptides of 10 to 15 contiguous amino acids. Because of the tolerance to self-proteins, T epitopes should be characterized by sequence motifs of low probability within the self-protein of a given species. This proposal has been tested and statistically validated using the human and mouse (the only species for which there is a sufficient amount of data) protein sequences available in data banks.[13] As a consequence, the search for T-cell epitopes can be now directed by the information profile of the putative immunogenic protein. Candidate T epitopes will be indicated by regions of the profile exhibiting the lowest probability of occurrence in the relevant (human or mouse) self-protein set.

Figure 3 presents the information profile of the human immunodeficiency virus *gag* polyprotein (HIV-1, BRU strain) which allowed the mapping of four T epitopes of the human cellular response (as detected in patients with acquired immunodeficiency syndrome). Here the sequence was analyzed in 4-tuples, using a catalog of 4-tuple frequencies computed from all available human protein sequences (with the exception of the immune-related proteins, the tolerance status of which is unclear: immunoglobulins, MHC proteins, etc.). The graph was smoothed over by an averaging window of seven positions. Thus, the profile reaches the baseline (average frequency 0) only when the sequence contains at least seven consecutive tetrapeptides, as yet never seen in any human sequence.

Discriminant Analysis Using Two *k*-Tuple Frequency Catalogs

Principle of Discriminant Analysis

With the use of two reference sets, the general methodology of *k*-tuple frequency analysis can be turned into a straightforward discriminant analysis. We thus now consider two mutually exclusive properties (or functions), such as being an intron or an exon, being a coding reading frame or not, then build two reference sets of sequences corresponding to each of the opposite properties. For a given *k*-tuple size, the two reference sets will be characterized by two *a priori* distinct frequency distributions, $F_1(s_k)$ and $F_2(s_k)$, where F_1 and F_2 are again two tables of 4^k (for nucleic acids) or 20^k (for proteins) frequencies corresponding to all possible *k*-tuples s_k.

[20] P. Kourilsky and J.-M. Claverie, *Avd. Immunol.* **45,** 107 (1989).

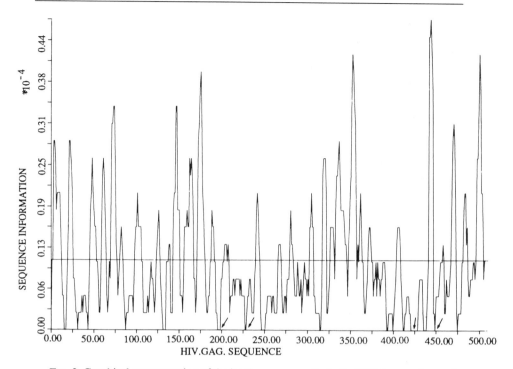

FIG. 3. Graphical representation of the local sequence rarity in the HIV-1 *gag* polyprotein sequence (BRU isolate). The Y axis shows the relative frequency of each 4-tuple in the reference set (here, human somatic proteins). The four peptides tested and confirmed as T-cell epitopes are indicated by arrows. (From Ref. 13.)

k-Tuples which are found at the same (relative) frequencies in the $F_1 F_2$ tables are, by definition, nondiscriminant between the two functions 1 and 2. On the contrary, k-tuples with frequencies that largely differ in the two reference sets are the most discriminant. This will naturally include bona fide specific functional signatures of size k, if they exist (for instance, k-tuples always present in sequences of type 1 and absent from sequences of type 2).

To analyze the test sequence, we now compute a local discriminant index, defined for each position in the sequence as

$$d(i) = F_1(s_{k,i})/[F_1(s_{k,i}) + F_2(s_{k,i})]$$

For a given k and the k-tuple frequency catalogs F_1 and F_2 computed from two reference sets (e.g., introns and exons) once and for all, this index varies in the test sequence from one position to the next as a function of the current k-tuple, $s_{k,i}$. This index takes its values in the range 0–1. It will be close to 0 for k-tuples occurring at a negligible frequency in the sequence set associated with function 1 and reach 1 for the converse situation. It will stay around 0.5 for all locations in the test sequence where the current k-tuple exhibits a similar frequency in both reference sets (whatever its precise high or low value). If the frequency of a given k-tuple happens to be 0 in both sets, the discriminant index is also set to 0.5 (this case is not supposed to occur frequently if the size of the reference sets are well matched to the total number of k-tuples). The results of this discriminant analysis is displayed along the sequence as a profile varying around 0.5 down to 0 and up to 1. A natural interpretation is thus that, at location i, the sequence appears to have a $d(i)$ probability of belonging to the reference set 1 (and a complementary probability of belonging to the other reference set). In this representation, the averaging of the primary profile with a sliding window becomes essential to connect domains of consistent tendencies (e.g., "1-like" regions) over k-tuples with no discriminant power. The profile then provides a heuristic but quantitative mesure of similarity with one or the other function specific reference sets (we like to speak of function related flavor or taste). Again, the choice of the value of k and of the averaging window span $(2w + 1)$ is empirical and must be optimized for each problem.

We present below a classic application of this method for exon–intron discrimination in genomic sequences.[12] However, we have used the same method for studying sequences from many other points of view. Thus, specific sequence signals can be tentatively assigned even when they do not fit into the scheme of two mutually exclusive properties. In this case, the second frequency catalog F_2 can be theoretically computed. For instance, searching for specific signals in promoters can be done by computing the actual k-tuple frequencies from a collection of promoter sequences sensitive to the same factors and using them against theoretical k-tuple frequencies computed from their overall nucleotide composition. This way, anomalous (positive or negative) regions of a test sequence will be revealed against a background of theoretically expected motifs. These techniques can be used to map function-specific subsequences in proteins as well.

Intron–Exon Discrimination in Unassigned Genomic Sequences

The development of automated techniques makes possible the determination of the sequence of large loci or even complete genomes. In most cases, very little information will be available on the organization of the genes prior to sequence determination. The boundaries of the transcription units, the splicing scheme, the location of the coding regions (if any), the function of the corresponding protein(s) are all problems which should be first taken up by the study of the raw sequence. Eukaryotic genomic sequences are characterized by the scarcity (a few percent) of "useful" sequences (exon) compared to "junk" DNA (intron). However, since most functions are related to proteins, most of the analysis will usually concentrate on the exons. Computerized tools to locate them are thus welcome.

Since exon and intron sequences have been submitted to very different evolutionary pressures, their k-tuple compositions are likely to be different, and the use of a dual frequency catalog discriminant analysis is very natural in this context. This is apparent from a two-dimensional rank analysis performed on the frequency distribution of all 6-tuples in exon versus intron sequences (Fig. 4).[21] Each point represents a given 6-tuple, the coordinates of which are its rank in the exon (X axis) or intron (Y axis) frequency distribution. Near the diagonal are found all the 6-tuples occurring at a similar frequency in both sets: they have no discriminant power. The farthest from the diagonal are the most discriminant ones and, indicated by arrows, the U1 snRNA homologous "consensus sequences." Unfortunately, these consensus sequences are not bona fide functional signatures because their presence suffers numerous exceptions.[10] The consideration of these sole motifs to locate intron boundaries leads to very unsatisfactory results. In contrast, our discriminant analysis, based on the whole 6-tuple frequency catalogs, takes advantage of all the information contained in this graph, from the most specific 6-tuples to those exhibiting only a slight statistical bias.

Figure 5 presents the analysis of the first 2500 nucleotides of the HLA-A3 gene encompassing the first five exons of the corresponding MHC class I protein. For intron–exon discrimination we determined that the frequencies of 6-tuples were the most discriminant and that large averaging windows (31–41) gave clearer profiles.[12] The F_1 and F_2 frequency catalogs were computed from all vertebrate exon and intron sequences available in GenBank (Release 54.00). For this we used a modification of the program we designed to generate the PGtrans[21] data bank (translated coding regions of GenBank). Note, however, that no consideration of reading phase is taken into account in the analysis (as overlapping 6-tuples) of the exon, intron, or test sequences. The notion of "codon usage" is thus not used in the present method (see below for phase-dependent analysis). Because of the arbitrary catalog assignment (F_1, exons; F_2, introns), regions of the gene exhibiting more similarity to exons are above the 0.5 line, while the intronlike regions are below.

In interpreting such graphs, one looks for uninterrupted domains of consistent tendency. The actual boundaries of the five first exons are

---

[21] J.-M. Claverie and I. Sauvaget, *Nature (London)* **318,** 19 (1986).

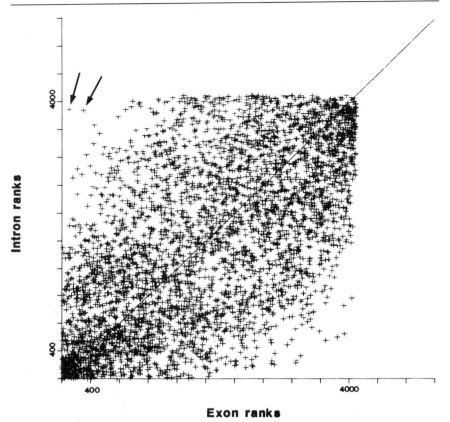

Exon ranks

FIG. 4. Two-dimensional rank distribution of 6-tuples in exons versus introns. Each 6-tuple is represented by a point, the coordinates of which are its frequency rank in each set of reference sequences, the less frequent being assigned rank 1 and the most frequent rank 4096. The 6-tuples appearing as a scattered cloud around the diagonal are those with similar frequencies in both reference sets. By contrast, the arrows point to the most discriminant 6-tuples (GTAAGT and GTGAGT) complementary to the U1 snRNA. (Adapted from Ref. 22.)

indicated by arrows. This figure truly represents a discriminant profile of average quality. Much sharper results are obtained in many cases.[12] Some features of this profile are consistently found: hence, the 5′ boundaries of exons are usually more accurately mapped than their 3′ counterparts. This correlates with our observation that the exon open reading frame often continues for a while in the next intron. The 5′ boundaries of the first coding region (indicated by E1) are normally situated downstream from the beginning of the exonlike domain owing to the presence of 5′ untrans-

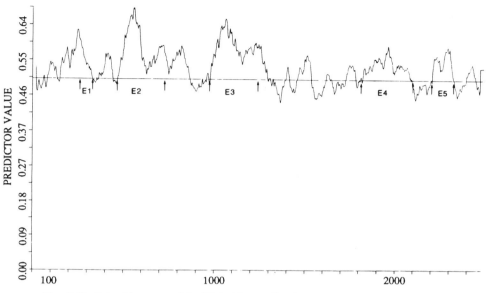

FIG. 5. Predictor histogram of the HLA-A3 gene (first 2500 bases). The sequence coded in 6-tuples is scanned against two reference catalogs (here, exons and introns), and the probability index computed as indicated in the text is noted on the *Y* axis. The reported positions of the exons are delineated by arrows. Parasite peaks, found in all MHC genes, may reflect an alternative splicing propensity.[10]

lated mRNA sequences in the exon reference set. The interpretation of such profiles is routinely complemented by the search for a 5′ GT box (U1 snRNA homologous sequences), a 3′ AG box, and 5′ GT box complementary sequences in the proximity of the predicted intron–exon boundaries.[10]

*Phase-Dependent Frequency Catalogs: Determination of
Protein Coding Frame*

Specific evolutionary constraints are selectively imposed on the coding regions of DNA sequences and are the source of statistical biases in their composition. Some are linked to a preferential codon usage (itself related to several causes) and others to better resistance to mutational drift (as revealed, e.g., by higher CpG content;[22] see also above). The *k*-tuple frequency analysis method offers a way to take advantage of any overall statistical bias without trying to identify its various components or under-

[22] L. Bougueleret, F. Tekaia, I. Sauvaget, and J.-M. Claverie, *Nucleic Acids Res.* **16,** 1729 (1988).

stand its origins. It is thus natural to apply it to the detection of protein coding regions (and reading frames) in eukaryote genes. We thus introduced the notion of phase in the building of the catalogs and computed phase-dependent k-tuple frequencies. Taking a reference set of all (e.g., vertebrate) available coding sequences, we successively computed the frequency distribution of the following:

6-tuples $F_{0,6}$ in phase with the coding frame:
    ATGCTAGATCGACATAGA . . .,
    ATGCTA
        CTAGAT
            GATCGA

$$\vdots$$

Different 6-tuple distribution $F_{1,6}$ with a shifted origin
    ATGCTAGATCGACATAGA . . .
     TGCTAG
        TAGATC
           ATCGAC.

$$\vdots$$

Third 6-tuple distribution $F_{2,6}$ shifted once more:
    ATGCTAGATCGACATAGA . . .,
      GCTAGA
         AGATCG
           TCGACA

$$\vdots$$

A priori, these three different 6-tuple frequency catalogs can be used in three different combinations to define a discriminant index as explained in the previous section. Each of these combinations is in turn used as a reference for the analysis of the three possible frames of the test sequence. In fact, we have determined in numerous cases that the most discriminant profile is always obtained with the F_0 versus F_1 catalogs, i.e., with the discriminant index defined as:

$$d(i) = F_0(s_{k,i})/[F_0(s_{k,i}) + F_1(s_{k,i})] \qquad i = i_0, i_0 + 3, i_0 + 6, i_0 + 9, \ldots$$

This seems to indicate that there is much less information (i.e., it is more random) in the F_2 (2 out of phase) k-tuple frequency distribution than in the F_0 (coding phase) and F_1 (one out of phase) distributions. As for intron–exon discrimination, we found that a value of $k = 6$ and large averaging window ($w = 20$) were optimal. Note that when phase-dependent catalogs are used, the test sequence is also analyzed in a phase-dependent way, with the profile being truly defined every three nucleotides, starting from position $i_0 = 1$, 2, or 3. For the purpose of combining different profiles, intervening positions are in fact filled in with the preceding value: $V_1 v_1 v_1 V_2 v_2 v_2 V_3 v_3 v_3$. . . .

To explore all possibilities of coding frames, three profiles of F_0 versus F_1 discriminant analysis are needed, each of them giving information on two reading frames at once. Figure 6 presents again the first 2500 nucleotides of the HLA-A3 gene, analyzed with the F_0 versus F_1 6-tuple frequency catalogs, starting at $i_0 = 2$. In this profile (referred to as P_2), putative coding regions (indicated by hatching) correspond to high-intensity peaks, either above or below the $p = 0.5$ baseline. Here, peaks above 0.5 denote coding regions in phase with i_0; peaks below 0.5 denote coding regions in phase with $i_0 - 1$. The leftover reading phase or noncoding regions would appear as fluctuations around $p = 0.5$. The profile presented in Fig. 6 already correlates well with the locations of the five exons indicated in Fig. 5. Proper arithmetic combinations with the two other profiles obtained for $i_0 = 1$ (P_1) and $i_0 = 3$ (P_3) allow a better mapping of the coding regions in the respective reading frames. Coding regions in phase with the $i_0 = 1$, $i_0 = 2$, and $i_0 = 3$ frames will appear as positive peaks in the $(P_1 - P_2)$,

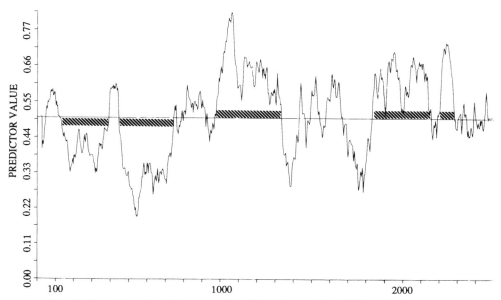

FIG. 6. Phase-dependent histogram of the HLA-A3 gene (first 2500 bases). The sequence has been coded in 6-tuples starting from position $i_0 = 2$, and the index noted on the Y axis has been computed from the two catalogs $F_{0,6}$ and $F_{1,6}$ as indicated in the text.

$(P_2 - P_3)$, and $(P_3 - P_1)$ combined profiles, respectively. Again, the output of this algorithm is routinely complemented by the search for initiation or stop codons or putative splice junctions.

Conclusion

The general methodology of k-tuple frequency analysis can be applied to different aspects of the interpretation of nucleotide or amino acid sequences of unknown function, in particular when no obvious homology is found with documented sequences. It has been used most successfully to locate exon–intron positions in genomic eukaryotic sequences and T-cell epitopes in immunogenic proteins. An important feature is that its application to a given problems does not require the prior recognition of consensus patterns, nor any knowledge of the structure–function relationship. It takes advantage, in a nonexplicit way, only of the statistical bias in k-tuple composition which often betrays the common evolutionary history of sequences endowed with a similar function. Its most important limitation, however, is to require the constitution of good quality reference sets of a large number of well-documented sequences. Obviously, not all function discrimination problems can meet this constraint. Finally, it must be noted that the poor quality of the current data banks does not allow sequence subsets to be directly taken out (e.g., by computer query on the feature table) to serve as reference sets without substantial "manual" checking of errors and redundancies.

[16] Splice Junctions, Branch Point Sites, and Exons: Sequence Statistics, Identification, and Applications to Genome Project

By Periannan Senapathy, Marvin B. Shapiro, and Nomi L. Harris

Introduction

Almost all the nuclear genes coding for proteins in eukaryotes are split into coding (exon) and noncoding (intron) sequences. The intron sequences are precisely spliced out of the initial gene transcript before the mRNA is transported to the cytoplasm for translation. The mechanism of splicing, by which introns are eliminated and exons are connected, is becoming understood in increasing detail.[1-3] It has been found that sequences immediately bordering splice junctions are strongly conserved in genes of a wide variety of eukaryotic organisms, ranging from yeast to humans.[4]

There exists a sequence of 8 nucleotides, highly conserved at the boundary between an exon and an intron, referred to as the 5' splice site (5' ss). The boundary between an intron and an exon, the 3' splice site (3' ss), also exhibits a highly conserved sequence of 4 nucleotides, preceded by a pyrimidine-rich region. The 5' and 3' splice sites are sometimes referred to as donor and acceptor sites. The first step in splicing is the assembly of a large ribonucleoprotein complex called a spliceosome on the pre-mRNA.[1] Before an intron is released, it forms an intermediate structure called a lariat. The final step of the splicing process occurs when the two exons are joined and the intron is released as lariat RNA.[5] A short 5–8 nucleotide sequence, containing the adenosine residue at which the lariat is formed, functions as the signal for lariat structure formation. This signal, which is called the branch point signal, lies within the intron, usually between 10 and 50 nucleotides upstream of the acceptor splice site.[6] These short conserved sequences are an essential part of the process of exon splicing and provide specific molecular signals by which the RNA splicing machinery can identify the precise splice points.

Breathnach and Chambon,[7] using a small number of published sequences, first noticed that a consensus sequence existed at each end of introns, and, soon after, Mount[8] made the first catalog of splice junction sequences, listing 269 (139 5' ss and 130 3' ss) sequences. More recently Shapiro and Senapathy[9] made a systematic analysis of the RNA splice junction sequences of eukaryotic protein coding genes using Release 46.0 of the GenBank data bank.[10] The accumulation of DNA sequences in the nucleic acid sequence data banks has continued at an even more rapid pace, and we report here the latest results of an analysis of splice sites in GenBank, which contains approximately two and one-half times as many such sites as last reported.

[1] P. A. Sharp, *Science* **235,** 766 (1987).
[2] P. Senapathy, *Proc. Natl. Acad. Sci. U.S.A.* **83,** 2133 (1986).
[3] M. R. Green, *Annu. Rev. Genet.* **20,** 671 (1986).
[4] R. Breathnach, C. Benoist, K. O'Hare, F. Gannon, and P. Chambon, *Proc. Natl. Acad. Sci. U.S.A.* **75,** 4853 (1978).
[5] J. W. S. Brown, G. Felix, and D. Frendeway, *EMBO J.* **5,** 2749 (1986).
[6] J. W. S. Brown, *Nucleic Acids Res.* **14,** 9549 (1986).
[7] R. Breathnach and P. Chambon, *Annu. Rev. Biochem.* **50,** 349 (1981).
[8] S. M. Mount, *Nucleic Acids Res.* **10,** 459 (1982).
[9] M. B. Shapiro and P. Senapathy, *Nucleic Acids Res.* **15,** 7155 (1986).
[10] H. S. Bilofsky and C. Burks, *Nucleic Acids Res.* **16,** 1861 (1986).

Nucleotide frequencies obtained for the highly conserved regions around splice sites for different categories of organisms closely resemble each other. A striking similarity among the rare splice junctions which do not contain AG at the 3′ splice site or GT at the 5′ splice site indicates the existence of special mechanisms to recognize them, and it suggests that these unique signals may be involved in crucial gene regulation events as well as in differentiation. We have also carried out a systematic analysis of the distribution and consensus of the branch point sequences in various regions of genes using the GenBank sequences.[11] We found that, except for a slightly higher frequency of the branch point sequence in the expected "lariat" region, branch point sequences were distributed nearly randomly in other regions of introns as well as in exons.

In order to interpret the deluge of sequence information generated by the human genome project, it is important to be able to identify genes in uncharacterized sequences. Without this capability, the sequence information is of little value. We have used the tabulated consensus scoring matrices to find the most probable splice sites in a given sequence. By comparing the actual splice sites in a number of known gene sequences with those predicted by our method, we find that the actual splice sites can be identified with a great degree of accuracy, indicating that this method can be used to find potential splice sites in uncharacterized DNA sequences. A method was developed to predict potential exons in an uncharacterized sequence based on splice site scores and by using other parameters of exons and eukaryotic coding sequences. However, although this method could identify some complete exons of a gene, it cannot identify all the exons of a gene completely. As we discuss below, gene signatures, severalfold more than those which circumscribe actual exons, exist in a gene sequence, and only some exons of a gene can be clearly identified. Thus, the problem of identifying complete genes is an order of magnitude more complex than finding individual exons, making it necessary to identify the rules which select the real gene attributes among the many randomly distributed ones. The use of multiple parameters to detect exons, which may yield more accurate and reliable results than the use of single parameters, is also discussed.

Methods

Release 57.0 of the GenBank, dated September 1, 1988, was used in this study. The data from seven GenBank files: PRIMATE, RODENT, MAMMAL, VERTEBRATE, INVERTEBRATE, PLANT, and VIRAL were used in the counts; separate tabulations were also made for immuno-

---

[11] N. Harris and P. Senapathy, *Nucleic Acids Res.* (submitted for publication).

globulin and yeast genes. The ORGANELLE category in GenBank was also looked at, but the count of consensus splice sites was too small to be included; most ORGANELLE splice sites do not conform to the eukaryotic consensus, and they are discussed below. A total of 3724 5′ splice sites and 3683 3′ splice sites were examined from the above seven categories.

For each sequence in GenBank, the splice sites were located using the information in the annotation preceding the sequence, a FEATURES line being followed by multiple lines indicating different features present in the gene. The feature used here was the IVS line, which indicates the first and last location of an intervening sequence (intron) between exons, thus giving the location of the G in GT at the 5′ ss and the G in AG at the 3′ ss for all sequences following the established consensus. Nucleotide counts were made for positions −7 to +40 for all 5′ ss with a GT at the beginning of the intron and for −60 to +5 for all 3′ ss having an AG at the end.

The 5′ splice sites not conforming to the GT or 3′ ss to the AG rule are called nonconsensus sites. The nonconsensus sites found in GenBank fall into one of the following three categories: (1) they are in pseudogenes or genes where there is uncertainty in the sequence; (2) the IVS reported appears to be an error (e.g., where the end location of an intron is off by one from the location where G in AG would normally be); or (3) splice sites that must be accepted as nonconsensus. Owing to the large number in category 2, only those sites in category 3 are reported here. Thus, the listing given here is conservative, erring in the direction of showing fewer nonconsensus splice sties than may actually exist. The number found for each category is 65 for category 1, over 100 for category 2, and these do not include a large group of organelle sequences and transfer RNAs, which are discussed separately below, and 49 for category 3, listed below in Tables III and IV.

The sites in category 2 represent errors either in reporting of results or in the GenBank entry of the data, or finally they may be in fact true nonconsensus sequences. Because laboratory sequencing is difficult and time consuming, and since data entry is done carefully by GenBank, most of the errors probably originate in the laboratory. Sequence data that contain nonconsensus splice site sequences should be carefully examined before being submitted to GenBank.

In the second part of the study, the tabulated nucleotide frequencies, obtained as described above, were used to find potential splice sites in a given sequence. The formulas for this are described in Shapiro and Senapathy.[9] In our analysis of the branch point sequences, we searched for their occurrence in several "windows" upstream and downstream of both 5′ and 3′ splice junctions. The search was based on a composite scoring matrix which we built from published data.[11] Sequences with a score of at least

97% were selected as branch point sequences in each window, and statistics for their occurrence in various windows as well as in different GenBank sequence categories were assembled.

Results and Discussion

Splice Junction Consensus Sequence from Different Classes of Organisms

Table I gives the percentage of each nucleotide at 5′ ss positions −3 to +6 and 3′ ss positions −14 to +1, for the seven GenBank categories, and Table II for immunoglobulin and yeast genes. The following rule was used in arriving at a consensus sequence for each location: If the highest percentage computed for a particular nucleotide site equals or exceeds 40, choose the corresponding nucleotide; choose also the nucleotide with the second highest percentage if it equals or exceeds 30 and is at least twice as large as the third highest percentage. The consensus nucleotides thus found are listed at the end of each row in Tables I and II. (The consensus nucleotides reflect the true percentages, not the rounded values in the table.)

For the complete data bank sequences (i.e., the total of the seven GenBank categories mentioned in Methods), the following consensus sequences can be established:

5′ splice site $\quad$ AGGT$_\text{G}^\text{A}$AGT

3′ splice site $\quad$ $_\text{CC}^\text{TT}$TT$_\text{CCCCCC}^\text{TTTTTT}$NCAGG

where N is any nucleotide and where the nucleotide above is more common among the two nucleotides at a given position. The pyrimidine-rich region preceding the 3′ ss actually extends well into the intron. Figure 1 gives a summary of the consensus splice junction sequences for the seven GenBank categories of organisms, the immunoglobulins, and the yeast sequences. The consensus sequences at the top (total of all seven categories) are identical to those reported earlier.[9]

All seven subtables in Table I (and the immunoglobulin listed in Table II) have very similar consensus sequences (although the invertebrate, plant, and viral genes have less G following the 5′ ss GT, and less C in the pyrimidine-rich region preceding the 3′ ss AG), and all are essentially the same as given previously.[9] The yeast subtable in Table II shows a quite different consensus, however, but this is inconclusive for two reasons: on one hand, the number of consensus sites counted is small, while on the other, the number of nonconsensus yeast sites is large (see below).

TABLE I
Nucleotide Percentages at Splice Junctions[a]

Location	Total 3724					Primate 1333					Rodent 1129					Other mammals 173				
	A	C	G	T	CN[b]	A	C	G	T	CN	A	C	G	T	CN	A	C	G	T	CN
5' splice sites																				
−3	32	37	18	13		28	40	17	14	C	33	38	15	14		36	36	18	9	
−2	60	13	12	15	A	59	14	13	14	A	63	11	12	14	A	59	13	16	13	A
−1	9	5	79	7	G	8	5	81	6	G	8	4	82	6	G	8	5	79	8	G
+1	0	0	100	0	G	0	0	100	0	G	0	0	100	0	G	0	0	100	0	G
+2	0	0	0	100	T	0	0	0	100	T	0	0	0	100	T	0	0	0	100	T
+3	59	3	35	3	AG	54	2	42	2	AG	58	4	38	3	AG	45	2	49	4	GA
+4	71	9	11	9	A	74	8	11	8	A	74	6	12	8	A	71	7	14	8	A
+5	7	6	82	6	G	5	6	85	4	G	8	5	81	6	G	6	3	88	2	G
+6	16	16	18	50	T	16	18	21	45	T	19	14	20	47	T	11	17	21	51	T
	3683					1293					1104					192				
3' splice sites																				
−14	11	29	14	46	T	10	31	14	44	TC	9	33	13	45	TC	6	33	9	53	TC
−13	11	33	12	44	TC	8	36	14	43	TC	8	34	10	48	TC	4	46	11	40	CT
−12	10	30	10	50	TC	6	34	12	48	TC	7	34	8	51	TC	4	35	8	53	TC
−11	8	30	10	52	TC	6	34	8	52	TC	5	31	10	54	TC	3	29	6	62	T
−10	11	32	9	48	TC	9	37	8	45	TC	5	33	9	50	TC	6	44	9	41	CT
−9	10	34	11	45	TC	9	38	10	44	TC	8	33	11	49	TC	11	28	11	50	T
−8	11	37	10	42	TC	8	44	9	40	TC	9	38	11	42	TC	5	49	10	36	CT
−7	11	38	9	43	TC	9	41	8	41	CT	11	39	6	44	TC	4	45	5	46	TC
−6	7	39	7	47	TC	6	44	8	45	TC	5	41	6	47	TC	9	53	6	32	TC
−5	8	36	6	51	TC	7	40	6	48	TC	7	38	5	49	TC	7	53	5	36	CT
−4	25	26	26	23		23	28	26	23		28	28	21	22		23	24	25	28	
−3	3	75	1	21	C	2	79	1	18	C	3	76	1	20	C	3	78	2	18	C
−2	100	0	0	0	A	100	0	0	0	A	100	0	0	0	A	100	0	0	0	A
−1	0	0	100	0	G	0	0	100	0	G	0	0	100	0	G	0	0	100	0	G
+1	27	14	49	10	G	28	14	47	11	G	25	14	52	9	G	16	22	55	6	G

(continued)

TABLE I (continued)

5' splice sites

Location	Other vertebrates 417					Invertebrate 288					Plant 219					Viral 165				
	A	C	G	T	CN	A	C	G	T	CN	A	C	G	T	CN	A	C	G	T	CN
−3	35	35	19	11		36	28	21	15		35	33	21	11		35	31	23	10	
−2	59	13	13	15	A	55	18	9	18	A	50	15	8	27	A	65	15	9	10	A
−1	10	5	76	9	G	12	10	71	7	G	8	6	73	13	G	22	8	67	3	G
+1	0	0	100	0	G	0	0	100	0	G	0	0	100	0	G	0	0	100	0	G
+2	0	0	0	100	T	0	0	0	100	T	0	0	0	100	T	0	0	0	100	T
+3	66	2	30	2	AG	75	0	21	4	A	68	8	16	9	A	71	4	21	1	A
+4	72	9	10	10	A	76	13	4	8	A	41	23	8	28	A	59	13	22	5	A
+5	9	5	81	5	G	6	9	81	4	G	12	6	71	12	G	9	7	73	10	G
+6	14	16	19	51	T	13	10	5	72	T	12	17	9	62	T	13	15	16	56	T

3' splice sites

Location	Other vertebrates 409					Invertebrate 294					Plant 219					Viral 172				
	A	C	G	T	CN	A	C	G	T	CN	A	C	G	T	CN	A	C	G	T	CN
−14	13	30	14	43	T	18	16	17	49	T	21	16	14	49	T	8	18	19	55	T
−13	8	33	10	49	TC	24	25	11	40	T	24	18	16	42	T	21	26	17	36	
−12	9	26	11	54	T	25	20	8	46	T	22	20	13	45	T	19	20	15	46	T
−11	7	31	9	54	TC	26	22	11	41	T	16	21	15	48	T	8	22	14	56	T
−10	8	27	9	56	T	27	18	9	46	T	23	20	18	39	T	16	19	6	59	T
−9	6	36	14	44	TC	21	40	6	33		21	21	18	40	T	6	16	19	59	T
−8	7	36	8	49	TC	33	19	9	39		23	18	16	42	T	11	18	14	57	T
−7	8	38	8	46	TC	21	34	10	35		19	15	22	44	T	10	29	18	43	T
−6	5	40	5	50	TC	17	26	6	51	T	16	20	18	47	T	6	27	6	60	T
−5	7	38	8	55	TC	13	26	3	58	T	15	15	11	59	T	10	18	5	67	T
−4	24	26	30	21	C	29	20	23	28		28	13	37	21	T	26	22	30	22	T
−3	5	74	0	21		3	66	2	30	C	4	63	0	32	CT	5	64	2	29	C
−2	100	0	0	0	A	100	0	0	0	A	100	0	0	0	A	100	0	0	0	A
−1	0	0	100	0	G	0	0	100	0	G	0	0	100	0	G	0	0	100	0	G
+1	35	27	14	50	G	47	13	29	11	A	21	11	56	12	G	20	12	57	12	G

[a] The percentages of nucleotides at 5' ss locations −3 to +6 and 3' ss locations −14 to +1 are given, for the total and for each of the seven Genbank categories making up the total. The count is at the top of each category.

[b] CN, Consensus nucleotides.

TABLE II
NUCLEOTIDE FREQUENCIES IN IMMUNOGLOBULIN AND YEAST SPLICE SITES

	Immunoglobulin 334					Yeast 24				
Location	A	C	G	T	CN[a]	A	C	G	T	CN[a]
5′ splice sites										
−3	25	53	9	13	C	46	21	17	17	A
−2	72	11	4	13	A	33	17	17	33	
−1	7	9	80	4	G	17	0	63	21	G
+1	0	0	100	0	G	0	0	100	0	G
+2	0	0	0	100	T	0	0	0	100	T
+3	60	5	32	4	AG	92	0	8	0	A
+4	82	2	13	3	A	29	8	13	50	T
+5	9	3	79	8	G	4	0	96	0	G
+6	24	14	20	41	T	17	4	0	79	T

	340					27				
	A	C	G	T	CN[a]	A	C	G	T	CN[a]
3′ splice sites										
−14	10	30	18	43	T	26	26	11	37	
−13	8	36	5	51	TC	30	19	19	33	
−12	6	31	8	56	TC	41	19	11	30	A
−11	1	29	3	67	T	26	30	7	37	
−10	10	47	5	38	CT	52	19	4	26	A
−9	8	39	7	47	TC	22	11	4	63	T
−8	14	42	13	31	CT	37	15	7	41	TA
−7	17	30	5	47	T	22	7	4	67	T
−6	2	51	2	45	CT	26	11	19	44	T
−5	6	47	1	46	CT	30	4	7	59	T
−4	27	18	27	28		37	19	15	30	
−3	1	90	0	9	C	26	11	0	63	T
−2	100	0	0	0	A	100	0	0	0	A
−1	0	0	100	0	G	0	0	100	0	G
+1	17	15	61	7	G	30	19	33	19	

[a] CN, consensus nucleotides.

Striking Similarities among Nonconforming Splice Sites

Of more than 7500 splice sites examined in the study (not counting pseudogenes), 49 are reported in Tables III and IV as most probably nonconsensus sites (i.e., not having a 5′ ss GT or 3′ ss AG). This is after discounting over 100 entries as probable errors and treating over 200 yeast and organelle nonconsensus sites separately (see below).

	5' Splice Site	count	Intron	3' Splice Site	count
Total	NAG$\boxed{GT}$$^A_G$AGT	3724	....	NTN^T_CT$^T_C$$^T_C$$^T_C$$^T_C$$^T_C$$^T_C$$^T_C$$^T_CNC\boxed{AG}$G	3683
Primate	CAGGT^A_GAGT	1333	....	NNN$^T_C$$^T_C$$^T_C$$^T_C$$^T_C$$^T_C$$^C_T$$^T_C$$^T_C$NCAGG	1293
Rodent	NAGGT^A_GAGT	1129	....	NTNN$^T_C$$^T_C$$^T_C$$^T_C$$^T_C$$^T_C$$^T_C$$^T_C$NCAGG	1104
Mammal	NAGGT^G_AAGT	173	....	NN^T_CN$^T_C$$^C_TT^C_TT^C_TT^C_T$$^C_T$NCAGG	192
Vert	NAGGT^A_GAGT	417	....	TT^T_CTT^T_CT^T_CT^T_CT$^T_C$$^T_C$$^T_C$$^T_C$NCAGG	409
Invert	NAGGTAAGT	288	....	TTTTTTTTTNNNTTNCAGA	294
Plant	NAGGTAAGT	219	....	NTNTTTTTNTTTTTN^C_TAGG	219
Viral	NAGGTAAGT	165	....	NTTNTNTTTTTTTTNCAGG	172
IG	CAGGT^A_GAGT	334	....	N^T_CN^T_CT$^T_C$$^C_TT^C_T$$^C_TT^C_T$$^C_T$NCAGG	340
Yeast	ANGGTATGT	24	....	T$^T_A$$^T_A$TNNANAT$^T_A$TTTNTAGN	27

FIG. 1. Consensus splice junction sequences for three groups of genes: the seven GenBank categories primate, rodent, mammal, vertebrate, invertebrate, plant, and viral, and their total; immunoglobulin genes; and yeast genes. The consensus was arrived at using the nucleotide percentages (given in Tables I and II) at the splice sites, which were computed as described in Ref. 9.

Of the 24 nonimmunoglobulin 5' ss 6 have almost the same sequence, AAG/GCAAGC. In our previous study,[9] of 8 nonconforming sites 5 had this sequence. This subset of nonconforming splice sites is found in genes coding for quite different kinds of proteins, suggesting a regulatory function for such unique sites. It is possible that genes containing such rare nonconforming splice sites may function as crucial nodes in the regulation of the genetic pathway of cell growth and differentiation. A particular example supporting this possibility is the splicing of *Drosophila* myosin genes. In *Drosophila* larvae, myosin light chain pre-mRNA is spliced by usage of a nonconsensus 3' splice site, in which the AG is preceded by a purine-rich as opposed to pyrimidine-rich sequence.[12] The myosin heavy chain pre-mRNA undergoes an alternative splice that also uses a purine-rich 3' splice site, but only in pupal and adult stages.[13,14]

There do appear to be some other common patterns among those listed in Tables III and IV. Most of the nonimmunoglobulin 5' entries have GC

[12] S. Falkenthal, V. P. Parker, and N. Davidson, *Proc. Natl. Acad. Sci. U.S.A.* **82**, 449 (1985).
[13] S. I. Benstein, C. J. Hansen, K. D. Becker, D. R. Wassenburg, E. S. Roche, J. J. Donady, and C. P. Emerson, Jr. *Mol. Cell Biol.* **6**, 2511 (1986).
[14] C. E. Rozek and N. Davidson, *Proc. Natl. Acad. Sci. U.S.A.* **83**, 2128 (1986).

TABLE III
NONCONFORMING 5′ SPLICE SITES

GenBank			
Locus	Acc. no.	Sequence	Description
CHIBRCB	X06039	GAG/TAAGTT	*Chironomus thummi* BRC
CHKHBADA1	J00853	AAG/GCAAGC	Chicken α-globin
CHKMYHE	J02714	CAG/GCAAGT	Chicken myosin
DROANTPG4	M14495	CAG/GAAAGT	*Drosophila melanogaster* ANTP
DROOPSB2	K02317	GAT/TGCCTA	*D. melanogaster* opsin
DUKHBAD	X01831	AAG/GCAAGC	Duck α_D-globin
HAMCRYAA1	X02950	AGG/CAAGTT	Syrian golden hamster
HUMACHRG6	X01720	AAG/GCAAGG	Human acetylcholine receptor
HUMCFXII3	M17466	CCG/GCGAGT	Human blood coagulation factor
HUMFVII	M13232	CAG/GCGGGG	Human factor VII
HUMMHSXA	M11591	ATG/GCACTG	Human MHC class II
HUMP45SC6	D00166	AAG/GCAAGC	Human cytochrome *P*-450
HUMSODG1	X01780	AAG/GCAAGG	Human superoxide dismutase
HUMTHB	M17262	CTG/GCAAGT	Human prothrombin
MUSAPRT	M11310	CAG/GCGAGT	Mouse APRT
MUSCRYA5A	J00375	AAG/GCAAGT	Mouse lens
MUSGFN1	M17295	GCA/TCGGTG	Mouse growth factor
MUSGFN3	M17297	TGC/AGAATT	Mouse growth factor
MUSHOX162	M20215	CAG/GGAAGG	Mouse HOX-1.6
MUSRP0II2	M12130	CAG/GCAAGA	Mouse polymerase
NEUQA1SRA	M13208	TAG/GCACGT	*Neurospora crassa*
PIGGH	M17704	CAG/GCAAGT	Pig growth hormône
RATHOXA	J02722	CAG/GCAAGC	Rat heme oxygenase
SOYNOD24H	M10595	AGG/GCAAGT	Soybean nodulin-24
HUMIGHBC	M18809	CTC/TGAGTC	Human Ig heavy chain
HUMIGHBD	M18810	TCA/GGAGAT	Human Ig heavy chain
HUMIGHBD	M18806	AGA/CACACA	Human Ig heavy chain germline
MUSIGHBC	K00608	GGG/AAGGGG	Mouse Ig active heavy chain
MUSIGKJC2	X00268	AAC/CTAAGT	Mouse Ig κ germline
RABIGKCA	K01360	AAC/CTAAGT	Rabbit Ig κ germline
RABIGKCC1	K01361	AAC/CTAAGT	Rabbit Ig κ germline

at the 5′ intron border and AAG or CAG in the adjacent exon. It is difficult to detect any patterns in the 3′ entries. In fact, some of these sequences are not pyrimidine-rich and are therefore suspect of being errors. There are two large classes on nonconforming splice sites not accounted for above. There are 123 such yeast gene splice sites, all of which are transfer RNAs, and about one-fourth have AA at the 5′ ss boundary and one-fourth have TA at the 3′ end. There is nothing close to a consensus in these 123. It is

TABLE IV
NONCONFORMING 3' SPLICE SITES

GenBank			
Locus	Acc. no.	Sequence	Description
ADBCG	J01917	GGGGTCGTGC/A	Adenovirus
CHIBRCB	X06039	CGAAAGCAAT/G	*Chironomus thummi* BRC
CHKMYHD	M13515	TCCTCTGTCA/A	Chicken myosin
HUMPLP4	M15029	GTTTGTGGGC/A	Human myelin proteolipid
HUMPLP5	M15031	CCTCTTTTCA/T	Human myelin proetolopid
MLAP53I	K02705	CTAGTCCCGC/T	Murine leukemia virus
MUSCD42	M17076	GGAGACCACC/A	Mouse T-cell antigen
MUSGSTYA2	M19251	AGTTGCTGCA/A	Mouse transferase
MUSLYT212	M12979	GACCTGGACC/T	Mouse T-cell antigen
RATTMA3	M16433	AAGAGTTGAA/A	Rat α-tropomyosin
SEHCRYAA1	M17247	CCATCAAGGC/A	Mole α_a-crystallin
HUMIGHBA	M18807	ACAATGTGAC/A	Human Ig heavy chain
HUMIGHBD	M18810	CTCCACCATC/A	Human Ig heavy chain
HUMIGHBd	M18810	ACAATGTGAA/C	Human Ig heavy chain
MUSIGHAB2	J00516	TCTACTATCG/G	Mouse Ig active heavy chain
MUSIGHEG	M14637	GTTAGTGACG/G	Mouse Ig γ chain
MUSIGKBB	K02157	CTACACTAAT/G	Mouse Ig κ V-region
MUSIGLAA2	J00579	TCATCCTGCG/G	Mouse Ig λ_1

well known that splice junctions in tRNA genes do not have any consensus,[7] which is consistent with the theory of the origin of splice junctions in eukaryotic protein coding genes.[2,15] There are more than 120 organelle splice sites, over 80% of which reside in the nine genes listed in Table V; they are primarily chloroplast and cytochrome structures.

Other Possible Signals Upstream of 3' Splice Junctions

Mount[8] gives statistics for the distance between the splice site AG at $-2, -1$ and the nearest AG 5' of it, with the average distance being 41.5 bases and the shortest 14 bases. We compiled similar statistics for the distance of each of the 16 dinucleotide pairs from the 3' ss AG. A histogram of these AG to nearest upstream AG distances is shown in Fig. 2. Only introns extending at least 96 bases in the 5' direction from the 3' ss were used, making Fig. 2 a true representation of the AG distances. Of the 1250 3' splice sites examined, there were 36 distances of 97 bases or greater

[15] P. Senapathy, *Proc. Natl. Acad. Sci. U.S.A.* **85,** 1129 (1988).

TABLE V
NINE ORGANELLE GENES CONTAINING NONCONSENSUS SPLICE SITES

| GenBank | | |
Locus	Acc. no.	Description
EGRCPPSBA	X00735	Green algae chloroplast
EGRCPRBCL	M12109	*Euglena gracilis* chloroplast
LEIKPMAX	M10126	Ribosomal RNA, cytochrome oxidase
MPOCPCG	X04465	Liverwort chloroplast, ribosomal RNA, cytochrome
MZECPRG16	X05422	Maize chloroplast, ribosomal RNA, transfer RNA
NEUMTND	K01181	*Neurospora crassa* mitochrondrial dehydrogenase
PANMTCOI	Y00403	Podospora cytochrome-*c* oxidase
TOBCPCG	X00616	Tobacco membrane protein, transfer RNA
TOBCPTGRG	J01452	Tobacco chloroplast, ribosomal RNA, transfer RNA

($<3\%$) (not shown in Fig. 2). The average distance for the 1214 distances of 96 or less in the current study is 36.6 (Fig. 2). This number would be increased slightly if the 36 distances greater than 96 were included.

With only 20 exceptions (2% of the sequences) there was no AG in the 10 nucleotides preceding the AG at the 3' ss. These results are consistent with those of Mount,[8] who suggested that the absence of AG close to the 3' ss may indicate a scanning mechanism for recognition of the splice site by the splicing machinery. Only one sequence (from the mouse immunoglob-ulin κ active *V*-region gene) has an AG immediately preceding the 3' ss AG, and it also has two AGs within 12 bases 5' of this.

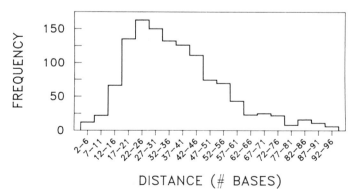

DISTANCE (# BASES)

FIG. 2. Histogram of counts of the number of nucleotides separating the AG at the 3' ss from the nearest AG in the intron. Only introns extending at least 96 bases from the AG were used. In total, 1250 sites were counted. Five distances are grouped together. Thirty-six distances of 97 bases or greater are not shown.

In addition to AG, we followed the frequency of the other 15 dinucleotides as a function of distance from the 3′ ss AG (Fig. 3). The clear peak in the frequency of occurrence of the upstream AG (Figs. 2 and 3) indicates that an AG, about 40 base pairs upstream from the 3′ ss, may be a molecular signal in the splicing process. The GA and GG dinucleotides have similar peaks, although slightly less pronounced than that of AG, indicating that these also may be molecular signals.

Analysis of Consensus and Distribution of Conserved Sequences around Branch Points in Eukaryotic Genes

Because the site of lariat formation is in the region 10–50 nucleotides upstream of the 3′ splice site, earlier investigators had looked for branch point sequences only in this region using computers. This approach fails to account for the possibility that this sequence is not uniquely confined to this particular region. If the branch point sequences were found only in this particular region with respect to the acceptor site, then one could presume that this information by itself could be used by the splicing process to

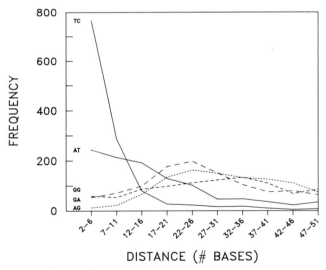

Fig. 3. Number of nucleotides separating the AG at the 3′ ss from the nearest XX in the intron, where XX is one of the 16 dinucleotide combinations. The dinucleotides are listed at the left, at the start of their curve. Dinucleotides AG, GA, and GG are shown with dashed lines. For clarity, only two other dinucleotide curves, TC and AT, are shown; the 11 omitted have similar shapes, that is, without peaks. Counts are for groups of five distances, and only distances less than 52 are shown.

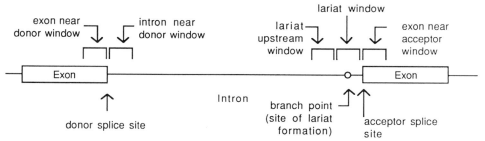

FIG. 4. DNA "windows" in introns and exons analyzed for branch point sequences. Different windows were analyzed for the frequency of occurrence of branch point-like sequences and for the nucleotide frequencies in these sequences. The windows we examined are (1) lariat window (−1 to −50 nucleotides upstream of the acceptor splice site in introns), (2) lariat upstream window (−51 to −100 nucleotides upstream of the acceptor splice site), (3) exon near acceptor window (1 to 50 nucleotides downstream of the acceptor splice site), (4) exon near donor window (−1 to −50 nucleotides upstream of the donor splice site), and (5) intron near donor window (1 to 50 nucleotides downstream of the donor splice site). Exons are indicated by boxes and the branch point site by a circle.

specify the site of lariat formation. However, if the occurrence of the branch point signal is not unique to this region, this would suggest that the sequence alone is not responsible for the recognition of the site of lariat formation.

In order to see if branch point sequences are found preferentially within 10–50 nucleotides upstream of the acceptor site, we computed the frequency of the best ranking branch point sequences in a number of different windows (as described in the section on Methods). Figure 4 illustrates the different windows we analyzed for branch point sequences. In order to obtain a consensus weight matrix, we computed the composite frequencies of each of the four nucleotides found at the consensus branch point sequences of plant, rat, human, chick, and *Drosophila,* based on data from Keller and Noon[16] and Brown.[6] The resulting standard composite data are given in Table VI.

Our aim was to compare the different windows (see Fig. 4) with respect to the frequency of the high-scoring branch point sequence having a score of at least 97%. The frequency (percentage) of windows containing branch point sequences with a score of at least 97% was determined for each window category described in Fig. 4, as well as for each of the seven GenBank categories of organisms. The results (Table VII) indicate that among the GenBank categories of the sequences analyzed, the lariat win-

[16] E. B. Keller and W. A. Noon *Proc. Natl. Acad. Sci. U.S.A.* **81,** 7417 (1984).

TABLE VI
STANDARD BRANCH POINT SEQUENCE WEIGHT
TABLE[a]

Nucleotide	Frequency of occurrence (%) at sequence position				
	−3	−2	−1	0	1
A	1	0	39	99	11
C	76	8	15	1	45
G	2	0	42	0	6
T	21	91	4	0	38

[a] Composite weight table derived from empirical data for plant, rat, human, chicken, and *Drosophila* DNA.[5,12] The values from these published tables were combined, and the percentage for each nucleotide occurring at each position is given here. This is used as the standard weight table to score potential branch point sequences in the analysis described in this chapter.

TABLE VII
BRANCH POINT SEQUENCE FREQUENCY IN DIFFERENT WINDOWS OF GENE SEQUENCES[a]

Category of organism	Window	Frequency of windows containing high-scoring branch point sequences (%)	
		Unconstrained middle nt	Constrained middle nt
Invertebrate	Exon near acceptor	21	4
	Lariat	48	30
	Lariat upstream	28	11
	Exon near donor	18	11
	Intron near donor	25	13
Mammal	Exon near acceptor	20	11
	Lariat	57	46
	Lariat upstream	35	14
	Exon near donor	25	13
	Intron near donor	34	19
Plant	Exon near acceptor	27	14
	Lariat	62	48
	Lariat upstream	32	18
	Exon near donor	30	14
	Intron near donor	37	24

TABLE VII (*continued*)
BRANCH POINT SEQUENCE FREQUENCY IN DIFFERENT WINDOWS OF GENE SEQUENCES[a]

Category of organism	Window	Frequency of windows containing high-scoring branch point sequences (%)	
		Unconstrained middle nt	Constrained middle nt
Primate	Exon near acceptor	24	11
	Lariat	46	24
	Lariat upstream	26	13
	Exon near donor	36	17
	Intron near donor	31	13
Rodent	Exon near acceptor	15	9
	Lariat	68	36
	Lariat upstream	24	4
	Exon near donor	34	15
	Intron near donor	31	12
Vertebrate	Exon near acceptor	24	19
	Lariat	47	28
	Lariat upstream	37	20
	Exon near donor	30	11
	Intron near donor	35	18
Viral	Exon near acceptor	31	9
	Lariat	39	14
	Lariat upstream	29	12
	Exon near donor	32	0
	Intron near donor	26	8
Random	(50 nt)	32	17

[a] The frequency of windows with high-scoring (97% or higher match) branch point sequence in genes of different categories of organisms was computed as follows. Potential branch point sequences were rated and scores were assigned according to how well they matched the standard (composite) consensus sequence derived from published data (see Table VI). The five-nucleotide sequence, among all possible five-nucleotide sequences in a window, that best matched the standard consensus was taken as the branch point sequence in that window. If the score of this branch point sequence was 97% or greater, then the window was counted as having a branch point sequence. The frequency of such windows containing a branch point is given for each window described in Fig. 4 and for each of the seven categories of organisms. The middle position of the branch point sequence was either constrained to match the standard consensus (G/A) or left unconstrained. When constrained, the branch point sequence should best match all five nucleotides in the composite, standard, table; when not constrained, the branch point sequence should best match all these positions except the fourth position, i.e., the A or G.

dow has the highest percentage of high-scoring branch point sequences. In order to compare these results with a window containing a purely random nucleotide sequence, we generated 200 50-nucleotide random sequences and performed the same analysis. The results (Table VII) show that in the case where the middle nucleotide was constrained to be a G/A as in the standard table (Table VI), the lariat window in invertebrate, mammal, plant, primate, rodent, vertebrate, and viral categories of genes contained, respectively, 30, 46, 48, 24, 36, 28, and 14% of sequences that matched 97% or better with the standard weight table, whereas the random sequence contained only 17% of windows that similarly matched. When the middle nucleotide was not constrained to be G/A, a higher percentage of sequences in each window category, as with the random sequence, matched the standard table. However, the overall pattern of frequencies with high-scoring branch point sequences in different window categories remained similar in all GenBank groups of organisms.

The percentages of lariat windows with high-scoring branch point sequences in all GenBank categories, with the exception of viruses, are higher than those of random sequences. A notable observation is that even the window which contains the second highest frequency of branch points among the five windows analyzed always had a lower frequency of high-scoring branch points than expected for a window with a random sequence. However, the values are still comparably closer to that for the windows with a random sequence. This suggests that, in evolution, there has been a slightly negative selection for the branch point sequence in regions other than the lariat region, where there has been a slightly positive selection for it. Furthermore, the low frequency (20–40%) of high-scoring branch point sequences even in the lariat window (when the middle nucleotide is constrained), indicates that the majority of real branch point sequences varies from the standard branch point sequence (CTRAY, R = purine, Y = pyrimidine) used in the current study. Thus, more than 60% of branch points must have sequences different from this consensus sequence. Further analysis is needed to identify the other branch point sequences.

Implications of Random Distribution of Branch Point Signal Sequences throughout Genes

The fact that branch point-like sequences are found in such abundance in eukaryotic genes implies that the splicing machinery must have some mechanism for recognizing the correct branch point site. The process of splicing seems to recognize exons and introns sequentially with a 5' to 3' scanning mechanism. The present results indicate that the scanner could not identify the donor splice site, branch point, and acceptor splice site within an intron in a sequential manner, because of the random occurrence of the branch point sequence in introns. Thus, even a scanning model, in which a donor site is identified first and then the intron scanned for a branch point site which helps determine the acceptor site, is unten-able.

We propose that the splicing machinery first locates the donor splice site for an intron. In a second step, it locates the first downstream acceptor site which also has a good branch point sequence within the first 50 nucleotides upstream of this site. In other words, neither a branch point site nor an acceptor sequence, occurring downstream of a 5' ss, can be a real site independent of the other. Both have to occur together in the 5' branch point–3' ss orientation (within 10–50 nucleotides) for both to be real sites. This argument is supported by the fact that there exist many nonfunctional 3' splice sites scoring higher than some of the true 3' splice sites in the introns (Table IX and Fig. 6). This hypothesis is also supported by the observation that, if an intron happens to have more than one sequence which could function as branch point, only the sequence closest to the acceptor splice site appears to serve this function in most cases.[17] When the branch point sequence in intron 1 of the human β-globin gene was removed, it was found that splicing was not prevented; rather, a cryptic branch point sequence upstream of the deleted one was activated. The observation that introns lacking a viable branch point fail to splice out normally confirms the importance of branch points in the splicing pro-cess.[16]

The nearly random occurrence of branch point sequences in genes may have some implications in understanding their evolution. This observation is consistent with the hypothesis that the very first genes evolved from random primordial sequences by a gene search mechanism evolving the split-gene architecture in the first genes;[2,15,18] the mechanism selected exons, introns, and splice signals from the preexisting, primordial, random sequences. Consistent with this is our observation that many good splice junction sequences (Table IX) and branch point sequences (Table VII) in genes do not function as real splice sites (Fig. 6). These findings suggest that real splicing signals may exist in a specific positional context in the gene sequence.

We tabulated the codon frequencies at each of the five nucleotide positions of the highest scoring branch point sequences of the lariat win-dow. When the middle position was constrained to be a G or A, we found that the codon at the third position was almost always a stop codon (data not shown). This is consistent with the hypothesis that branch point se-quences evolved from stop codons.[2,15] This hypothesis suggests that the splicing mechanism for removing introns was developed in order to over-come the problem of randomly distributed stop codons. A stop codon scanning mechanism may have been responsible for the evolution of the

---

[17] B. Ruskin, A. R. Krainer, T. Maniatis, and M. R. Green, *Cell* **38**, 317 (1984).
[18] P. Senapathy, *Mol. Genet. (Life Sci. Adv.)* **7**, 53 (1988).

splice junction signals, as well as the branch point signal, from stop codons. Thus, how a particular consensus sequence came to serve this special function of signaling lariat formation may be explained by its possible mode of evolution.

Identifying Potential Splice Sites

Considering efforts to sequence the human genome, it seems inevitable that large amounts of uncharacterized DNA sequences will emerge, without knowledge of the mRNA and protein products they encode. Predicting potential coding genes in these sequences is imperative and will be a major task involving several steps, such as identifying potential promoter sites, transcription and translation start sites, 5' and 3' splice sites, polyadenylation sites, and potential exons and introns in the right context. Undoubtedly, computer methodologies will play an important role in this task. The existing information on sequence characteristics of these different architectural elements of genes will have to be employed.

We have used the tabulated information on the splice sites to identify potential splice sites in a given sequence.[9] Previous work in locating sequence signals such as splice sites have used weight matrices,[19-22] pattern recognition methods,[23] and perceptron.[24,25] As with the other scoring matrix methods, our procedure scores every possible subsequence of consensus sequence length in a given sequence, to see which subsequences are most closely related to the consensus sequence. Using the highest scores for both the 5' and 3' splice sites, we have formulated a method of identifying potential exons.

In order to verify the validity of our method, we scored known gene sequences for 5' and 3' sites and assessed the ranks of the known splice sites in the list of ranks generated by our procedure. The sequence of the gene for tumor necrosis factor,[26] 3200 bases in length, was searched in this way for potential splice sites. Table VIII lists the top 20 ranked 5' and 3' splice sites and their scores. Note that Table VIII contains all three actual donor (852, rank 1; 1356, rank 13; 1564, rank 2) and acceptor sites (1300, rank 5; 1515, rank 6; 1853, rank 12) of the gene within the top 13 (top 0.4%) ranks, from a total of approximately 3200 ranks.

---

[19] R. Harr, M. Haggstrom, and P. Gustafsson, *Nucleic Acids Res.* **11**, 2943 (1983).

[20] R. Staden, *Nucleic Acids Res.* **12**, 505 (1983).

[21] M. E. Mulligan and W. McClure, *Nucleic Acids Res.* **14**, 109 (1986).

[22] Y. Ohshima and Y. J. Gotoh, *Mol. Biol.* **195**, 247 (1988).

[23] J. Quinqueton and J. Moreau, *Biochimie* **67**, 541 (1985).

[24] K. Nakata, M. Kanehisa, and C. Delisi, *Nucleic Acids Res.* **13**, 5327 (1985).

[25] Y. Iida, *CABIOS* **3**, 93 (1987).

[26] H. Ito, T. Shirai, S. Yamamoto, M. Akira, S. Kawahara, C. Todd, and R. B. Wallace, *DNA* **5**, 157 (1986).

TABLE VIII
COMPUTER-PREDICTED SPLICE SITE RANK TABLE FOR RABBIT TUMOR NECROSIS
FACTOR GENE SEQUENCE[a]

Donors				Acceptors			
Rank	Score	Position	Sequence	Rank	Score	Position	Sequence
7	80.0	167	TGGTGGGGG	20	81.4	59	TCTCCCGCTCAAGGT
12	78.7	219	TGGTAACTG	7	87.4	107	CTACCCCCCACAGGA
3	87.8	262	GGGTGAGGA	17	82.1	158	TCGGTCCCCAAAGTC
14	77.2	378	AGATGAGCT	30	77.2	260	TCCGGGTCGGGAGGG
5	81.2	419	GGGTGAATG	9	86.1	305	CAACTTTCCAAAGCC
26	74.0	471	AGCTGTGTG	29	77.6	377	CCGCTTCCTCCAGAT
30	72.5	504	CAGTGAGGA	1	92.5	403	TTTCTCCACCAAGGA
15	77.2	601	CGGTGTGCC	27	79.0	456	CCTCTCGCCCGCGGG
27	74.0	828	GGGTGATCG	26	79.0	629	TCTCACGCACCAGCC
1	**100.0**	**852**	**AGGTGAGTG**	11	84.5	657	CCTCCCTCTGGAGAG
17	76.2	954	AGGAGGGTG	28	78.6	795	CTCGTGGCTGGAGCC
24	74.5	1042	AGATAAGGA	18	81.8	826	TGCTGCACTTCAGGG
28	74.0	1078	GGGTGAAGA	23	79.9	842	GATCGGCCCTCAGGA
25	74.5	1181	AGGGAAGGA	19	81.5	1120	AACCTCCCGAAAGAG
19	76.0	1193	CGATGAGTG	4	89.7	1260	TTGCTCCTTTGAGCG
13	**78.0**	**1356**	**CAGTAAGTG**	**5**	**88.1**	**1300**	**TTTTTTCTCCCAGCA**
16	77.2	1401	AGGGGAGCT	22	80.8	1462	TGGGTTTTGGGAGGG
11	79.0	1482	GGGTAGGGG	2	92.4	1480	ATCATCTCCTGAGGG
31	72.5	1560	TAGTAGGTA	**6**	**87.8**	**1515**	**TTCTTCTTCTCAGGA**
2	**91.2**	**1564**	**AGGTAAGAG**	25	79.1	1537	CTCGGGCCCTGAGTG
22	74.8	1653	GAGTGAGAA	**12**	**84.3**	**1853**	**TCTTTTCTCCAGCA**
18	76.2	1677	AGGGGAGGG	8	86.3	1984	CATCTACTCCCAGGT
8	79.5	1762	ATGTGAGGG	21	81.0	2092	CTCTGCCATCAAGAG
9	79.5	1793	ATGTGAGGG	10	84.8	2273	ATTGCCCTGTGAGGG
21	75.0	1799	GGGTGAACC	14	84.1	2310	TCTCCCACCCCAGCC
20	75.5	2140	TGGTACGAG	24	79.8	2354	CATCCTCCTCTAGCT
23	74.8	2184	GGGTGACCG	16	82.7	2641	TTCCTCTCCTCAGAT
10	79.5	2720	TGGTGATTA	13	84.1	2689	GCCCCCCGCCCAGGG
6	80.3	2965	TGGTGACTG	31	76.6	2711	CCTCTATTTATAGTT
29	74.0	3012	AGTTGTGTT	15	83.0	2941	TTTATCTGATCAGGT
32	72.5	3020	TGGTAACCG	3	91.1	3007	TCTGCTCCCTAAGGG
4	85.7	3035	TGGTCAGTG	32	76.1	3165	CTTATCCCATAATGC

[a] Each subsequence in a given gDNA sequence is weighted based on the known frequencies of nucleotides at the donor and the acceptor splice sites and a score obtained.[8] The sequence locations are sorted by the scores and arranged in a decreasing order of ranks. The top 2% of the ranks are then sorted by sequence position. The 3200-nt tumor necrosis factor gene sequence has 3192 donor and 3190 acceptor splice site scores and ranks (the conserved donor sequence is 8 nt long and the conserved acceptor sequence is 10 nt long). Only the top 1% of these is shown here, which happens to contain all the real splice sites of the gene (shown in bold-face type).

There are three possible explanations for these results. (1) If high scores indicate valid splice sites, there may be proteins other than those found experimentally that are manufactured by this gene. Thus, it is possible that some of the top ranked spliced sites in genes may actually be used, generating very low quantities of essential mRNAs that are difficult to identify in the laboratory owing to the lack of experimental sensitivity. (2) If actual splice sites have lower scores than some which appear to be nonfunctional as splice sites, there must be other sequence signals in addition to the splice junction sequence signal that directs the splicing. (3) Among the many randomly distributed sites, the correct 5' ss, 3' ss, and the branch point sites are identified by a specific positional context.

Table IX shows the results of scoring the potential splice sites of six sequences, selected from GenBank at random from those having at least seven exons. The scores and ranks of the splice sites found experimentally are given for each of the sequences. The scores for each gene were computed using the appropriate subtable of Table I, e.g., the primate percentages were used to score the human HLA gene. The results show that most of the splice junction sequences found in a gene in the laboratory are ranked at or near the top of the list. While Table IX admittedly represents a small sample, the results indicate that the splice junction sequence itself is, in fact, an important requirement (and that our scoring system is valid). However, some sites are ranked fairly low, e.g., the 5' ss for intron six and 3' ss for intron three of the rat chymotrypsin gene, and the intron two 3' ss for rabbit MHC, with almost no pyrimidine content upstream of the 3' ss. This result indicates that sequence elements other than the junction sequence itself (such as the lariate sequence) could also be involved in locating splice junctions. Scoring formulas will have to use such additional information as it becomes known.

In all six genes in Table IX, the top ranked 5' splice sites found by the computer are actual splice sites, but three of the computer top ranked 3' splice sites are not known to be actual splice sites. At least some of the top-ranked potential splice junction sequences that have not been identified in the laboratory may be real sites possibly used in alternative splicing situations. About two-thirds of the actual splice sites in Table IX are ranked at the top (rank 1 to 8), one-sixth are medium ranked (9 to 20), and the remainder are lower. It may be possible that different components of the splicing machinery are involved in recognizing splice sites with infrequently occurring bases, as opposed to those which scored high.

It is striking that out of the thousands of locations scored for each gene, the true splice sites are ranked so high, e.g., four of the six human HLA 5' splice sites are in the top 0.1% of all ranks (4 of 4123). It is also interesting to note that there is some pattern in the distribution of the ranks: (1) all six

TABLE IX
SPLICE SITE SCORES AND RANKS FOR SIX GENES

	5' splice site			3' splice site		
Description[a]	Rank	Score	Sequence	Rank	Score	Sequence
Human HLA gene, 4123 bp, primate						
Intron 1	12	75.1	GC/GTGAGT	7	91.4	CTCCTCGCTCCCAG/G
Intron 2	2	88.4	CG/GTGAGT	>100	61.7	CGGGGGCGGGCCAG/G
Intron 3	>100	60.8	GG/GTACCA	8	91.0	CTCTTTCCCGTCAG/A
Intron 4	3	87.6	GG/GTAAGG	11	87.6	CCCCCTTTTCCCAG/A
Intron 5	1	95.4	AG/GTAAGG	4	93.8	TTTTCTTCCCACAG/A
Intron 6	4	87.6	AA/GTAAGT	3	95.8	TTATTCTACTCCAG/G
Rat chymotrypsin, 5809 bp, rodent						
Intron 1	1	92.8	TG/GTAAGT	3	94.3	TATTCCCTTCCAG/G
Intron 2	2	89.5	AG/GTGAGA	6	92.4	TTCGTCCTCCCAG/G
Intron 3	5	84.4	AA/GTGAGT	>100	72.5	GTCTCAATACCTAG/G
Intron 4	71	67.2	AG/GTACAC	1	95.9	TCTTGTCCCCACAG/G
Intron 5	3	89.5	TG/GTGAGT	14	87.6	CTGTATCCCTGCAG/C
Intron 6	>100	68.3	TG/GTACAG	2	95.6	ACCTTTCTCTGCAG/G
Rabbit MHC, 3881 bp, other mammal						
Intron 1	2	94.8	GG/GTGAGT	8	87.3	CCTGTCGCCCGCAG/G
Intron 2	5	92.3	CG/GTGAGT	>100	59.6	CGGGGCGGGGCCAG/G
Intron 3	30	75.1	AG/GTTCGC	7	86.9	AATCTTTCCCTCAG/A
Intron 4	>100	63.6	GG/GTAGAG	11	86.2	CCTTCCCTTCTCAG/A
Intron 5	12	83.3	AG/GTAGGG	2	92.7	TTTTCTTCTCACAG/A
Intron 6	1	98.3	AG/GTAAGT	1	96.0	TTTGTTCACCCCAG/G
Chicken thymidine kinase, 3008 bp, other vertebrate						
Intron 1	19	71.5	AG/GTACCG	7	84.9	GGCTCTGCCTCCAG/G
Intron 2	1	95.7	AG/GTAAGG	6	86.7	TCCTGCTCTTGCAG/C
Intron 3	10	76.0	AG/GTGGGC	5	88.0	TCCCTTCTCCTCAG/G
Intron 4	6	80.4	TT/GTAAGT	25	75.8	TGTACCTCCTGTAG/T
Intron 5	5	80.8	AG/GTAAAA	2	91.0	CTGTTTTGGTGCAG/G
Intron 6	3	93.1	AG/GTGAGT	11	82.5	GATTTCCTTGATAG/G
Clostridium elegans myosin isozyme, 9000 bp, invertebrate						
Intron 1	2	91.2	AG/GTGAGT	8	90.0	TATTGTAATTTCAG/G
Intron 2	9	77.8	AG/GTTCGT	>100	70.6	GGTTTTCAATGTAG/T
Intron 3	5	87.8	AC/GTAAGT	1	94.5	GTTTCCTATTTCAG/A
Intron 4	24	71.6	TT/GTGAGT	3	92.9	TATAAATTTTGCAG/A
Intron 5	1	92.0	AG/GTAAGA	2	93.7	TTCTTATTTTTCAG/C
Intron 6	4	88.3	AT/GTAAGT	6	91.3	TGTGAATCTTTCAG/G
Intron 7	8	79.8	AG/GTAATA	7	90.8	TTCATGGTTTTCAG/G
Intron 8	6	80.1	CT/GTAAGT	4	91.5	CTTATTAATTTCAG/G
Maize alcohol dehydrogenase, 3495 bp, plant						
Intron 1	22	76.5	AG/GTCCGC	>100	61.9	CTGGACCCGTGCAG/C
Intron 2	5	87.2	AG/GTATCT	7	82.8	CTTATCTGTCTCAG/G
Intron 3	2	92.8	GG/GTATGT	11	79.1	CTTGATTTTGCCAG/T
Intron 4	1	96.7	TG/GTAAGT	4	84.8	CTTTCTCTGTTTAG/G
Intron 5	16	79.0	CC/GTAAGT	76	66.5	TGCGCATGGTTAAG/G
Intron 6	40	73.8	AG/GTACAG	14	78.3	GTCCCATTTTTCAG/C
Intron 7	3	88.9	AG/GTCTGT	9	79.9	CCTTTATGGTCTAG/G
Intron 8	6	86.8	AT/GTAAGT	23	74.7	CCCTCGTGATCCAG/G
Intron 9	4	87.8	AG/GTAAAT	37	70.9	GCAATTCTGCACAG/G

[a] Includes the gene length and GenBank category.

genes contain the top ranked 5' ss, (2) the top ranked 5' ss is, in all cases, paired with a high-ranked 3' ss, and (3) all but one gene contains at least one very low-ranked splice site. The association pairs, while the splicing signals are distributed nearly randomly in genes, suggests a complex scanning mechanism to identify such pairs.

Predicting Potential Exons

In order to identify potential exons of a gene in a given sequence, we first computed the 10 top ranked 5' and 3' splice sites in the sequence. Then we identified the potential exons as follows. (1) Mark the sites depicting all three reading frames and the open reading frames of the gene sequence. (2) Mark the region between a 3' ss and its closest downstream 5' ss as a potential exon if this region would fall within a reading frame (i.e., uninterrupted by a stop codon) in any of the three reading phases. (3) Repeat this until the end of the sequence, in order to identify all potential exons resulting from the 10 top ranked splice sites.

Next we determine if each of the potential exons is a part of the gene in question by the following rules of assignment of exons to different reading frames, using the procedure described earlier.[2] The reading frame of the first exon, RF_1, is L_1 mod 3, where L_1 is the starting location of the ATG initiator codon, and mod 3 indicates the remainder when divided by 3. A zero remainder specifies RF_3. The RF of the ith exon is $RF_i = (RF_i - 1 + I_i - 1)$ mod 3, where I_i is the length of the ith intron in the sequence.

Almost all the actual exons from the six genes examined (Table IX) were found by this method. However, some additional exons were found, using this method, that result from highly ranked splice sites. These additional exons may actually be used in the cell to generate alternatively spliced mRNAs, or they may represent artifacts.

Figure 5 illustrates the method for detecting exons in a human HLA gene,[27] the first gene in Table IX. An alternative fourth exon, 4', was found to the left of exon 4. Exon 4' is located between the top ranked 3' ss and the fifth ranked 5' ss, and it is in the first reading frame. If the combined ranks of 5' and 3' splice sites are taken to be an exon score, the exon labeled 4' would be top ranked. Continuations from exon 4', yielding alternative mRNAs, can be achieved by splicing it to existing or newly found exons downstream.

Problem of Identifying Complete Genes

An important finding in our studies is that the splice sites and branch point sites are almost randomly distributed in gene sequences. When we

---

[27] M. Malissen, B. Malissen, and B. R. Jordon *Proc. Natl. Acad. Sci. U.S.A.* **79**, 893 (1982).

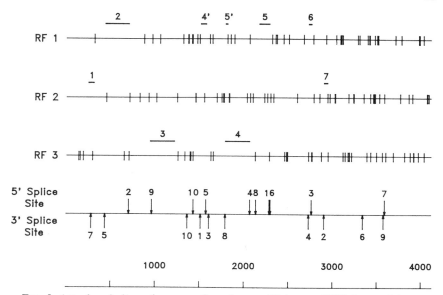

FIG. 5. Actual and alternative exons for a human HLA gene (4123 base pairs). Exon numbers appear over the lines indicating exon locations. Exon 4' indicates an alternate exon found using our scoring system (see text). Tick marks indicate the location of stop codons in the three reading frames (labeled RF1, RF2, RF3), under which the location and rank of the top 10 5' and 3' splice sites are shown. Some of the actual splice sites are not marked here, as they ranked below 10. See Table IX for the ranks of all the actual splice sites.

plotted the very top ranking splice sites and branch point sites, it was clear that they were almost randomly distributed in both exons and introns (Fig. 6). This indicates that the problem of "selecting" the right sites that circumscribe the exons is highly complex. Based on the above method (illustrated in Fig. 5), we can identify some exons of a gene. By using a larger number of top ranked splice sites and branch point sites (see below), more real exons can be identified. By the same token, however, more artifactual exons will also be predicted. The major task, then, is to isolate real exons from artifactual ones. Thus, after identifying the most probable exons constituting a probable gene in a given sequence, the first problem is to find if each one of the exons identified as above belongs to the gene in the region or is an artifact. (It may sometimes represent an alternatively spliced exon.) The second problem is to find if any exon of a gene is missing from the set of identified exons. We are using additional rules and combinatorial analysis in order to do these, some of which are discussed below.

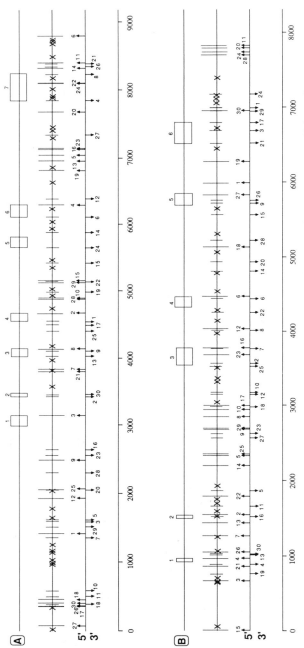

Fig. 6. The nearly random distribution of splice sites and branch point sites in eukaryotic gene sequences. Eukaryotic gene sequences containing at least five exons were chosen randomly to analyze the distribution of splice signals in them. The sequences were analyzed by the computer program RATE, which ranks each sequence location based on the splice site scoring matrix (as described in Ref. 9). In each sequence the top 30 ranking locations are shown (| for 3′ ss and | for 5′ ss). Similarly, we found (with the BRANCH program) the locations matching with the branch point sequences (X). The distribution of the 5′ ss, 3′ ss, and branch point sites are shown for two genes: (A) chicken ovalbumin gene and (B) human interleukin 1 β gene. The exon locations are indicated by numbered boxes on the first line of each gene. The second line shows the distribution of splice sites and branch point sites. On the third line, the up arrows indicate the 5′ ss (with their corresponding ranks), while the 3′ ss are indicated by the down arrows (along with their ranks). The fourth line is a reference scale, giving number of nucleotides.

Conclusion and Future Directions

This study provides a 2.5-fold increase over the number of splice sites last reported.[9] The results for the four mammal categories (primates, rodents, other mammals, and other vertebrates) are basically in agreement with previous findings. The availability of statistics for a substantial number of splice sites makes it possible to rate each location in an uncharacterized sequence for its potential as a splice site. A conspicuous pattern of the ranks of the splice sites from known genes gives possible clues to the workings of the splicing machinery. Thus, computer analysis of splice junction sequences can lead to a better understanding of the general mechanism by which splice sites are precisely recognized.

The ratings of sequences based on splice-site scoring matrices help in devising a method in order to identify potential exons. Because this method is able to identify some of the actual exons in known gene sequences, it should be possible with such methods to identify exons in uncharacterized sequences whose gene content is unknown. This approach will be very useful in the human genome sequencing project, in which large quantities of uncharacterized DNA sequence information would be generated with no knowledge of exon–intron content.

Although our efforts have focused on developing rules for the detection of individual gene attributes, our current effort is centered around the question of defining rules that detect the real sites out of the nearly randomly distributed individual gene attributes and that circumscribe the real exons which form the complete gene. The bottom line plan is to enhance the reliability of the exon detection scheme by using multiple, imperfect, and independent search criteria, relying on the assumption that functional exons will be detected by the convergence of multiple scan parameters. Further, we first identify exons and then position the upstream and downstream regulatory sequences based on the exon information. The properties of the first exon and the last exon will vary from those of the internal exons because they lack the 5′ ss and the 3′ ss, respectively. Special algorithms to identify them will be necessary. Our preliminary results using branch point sites to identify the real 3′ splice sites are very encouraging (P. Senapathy, unpublished). By using branch point sequence scores in a positional context with both 5′ ss and 3′ ss scores, the reliability of exon identification improves by reducing the score of artifactual exons and at the same time improving the scores of real exons. However, there still seem to be many high-scoring artifactual exons detected using the above rules. We are using further rules, such as lack of stop codons in the exons, a high RNY periodicity, and a high codon-bias statistic, to improve and isolate the real exons from the artifactual. It is noteworthy that there seems to be a hierarchy which must be followed while using the rules to produce satisfactory results. We have investigated the classification and regression tree (CART) methods[28] to find the hierarchy of the various parameters (e.g., the consensus sequence around the start codon, the length of the open reading frame, and the 5′ splice site score around the first stop codon in an exon) and their scoring thresholds in identifying the first exon (M. Shapiro and P. Senapathy, unpublished).

Even after all this, although most of the real exons may have high scores, some may have considerably lower scores, because many artifactual exons may have much higher combined scores than some real exons. New rules must be found to detect these low-scoring, "problematic" exons. Owing to the rapid influx of sequence information, a large amount of complete gene sequences with multiple exons will accumulate very soon, and new rules will eventually be found. The evidence so far is convincing that complete genes can be found in uncharacterized sequences using the computer alone, which will make reading and interpreting the instructions in a complex genome, such as the human genome, more meaningful and reliable.

[28] L. Breiman, J. Freidman, R. Olshen, and C. Stone, "Classification and Regression Trees." Wadsworth International Group, 1984.

Section IV

Predicting RNA Secondary Structures

[17] Predicting Optimal and Suboptimal Secondary Structure for RNA

By JOHN A. JAEGER, DOUGLAS H. TURNER, and MICHAEL ZUKER

Introduction[1]

The secondary structure of RNA is important for many functions, including regulation of transcription and translation, catalysis, and transport of proteins across membranes. Where once RNA was considered only a structural framework for proteins, it is now known to play an active role in several enzymes. Sequencing RNA is straightforward, and thousands of sequences are known. In addition to the genetic code, a "structural code" must exist in the sequence as well.

RNA forms both secondary and tertiary structure, as shown by the crystal structure of tRNA[Phe].[2] We currently have little idea of the tertiary interactions, but studies on tRNA indicate these are weaker than secondary structure interactions.[3] Thus, secondary structure is a good place to start determining RNA structure.

There are two methods of predicting secondary structure: phylogeny and energy minimization. Phylogeny relies on alignment and subsequent folding of several sequences into similar structures for functionally analogous RNA.[4] Energy minimization relies on thermodynamic parameters and computer algorithms to determine the minimum and near minimum free energy foldings of an RNA.[5] Among the advantages of energy minimization are (1) only one sequence is required, allowing predictions for unique sequences, such as mutations and synthetic genes; and (2) insights are obtained into the reasons a structure is formed — the structural code — not only the structure itself. While much of the structural code remains to be discovered, current knowledge is adequate to correctly predict about 70% of the secondary structure from a single sequence.[6]

---

[1] National Research Council of Canada Publication No. 30748.

[2] S. H. Kim, F. L. Suddath, G. J. Quigley, A. McPherson, J. L. Sussman, A. H. J. Wang, N. C. Seeman ,and A. Rich, *Science* **185,** 435 (1974).

[3] D. M. Crothers and P. E. Cole *in* "Transfer RNA" (S. Altman, ed.), p. 196. MIT Press, Cambridge, Massachusetts, 1978; D. Riesner and R. Romer *in* "Physicochemical Properties of Nucleic Acids" (J. Duchesne, ed.), Vol 2, p. 237. Academic Press, New York, 1973.

[4] For example, C. R. Woese, R. Gutel, R. Gupta, and H. F. Noller, *Microbiol. Rev.* **47,** 621 (1983).

[5] I. Tinoco, Jr., O. C. Uhlenbeck, and M. D. Levine, *Nature (London)* **230,** 362 (1971).

[6] J. A. Jaeger, D. H. Turner, and M. Zuker, *Proc. Natl. Acad. Sci. U.S.A.* (in press) (1989).

METHODS IN ENZYMOLOGY, VOL. 183

Program

The program is the work of M. Zuker.[7] It extends the recursive method[8,9] to predict both optimal and suboptimal secondary structures and to include favorable stability increments from single-stranded regions. All suboptimal structures are not predicted. Instead, a representative set of structures using all possible local motifs are generated. An example is shown in Fig. 1. Since all motifs are predicted, but not all combinations, the user must look at the results and not expect the computer to find all structures. Analysis of suboptimal structures is aided by two new methods of looking at RNA folding: (1) the energy dot plot and (2) a plot of the number of possible different base pairs versus nucleotide position in the sequence (P-Num graphs).

There are two versions of the program: LRNA folds linear sequences, and CRNA folds circular sequences. LRNA and CRNA have the same user interface. Thus, all information presented here applies to CRNA as well as LRNA.

The program, written in FORTRAN 77 with DEC extensions, runs on a VAX computer under the VMS operating system. A C translation has been created to facilitate porting the program to UNIX environments. The graphics section requires a Tektronics-compatible terminal with graphics cross hairs. Currently supported graphics terminals are the VISUAL 600 series and the Tektronics 4105. The distribution copy includes a precompiled version for the VAX. All examples in this chapter were run using the VAX FORTRAN version of the program.

Input Files

The program uses as input the sequence of the RNA, various constraints on the foldings, and a set of thermodynamic parameters. Each is discussed below.

Sequence Files

The program accepts sequences in any number of formats: Stanford[10] (IntelliGenetics), GenBank,[11] Protein Identification Resource,[12] and

[7] M. Zuker, *Science* **244**, 48 (1989).

[8] M. Zuker and P. Steigler, *Nucleic Acids Res.* **9**, 133 (1981).

[9] R. Nussinov, G. Pieczenik, J. R. Griggs, and D. J. Kleitman, *SIAM J. Appl. Math.* **35**, 68 (1978).

[10] D. Roode, R. Liebschutz, S. Maulik, T. Friedemann, D. Benton, and D. Kristofferson, *Nucleic Acids Res.* **16**, 1857 (1988).

[11] H. S. Bilofsky and C. Burks, *Nucleic Acids Res.* **16**, 1861 (1988).

[12] K. E. Sidman, D. G. George, W. C. Barker, and L. T. Hunt, *Nucleic Acids Res.* **16**, 1869 (1988).

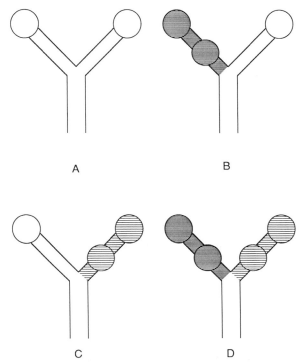

FIG. 1. Representative set of motifs found by the program. If A is the optimal predicted structure, and the motifs shown in B and C are found in suboptimal structures, then the D structure is *not* found, since it contains two independent suboptimal regions.

EMBL,[13] as well as the original RNAFold format;[8] the simplest is the Stanford format (see Appendix 1). Each file can contain one or more sequences. LRNA will ask the user to indicate which sequence to fold as well as the start and end of the fragment. The distributed copy accepts only 750 bases, a practical folding limit for a VAX 750. This limit is easily changed.

Constraints on Folding

Information about the structure helps the program fold the sequence into a better secondary structure. The locations of single-stranded sites, double-stranded sites, or known helixes are easily entered in LRNA. These locations are often available from nuclease mapping and chemical modifi-

[13] G. N. Cameron, *Nucleic Acids Res.* **16,** 1865 (1988).

cation data.[14] Although it is tempting to put in all constraints, erroneous data hinder the program. Thus, only the most certain single- and double-stranded sites should be used.

There are two ways to enter single-stranded positions directly in the sequence file. Rewriting individual bases in lowercase letters will force them to be single stranded; they will, however, contribute no terminal stack or dangling end stabilization. An alternative method is to replace bases susceptible to nuclease or chemical modification by an alternate letter: B, Z, H, V, and W replacing A, C, G, U, and T, respectively. A base marked this way can pair only if the base 3′ to it is single stranded, otherwise the marked base is forced to be single stranded (e.g., if the sequence is . . . AGHUC . . . then H will be single stranded if U is paired, and H may be paired or single stranded if U is single stranded). This allows for chemical modification or nuclease susceptibility of base-paired mucleotides adjacent to single-stranded regions.

It is also possible to force secondary structure motifs using the Force/Excision menu in the program. In particular, two sections can be forced to pair together, one section forced to pair anywhere, one section forced single stranded, or two sections not allowed to pair together. More information on these topics is given in the Method section.

Energy Files and Parameters

A compilation of thermodynamic parameters at 37° are distributed with the program.[6] These are largely based on measurements. [15,16] Parameters not measured were extrapolated with theory.[17] The only two exceptions to this rule were multibranched loops and certain hairpins of four nucleotides ("tetraloops"). Multibranched loops were approximated as $\Delta G^0 = 4.6 + 0.4f + 0.1h$, where f is the number of single-stranded bases in the multibranched loop and h is the number of helixes. This linear approximation was optimized with a set of 210 phylogenetic structures.[6] The linear approximation speeds up the folding algorithm. Recently, tetraloops were shown to have unusual stability.[18,19] Eight of these loops are given an

[14] C. Ehresmann, F. Baudin, M. Mougel, P. Romby, J. P. Ebel, and B. Ehresmann, *Nucleic Acids Res.* **15**, 9109 (1987).

[15] D. H. Turner, N. Sugimoto, and S. M. Freier, *Annu. Rev. Biophys. Biophys. Chem.* **17**, 167 (1988).

[16] S. M. Freier, R. Kierzek, J. A. Jaeger, N. Sugimoto, M. H. Caruthers, T. Neilson, and D. H. Turner, *Proc. Natl. Acad. Sci. U.S.A.* **83**, 9373 (1986).

[17] H. Jacobson and W. H. Stockmayer, *J. Chem. Phys.* **18**, 1600 (1950).

[18] C. Tuerk, P. Gauss, C. Thermes, D. R. Grobe, M. Gayle, N. Guild, G. Stormo, Y. d'Aubenton-Carafa, O. C. Uhlenbeck, I. Tinoco, Jr., E. N. Brody, and L. Gold, *Proc. Natl. Acad. Sci. U.S.A.* **85**, 1364 (1988).

[19] C. R. Woese, *Microbiol. Rev.* **51**, 221 (1987).

FOLDING BASES 1 TO 76 OF #1 A570? YEAST
ENERGY = -21.8

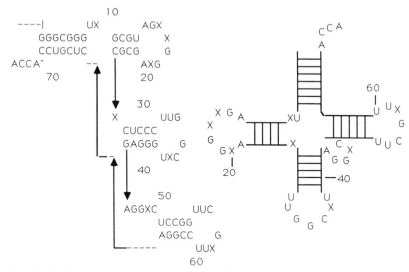

FIG. 2. Reading printer secondary structures. The structure at left is the output for yeast glycine (GGC) tRNA (arrows added for clarity), while the structure at right is the common cloverleaf representation. X represents modified nucleotides that are not able to base pair.

extra stability of −2.0 kcal/mol. This is a modest increment, equivalent to a strong hydrogen bond or stacking interaction.[16,20]

Output Files

As with the previous program, RNAFold,[8,21] there are three types of output files: structural, CT, and Region. The structural file is a secondary structure in line printer output, not suitable for publication. An example is shown in Fig. 2, along with a more conventional representation. The arrows, which are not usually included, show how to read a multibranched loop. There are two rules that help in reading these structures: (1) the numbers on nucleotides help trace out the structure, and (2) a space on the 3′ side of a helix means drop down to the next helix (e.g., the space between C25 and X26 in Fig. 2).

CT (coordinate table) files and Region files produce output for other graphics programs such as SQUIGGLES by the University of Wisconsin

[20] D. H. Turner, N. Sugimoto, R. Kierzek, and S. D. Dreiker, *J. Am. Chem. Soc.* **109**, 3783 (1987).
[21] M. Zuker, this series, Vol. 180, p. 262.

Genetics Computer Group (GCG),[22] DRAW by Shapiro *et al.*,[23] and NAVIEW by Bruccoleri and Heinrich.[24] These programs provide publication-quality representations of secondary structure. Since these files represent the structure in numeric form, simple statistical information is easily accrued (see the Examples section).

Information displayed using the graphic section of LRNA can be dumped to data files. These files can be translated to drive the GCG FIGURE program,[22] using two programs distributed with LRNA.

Method

The recursive algorithm[7,8] has two sections: Fill and Traceback. The Fill step creates a matrix with information of the structures, which can be stored in a Save file. This is a time-consuming step, because the program determines the minimum energy structure for each fragment of the sequence. Traceback examines the matrix created by Fill to produce structures, suboptimal plots, or P-Num graphs. This step is much quicker than the Fill step.

Typically, the user executes the Fill algorithm in batch mode, creating a Save file. A continuation run reads this Save file and executes the traceback, finding structures using either the graphical or "N best" method. Although this two-step method may seem clumsy, generating a Save file has two advantages: the Save file allows the user to save all the folding information in an easily accessible form, and creating the Save file in batch mode is often cheaper than running in real time. This last point is significant, since for an RNA of N bases, the program will take about $5.2 \times 10^{-8} \times N^{3.7}$ min to generate a Save file (~ 8.3 hr on a VAX 750 for 500 bases, depending on system parameters). The Save file, however, takes up a lot of space. An N-base sequence will create a Save file of approximately $6.1 \times N^2$ bytes (about 1.5×10^6 bytes for 500 bases) using LRNA; CRNA requires approximately $4.1 \times N^2$ bytes.

In batch mode, a sequence of commands in a batch file is run noninteractively, usually in nonprime time. The user can type in the batch file using an editor or the interactive program BATGEN. The use of BATGEN is described below and illustrated in Appendix 2.

BATGEN's questions are similar to those asked by LRNA, but answers are written to a file rather than executed directly. Different queries are

[22] J. Devereux, P. Haeberli, and O. Smithies, *Nucleic Acids Res.* **12,** 387 (1984).

[23] B. A. Shapiro, J. Maizel, L. E. Lipkin, K. Currey, and C. Whitney, *Nucleic Acids Res.* **12,** 75 (1984).

[24] R. E. Bruccoleri and G. Heinrich, *CABIOS* **4,** 167 (1988).

usually shortcuts around certain sections. BATGEN presents the user with a default option for many questions; hitting return chooses the default.

BATGEN first asks for the batch file name and the header file name. The header file contains information placed at the top of the batch file, such as a system command to change directories. The next prompt asks which type of folding program will be run: linear (LRNA) or circular (CRNA). BATGEN then asks for the run type: regular, save, or continuation. The save option completes the Fill part of the program, saving the results in a file. The continuation option generates structures or graphic output, using the file generated in a save run. A regular run is a Fill followed by a Traceback, but without generation of a Save file. As mentioned before, the two-step approach is preferred.

After the save option is chosen from the run-type menu, the program asks for the Save file name and the sequence file name. The program displays all sequences; the user selects one by name or number. Only the first 5000 bases of a sequence can be accessed in the distributed version of the program, but this number is easily increased. The structure prediction program is limited to a 750-base fragment within the sequence, not counting excised regions.

The next series of prompts asks for energy files, force or excision information, and energy parameters. Users can choose their own energy files, in which case the program will prompt for each file, in the same way it asks for the batch file name. The program asks for Force or Excision data next, in a method similar to LRNA.

The Force/Excision menu has nine options: two force, two prohibit, two excisions, and three controls. The force and prohibit options refer to helixes; thus, a double force causes the specified helix to form, and a double prohibit prevents a specified helix from forming. A single force causes a specified fragment to be double stranded, while a single prohibit causes the specified fragment to be single stranded. An excision removes fragments from a sequence and folds what remains. This feature is useful when a partial structure is known or when removing introns from a sequence. A closed excision forces the first and last bases of the designated fragment to be paired and excises the intervening bases, while an open excision removes the entire designated fragment and splices the ends together. Finally, there are three control choices: (1) show the current force/excision data; (2) clear all force or excision data; and (3) exit the menu. BATGEN then gives the user the option of changing certain, easily adjustable energy parameters.

There are ten options in the Energy Parameter menu. These allow easy modifications of some of the energy parameters and constraints. They are mostly self-explanatory, save three important points. A negative energy favors something, while a positive energy discourages something, and the parameters are in tenths. For example, the force parameter is -500, which is -50.0 kcal/mol. Making this value more positive will decrease the chances that a structure forced at the Force/Excision menu will form. Finally, the parameters "internal loop size" and "internal loop asymmetry" set an upper size limit on the program's search for internal loops. These parameters affect not only the resultant structure but also the Fill time. The default value of 30 is usually adequate for folding at 37°.

At this point, the user has the option to arrange for additional runs in the batch file. For example, the user could fold another molecule, or continue with the Traceback, as described next. Usually though, the user will stop here, submit the batch file just generated with BATGEN, and make a Save file. An example of using BATGEN is shown in Appendix 2.

Once the Save file has been generated, the program is run in continuation (or Traceback) mode to find secondary structures from the Save file. There are two methods of finding structures in continuation mode: N best and suboptimal plot. The N best traceback can be run from a batch file or interactively. In contrast, the suboptimal plot must be generated interactively at a graphics terminal.

N best Traceback

As mentioned above, the N best traceback can be run from batch or interactively. The text below applies to BATGEN as well as running the program interactively, since BATGEN's queries are similar to those of the program. To start the N best traceback, the user chooses the continuation mode from the run-type menu and N best from the run mode menu. The program then asks for the three N best parameters, which limit the number of structures LRNA finds.

The three parameters are denoted P, d, and N. P which is zero or a positive integer, limits the program to finding structures within $P\%$ of the minimum energy. This group of structures are called P-optimal. For example, 10-optimal is the set of structures within 10% of the minimum free energy. The distance or window parameter, d, is a measure of structure similarity. It is defined and illustrated in Fig. 3; d is a nonnegative integer. N is an upper bound on the number of structures found. The total number of structures generated depends on these parameters, as well as the sequence length and composition. Typical values range from less than 10 structures for 76 bases when $P = 10\%$, $d = 1$, to roughly 100 for 425 bases when $P = 8\%$, $d = 10$.

After choosing these parameters, LRNA asks the user for the Save file name and a continuation dump file name. The continuation dump file is a

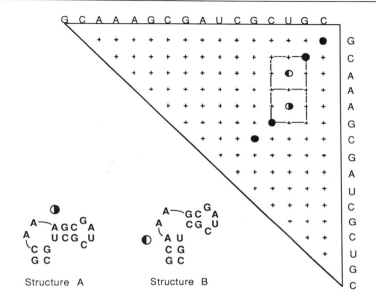

●   Common Base Pairs of A and B

FIG. 3. Distance, d, is a measure of dissimilar base pairs between two structures. This difference is easily seen on a dot plot. If d is the distance between two foldings, then a square of side $2d$ centered on a base pair from one form will always contain a base pair from the other folding. No smaller square has this property. The maximum distance between structures of a linear sequence of N bases is $N - 5$. The distance between the structures shown is 1.

list of the current energy parameter menu and what kind of forcing information is included in the fold. The usual response to this query is ⟨return⟩, which will show these parameters on the screen. After asking whether the energy parameters should be displayed, the program asks what kind of output files the user wants. Usually, this is at least a structure output file. After asking about CT and region files, the program finds the structures and then stops. If more than one structure is found, each output file will contain multiple structure information.

Graphic Methods: Suboptimal and P-Num Plots

The suboptimal plot mode is an interactive way of choosing structures. This method uses a graphics terminal with the cross hairs feature to display all structures within a specified increment of the computed minimum energy. The display is called an energy dot plot. A base pair between the ith and jth nucleotides ($i < j$) is displayed as a point in the ith row and jth column of a triangular grid. The user can bring out the cross-hairs and move them so that they intersect at any base pair displayed on the screen. By pressing the appropriate key (See Table I), the user can display the base numbers of the selected pair and the minimum energy of a structure containing that base pair, or cause the generation of that structure. The generated structure is output as any of the three types of output files. As many base pairs as desired can be selected.

As with the N best mode, the user first chooses continuation run, but then picks suboptimal plot from the run mode menu. The user chooses the type of graphics terminal from the next menu, and then is asked for the minimum vector size to plot (INAROW). A base pair will not be shown unless there are at least INAROW base pairs stacked on one another in a suboptimal structure within the given energy increment. Thus, to see all base pairs, set INAROW to 1. The next parameter is Window, which is similar to d, the distance between structures. The Window parameter ensures that only essentially different structures are found in the N best mode, and it controls what base pairs are displayed in the suboptimal plot mode. As before, the program will then ask for the Save file name, for a continuation dump, whether the energy parameters should be displayed, and what kind of output files are desired. The screen clears, a triangle is drawn, and the user is prompted for an energy increment. The triangle is then filled with dots representing all base pairs which can exist in structures with energies within the given increment from the minimum. Examples are given in Figs. 5 and 9.

Initially, typing 0 displays the base pairs for the minimum energy structure(s). Typing a positive number (in tenths of a kcal/mol: 17 is actually 1.7 kcal/mol) plots the unique base pairs of foldings within this energy increment, subject to the constraints of INAROW and WINDOW. A negative number clears the screen, redraws the triangle, and prompts the user for an energy increment.

Typing / or hitting return at this point brings out the cross hairs. Moving the cursor to a specific dot (base pair) and typing q generates the optimal structure containing that base pair. Typing other letters produces other output, as described in Table I. The clear and redraw option (r) is similar to typing a negative number for the energy increment (the screen is cleared, etc. . . .). When an increment is typed in, the base pairs of structures already picked are not displayed. In addition, all base pairs within a distance WINDOW of ones which have already occurred in structures are not displayed. This feature helps the user pick only new and essentially different structures.

The graph mode, or P-Num plot, plots base number (x) versus the total number of different pairs that base can make (y) for all P-optimal foldings.

TABLE I
ENERGY DOT-PLOT COMMANDS[a]

Key	Option
g	Enter *P*-Num graph mode
p	Dump the current dot plot to an ASCII file; all admissible base pairs are included (i.e., WINDOW and INAROW have no effect)
q	Generate an optimal folding containing the selected base pair
r	Clear screen and redraw grid outline
t	Terminate, leave program
Other	Display the $i-j$ coordinates of the selected point and the energy of the structure containing that pair

[a] When graphics cross hairs are displayed.

This plot shows how well determined pairing is over the entire sequence. A large value means that base can pair with many other bases, whereas a small value suggests the pairing for that base is well determined or that the base is usually single stranded.

As with the energy dot plot, typing / brings out the graphics cursor. The horizontal position of the cursor picks a specific nucleotide. When a space is typed, the number of potential pairs for that nucleotide is displayed. Other commands are found in Table II.

Another useful, but completely separate, program is SNUM. *S*-Num is a measure of how many times a base is found single stranded in a series of structures. This is an example of simple statistics derived from a CT or

TABLE II
ENERGY MODE COMMANDS[a]

Key	Option
p	Dump the current *P*-Num plot to a file
q	Return to energy dot-plot mode
r	Clear and redraw graph
t	Terminate, leave program
Other	Display base position and number of possible pairs for a given increment

[a] When graphics cursor is displayed.

region file. The program SNUM first prompts the user for a multiple CT file generated with an N-best traceback, and then asks for an output file for the S-Num data. The program counts the number of times a base is single stranded and outputs the data to the output file. The data are in simple $X-Y$ format and can be plotted.

Other Folding Methods

In addition to the save and continuation method, the user can choose to execute these in a single step, the regular run. There are two types of regular runs: single and multiple. The single run fills the folding matrix, and immediately performs a traceback, in either N best or suboptimal plot mode. The multiple run predicts only the optimal structure for a series of sequences in a single file. Both of these methods suffer from the same problem: after leaving the program, any folding information (besides the structure, CT, or region files generated) is lost.

Performance of Predictions

The performance of the program has been tested on 141 tRNAs, 67 5 S rRNAs, *Escherichia coli* 16 S rRNA (separated into four domains), and the Group Ib intron of the 26 S rRNA of *Tetrahymena thermophila.*[6] On average, 70% of the phylogenetically deduced helixes are predicted for a given structural type. A helix is considered correct if it does not lack more than two base pairs in the phylogenetic structure. Better results are obtained by looking at the suboptimal predictions. For example, the optimal and best 5-Num structures for *E. coli* 16 S rRNA include 63 and 94% of the phylogenetic helixes, respectively Further improvements in prediction are expected as more of the structural code is discovered, since it is relatively simple to change the energy model in LRNA.

Examples

Two sample analyses are shown. The first is the Group Ib self-splicing intron from the 26 S rRNA from *T. thermophila* (IVS, Fig. 4a),[25,26] and the second is the suggested 5' end of the human c-*fes* gene (c-*fes*).[27] The general strategy is analysis of the optimal structure then generation of other possible structures with N best and energy dot-plot modules. Finally, two kinds of plots, P-Num and S-Num, show how well the pairing and single-

[25] F. Michel and B. Dujon, *EMBO J.* **2**, 33 (1983).
[26] J. M. Burke, M. Belfort, T. R. Cech, R. W. Davies, R. J. Schweyen, D. A. Shub, J. W. Szostak, and H. F. Tabak, *Nucleic Acids Res.* **15**, 7217 (1987).

a

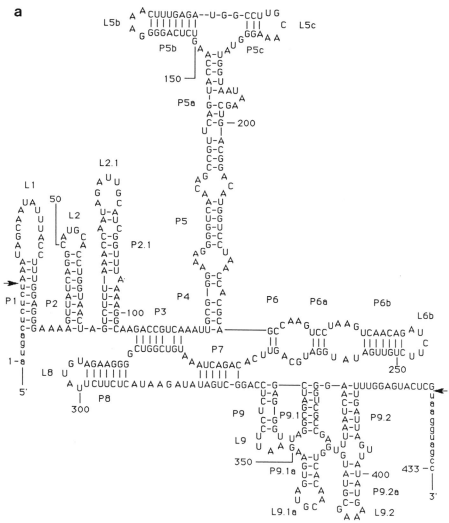

Fig. 4. (a) Standard structure of the Group I intron from *Tetrahymena thermophila* 26 S rRNA (IVS) [J. M. Burke, M. Belfort, T. R. Cech, R. W. Davies, R. J. Schweyen, D. A. Shub, J. W. Szostak, and H. F. Tabak, *Nucleic Acids Res.* **15**, 7217 (1987)]. (b) An optimal calculated structure of the same region ($\Delta G^0 = -115.9$ kcal/mol). Lowercase letters represent portions of the intron included in the folded sequence. Arrows show exon–intron boundries. Numbering is from the 5′ end of the sequence. This sequence corresponds to bases -10 to -1, 2 to 424 of Cech *et al.* [T. R. Cech, N. K. Tanner, I. Tinoco, Jr., B. R. Weir, M. Zuker, and P. S. Perlman, *Proc. Natl. Acad. Sci. U.S.A.* **80**, 3903 (1983)].

b

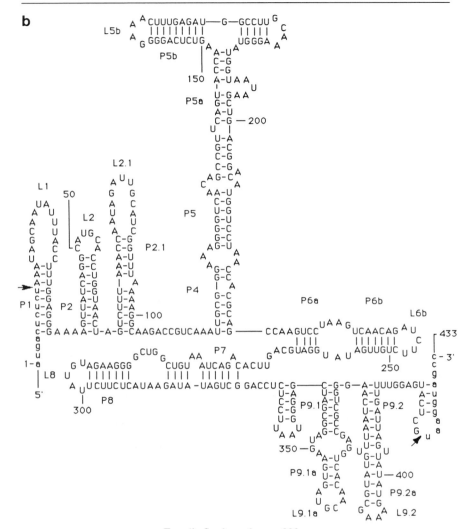

FIG. 4b. See legend on p. 293.

stranded regions are defined. In both cases, no data besides the sequence and the default energy files were used.

IVS

The intron from *T. thermophila* provides an interesting case because it has a convenient size for folding and a combination of phylogenetically conserved and variable regions. The sequence folded is the intron (414 bases) plus 10 bases of exon on both the 5' and 3' sides.

There are three structures at the minimum of $\Delta G^0 = -115.9$ kcal/mol. One is shown in Fig. 4b, and the others differ only slightly. The structure shown has most of the features of the phylogenetic model described by Burke et al.[26] (Fig. 4a). The calculated structure is also consistent with modification data.[28] The major difference between the model shown and that of Burke et al. is that the P3 helix is not predicted; P3 and P7 form a pseudoknot, a structure that cannot be predicted by this algorithm. Other minor differences include "slipped" P5c and P9 helices. The exact pairing of P9 is not proven by phylogeny, and the calculated structure of P5c is possible from phylogenetic data.[29]

All the helixes of Burke et al. are found within the set of suboptimal foldings within 5% of the minimum free energy (5-optimal foldings). The P3 helix, for example, is found 1.8 kcal/mol from the minimum; this prevents P7 from forming the pseudoknot. The published structure for P9 is found 2.4 kcal/mol away from the minimum.

Energy dot plots of 0-, 5-, and 10-optimal structure sets are shown in Fig. 5. Three folding domains are seen in the 5-optimal plot, as shown by the lines in Fig. 5b. These lines show regions of the molecule which fold largely independently of each other. These domains are obscured at the 10-optimal level. Certain regions, however, show unique pairing, even at the 10-optimal level. The stem-loop structure at 150–171 is the most well defined of such regions, followed by regions 236–256 and 378–411.

The 5- and 10-Num plots are shown in Fig. 6. They indicate how variable the possible pairing is throughout the molecule. The three regions mentioned above are shown as deep valleys in this plot, indicating either well-defined pairing or single strandedness. Other relatively well-defined pairing regions include 40–65, 289–307, and 341–376. Interestingly enough, three of the valley regions (236–256, 289–307, and 378–411) show compensating structure across the *Tetrahymena* family.[30]

To remove some of the ambiguity of the *P*-Num plot, the *S*-Num and *P*-Num plots for the region 140–240 are shown together in Fig. 7. A comparison indicates that part of the valley in 150–171 is single stranded, while the remainder is double stranded. The region 210–230 indicates potential fluxional structures, where the sequence can fold into both single-, and double-stranded regions.

[27] W. Van de Ven, A. Roebroek, and J. Schalken, Eur. Patent 87,200,940.2 (1987).
[28] T. Inoue and T. Cech, *Proc. Natl. Acid. Sci. U.S.A.* **82**, 648 (1985).
[29] F. Michel, personal communication (1988).
[30] T. R. Cech, N. K. Tanner, I. Tinoco, Jr., B. R. Weir, M. Zuker, and P. S. Perlman, *Proc. Natl. Acad. Sci. U.S.A.* **80**, 3903 (1983).

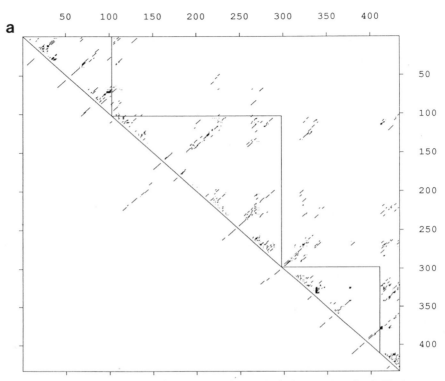

FIG. 5. IVS energy dot plots: (a) 0- and 5-optimal and (b) 0- and 10-optimal. The bottom left plots represent the minimum free energy structures.

c-fes

Another application of the folding program is the prediction of structure in mRNA. The secondary structure of mRNA is important in at least two regards: start codon or Shine–Dalgarno sequence sequestering and the design of antisense nucleic acids. In the former case, when the start codon or Shine–Dalgarno sequences are paired in a strong helix, translation may be inhibited. It is important to avoid this when constructing a gene based on a known protein sequence using the most preferred codons.[31] Conversely, gene expression can be slowed or shut down using antisense RNA

[31] For example, G. Gröger, F. Ramalho-Ortigao, H. Steil, and H. Seliger, *Nucleic Acids Res.* **16,** 7763 (1988).

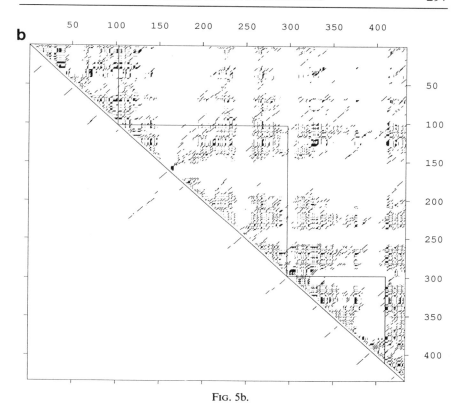

FIG. 5b.

or DNA pairing to single-stranded regions.[32] In this case, prediction of these target regions can facilitate design of the antisense sequence. One example is the human c-*fes* protooncogene.

Much is unknown about c-*fes*, including the transcription start site. A good guess for this site is the AUG near the beginning of exon 2.[27,33] The sequence folded for this example includes exon 1 and 396 bases of exon 2. There are three optimal structures that differ little. One is shown in Fig. 8. Although representations of mRNA structure usually show only stem-loops, the predicted c-*fes* structure shows multibranched loops, bulges, and internal loops also.

---

[32] For example, E. V. Prochownik, J. Kukowska, and C. Rodgers, *Mol. Cell. Biol.* **8,** 3683 (1988).

[33] G. Skuse and P. Rowley, personal communication (1988).

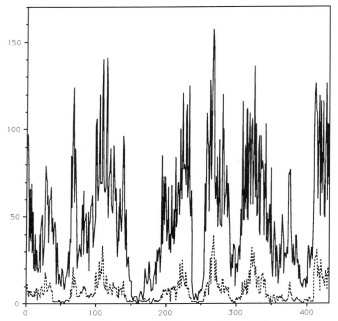

FIG. 6. 5- and 10-Num plots of IVS. · · · , 5%;—, 10%.

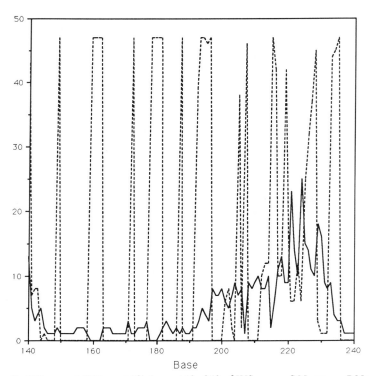

FIG. 7. P-Num and S-Num at 5%, bases 140–240 of IVS. · · · , S-Num;—, P-Num.

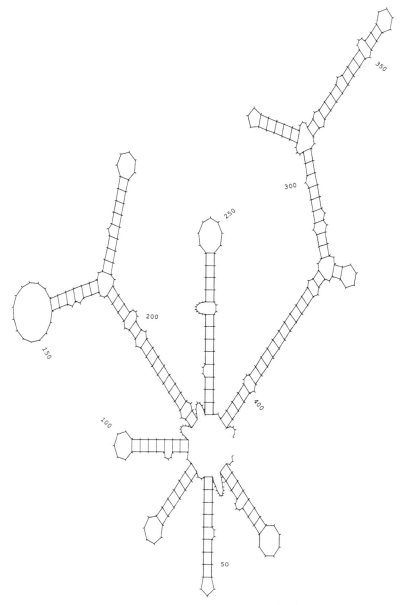

FIG. 8. Minimum free energy structure of c-*fes*.

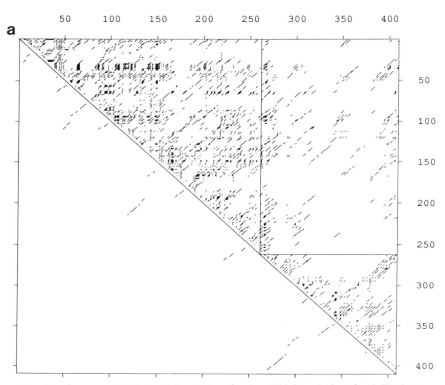

Fig. 9. c-*fes* energy dot plots: (a) 0- and 5-optimal and (b) 0- and 10-optimal. The bottom left plots represent the minimum free energy structures.

The 5-optimal dotplot (Fig. 9a) suggests two folding domains: 1 to 269 and 270 to 409. These domains are separated by the strong helix which starts at C^{277}–G^{398} and ends at C^{290}–G^{385}.

The domains are also seen in the 5-Num plot (Fig. 10a), as the variability in pairing changes around 270. The 5- and 10-Num plots of c-*fes* (Fig. 10b) are quite different from the *P*-Num plots of IVS (Fig. 6). Instead of well-defined and variable pairing regions (valleys and peaks, respectively), only variable regions are observed, especially at the 10% level.

As mentioned at the beginning of this section, one application of this program is to find regions of RNA that are single stranded in the predicted structure. These regions are likely to be available for binding by antisense nucleic acid. The *S*-Num plot shown in Fig. 11 illustrates this for c-*fes*. For example, region 100–120 provides a suitable target. The area around

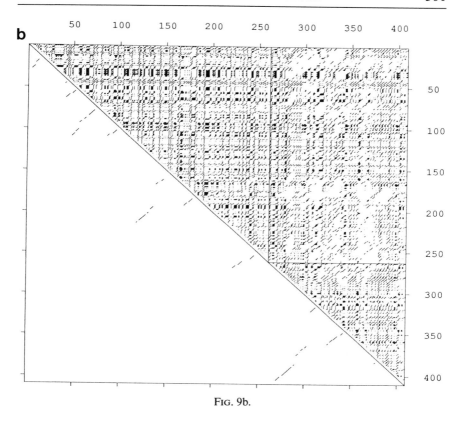

Fig. 9b.

110–120 seems to be a core of single-stranded structure (high value of single-stranded count, S-Num) surrounded by some highly variable pairing regions (high value of different base pair count, P-Num).

Conclusion

The programs described provide a convenient way to fold an RNA sequence and examine suboptimal as well as optimal structures. On average, structures predicted from single sequences are roughly 70% correct.[6] Experimental data defining single and double stranded regions can be used to constrain the prediction and presumably increase accuracy. The output files present structures in both human- and machine-readable forms. The new graphics routines provide a convenient way to examine large numbers of structures to look for well-defined motifs and variable regions. Thus, the

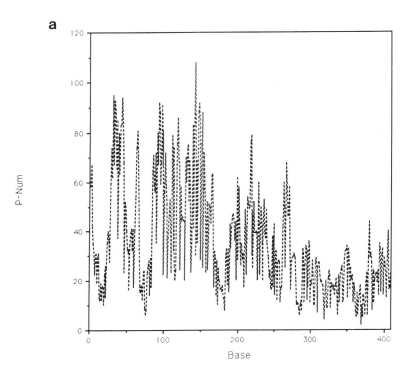

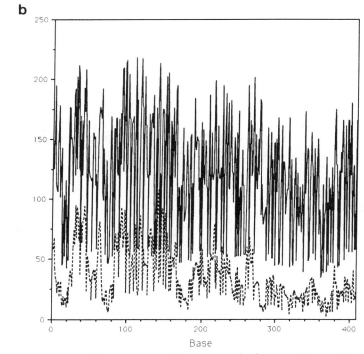

FIG. 10. (a) 5-Num and (b) 5-Num and 10-Num plots of c-*fes*. · · ·, 5%; —, 10%.

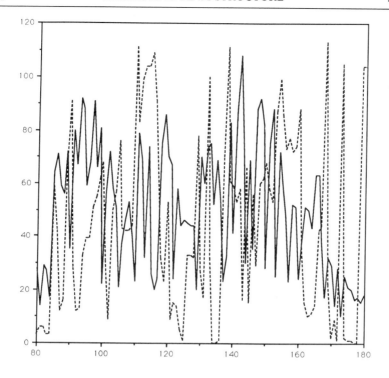

Base

FIG. 11. *P*-Num and *S*-Num at 5%, bases 80–180 of c-*fes*. · · · , *S*-Num;—, *P*-Num.

program seems well suited to help in the design of experiments involving RNA. Finally, the program can be modified to include new energy rules. Thus, its performance should improve as more of the "structural code" is discovered.

Appendix 1

One of the simplest standard sequence formats is Stanford or IntelliGenetics (IG):

1. At least one line starting with a semicolon, followed by nothing or optional text. Multiple lines of this kind are fine and are actually preferred to document the sequence.
2. A line of text. This text will appear as a label before the structures generated with the folding program. Only the first 30 characters are read.

3. The sequence, having the following characteristics:
   (a) it must be in upper case (otherwise the folding program will not recognize it),
   (b) can have any number of bases per line,
   (c) accepts T as well as U,
   (d) can be up to 5000 nucleotides (only a subset can be folded at any given time), and
   (e) ends with the number 1, to signify the end of the sequence.
4. After the sequence, one can optionally put in more sequences. Each sequence must be separated by at least one line starting with a semicolon.

Example:
```
; tRNA sequence
#1 AGC YEAST
GGGCGGGUXGCGUAGXXGGXAGCGCXCUCCCUUGGCXUGGGAGAGGX
CUC
CGGUUCGXUUCCGGACUCGUCCACCA1
```

Appendix 2

An example of using BATGEN to set up a batch file to fold the tRNA shown in Fig. 2. The output of the program is in plain text, while the user's responses are in **boldface** type. When a prompt ($\rightarrow$) has no response after it, the default option was chosen by hitting return. After running BATGEN, the user lists the batch file just generated.

The program generates batch files for the RNA folding programs LRNA and CRNA. Start-up information (like changing directories) is put into HEADER.DAT. At each prompt, one may type in a response or hit return to use the default,

```
$r batgen
Batch file name? Default:Fold.Com
→ trna1.com
Header file name? Default:Header.Dat
→
Enter folding program type
   1 Linear
   2 Circular
Default = 1 → 1
Enter run type
   0 Regular run
   1 Save run
   2 Continuation run
Default = 1 → 1
What is the save file name? Default:Fold.Sav
trna1.sav
   INPUT SEQUENCE FILE NAME (CTRL-Z TO end):
[jaeger.fold.seq]trnas.seq
AVAILABLE SEQUENCES IN [jaeger.fold.seq]trnas.seq
```

1. #1 A570?	GGC YEAST — use align	2. #2 I580	AAU TORULOPSIS UTIL
3. #3 L580	CAA TORULOPSIS UTIL	4. #4 X580	CAU TORULOPSIS UTIL
5. #5 P580	UGG TORULOPSIS UTIL	6. #6 V580	AAC TORULOPSIS UTIL
7. #7 Y580	GUA TORULOPSIS UTIL	8. #8 F590	GAA YEAST
9. #9 E575	UUC SCHIZOSACCHA. P	10. #10 F575	GAA SCHIZOSACCHA.

CHOOSE SEQUENCE by NUMBER or NAME, or ? for RELIST;
⟨CR⟩ DEFAULTS TO THE FIRST ONE, CTRL–Z for NEW file.
1
5′ END?
Default = 1 →
3′ end?
Default = 76 →
Use default energy files?
(Y,N) Default:Y →
Do you have Force or Excision information?
(Y,N) Default:N → **n**
Change any of the energy parameters?
(Y,N) Default:N → **n**
Run the folding program again in the batch file?
(Y,N) Default:N → **y**
Enter run type
   0 Regular run
   1 Save run
   2 Continuation run
Default = 1 → **2**
Within how many percent of the minimum?
Default = 0 → **0**
Maximum number of structures found?
Default = 1 → **10**
Minimum distance between structures?
Default = 0 → **0**
What is the save file name? Default:trna1.sav
→
That save file doesn't exist.
Continue anyway.
(Y,N) Default:N → **y**
Make a structure file?
(Y,N) Default:Y → **y**
Structure file name? Default:Fold.Out
→ **trna1.out**
Printer width? Default: 80
Default = 80 → **120**
Make a CT file?
(Y,N) Default:N → **y**
CT file name? Default:Fold.CT
→ **trna1.ct**
Make a REGION file
(Y,N) Default:N → **n**
Run the folding program again in the batch file?
(Y,N) Default:N → **n**
$type trna1.com
$set default [turner.fold]
$R LRNA
1
trna1.sav

[jaeger.fold.seq]trnas.seq
  1
  1
  76
Dangle.Dat
Loop.Dat
Stack.Dat
TStack.Dat
TLoop.Dat
8
$R LRNA
2
1
   10
   100
   0
trna1.sav
N
Y
N
trna.out
120
Y
trna1.ct
N
$

Acknowledgments

This work was supported by National Institutes of Health Grant GM22939. Michael Zuker is a Fellow of the Canadian Institute for Advanced Research.

[18] Detecting Pseudoknots and Other Local Base-Pairing Structures in RNA Sequences

By Hugo M. Martinez

Introduction

Martinez[1] describes a suite of programs, RNAFOLD, comprising a workbench for studying RNA secondary structures. We here describe a

---

[1] H. M. Martinez, *Nucleic Acids Res.* **16**, 1789 (1988).

number of extensions to this workbench aimed at finding pseudoknots and hairpins in both single and multiple sequences.

As originally described, RNAFOLD consists of four principal programs: DYNPRO for finding global structures of the orthodox type using a dynamic programming algorithm; MONTECARLO for generating a set of orthodox, global structures using an algorithm intended to simulate the folding process; STEMS for finding hairpins; and, finally, DRAW-STRUCT for the drawing of a structure when presented as a base-pairing list. The extensions apply to the first three of these programs. Specifically, (1) an option has been added to STEMS for finding pseudoknots, and (2) all three programs can now deal with more than one sequence at a time to obtain individual and common structures. Thus, given an input set S of sequences, DYNPRO finds a global structure for each member of S and then hairpins common to these global structures. MONTECARLO obtains a set of structures for each member of S, finds a common hairpin list for each such set and then finds the hairpins common to these lists. STEMS generates a set of hairpins for each member of S as a set of local structures and then finds the hairpins common to these sets. When used in the pseudoknot mode, STEMS instead deals only with pseudoknots, finding a set for each member of S and then finding the common ones. In the following these extensions are described in detail, giving the rationale for the extensions and some results of their application to particular sets of RNA sequences.

STEMS Extensions

The pseudoknot option was prompted by two reports: that of Puglisi *et al.*[2] regarding the experimental confirmation of a pseudoknotted RNA oligonucleotide and that of McPheeters *et al.*[3] implicating a pseudoknot structure as an autogenous regulatory site on the bacteriophage gene 32 messenger RNA. Both of these reports strongly support the thesis of the potential importance of pseudoknots as earlier advocated by Pleij *et al.*[4]

In trying to deal with this new kind of base-pairing structure which partially violates the "orthodox" constraint of bases within a hairpin loop not pairing with bases outside the loop, the decision was made to only consider them as independent of a global structure, just as is done when finding hairpins independent of global structure considerations. In part this strategy was a matter of expediency because including pseudoknots as parts

[2] J. D. Puglisi, J. R. Wyatt, and I. Tinoco, Jr., *Nature (London)* **331**, 283 (1988).
[3] D. S. McPheeters, G. D. Stormo, and L. Gold, *J. Mol. Biol.* **201**, 517 (1988).
[4] W. A. Pleij, K. Rietveld, and L. Bosch, *Nucleic Acids Res.* **13**, 1717 (1985).

of a global structure presents some as yet unresolved technical difficulties and also because of the increasing reliance on comparative methods which base the significance of a local structure on conservation criteria.

An example of a pseudoknot is shown in Fig. 1 in two different ways. In the first, the base-pairing sections, identified as stems 1, 2, and 3, are drawn in a noncoaxial manner, while in the second a coaxial arrangement has been imposed to emphasize the stacking of stem 2 to stem 1 and the stacking of stem 3 to stem 2. It is this stacking which gives the coaxial extension of stem 1 and of stem 2 to form a larger, single stem comprising the pseudoknot complex. A simpler pseudoknot would consist of just stems 1 and 2 or of just stems 2 and 3.

Identifying a pseudoknot as stacked stems readily suggests how to go about finding them. Thus, since stems 1 and 3 are the stems of two hairpins with stem 2 linking these two hairpins together, one need only generate a list of hairpins and determine which pairs of them can be linked together by an intermediate stem.

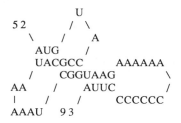

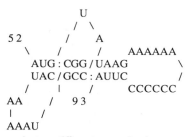

FIG. 1. Pseudoknot shown in two different ways: (top) noncoaxial manner and (bottom) coaxial manner. The slashes indicate extended bonds between successive bases. Paired bases are shown as letters in the same column. As referred to in the text, stem 1 is AUG paired with CAU, stem 2 is GCC paired with GGC, and stem 3 is UAAG paired with CUUA.

The pseudoknot mode in STEMS is selected by setting the pseudoknot option parameter to 1. In this mode STEMS generates the hairpins (with their stems constrained to have no bulges or inner loops) for all members of the sequence set S but does not report them. Instead, each hairpin list (one list for each sequence) is first sorted in increasing order relative to the 5′ starting position of a hairpin and then scanned in the following manner. For a given hairpin in the list, successive ones are examined for linking to it via base pairing in the form of an intermediate stem that forms the actual pseudoknot (a stem 2). The first one found such that the complex has a negative free energy is kept, and no further possibilities are examined. If no such complex can be formed for the given hairpin, the sequence of bases is examined for a stem 2 that can give the simpler form of a pseudoknot complex. Regarding the energy of a pseudoknot complex, this is calculated using the same base-pairing, stacking, and destabilization energies as for orthodox stems and hairpins.

Once a pseudoknot list is formed for each member of the sequence set S, a search is made for pseudoknots which are common to those of the first sequence. That two pseudoknots are to be regarded as the same is subject to the following six user-adjustable parameters:

1. Min-pseudoknot-loop-size, which specifies the least number of free bases in the hairpin loop of stem 1 or of stem 2 (default is 2)
2. Max-pseudoknot-loop-size, which specifies the maximum number of free bases in the hairpin of stem 1 or of stem 2 (default is 30)
3. Max-pseudoknot-length, which specifies the maximum number of base pairs in stem 2 (default is 9)
4. Reg-fact, which concerns the allowed difference in position of the pseudoknots in their respective sequences as a fraction of the smaller sequence length (default is 0.3)
5. Len-fact, which concerns the difference in length of two pseudo-knots as a fraction of the smaller length (default is 0.3)
6. Align-fact, which concerns the alignment score of the subsequences comprising the pseudoknots as a fraction of the maximum possible (default is 0.6)

For each pseudoknot of the first sequence a list is made of the pseudoknots in the other sequences which are identical to it, according to the above criteria. The corresponding list of subsequences is then passed on to a multiple sequence alignment program for an assessment of what is conserved among them.

Figures 2 and 3 show the output of STEMS in the pseudoknot mode for the two sequences investigated by McPheeters et al.[3] This output has been edited to show only the common pseudoknots. There are two common

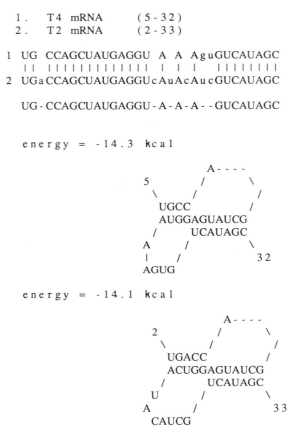

```
1 .    T4  mRNA      ( 5 - 3 2 )
2 .    T2  mRNA      ( 2 - 3 3 )

1  UG  CCAGCUAUGAGGU  A  A  A g u GUCAUAGC
   | |  | | | | | | | | | | | | |  | |  |   | | | | | | |
2  UGa CCAGCUAUGAGGUc AuAcAu cGUCAUAGC

UG - CCAGCUAUGAGGU - A - A - A - - GUCAUAGC


energy  =  - 1 4 . 3   kc a l
```

```
                          A - - - -
              5          /          \
               \        /            /
                UGCC              /
                AUGGAGUAUCG
                /       UCAUAGC
              A     /              \
              |    /                3 2
              AGUG
```

```
energy  =  - 1 4 . 1   kc a l
```

```
                          A - - - -
              2          /          \
               \        /            /
                UGACC             /
                ACUGGAGUAUCG
                /       UCAUAGC
              U     /              \
              A    /                3 3
              CAUCG
```

FIG. 2. Output of STEMS in pseudoknot mode showing one common pseudoknot pair of the sequences of T4 mRNA and T2 mRNA and the alignment of the corresponding subsequences.

pseudoknots, and both are of the simple type in that a stem 3 is missing. The first pair of pseudoknots (Fig. 2) are those reported by McPheeters *et al.*[3] The alignment of the corresponding subsequences and the structures themselves clearly show what is conserved.

When the pseudoknot option is not invoked, STEMS proceeds to find the hairpins in all the members of S subject to the normal constraints of max-bulge-size, max-inner-loop-size, and max-hairpin-loop-size. As for pseudoknots, it then proceeds to find the hairpins which are common to those of the first sequence. The criteria of when two hairpins are the same are just the last three for pseudoknots. The output shown in Fig. 4 is also

```
1.   T4  mRNA      ( 1 0 - 3 6 )
2.   T2  mRNA      ( 8 - 3 5 )

1  GCUAUGAGGU  A  A  A g u GUCAUAGCAC c a
   | | | | | | | | | |   |   |   |     | | | | | | | | | |
2  GCUAUGAGGU c A u A c A u c GUCAUAGCAC

   GCUAUGAGGU - A - A - A - - GUCAUAGCAC - -
```

energy =   - 2 . 2 k c a l

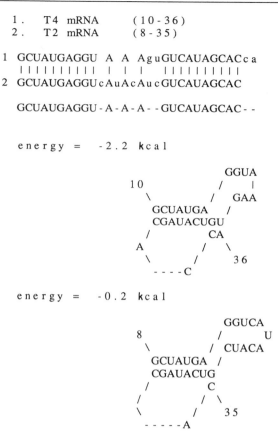

energy =   - 0 . 2 k c a l

FIG. 3. Output of STEMS in pseudoknot mode showing another common pseudoknot of the sequences of T4 mRNA and T2 mRNA and the alignment of the corresponding subsequences.

edited to show only common hairpins, and it also applies to the two sequences treated with the pseudoknot option. The max-bulge-size parameter has been set to 0 in order to concentrate on hairpins with perfect stems. Note that the common hairpins are just those found in the common pseudoknots.

MONTECARLO Extensions

The Monte Carlo method which we use for folding an RNA sequence is an iterative one. First the most energetic hairpins are found, up to a

```
1.   T4  mRNA       (10-32)
2.   T2  mRNA       (8-33)


1  GCUAUGAGGU  A  A  AguGUCAUAGC
   |||||||||| | | |   |||||||||
2  GCUAUGAGGUcAuAcAucGUCAUAGC


   GCUAUGAGGU - A - A - A - - GUCAUAGC


energy =   -7.3  kcal
                 ggua
10      GCUAUGA        a
32      CGAUACU        /
                 cugu


energy =   -7.1  kcal
                 ggucau
8       GCUAUGA           \
33      CGAUACU           /
                 cugcua


   1.   T4 - mRNA      (5-20)
   2.   T2 - mRNA      (2-19)


1  UG  CCAGCUAUGAGGU  A
   ||  ||||||||||||| |
2  UGaCCAGCUAUGAGGUcA


   UG - CCAGCUAUGAGGU - A


energy =   -1.2  kcal
                 agcu
5       UGCC         \
20      AUGG         /
                 ggag


energy =   -6.5  kcal
                 agcu
2       UGACC        \
19      ACUGG        /
                 ggag
```

FIG. 4. Output of STEMS in hairpin mode. Only common hairpins are shown.

user-specifiable maximum number. Such a list, called L0, serves as the initial one for all the foldings. Regarding L0 as a population of stems competing to form, members from L0 are selected with a probability that is a function of the free energy relative to the competing population. Once a hairpin is selected for adding to the current structure, the remaining population is screened so that only those which are compatible with the new structure can compete. The sequential selection and readjustment of the competing population continues until L0 is exhausted of members compatible with the current structure. On exhaustion of L0 a new hairpin list L1 is constructed using the unpaired bases of the structure obtained from L0; L1 is then used for the competition until it is exhausted, etc.

The structure resulting from this kind of folding is encoded as a list of base pairs so that it is an easy matter to compare the structures resulting from separate foldings and thus tally the frequency with which a particular structure occurs. Only the distinct base pair lists are retained, and from each a hairpin list is constructed consisting only of hairpins whose loops are free of additional hairpins (do not bifurcate). It is these kinds of local structures which are compared from the various foldings using the same comparison criteria as for the Hairpin mode of STEMS. What is reported are the hairpins common to the base pair list with the largest frequency of occurrence.

In the case of several sequences, each is folded the prescribed number of times, and for each the corresponding common hairpin list is found. Then these common hairpin lists are compared, and the hairpins common to those of the first sequence are saved.

Figure 5 shows the MONTECARLO folding of three small nuclear RNA sequences obtained from the GenBank database: *Drosophila* (locus drour2b, 99 bp), pheasant (locus phsur2a, 95 bp), and pea (locus peanur2, 96 bp). The foldings (10 of each) have been done with default parameter settings, and in order to conserve space the output has been edited to show only the principal features of the foldings. Thus, in 10 foldings the sequence drour2b yields 5 distinct structures with the corresponding occurrence frequencies and energies shown, and for these 5 distinct structures the common hairpins are shown. The same kind of output applies to the other two sequences, and finally the hairpins common to those of the first sequence drour2b are shown.

DYNPRO Extensions

With DYNPRO a sequence is folded using an adaptation of the dynamic programming algorithm of Zuker and Stiegler.[5] The corresponding minimum free energy structure is in the form of a base pair list which is

---

[5] M. Zuker and P. Stiegler, *Nucleic Acids Res.* **9**, 133 (1981).

*Sequence 'drour2b' foldings (5 structures):

    struct #0:   frequency = 4/10, energy = -30.7 kcal

    struct #1:   frequency = 2/10, energy = -28.0 kcal

    struct #2:   frequency = 2/10, energy = -30.7 kcal

    struct #3:   frequency = 1/10, energy = -29.0 kcal

    struct #4:   frequency = 1/10, energy = -29.0 kcal

  ** Hairpins common to the foldings of 'drour2b' **

    6   UGAUUuuu
    20   ACUAAgg

    22   GACGGAGuGcuUG::Gagc
    54   CUGUCUC:C::ACcuCguu

    57   GGGUUGg:::CCCGGuauugca
    96   CCCGGCuuuaGGGCCgccaug

*Sequence 'peaur2' foldings (3 structures):

    struct #0:   frequency = 7/10, energy = -30.1 kcal

    struct #1:   frequency = 2/10, energy = -29.0 kcal

    struct #2:   frequency = 1/10, energy = -27.6 kcal

  ** Hairpins common to the foldings of 'peaur2' **

  10 GGGG:GaagaGUCacacaguAGC::UUGCUAuUGG:GucucuuCGCgugucg
  94 CCCCaCg:::CGG:::::::UCGuuAACGAU:AUCaCguu:::GCGuuuuc

*Sequence 'phsur2a' foldings (1 structure):

    struct #0:   frequency = 10/10, energy = -51.7 kcal

  ** Hairpins common to the foldings of 'phsur2a' **

    4   CGGAuuuuugggcGCgGGAGuuGGA:Cccgga
    53   GCCU:::::::::CG:CCUC::CCUcGuucg

    55   GCAUCGU:CCCGGuauGGcag
    92   CGUGGCAcGGGCCu::CCau

FIG. 5. Output of MONTECARLO.

```
** HAIRPINS COMMON TO THOSE OF SEQUENCE 'drour2b' **

        1.  drour2b      (57-96)
        2.  phsur2a      (55-92)


    1  GggTtGgCCCGGTATtGCAGTACCgCCGGGatttCGGccC
       |   |  |  |||||||||| |||||||| |||||    ||| |
    2  GcaTcGtCCCGGTATgGCAGTACCtCCGGG    caCGGtgC

       G- - T-G-CCCGGTAT-GCAGTACC-CCGGG- - - -CGG- -C


 energy = -21.5 kcal

 57    GGGUUGg : : : CCCGGuauugca
 96    CCCGGCuuuaGGGCCgccaug


 energy = -28.3 kcal

 55    GCAUCGU : CCCGGuauGGcag
 92    CGUGGCAcGGGCCu : : CCau
```

Fig. 5. (*Continued*)

then converted to a list of hairpins. In the case of several sequences these lists serve to identify the nonbifurcating hairpins common to those of the first sequence just as is done within MONTECARLO. Figure 6 shows the output of DYNPRO for the same set of sequences used by MONTE-CARLO (Fig. 5). Although edited, the actual structures have been retained rather than just their energies being shown.

It will be noticed that the commonality output here differs from that of MONTECARLO. As is often the case, the results agree in certain areas, such as the commonality of hairpins belonging to the same pair of sequences, but differ in which hairpins are common. The difference resides in the different global structures produced, which generally give different sets of hairpins to compare.

Summary

The current version of RNAFOLD now has the capability for conducting comparative studies relative to nonbifurcating hairpins contained in the global structures produced by the dynamic programming and Monte Carlo methods. It also has the capability of both finding and comparing

```
*Sequence  'drour2b',  energy  =  -37.2  kcal

    >  stem  1:  a  hairpin

    6          UGAUUuuu
    20         ACUAAgg


    >  stem  2:  a  hairpin

    22         GACGGAGuGcuUGGAGcu
    54         CUGUCUC:C::ACCUCgu


    >  stem  3:  a  hairpin

    57         GGGUUGg:::CCCGGuauugca
    96         CCCGGCuuuaGGGCCgccaug


*Sequence  'peaur2',  energy  =  -43.9  kcal

    >  stem  1:  bifurcates  into  stems  2  3

    7          UGaGGG(2)g(3)
    95         AC:CCC


    >  stem  2:  a  hairpin

    13         G:GAAGAGuCaCaCAGUAGcu
    51         CgCUUCUCuG:G:GUUAUCgu


    >  stem  3:  a  hairpin

    53         UGuCGCuuuuGCg:UUGCacu
    90         AC:GCGgu::CGuuAACGaua
```

FIG. 6. Output of DYNPRO.

hairpins and pseudoknots independently of global structures. The efficacy of these increased capabilities has been tested for select families of sequences, and the results thus far indicate favorable utility. Under consideration is a further extension designed to incorporate pseudoknots within global structures. Written in the C language, RNAFOLD and its companion program, GENALIGN, for doing multiple alignments in the comparison of pseudoknots and hairpins, is available for UNIX systems on standard ½-inch, 9-track tape or on SUN tape cartridges.

```
*Sequence 'phsur2a', energy = -58.1 kcal

  > stem 1: bifurcates into stems 2 3

  13     GG:GC(2)(3)
  95     CCaCG


  > stem 2: a hairpin

  17     GCG:GGAGuuGGAcccGGAGcu
  56     CGCgCCUCg:CCUc::CCUCgu


  > stem 3: a hairpin

  57     AUCGU:CCCGGuAuGGcag
  90     UGGCAcGGGCC:U:CCau


 ** HAIRPINS COMMON TO THOSE OF SEQUENCE 'drour2b' **

     1.   drour2b     (22-54)
     2.   phsur2a     (17-56)


  1 GacGGAG T G CttGGAGCTTGCT CC aC CTCtG tC
    |  |||| | | |   |||||||||| || | ||| | |
  2 GcgGGAGtTgGaCccGGAGCTTGCTcCCtcCgCTCcGcgC

  G--GGAG-T-G-C--GGAGCTTGCT-CC--C-CTC-G--C


energy = -15.4 kcal

22     GACGGAGuGcuUGGAGcu
54     CUGUCUC:C::ACCUCgu


energy = -25.0 kcal

17     GCG:GGAGuuGGAcccGGAGcu
56     CGCgCCUCg:CCUc::CCUCgu
```

FIG. 6. (*Continued*)

Acknowledgments

The work reported here was undertaken while the author was visiting the Mathematical Biology Laboratory at the Frederick Cancer Research Facility, Frederick, Maryland, during 1988. I wish to thank the director, Dr. Jacob Maizel, and staff for many useful discussions and for providing excellent computational facilities.

[19] Computer Modeling and Display of RNA Secondary and Tertiary Structures

By Daniel Gautheret, François Major, and Robert Cedergren

Introduction

The evaluation of all possible conformations of a nucleic acid at the atomic level is an extremely complex computational problem: with six rotation angles per nucleotide, each nucleotide may assume 360^6 possible conformations allowing only integer values for each angle. Even the evaluation of different conformations for a single-stranded loop of, say, seven nucleotides as in the anticodon loop of tRNA is an arduous task. On the other hand, knowledge of RNA secondary and tertiary structure is essential to the comprehension of RNA function. It is therefore understandable that a major research effort has been devoted to learning the rules by which RNA chains fold and interact with themselves. And even though the prediction of RNA secondary structures is still not fully reliable, hypothetical folding patterns can be extremely useful in the characterization of an RNA or RNA gene sequence.

We present here a computer program, RNASE (ribonucleic acid structure editor), for the construction and representation of RNA secondary and tertiary structures.[1] The program can be used to predict and display helical regions of RNA secondary structures. After selection of helices, a three-dimensional image of the molecule can be obtained. In addition, several graphic functions are available that allow the manipulation of the two- and three-dimensional representation of the molecule. In order to implement these features in a microcomputer environment, we choose to represent the polynucleotide chain by a skeletal-type model, a representation which allows two degrees of rotational freedom per nucleotide.[2] This skeletal model facilitates the evaluation of conformational change and the graphic display of RNA chains, i.e., conformational changes can be calculated and displayed in virtually real time.[3]

The RNASE program is an attempt to address certain weaknesses in other structural prediction programs and to extend display graphics into

[1] R. Cedergren, D. Gautheret, G. Lapalme, and F. Major, *CABIOS* **4**, 143 (1988).

[2] R. J. Fletterick, T. Schroer and R. J. Matela, *in* "Molecular Structure: Macromolecules in Three Dimensions," p. 99. Blackwell, Oxford, 1985.

[3] F. Major, Master's thesis, Technical Report 190, Département d'Informatique et de Recherche Opérationnelle, Université de Montréal, 1987.

the relatively unexploited realm of tertiary structure. In particular, the program allows the display of user-selected helices, and simple knotted structure calculations (pseudoknots or other base pair interactions definable in the tertiary structure) are incorporated. Also, we have implemented the recent energy values of Freier *et al.*[4] that take into account the effect of nucleotides adjacent to helical regions.

RNASE itself has been designed to provide an integrated environment for the structural study of nucleic acid sequences; therefore, different levels of computation, from sequence input to simple tertiary modeling, have been coupled in such a way that no complex user interaction nor program swaps are necessary. Changes in helix selection (see below) automatically modify the secondary and tertiary structure displays. The data flow diagram for RNASE is shown in Fig. 1. The program is written in Turbo Pascal and runs on IBM-PC or compatible computers (640 kilobyte memory and Hercules graphic board recommended) and is available from the authors. The main characteristics of the program are described herein, and the use of the program is demonstrated in the three-dimensional modeling of the so-called hammerhead or self-cleaving RNA molecule.

Program Description

Menus. The menus used in RNASE are shown in Fig. 2. These are referred to occasionally in the following discussion and other figures.

Secondary Structure Prediction

The secondary structure prediction utility is based on the generation/selection method first described by Pipas and McMahon.[5] Although energetically optimal structures are not always obtained, this approach has the advantage of producing a set of suboptimal structures and of permitting knotted structures, both of which are usually rejected in dynamic programming algorithms.

Helix List Generation. For helix generation and energy calculations, a simple folding algorithm (of order n^2, where n is the length of the sequence) generates a list of all possible helical regions (Fig. 3). The free energy of each double helix is calculated by the addition of stacking and base-pairing thermodynamic parameters obtained from Freier *et al.*[4] These values take into account recent energy measurements on oligonucleotides, which demonstrate the importance of dangling ends and terminal mismatches in the

[4] S. M. Freier, R. Kierzek, J. A. Jaeger, N. Sugimoto, M. H. Caruthers, T. Neilson, and D. H. Turner, *Proc. Natl. Acad. Sci. U.S.A.* **83**, 9373 (1986)
[5] J. M. Pipas and J. E. McMahon, *Proc. Natl. Acad. Sci. U.S.A.* **72**, 2017 (1975).

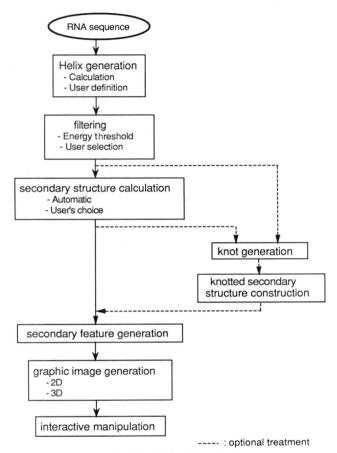

----- : optional treatment

FIG. 1. RNASE data flow diagram.

stabilization of RNA helices. Our program considers as well the initiation and propagation energies described by Freier *et al.*[4] Bulges and mismatches other than G–U pairs are not permitted inside paired regions. In the case of long sequences, the number of potential helical regions could exceed the memory capacity of a microcomputer, therefore the program retains a constant number of helices, 300 in the present version.

Selection. The complex problem of selecting the most favorable set of compatible helices describing the secondary structure for a given RNA sequence has been investigated by Ninio and co-workers.[6,7] Using their

---

[6] J. P. Dumas and J. Ninio, *Nucleic Acids Res.* **10,** 197 (1982).
[7] M. Gouy, P. Marlière, C. Papanicolaou, and J. Ninio, *Biochimie* **67,** 523 (1985).

```
                                    HELIX MENU

                    A  - Add helix
                    H  - Hide helix
                    R  - Restore helix
                    S  - Select helix for drawing
          MAIN MENU                T  - change energy Threshold
                                   P  - Print helices
  L  - Load file                   L  - sort helix List
  S  - Save file
  H  - Helix editor      Home, end, PgUp, PgDn to Scroll
  2  - 2D structure      <esc>  - Main menu
  3  - 3D structure      <F1>   - show menu
  <esc> - Quit progr
                      a   b
```

```
                                   3D structure editor

                              N  - Nucleotide number
                         c    F  - select Fragmnt to be rotated
                              S  - Stereo on/off
                              D  - adjust Distance for stereo
         2D structure editor  A  - change rotation Angle
                              O  - display Options
  N  - Nucleotide number      P  - Print image
  S  - rotate Stem            C  - Center image
  W  - rotate Whole molecule  <F9>  - next nucleotide
  O  - display options        <F10> - previous nucleotide
  A  - change rotation Angle  J     - Jump to paired nuc.
  P  - Print image            G     - Go to nucleotide

  +, →           - spin on Y axis    +, →          - rot. on Y axis
  +, +           - spin on X axis    +, +          - rot. on X axis
  PgUp, PgDn     - rot. on Z axis    PgUp, PgDn    - rot. on Z axis
  +, -           - enlarge or reduce +, -          - enlarge or reduce

  <esc>          - Main menu         <esc>         - Main menu
  <F1>           - show menu         <F1>          - show menu
```

FIG. 2. (a) Main menu. Option L loads the RNA sequence from a disk. Various formats are supported, including IntelliGenetics, GenBank/Fastn, and a free format containing ASCII sequence characters along with any blank characters. Modified bases are considered as their unmodified homolog (U for D) but are displayed on screen in the original notation. Option S saves the helix list and graphics on a disk. Options 2, 3, and H call other menus. (b) Helix editor menu. Option A adds user-defined helices to the list. Option S allows helix selection for secondary and tertiary structure construction. Helices may be selected automatically or by the user. Options T, H, and R remove or recall helices from the helix list. Removed helices are not considered in the selection procedure. (c and d) Graphic menus. Common features include option O, which sets up display options. Base characters, backbone line, hydrogen bonds, and knotted regions may be visualized or hidden. Arrows and PgUp/PgDn keys rotate the molecule or a fragment in space on the X, Y, or Z axis. Plus and minus enlarge or reduce screen image size. In option P, screen structures may be sent directly to a printer (Epson dot printer or Hewlett Packard laser printer). Option N displays given nucleotide positions. For the 2D graphic menu, after a paired nucleotide has been selected with option S, the stem may be rotated around a pivot point, which may be used to untangle long sequences. For the 3D graphic menu, options <F9>, <F10>, J, and G change the pivot nucleotide for segment rotations. Option F selects whether a segment or the whole molecule is rotated; S sets stereo display on/off.

(1) -8.3Kc	(10-3) -7.8Kc	(22-3) -7.4Kc	(27-2) -7.3Kc
1-GCGGAUUu7	10aGCUCa13	22aGAGCGCc27	27gCCAGAc31
72-CGCUUAAg66	25gCGAGa22	56gCUTGUGu51	43aGGUCUa39
(12-3) -6.5Kc	(34-3) -6.2Kc	(49) -6.1Kc	(3-3) -6.1Kc
12cUCAGUu16	34uGAACAu38	49cCUGUGt53	3cGGAUUUa8
35aAGUCAg31	56gCUTGUGg52	65aGACACc61	49uCCUGGAg44
(3-2) -5.7Kc	(41-3) -5.1Kc	(2-4) -4.8Kc	(39) -4.8Kc
3cGGAUu6	41cUGGAg44	2gCGGAUu6	39aUCUGg42
61aCCUAg31	62cACCUa59	42gGUCUAc38	66aAGACa63
(19-5) -4.7Kc	(32) -4.7Kc	(19-3) -4.5Kc	(24) -4.4Kc
19gGGAg21	32aCUGa34	19gGGAg21	24aGCGc26
49uCCUg47	65aGACa63	61aCCUa59	72aCGCu70
(1-2) -4.2Kc	(45) -4.1Kc	(11-3) -3.9Kc	(41) -3.9Kc
1-GCGg3	45aGGUc47	11gCUCa13	41cUGGa43
27cCGCg25	75aCCAc73	45gGAGg43	76-ACCa74

FIG. 3. Example of a helix list with a tRNA sequence. Numbers on stems refer to nucleotide positions. Each helical dangling end or terminal mismatch (shown in lowercase letters) is considered in the stem energy calculation. (Kc, Kilocalories.)

algorithm, secondary structures consisting of compatible helical regions can be generated in a reasonable time for sequences up to 200 bases in spite of the exponential nature of the problem. Their procedure first groups incompatible helical regions together in "islets" or sets, then constructs a secondary structure by picking one compatible helical region from each islet. At each stage of the selection process, a test is done to determine whether a more energetically favored structure can be found using the best helix of each remaining islet. If not, there is no reason to continue this pathway, and control is returned recursively to the previous level. The system retains a given number of the best structures.

Knots. The structures generated in the previous section involve only nested base-paired regions. However, recent research has documented the importance of less orthodox structures in RNA molecules called pseudo-knots (Fig. 4). Most secondary structure prediction algorithms do not consider such conformations. Since our system generates all possible helices, we thought it would be useful to incorporate pseudoknots (or tertiary interactions) in the search and display functions of RNASE. Ide-ally, the pseudoknot prediction would be best accomplished at the same time as the helix selection above, since less stable helices, rejected in the selection process, could be stabilized by pseudoknot or other tertiary inter-actions. However, even though the approach used here performs the energy calculations for pseudoknots after rather than during the helix selection

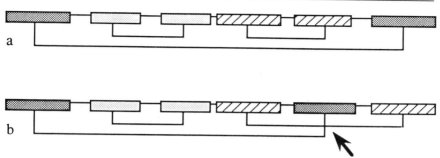

FIG. 4. (a) Canonical secondary structure. (b) Knotted secondary structure.

process, this feature does give the user at least some insight into the energetics and spatial organization of these interactions.

Knot search and selection is done by starting with the list of all possible helices and the particular combination selected for the secondary structure from above. Then, a list of potential knots is made from the complex helix list by removing all helices which contain nucleotides common to helices selected for the secondary structure. The selection procedure used above to choose the helices of the secondary structure is then applied to generate optimal combinations of knot-forming helices. Owing to the lack of knowledge about the energetic contribution of knots, we do not use tertiary interactions to calculate the overall stabilization energy of a given conformation.

User-Defined Helices. Along with automatic generation and selection procedures, the RNASE program provides tools for the definition of arbitrarily defined helical regions and their incorporation into the final structure (see Fig. 2).

Graphical Display Algorithm

The graphical display algorithm first analyzes the selected set of secondary structure helices by converting the "sequence" set, the linear sequence of the molecule, to a "pairing" set, a list of paired regions.[1] Subsequently, a "forms" set (the list F below) is established which describes the secondary structural feature of each region, i.e., whether the region is a bulge, an internal loop, a hairpin loop, or a helical region. In the case of the yeast tRNA^{Phe} structure, the form set is:

$$F = \{\text{helix}_{1-7}, \text{intern\_loop}_{8-9}, \text{helix}_{10-13}, \text{hairpin\_loop}_{14-21}, \text{helix}_{22-25},$$
$$\text{intern\_loop}_{26-26}, \text{helix}_{27-31}, \text{hairpin\_loop}_{32-38}, \text{helix}_{39-43},$$
$$\text{intern\_loop}_{44-48}, \text{helix}_{49-53}, \text{hairpin\_loop}_{54-60}, \text{helix}_{61-65},$$
$$\text{helix}_{66-72}, \text{unstructured}_{72-76}\}$$

TABLE I
ANGLE ASSIGNMENT[a]

Form	Secondary structure (θ, τ)	Tertiary structure (θ, τ)
Helix	$(0°, 0°)$	$(29°, 195°)$
Hairpin loop	First and last nucleotide: $(270° + 360°/n, 0°)$	Last link: see text
		All others $(29°, 195°)$
	Other nucleotides: $(360°/n, 0°)$	
Other loops	*Idem*	$(0°, 0°)$

[a] n, Number of nucleotides in complete circle.

Once the process of form assignment has been completed, the secondary or tertiary image may be calculated. For the ribophosphate backbone representation used, only two angles need to be defined for each nucleotide: a torsional angle (τ) and a bend angle (θ). The values of τ and θ for each form are given in Table I. Since the secondary structure representation is planar, the torsion angle τ is zero for nucleotides of each form (Table I and Fig. 5). The values of θ in loop regions are calculated by the expression: $360°/n$, where n is the number of nucleotides in the same graphic circle.[8] Once the figure is displayed on the screen, a series of manipulative procedures are available to the user. These include rotations of the structure in the x, y, and z axes and modifications of the angle that the stem region makes with the circle branch point. This latter feature is a common method to untangle overlapping stem regions of a molecule. Finally, the molecule can be displayed in the form of a sequence, a backbone tracing, with or without hydrogen bonds (see Fig. 2). Displays can be magnified or diminished in size by using the "+" or "−" options.

The three-dimensional representation is constructed and displayed using the angles of Table I, some of which are taken from Fletterick *et al.*[2] and assume an RNA A-type helix. The three-dimensional structure of loops is much more complicated to construct, since the number of combinations of angles θ and τ which would close the loop is extremely large. Also, the number of nucleotides in a loop is not a constant. To simplify this problem, we use the tertiary structure of the anticodon loop of tRNA as a model;[9] in other words, the algorithm first extends the 3′ helix of the stem in the 5′ direction using the helical angles of Table I. The last two nucleotides at the 5′ terminus of the loop are then placed on a straight line joining the 3′ extended helix strand to the 5′ strand of the stem. This structure is

[8] G. Lapalme, R. J. Cedergren, and D. Sankoff, *Nucleic Acids Res.* **10,** 8351 (1982).
[9] A. Rich and U. L. RajBhandary, *Annu. Rev. Biochem.* **45,** 805 (1976).

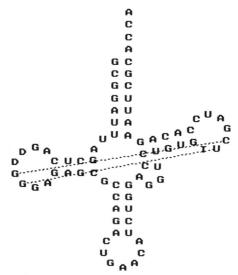

F IG. 5. Representation of knot-forming base pairs in the tRNA cloverleaf, secondary structure as produced by RNASE.

consistent with the RNA hairpin model described by Haasnoot *et al.*[10] and with the crystal structure of the tRNA anticodon loop.[9] It should be noted, however, that this model fits medium sized hairpin loops (five to eight nucleotides) better than others.

For single-stranded regions (bulges and internal loops) θ and τ are set to $0°$, which results in the placing of these nucleotides on a straight line. These values are completely arbitrary and therefore do not correctly juxtapose two nonadjacent strands of the same stem in the screen display. However, this disadvantage is turned into an advantage, since the angle values of single-stranded regions can be modified by the user. Using the modeling tools of RNASE, two nonadjacent strands of a helical region can be brought together, or the three-dimensional proximity of any feature (helices, bulges, or loops) of the molecule can be redefined. The interactive manipulation of single-stranded regions allows virtually unlimited modeling of a given molecule.

[10] C. A. G. Haasnoot, C. W. Hilbers, G. A. van der Marel, J. H. van Boom, U. C. Singh, N. Pattabiraman, and P. A. Kollman, *J. Biomol. Struct. Dynamics* **3**, 843 (1986).

Application: Three-Dimensional Modeling of Small Self-Cleaving
RNA

Catalytic RNA and Hammerhead Structure

The recent discovery of the catalytic properties of RNA molecules has
opened a new and rapidly expanding field in nucleic acid research. The
catalytic RNAs, called ribozymes, are mostly involved in RNA cleavage
reactions, for which two cleavage mechanisms have been proposed.[11] Currently, little conformational information is available on these molecules or
the reaction intermediates, even though the three-dimensional arrangement of the RNA chain (particularly the active site) is a determining factor
in the molecular mechanism of catalysis. Analysis, design, and comparison
of putative tertiary structures could thus be of great help in the elucidation
of these mechanisms.

One of the frequently cited ribozymes, the hammerhead structure, is
found in circular satellite RNA associated to plant viruses.[12] Replication of
this RNA by the rolling circle model produces long transcripts which
undergo self-cleavage to release unit-length RNA molecules. A 40- to
50-nucleotide structure in this molecule has been identified as necessary
and sufficient for the cleavage activity. Although this structure is composed
of two segments, which may be separated by several hundred nucleotides
in the original molecule, Uhlenbeck has been able to recreate the cleaving
activity by using two synthetic oligoribonucleotides having lengths of 24
and 19 nucleotides.[13] Analysis of a number of naturally occurring sequences and mutants has established a consensus primary and secondary
structure required for ribozyme activity (Fig. 6).

Three-Dimensional Design

To facilitate treatment by RNASE, we have used the sTobRV self-
cleaving structure, in which the catalytic and substrate segments are linked
through a seven-residue hairpin loop.[12] The secondary structure algorithm
in RNASE correctly finds the three presumed helices (I, II, and III in Fig.
6), and the secondary structure containing these three stems is automatically selected as the best helix combination (free energy of -8.1 kcal/mol).
The first tertiary output is shown in Fig. 7a. Note that helices II and III are
folded according to the standard angles in Table I, and the two hairpin
loops are constructed after the tRNA anticodon loop model.[9] However, the
two non-adjacent strands forming stem I are not base-paired as discussed

---

[11] R. Cedergren, B. F. Lang, and D. Gravel, *FEBS Lett.* **226**, 63 (1987).
[12] J. Haseloff and W. L. Gerlach, *Nature (London)* **334**, 585 (1988).
[13] O. C. Uhlenbeck, *Nature (London)* **328**, 596 (1988).

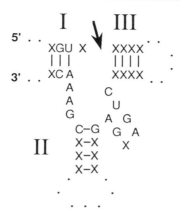

FIG. 6. Conserved primary and secondary structure in hammerhead self-cleaving domains. The cleavage site is indicated by an arrow, and the three helical regions are labeled I, II, and III.[12]

above. The presence of a single nucleotide between the 5' strand of helix I and helix III allows the program to stack the strand on the axis of helix III. We then used the three-dimensional manipulation tools of the program to correctly fold the two strands of helix I by modifying the torsion angles in the single-stranded regions which link helix II to the two other helices. The result of this modeling is presented in Fig. 7b.

Model of Hammerhead and Mechanism of Cleavage

The products of the self-cleaving reaction are 5'-hydroxyl and a 2',3'-cyclic phosphate; therefore, the reaction must proceed by nucleophilic attack of the adjacent 2'-hydroxyl on the phosphodiester bond. In the case of Pb^{2+}-catalyzed hydrolysis of tRNA, a metal binding site positions the lead ion so that the nucleophilic nature of the adjacent 2'-hydroxyl group is greatly enhanced.[14] In like manner, the self-cleaving RNA is dependent on the presence of metal ions (Mg^{2+}, Mn^{2+}), and the structure of Fig. 7b suggests the presence of a binding cleft at the center of the molecule near the base of the three helical regions. Also, this central core encompassing the putative cleft is composed of most of the nucleotides that have been shown to be strictly required for the reaction. However, the nucleotides at the base of helix II are clearly more distant from the catalytic site (shown by the arrow), so that the reason for their requirement is not obvious. It may be that these nucleotides are required for tertiary interactions which

[14] R. S. Brown, J. C. Dewan, and A. Klug, *Biochemistry* **24,** 4785 (1985).

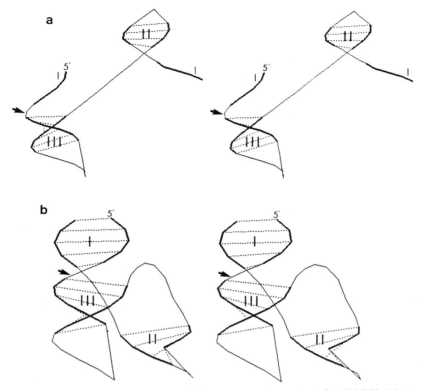

FIG. 7. (a) RNASE stereo output of the structure determined for TobRV (+)-strand satellite RNA. Paired regions, which are represented by thick lines, assume a typical A-type helix shape. The cleavage site is shown by the arrow. Helix I is not formed although the two stems which make up the double helix are shown in a helical conformation. (b) Modeled, three-dimensional structure obtained after manipulation of single-strand regions.

could involve the formation of base triples as found in the tRNA tertiary structure[9] or that the long single-stranded region of mainly required nucleotides between helices II and III are able to fold back on one another to create a structure which is much more tightly folded around the catalytic site.

Discussion

Secondary structure prediction algorithms have improved considerably since the mid-1970s. Although they do take advantage of sophisticated mathematical methods and accurate energy parameters, they face the major drawback of not considering tertiary interactions. Optimal secondary structures according to canonical nonknotting pairings are often different from generally reliable structures inferred from phylogeny. Moreover, tertiary interactions probably play a considerable role during the folding of an RNA molecule in solution: helices normally rejected by the programs based on maximizing stabilization energies determined in simple model compounds could easily be much more stable in the three-dimensional conformation. The D loop stem in tRNA is often composed of only three base pairs, which can present problems for prediction algorithms. It is likely that the precise geometry of the interaction of the D loop with the T loop and with the uridine at position 8 stabilizes the stem region to the extent that ultimately it participates in the tertiary structure.[9] Consideration of tertiary interactions can be essential in determining correct secondary structures.

In the approach to RNA folding and modeling presented here, we do not depend only on computationally complicated energy minimizations, since many user-defined options are permitted and some attempt is made to consider tertiary (pseudoknot) interactions. It is quite clear that pseudoknot structures in a folded molecule are difficult to evaluate fully, since their presence must depend on the presence and the orientation of other helices; therefore, our use of this feature in RNASE is not completely rigorous. Pseudoknots can be inferred at the secondary structural level, but a spatial representation would be necessary to decide whether distances and base orientations would permit such interactions. However, we feel that the implementation of this option can be quite useful, especially when coupled to the ability of the user to carry out complex secondary and tertiary structure manipulations of the molecule.

RNASE does not as yet take advantage of all pertinent knowledge on RNA structures. For example, stacking, bulged-out, and turn-permitting nucleotides could be partially predicted from the statistical analysis of known sequences;[15] such information could reduce considerably the rotational freedom of single-stranded and loop regions. In the case of tRNA, only four nucleotides have a "sharp turn" conformation whereas the others generally stack in approximately helical conformations. The ability to predict the occurrence of turn-permitting residues and to translate all helical constraints in terms of spatial coordinates could greatly improve prospects for the prediction of single-stranded conformations particularly in loop regions. Likely structures could then be tested for stabilizing ter-

---

[15] C. Tuerk, P. Gauss, C. Thermes, D. R. Groebe, M. Gayle, N. Guild, G. Stormo, Y. D'Aubenton-Carafa, O. C. Uhlenbeck, I. Tinoco, E. N. Brody, and L. Gold, *Proc. Natl. Acad. Sci. U.S.A.* **85**, 1364 (1988).

tiary interactions and finally refined at the atomic level by an available energy minimization algorithm. Except for this last step, no complicated representation would be required, other than improving the backbone model used here so that base orientations could be graphically represented.

Acknowledgments

This work was supported by funds from the Medical Research Council of Canada. François Major holds a studentship from the NSERC and was previously supported on a grant from NSERC. Robert Cedergren is a Fellow of the Canadian Institute of Advanced Research.

Section V

Aligning Protein and Nucleic Acid Sequences

[20] Mutation Data Matrix and Its Uses

By DAVID G. GEORGE, WINONA C. BARKER, and LOIS T. HUNT

Introduction

More than one-half of the chapters in this volume are concerned with methods of comparing protein sequences. Since the 1960s, as sequence data have accumulated, sequence comparison techniques have proved to be extremely effective in studying the structural and functional properties of proteins and how these, and the organisms in which they are expressed, have evolved. It is important to realize at the outset that these methods are approximate and have many inherent limitations. The fundamental problem is to judge whether two or more sequences exhibit similar three-dimensional structure and/or function based on similarities observed in their amino acid sequences. Unfortunately, present biological understanding of the processes involved in determining protein structure and/or function is not sufficient to allow an adequate assessment of the sequence characteristics important for these processes. As a result, all sequence comparison algorithms are limited by the underlying model of sequence similarity and hence are not guaranteed to produce biologically meaningful results. In this sense, the results of these methods must be evaluated in the same manner as results are analyzed in classic comparative biology. No single method is superior to all others; evidence is drawn by comparing the results of several approaches, and reasoning based on other biological knowledge must be incorporated. The interpretation of evolutionary histories derived from comparisons of protein (and/or nucleic acid) sequences suffers similar problems in that evolutionary processes are also not fully understood.

Basic to all sequence comparison is the concept of an alignment that defines the relationship between sequences on a residue-by-residue basis. Aligned residues are presumed to be related in an evolutionary and/or functional sense; residues occupying equivalent positions are believed to share common ancestors and/or to have equivalent biological roles. There is a wealth of genetic evidence supporting the role of insertions and deletions in the evolution of macromolecules, and it is customary to allow for the presence of unrelated segments reflecting these events in the alignment. A series of dashes is often used to indicate the "gaps" in sequences resulting from these events. Unfortunately, the number of identities or similarities seen between sequences that may not be related can also be greatly in-

METHODS IN ENZYMOLOGY, VOL. 183

creased by allowing gaps in the alignment. Even when it is clear that gaps must be inserted to align related sequences, the proper positioning of the gaps is often not obvious; therefore, great care must be taken not to overinterpret the significance of identities observed after gaps have been inserted. Sequence alignment algorithms attempt to place these gaps based on some model of sequence similarity.

In any sequence comparison application, whether it be database searching, pattern recognition, or sequence alignment, the best solution is selected from all other possible solutions based on a set of scoring rules that define the similarity model. The optimization generally involves assigning a score to each possible alignment or subalignment and selecting those with the "best" scores. The results of the analysis are dependent on the choice of scoring rules; thus, it is customary to vary these according to the type of relationship being analyzed. For example, in examining possible evolutionary relationships between sequences, it is useful to base the scoring matrix on observed exchange frequencies, whereas when investigating possible structural similarities in proteins, it may be more appropriate to match residues of similar size and hydrophobicity.

Similarity Scoring Matrices

There are two distinct types of scoring systems: similarity scoring systems and difference scoring systems. Difference scoring systems have been introduced primarily for derivation of evolutionary trees; the evolutionary distance between proteins or nucleic acids can be inferred from differences observed between their sequences. However, inasmuch as one is generally more interested in the similar regions within sequences, it is more natural to use similarity measures for most other sequence comparison purposes.

Sequence comparison methods use a scoring matrix that assigns a value to each possible pair of aligned amino acids. In a similarity scoring matrix, higher values are assigned to more similar pairs and lower values to dissimilar pairs. The simplest similarity matrix, which we call the unitary matrix (UM), assigns values of +1 for identities and 0 for nonidentities. A scoring matrix based on the genetic code (GCM) reflects the maximum numbers of nucleotides that the codons for two amino acids may have in common, for example, scoring +3 for identities, +2 for amino acids whose codons must differ by at least one nucleotide, +1 for those whose codons must differ by two nucleotides, and 0 for those whose codons cannot have any nucleotides in common.

Similarity scoring matrices can be based on any property of amino acids that can be expressed numerically. A similarity matrix can be computed from the amino acid property values in the following way. A matrix whose elements are the absolute values of differences between each of the

TABLE I
AMINO ACID HYDROPHILICITY VALUES[a]

Symbol	Amino acid	Hydrophilicity	Symbol	Amino acid	Hydrophilicity
A	Ala	−0.50	M	Met	−1.30
R	Arg	3.00	F	Phe	−2.50
N	Asn	0.20	P	Pro	−1.40
D	Asp	2.50	S	Ser	0.30
C	Cys	−1.00	T	Thr	−0.40
Q	Gln	0.20	W	Trp	−3.40
E	Glu	2.50	Y	Tyr	−2.30
G	Gly	0.00	V	Val	−1.50
H	His	−0.50	B	Asx	1.40
I	Ile	−1.80	Z	Glx	1.40
L	Leu	−1.80	X	Unk	−0.40
K	Lys	3.00			

[a] From Levitt.[1]

property values is constructed. The maximum matrix element is determined, and each matrix element is divided by this value. The resultant matrix is a dissimilarity matrix with values between 0 and 1. This matrix can be converted to a similarity matrix with values between 1 and 0 by subtracting each element from 1. The similarity matrix elements are converted to integer values by multiplying each element by a scaling factor and then rounding off to the nearest integer. Table I shows the amino acid hydrophilicity data of Levitt.[1] Figure 1 shows the similarity scoring matrix derived from these data; a scaling factor of 10 was used in the derivation. In 1971, McLachlan[2] published a scoring matrix (AAAM) based on the alternative amino acids found at corresponding positions in alignments of groups of related sequences. Relative substitution frequencies were converted to integers that range from 1 to 6, identities were assigned scores of 8 or 9, and gaps were given a score of zero. A score of 3 represents a neutral substitution.

Often the evolutionary histories of two sequences being compared are not known. Consequently, if at some site the first sequence contains amino acid A and the second amino acid B, it is not known whether the difference has resulted from amino acid A changing to amino acid B or from amino acid B changing to amino acid A. Hence, comparison methods usually require that similarity scoring matrices be symmetric; all of the matrices described above satisfy this criterion.

[1] M. Levitt, *J. Mol. Biol.* **104,** 59 (1976).
[2] A. D. McLachlan, *J. Mol. Biol.* **61,** 409 (1971).

	R	K	D	E	B	Z	S	N	Q	G	X	T	H	A	C	M	P	V	L	I	Y	F	W
R	10	10	9	9	8	8	6	6	6	5	5	5	5	5	4	3	3	3	3	3	2	1	0
K	10	10	9	9	8	8	6	6	6	5	5	5	5	5	4	3	3	3	3	3	2	1	0
D	9	9	10	10	8	8	7	6	6	6	5	5	5	5	5	4	4	4	3	3	3	2	1
E	9	9	10	10	8	8	7	6	6	6	5	5	5	5	5	4	4	4	3	3	3	2	1
B	8	8	8	8	10	10	8	8	8	8	7	7	7	7	6	6	6	5	5	5	4	4	3
Z	8	8	8	8	10	10	8	8	8	8	7	7	7	7	6	6	6	5	5	5	4	4	3
S	6	6	7	7	8	8	10	10	10	10	9	9	9	9	8	8	7	7	7	7	6	6	4
N	6	6	6	6	8	8	10	10	10	10	9	9	9	9	8	8	8	7	7	7	6	6	4
Q	6	6	6	6	8	8	10	10	10	10	9	9	9	9	8	8	8	7	7	7	6	6	4
G	5	5	6	6	8	8	10	10	10	10	9	9	9	9	8	8	8	8	7	7	6	6	5
X	5	5	5	5	7	7	9	9	9	9	10	10	10	10	9	9	8	8	8	8	7	7	5
T	5	5	5	5	7	7	9	9	9	9	10	10	10	10	9	9	8	8	8	8	7	7	5
H	5	5	5	5	7	7	9	9	9	9	10	10	10	10	9	9	9	8	8	8	7	7	5
A	5	5	5	5	7	7	9	9	9	9	10	10	10	10	9	9	9	8	8	8	7	7	5
C	4	4	5	5	6	6	8	8	8	8	9	9	9	9	10	10	9	9	9	9	8	8	6
M	3	3	4	4	6	6	8	8	8	8	9	9	9	9	10	10	10	10	9	9	8	8	7
P	3	3	4	4	6	6	7	8	8	8	9	9	9	9	10	10	10	9	9	9	8	8	7
V	3	3	4	4	5	5	7	7	7	8	8	8	8	8	9	10	10	10	10	10	9	8	7
L	3	3	3	3	5	5	7	7	7	7	8	8	8	8	9	9	9	10	10	10	9	9	8
I	3	3	3	3	5	5	7	7	7	7	8	8	8	8	9	9	9	10	10	10	9	9	8
Y	2	2	3	3	4	4	6	6	6	6	7	7	7	7	8	8	9	9	9	9	10	10	8
F	1	1	2	2	4	4	6	6	6	6	7	7	7	7	8	8	8	9	9	9	10	10	9
W	0	0	1	1	3	3	4	4	4	5	5	5	5	5	6	7	7	7	8	8	8	9	10

FIG. 1. Hydrophobicity scoring matrix constructed from the hydrophilicity data of Levitt,[1] as described in the text. The matrix can be used in comparison of two sequences to generate dot matrix graphic plots, using, for example, a window size (segment length) of 25 and a minimum score in the range of 200 to 210. When using the matrix with the ALIGN program, the matrix bias and the gap penalty should be about 24 rather than the default values of 6 and 6 used with the MDM (see text).

Dayhoff Mutation Data Matrix

One of the most widely used similarity measures is the mutation data matrix (MDM) developed by Dayhoff and colleagues. The first MDM, published in 1968, was derived from over 400 accepted point mutations (evolutionary replacements of one amino acid for another at homologous positions) between present-day sequences and inferred ancestral sequences.[3] The relative frequency of exposure of each type of amino acid to mutational change and relative mutability were also taken into account, as described below. The MDM used extensively in the 1980s was calculated on the basis of nearly 1600 accepted point mutations in 71 groups of closely related proteins (< 15% different).[4]

---

[3] M. O. Dayhoff and R. V. Eck, *in* "Atlas of Protein Sequence and Structure" (M. O. Dayhoff and R. V. Eck, eds.) Vol. 3, p. 33. National Biomedical Research Foundation, Silver Spring, Maryland, 1968.

[4] M. O. Dayhoff, R. M. Schwartz, and B. C. Orcutt, *in* "Atlas of Protein Sequence and Structure" (M. O. Dayhoff, ed.), Vol. 5, Suppl. 3, p. 345. National Biomedical Research Foundation, Washington, D.C. 1979.

The original Dayhoff model of amino acid mutation assumes a Markovian model of amino acid substitution, i.e., the model assumes that the probability of a mutation at any site within a protein is independent of its previous history. For example, the probability of cysteine being transformed to serine by mutational processes is the same for all cysteine residues irrespective of the identity of their immediate ancestors.

Within the Markovian model, the MDM is derived from a transition probability matrix in which each matrix element gives the probability that amino acid A will be replaced by amino acid B in one unit of evolutionary change. The diagonal elements give the probabilities that the amino acids will remain unchanged. The sum of the diagonal elements gives the probability that there will be no change during the represented evolutionary interval. In the Dayhoff derivation, the probability matrix was normalized such that this sum corresponded to a chance of 99 out of 100. Thus, the unit of evolution represented by the probability matrix corresponds to one accepted amino acid substitution per hundred sites (1 PAM unit). Note that time is not explicitly taken into account in this derivation. The transition probabilities are assumed to be constant with respect to a unit of amino acid substitution. They are not functions of the length of time required for the substitution to occur, and the time intervals may vary. Hence, within the limitations of this approximation, it is valid to apply this model to proteins that exhibit widely varying rates of evolutionary change.

Each element of the mutation probability matrix was calculated as the product of the conditional probability of amino acid A being replaced by amino acid B, given that amino acid A is replaced, and the probability that amino acid A is replaced. The conditional probabilities were calculated as the observed frequencies of exchange between the amino acids; these frequencies were derived in the following way. Groups of closely related sequences ($< 15\%$ different) were aligned, and, based on a given unrooted evolutionary topology, ancestral sequences were generated for each node in the topology. The data were compiled from groups of closely related sequences to reduce the chance of the observed differences being the result of superimposed mutations, i.e., amino acid A changing to amino acid B, which then changes to amino acid C. The exchanges between each sequence and its immediate ancestor were compiled and summed for all sequence pairs. Exchanges involving positions in the ancestral sequences where an amino acid could not be unambiguously assigned were treated statistically by giving fractional scores to each alternative replacement. In the derivation it was assumed that the exchanges are symmetric; hence, an exchange of amino acid A for amino acid B was also counted as an exchange of amino acid B for amino acid A. This assumption is reasonable because the conditional replacement probability is expected to depend primarily on the chemical and physical similarity of the two amino acids.

The probability of an amino acid being replaced is estimated as its relative mutability, which is calculated as the ratio of the number of observed changes of an amino acid to its total exposure to change. The total number of observed changes was computed as the total number of times the amino acid was observed to change in the sequences examined. For each pair of sequences examined, the exposure to change of each amino acid was calculated as the frequency of occurrence of the amino acid multiplied by the total number of all amino acid changes observed for that sequence pair per hundred sites. The last factor normalizes the values to equal sequence length and equal evolutionary distance. These values were summed for all sequence pairs to compute the total exposure.

In accordance with the Markovian model, mutation probability matrices corresponding to larger intervals of evolutionary distance can be obtained by repeatedly multiplying the original matrix by itself, i.e., the 2-PAM matrix corresponds to the square of the 1-PAM matrix, the 3-PAM to the cube, etc. For each evolutionary distance, the sum of the diagonal elements gives a measure of the expected number of observed amino acid changes. A plot of these data (Fig. 2) clearly shows a saturation phenomenon. As the number of mutations increases, the probability of superimposed mutations at individual sites increases. Inasmuch as these multiple mutations are indistinguishable from single mutations, the observed number of changes asymptotically approaches a constant value. For the 1978

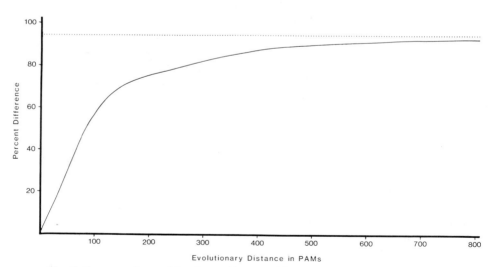

FIG. 2. Correspondence of the observed percent difference and the estimated evolutionary distance between two amino acid sequences.

mutation data, this value corresponds to 94% difference. At large evolutionary distances, the matrix values themselves exhibit an asymptotic behavior (this is a general property of Markovian models). Transition to any amino acid becomes equal to its frequency of occurrence irrespective of the identity of the initial amino acid. The equilibrium frequencies are implicitly determined by the data set from which the matrix was derived. This same effect is observed in simulation studies. Repeated applications of the 1-PAM probability matrix will produce sequences that tend toward the equilibrium composition regardless of the composition of the initial sequence.

Although the PAM values represent estimates rather than observed quantities, they provide a more realistic measure of evolutionary distance than do the numbers of percentages of observed differences between pairs of sequences. When calculating topologies by the least-squares matrix method, it has been found that correction of the observed percent difference values for inferred superimposed mutations, using the data depicted in Fig. 2, gives a more realistic reconstruction of the evolutionary topologies.[5]

For generalized sequence comparisons it is more useful to employ a similarity matrix whose elements reflect the ratio of the probability of an amino acid exchange to the probabilities of the two amino acids occurring at random. These ratios are given by the elements of the relatedness odds matrix. This matrix was derived by dividing the elements of the mutation probability matrix by the normalized frequencies of occurrence of the replacement amino acids. Each element gives the probability of replacement of amino acid A with B per occurrence of A per occurrence of B. The matrix has the desired attribute that it is symmetric. McLachlan's scoring matrix[2] reflects a similar ratio, but it was derived by a different method.

When one protein is compared with another, position by position, one should multiply the odds for each position to calculate an odds for the comparison of the entire sequences. However, it is more convenient to add the logarithms of the odds. Thus, the MDM scoring matrix contains the logarithms of the elements of the 250-PAM odds matrix; to allow more rapid computation by avoiding floating point operations, the elements of the log odds matrix are multiplied by 10 and rounded to the nearest integer. The neutral score is zero. A score of $+10$ indicates that a pair of amino acids is expected to occur 10 times as frequently in related sequences (after 250 PAMs of evolutionary change) as would occur by

[5] W. C. Barker, L. T. Hunt, and D. G. George, in "Computer Simulation of Carcinogenic Processes" (B. D. Silverman, ed.) p. 1. CRC Press, Boca Ratan, Florida, 1988.

chance. The elements of the matrix range from -8 (for Trp/Cys) to $+17$ (for Trp/Trp).

Simulation studies have indicated that for distantly related sequences (between 73 and 86% difference) the MDM for 250 PAMs is optimal for distinguishing between related proteins and those whose observed similarity is due to chance.[6] Hence, the matrix corresponding to 250 PAMs has become the standard matrix for sequence comparison studies. It has been reported[7] that in sequence comparison it is more effective to use a matrix corresponding more closely to the actual evolutionary distance between the sequences being compared, and a computer program that facilitates this approach has been developed.[8] These results are not unexpected; however, this approach is limited because it requires *a priori* knowledge of the approximate evolutionary distance between the sequences being compared.

The MDM (shown in Fig. 3) clearly reflects the physicochemical properties of the amino acids. Figure 3 has been arranged to designate groups of chemically similar amino acids: the aromatic group (tryptophan, tyrosine, phenylalanine); the hydrophobic group (valine, leucine, isoleucine, methionine); the basic group (lysine, arginine, histidine); the acid–acid amide group (glutamine, glutamic acid, aspartic acid, asparagine); the group of amino acids that are small and not strongly hydrophilic or hydrophobic (glycine, alanine, proline, threonine, serine); and cysteine. Some groups overlap: the basic and the acid–acid amide groups tend to replace one another to some extent, and phenylalanine interchanges with the hydrophobic amino acids more often than chance expectation predicts. These patterns are imposed principally by natural selection and only secondarily by constraints of the genetic code; they reflect the similarity of the functions of the amino acids in their interactions with one another in the three-dimensional conformation of proteins. Some of the properties of an amino acid residue that determine these interactions are size, shape, and local concentrations of electric charge; conformation of van der Waals surface; and ability to form salt bonds, hydrophobic bonds, and hydrogen bonds. For specific applications, it is possible to derive scoring matrices that reflect one or another of these particular properties. The values in the MDM correlate particularly well with similarity of hydrophobicity and molecular bulk of the side chains of the amino acids[9,10] as well as with secondary structure-forming propensity.[9]

[6] R. M. Schwartz and M. O. Dayhoff, *in* "Atlas of Protein Sequence and Structure" (M. O. Dayhoff, ed.), Vol. 5, Suppl. 3, p. 353. National Biomedical Research Foundation, Washington, D.C., 1979.

[7] J. F. Collins, A. F. W. Coulson, and A. Lyall, *CABIOS* **4,** 67 (1988).

[8] A. H. Reisner and C. A. Bucholtz, *Nucleic Acids Res.* **14,** 233 (1986).

[9] S. French and B. Robson, *J. Mol. Evol.* **19,** 171 (1983).

	C	S	T	P	A	G	N	D	E	Q	H	R	K	M	I	L	V	F	Y	W
C Cys	12																			
S Ser	0	2																		
T Thr	-2	1	3																	
P Pro	-3	1	0	6																
A Ala	-2	1	1	1	2															
G Gly	-3	1	0	-1	1	5														
N Asn	-4	1	0	-1	0	0	2													
D Asp	-5	0	0	-1	0	1	2	4												
E Glu	-5	0	0	-1	0	0	1	3	4											
Q Gln	-5	-1	-1	0	0	-1	1	2	2	4										
H His	-3	-1	-1	0	-1	-2	2	1	1	3	6									
R Arg	-4	0	-1	0	-2	-3	0	-1	-1	1	2	6								
K Lys	-5	0	0	-1	-1	-2	1	0	0	1	0	3	5							
M Met	-5	-2	-1	-2	-1	-3	-2	-3	-2	-1	-2	0	0	6						
I Ile	-2	-1	0	-2	-1	-3	-2	-2	-2	-2	-2	-2	-2	2	5					
L Leu	-6	-3	-2	-3	-2	-4	-3	-4	-3	-2	-2	-3	-3	4	2	6				
V Val	-2	-1	0	-1	0	-1	-2	-2	-2	-2	-2	-2	-2	2	4	2	4			
F Phe	-4	-3	-3	-5	-4	-5	-4	-6	-5	-5	-2	-4	-5	0	1	2	-1	9		
Y Tyr	0	-3	-3	-5	-3	-5	-2	-4	-4	-4	0	-4	-4	-2	-1	-1	-2	7	10	
W Trp	-8	-2	-5	-6	-6	-7	-4	-7	-7	-5	-3	2	-3	-4	-5	-2	-6	0	0	17
	C	S	T	P	A	G	N	D	E	Q	H	R	K	M	I	L	V	F	Y	W
	Cys	Ser	Thr	Pro	Ala	Gly	Asn	Asp	Glu	Gln	His	Arg	Lys	Met	Ile	Leu	Val	Phe	Tyr	Trp

Cluster labels within the matrix:
C – sulfhydryl
S – small hydrophilic
N – acid, acid amide, hydrophilic
H – basic
V – small hydrophobic
F – aromatic

FIG. 3. Mutation data matrix for 250 PAMs (amino acid mutations per 100 residues). The amino acids were arranged by assuming that positive values represent evolutionarily conservative replacements; the clusters correspond to groupings based on the physicochemical properties of the amino acids. [Reproduced, with permission, from D. G. George, L. T. Hunt, and W. C. Barker, in "Macromolecular Sequencing and Synthesis" (D. H. Schlesinger, ed.), p. 127. Alan R. Liss, New York, 1988.]

Limitations of Model

It was known at the time of original publication of the matrix that the observed frequencies of amino acid exchanges that require two or more nucleotide replacements were significantly greater than that expected, based on the frequencies of amino acid replacements requiring single nucleotide exchanges.[3] As was correctly pointed out by Wilbur,[11] this is

---

[10] W. R. Taylor, *J. Theor. Biol.* **119**, 205 (1986).
[11] W. J. Wilbur, *Mol. Biol. Evol.* **2**, 434 (1985).

inconsistent with a simple Markovian model of nucleotide replacement. It is important to realize that the Dayhoff model makes no assumptions about the genetic mechanisms involved in point mutations on the DNA level, and non-Markovian behavior on the nucleotide level does not necessarily imply that amino acid exchanges cannot be treated in Markovian fashion. The two-step mutation model described by Wilbur[11] implies that at each step (each nucleotide replacement in a multiple nucleotide exchange), the mutation must be fixed in the population before the next step may begin. It has never been demonstrated that this is required, and the data seem to suggest otherwise. Hence, one may justifiably ignore events at the nucleotide level and model amino acid mutation based entirely on observed amino acid exchange frequencies.

More importantly, the model makes the assumptions that there are no correlations in exchange frequencies between neighboring sites and that the exchange frequencies are the same regardless of the position of the site within the protein sequence. Molecular modeling studies have indicated that correlations between neighboring sites play a significant role in structure formation. Moreover, it is well known that different sites within proteins show dramatically different levels of variability. These two approximations limit the utility of this matrix as a measure of biological similarity.

These restrictions apply equally to virtually all currently employed sequence comparison methods; therefore, using this matrix in these procedures does not introduce any additional approximations. In general, such methodologies do not provide a true measure of biological similarity.[12] Several new approaches to the sequence comparison problem have been reported that attempt to rectify these problems,[13,14] but these approaches introduce their own approximations and generally require *a priori* knowledge concerning the class of proteins in questions.

Three other criticisms of the methodology have repeatedly appeared in the literature. (1) The derivation of the matrix employs circular logic, i.e., a matrix is required to construct the sequence alignments, and these alignments are used in the derivation of the matrix. All the sequence alignments examined in these studies were between sequences less than 15% different. Within such alignments, very few gaps (insertion/deletions) are observed, and very few decisions are required for their construction. As a result, any

[12] D. G. George, L. T. Hunt, and W. C. Barker, *in* "Macromolecular Sequencing and Synthesis" (D. H. Schlesinger, ed.), p. 127. Alan R. Liss, New York, 1988.
[13] M. Gribskov, A. D. McLachlan, and D. Eisenberg, *Proc. Natl. Acad. Sci. U.S.A.* **84,** 4355 (1987).
[14] M. Gribskov, M. Homyak, J. Edenfield, and D. Eisenberg, *CABIOS* **4,** 61 (1988).

realistic method of sequence alignment (including alignment by eye) and any similarity scoring matrix (including the most simple matrix, which assigns one score for matches and one score for mismatches) will virtually always produce the same alignment. (2) The methodology suffers because it depends heavily on ancestral sequence methods, which are known to have significant limitations. The limitations of these methods involve the problems associated with multiple mutation events. Among closely related sequences, these problems have very little effect. (3) The methodology depends heavily on the selection of the correct topological relationships among the sequences in each group. The effect of this limitation is more difficult to assess as there is no rigorous measure for defining correct topology. The chief criterion for selecting the topologies to be used was minimum overall length of the topology (total number of mutations). When there was a choice between several nearly minimal trees, the one that was consistent with generally accepted phylogeny or with topologies derived from other sequences was chosen, if such information was available. It is noteworthy that the values of the matrix elements compiled in 1978 did not significantly differ from those compiled in 1968, although during this time there was a doubling of the amount of sequence information examined, new groups of sequences previously unrepresented in the data set were introduced, and as a result the topologies corresponding to many of the previously examined groups substantially changed. Hence, this does not appear to be a severe restriction on the methodology.

There is a valid question concerning how well this matrix represents the currently available sequence data, however. Since 1978, the amount of protein sequence data has increased 10-fold. More importantly, data are now available from groups of sequences that were underrepresented or not represented at all in the 1978 data set. Some of the groups, such as the nonaqueous soluble (hydrophobic) proteins and the viral proteins, exhibit properties dramatically different from those represented in the 1978 data set. Thus, the lack of significant change between the 1968 and 1978 derivations cannot be taken to indicate that no appreciable changes would be observed in a new derivation.

Computer Applications Using MDM

Many types of sequence comparison and searching applications use scoring matrices in order to enhance their sensitivity. Among these are the dynamic programming approaches to aligning sequences, the sliding-window segment comparison method and its graphic implementations, various database searching methods, and methods for making alignments of three or more related sequences. We[6,15] and others have found that incorporating the MDM into such methods markedly increases their sensitivity for illuminating even very distant relationships. In the discussion that follows, we draw examples from several of these methods; there are many examples, some of which are described in other chapters of this volume.

The ALIGN program[12,16] is a global alignment program based on the algorithm of Needleman and Wunsch;[17] it is most sensitive in comparing sequences of similar size and architecture that are assumed to be related along their entire lengths. It determines an optimal alignment (including gaps) of a pair of sequences by dynamic programming. To obtain realistic alignments, two types of gap penalty are employed: one is a penalty applied every time a break (gap) is inserted in either sequence (the break penalty), regardless of the length of the break; the other assesses a penalty based on the size of the break. These parameters can be shown to be equivalent to those of the optimal sequence alignment method of Smith and colleagues.[18] The ALIGN program exacts the second penalty by giving a bonus to every score for an amino acid matching another amino acid, whereas the score for an amino acid aligned with a gap is always zero. This bonus is called a matrix bias because it is achieved by adding the value to every element in the scoring matrix rather than by modifying the algorithm. We most often set both the break penalty and the matrix bias at 6. This value for the matrix bias ensures that almost all replacements of one amino acid for another receive a higher score than an insertion or deletion. If 8 is used, all replacements except Trp/Cys receive scores higher than 0.

Schwartz and Dayhoff[6] tested eight pairs of distantly related sequences using ALIGN and four different scoring matrices (Table II). The average score using the MDM was twice as high as that using the unitary matrix (UM), with the genetic code matrix (GCM) and the matrix of McLachlan[2] (AAAM) giving intermediate scores. The MDM gave the highest score for six of the eight comparisons, with UM and AAAM each giving the highest score in one case. Similar results were reported by Feng and Doolittle[19] who used a different global alignment program. If the sequences being compared are of markedly different lengths or if only portions of the sequences being compared are related, the ALIGN program may not produce a reasonable alignment or score; in such cases, using a high break

[15] W. C. Barker and M. O. Dayhoff, *in* "Atlas of Protein Sequence and Structure" (M. O. Dayhoff, ed.), Vol. 5, p. 101. National Biomedical Research Foundation, Washington, D.C., 1972.

[16] M. O. Dayhoff, W. C. Barker, and L. T. Hunt, this series, Vol. 91, p. 524.

[17] S. B. Needleman and C. D. Wunsch, *J. Mol. Biol.* **48,** 443 (1970).

[18] W. M. Fitch and T. F. Smith, *Proc. Natl. Acad. Sci. U.S.A.* **80,** 1382 (1983).

[19] D.-F. Feng and R. F. Doolittle, *J. Mol. Evol.* **25,** 351 (1987).

TABLE II
SEQUENCE COMPARISON SCORES OBTAINED WITH
VARIOUS SCORING MATRICES[a]

| Matrix | ALIGN Scores | | RELATE Scores | |
	Range (SD)	Mean (SD)	Range (SD)	Mean (SD)
UM	0.1–5.8	3.0	0.5–8.0	4.9
GCM	0.4–9.0	3.9	1.3–9.3	4.6
AAAM	0.4–9.9	4.9	0.7–8.0	4.6
MDM	2.9–12.1	5.9	3.5–15.2	7.4

[a] Adapted from Schwartz and Dayhoff.[6] Eight pairs of distantly related proteins were compared using the program ALIGN; five pairs of distantly related proteins and four proteins with internal duplication were tested with program RELATE.

penalty may give satisfactory results. Often, however, regions of similarity must be preselected to obtain reasonable results.

Computer methods that locate and compare subregions within proteins are collectively known as local similarity methods. the RELATE program,[16] based on a method first introduced by Fitch,[20] was one of the earliest of such methods. It compares all segments of a given length from one sequence with all segments of the same length from the second sequence. This method is useful when the correspondence of segments is not known or when looking for regions of similarity in otherwise unrelated sequences. The segment score is the sum of the scores for each pair of amino acids occupying corresponding positions within the two segments. If the proteins are related, there will be two populations of scores: a large population from unrelated segments and a very small population of higher scores from the related segments. The program calculates the expected number of scores in this second population and computes the average of that same number of top scores from the upper tail of the distribution. The sequences are permuted, and a distribution of average top scores is generated and used to calculate a Z value, the RELATE score (in SD units). Using the RELATE program and four scoring matrices (Table II), Schwartz and Dayhoff[6] tested five pairs of distantly related sequences and four internally duplicated sequences. The MDM gave the highest score for seven of the nine tests, with UM giving higher scores in two cases.

[20] W. M. Fitch, *J. Mol. Biol.* **16**, 9 (1966).

The DOTMATRIX graphic comparison program[12,21] is a convenient method for locating and visualizing regions of similarity between two sequences, especially when the lengths are different, and for finding regions of repetitive structure within a sequence. The plot is a graphic equivalent of the list of highest segment comparison scores obtained with RELATE. It is derived by computing a comparison matrix whose elements reflect the similarity of the segments being compared. The sequences are represented on the axes; dots are placed within a rectangular grid at positions corresponding to the central residues in the segments for which the comparison scores exceed a specified minimum value. Regions of sequence similarity are visualized as diagonal lines composed of contiguous dots; an offset in a diagonal represents a break (insertion or deletion) in one sequence with respect to the other in the region of similarity. Shorter diagonals above and/or below a main diagonal may represent repeats in one or both sequences.

Varying the parameters (scoring matrix, window size, minimum score) of the dot matrix comparison allows more subtle similarities to be resolved. A low minimum score brings out weaker internal similarities, although the background noise is much higher. A larger window (segment length) favors longer repeats and may be used to visualize these in a highly periodic sequence.[21] Figure 4 shows four DOTMATRIX comparisons of troponin C^{22} on the horizontal axes and myosin L2 chain[23] on the vertical axes; both proteins are from rabbit skeletal muscle. The plot using the MDM (Fig. 4a) has a very prominent main diagonal showing the similarity of the two proteins along their entire lengths; their amino halves have nearly twice as many identities as their carboxyl halves. In addition, three parallel broken diagonals can be seen above and below the main diagonal. These reflect the comparisons of the more well-conserved regions of myosin L2 with the four homologous calcium-binding domains of troponin C. The plot with the hydrophobicity matrix (Fig. 4b) shows the main diagonal and the homology of both halves of troponin C to the first half of myosin L2, but only some short segments of the off-diagonals can be seen. Many of these are present in the plot made using the unitary matrix and a minimum score of 5 identities in a 25-residue span (Fig. 4c), but the plot is very noisy, as many nonhomologous pairs of segments also show this much identity. Raising the threshold to 7 (Fig. 4d) eliminates both the noise and most of the diagonals representing homologous comparisons.

[21] W. C. Barker, L. T. Hunt, and D. G. George, *Protein Seq. Data Anal.* **1,** 363 (1988).
[22] J. H. Collins, M. L. Greaser, J. D. Potter, and M. J. Horn, *J. Biol. Chem.* **252,** 6356 (1977).
[23] G. Matsuda, T. Maita, Y. Suzuyama, M. Setoguchi, and T. Umegane, *Hoppe-Seyler's Z. Physiol. Chem.* **359,** 629(1978).

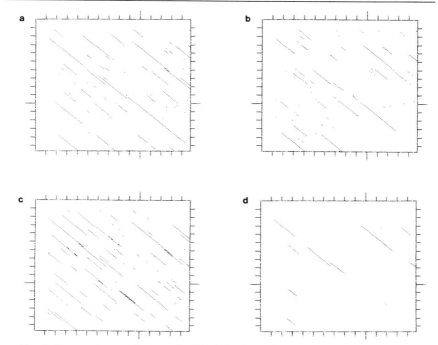

FIG. 4. Dot matrix graphic plots of rabbit skeletal muscle troponin C (horizontal axes) and myosin L2 (DTNB) regulatory light chain (vertical axes) with different scoring matrices. The segment length (window size) is 25 in all plots. (a) MDM with minimum score of 15; (b) hydrophobicity matrix with minimum score of 195; (c) unitary matrix with minimum score of 5; (d) unitary matrix with minimum score of 7.

Several other approaches to the local similarity problem, in which the MDM was among the scoring matrices employed, have recently been reported. Hall and Myers developed a program[24] that can find locally optimal alignments between two sequences even when they are separated by long nonsimilar regions or when the sequences have little overall similarity. Lawrence and Goldman employed an alternative approach[25] to dynamic programming for detecting and evaluating similarities. Based on a nonlinear similarity score and their modification of the DD algorithm of Altschul and Erickson,[26,27] and using cost matrices such as one they derive from the MDM, this method finds the boundaries of so-called homology

[24] J. D. Hall and E. W. Myers, *CABIOS* **4,** 35 (1988).
[25] C. B. Lawrence and D. A. Goldman, *CABIOS* **4,** 25 (1988).
[26] S. F. Altschul and B. W. Erickson, *Bull. Math. Biol.* **48,** 617 (1986).
[27] S. F. Altschul and B. W. Erickson, *Bull. Math. Biol.* **48,** 633 (1986).

domains. An aggregate score for all such domains found in a comparison can be calculated by a method similar to the derivation of an ALIGN score.

By far the most effective method for large-scale database searching, which was introduced by Wilbur and Lipman,[28] uses a fast algorithm for locating the maximum diagonals in a dot matrix-type comparison of two sequences. Lipman and Pearson[29] developed FASTP, a version of the program specifically designed for rapid searching of protein sequence databases and incorporating the MDM into the scoring procedure. As in the original method, a maximum diagonal is selected to represent each sequence comparison. In this implementation, however, the five best scoring diagonals from each sequence comparison are first selected, based on the relative number of matches versus mismatches. The selected diagonals are then rescored using the MDM, and, based on these scores, one representative diagonal is selected. The program then proceeds as before except that the optimized score is computed using the MDM as the similarity matrix. This implementation incorporates other changes, including an improvement to the methodology used for constructing the maximum diagonals, that result in an overall increase in speed of the program, but the major advance is the dramatic increase in the sensitivity resulting from the introduction of a similarity scoring matrix.

Pearson and Lipman[30] have recently introduced a refinement of this method. In the new program, FASTA, the unknown sequence is compared to database sequences in a four-step process that can combine multiple sequential regions of similarity to produce a longer diagonal than in FASTP (in which only one diagonal is selected) for final comparison of each pair of sequences. FASTA is a modification for detecting local similarities, including repetitive structure; it saves and displays, as alignments or dot matrix plots, all diagonals (not just the top 10) with similarity scores above a threshold value.

FASTP and FASTA are not as effective as some other searching methods, however, such as the SEARCH program of the PIR, in finding similar segments of short length (less than 30–40 residues) or all copies of repetitive segments. The SEARCH program[12,16,31] compares a sequence of specified length with all other sequences of the same length in the database and computes the score based on a specified scoring matrix. Typically, for

[28] W. J. Wilbur and D. J. Lipman, *Proc. Natl. Acad. Sci. U.S.A.* **80**, 726 (1983).

[29] D. J. Lipman and W. R. Pearson, *Science* **227**, 1435 (1985).

[30] W. R. Pearson and D. J. Lipman, *Proc. Natl. Acad. Sci. U.S.A.* **85**, 2444 (1988).

[31] W. C. Barker, D. G. George and L. T. Hunt *in* "Computer Analysis for Life Science" (C. Kawabata and A. R. Bishop, eds.), p. 194. OHMSHA Ltd., Tokyo, 1986.

a 25-residue piece, the corresponding segments from all closely related sequences (> 50% identical overall), and many from more distantly related sequences, will appear above the distribution of segments from unrelated proteins. When very little sequence data are available, we frequently perform the search using both MDM and UM. The former is usually more effective in locating segments from more distantly related proteins; the latter may be more effective when several residues must be absolutely conserved to retain an activity or to locate possible cross-reactive epitopes in unrelated sequences.

The MDM has been found to be useful in several methods of making consensus sequence alignments. These methods allow one to base comparisons on genetic and other data in addition to sequence similarities. If the sequence alignments can be improved, then the evolutionary trees derived from them will have more accurate branch lengths and branching order. Furthermore, it may be possible to detect and establish extremely distant relationships, thus extending the information on protein and species evolution. The ALIGN program is capable of revealing similarity (and suggesting possible relationships) between sequences 65–80% different, whereas profile analysis[13] may allow correspondence to be established between sequences 80–90% different.

Profile analysis is a two-step, two-program method that defines a consensus sequence, or profile, from an alignment and matches this profile against other sequences. First an alignment of structurally and functionally similar sequences, either entire proteins or domains, is used to construct a sequence position-specific scoring matrix. Using a comparison table based on the MDM, the first program (PROFMAKE) then generates a consensus sequence or profile, in which position-dependent gap penalties are included. The profile is thus a pattern that can simultaneously embody the aggregate of structural, functional, and genetic information about a specific group of proteins. In the second step of the analysis, the profile is compared either with a database to find any similar sequences (PROFANALDB) or with a single sequence (PROFANAL) to learn if it belongs to a known family (or superfamily) and, if so, to obtain the best alignments. Once a library of such profiles has been accumulated, it can be used as a database against which a target sequence can be searched; Gribskov et al.[14] have developed program PROFILESCAN, which incorporates a modified form of the MDM, to do this.

Several other approaches that employ a consensus sequence concept (reviewed by Taylor[32,33]) have used the MDM as the scoring matrix at some

[32] W. R. Taylor, Protein Eng. **2**, 77 (1988).
[33] W. R. Taylor, *J. Mol. Evol.* **28**, 161 (1988).

point. The consensus sequence is derived, as in profile analysis, from alignments of known related sequences. Usually the alignments are constructed by beginning with the most similar or recently diverged sequences (two to several), deriving a consensus, then adding more sequences in repeated pairwise comparisons of new sequence to consensus sequence.[19,33,34] A variation is to derive a consensus from each of several groups of closely related sequences and then to compare the consensus sequences.[34] The MDM, or a modification of it, is the similarity scoring matrix most often used in the sequence comparisons,[19,33,34] as the results appear to fit best, compared with global optimization methods, with other biological information.[19] Again, all available information is used to establish gaps; the progressive alignment procedure of Feng and Doolittle retains any gaps once established, as more sequences are added.[19] Patthy was particularly concerned with deriving patterns (of identities, similarities, variable regions, gaps) in order to identify very distantly related sequences and employed both a modified MDM and the unitary matrix in the protocol.[34]

Taylor[10] and Risler et al.[35] among others, have examined various scoring matrices, including the MDM, to determine which properties are important in conservation of amino acids in structurally related proteins. Taylor[10] briefly reviewed previous efforts to apply multidimensional scaling techniques to the MDM; he developed two-dimensional Venn diagrams to represent properties of the amino acids. Size and hydrophobicity were the properties found to correlate with the values in the MDM. He also pointed out a type of situation for which the MDM may not be the best measure of conservation. Risler et al.[35] developed a matrix based on superposition of three-dimensional structures of related proteins, using observed replacements where α-carbon atoms at corresponding positions are less than 1.2 Å apart. Comparison of several scoring matrices in aligning sequences indicates that the MDM may weight identities too heavily. It is also pointed out[20,32,35] that the MDM may be inadequate for optimal alignment of very distant sequences because it does not give sufficient information about the local structural environment.

New Similarity Matrices

We are in the process of deriving new matrices based on much more data than were previously available. Although the MDM derived in 1978 showed only little change from the matrix published in 1969,[36] the amount

[34] L. Patthy, *J. Mol. Biol.* **198,** 567 (1987).
[35] J. L. Risler, M. O. Delorme, H. Delacroix, and A. Henaut, *J. Mol. Biol.* **204,** 1019 (1988).

of sequence data had only doubled and most of the groups of proteins present in the 1978 data were also represented in the 1969 data. There has been a nearly 10-fold increase of protein sequences in the family and subfamily categories since the MDM was last compiled. Nearly 70% of the entries in the current sequence database contain protein sequences obtained by inference from the elucidation of the corresponding nucleic acid sequences. Prior to 1978 the database was strongly biased toward soluble proteins that occur naturally in large quantities, as these types can be most easily sequenced by classic methods. Entire classes of proteins were previously absent, the most prominent being the membrane proteins, whose hydrophobic nature markedly distinguish them from other groups. In addition, many of the groups of proteins now represented in the database previously contained either no sequences or too few sequences to make any contribution to the compiled amino acid replacement data. Thus, even if the new derivation does not yield a different set of matrix values, the reliability of these values will be dramatically increased, and this recompilation will provide a more sensitive probe for the understanding of protein relationships and will have a direct influence on the investigation of many evolutionary questions.

As more sequence data have become available, it has become increasingly apparent that sequence similarity between proteins has profound biochemical implications irrespective of evolutionary relationships and that general evolutionary constraints may differ from those imposed by purely functional requirements. Furthermore, it is likely that acceptable replacement frequencies may vary among functionally distinct classes of proteins. Replacement matrices derived explicitly from specific functional groups of proteins are likely to be much more sensitive for examining questions of functional relatedness. Clearly, a generalization of the methods introduced by Dayhoff that allows the compilation of amino acid replacement data, independent of evolutionary considerations, is warranted. As more sequence data become available from specific classes of proteins, the techniques that we are now developing will lead to a wide variety of empirical matrix probes designed to study specific biological questions.

Acknowledgments

This work was supported by National Institutes of Health Grant GM37273 from the National Institute of General Medical Sciences. We wish to thank James K. Bair for editorial and technical support in the preparation of the manuscript and illustrations.

---

[36] M. O. Dayhoff, R. V. Eck, and C. M. Park, in "Atlas of Protein Sequence and Structure" (M. O. Dayhoff, ed.), Vol. 4, p. 75. National Biomedical Research Foundation, Silver Spring, Maryland, 1969.

[21] Sensitivity Comparison of Protein Amino Acid Sequences

By Patrick Argos and Martin Vingron

Introduction

Many procedures have been developed to align protein amino acid sequences; this volume is proof alone. To compare two protein primary structures, a classic technique involves positioning the two sequences in single-letter code and with lengths m and n such that one is vertical along a column (m) and the other horizontal along a row (n). Then, in matrix fashion, scores are inserted at each grid point $(m \times n)$ corresponding to every pairwise comparison of the amino acids in each sequence. The scores can simply be 0 (nonidentity) or 1 (identify) on up to the Dayhoff mutation matrix (20×20) expressing relative position weights with which amino acids prefer to exchange in aligned sequence families as c cytochromes or globins.[1] To achieve the actual alignment, all possible paths are traced through the sequence comparison (search) matrix with the residues consecutively aligned from the amino to the carboxy termini, albeit with deletions and insertions. For each path, the sum of positive Dayhoff values along the trace is determined with gap penalties subtracted each time an insertion/deletion is initiated and extended by one position in length (see Refs. 2, 3, and 4 for details and reviews). There are fast ways[5] to score every possible path to find the optimal one (maximum).

The classic procedure is flawed, especially when alignments result in amino acid identity at 35% or less of the matched positions. The technique relies on extrinsic choices for the (20×20) scoring matrix as well as for the values of the initiating and extending gap penalties. Controls using "standard-of-truth" alignments from tertiary structural superpositions[6] show that optimal matches are achieved for different sequence pairs (e.g., serine

[1] M. O. Dayhoff, W. C. Barker, and L. T. Hunt, this series, Vol. 91, p. 524.

[2] P. Argos and P. McCaldon, *in* "Genetic Engineering, Principles and Methods" (J. K. Setlow, ed.), Vol. 10, p. 21. Plenum, New York and London, 1988.

[3] G. von Heijne, "Sequence Analysis in Molecular Biology." Academic Press, New York, 1988.

[4] R. F. Doolittle, "Of URFS and ORFS: a Primer on How to Analyze Derived Amino Acid Sequences." University Science Books, Mill Valley, California, 1986.

[5] S. B. Needleman and C. D. Wunsch, *J. Mol. Biol.* **48**, 443 (1970).

[6] B. W. Matthews and M. G. Rossmann, this series, Vol. 115, p. 397.

proteases or globins) with different gap penalty choices. Gap penalties should even be varied within regions of the sequence alignment according to the matched secondary structures, such as β strands or exposed loops. Furthermore, local sequence dependency is not considered, as in comparing sequence fragments (oligopeptides); instead, residues are examined only singly in the search matrix.

This chapter describes a procedure for pairwise sequence alignment that attempts to overcome the aforementioned difficulties; the technique is correspondingly described as sensitive. Next, a multiple sequence alignment procedure is described that preserves both sensitivity and speed of execution. The significance or credibility of the resultant alignments is subsequently considered. Finally, computer programs based on the procedures and available from the authors are discussed.

Sensitive Pairwise Sequence Comparison

Alignments considered "standards of truth" can be had by spatial superposition of α-carbon backbone folds known from X-ray diffraction analysis.[7] The distribution of residue match and conservation is often not uniform. There may be 10 amino acids where half are identical and one is a cysteine essential for function, then an insertion/deletion of 8 residues, the next 30 residues with only 10% identity but strong hydrophobicity conservation to preserve a buried β sheet, another insertion/deletion, then 12 amino acids with three identically conserved glycine residues essential in turns, then 5 amino acids with no perceivable conservation because their exposure allows extreme variation, and so forth. The variable pattern is a result of ever-changing structural and functional requirements. So, in any sensitive sequence comparison technique, it is essential to consider sequence fragments of varying length as well as the physical and chemical characteristics of the 20 amino acids. Such a procedure has been developed and described in detail.[8]

The alignments of several sequences that relied on tertiary structures were collected; the identity level was only about 18%, on the average. The individual alignments were strung together such that over 11,000 amino acids were paired. Correlation coefficients were then calculated using about 100 different amino acid characteristics. For example, two strings of paired hydrophobicity values, each with 11,000 elements, are generated from the extended sequence alignment. The correlation between the two hydrophobicity series is determined. The following five characteristics proved to be

---

[7] M. G. Rossman and P. Argos, *J. Mol. Biol.* **109,** 99 (1977).

[8] P. Argos, *J. Mol. Biol.* **193,** 385 (1987).

most sensitive: (1) hydrophobicity, (2) residue bulk or shape, (3) preference of a residue to be in the turn structural conformation, (4) antiparallel β-strand conformational potential, and (5) amino acid refractivity index, which correlates strongly with molecular weight. The importance of these parameters seems obvious in conserving protein tertiary folds: hydrophobicity to maintain the interior core, turn preference to ensure the chain returns to form the core, β preference as strand folds often comprise protein interiors, and side chain size and shape for close-packed structures.

A window length (e.g., 25 residues) must be defined. When aligning every 25-residue span of one protein sequence with every such span in the other sequence, two scoring criteria are used. The sum of the Dayhoff mutation values for each of the residue pairs is determined. The correlation coefficients for each of the five aforementioned characteristics are calculated for the two aligned sequence fragments and averaged. Calculating correlations insists that the two fragments show similarly varying characteristic patterns without the necessity of similar absolute characteristic values. Assigning weights *a priori* to the parameters is next to impossible, because accurate predictions cannot be made of the tertiary structure of the sequence fragment. So, each parameter is given unit weight with the expectation that the locally most important will dominate the average.

These processes are repeated for all possible fragment comparisons in the two sequences. The Dayhoff score and mean correlation are then added and scaled such that each contributes equally, on average, to the combined score. Considering only those oligopeptide matches where mutation and structure agree reduces greatly the noise level in the search matrix.[8] The standard deviation and mean of the combined scores are calculated, and then the combined scores are expressed as the number of standard deviations (v) above the mean. The scores are entered into a search matrix bordered by the two sequences if they are greater than some threshold (usually $3.0v$). The v values are entered over the entire probe length, and overlaps are resolved by maintaining the largest values. For example, if positions 11 to 20 in one sequence (window length 10) were compared with 41 to 50 in the other sequence with a combined and normalized score of $3.1v$, then this value is placed at matrix sites (11,41), (12,42), . . . , (20,50). Now, when 12 to 21 is compared with 42 to 51 at $5.2v$, then $5.2v$ is placed at (12,42), (13,43), . . . , (21,51); $5.2v$ overlaps $3.1v$ in 9 positions and replaces the lower values.

The combined scores for several probe lengths are collected into one search matrix by merely repeating the above procedures for each window length, the highest standard deviation score always resolving any overlap. Usually it is sufficient to use window lengths of 5 to 35 in steps of 1, 2, or 3 (depending on available computer time and noting that insertions/deletions invariably occur after 35 or fewer residues have been consecutively matched). The use of the number of standard deviations as a score allows an equivalent comparison of the results from each probe where the noise or background is taken from the search at a given window length. This approach, along with plotting over the entire probe length, literally "fills in" the homology. An example is given in Fig. 1. Usually only those peaks above a certain threshold are placed in the search matrix ($3.0v$ for sequences less than 500 residues and $3.5v$ for those longer); the remaining positions are set to $0.0v$.

A computer program has been developed[9] that allows the user to select, through a "mouse" device, an alignment path in the search matrix directly on a graphics screen. Gap penalties are thus avoided, and lower valued or suboptimal peaks consistent with the overall alignment trend can be easily selected. Alternatively, a path can be traced automatically by considering all possible paths with an efficient algorithm.[5] Plotting over the entire probe length and selection of a minimum v value for display force the automated procedure to maintain a particular high-peak trace not achieved when only single residues are compared.

Four approaches to alignment were allowed to compete[9] in order to achieve a standard-of-truth alignment between chymotrypsin and α-lytic protease (176 matched positions, 20.5% identity, oligopeptide GDSGG conserved exactly); lupin and human hemoglobins (139 matches, 16.4% identity); southern bean mosaic and tomato bushy stunt virus amino-terminal region (122 alignment sites, 29.5% identity) as well as a carboxy-terminal region (60 residues, 26.7% identity); and valyl- and leucyl-tRNA synthetases (536 matches, 25.4% identity, oligopeptide TTRPETL shared) for which an unambiguous trace path could be visualized in the sensitive search matrix. Table I shows the number of errors for the five cases over four procedures using different gap penalties when appropriate. The acronyms for the techniques are ISSC (user interaction with the sensitive search matrix);[9] AUTO (automated trace path through the sensitive search matrix);[9] BESTFIT (automated alignment procedure relying on a modified Dayhoff matrix and the algorithm of Smith and Waterman[10] and found in the UWGCG program package);[11] and ALIGN (developed by Dayhoff and colleagues using the Dayhoff matrix and automated alignment.)[1]

It is clear from Table I that the sensitive search matrix used in the interactive or automated (lacking the advantage of various gap penalties)

[9] R. Rechid, M. Vingron, and P. Argos, *CABOIS* 5, 107 (1989).
[10] T. F. Smith and M. S. Waterman, *Adv. Appl. Math.* 2, 482 (1981).
[11] J. Devereux, P. Haeberli, and O. Smithies, *Nucleic Acids Res.* 12, 387 (1984).

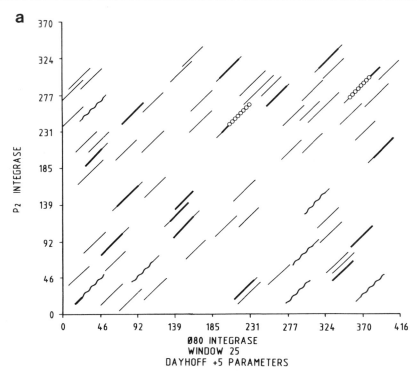

FIG. 1. (a) Search matrix, based on the Dayhoff and five parameter scores, for comparison of integrases from bacteriophages P2 and ϕ80. A constant search window length of 25 residues was used. The symbols indicate the number of standard deviations above the matrix mean for the search peak heights (S) according to the following scheme: $3.0v \leq S < 3.3v$ (thin lines); $3.3v \leq S < 3.7v$ (thick lines); $3.7v \leq S < 4.0v$ (wavy lines); $4.0v \leq S \leq 4.2v$ (circles). (b) As (a), except using windows 7 to 25 in steps 1 with symbol ranges of $3.7v \leq S < 4.0v$ (thin lines); $4.0v \leq S < 4.2v$ (thick lines); $4.2v \leq S \leq 4.6v$ (circles). It is clear that using many windows facilitates delineation of the alignment path by literally "filling it in."

modes yields alignments closest to those resulting from tertiary structural superposition. In the case of the synthetases, where the sensitive search matrix path was unambiguous, ALIGN and BESTFIT were substantially different. Functionally important and exactly conserved oligopeptides are not missed by the sensitive procedures while the other methods may or may not delineate them depending on the gap penalties (see Fig. 2 for an illustration). ALIGN, under all conditions, never discovered the TTREPTL of the synthetases. Furthermore, different gap penalties had to be utilized to achieve optimal alignments and to delineate the well-con-

b

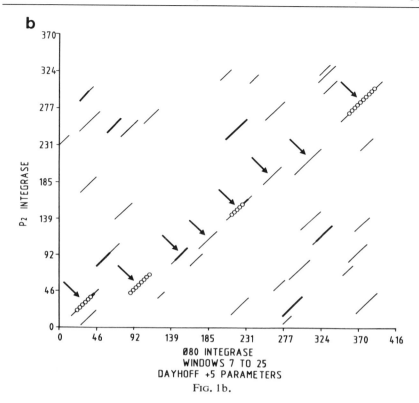

Fig. 1b.

served oligopeptides for the various examples. Choosing the appropriate values *a priori* is difficult without knowledge of tertiary structure, which is almost always not the case.

Multiple Sequence Alignment

Weak homologies may become visible with simultaneous alignments of several related sequences. This is accomplished by a multiple sequence alignment algorithm.

One natural approach to the multiple alignment problem is the generalization of the classic procedure described previously. A multidimensional search matrix is created with one sequence along each dimension and the Dayhoff mutation values entered at each matrix coordinate according to the two amino acids compared. To trace all possible paths which include

TABLE I
ERROR RATES FOR GIVEN SEQUENCE COMPARISONS[a]

Method	Proteases	Globins	Virus amino terminus	Virus amino terminus	tRNA synthetases
ISSC	111(Y)	8	12	12	0(Y)
AUTO *(GA, GB)*					
10, 1	122(Y)	8	19	60	106(Y)
BESTFIT *(GA, GB)*					
6.0, 0.4	162(N)	24	12	60	233(N)
5.0, 0.3	162(N)	24	12	60	168(N)
4.0, 0.2	123(N)	21	12	60	139(Y)
3.0, 0.1	112(N)	21	17	60	146(Y)
2.0, 0.0	123(Y)	21	14	60	194(Y)
ALIGN *(K, GA)*					
+2, 0	148(Y)	133	65	60	493(N)
+2, 10	127(Y)	19	15	60	200(N)
+2, 20	149(N)	18	12	60	179(N)
+10, 0	154(N)	54	33	60	430(N)
+10, 10	139(N)	22	16	60	474(N)
+10, 20	131(N)	25	12	60	464(N)
+20, 00	156(N)	40	59	60	470(N)
+20, 10	139(N)	22	12	60	468(N)
+20, 20	134(N)	25	12	60	454(N)

[a] K is the constant added to the Dayhoff matrix as allowed by the ALIGN program while GA and GB are the gap penalties to initiate and extend, respectively, an insertion/deletion. For proteases and tRNA synthetases, a pentapeptide (GDSGG) and heptapeptide (TTREPTL), respectively, were identical in the two compared sequences. An (N) next to an error rate indicates that the sequence identity was NOT recognized by the alignment technique at the given parameter settings while a (Y) indicates recognition. The error count was determined by subtracting the number of correct matches produced by the alignment scheme from the total number of alignment positions in the standard-of-truth alignment with gap sites not counted.

insertions and deletions becomes practically impossible, even with shortcuts (except for the three-sequence case[12]), on present-day computers.

Another approach builds multiple alignments from only pairwise comparisons of all the sequences.[13] In this case, the alignment of the two most related sequences is accepted using selected gap penalties. A third sequence, known to be most related to the previous sequences from all the

[12] M. Murata, J. S. Richardson, and J. L. Sussman, *Proc. Natl. Acad. Sci. U.S.A.* **82**, 3073 (1985).
[13] G. J. Barton and M. J. E. Sternberg, *J. Mol. Biol.* **191**, 153 (1987).

```
LeuSyn    (238)   WIGESIGAELVFKVADSKLENLIVFTTRPETLFAV
ValSyn1   (241)   RYKDLIGKYVILPLVNRRIP--IVGDEHADMEKGT
ValSyn2   (204)   LADGAKTADGK-------DYLVVATTRPETLLGD
ValSyn3   (201)   RYPLADGA----KTADGK-DYLVVATTRPETLLGD


LeuSyn            YVALALDHPIVQKYSE
ValSyn1           CVKITPAHDF------   (5.0,0.3)
ValSyn2           GVAVNPEDPRYKDLIG   (5.0,0.2)
ValSyn3           GVAVNPEDPRYKDLIG   (4.0,0.2)
```

FIG. 2. Alignment of particular regions in bacterial leucyl-(LeuSyn) and valyl-tRNA synthetases (ValSyn). The number in parentheses indicates the starting sequence position. The number following ValSyn indicates the respective alignments resulting from the program BESTFIT, with initiating (I) and extending (E) gap penalties given as (I,E) at the end of each line. It is clear that only a slight change in gap penalty yields very different alignments, even to the extent of missing seven consecutive and identical amino acids (TTRPETL).

pairwise comparisons, is aligned to the first two by the profile technique.[14] The set of previously aligned sequences is represented by a matrix, each column of which corresponds to a position in the underlying alignment. The number of rows is the number (20) of amino acids plus one added for the gap possibility. Every column of the profile matrix contains the distribution of the amino acids at the corresponding alignment position. A sequence is aligned to a set of matched sequences by creating a search matrix bounded by the sequence to be aligned and by the profile columns corresponding to the number of alignment positions in the sequence cluster. The score for matching a profile column with an amino acid from the sequence is calculated as the weighted sum over the Dayhoff values for the amino acid of the one sequence versus those occurring in the alignment position to be matched. Every search matrix value is weighted by the frequency of the residue as given in the profile column. All possible paths under given gap penalties can now be traced through the profile versus the one-sequence search matrix and the optimal one found. A new profile matrix is then calculated from the three aligned sequences, and then a fourth sequence can be compared. The procedure is repeated until the multiple alignment is effected.

Though the iterative pairwise alignments and profile analyses are clever and useful, they are flawed. Different gap penalty choices once gain result in different alignments, which for weaker homologies are difficult to choose among. The final alignment is dependent on sequence order; for

---

[14] M. Gribskov, A. D. McLachlan, and D. Eisenberg, *Proc. Natl. Acad. Sci. U.S.A.* **84,** 4355 (1987).

example, once gaps are introduced from the first few sequences, they can never be eliminated or modified with the addition of further sequences. Furthermore, there is no assurance that the early matched sequences will be correctly aligned in all local regions.

The authors have developed a multiple sequence alignment technique that avoids or mitigates some of these problems.[15] The dipeptides contained in each sequence are distinguished and counted. Those sequences that show a high count of the same 2-mers are clustered; they should be closely related. Within each cluster the sequences are aligned so that the largest number of dipeptides are brought into multiple coincidence for all the clustered subsets with preservation of alignment progression. For example, suppose the sequences MLRKENRKSELAGYFGG, RKRTTE-LYDGG, and MISTRKEMGGSTELLGFGG are to be aligned. It is clear that dipeptides RK, EL, and GG will allow the best progressive match of the sequences:

$$ML[RK]ENRKS-[EL]AGYF[GG]$$
$$MIST[RK]EMGGST[EL]LG-F[GG]$$
$$[RK]---RTT[EL]--YD[GG]$$

Aligning the carboxyl-terminal [GG] of the first sequence with the first-occurring (amino-terminal) one in the second sequence would of course not be performed. The regions bordered by the 2-mers are then matched from longest to shortest (e.g., EMGGST with ENRKS with RTT); this minimizes gaps. The first two (EMGGST and ENRKS) are matched by considering all possible paths in a classic search matrix dependent on Dayhoff mutation values and chosen gap penalties. The third is fitted to the profile of the first two, and so forth. The procedure is repeated for all intervals within a given cluster and then for all clusters.

The two clusters sharing the greatest number of dipeptides are joined by aligning each of their profiles. Two profiles are compared by once again calculating a search matrix where the columns correspond to the alignment positions of one sequence set and consist of their amino acid compositional distribution, and the rows are the similar values for the other sequence cluster. The score at each search matrix coordinate is the sum, over all possible amino acid comparisons, of the product of the residue compositions in each sequence cluster, the Dayhoff value for the residues compared, and the weights associated with each sequence in their respective clusters. All possible paths are traced in the resultant search matrix and the two profiles aligned. A new profile is then determined from the alignment of the two previous profiles. The next sequence cluster is added according to the dipeptide count, and its profile is aligned with the just created one.

[15] M. Vingron and P. Argos, *CABIOS* 5, 115 (1989).

Then a more extensive profile is calculated from the three clusters and the procedure repeated until all sequences are matched. The process can be performed even if a subset contains only one sequence.

Weights are attached to each sequence in a given alignment and can be used when calculating scores in the comparison of two profiles. A comparatively high number of similar sequences would normally bias a profile. A single sequence distantly related to the rest would contribute only marginally. The need to correct for underrepresentation of sequences arises. The total number of mismatches between one sequence and the remaining sequences at each alignment position is a useful indicator. A high number of mismatches means that the sequence is distant from the average sequence of the profile, while mismatch counts similar for all the sequences indicate a homogeneous set. After counting the mismatches for each sequence, weights are determined through division of each of the counts by the smallest one. For instance, if an alignment cluster consisted of two identical sequences and a very different one, the corrected profile would attach weights 1, 1, and 2 for the three sequences.

This approach is not so order dependent as previous methods owing to the preclustering of the sequences. Furthermore, aligning profiles is more accurate than aligning single sequences or one sequence to a profile. In a comparison of 9 azurins with 15 plastocyanins, where the standard-of-truth match contained 94 match positions not counting gaps, only 36 errors were encountered in comparing the two profiles, while relating the two single sequences with known tertiary structure yielded, under the same gap penalties, 62 errors. Even if the single sequence comparisons were allowed several possible gap penalties, the best error rate was 43. Profile comparison has also been shown to recognize more easily conserved catalytic regions despite the use of different gap weights. In this case the reinforcement of multiple sequences containing a conserved pattern allows easy recognition by profile fitting between sequence sets. Weighting of the residue count in a given profile column according to the closeness of the sequences also improves sensitivity. For the azurin–plastocyanin comparison, a 30% improvement was observed over the unweighted procedure.[15] Aligning only the intervals between dipeptide anchors increases greatly the speed of the method. Nonetheless, it still suffers from the required setting of gap penalties and certain cut-off scores, such as those used for clustering.

Significance of Alignment

If the identity level is above about 30% or so with uniform distribution over the sequence alignment, and there are no long insertions or deletions, meaningfulness is obvious to the eye and by several mathematical criteria.

For most other cases, significance estimation is not straightforward. Certain regions of the protein sequences may be aligned well, while others can hardly be matched. An alignment may be reasonable over one relatively short stretch (25–50 residues) and yet cannot be extended. Sometimes four or more amino acids show a perfect or near-perfect consecutive match, and yet the remaining regions do not align well. In other cases, the entire sequences can be aligned, but with only 15–30% identity.

Three approaches have generally been used to assess alignment quality: (1) Monte Carlo methods based on random shuffling of the two sequences and matching by the given procedure, (2) analytical formulas derived for particular alignment algorithms to yield expected match quality in shuffled sequences, and (3) use of real but unrelated protein sequences in large reference samples for analysis. In all cases a mean assessment and standard deviation (v) are calculated from the controls. For longer sequence alignments (say, greater than 50 residues), the literature consensus seems to state that greater than $3v$ is necessary for consideration of the match as "possible." Around $5v$ or $6v$ it becomes probable, and above $10v$, "certain." It has been found that scores greater than $6v$ yield 75% correct alignments in helices and β strands of known tertiary folds.[13] Application of biological knowledge is also important. If the proteins share some function, a sequence relationship is more likely.

Of course, no protein would use a sequence of randomly selected letters; careful engineering and bias prevail.[16] Thus, significance estimates relying on real sequences are more trustworthy. The authors, along with Rechid,[9] have run several unrelated sequence pairs through the sensitive technique, produced search matrices and associated automated alignments using weaker gap penalties to emphasize the use of large search scores, and determined that the average correlation coefficient for five residue properties, over all alignment positions not including gaps, is $+0.282$ (not 0.000 as random shuffling would yield). This value is subtracted from the mean correlation of the alignment to be tested and the result divided by the standard deviation of the trial scores, which depends on the alignment length. Then, in a strict sense, if the answer is greater than $2v$ (the 95% confidence level), the alignment may well be significant. If the biological connection is compelling, comparisons at a significance level less than $2v$ may even be considered. Obviously, the need to be greater than $3v$ for the typical control test is a result of the artificial shuffling of the sequences.

With any questionable homology it is always important to examine visually the search matrix calculated over several window sizes. The authors have encountered cases where the same percent identity and the same

[16] P. McCaldon and P. Argos, *Proteins* **4**, 99 (1988).

standard deviation significance have been achieved, yet the search matrices do not prove equally convincing.[17] With mere credibility scores, the distribution of high peaks along the entire path cannot be seen and the number and size of noise peaks relative to match peak cannot be assessed; the presence of weak peaks that fill a homology path well and that are unlikely to occur by chance must be visualized. The high credibility score may result from only a small fraction of the entire alignment region, leaving most of the matches in doubt. Examination of the search matrix can alleviate these pitfalls. Perhaps the best and only significance test is the observance of a good distribution of peaks collinear with the search matrix diagonal.

Despite biological and statistical correlations, mistakes can still be made. Two nucleotide binding domain sequences from alcohol and lactate dehydrogenases were compared.[8] A reasonable search matrix path resulted. Both sequence regions were known to bind NAD, yet the alignment (Fig. 3) was nearly two-thirds incorrect relative to the tertiary structural superposition, which achieved only 12% identity, half that from the computer results with about the same number of insertions/deletions.

Available Computer Programs

Program packages executing the sensitive procedure in comparing two sequences and the multiple sequence alignment technique are available from the authors.[8,9,15] Interested researchers can write to the authors; send a FAX request (0049-6221-387306); or relay a computer network message (ARGOS@EMBL for the sensitive procedure and VINGRON@EMBL for the multiple alignment method, both on BITNET). A magnetic tape containing the source code and executable files will be sent to academic institutions.

The sensitive protein sequence alignment package[9] consists of a main program (in C or Fortran languages) that calculates search matrix values for one or more sequence pairs. The user can choose the sequences, comparison matrix (Dayhoff or identity or whatever), residue characteristics, and probe lengths. The resultant files can be fed to various satellite programs that utilize the search matrix. The user can visually interact with the matrix by choosing peak lines and alignment paths with a "mouse" device. Sequence intervals not matched by an observable path can be automatically aligned. The alignment, its significance assessment, and the size of the matrix peaks utilized are all shown. The interactive programs are written for SUN[2,3,4] workstations or microVAXs (II or 3000 series)

---

[17] P. Argos and S. D. Fuller, *EMBO J.* **7**, 819 (1988).

```
LADH     Q G S T C A V F G L G G V G L S V I M G C K A A G - A A R I I G V D I
LDH(1)   S Y N K I T V V G V G A V G M A C A I S I L M K D L A D E V A L V D V
LDH(2)   S Y N K I T V V G V G A V G M A C A I S I L M K D L A D E V A L V D V


LADH     N K D K - - - - - - - - - - F A K A K E V G A T E C V N P Q D Y K K P
LDH(1)   M E D K L K G E M M D L - - Q H G S L F L H T A K I V S G K D Y S V S
LDH(2)   M E D K L K G E M M D L Q H G S L F L H T A K I V S G K D Y S V - - -


LADH     I Q E V L T E M S N G G V D F S F E V I G - - - - - - - - - - - - - -
LDH(1)   A G S K L V V I T A G A R Q Q E G E - - S - - - - - - - - - - - - - -
LDH(2)   - - - - - - - - - S A G S K L V V I T A G A R Q Q E G E S R L N L V Q


LADH     - - - R L D T M V T A L S C C Q E A Y G V S V I V G V P P D S Q N L S
LDH(1)   - - - R L N L V Q R N V N I F K F I F I I P N I V K H S P D C I I L V
LDH(2)   R N V N I F K F I I P N I V K H S P D C I I L V V S N P V D V L T Y V


LADH     M - N P M L L L S G R T W K G A I - - - - - - - - - -
LDH(1)   V S N P V D V L T Y V A W K L S G L P M H R I I G S S
LDH(2)   A - W K L S G L P M H R I I G S S
```

FIG. 3. Alignment of the NAD-binding domain from horse liver alcohol dehydrogenase (LADH) with that from dogfish lactate dehydrogenase (LDH) as determined by the tertiary structural superposition [LDH(2)] and by the sensitive search matrix [LDH(1)]. The bracketed arrows show the initiation site where the two alignments do not agree. Residues are boxed in LDH(1) and LDH(2) if they are conserved or identical with those in LADH. The computer search yielded 18 and 34 identical and conserved residues, respectively, while the structural superposition ("standard of truth") resulted in only 7 and 16.

which have graphics capability; the computer language used is C. If interaction (recommended) is not possible, then routines exist (variously written in C or Fortran) that provide a laser writer output (postscript format) of the search matrix, an automatic sequence alignment based on the sensitive search matrix and selected gap penalties, and alignment assessment. The main program requires some computer time and should be used to discover weak homologies (< 35% identity), to extend relationships limited in sequence alignment length, or to delineate matches in sequence intervals.

The multiple sequence alignment package,[15] written in the C language, consists of two main programs. The first simply accepts whatever sequences are given and automatically clusters them (if desired), determines the optimal set of dipeptide anchors for each cluster, then aligns intervals between anchors for each cluster, and finally outputs each of the aligned clusters successively according to their closeness. The output files containing the alignments of the first two sequence sets are presented to the second program, which aligns them by the profile method. The resultant output and the third sequence set are then given to the same program, yielding the three-cluster match. The process is repeated as necessary, even if a single sequence constitutes a cluster. The user may select values for certain parameters: (1) the two gap penalties for initiation and extension of insertions/deletions; (2) ON/OFF end gap setting if the sequences vary greatly in length; (3) the scoring matrix; (4) the oligopeptide length for determining anchor points and clustering (dipeptides for most cases and tripeptides if the sequences are closely related); (5) threshold oligopeptide match count to determine the clusters; (6) ON/OFF use of the weighting scheme in the profile analysis to avoid compositional bias in closely matched sequences; and (7) maximum difference between sequence positions within a matching set of oligopeptide anchors. The programs work quickly unless the sequences are particularly long and numerous.

[22] Three-Way Needleman – Wunsch Algorithm

By Mitsuo Murata

Introduction

In 1970 Needleman and Wunsch developed an elegant algorithm for aligning two protein sequences.[1] The algorithm finds the optimal alignment among all possible alignments that can be generated from the two sequences. Since then, the method has been a valuable tool in molecular biology. Properly aligned sequences, for example, often reveal structurally and functionally important regions of proteins such as catalytic sites and ligand binding sites. To find structurally equivalent residues from the best aligned sequences is the first essential step in predicting the tertiary structure of a protein based on a homologous protein whose three-dimensional structure is known.

Obviously, information from sequence comparisons becomes more reliable as we increase the number of sequences to be compared. To align more than two sequences is easy, if they are from closely related proteins; they may be so similar that a good alignment can be obtained visually or by manually adjusting the alignments from pairwise comparisons. When sequence similarity is less obvious, however, the manual approach becomes very tedious and the results more subjective. Such a situation was encountered when three copper-containing proteins, cucumber basic blue protein, stellacyanin, and plastocyanin, were compared. The difficulty of obtaining a consistent alignment of these sequences by pairwise comparisons moti-

---

[1] S. B. Needleman and C. D. Wunsch, *J. Mol. Biol.* **48**, 443 (1970).

vated us to extend the method of Needleman and Wunsch so that three sequences could be compared simultaneously.[2]

This chapter concerns specifically the algorithm of Needleman and Wunsch and its extension to a practical three-way comparison method. For other multisequence comparison methods, readers are referred to the articles (and references therein) by Sankoff and Cedergren[3] and by Waterman[4] and to the relevant chapters in this volume. Our original program compared three relatively short protein sequences. Modifications to the program for the comparison of longer sequences are described in this chapter. The modified version of the program called ALIGN3, written in Digital Equipment Corporation's (DEC) VAX-11 FORTRAN, is available from the author on request.

Needleman–Wunsch Algorithm for Pairwise Comparisons[1]

Let A and B be the two sequences with lengths m and n, and denote by $A(i)$ and $B(i)$ the ith residues in the respective sequences. Suppose that we assign to every possible residue pair from the two sequences a weight, wt, which reflects the similarity between the two residues:

$$wt(i,j) = \text{weight}[A(i),B(j)]$$

It is apparent that, in the optimal alignment, the sum of the weights which are assigned to the paired residues is maximum. The weights are taken from a suitable 20×20 weight matrix such as the one based on the genetic code, the similarity matrix of McLachlan,[5] or the mutation data matrix (MDM78) of Dayhoff et al.[6] To find the maximum sum of weights, the Needleman–Wunsch algorithm constructs a two-dimensional array, here denoted by L, in which the cell (i,j) corresponds to the residue pair $A(i),B(j)$ for $i = 1, \ldots , m, j = 1, \ldots , n$. The goal is to find a path in L along which the sum of wt values assigned to the cells in the path is maximum; Needleman and Wunsch called such a path the maximum

[2] M. Murata, J. S. Richardson, and J. L. Sussman, *Proc. Natl. Acad. Sci. U.S.A.* **82,** 3073 (1985).

[3] D. Sankoff and R. J. Cedergren, *in* "Time Warps, String Edits, and Macromolecules: The Theory and Practice of Sequence Comparison" (D. Sankoff and J. Kruskal, ed.), p. 253. Addison-Wesley, London, 1983.

[4] M. S. Waterman, *Bull. Math. Biol.* **46,** 473 (1984).

[5] A. D. McLachlan, *J. Mol. Biol.* **61,** 409 (1971).

[6] M. O. Dayhoff, R. M. Schwartz, and B. C. Orcutt, *in* "Atlas of Protein Sequence and Structure" (M. O. Dayhoff, ed.), p. 345. National Biomedical Research Foundation, Washington, D.C., 1978.

match pathway. We note that the two consecutive cells in a path are (i,j) and $(x,j+1)$ for $i < x \le m$ or (i,j) and $(i+1,y)$ for $j < y \le n$.

The summation of weights is carried out backward from the last cell (m,n), either row by row or column by column, in the following manner:

$$L(i,j) = wt(i,j) + \max\{L(i+1,j+1), L(x,j+1) - g, L(i+1,y) - g\}$$
$$(i < x \le m, j < y \le n)$$

The value of $L(i,j)$ is the sum of $wt(i,j)$ and the maximum value of already calculated cells that can be next to cell (i,j). In order to avoid an excess number of gaps in the final alignment, a penalty, g, is subtracted from the maximum value contribution unless it is from the cell $(i+1,j+1)$. Calculated in this way, $L(i,j)$ contains the maximum sum of weights for all possible partial sequence alignments of A and B starting from i and j, and ending at m and n. Once the summation is completed on all the cells, optimally aligned residues can be found by tracing back the path starting at the cell with the maximum value in the entire L. Because of the above condition for two consecutive cells in a path, the path so traced is the maximum match pathway.

Extension of Needleman–Wunsch Algorithm to Three-Way Method[2]

To extend the above Needleman–Wunsch algorithm to align three sequences, we introduce the third sequence, C, with p residues. For the value of wt, the sum of the pairwise weights is used:

$$wt(i,j,k) = \text{weight}[A(i),B(j)] + \text{weight}[B(j),C(k)] + \text{weight}[A(i),C(\text{k})]$$

The summation of weights is carried out in a three-dimensional array, again denoted by L, in which each cell is represented by (i,j,k) for $i = 1, \ldots, m, j = 1, \ldots, n, k = 1, \ldots, p$. An important aspect in the extension to a three-way method is that the cell consecutive to (i,j,k) now resides in one of the three planar regions: $(x,y,k+1)$ for $i < x \le m, j < y \le n$; $(x,j+1,z)$ for $i < x \le m, k < z \le p$; and $(i+1,y,z)$ with $j < y \le n, k < z \le p$. We used a term "subshell" to indicate these regions. Let the subshell of (i,j,k) be the set of cells in the three planar regions

$$\{(x,y,k); i \le x \le m, j \le y \le n\}$$
$$\{(x,j,z); i \le x \le m, k \le z \le p\}$$
$$\{(i,y,z); j \le y \le n, k \le z \le p\}$$

Then the above cell consecutive to (i,j,k) is in the subshell of $(i+1,j+1,k+1)$.

The three-way version of the calculation of L is

$$L(i,j,k) = wt(i,j,k) + \max\{L(i+1,j+1,k+1), L(x,y,k+1)$$
$$-g, L(x,j+1,z) - g, L(i+1,y,z) - g\}$$
$$(i < x \le m, j < y \le n, k < z \le p)$$

As in the pairwise comparison, the calculation of maximum value proceeds from the last cell (m,n,p) to the origin $(1,1,1)$. Again, a gap penalty, g, is subtracted from the maximum value contribution unless it is from the cell which is at the position diagonal to (k,j,k). This straight extension of the pairwise method implies that for the calculation of each $L(i,j,k)$ we must scan the subshell of $(i+1,j+1,k+1)$ to find the maximum value. The scanning of the subshells is, however, very time consuming, and the scheme is impractical even for the comparison of short (~ 100 residues) sequences.

One way to reduce the amount of computing time is to store the maximum values of the subshells in another three-dimensional array, denoted here by Q. Briefly, let $Q(i,j,k)$ contain the maximum value of the subshell of (i,j,k). Then $L(i,j,k)$ can be calculated from $wt(i,j,k)$ and $Q(i+1,j+1,k+1)$, and as $L(i,j,k)$ is calculated, $Q(i,j,k)$ is successively updated from the maximum of $L(i,j,k)$, $Q(i+1,j,k)$, $Q(i,j+1,k)$, and $Q(i,j,k+1)$. The summation process according to this scheme will take time proportional to l^3 instead of l^5 of the scanning method, where l is the geometric mean of m, n, and p.

The drawback of this method is, however, the use of another three-dimensional array, which is too expensive in terms of computer memory usage. Note, however, that Q serves only as a temporary storage. Therefore, it is not necessary to keep all values of Q all the time. Our solution to the memory problem is to reduce Q to a two-dimensional array, $Q1$, as in the following algorithm.

Algorithm 1:
```
for i := m downto 1 do
for j := n downto 1 do
for k := p downto 1 do
begin
    L(i,j,k) := wt(i,j,k) + max( L(i + 1,j + 1,k + 1),
                    Q1(j + 1,k + 1) − g );
    if j < n and k < p then
        Q1(j + 1,k + 1) := max( L(i,j + 1,k + 1), Q1(j + 1,k + 1),
                    Q1(j + 2,k + 1), Q1(j + 1,k + 2) )
end.
```

The values of $L(i,j,k)$ and $Q1(j,k)$ are initially set to 0 for $i > m$, $j > n$, and

$k > p$. Here, $Q1(j + 1,k + 1)$ holds successively the maximum value in the subshell of (i,j,k). The order by which $i, j,$ and k are operated and the inclusion of the ith term (instead of $i + 1$) in $L(i,j + 1,k + 1)$ in the calculation of a new $Q1(j + 1,k + 1)$ ensure that the value of $Q1(j + 1,k + 1)$ is continuously updated. Figure 1 explains the algorithm in more detail. The space and time required by Algorithm 1 are proportional to $l^2(l + 1)$ and l^3, respectively.

The tracing of the best path in L to obtain the optimal alignment is carried out as follows:

Algorithm 2:
    $(r,s,t) := (1,1,1)$;
    while $r \leq m$ and $s \leq n$ and $t \leq p$ do
    begin
        $(x,y,z) := (r,s,t)$;   {best location yet}
        $b := L(r,s,t)$;   {best value so far}
        $d := 0$;   {shortest distance}
        for each (i,j,k) in subshell of (r,s,t) do
            if $L(i,j,k) - g > b$ or $L(i,j,k) - g = b$
                and distance from (r,s,t) to $(i,j,k) < d$) then
            begin
                $(x,y,z) := (i,j,k)$;
                $b := L(i,j,k) - g$;
                $d :=$ distance from (r,s,t) to (i,j,k)
            end;
        print (x,y,z);
        $(r,s,t) := (x + 1,y + 1,z + 1)$
    end.

In this tracing algorithm a penalty is imposed for a gap occurring at the beginning of the path but not for one at the end of the path. Since it is desirable to allow the residues at both terminals to be aligned without penalty, the algorithm is modified in the actual program by giving $g = 0$ if $r = 1$ or $s = 1$ or $t = 1$.

Implementation of Gap Penalty

In the above description of the algorithms, a constant gap penalty is used regardless of the size of gaps. Fitch and Smith advocated, from the study of short chicken hemoglobin sequences, the use of a length-dependent gap penalty in the form of $g + rn$, where g is the penalty for having the gap, r is the weight for a single position in the gap, and n is the number of

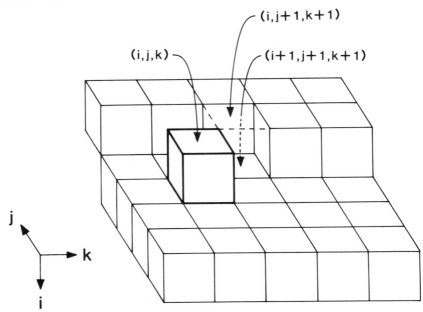

FIG. 1. Illustration of Algorithm 1 at the point where the value of L for the cell (i,j,k) (shown with heavily outlined edges) is calculated. Cells with lightly outlined edges are those for which the array $Q1$ holds the current value of Q just before the assignment to $Q1(j + 1,k + 1)$. After the assignment, $Q1(j + 1,k + 1)$ holds the value of Q for the cell $(i,j + 1,k + 1)$, indicated by dashed edges. As the depiction suggests, one may think of the array $Q1$ as holding the values of Q for a layer of cells with a moving ridge or fault line in it: the dashed cell is about to replace the one below it as part of this layer. (Reproduced from Ref. 2.)

residues in the gap.[7] They concluded that the correct alignment can be obtained only when a length-weighted gap penalty is used. In contrast, the Needleman–Wunsch algorithm with a constant gap penalty can produce the correct alignment of the same two hemoglobin sequences, as shown in Fig. 2a.[8] Barton and Sternberg also observed that the use of a length-dependent gap penalty is not necessary.[9] In fact the use of a length-dependent penalty may not always produce the optimal alignment, particularly when a very large gap is involved. Figure 2b shows the alignment of the α subunit of casein kinase II from *Drosophila melanogaster*[10] with that of

---

[7] W. M. Fitch and T. F. Smith, *Proc. Natl. Acad. Sci. U.S.A.* **80**, 1382 (1983).
[8] The pairwise comparison program used here, which is based on the Needleman–Wunsch algorithm, is available from the author.
[9] G. J. Barton and M. J. E. Sternberg, *Protein Eng.* **1**, 89 (1987).
[10] A. Saxena, R. Padmanabha, and C. V. C. Glover, *Mol. Cell. Biol.* **7**, 3409 (1987).

a

```
Weighting System : MDM78
Gap Penalty = 6

    1    FASFGNLSSPTAILGNPMV        19
    1    FPHF DLSH       GSAQI      13
         *   *   **        *

No. of aligned residues = 13
No. of matched residues = 5
Alignment Score =    129

sample size =    100
mean =    112.79
standard deviation =        6.59
standardized score =        2.46
```

b

```
  61    DSKVKIVIKMLKPVKKKKIKREIKILTDLSNEKVPPTTLPFQKDQYYTNQKEDVLKFIRP    120
  56    TTTEKCVVKILKPVKKKKIKREIKILENLR                                  87
        *  *  *  ****************  *

 121    YIFDQPHNGHANIIHLFDIIKDPI    144
  88            GGTNIITLLAVVKDPV    103
        *  ***  *       ***
```

FIG. 2. Sequence alignments obtained by the Needleman–Wunsch algorithm using length-independent gap penalties. (a) Alignment of chicken β- (upper) and α- (lower) hemoglobin segments. The same optimal alignment can be also produced with gap penalties of 4 and 5. (b) A portion of aligned sequences from the α subunit of casein kinase II from yeast (upper) and that of *D. melanogaster*. The gap penalty used is 8.

yeast[11] using the Needleman–Wunsch algorithm with a constant gap penalty of 8. The same kinase sequences were compared using length-dependent gap penalties derived from all the combinations of g and r values 1, . . . , 10, but none of the alignments produced the long deletion shown in Fig. 2b. Obviously, the constraint imposed by even the smallest penalty factors, $g = 1.0$, $r = 1.0$, becomes too great to create the long gap which allows the alignment of the two conserved segments on both sides of the gap.

The use of a gap penalty weighted for the residues involved in the secondary structure of one of the proteins seems to improve the quality of alignments.[9,12] The problem here, of course, is that accurate information on the secondary structure is not always available. When such information is available, however, the use of this method is recommended, as it is easily implemented in computer programs.

[11] J. L.-P. Chen-Wu, R. Padmanabha, and C. V. C. Glover, *Mol. Cell. Biol.* **8**, 4981 (1988).
[12] A. M. Lesk, M. Levitt, and C. Chothia, *Protein Eng.* **1**, 77 (1986).

Similarly, weights in the scoring system can be modified according to the importance of individual amino acids in the proteins in question. In the comparison of blue copper proteins mentioned above, for example, the quality of alignment was improved by increasing the weights for the metal-binding amino acids, namely, Cys, His, and Met.[2]

Programming Considerations

The first step in implementing the above three-way algorithm in a computer program is to convert each amino acid sequence to a sequence of integers which correspond to the entry positions of 20 amino acids in the weight matrix used. In the MDM78 matrix, for example, the weights are entered according to the following order of amino acids: C,S,T,P,A,G,N,D,E,Q,H,R,K,M,I,L,V,F,Y,W. Therefore, if sequence A is ACD . . . , the corresponding integer sequence, A', is 5,1,8,. . . . The calculation of wt in Algorithm 1 is carried out cheaply each time when needed from the weight matrix; for example, the following FORTRAN statement function can be used:

$$wt(i,j,k) = weight(A'(i),B'(j)) + weight(B'(j),C'(k)) + weight(A'(i),C'(k))$$

Thus, no storage is required for wt.

The tracing process described in Algorithm 2 takes time proportional to l^3. There is an alternative that takes much less time. As Needleman and Wunsch suggest in their original paper (Ref. 1, p. 446), if the location of the cells which contributed the maximum values in the calculation of L in Algorithm 1 is recorded in another storage, we could trace the overall best path in time proportional to l. A similar method has been adopted by Gotoh.[13] However, because of the additional storage requirement, we have decided not to use this approach; instead, the same matrix L is used to trace the maximum path.

As pointed out by Johnson and Doolittle,[14] the path tracing described in Algorithm 2 prints out only triply aligned residues and does not list the two residue alignments opposite to a gap. In our original program, doubly aligned residues were inserted manually in the final alignment. It is, of course, possible, with a small expense of computing time, to trace L so that doubly aligned residues are automatically printed out. Instead of tracing L, however, the current version of the program uses a subroutine program for pairwise comparisons to maximize the alignment of the residues in these segments.

[13] O. Gotoh, *J. Theor. Biol.* **121**, 327 (1986).
[14] M. S. Johnson and R. F. Doolittle, *J. Mol. Evol.* **23**, 267 (1986).

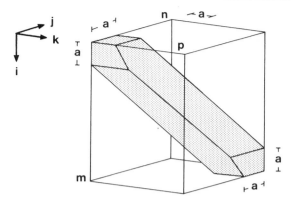

Fig. 3. The cells in the dotted region are included in the matrix calculations in ALIGN3. The region indicated can be imagined as being created by cutting corners of L with two pairs of parallel planes, one perpendicular to the $i-j$ face and the other perpendicular to the $i-k$ face. The upper and lower bounds of j with respect to i are $\min\{\lfloor u(i-1)+a\rfloor,n\}$ and $\max\{\lceil u(i-a)+1\rceil,1\}$, respectively, where $u=(n-a)/(m-a)$. Similarly, the upper and lower bounds of k with respect to i can be obtained by replacing n with p in the formulas.

The Needleman–Wunsch algorithm for the comparison of more than two sequences requires a large amount of computer memory. The three-way method described above can compare relatively small proteins with commonly used computers; three sequences each up to 190 residues can be compared on our computer, the DEC VAX11/750 allocated with 4096 working set pages of main memory (2 megabytes) and with 30,000 pages of virtual memory (15.36 megabytes). The same algorithm can be employed for the comparison of longer sequences by carrying out the matrix calculations for a limited number of cells in L. A shortcut method similar to the one described by Kruskal and Sankoff can be used.[15] Figure 3 shows the cells that are included in the matrix calculations in ALIGN3. In the actual program, the reduction in the number of cells is effected simply by setting boundaries for the j- and k-loop operations in Algorithm 1. In ALIGN3, a is set to 40, and the values of m, n, and p are chosen such that $m \geq n$, $m \geq p$, thus a maximum of 80×80 cells are calculated on each ith plane. The amount of memory saved depends on the size of a. With the above memory configuration of the computer we use, ALIGN3 compares three sequences, each up to 900 residues. Note that, although not used in ALIGN3, a further small saving in memory is possible if we similarly cut L

[15] J. B. Kruskal and D. Sankoff, *in* "Time Warps, String Edits, and Macromolecules: The Theory and Practice of Sequence Comparison" (D. Sankoff and J. Kruskal, ed.), p. 276. Addison-Wesley, London, 1983.

```
(Top)       ATPASE BETA CHAIN - ESCHERICHIA COLI
(Middle)    ATPASE BETA CHAIN - BOVINE MITOCHONDRIA
(Bottom)    ATPASE ALPHA CHAIN - BOVINE MITOCHONDRIA

            Gap penalty used:    16
            Maximum score:   13384


                 +                      +            +
   1        MATGKIVQ  VIGAVVDVEFP        QDAVPRVYDALEVQNGN    E    37
   1   QASPSPKAGATTGRIVA  VIGAVVDVQF     DEGLPPILNALEVQGRE    T    45
   1   EKTGT    AEVSSILEERILGADTSVDLEETGRVLSIGDGIARVHGLRNVQAEEMVEFS     56
            . ...  *.  ..**...* .         .. ...  ...**  .


             +          +          +          +          +          +
  38   RLVLEVQQQLGGGIVRTIAMGSSDGLRRGLDVKDLEHPIEVPVGKATLGRIMNVLGEPVD     97
  46   RLVLEVAQHLGESTVRTIAMDGTEGLVRGQKVLDSGAPIRIPVGPETLGRIMNVIGEPID    105
  57   SGLKGMSLNLEPDNVGVVVFGNDKLIKEGDIVKRTGAIVDVPVGEELLGRVVDALGNAID    116
       ....... . *.   *......   .. ..* *... .... .*** ..***.....*...*


             +          +          +          +          +          +
  98   MKGEIGEEERWAIHRAAPSYEELSNSQELLETGIKVIDLMCPFAKGGKVGLFGGAGVGKT    157
 106   ERGPIKTKQFAAIHAEAPEFVEMSVEQELLVTGDKVVDLLAPYAKGGKIGLFGGAGVGKT    165
 117   GKGPIGSKARRRVGLKAPGIIPRISVREPMQTGIKAVDSLVPIGRGQRELIIGDRQTGKT    176
       .*.*. .    ** .     .. .*.. **.*..*.. * ..*.....*....***


             +          +               +          +          +
 158   VNMMELIRNIAIEHSG        YSVFAGVGERTREGNDFYHEMTDSNVID       KVS   204
 166   VFIMELINNVAKAHGG        YSVFAGVGERTREGNDLYHEMIESGVINLKDATSKVA    218
 177   SIAIDTIINQKRFNDGTDEKKKLYCIYVAIGQKRSTVAQLVKRLTDADAMK       YTI   230
       . ...** * . .*         *......*.................  ..       ..


          +          +          +          +          +          +
 205   LVYGQMNEPPGNRLRVALTGLTMAEKFRD EGRDVLLFVDNIYRYTLAGTEVSALLGRMP    263
 219   LVYGQMNQPPGARARVALTGLTVAEYFRDQEGQDVLLFIDNIFRFTQAGSEVSALLGRIP    278
 231   VVSATASDAAPLQYLAPYSGCSMGEYFRD NGKHALIIYDDLSKQAVAYRQMSLLLRRPP    289
       .*.....  ...  ....*....*.***  .* ..*.. *...  . *. ..*.**.* *


             +          +               +          +          +
 264   SAVGYQPTLAEEMGVLQERITSTKT       GSITSVQAVYVPADDLTDPSPATTFAHLDAT    319
 279   SAVGYQPTLATNMGTMQERITTTKK       GSITSVQAIYVPADDLTDPAPAVTFAHLDAT    334
 290   GREAYPGDVFYLHSRLLERAAKMNDAFGGGSLTALPVIETQAGDVSAYIPTNVISITDGQ    349
       ....*......  .. ..**.. ..        **.*.........*.*..... *. .....*..


             +          +          +          +          +          +
 320   VVLSRQIASLGIYPAVDPLDSTSRQLDPLVVGQEHYDTARGVQSILQRYQELKDIIAILG    379
 335   TVLSRAIAELGIYPAVDPLDSTSRIMNPNIVGSEHYDVARGVQKILQDYKSLQDIIAILG    394
 350   IFLETELFYKGIRPAINVGLSVSR VGSAAQTRAMKQVAGTMKLELAQYREVAAFAQFGS    408
       .*.. .. .**.**.....*.**  .. ...*.....*. * ........


             +          +          +          +          +          +
 380   MDELSEEDKLV  VARARKIQRFLSQPF  FVAEVFTGSPGKYVSLKDT IRGFKGIM    432
 395   MDELSEEDKLT  VSRARKIQRFLSQPF  QVAEVFTGHLGKLVPLKET IKGFQQIL    447
 409   DLDAATQQLLSRGVRLTELLKQGQYSPMAIEEQVAVIYAGVRGYLDKLEPSKITKFENAF    468
       .........*    . ....*.........  .**.....* *....*.. * .*..*.  .


             +          +          +
 433   EGEYDHLPEQAFYMV   GSI EEAVEKAKKL           460
 448   AGEYDHLPEQAFYMV   GPI EEAVAKADKLAEEHS       480
 469   LSHVISQHQALLGKIRTDGKISEESDAKLKEIVTNFLAGFE   509
       ..............  * * **...*....
```

FIG. 4. Alignment of three ATPase sequences.

in Fig. 3 by a pair of parallel planes perpendicular to the $j-k$ face. In this case, however, the cells included within the j and k boundaries for each ith plane are not always confined in a rectangular region. Consequently, the tracing process of Algorithm 2 becomes more elaborate and time consuming.

Application Example

As an example of the application of the above three-way algorithm, an alignment of three ATPases produced by ALIGN3 is shown in Fig. 4. The sequences used were the α and β subunits of bovine mitochondria[16] and the β subunit from *Escherichia coli.*[17] The MDM78 weighting system was used with a constant 8 added to make all weights nonnegative. The CPU time required for the matrix summation was 5 min 4 sec and that for the tracing process was 45 sec on the DEC VAX11/750.

Acknowledgments

This work was supported in part by a grant from the National Institutes of Health, GM28139 (to J. W. Lee).

[16] J. E. Walker, I. M. Fearnley, N. J. Gay, B. W. Gibson, F. D. Northrop, S. J. Powell, M. J. Runswick, M. Saraste, and V. L. J. Tybulewicz, *J. Mol. Biol.* **184,** 677 (1985).
[17] J. E. Walker, M. Saraste, M. J. Runswick, and N. J. Gay, *EMBO J.* **1,** 945 (1982).

[23] Progressive Alignment and Phylogenetic Tree Construction of Protein Sequences

By DA-FEI FENG and RUSSELL F. DOOLITTLE

Introduction

The relationship of a set of related protein sequences can be expressed quantitatively in terms of a phylogenetic tree. The topology of the tree gives an indication of how the sequences should be grouped; the branch lengths provide some sense of the true evolutionary distances. The accuracy of the tree naturally depends on the alignment of the sequences. When there are more than two sequences, the problem lies in the gaps which must be introduced to align sequences optimally. For the simple case of three sequences A, B, and C, the locations of the gaps in a three-way alignment may be very different from those between any pair of sequences.

The scheme discussed here, the progressive alignment method, produces a multiple alignment for a set of protein sequences by iteratively acting on the sequences.[1] The essence of the method is based on the simple rule, "once a gap, always a gap." Consequently, the order in which the sequences are arranged is crucial. In this regard, an approximate phylogenetic order of the sequences is first determined by a series of pairwise alignments by the Needleman and Wunsch method.[2] This provides the most closely related pair, which is the starting point for the progressive alignment procedure. The preliminary set of pairwise measurements also reveals any subclusters that may exist in the set. These subclusters can be treated as units during the alignment process, ensuring that the relative positions of the residues within the cluster will not be altered. Thus, all subclusters are prealigned before the final alignment is undertaken.

The progressive alignment method utilizes the minimum mutation matrix of Dayhoff,[3] thereby permitting the matching of not only identical residues but also those that are similar to each other. To further increase the chances of maximizing the matching, the characters from the last added sequence, or set of sequences, are averaged across all previously aligned residues.

Methods

All the programs described here are written in the C language and run on a VAX computer with the UNIX (Berkeley 4.3) operating system. Other users have run these programs successfully with the VMS operating system, although some minor adjustments are usually necessary. The ensemble of programs dealing with sequence alignment and tree building can be obtained by sending a blank magnetic tape to the authors.

Definitions. For purpose of description, we distinguish between simple and compound trees. Simple trees are those in which the branching order follows the simple clustering pattern $((((AB)C)D)E)$, etc., whereas compound trees have subclusters, as in $(((AB)(CD))E)$, etc. Neutral elements are simple characters (Xs or Js) that are filled into sequences when gaps occur. They are neutral in the sense that they are invisible to the scoring system used to establish subsequent alignments, which is to say, when X is

---

[1] D.-F. Feng and R. F. Doolittle, *J. Mol. Evol.* **25**, 351 (1987).

[2] S. B. Needleman and C. D. Wunsch, *J. Mol. Biol.* **48**, 443 (1970).

[3] M. O. Dayhoff, "Atlas of Protein Sequence and Structure," Vol. 5, Suppl. 3, National Biomedical Research Foundation, Washington, D.C., 1978.

matched with another residue, the value is equal to zero. Negative segments are those internodal connecting distances with negative values that occasionally emerge from Fitch–Margoliash[4] trees when data scatter confounds the segment averaging (or least-squares treatment). Percent identity is taken as the number of identities per 100 aligned residues.

Pairwise Alignments. The Needleman and Wunsch algorithm[2] was used in a three-matrix form[5] and utilized the minimum mutation matrix of Dayhoff[3] in its scoring. The similarity scores obtained from the alignments are converted to difference scores by the relationship

$$D = -\ln S_{\text{eff}} \times 100 = -\ln(S_{\text{real}} - S_{\text{rand}})/(S_{\text{ident}} - S_{\text{rand}}) \times 100$$

where S_{real} is the alignment score itself, S_{rand} the score obtained with random sequences of the same lengths and compositions, and S_{ident} the average score of the two sequences being compared when each is aligned with itself. Based on previous comparisons of many different kinds of sequences,[6] a value of 770 determined as an average random score is used for S_{rand} in these initial comparisons. In the event that the binary results indicate a compound tree, the sequences forming the subclusters are first prealigned.

Multiple Alignments. The Needleman–Wunsch algorithm[2] is used successively on pairs of sequences presented in an appropriate order of relatedness as determined by the preliminary comparisons. Gaps are concurrently filled with neutral elements, and a limited number of nearest alternative sequences is tried in search of better alignments as determined by higher similarity scores.

Tree building. When the alignment is completed, random scores (S_{rand}) are generated by jumbling the actual sequences under comparison. This is necessary to offset the effect of unusual amino acid combinations that can give high (or low) scores with the minimum mutation matrix.[3] S_{rand} values are also calculated for each pair of sequences. Then, using the above equation, all the pairwise similarity scores S_{real} are converted to a distance matrix. The conventional method of Fitch and Margoliash[4] is used to determine the branching order of the sequences. Combining the branching order with the distance matrix, a least-squares approach as described by Klotz and Blanken[7] is used to determine the branch lengths. At this point, branches around the shortest interior segments may be interchanged in search of better trees as measured in terms of lower percent standard

[4] W. M. Fitch and E. Margoliash, *Science* **155**, 279 (1967).
[5] M. L. Fredman, *Bull. Math. Biol.* **46**, 553 (1984).
[6] D.-F. Feng, M. S. Johnson, and R. F. Doolittle, *J. Mol. Evol.* **21**, 112 (1985).
[7] L. C. Klotz and R. L. Blanken, *J. Theor. Biol.* **91**, 261 (1981).

TABLE I

ENSEMBLE OF PROGRAMS FOR PROGRESSIVE ALIGNMENT AND PHYLOGENETIC TREE
CONSTRUCTION

Program	Objective
SCORE	Initial pairwise alignment of sequences to be studied; yields a preliminary branching order
PREALIGN	Align subcluster of sequences
TREE	Progressive alignment of properly ordered sequences (and subsets of sequences); yields branching order and branch lengths of the phylogenetic tree
BLEN	Determines the branch lengths from a given topology combined with a distance matrix
ALIGN	Progressive alignment of properly ordered sequences (and subset of sequences); yields multiple alignment only
FORMAT	Converts sequences recorded in other formats to the "Old Atlas" format so that it can be recognized by this set of programs
TREEPLOT	A customized plotting routine designed to display the phylogenetic tree data in the form of a dendrogram on a Zeta plotter
MULPUB	Format progressively aligned sequences

deviation. Finally, the phylogenetic tree of the sequences in the form of a dendrogram is plotted on a plotter or, if an automatic plotter is not available, drawn by hand.

Outline of the Progressive Method

In the time since the original publication of the progressive alignment method,[1] a number of improvements regarding execution have been made. Several of the programs have been combined so that they will run consecutively without interruption (Table I). The following describes the current status of the procedure.

Initial Branching Order. The program SCORE is written so that the Needleman and Wunsch pairwise alignment procedure[2] can be applied directly when a set of n sequences is provided in a single file. At this stage, the order in which the sequences are assembled is unimportant; the program SCORE performs all $n(n-1)/2$ alignments and sorts out a roughly proper order. Furthermore, SCORE will reveal whether the starting set contains any clusters. If one or more clusters exist, those sequences should be prealigned with the program PREALIGN. As a hypothetical example, assume that the result of the binary alignments conducted by SCORE yields a branching order $((((AB)C)D)(EF))$. The one-line branching order notation[8] shows that sequences A and B represent the most closely related pair. They are followed by C and then D and then a cluster EF. PRE-ALIGN aligns sequences E and F, filling gaps with neutral elements (Js). From here on, E and F are treated as a unit, EF. All sequences, A through F, are then entered into a single file in the order provided by the SCORE results. This becomes the input file for the programs ALIGN or TREE, depending on whether the user is interested in only getting a multiple alignment or also obtaining a phylogenetic tree.

Progressive Alignment. The program TREE (or ALIGN), which is the heart of the procedure, is now used to generate the multiple alignment for sequences A through F. It begins by inserting neutral elements (Xs) in gaps that occur in the aligned pair of A and B, now designated as AB. Next, the first ternary case $(AB)C$ is considered; that is, when C is brought in, a new alignment is made and a score determined. The key to this alignment is that new gaps can be incorporated into either C or B. If one or more gaps are inserted into B, the same gaps must be inserted into A (since the AB alignment must be conserved). The result is compared with the alternative $(BA)C$. Assuming that the former has a higher score, it will then set the path for the next alignment. When sequence D is brought in, the arrangements $((AB)C)D$ and $((AB)D)C$ are compared. Again, assuming that $((AB)C)D$ wins, the next two configurations to be examined will be $(((AB)C)D)(EF)$ and $(((AB)C)(EF))D$. This fine tuning is applied only on the nearest neighbors for the simple reason that it is impractical as well as unnecessary to make an exhaustive trial. The procedure is continued iteratively until all sequences have been incorporated.

To carry out the above operations, one has a choice of using either the programs ALIGN or TREE. The former will provide only the multiple alignment and therefore takes much less time to complete the task. The latter is more time consuming, since TREE must calculate all the pairwise S_{rand} values determined from shuffled sequences. Each of the S_{rand} values is based on 50 jumbles. Based on the equation noted above, the similarity scores, S_{real}, are converted to distance values. In line with the method of Fitch and Margoliash,[4] the pair of sequences separated by the smallest distance is grouped together first. The matrix is then reduced by one, and all distance values related to this pair are averaged in a new matrix. This procedure is repeated until the matrix is reduced to a single dimension. The end result is a branching order with the associated branch lengths. The branching order may or may not be the same as the one determined from binary alignment alone; the branch lengths may contain some negative

[8] W. M. Fitch, *Am. Nat.* **111**, 223 (1977).

values. The program saves the branching order and the distance scores in a file called ALBR which the user can modify so that alternative topologies can be tried when it is used in conjunction with the program BLEN.

Negative Branch Lengths. Occasionally, negative branch lengths may occur. When that happens, alternative trees are chosen with the branches on either side of the negative segment reversed, and a new set of branch lengths is calculated. The amended branching order serves as the input file for the BLEN program, and the new tree can be calculated with very little effort. Experience tells us that only a small number of trials need to be undertaken to eliminate the negative segments. In general, only the lengths in the immediate neighborhood of the reversed segments are affected in the recalculation. Even if no negative values appear in the initial tree, it is prudent to explore alternative topologies to see if better trees are generated. Ordinarily, this involves swapping the branches around the shortest interior segments. The criterion for "better" in this case is lower percent standard deviation.

Automatic Plotting. The program TREEPLOT is used to generate the dendrogram. TREEPLOT is designed specifically to be run on a Zeta plotter; the library functions it uses are not generally available, and users will have to make their own arrangements for plotting details. A schematic outline of the progressive alignment method from start to finish is presented in Fig. 1.

Example

In the following example, all the computer execution statements assume that the user is operating in the UNIX environment. When working under VMS or some other operating system, some of these statements may have to be modified. It is essential that all sequences be in the "Old Atlas" format. Sequences written in any other format can be converted to this arrangement by using the program FORMAT. Figure 2 shows an example of such a sequence.

Illustration Sequences. Seven globin sequences have been chosen to illustrate progressive alignment and phylogenetic tree construction. They are human hemoglobin α chain (hbah); human hemoglobin β chain (hbbh); human hemoglobin γ chain (hbgh); hagfish hemoglobin (heha); lamprey hemoglobin (hbrl); human myoglobin (myoh); gastropod myoglobin (mycr).

Binary Order. To begin with, one concatenates all the sequences (any order) into a file called "test0",

    cat *hbah hbbh hbgh heha hbrl myoh mycr* > test0

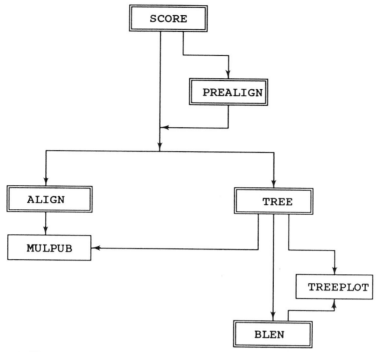

FIG. 1. Flow chart of progressive alignment procedure. Program names are shown in boxes: the primary programs used for progressive alignment are enclosed in double boxes; other utility programs are shown in single boxes.

("cat" is a UNIX command). The next step is to use the SCORE program to perform all the pairwise comparisons,

score test0 test1 > test2

At completion, "test2" contains $n(n - 1)/2$ pairwise alignments. All the results calculated based on binary alignments and, in particular, the

```
THBRL  HEMOGLOBIN, RIVER  LAMPREY (LAMPETRA FLUVIATILIS)
PHBRL    1 P I V D S G S V A P L S A A E K T K I R S A W A P V Y S N Y
PHBRL   31 E T S G V D I L V K F F T S T P A A Q E F F P K F K G M T S
PHBRL   61 A D Q L K K S A D V R W H A E R I I N A V N D A V A S M D D
PHBRL   91 T E K M S M K L R D L S G K H A K S F Q V D P Q Y F K V L A
PHBRL  121 A V I A D T V A A G D A G F E K L M S M I C I L L R S A Y *
```

FIG. 2. The "Old Atlas" format for protein sequence data. The first column of every line specifies either T (title) or P (protein data). The next four columns are devoted to a four-letter identification code. The sequence is arranged 30 residues across and terminates with an asterisk. The programs mentioned in the text recognize any sequence in this format.

branching order are in "test1." For example, the preliminary branching order of these sequences should be *hbgh, hbbh, hbah, myoh, mycr, heha, hbrl,* with *heha* and *hbrl* forming a subcluster. The output reads

Branching order: (HBGH,HBBH)HBAH)MYOH)MYCR)HEHA)-HBRL)

Clusters: (HEHA,HBRL)

Prealigning Sequences. Concatenate *heha* and *hbrl* into a file called "test3,"

cat *heha hbrl* > test3

Then, prealign these two sequences,

prealign test 3 test 4 > test5

The write-file "test4" will contain the prealigned sequences in the "Old Atlas" format with neutral elements (Js) occupying all gap positions. The file "test5" contains the equivalent alignment without Js and also general information about the degree of similarity. If more than two sequences are to be prealigned, the same procedure is used. In the event that nesting of clusters occurs, the deeper group should be prealigned first; then systematically work back through the entire cluster.

Progressive Alignment. The prealigned sequences in the file "test4" are combined with the other sequences in a single file called "sample," but the order must now be in tune with the preliminary branching order. The concatenation should therefore be performed in the following order,

cat *hbgh hbbh hbah myoh mycr test4* > sample

If only a multiple alignment is desired, the program ALIGN should be used, since it does not require the detailed distances and is much faster.

align sample sample1 > sample2

The file "sample1" will contain all the aligned sequences with gaps occupied by Xs in the "Old Atlas" format. It can be used as the input file for another program, MULPUB, which reprints the alignment in a closepack format without Xs (Fig. 3). A multiple alignment without Xs, and results such as the similarity scores and percent identity between each pair of sequences after multiple alignment, will be given in the file "sample2."

If a phylogenetic tree is needed as well, then one should use the program TREE,

tree sample sample1 > sample3

```
                                      *                      *      *  *
hbgh            GHFTEEDKATITSLW    GKV   NVEDAGGETLGRLLVVYPWTQRFFDSFGNLSSASAIMGNPKVKAHGKKVLTSLG
hbbh            VHLTPEEKSAVTALW    GKV   NVDEVGGEALGRLLVVYPWTQRFFESFGDLSTPDAVMGNPKVKAHGKKVLGAFS
hbah            VLSPADKTNVKAAW    GKVGAHAGEYGAEALERMFLSFPTTKTYFPHF DLSH      GSAQVKGHGKKVADALT
heha   PITDHGQPPTLSEGDKKAIRESW    PQIYKNFEQNSLAVLLEFLKKFPKAQDSFPKFSAKKS  HLEQDPAVKLQAEVIINAVN
hbrl   PIVDSGSVAPLSAAEKTKIRSAW    APVYSNYETSGVDILVKFFTSTPAAQEFFPKFKGMTSADQLKKSADVRWHAERIINAVN
myoh            GLSDGEWQLVLNVW    GKVEADIPGHGQEVLIRLFKGHPETLEKFDKFKHLKSEDEMKASEDLKKHGATVLTALG
mycr   SLQPASKSALASSWKTLAKDAATIQNNGATLFSLLFKQFPDTRNYFTHFGNMSDA EMKTTGVGKAHSMAVFAGIG

                                      *   *
hbgh   DAIKHLDD   LKGTFAQLSELHCDKLHVDPENFKLLGNVLVTVLAIHFGKEFTPEVQASWQKMVTGVASALSSRYH
hbbh   DGLAHLDN   LKGTFATLSELHCDKLHVDPENFRLLGNVLVCVLAHHFGKEFTPPVQAAYQKVVAGVANALAHKYH
hbah   NAVAHVDD   MPNALSALSDLHAHKLRVDPVNFKLLSHCLLVTLAAHLPAEFTPAVHASLDKFLASVSTVLTSKYR
heha   HTIGLMDKEAAMKKYLKDLSTKHSTEFQVNPDMFKELSAVFVSTM       GGKAAYEKLFSIIATLLRSTYDA
hbrl   DAVASMDDTEKMSMKLRDLSGKHAKSFQVDPQYFKVLAAVIADTV       AAGDAGFEKLMSMICILLRSAY
myoh   GILKKKGH   HEAEIKPLAQSHATKHKIPVKYLEFISECIIQVLQSKHPGDFGADAQGAMNKALELFRKDMASNYKELGFQG
mycr   SMIDSMDDADCMNGLALKLSRNHIQR KIGASRFGEMRQVFPNFLDEALGGGASGDVKGAWDALLAYLQDNKQAQAL
```

FIG. 3. Multiple alignment of seven hemoglobin sequences. Asterisks denote locations where all seven residues are identical. The four-letter designations are as follows: hbgh, human globin γ chain; hbbh, human globin β chain; hbah, human globin α chain; heha, hagfish hemoglobin; hbrl, lamprey hemoglobin; myoh, human myoglobin; mycr, gastropod myoglobin.

The file "sample3" will end up not only with the same calculated results as were given in "sample2" (from the ALIGN program), but it will also contain the difference scores, branching order, and branch lengths.

The multiple alignment of the seven globin sequences is shown in Fig. 3. At this point, it is useful to draw on paper a tree based on the branching order provided in the file "sample3." It is important to note how the segments are numbered so that one can identify which branch length in "sample3" corresponds to which segment on the tree (see Fig. 4). As a general rule, tip segments are numbered first, followed by the nearest interconnecting segment; the very first segment is always located at the left-hand side of the closest pair. Applying this rule to the globin case, the appropriately numbered segments start at tip *hbgh,* then *hbbh,* then proceed to the nearest internal segment, then *hbah,* etc. (see Fig. 4).

The globin tree that emerged directly from the initial alignment is shown as a dendrogram in Fig. 5 (top). It should be noted that one of the branch lengths is small and negative; this is an indication that an alternative topology should be tried with the branches adjacent to the negative segment reversed.

Exploring Alternative Topologies. To explore other topologies, one should note that after the completion of the run using TREE, there is a hidden file called ALBR (see Table II) in the user computer work space. On examination, one will find that it contains at least one line of letters, in alphabetical order, beginning with upper case. If there are more than 26 sequences, the lower case alphabet will be invoked. The file also contains a set of triangularly arranged numbers representing the distance matrix, in the same order as the letters. The letters refer to the sequences; for example, *A* is *hbgh, B* is *hbbh, C* is *hbah, D* is *myoh, E* is *heha, F* is *hbrl,* and *G* is

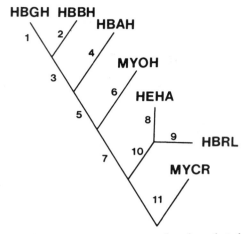

FIG. 4. Example showing how segments are numbered so that the calculated branch lengths can be matched with the phylogenetic tree. The segments in this diagram are not drawn proportional to the actual calculated lengths. See the legend to Fig. 3 for four-letter designations.

mycr. When there is more than one line of letters, clusters exist. In "sample3," the second line is *EF* representing *heha* and *hbrl.* The branching order and the distance matrix are always separated by an asterisk.

The most direct way to investigate the negative segment is by interchanging *myoh* and the cluster (*heha,hbrl*). This involves modifying the ALBR file (see Table II). Let the new file be "sample4" and use it as input file for BLEN,

     blen sample4

The new topology and the corresponding branch lengths are given in Fig. 5 (bottom). Note that the negative branch segment disappeared and that the goodness of fit, as measured by the percent standard deviation, improved.

Discussion

Since the progressive alignment method was first published in 1987, it has been applied to a wide range of sequences with great success. Although the basic method has remained intact, the computing procedures have been significantly updated. In the initial version, a number of steps had to be carried out separately in order to arrive at the multiple alignment and phylogenetic tree for a set of protein sequences. For example, merely to obtain the preliminary binary branching order and prepare the sequences

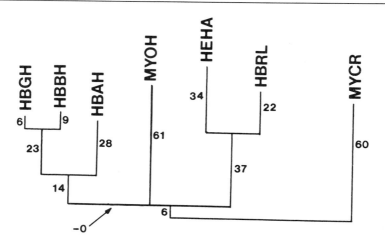

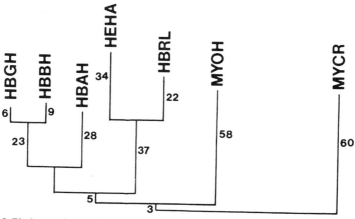

FIG. 5. Phylogenetic trees for the seven hemoglobins determined directly from the progressive alignment method. In the top tree, the segments corresponding to number 7 in Fig. 4 is −0.08. The percent standard deviation (%SD)[4] (calculated from the actual distance matrix and the distance matrix derived from the tree) is 3.45. The bottom tree was obtained by switching *myoh* and the cluster *heha – hbrl.* The %SD of this tree is 3.18. See the legend to Fig. 3 for four-letter designations.

TABLE II
Two Examples of the Output File ALBR from the
TREE Program[a]

ABCDEFG
EF
*

15.43	59.63	104.67	112.51	100.49	107.18
57.41	111.48	113.38	103.86	115.66	
98.72	115.28	97.23	111.01		
139.39	118.09	121.07			
55.52	129.51				
131.07					

ABCDEFG
EF
*

15.43	59.63	104.67	112.51	100.49	107.18
57.41	111.48	113.38	103.86	115.66	
98.72	115.28	97.23	111.01		
139.39	118.09	121.07			
55.52	129.51				
131.07					

[a] The first example gives the file that produces the branch lengths shown in Fig. 5 (top) when used with the program BLEN. Above the asterisk is the branching order and below it is the distance matrix. The capital letters on the first line denote the seven globin sequences. The second line indicates that sequences E and F form a cluster. The distance matrix is arranged in the same oder as the letters on the first line. The entry 129.51 refers to the distance between sequences E and G. The second example shows the modified letter order after interchanging sequence D and cluster EF. Note that the distances are not modified. The latter was the input file for BLEN, which gives the branch lengths shown in Fig. 5 (bottom).

for progressive alignment, one had to use three different programs: SCORE, BORD, and PREALIGN. Similarly, in the original progressive mode, four programs, DFALIGN, SHUFFLE, BORD, BLEN, were needed to obtain a phylogenetic tree. Presently, only two programs (SCORE and PREALIGN) are necessary in the preprogressive alignment stage; even better, the four programs previously needed for the progressive mode have been combined into a single program called TREE.

Aside from the fact that fewer programs are now involved, the single most important improvement is in the TREE program. Formerly, in order to get the branch lengths of a phylogenetic tree, one had to manually construct a connectivity table, which is not only cumbersome but also error prone, especially when working with a large number of sequences. The automation of that step eliminated a great deal of manual work and allows one to start with a set of properly ordered sequences and to obtain the phylogenetic tree in a single operation. Furthermore, it is now a simple matter to try various alternative topologies in order to gain some sense of reliability of the tree. In this respect, it is worth noting how the branching order for the hemoglobin sequences changed during the various operations. The preliminary binary order from SCORE was *hbgh, hbbh, hbah, myoh, mycr, heha, hbrl,* and the initial progressive changed it to *hbgh, hbbh, hbah, myoh, heha, hbrl, mycr.* The appearance of a negative segment between *myoh* and the cluster *heha* and *hbrl* prompted a further exploration, and the alternative branching order with *myoh* and the cluster group reversed yielded a tree with no negative segment and a lower percent standard deviation. The final order is *hbgh, hbbh, hbah, heha, hbrl, myoh, mycr,* which also makes the most evolutionary sense.

Proper sequence alignment is essential to any scheme of phylogenetic tree construction based on sequences. The proper alignment, however, is not necessarily the mathematically optimal alignment, and we have made the point before that it is a historical alignment that warrants the attention of the evolutionist.[9] The progressive alignment procedure we have developed puts more stock in the relationship of recently diverged sequences than it does in distant relationships. Thus, it seems to us foolish to omit or move a gap that occurs when two similar sequences are aligned just because an additional match might be made with some very distantly related sequence, as is often the case when mathematically optimized alignments are generated.

But if we are going to imply that parsimony is not an inviolate rule, how are we ever to choose the best tree? There are empirical approaches that one can adopt in addition to statistical measures like the lowest percent standard deviation. For example, one can examine sequences from more than one set of gene products from a given set of organisms to see if they all yield the same phylogeny. Or, if only one type of sequence is involved, the sequences can be cut into halves (amino-terminal halves and carboxy-terminal halves) and two separate trees made. If the data are reliable, the two trees ought to have the same topology. Recently, we developed another check on the reliability of protein sequence-based trees. It is a simple character-based test that we call PAPA (parsimony after progressive alignment). Its features are discussed in another chapter in this volume.[10]

---

[9] R. F. Doolittle, "URFs and ORFs: A Primer on How to Analyze Derived Amino Acid Sequences," Chap. 3. University Science Books, Mill Valley, California, 1986.

[10] R. F. Doolittle and D.-F. Feng, this volume, [41].

[24] Identification of Significant Sequence Patterns in Proteins

By Samuel Karlin, B. Edwin Blaisdell, and Volker Brendel

Introduction

The three-dimensional structure of crystalline proteins has been determined to high resolution for only about 70 distinct, mostly globular, proteins. These structures may not be a good representative sample of the protein universe, and new structures are being determined at the rather slow rate of at most 10–20 per year.[1] Moreover, the crystal structure of a protein may be different from its natural conformation under physiological conditions. The three-dimensional structure of proteins in solution has been determined by NMR methods for only a few small oligopeptides.[2] To date, prediction of secondary and tertiary structure from amino acid sequence is generally considered inadequate. An acceptable solution of the protein folding problem appears remote. At present it may be more productive to attempt to find correlations between amino acid sequence features and protein function without knowledge of the precise molecular structures involved.

Support for this approach may be found in the recent observation that, in yeast transcription activators GAL4 and GCN4, activating domains are associated with nonspecific acidic regions.[3,4] In other cases, regions rich in other residue types are implicated in activation, e.g., glutamine-rich regions in the human transcription factor Sp1.[5] The DNA-binding domains of these proteins generally correspond to basic regions. Zinc finger and homeo DNA-binding sequences are regions of high mixed charge with a predominance of positive charge.[6]

The methods described in this chapter identify statistically significant amino acid sequence configurations of many kinds. Our objective is to identify diagnostic sequence features that might provide insights into

[1] M. J. Rooman and S. J. Wodak, *Nature (London)* **335,** 45 (1988).

[2] K. Wüthrich, *Science* **243,** 45 (1989).

[3] I. A. Hope and K. Struhl, *Cell* **46,** 885 (1986).

[4] J. Ma and M. Ptashne, *Cell* **48,** 847 (1987).

[5] A. J. Courey and R. Tjian, *Cell* **55,** 887 (1988).

[6] V. Brendel and S. Karlin, *Proc. Natl. Acad. Sci. U.S.A.* **86,** 5698 (1989).

METHODS IN ENZYMOLOGY, VOL. 183

protein function and structure and ways of protein classification. The cited examples come from use of the charge alphabet (amino acids are denoted in the one-letter code) $+ = \{H,K,R\}$, $- = \{D,E\}$, and $0 = \{A,C,F,G,I,L,M,N,P,Q,S,T,V,W,Y\}$. Configurations of other groupings of amino acids into classes similar in some other way may be tested for significance with the same methods described below. For example, the classification internal, external, ambivalent ($i = \{F,I,L,M,V\}$, $e = \{D,E,H,K,N,Q,R\}$, and $a = \{A,C,G,P,S,T,W,Y\}$) is a good alphabet for distinguishing levels of hydrophobicity.

Our focus here is to identify statistically significant clusters, runs, and periodic patterns of charge. A charge cluster in our notation refers to a short protein segment (25–75 residues) with significantly high specific charge content relative to the charge composition of the whole protein; a positive charge cluster is a segment with high positive net charge, a negative charge cluster is a segment with high negative net charge, and a mixed charge cluster is a segment high in charged residues of both signs. A charge run of length n denotes n consecutive charged residues; intermittent uncharged residues will be counted as errors. Finally, periodic charge patterns are sequences like $(+,0)_n$ or $(-,0,0)_n$, where n is the repeat count of the unit pattern. All of these configurations do occur in natural protein sequences and in some cases have been implicated in protein function. For example, charge clusters are associated with transcriptional activation as mentioned above, positive charge runs are a motif in nuclear location signals,[7] and $(+,0,0)_{5-8}$ sequences are conspicuously conserved in voltage-gated ion channel proteins.[8]

Ordinarily, any protein will contain some stretches that are higher in net charge than others and will contain some short runs of charge. These would also occur in random sequences simply because of chance fluctuations. We describe how to evaluate the statistical significance of clusters, runs, and periodic patterns by giving formulas that estimate the probability of observing a cluster of ascertained concentration of charge or a run or periodic pattern of a certain length in a random sequence of the observed composition of the protein. These formulas allow us to single out highly unlikely charge configurations. Several examples are given (Table I, Fig. 1). At the end of the chapter we discuss the merits and limitations of this approach.

[7] H. Siomi, H. Shida, S. H. Nam, T. Nosaka, M. Maki, and M. Hatanaka, *Cell* **55**, 197 (1988).
[8] T. Tanabe, H. Takeshima, A. Mikami, V. Flockerzi, H. Takahashi, K. Kangawa, M. Kojima, H. Matsuo, T. Hirose, and S. Numa, *Nature (London)* **328**, 313 (1987).

TABLE I

SIGNIFICANT RUNS AND PERIODIC PATTERNS OF CHARGE[a]

Protein	Length	f_+	f_-	Pattern	Errors	Sequence	L_{min}	Eq.
hα₁AR	515	12.2%	7.4%	$(+)_8$ at 371–378	0	RRRRRRRR	5.1	(3)
yRAD6	172	11.6%	23.3%	$(-)_{19}$ at 150–170	2	EDDMDDMDDDDDDDDDDDDDE	9.4	(4), (5)
yCDC34	295	11.5%	24.7%	$(-)_{15}$ at 245–252	3	DDDDERIEFEDDDDDDDD	11.6	(4), (5)
yMAK16	306	19.3%	21.6%	$(-)_{10}$ at 201–211	1	DEEEDWDEEEE	8.3	(4), (5)
				$(0, -)_7$ at 251–264	0	SESDSESESDSDSD	10.5	(6)
tonB	244	13.9%	9.8%	$(-, 0)_8$ at 71–86	2	EPVVEPEPEPEPIPEP	11.7	(7), (8)
				$(+, 0)_6$ at 96–107	0	KPKPKPKPKP	9.7	(6)
Prepromelittin	70	8.6%	12.9%	$(-, 0)_{10}$ at 24–43	1	EPEPAPEPEAEADAEADPEA	9.3	(7), (8)
Bicoid	493	14.2%	8.3%	$(+, 0)_{14}$ at 13–41	2	HPLPHTHTHPHPHSHPHSHPHPHHQHP	11.2	(7), (8)
EBNA-3	839	11.0%	10.5%	$(-, 0, 0)_6$ at 739–756	0	ESGEGSDTSEPCEALDLS	11.8	(6), (9)
MBK1	495	11.5%	13.3%	$(+, 0, 0)_7$ at 292–312	1	RVIRLVRFRIFKLSRHSKGL	13.7	(7)–(9)

[a] For each protein the name, length, percentage of positively and negatively charged residues, the charge patterns, their locations and numbers of errors, the minimum length (not counting errors) that is significant at the 1% level, and the equations in the text used to determine L_{min} are given. Sequences are displayed in the standard one-letter code; errors are indicated by underlining. All examples are discussed in the text. hα₁AR, Hamster α₁-adrenergic receptor; yRAD6, *Saccharomyces cerevisiae* RAD6 protein; yCDC34, *Saccharomyces cerevisiae* CDC34 protein; yMAK16, *Saccharomyces cerevisiae* MAK16 protein; tonB, *Escherichia coli* tonB protein; EBNA-3, Epstein–Barr virus nuclear antigen 3; MBK1, mouse MBK1 protein.

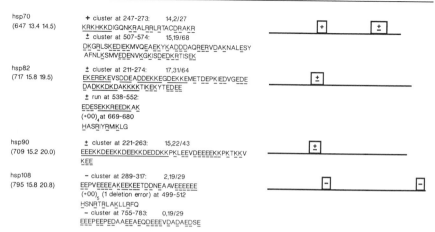

hsp70
(647 13.4 14.5)

+ cluster at 247-273:   14,2/27
KRKHKKDIGQNKRALRRLRTACDRAKR
± cluster at 507-574:   15,19/68
DKGRLSKEDIEKMVQEAEKYKADDDAQRERVDAKNALESY
AFNLKSMVEDENVKGKISDEDKRTISEK

hsp82
(717 15.8 19.5)

± cluster at 211-274:   17,31/64
EKEREKEVSDDEADDEKKEGDEKKEMETDEPKIEDVGEDE
DADKKDKDAKKKKTIKEKYTEDEE
± run at 538-552:
EDESEKKREEDKAK
(+00)₄ at 669-680
HASRIYRMIKLG

hsp90
(709 15.2 20.0)

± cluster at 221-263:   15,22/43
EEEKKDEEKKDEEKKDEDDKKPKLEEVDEEEEKKPKTKKV
KEE

hsp108
(795 15.8 20.8)

− cluster at 289-317:   2,19/29
EEPVEEEEAKEEKEETDDNEAAVEEEEEE
(+00)₅ (1 deletion error) at 499-512
HSNRTRLAKLLRFQ
− cluster at 755-783:   0,19/29
EEEPEEPEDAAEEAEQDEEEVDADAEDSE

FIG. 1. Significant charge configurations in high molecular weight heat-shock proteins. Below the name of each protein are given the number of residues and the percentage of positively and negatively charged residues in the protein. Each charge cluster (of positive, negative, or unspecified sign) is identified by its coordinates in the protein, the number of positive and negative charges, and the length of the segment. Sequences are displayed in the standard one-letter code; charged residues are emphasized by underlining. Schematic representations of the proteins are drawn approximately to scale. References for the sequences are given in the text.

Statistical Evaluation of Charge Configurations

The distribution of specified residues within a protein may be analyzed both with respect to its primary structure and, if available, with respect to its three-dimensional structure. First, we discuss significance estimates for clusters, runs, and periodic patterns in the amino acid sequence, and then we outline some approaches that may be used to evaluate clusters in three dimensions. Proteins will be represented in the charge alphabet $+,-,0$. Let f_+, f_-, and f_0 denote the fractions of positive, negative, and uncharged amino acids within the protein (or, possibly for small proteins, within a grouping of similar proteins, e.g., ribosomal proteins or histones). We estimate the statistical significance of the observed residue distribution relative to the probability of finding a cluster, run, or periodic pattern of specified content in a random sequence generated by independent sampling of $+$, $-$, and 0 with probabilities f_+, f_-, and f_0, respectively. Models assuming neighbor dependencies have also been studied,[9] but in practice the corresponding more complex formulas yield results which differ little from those obtained with the independence model. Analysis of randomly

[9] S. Karlin and F. Ost, *Adv. Appl. Prop.* **19L** 293 (1987).

shuffled counterparts of natural sequences produced good agreement with the analytical significance estimates (data not shown).

Clusters

If we had just one **window** of length W residues to evaluate, the **count** c of charged residues (positive, negative, or both charges) in this window would follow a binomial distribution with mean Wf and variance $Wf(1 - f)$, where f denotes f_+, f_-, or $(f_+ + f_-)$, for positive, negative, or mixed charge clusters, respectively. Since we are interested in all possible windows of lengths 25–75 residues, the problem of multiple tests arises. We deal with this conservatively by choosing a low value (.0001) of the tail probability in the normal approximation to the individual binomial test. For a protein of typical length 150–800 residues, for which we might look at the order of 100 independent windows, this procedure would yield an overall significance level of about 1%.

In our programmed procedure we first calculate the minimum significant count c_W for window sizes $W = 30$, 45, and 60 according to Eq. (1):

$$c_W = \text{least integer equal to or greater than } Wf + 4\sqrt{Wf(1-f)} \quad (1)$$

The factor 4 makes an approximate adjustment of the significance level as outlined above; for proteins exceeding 1,000 residues we use the factor 4.5. The chosen window sizes are long enough to avoid most chance fluctuations and to assure reasonable accuracy of the normal approximation, and they are also short enough not to sacrifice the concept of local clustering (as distinguished from net charge of the entire protein). From a molecular standpoint, they correspond to the lengths of typical exons.[10] As the protein is read by the program, a one-dimensional array is created which contains the positions of the charged residues; for negatively charged residues the positions are saved as negative values. This array contains all the information required for subsequent analysis.

For illustration consider identification of significant mixed charge clusters. Beginning at the amino terminus, each successive charge defines the starting point of a window. For the ith charge the program determines (by looking at the array) whether the $(i + c_W - 1)$th charge occurs within distance W, where W loops through 30, 45, 60. When a success is scored the program marks i as a possible starting point of a significant charge cluster. The cluster is then optimized in length by maximizing

$$t = \frac{c - Wf}{\sqrt{Wf(1 - f)}} \quad (2)$$

---

[10] T. W. Trout, *Proc. Natl. Acad. Sci. U.S.A.* **85**, 2944 (1988).

with respect to W, where c is the observed count in the window. Only nonoverlapping clusters (unless of different types) of length at least 20 residues are printed; the significance of shorter clusters is evaluated more reliably by considering these clusters as runs with errors (see next section). The search for another cluster is resumed after the end of the extended optimized cluster. The identification of positive and negative charge clusters is analogous to the procedure for detecting mixed charge clusters. However, as another conservative measure, we routinely count c as net charge. Thus, for a positive charge cluster, for example, we require a net count of c_W positive charges to occur within a segment of length less than W. It should be noted that the procedure outlined here is probabilistically essentially the same as scanning with a fixed window size W.[11]

For example, the *Drosophila cut* protein[12] contains 11.8% basic residues in its amino acid sequence of length 2175 residues. According to Eq. (1) (with factor 4.5), $c_{30} = 11$. A net count of 11 positive charges occurs in segment 665–692 of length $28 < 30$. Thus, 665 marks the beginning of a significant positive charge cluster; this cluster extends to length 71 (residues 665–735), containing 26 basic residues and only one acidic residue.

Runs

Let N be the length of the sequence. For the independence model the probability of observing a run of a particular letter of length exceeding $L = \log N/(-\log \lambda) + z$, where λ is the fraction of the prescribed letter in the sequence, is bounded (for large N; $N \geq 100$ suffices) between

$$P_- = 1 - \exp\{-(1 - \lambda)\lambda^{z+1}\} \quad \text{and} \quad P_+ = 1 - \exp\{-(1 - \lambda)\lambda^z\} \quad (3)$$

Note that these bounds are very narrow (corresponding to runs differing by only one residue in length); for a detailed discussion of these formulas see Refs. 9, 13, and 14.

In practice, we determine z so that $P_+ = 0.01$. For example, the *Drosophila cut* protein[12] is of length 2175 residues, with a 12.1% proportion of acidic residues (i.e., $\lambda = 0.121$). Thus $L = 3.64 + 2.12. = 5.76$, and an acidic run of length 6 or longer would be significant at the 1% level. An acidic run of length 18 occurs at positions 276–293.

Criteria are also available for runs containing insertions of a nonmatching residue type. Let $K(\mathbf{n})$, $\mathbf{n} = (n_1, \ldots, n_e)$, denote the cumulative length of the longest run of the prescribed letter type interrupted by at most

[11] S. Karlin, B. E. Blaisdell, E. S. Mocarski, and V. Brendel, *J. Mol. Biol.* **205**, 165 (1989).
[12] K. Blochlinger, R. Bodmer, J. Jack, L. Y. Jan, and Y. N. Jan, *Nature (London)* **333**, 629 (1988).
[13] D. E. Foulser and S. Karlin, *Stochastic Processes Appl.* **24**, 203 (1987).

n_1 singlet errors, at most n_2 doublet errors, . . . , and at most n_e e-length errors. Then $K(\mathbf{n})$ is of asymptotic order

$$K = \frac{\log N + n^*[\log \log N - \log(-\log \lambda)]}{-\log \lambda} \qquad (4)$$

where $n^* = \sum_{\mu=1}^{e} n_\mu$ counts the number of error blocks. The probability of occurrence of an extended run of length at least $K(\mathbf{n})$ with errors conforming to the pattern $\mathbf{n} = (n_1, \ldots, n_e)$ can conservatively be approximated by

$$Pr\{K(\mathbf{n}) \geq K + z\} \approx 1 - \exp\left[-\frac{(1-\lambda)\lambda^z}{n_1! \ldots n_e!} \left(\frac{1-\lambda}{\lambda}\right)^s \right] \qquad (5)$$

where $s = \sum_{\mu=1}^{e} \mu n_\mu$ is the total number of characters spanned by all the error blocks (see Ref. 9). In assessing statistical significance we determine z so that the right-hand term of Eq. (5) equals 0.01.

The fraction of uncharged residues in the *Drosophila cut* protein[12] is 0.761. Thus by Eqs. (4) and (5) the minimum length of a 1% significant perfect uncharged run is $28.14 + 11.60 = 39.74$, and the minimum length of a 1% significant uncharged run interrupted by a single charged residue is $40.36 + 7.36 = 47.72$. Significant uncharged runs with no errors occur at positions $740-782$ and $1231-1284$; significant uncharged runs with one single error occur at positions $120-179$, $1847-1894$, and $2123-2175$. Other examples are given in Table I and are discussed below in the section on Applications.

Periodic Patterns

Consider a random sequence composed from three letters A, B, and C, occurring with probabilities α, β, and γ, respectively ($\alpha + \beta + \gamma = 1$). We assess the significance of either of the alternating forms $\{ABAB \ldots \}$ or $\{BABA \ldots \}$ in a sequence of length N. Here set $\lambda = \sqrt{\alpha\beta}$. Then an alternating run as prescribed of length exceeding $\log N/(-\log \lambda) + z$ has asymptotic probability bounded between

$$P_- = 1 - \exp\{-\tau\lambda^z\} \quad \text{and} \quad P_+ = 1 - \exp\{-\rho\lambda^z\} \qquad (6)$$

where $\tau = \sqrt{\alpha\beta}(2 - \alpha - \beta)$ and $\rho = (\alpha + \beta - 2\alpha\beta)/\sqrt{\alpha\beta}$.[14] In practice, we determine z so that $P_+ = 0.01$. For example, for the *Drosophila cut* pro-

---

[14] S. Karlin, F. Ost, and B. E. Blaisdell, *in* "Mathematical Methods for DNA Sequences" (M. Waterman, ed.), p. 133. CRC Press, Boca Raton, Florida, 1989.

tein[12] $\lambda = \sqrt{f_+ f_-} = 0.119$ and $\rho = 1.76$. Thus a $(+,-)_n$ repeat would be significant at the 1% level if $2n$ exceeded $L = 3.61 + 2.43 = 6.04$. A $(+,-)_3$ occurs at positions 1063–1068; this pattern extends to a significant pattern with one error (see below).

Allowing for an error pattern $\mathbf{n} = (n_1, \ldots, n_e)$ (see above), we now determine $\tilde{z}$ to satisfy

$$1 - \exp\left[-\frac{\rho\lambda^z}{n_1! n_2! \ldots n_e!}\left(\frac{1-\lambda}{\lambda}\right)^s\right] = 0.01 \tag{7}$$

Then statistical significance (at the 1% level) for the periodic alternating run $(AB)_n$ with error pattern $\mathbf{n}$ requires a length of pattern agreement (i.e., counting matches only) exceeding

$$\frac{\log N + n^*[\log\log N - \log(-\log\lambda)]}{-\log\lambda} + \tilde{z} \tag{8}$$

In Eqs. (7) and (8), as before, $n^* = \sum_{\mu=1}^{e} n_\mu$ and $s = \sum_{\mu=1}^{e} \mu n_\mu$.

The alternating pattern $(+,-)_3$ in the *Drosophila cut* protein continues with $0(+,-)_2$ to a total length of 10 matching residues and one singlet error, exceeding the critical number $4.21 + 3.37 = 7.58$ determined from Eq. (8). Actually the pattern is the highly restricted peptide $(RE)_3Q(RE)_2$, rendering it even more striking.

As another illustration of a periodic letter pattern we consider the longest run of the unit BAA or ABA or AAB. In this case, we prescribe $\lambda = (\alpha^2\beta)^{1/3}$ and

$$\rho^* = \frac{(1-\alpha^2\beta)\alpha[\alpha^2 + 2\beta^2 + \gamma(\alpha + 2\beta)]}{\alpha^3 + \alpha\beta + \beta + \gamma(1 + \alpha + \alpha^2)}$$

For a pattern of errors $\mathbf{n} = (n_1, n_2, \ldots, n_e)$ we determine $\tilde{z}$ from Eq. (7) with ρ^* replacing ρ, and the critical length is again given by Eq. (8). Examples (see Table I) are discussed below.

Clusters, Runs, and Periodic Patterns in Three Dimensions

In this section we discuss how the concept of clusters in primary structure can be extended to encompass clusters in three dimensions. While limited in application by the small number of proteins for which the crystal structure has been resolved, such studies may nevertheless provide novel perspectives on attributes of protein structure as well as confirmation for the sequence studies.

In order to investigate the spatial distribution of charged residues in a protein it suffices to have the (x,y,z) coordinates of the α-carbons of the backbone and the identity of the associated residues in the chosen alphabet. Alternatively, one could choose the centroid of the side chain or the β-carbon to identify each residue. Given some measure of intramolecular flexibility the particular choice adopted is immaterial, however, and for the present discussion we abide with the more accessible α-carbon coordinates.

As discussed in an earlier section, identification of charge clusters in an amino acid sequence involves counting the number of charges in a window of specified length W, extending from the ith residue toward the carboxy terminus; if the count exceeds the critical value c_W [Eq. (1)], the given window is the core of a significant cluster, the extent of which is found by the optimization procedure described above. For three-dimensional analysis we may apply the same procedure and statistics if we define c_i^α as the reference point of some volume (rather than window). Increasing the volume of a specified type (see below) until it includes W α-carbons takes the place of increasing the window size. If c_W of the α-carbons specify charged residues, the given volume is the core of a significant cluster in three dimensions.

Several different spatial shapes may be used in the general scheme to emphasize different notions of clustering. In the simplest case we may want to apply spheres centered on c_i^α. Computationally we would determine the Euclidean distances for all pairs (c_i^α, c_j^α) and determine the number of charged residues within the W closest neighbors of the ith residue. In order to more strongly emphasize surface contours, cylindrical shapes may be more appropriate. For example, one might define two cylinders of length L and radius R with axes along the c_i^α to c_{i-1}^α and c_i^α to c_{i+1}^α lines, respectively, and end point c_i^α. This volume may be increased by letting L and R grow in some specified ratio. A simple way of capturing known secondary structures with the previous approach would be to define the axes by the lines $c_i^\alpha - c_{i-2}^\alpha$ and $c_i^\alpha - c_{i+2}^\alpha$, appropriate for 2-period β strands, or by the lines $c_i^\alpha - c_{i-7}^\alpha$ and $c_i^\alpha - c_{i+7}^\alpha$, appropriate for 3.5-period α helices. Other shapes may be contemplated involving, for example, right circular conical forms or elliptical cross sections that capture other structural attributes of specific proteins. All shapes allow use of the same statistics described above.

We can also extend the notion of runs and periodic patterns to three dimensions as follows. For a given α-carbon find the α-carbon closest to it (say in the sense of Euclidian distance). Then find the α-carbon closest to the second and different from the preceding one, and so on. This process produces a chain of α-carbons, and in the case that all of the residues so defined are basic, for example, we would speak of a positive charge run, the statistical significance of which can be evaluated as in the case of runs in the primary structure. Note that adjacent α-carbons in the chain need not be close to each other in amino acid sequence. For example, if the $(EP)_6$

and $(KP)_6$ patterns in tonB (see below) were to interlock in the three-dimensional structure, the resulting chain could be a three-dimensional (EK) run. Obviously, the given definition of three-dimensional chains can also accommodate periodic patterns and runs and patterns with errors.

Applications

In this section we illustrate use of the methods. First, we discuss various examples of charge runs and periodic patterns in proteins (Table I). Next, a discussion is given of charge clusters in high molecular weight heat-shock proteins (hsp) (Fig. 1).

Examples of Charge Runs and Periodic Patterns in Proteins

Table I displays a variety of distinctive charge runs and periodic patterns. We comment briefly on these examples.

The hamster α_1-adrenergic receptor[15] carries an unusual run of eight arginines, encoded from the two codons CGT and CGC. In a previous survey of more than 2500 protein sequences we have encountered no uninterrupted positive charge run of length exceeding 9 residues; there were only four of length 8 and one of length 9. In sharp contrast, there are more than a dozen cases of acidic runs of 10 residues or longer. This contrast is intriguing in view of the preponderance of protein sequences of net positive charge. The arginine run in the α_1-adrenergic receptor, located in the cytoplasmic portion of this membrane protein, is poorly conserved in the β_1-, and not conserved in the α_2-, and β_2-adrenergic receptors.[15] The adrenergic receptors are coupled to G proteins, and it is of note that several ras proteins (which are similar to G proteins) also contain carboxy-terminal basic charge runs.[16]

The Saccharomyces cerevisiae RAD6 gene encodes a ubiquitin-conjugating enzyme that is required for a variety of cellular functions, including DNA repair.[17] The RAD6 gene product contains a highly significant $(-)_{19}$ charge run interrupted by two separated methionine residues at its carboxy terminus. The yeast cell cycle gene CDC34 encodes a ubiquitin-conjugating enzyme similar to RAD6.[18] This protein contains a run of 8 aspartate residues, encoded from both its possible codons, in a significant

[15] S. Cotecchia, D. A. Schwinn, R. R. Randall, R. J. Lefkowitz, M. G. Caron, and B. K. Kobilka, *Proc. Natl. Acad. Sci. U.S.A.* **85,** 7159 (1988).

[16] S. Karlin and V. Brendel, *Oncogene,* in press (1989).

[17] S. Jentsch, J. P. McGrath, and A. Varshavsky, *Nature (London)* **329,** 131 (1987).

[18] M. G. Goebl, J. Yochem, S. Jentsch, J. P. McGrath, A. Varshavsky, and B. Byers, *Science* **241** 1331 (1988).

charge run of 15 acidic residues, with two errors. This run is central in an acidic charge cluster (positions 234–267, containing 1 basic and 22 acidic residues in a segment of 34 residues). The acidic regions in these two proteins may well play a role in their specificity for substrates, which include basic proteins such as histones.

The yeast MAK16 gene, presumed to encode a nuclear cell cycle regulator,[19] displays the striking charge pattern $(S, D/E)_7$ at positions 251–264, preceded by SASQ. This peptide is considered a good candidate for phosphorylation by casein kinases that phosphorylate serine residues present in an acidic environment.[19]

The tonB membrane protein of *Escherichia coli,* involved in iron uptake and in the energy-dependent phase of vitamin B-12 transport,[20] contains two striking charge patterns toward the amino terminus, $PPPEPVV(EP)_4IPEPP$ (positions 68–87), where the (EP) iteration is prominent, and nearby downstream $(KP)_6VKK$ (positions 96–110), where the (KP) iteration is prominent. The apparent molecular weight of the tonB protein by polyacrylamide gel electrophoresis is much larger than the calculated molecular weight, and this effect is largely due to properties of the amino-terminal one-third of the protein.[20] A possible explanation for this may be some dimerization of the protein facilitated by charge interactions between the (EP) and (KP) repeats on two molecules. Such interactions might also conceivably occur within one molecule, thus forming an alternating charge chain in three dimensions.

The leader segment of promelittin, the lytic peptide in bee venom,[21] displays the sequence $(A/P, -)_{10}$ with one mismatch. Processing entails sequential digestion of promelittin to melittin via the type-4 dipeptidylaminopeptidase, the target site of which is the bond of the dipeptides XP or XA. The intervening acidic amino acids X are not relevant to the digestion but may serve other purposes. We may speculate that in promelittin a conformation is established in such a manner that the negative charge of the leader sequence binds to the positively charged region of melittin, thereby protecting the protein from premature cleavage.

The histidine periodic alternation $(H, 0)_n$, $n \geq 9$, occurs in several developmental control proteins of *Drosophila,* including *bicoid,*[22] *e74, paired,* and *spalt.* The major nuclear proteins (EBNA1, 2, and 3) of the latent infectious state of Epstein–Barr virus all contain significant periodic

[19] R. B. Wickner, *Proc. Natl. Acad. Sci. U.S.A.* **85,** 6007 (1988).

[20] K. Postle and R. F. Good, *Proc. Natl. Acad. Sci. U.S.A.* **80,** 5235 (1983).

[21] G. Suchanek, G. Kreil, and M. A. Hermodson, *Proc. Natl. Acad. Sci. U.S.A.* 75, 701 (1987).

[22] W. Driever and C. Nüsslein-Volhard, *Cell* **54,** 83 (1988).

charge patterns and separate charge clusters.[23] The $(-, 0, 0)_6$ pattern of EBNA3 involves several intervening serine residues, a charge structure possibly conducive to serine phosphorylation. MBK-1, a probable mouse potassium channel protein,[24] exemplifies the amphipathic helical S4 segment $(+, 0, 0)_{4-8}$, conserved in all voltage-gated ion channel proteins.

Charge Clusters in High Molecular Weight Heat-Shock Proteins

It is of interest to analyze families of functionally related proteins and to determine whether significant charge configurations are conserved among members of the family. Here we show that all sequenced eukaryotic high molecular weight heat-shock proteins abound with significant charge clusters (see Fig. 1). In both eukaryotic and prokaryotic cells these proteins are synthesized on exposure to raised temperature or to various other stresses. In some cases they are also subject to developmental regulation. While the functions of heat-shock proteins, which are closely related in different species, remain mostly speculative,[25] these proteins are of particular interest because of their interactions with transforming proteins, viral transactivators, and steroid hormone receptors.

The frog hsp70 protein contains two distinct charge clusters (Fig. 1). The corresponding hsp70 genes appear to be constitutively active during oogenesis and become heat-inducible in early embryos.[26] The level of expression of hsp70 is elevated in several tumor cell lines, and its expression is stimulated by both the *c-myc* product[27] and by E1A.[28] The positive charge cluster (but not the mixed charge cluster) of the frog protein (Fig. 1) is conserved in *Drosophila* hsp70.[29] The 70K heat shock protein dnaK of *E. coli*,[30] however, does not contain any charge clusters (not shown), consistent with our previous observation that prokaryotic proteins in the main do not carry significant charge clusters.[31]

Drosophila heat-shock protein 82 is a common protein in early embryos kept at normal temperatures.[32] Its mixed charge cluster (see Fig. 1) is conserved among various *Drosophila* species.[32] Yeast hsp90,[33] which is

[23] B. E. Blaisdell and S. Karlin, *Proc. Natl. Acad. Sci. U.S.A.* **85**, 6637 (1988).
[24] B. L. Tempel, Y. N. Jan, and L. Y. Jan, *Nature (London)* **332**, 837 (1988).
[25] H. R. B. Pelham, *Cell* **46**, 959 (1986).
[26] M. Bienz, *EMBO J.* **3**, 2477 (1984).
[27] R. E. Kingston, A. S. Baldwin, and P. A. Sharp, *Nature (London)* **312**, 280 (1984).
[28] J. R. Nevins, *Cell* **29**, 913 (1982).
[29] T. D. Ingolia, E. A. Craig, and B. J. McCarthy, *Cell* **21**, 669 (1980).
[30] J. C. A. Bardwell and E. A. Craig, *Proc. Natl. Acad. Sci. U.S.A.* **81**, 848 (1984).
[31] S. Karlin and V. Brendel, *Proc. Natl. Acad. Sci. U.S.A.* **85**, 9396 (1988).
[32] R. K. Blackman and M. Meselson, *J. Mol. Biol.* **188**, 499 (1986).
[33] F. W. Farrelly and D. B. Finkelstein, *J. Biol. Chem.* **259**, 5745 (1984).

60% identical to hsp82 of *D. melanogaster,*[32] also contains a very highly charged region. Within this charge cluster are five repeats of three consecutive aspartates or glutamates followed by two lysines (Fig. 1). Interestingly, in *Drosophila* and yeast this region is poorly conserved at the amino acid level, perhaps suggesting that charge concentration rather than other residue attributes is important.

hsp108 of chicken[34] is constitutively expressed in many chicken tissues. In addition to being heat-inducible it is also steroid-inducible. Its sequence has a high degree of similarity to the sequences of hsp82 and hsp90, but rather than mixed charged clusters two highly acidic clusters stand out in this protein (see Fig. 1). The highly acidic carboxy terminus is conserved in the related hamster glucose-regulated protein grp94,[35] where it is speculated to have a possible calcium ion binding function in the endoplasmic reticulum. Both hsp82 and hsp108 contain a significant $(+, 0, 0)$ pattern (Fig. 1). In view of the prominence of charge clusters in transcription factors[3-6], one may speculate that at least some of the heat shock proteins participate via their charge clusters in transcriptional responses to stress.

Merits and Limitations

We have described methods for identifying statistically significant sequence patterns in proteins, exemplified by distinctive charge configurations. Other sequence features that can be investigated similarly by our algorithms and statistical analyses include patterns of amino acids grouped by size or hydrophobicity, and the distribution of specific amino acids (e.g., cysteine) or specific peptides or phosphorylation and glycosylation sites. These approaches can provide insights into the function and classification of proteins.

The length of the protein is rarely a restriction in applying the formulas for discerning significant charge runs and periodic patterns (100 residues suffice). However, charge clusters for segments 25–75 residues in length imply a high concentration of charge relative to the charge composition of the whole protein. It follows that the cluster significance criteria are intrinsically more demanding for establishing a charge cluster in small proteins (<200 residues). We can overcome this difficulty by joining a small protein to a natural group of similar small proteins, concatenating them, and evaluating the extended protein collection for charge clusters. To illustrate this procedure, the human histones H2A, H2B, H3, and H4 were concate-

[34] M. S. Kulomaa, N. L. Weigel, D. A. Kleinsek, W. G. Beattie, O. M. Conneely, C. March, T. Zarucki-Schulz, W. T. Schrader, and B. W. O'Malley, *Biochemistry* **25,** 6244 (1986).
[35] P. K. Sorger and H. R. B. Pelham, *J. Mol. Biol.* **194,** 341 (1987).

nated, yielding an extended protein of length 489 residues having $f_+ = 0.239$ and $f_- = 0.076$. It contained no significant charge clusters. Other natural functional groupings of relatively small proteins could include ribosomal proteins, the globin superfamily, and the *ras* family. We have found no correlation between length and the presence or absence of charge clusters for proteins at least 300 residues in length (the data collection included more than 2500 protein sequences).

Statistical methods, by their nature, cannot identify all peptides that convey important functional activities. For example, the well-established nuclear location signals PKKKRKV in the large T-antigen of SV40, or SRKRPRP in polyomavirus large T,[36] are not, according to our stringent criteria, long enough to qualify as statistically significant positive charge runs. Similarly, the acid concentration (18 acidic amino acids, 2 basic, stretching over 60 residues) in the third quartile of the GCN4 protein[3] is not sufficiently high for this region to qualify as a negative charge cluster. Were the conditions of statistical significance relaxed to accommodate these cases, a deluge of charge configurations would generally be found, and most of these would be due to sampling fluctuations. We opted for our more conservative procedures at the cost of possibly missing some biologically important configurations.

Studies on the charge distribution of a large number of viral proteins revealed a richness of significant charge configurations mainly in eukaryotic DNA viruses.[11,31] A striking example[23] is the Epstein–Barr virus (EBV) nuclear antigen EBNA1, a protein required for replication and maintenance of viral genomes in latently infected cells. This protein carries four separate statistically significant charge clusters, two acidic and two basic. Moreover, the positive charge clusters each contain the periodic patterns $(RG)_4$ that are extended to $(+, 0)_8$ and $(+, 0)_{15}$, respectively. Significant charge configurations further occur in the other major nuclear antigens of EBV expressed in the latent state (EBNA2, 3, 4, and 6), in the latent plasma membrane antigen LMP, and in the principal transactivator of the lytic cycle (BMLF1),[23] as well as in proteins of other eukaryotic viruses, including the immediate early transactivators ICP0 and ICP4 of herpes simplex virus, p62 and p63 of varicella zoster virus, and IE1 of cytomegalovirus;[11] the E1A-32K protein of adenovirus; the middle T-antigen of polyomavirus; and E1 and E7 of papillomavirus.[31] All these proteins are involved in regulation of early viral transcription or of viral replication.

The numbers of significant charge configurations are very different in proteins of DNA versus RNA eukaryotic viruses.[31] From 20 to 35% of the proteins in double-stranded DNA mammalian viruses, independent of

[36] W. D. Richardson, B. L. Roberts, and A. E. Smith, *Cell* **44**, 77 (1986).

genome size, contain some significant charge clusters compared to less than about 3% of RNA viral proteins. Many capsid and core proteins in DNA viral genomes carry a positive charge cluster or run, generally near the carboxy terminus; many membrane-associated and glycoproteins carry a negative charge cluster,[31] and many transactivating proteins carry multiple significant charge configurations.[23] Bacteriophage proteins carry few charge structures compared with many in the proteins of eukaryotic DNA viruses.[31]

In cellular proteins significant charge clusters are prevalent among nuclear transcription and replication factors,[6] nuclear oncogene products and transforming proteins,[16] steroid and thyroid hormone receptors,[6] the opsin superfamily, and homeobox proteins. However, charge clusters are lacking in the bulk of cytoplasmic enzymes and housekeeping proteins. Multiple charge clusters are found in less than 3% of protein sequences but are prominent in voltage-gated ion channel proteins, major developmental control proteins, and high molecular weight heat-shock proteins. Specific association of sequence attributes with functional properties encourages the use of the methods presented here to predict functional domains of proteins. The methods suggest experimental designs including deletions, attenuations, relocations, site-specific mutagenesis, and fusion constructions. Insights into protein structure might also result by synthesizing specific natural peptides, like the unusual charge patterns given in Table I, and examining their conformation under various solvent conditions.

Acknowledgments

This work was supported in part by National Institutes of Health Grants GM10452-26 and GM39907-01 and National Science Foundation Grant MCS82-15131 to S.K., and by Sloan Foundation Grant B1987-2 to V.B.

[25] Protein Multiple Sequence Alignment and Flexible Pattern Matching

By Geoffrey J. Barton

Introduction

The alignment of two or more protein sequences can provide a wealth of information to guide further experimentation, particularly if one of the aligned proteins has been biochemically or crystallographically well characterized. However, any inference from the alignment is crucially dependent on its accuracy. Thus, in this chapter, alignments obtained from comparison of the protein three-dimensional structures are used as a standard against which to test an automatic method for the pairwise alignment of protein sequences. The accuracy of the resulting alignments is shown to improve when additional nonsequence information is incorporated into the algorithm.

Rigorous methods for the alignment of two protein sequences have long been known;[1] however, the calculation of an optimal alignment of four or more sequences is beyond the capabilities of even the most powerful computers. Alignment of multiple sequences by eye is at best a tedious and time-consuming operation; at worst it is unsystematic and leads to an alignment about which no degree of confidence may be expressed. In this chapter, a practical strategy for the rapid multiple alignment of protein sequences is described. Although not guaranteed to give the mathematically optimal alignment, the algorithm is able to cope with large numbers of sequences. It is also a fast procedure that gives alignments generally as good or better than those obtained by pairwise methods.

When sequence similarity is weak, conventional alignment procedures can fail to identify biologically significant relationships against the background of all known sequences. The sensitivity and selectivity of alignment methods that exploit information from single or multiple sequences with and without additional nonsequence information are also evaluated. Furthermore, a technique that relies on the systematic derivation of *flexible patterns* is shown to be superior to all these methods when applied to the globin family of proteins.

---

[1] S. B. Needleman and C. D. Wunsch, *J. Mol. Biol.* **48**, 433 (1970).

Protein Sequence Comparison

Broadly, there are three categories of methods for the comparison of protein sequences. Segment methods compare all overlapping segments of a predetermined length (e.g., 10 amino acids) from one protein with all segments from the other. The distribution of scores obtained for all segment pairs can be used directly to infer homology.[2] Alternatively, the segment scores may be plotted graphically as a "comparison matrix."[3,4] Segment methods have the advantage of simplicity; however, they do not cater explicitly for insertions and deletions (gaps).

Optimal global alignment methods allow the best overall score for the comparison of the two sequences to be obtained including a consideration of gaps. The Needleman and Wunsch algorithm[1] was the first description of a global alignment method applied to protein sequences, but variants of the basic dynamic programming algorithm have been independently developed and applied in many fields (see Sankoff and Kruskal[5] for review). The advantage of these techniques is that they are guaranteed to find the best overall score for the comparison of the sequences including a consideration of gaps. Furthermore, these methods can also produce one or more alignments consistent with this best score. As a consequence, computer programs based on this method (e.g., NBRF, ALIGN) have been widely used for biological sequence comparison. For these reasons, the algorithm of Needleman and Wunsch[1] forms the nucleus of the techniques for pairwise and multiple sequence alignment described in the following sections.

Finally, *Optimal local alignment* algorithms seek to identify the best *local* similarities between two sequences but, unlike segment methods, include explicit consideration of gaps. The methods are based on modified Needleman and Wunsch-style algorithms (e.g., see Refs. 6 and 7) and represent an important class of comparison algorithm, particularly for the location of significantly similar regions between long sequences.

All the methods require a *scoring scheme* for the matching of each of the 210 possible pairs of amino acids (i.e., 190 pairs of different amino acids plus 20 pairs of identical amino acids). For example, the simple identity scoring scheme gives a score of 1 to identical pairs and 0 to all others. More sophisticated schemes can incorporate knowledge about the

[2] W. M. Fitch, *J. Mol. Biol.* **16**, 9 (1966).

[3] A. D. McLachlan, *J. Mol. Biol.* **61**, 409 (1971).

[4] P. Argos, *J. Mol. Biol.* **193**, 385 (1987).

[5] D. Sankoff and J. B. Kruskal (eds.) "Time Warps, String Edits, and Macromelecules: The Theory and Practice of Sequence Comparison. Addison Wesley, Reading, Massachusetts, 1983.

[6] T. F. Smith and M. S. Waterman, *J. Mol. Biol.* **147**, 195 (1981).

[7] D. R. Bowsell and A. D. McLachlan, *Nucleic Acids Res.* **12**, 457 (1984).

physical properties of the amino acids,[3] minimum allowed base changes,[2] or observed substitutions[8] to give a symmetrical 20×20 matrix of scores.

Gap Penalties and Needleman–Wunsch Algorithm

The Needleman–Wunsch algorithm is an elegant procedure that allows the best alignment of two sequences of length N to be calculated in N^2 steps. When comparing two sequences the algorithm seeks to model the real (possibly evolutionary) processes involved in converting one sequence to the other. The scoring scheme that dictates the weight for aligning one type of amino acid with another is part of this model. For any chosen scoring scheme, the Needleman–Wunsch algorithm will find the maximum possible score for the comparison of the two sequences. However, this optimal alignment may require the insertion of an unrealistically large number of gaps (residues aligned with blanks).

In order to overcome this problem and limit the total number of gaps created, an additional factor is introduced into the model. This takes the form of a gap penalty which is subtracted during the process of calculating the best alignment whenever a gap is allowed. One of the most commonly used gap-penalty functions takes the form:

$$P = G_1 L + G_2 \tag{1}$$

where L is the length of the gap while G_1 and G_2 are user-defined constants. This form of penalty has both length-independent (G_2) and length-dependent ($G_1 L$) terms that are sometimes known as penalties for creation of a gap and extension of a gap, respectively.

Criteria for Assessing Quality of Alignment

Given the Needleman–Wunsch algorithm, a scoring scheme, and gap-penalty function, we have a system that can optimally align any two protein sequences. It is important to bear in mind, however, that this alignment is optimal only with regard to the chosen model; changing the model, either by using a different scoring scheme or a modified gap penalty, can lead to completely different alignments. While all of these alignments will be mathematically optimal it is possible that none of them illustrate genuine, biologically significant equivalences. Thus, when applying and interpreting automatically obtained alignments three questions need to be answered: What is a good protein sequence alignment? How

---

[8] M. O. Dayhoff, R. M. Schwartz, and B. C. Orcutt, *in "Atlas of Protein Sequence and Structure"* (M. O. Dayhoff, ed.), Vol. 5, p. 345. National Biomedical Research Foundation, Washington, D.C., 1978.

closely can automatic alignment procedures reproduce a good alignment? Can we estimate the likely quality of an alignment from sequence information alone?

What Is Good Protein Sequence Alignment?

The protein three-dimensional structure determines its biological activity. It is therefore of crucial importance when two or more protein sequences are aligned that those residues defining a common tertiary fold, together with any common catalytic and binding residues, are correctly equivalenced. It follows that a *good* alignment of two globular proteins is one which faithfully reflects any similarities in three-dimensional structure.

However, although the overall fold may be conserved within a protein family, there is usually considerable variation in the details of structure for individual family members. In particular, although core secondary structures (α helices and β strands) may exhibit similar conformations and relative positions in three dimensions, the loop regions linking these structures may not. Thus, even when the proteins to be aligned have crystallographically determined structures known to high resolution, it can be difficult or impossible to obtain a consistent alignment over the whole length of *all* the family members.

In order to be useful, a protein sequence alignment method must at least align those residues that are performing equivalent structural roles in the two proteins. For this reason when assessing an alignment algorithm for two or more protein sequences, *test zones*[9] are selected that correspond to equivalent secondary structural regions in the proteins. The accuracy of any alignment can then be expressed in terms of the number of residues within the test zones that are equivalenced in the same way as expected from three dimensional structure comparison.

How Well Do Automatic Alignment Methods Perform?

The option of using different scoring scheme and gap-penalty combinations complicates the evaluation of alignment methods. Studies performed by Dayhoff *et al.*[8] and more recently by Feng *et al.*[10] indicate that, on average, the Dayhoff mutation data matrix (MDM) is more effective than the simple identify matrix, genetic code, or physical property scoring schemes at detecting homology between distantly related proteins. Using Dayhoff's matrix, Barton and Sternberg[9] considered a range of gap-penalty constants ($G_1 = 0-10, G_2 = 0-10$ in integer steps) for five polypeptide

---

[9] G. J. Barton and M. J. E. Sternberg, *Protein Eng.* **1**, 89 (1987).

[10] D. F. Feng, M. S. Johnson, and R. F. Doolittle, *J. Mol. Evol.* **21**, 112 (1985).

pairs for which alignments based on three-dimensional structure were known. For each pair of sequences, test zones were selected from the common core secondary structural regions and mean, maximum, and minimum accuracies of alignment obtained over the 121 comparisons (Fig. 1). Clearly, some protein pairs align very well (1, 4 and 5), whereas others give poor alignments (2, 3). Furthermore, the best alignment obtained for each pair did not require a length-dependent gap penalty (i.e, $G_1 = 0$).

Use of Significance Scores to Estimate Likely Quality of Alignment

With the increase in the number of protein sequences derived directly from cDNA one frequently wishes to align two sequences for which there is little or no additional nonsequence information available to guide the alignment (e.g., three-dimensional structure, catalytic residues). Under

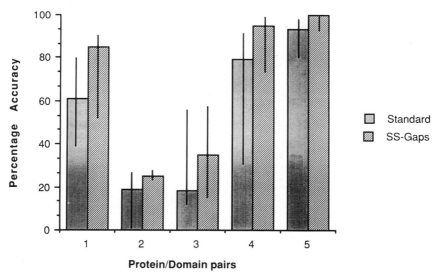

FIG. 1. Accuracy of pairwise sequence alignments by the Needleman–Wunsch method by comparison with tertiary structural alignments. Values of G_1 and G_2 from 0 to 10 in integer steps were used with MDM (+8 to remove negative elements). 1, Immunoglobulin light chain variable region (FABVL) versus heavy chain variable region (FABVH); 2, FABVH versus light chain constant region (FABCL); 3, plastocyanin versus azurin; 4, human α-hemoglobin versus leghemoglobin; 5, trypsin versus elastase. Standard: Mean values for 121 comparisons using the conventional Needleman–Wunsch algorithm. SS-Gaps: Effect of including secondary structure-dependent gap penalties. Upper and lower extremities of vertical bars indicate the best and worst alignments obtained.

these circumstances it is invaluable to have an indication of the likely accuracy of any automatic alignment performed.

A commonly used method of assessing the similarity between two sequences proceeds as follows. First, the best score V for the comparison of the native sequences is obtained. The sequences are then randomized a number of times (typically 100) to give artificial sequences with the same length and composition as the native. The best score for aligning each pair of randomized sequences is then obtained, and the mean m and standard deviation (S.D.) of this distribution of random scores are calculated. The similarity of the native sequences is then expressed in terms of the number of standard deviation units away from the mean of the random distribution [i.e., Score $= (V - m/\text{S.D.}]$.

A study was performed[11] which considered all pairwise comparisons within seven globin sequences and eight immunoglobulin domains (49 unique pairs in all). The results shown in Fig. 2 illustrate that alignments scoring above 15.0 S.D. (seven examples) give at or near 100% agreement with the reference alignment. Those scoring between 5.0 and 15.0 S.D. (25 examples) give better than 70% agreement with the reference alignment, whereas scores below 5.0 SD (17 examples) show a sharp rise in alignment accuracy correlated with significance score and ranging from 0% (0.57 S.D.; FABCH1 versus FB4VH) to 84% (2.4 S.D.; FABVL versus FABCL). Above 5.0 S.D. there are *no* really poor alignments; however, in the lower standard deviation range small changes in observed significance score can indicate a considerable difference in alignment accuracy.

These studies provide guidelines for the quality of a protein sequence alignment. Clearly, a near ideal alignment is indicated by significance scores above 15.0 S.D. Scores above 5.0 S.D. suggest a "good" alignment, whereas an alignment giving a score below 5.0 S.D., although possibly good, must be regarded with greater caution.

Improving Sequence Alignments by Using Secondary Structure-Dependent Gap Penalties

The alignment model consisting of a scoring scheme and gap-penalty function can produce good alignments, however, the gap-penalty function [Eq. (1)] acts equally over the entire sequence length. This feature does not reflect observations on families of known protein three-dimensional structures where there is a clear preference for insertions/deletions to occur in loop regions linking the core secondary structures (e.g., see Refs. 12 and

[11] G. J. Barton and M. J. E. Sternberg, *J. Mol. Biol.* **198,** 327 (1987).
[12] M. F. Perutz, J. C. Kendrew, and H. C. Watson, *J. Mol. Biol.* **104,** 59 (1965).

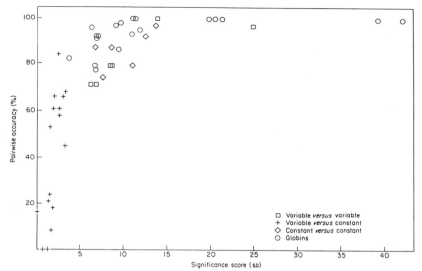

FIG. 2. Pairwise alignment accuracy versus significance score (100 randomizations) for seven globins [human α-hemoglobin (HAHU), human β-hemoglobin (HBHU), horse α-hemoglobin (HAHO), horse β-hemoglobin (HBHO), myoglobin (MYWHP), lamprey globin (P1LHB), leghemoglobin (LGHB)] and eight immunoglobulin domains [consisting of one light chain constant domain (FABCL), three heavy chain constant domains (FABCH, FCCH2, FCCH3), two light chain variable domains (FABVL, FB4VL), and two heavy chain variable domains (FABVH, FB4VH)]. Dayhoff's matrix was used: (250 PAM) + 8, $G_2 = 8$, $G_1 = 0$.

13). Indeed, errors in sequence alignment can frequently be attributed to the misplacing of a gap in a core secondary structural region.

The alignment model may be improved to better match the observed pattern of insertions by using a modified gap-penalty function:

$$P_{ss} = Q\,(G_1 L + G_2) \qquad (2)$$

where $0 \le Q \le 1$ and the subscript ss denotes the inclusion of secondary structural information. This change has the effect of reducing the penalty for a gap in loop regions relative to secondary structural regions.

In its simplest form, Q takes a value of 1.0 for regions of secondary structure and a value of less than 1.0 for loop regions. The effect of applying this type of penalty is illustrated in Fig. 1. All five protein pairs show improvements in mean accuracy and improvements in the worst alignment obtained, and, with the exception of FABVH *versus* FABCL, the best alignment obtained also gives a higher accuracy.

[13] A. M. Lesk and C. Chothia, *J. Mol. Biol.* **136**, 225 (1980).

In its most general form, Q may be derived from a property of the sequence which exhibits a *maximum* for regions likely to be involved in secondary structures or other conserved regions and a minimum for regions likely to be subject to greater variability. Q might therefore be derived from a secondary structure prediction profile,[14,15] a smoothed profile based on hydrophobicity,[16] or a profile of likely buried residues.[17] Unfortunately, none of these methods predict the location of secondary structural elements with sufficient accuracy to improve the alignment quality. Indeed, in studies using several alternative predictive schemes to derive values of Q over the sequences, the overall accuracy of alignment actually decreased.

This observation clearly limits the applicability of the modified gap-penalty function to systems where one of the proteins has a known X-ray structure. For such systems the improvement in accuracy obtained justifies the inclusion of secondary structural information into the alignment and is of particular use when the alignment is to be used for subsequent building of a three-dimensional model by homology (e.g., see Ref. 18). Lesk and co-workers[19] described a similar technique and showed that it improves the alignment of sequences within the globin and serine proteinase families.

Simultaneous Alignment of More than Two Sequences (Multiple Alignment)

Needleman and Wunsch[1] suggested that their dynamic programming algorithm could be extended to the simultaneous comparison of many sequences. Waterman *et al.*[20] also described how dynamic programming could be used to align more than two sequences. In practice, however, the need to store an N-dimensional array (where N is the number of sequences) limits these extensions to three-sequence applications (e.g., see Ref. 21). In addition, the time required to perform the comparison of even three sequences is proportional to N^5. Murata *et al.*[22] described a modification of the Needleman–Wunsch procedure for three sequences which ran in time proportional to N^3; unfortunately, this approach required an additional

[14] J. Garnier, D. J. Osguthorpe, and B. Robson, *J. Mol. Biol.* **120**, 97 (1978).
[15] P. Y. Chou and G. D. Fasman, *Adv. Enzymol.* **47**, 45 (1978).
[16] M. Levitt, *J. Mol. Biol.* **104**, 59 (1976).
[17] J. Janin, *Nature (London)* **277**, 491 (1979).
[18] T. L. Blundell, B. L. Sibanda, and L. Pearl, *Nature (London)* **304**, 273 (1983).
[19] A. M. Lesk, M. Levitt, and C. Chothia, *Protein Eng.* **1**, 77 (1986).
[20] M. S. Waterman, T. F. Smith, and W. A. Beyer, *Adv. Math.* **20**, 367 (1976).
[21] R. A. Jue, N. W. Woodbury, and R. F. Doolittle, *J. Mol. Evol.* **15**, 129 (1980).
[22] M. Murata, J. S. Richardson, and J. L. Sussman, *Proc. Natl. Acad. Sci. U.S.A.* **82**, 3073 (1985).

three-dimensional array, thus further limiting its application to short sequences.

The multiple alignment of four or more sequences cannot in practice be performed by a rigorous method since even when gaps are not explicitly considered, the number of segment comparisons that must be made is of the order of the product of the sequence lengths. Algorithms for multiple sequence alignment therefore seek to identify an optimum alignment by considering only a small number of the total possible residue or segment comparisons. Several authors have described multiple alignment algorithms; however, they either do not give an overall alignment,[23] are restricted to relatively few sequences,[24] or are specifically intended for aligning nucleic acid sequences and do not allow the flexibility in scoring scheme that is useful for protein sequence comparison.[25,26] In the following section a method that permits large numbers of protein sequences to be aligned quickly is described and tested by comparison with alignments obtained from the comparison of protein three-dimensional structures.

Effective and Rapid Strategy for Multiple Protein Sequence Alignment[11]

The alignment algorithm described here reduces the multiple alignment of N sequences to a set of $N - 1$ pairwise alignments and is summarized in Fig. 3. (1) Sequences A and B are optimally aligned by the Needleman–Wunsch algorithm. (2) The third sequence is optimally aligned with the *alignment* resulting from Step 1. Average scores are used when comparing a residue in sequence C to an aligned position in the result of Step 1. For example, the score for matching the alignment of Ala and Val with Ala would be given by the score for (Ala versus Ala) plus (Val versus Ala) divided by 2. Gaps that are already present in the alignment from Step 1 are maintained, and a low score is assigned to matching an amino acid in sequence C with any such gap. This score is used when calculating the average score at the aligned position. For example, if the aligned position is AlaGap then the score for matching Ala would be given by (Ala versus Ala) plus (Gap versus Ala) dived by 2. (3) The multiple alignment of sequences A, B, and C obtained in Step 2 is now optimally aligned with sequence D using a procedure similar to Step 2. (4) Step 3 is repeated until all sequences have been added to the alignment. (5) The alignment from Step 4 may optionally be refined by reoptimizing the alignment of each sequence with the completed alignment less that sequence.

---

[23] D. J. Bacon and W. F. Anderson, *J. Mol. Biol.* **191**, 153 (1986).
[24] M. S. Johnson and R. F. Doolittle, *J. Mol. Evol.* **23**, 267 (1986).
[25] E. Sobel and H. M. Martinez, *Nucleic Acids Res.* **14**, 363 (1986).
[26] W. Bains, *Nucleic Acids Res.* **14**, 159 (1986).

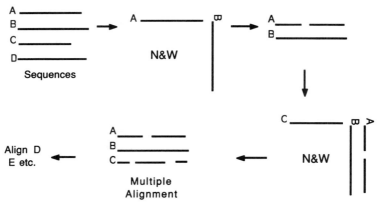

FIG. 3. Summary of the multiple alignment process.

Order of Alignment

Since the multiple algorithm shares the Needleman–Wunsch procedure with pairwise methods, the alignment will be dependent on both the scoring scheme and gap penalty. In addition, there are $N!$ alternative orders in which the sequences could be aligned. A systematic procedure for determining the alignment order must therefore be applied.

The pairwise comparison tests shown in Fig. 2 demonstrate that the accuracy of alignment is correlated with the significance score. The single alignment order may therefore be determined by first calculating significance scores for all unique *pairwise* comparisons within the sequence set. Then, when generating the multiple alignment, the pair of sequences that gives the highest significance score is aligned first. Of the remaining sequences, the one which gives the highest score when compared to A or B is then aligned. The process is repeated for all remaining sequences, where every *ith* sequence being added to the alignment is the one that gives the highest pairwise significance score with the $i - 1$ sequences already aligned.

Cluster Analysis

A useful method of visualizing the pairwise comparison data is to apply the technique of single linkage cluster analysis. This provides a convenient representation in the form of a dendrogram that can illustrate some of the interrelationships between the members of a sequence group.

The dendrograms illustrated in Fig. 4a,b for the seven globins and eight immunoglobulins used to evaluate pairwise methods clearly show the

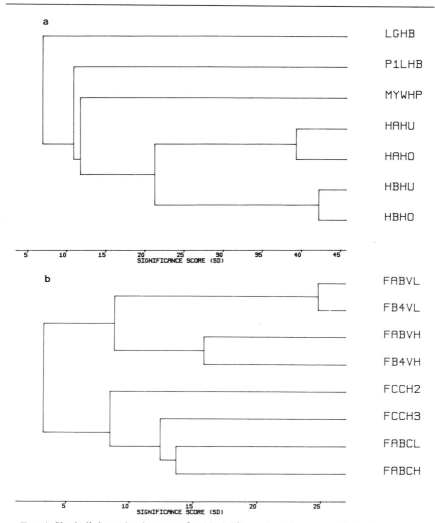

FIG. 4. Single-linkage dendrograms for (a) globins and (b) immunoglobulin domains.

sequences grouped by similarity. The maximum level of similarity between the groups is also readily apparent. Given that the relationship between significance score and alignment accuracy is known, the dendrogram in conjunction with pairwise scores can help to identify quickly which pairs of sequences may align to high accuracy. Furthermore, the high scoring clusters may indicate groups of sequences that will also align well by the multiple alignment algorithm.

Reduction of Calculation Time

Before a group of sequences can be ordered, or cluster analysis performed, it is necessary to calculate scores for all sequence pairs. This is an expensive procedure since if M randomizations are performed, $N(N - 1)M/2$ alignments must be generated. Feng *et al.*[10] considered how many randomizations need be performed on a pair of sequences before consistent significance values are obtained. On the basis of 4 pairs of sequences they suggested that as few as 25 could produce a genuinely reflective score. A related study using a larger dataset (47 protein pairs)[11] indicated that instabilities in significance score do not damp out until at least 60 randomizations are performed. It is therefore impractical to use a randomization procedure to establish the order when large numbers of long sequences are to be aligned. However, it is possible to derive a normalized alignment score (NAS) directly from the match score V without the need for randomization.[10,11] Scores of this type correlate well with the significance score, suggesting that when central processing unit (CPU) time would otherwise by prohibitive NAS values can be used to establish an alignment order and reduce the number of comparisons that need be made by a factor of at least 60.

Evaluation of Multiple Alignment Algorithm: Comparison with Pairwise

The alignment procedure described above is able to produce a multiple alignment for *any* set of sequences. However, as with pairwise methods, it is of vital importance that the properties and limitations of the method are well understood so that its best features can be exploited when it is applied to new systems.

The seven globin and eight immunoglobulin sequences used to evaluate pairwise methods also provide a good test system for the multiple algorithm. In common with the pairwise method, the gap penalty and scoring scheme may be varied; however, there are three additional factors to be considered: (1) Which sequences should be included in a multiple alignment? (2) Does the order of alignment have a serious effect? (3) Can alignments be improved by iteration (Step 5 above)?

Point (1) was considered by multiply aligning four groups of sequences derived from the globins and immunoglobulins: Alignment 1, the seven globin sequences HBHU, HBHO, HAHU, HAHO, MYWHP, P1LHB, and LGHB (see Fig. 5); Alignment 2, the four constant immunoglobulin domains FABCL, FABCH, FCCH3, AND FCCH2; Alignment 3, the four variable domains FABVL, FB4VL, FB4VH, AND FABVH; and Alignment 4, the eight immunoglobulin domains used in Alignments 2 and 3.

The effect of order was addressed by considering alternative alignment

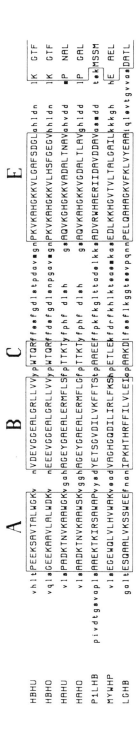

FIG. 5. Multiple alignment of seven globin sequences. Boxed regions with capital letters refer to test zones. Regions A, B, C, E, F, G, and H are all α helical in the known protein structures.

orders for Alignments 1 and 4, and the effect of applying up to four iterations was investigated for all four alignments. Figure 6 shows the accuracy of alignment obtained for pairs of sequences within the four multiple alignments compared to the accuracy obtained when the sequences are aligned pairwise. Points above the diagonal represent an improvement in alignment when the multiple algorithm is applied. The globin multiple alignment (1) gives an overall improvement from 90 to 99% accuracy, with the largest improvement for the comparison of leghemoglobin with human β-hemoglobin (77 to 99%). Alignments 2 and 3 also show an overall improvement in accuracy; for the constant domains this is from 86 to 90%, while the variable domains improve from 83 to 84%. The

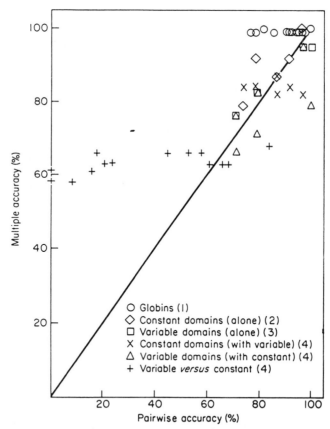

FIG. 6. Accuracy of alignment obtained by the multiple method versus the conventional pairwise method.

most striking improvement is for variable versus constant domains within Alignment 4. Some sequence pairs which were completely misaligned by the pairwise procedure gave around 60% accuracy when multiple aligned (e.g., FB4VH versus FABCH). However, this improvement was obtained at the expense of a slight degradation of variable *versus* variable and constant *versus* constant alignments.

Although Alignment 4 shows a large improvement in accuracy for low scoring sequence pairs, further studies (results not shown) suggest that improvements in accuracy can be very variable for this type of sequence. Furthermore, when ten alternative alignment orders were generated for the seven globin and eight immunoglobulin alignments (1 and 4), the alternative orders had very little effect on the globin accuracy ($< 1\%$), but for the immunoglobulin most alternative orders gave poorer alignments (mean of 57.6% compared to 70.8% for the order based on S.D. score). These findings are consistent with the observed variation in alignment accuracy below significance scores of 5.0 S.D. for pairwise comparisons (Fig. 2).

The use of up to four iterations to refine the initial multiple alignment showed that in general there was no benefit in performing more than two iterations. Alignments 1–3 improved by approximately 1% over an alignment with no iterations, although, once again, Alignment 4 proved to be atypical with an improvement of around 9%.

In summary, this evaluation suggests that for sequence groups that on pairwise comparison cluster above 5.0 S.D. (e.g., Alignments 1, 2, and 3) the resulting multiple alignment is likely to be as good or better than corresponding pairwise alignments. In common with the findings for pairwise methods, the multiple alignment obtained for sequences that cluster below 5.0 S.D. (e.g., Alignment 4) is likely to be unpredictable in quality. Furthermore, time can be saved when large numbers of sequences are to be aligned by using the minimum calculation of normalized alignment scores to establish the alignment order rather than significance scores. The observation that alignment order has little effect on the result for some groups of sequences (e.g., the seven globins) suggests that an arbitrary alignment order may often be acceptable, thus removing the time-consuming need to perform all pairwise comparisons prior to multiple alignment.

Tree-Based Multiple Alignment

The algorithm evaluated in the previous section considers the sequences in a single linear order. However, the dendrogram representation of the pairwise comparison data (Fig. 4a,b) suggests an alternative way in which to order the multiple alignment process. Rather than starting with the most similar pair and adding successively to that alignment, the dendrogram or tree is followed exactly from its branches to the root. For example, with the seven-globin alignment the following series of alignments are performed: (1) Align the most similar pair of sequences, HBHU and HBHO, to give *alignment* HBHU:HBHO. (2) Align the next most similar pair, HAHU and HAHO, to give alignment HAHU:HAHO. (3) Now align the two *alignments* HBHU:HBHO and HAHU:HAHO to give the four-sequence alignment HBHU:HBHO:HAHU:HAHO. (4) Align MYWHP to the four-sequence alignment obtained in Step 3, then P1LHB to the resulting five-sequence alignment, and finally LGHB to the six-sequence alignment to give the final seven-sequence alignment.

The only new operation involved in this process is the alignment of two *alignments* shown in Step 3. This step is essentially the same as adding a single sequence to an alignment, only now it is necessary to calculate mean scores over all unique pairs of residues at each position. As before, gaps that already exist in either alignment are maintained, and the score for matching two such gaps is given a low score.

The tree-based approach is intuitively better than a single order alignment method. However, for sequences that all cluster at high scores, the differences in alignment are only slight, and for the seven globins the end result is identical. Where there are two or more distinct high scoring clusters that do not form a single high scoring cluster (as for the eight immunoglobulins), then a tree-based alignment will give better results within the high scoring clusters. However, the problem of variable alignment quality when low scoring sequence pairs are compared still remains. Thus the eight-immunoglobulin alignment when performed tree-wise gives better accuracy for constant *versus* constant and variable *versus* variable domains but equally unpredictable results for variable *versus* constant domains. The tree-based, or *progressive,* multiple alignment method has been described by Feng and Doolittle[27] (also this volume,[23]), who demonstrated that better phylogenetic trees could be obtained from its application.

Speed of Multiple Alignment and Applications

An advantage of the multiple alignment algorithm described here is its speed. For example, the complete seven-sequence globin alignment required only 65 CPU sec (2 iterations) on a VAX 11/750. Pairwise comparisons to establish the order without randomization required 44 sec, giving a total time of 109 sec. If an arbitrary order with no iterations had been used, the total time required would be approximately 20 sec. This compares

[27] D. F. Feng and R. F. Doolittle, *J. Mol. Evol.* **25**, 351 (1987).

favorably with the algorithm of Johnson and Doolittle[24] which requires 60 *min* of CPU time to align five sequences of less than 50 residues in length and cannot easily be extended to cope with large numbers of sequences.

Aligning large numbers of medium length sequences (150–300 residues) by single order or tree methods is therefore a matter of routine. For example, the alignment of 128 globin sequences including α and β hemoglobin, myoglobin, and leghemoglobin from a wide range of species required only 8.5 min of CPU time to produce an alignment prior to refinement by iteration.[11] The alignment of longer sequences is also practical. This is shown in one application of the algorithm to the prediction of potential T and B lymphocyte-defined epitopes on the *env, gag,* and *pol* viral polyproteins of the human immunodeficiency virus (HIV). Four viral isolates were aligned (500–1000 amino acids in length), and analysis of residue conservation in combination with structure prediction methods allowed potential epitopes to be identified.[28,29]

The speed and accuracy of the alignment method have also permitted an improved secondary structure prediction method to be developed.[30] The prediction algorithm combines Robson prediction values[14] averaged over all aligned sequences with a measure of the residue conservation at each aligned position. The use of a conservation value has the effect of reducing the likelihood of predicting secondary structure (α helix or β strand) in regions where gaps have been inserted in the alignment. The overall improvement in accuracy over the standard Robson method was 8.5% obtained for 11 protein families representative of the most common structural classes (α/α, β/β, α/β, and $\alpha + \beta$).

Refinements to Improve Speed and Sensitivity

Since the multiple algorithm is built from successive applications of a pairwise technique, any refinements available to pairwise methods may also be incorporated into the multiple alignment procedure. For example, the time required to perform a Needleman–Wunsch alignment can be reduced by "cutting corners" during calculation of the best alignment.[31] Fast but approximate pairwise methods (e.g., the algorithm of Lipman and

[28] A. R. M. Coates, J. Cookson, G. J. Barton, M. J. Zvelebil, and M. J. E. Sternberg, *Nature (London)* **326**, 549 (1987).

[29] M. J. E. Sternberg, G. J. Barton, M. J. J. Zvelebil, J. Cookson, and A. R. M. Coates, *FEBS Lett.* **281**, 231 (1987).

[30] M. J. J. Zvelebil, G. J. Barton, W. R. Taylor, and M. J. E. Sternberg, *J. Mol. Biol.* **195**, 957 (1987).

[31] J. B. Kruskal and D. Sankoff, *in* "Time Warps, String Edits, and Macromolecules: The Theory and Practice of Sequence" (D. Sankoff and J. Kruskal, eds). p 265. Addison-Wesley, Reading, Massachusetts, 1983.

Pearson[32] may also be used in place of the rigorous Needleman–Wunsch algorithm. Both these methods are most useful when long sequences that have relatively few differences are to be aligned.

Nonsequence information may also be incorporated to guide multiple alignments. Secondary structure-dependent gap penalties as described earlier may be incorporated when at least one sequence has a known three-dimensional structure. Furthermore, position-specific weights may be assigned to residues of known importance (e.g., catalytic amino acids) to increase their likelihood of aligning with similar amino acids.

Guidelines for Performing Multiple Protein Sequence Alignments and Assessing Accuracy

Given a group of sequences to multiply align the following steps may be followed: (1) Ideally all pairwise comparisons for the sequences should be performed using at least 60 randomizations to establish significance scores. (2) Cluster analysis may then be applied to the pairwise data resulting from Step 1 and a dendrogram drawn to represent the results. (3) The dendrogram should be inspected to locate any high scoring clusters of sequences. The sequences that cluster above 5.0 S.D. can be multiply aligned with a high degree of confidence. Outlying sequences (those that do not belong to high scoring clusters) should be removed for possible incorporation in Step 5. (4) The sequences within the high scoring clusters identified in Step 3 should be multiply aligned following the order suggested by the pairwise significance scores. (5) Steps 1–4 will produce one or more "core" alignments that are largely correct. The next step is to align the remaining weakly similar sequences to one or more cores, making use of additional nonsequence information where possible, for example, the location of known catalytic or structural regions common to both sequence groups. A flexible pattern derived from the core alignments may be used to assist in this procedure (see following section). (6) The final alignment(s) must *always* be discussed in the light of the likely error rates implied by the pairwise significance scores and any assumptions made in combining core alignments.

Flexible Patterns: Sensitive Method to Detect Weak Structural Similarities

The score obtained when two sequences are optimally aligned by the Needleman–Wunsch algorithm tells us how similar the sequences are

---

[32] D. J. Lipman and W. R. Pearson, *Science* **227**, 1435 (1985).

according to the model of evolutionary change implied by the scoring scheme and gap penalty. However, this scheme can give lower scores for protein pairs that are known to have similar tertiary structures than for either random sequences of the same length and composition or an arbitrary pair of unrelated protein sequences. In other words, the similarity between the proteins may be hidden in the noise generated by chance high scoring alignments.

The overall improvement in alignment accuracy observed when multiple rather than pairwise alignments are used suggests one route by which the sensitivity of an alignment procedure may be improved. Reliably aligned protein families clearly contain information that is not available from a single sequence, for example, the importance of conservation at particular residue positions and the disposition of gaps. Multiple alignment information of this type has been exploited to improve the sensitivity of comparison between families of aligned proteins (e.g., see Refs. 4 and 33), and, as suggested in the previous section, they can also be used in conjunction with additional nonsequence information (e.g., additional weights for important residues, secondary structure-dependent gap penalties).

Another route by which greater alignment sensitivity has been achieved is to abstract a pattern of allowed residues that represents a particular protein fold then use this pattern rather than the complete sequence to identify the fold in another protein (e.g., ADP-binding proteins[34]). The flexible pattern method[35] allows patterns of this type to be readily expressed and compared to any number of protein sequences. Briefly, a flexible pattern is defined in terms of a series of n *elements*, E_i, and $n - 1$ *gaps*, F_j, where each pattern starts and ends with an element (e.g., E_1, $F_1, E_2, F_2, E_3, F_3, E_4$). In its most general form an element is a place marker defined in terms of its position and the score obtained when it is aligned with each amino acid type. This definition allows all conventional scoring systems to be accommodated. *Gaps* are defined with a specific length range ≥ 0. For example F_1 might be set to $0, F_2$ to a value of $5 \leq F_2 \leq 12$. This definition implies that deletions within the pattern are not allowed, although deletions from the ends (where gap lengths are not explicitly stated) may occur. Figure 7 illustrates a hypothetical pattern derived from a number of different information sources. A modified Needleman–Wunsch algorithm was developed to allow the best alignment between a pattern and a sequence to be determined. This algorithm also allows any repeats of the pattern to be located within the sequence.[36]

[33] W. M. Fitch, *J. Mol. Biol.* **49,** 1 (1970).
[34] R. K. Wierenga, P. Terpstra, and W. G. J. Hol, *J. Mol. Biol.* **187,** 101 (1986).
[35] G. J. Barton, *Ph.D. Thesis,* University of London, 1987.
[36] G. J. Barton and M. J. E. Sternberg, *J. Mol. Biol.* submitted (1989).

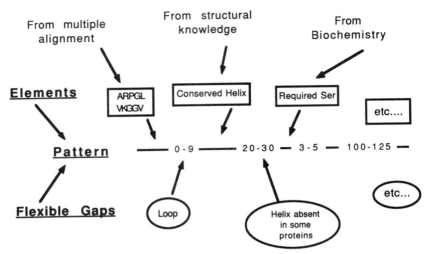

Fig. 7. Generalized flexible pattern consisting of pattern elements, which may be derived from a variety of sources defining alternative scoring schemes, and flexible gaps, which permit an allowed range of insertions but no others.

Alignment Methods Considered

If we accept that using multiple sequence data or patterns is an improvement on using a single sequence to identify similarity, it remains to decide which is the best approach and what are the limits of its sensitivity. The following approaches may be used: FASTP, the database scanning procedure of Lipman and Pearson, which uses a single query sequence and identity scoring scheme and treats gaps uniformly; NW, the Needleman–Wunsch rigorous pairwise alignment procedure, which uses a single query sequence and treats gaps uniformly; NW–SS, the NW method but with secondary structure-dependent gap penalties; BS, the multiple alignment procedure of Barton and Sternberg, in which an *alignment* of two or more sequences is optimally aligned with each entry in the database in turn and which treats gaps uniformly; BS–SS, the BS method but with secondary structure-dependent gap penalties, a procedure similar to that of Gribskov et al.;[37] and FP, the flexible pattern approach (with patterns derived from one sequence or a sequence alignment), which may include secondary or tertiary structural information and for which gap ranges between pattern elements are explicitly defined.

[37] M. Gribskov, A. D. McLachlan, and D. Eisenberg, *Proc Natl. Acad. Sci. U.S.A.* **84**, 4355 (1987).

Evaluation of Alignment Methods by Database Scanning

A convenient method to assess the sensitivity and selectivity of an alignment procedure is to test its ability to identify known members of a protein family from the database of all known sequences. The evaluation procedure consists of optimally aligning the query sequence(s) or pattern with each sequence in the database then rank ordering the scores. The selectivity of the method is then estimated by counting how many of the known family members have higher scores than the first nonfamily protein. The sensitivity of the procedure is shown by the overall profile of scores given for family members.

In the studies described here the globin family was used as a test system since there are a large number of entries in the database (345 complete sequences as well as 17 fragments in PIR Release 14), which vary in biological source (representatives from mammals, plants, and bacteria). Furthermore, the globins exhibit very similar protein folds, and several globin structures have been determined to high resolution by X-ray crystallography. For each scan performed, three values were determined: (1) the number of whole globins giving higher scores than the first nonglobin, (2) the number of whole globins not in group 1 but still present in the top 500 scoring sequences, and (3) the number of whole globins not in groups 1 or 2. The three-dimensional structure-based alignment of seven globins performed by Bashford *et al.*[38] was used as the information source for those algorithms that require multiple alignment data and/or secondary structural information (see Fig. 8).

Comparison of Alignment Methods

The result of the scans performed is summarized in Fig. 9. Scans 1, 2, and 3 used the complete human α-hemoglobin (HAHU) sequence to query the database. The NW procedure (scan 2) performed better than the widely used FASTP program (scan 1) both in terms of selectivity (NW placed 306 globins before first the nonglobin, whereas FASTP only 297) and sensitivity (only 31 globins not in the top 500 scores versus 41). The inclusion of secondary structural information in the form of modified gap penalties (scan 3) gave a further improvement; however, there were still 34 globins that gave alignment scores below that of a nonglobin, 25 of which were not in the top 500 sequences.

Scans 4, 5, and 6 applied multiple sequence information from the structural alignment of seven globin sequences. As expected, this additional information makes the BS algorithm (scan 4) more selective and

---

[38] D. Bashford, D. C. Chothia, and A. M. Lesk, *J. Mol. Biol.* **196**, 199 (1987).

Secondary Structure	Alignment							Elements of Flexible Patterns 1 2 3 4 5 6 7 8
.	.	.	.	.	.	.	.	
.	.	.	.	.	.	.	.	
		V				A	G	
	V	H	V		P	A	G	
	L	L	L	L	L	L	L	
A 1	S	T	S	S	S	T	S	| | | | | | |
A 2	P	P	E	A	A	E	A	| |
A 3	A	E	G	D	A	S	A	| | |
A 4	D	E	E	Q	E	Q	Q	| | | | | |
A 5	K	K	W	I	K	A	R	|
A 6	T	S	Q	S	T	A	Q	| | | | |
A 7	N	A	L	T	K	L	V	|
A 8	V	V	V	V	I	V	I	| | | | | | | |
A 9	K	T	L	Q	R	K	A	| |
A10	A	A	H	A	S	S	A	|
A11	A	L	V	S	A	S	T	| | |
A12	W	W	W	F	W	W	W	| | | | | | | |
A13	G	G	A	D	A	E	K	|
A14	K	K	K	K	P	E	D	|
A15	V	V	V	V	V	F	I	| | | | | | |
A16	G		E	K	Y	N	A	
	A		A	G	S	A	G	**Flexible Gap**
							N	Explicit gap range
							D	for Pattern 1
B 1	H	N	D		T	N	N	
B 2	A	V	V		Y	I	G	0 - 12 residues
B 3	G	D	A		E	P	A	
B 4	E	E	G		T	K	G	
B 5	Y	V	H	D	S	H	V	|
B 6	G	G	G	P	G	T	G	| | | | |
B 7	A	G	Q	V	V	H	K	|
B 8	E	E	D	G	D	R	D	| |
B 9	A	A	I	I	I	F	C	| | | | | |
B10	L	L	L	L	L	F	L	| | | | | | |
B11	E	G	I	Y	V	I	I	|
B12	R	R	R	A	K	L	K	| |
B13	M	L	L	V	F	V	H	| | |
B14	F	L	F	F	F	L	L	| | | | | | |
B15	L	V	K	K	T	E	S	|
B16	S	V	S	A	S	I	A	| | | |
.	.	.	.	.	.	.	.	
.	.	.	.	.	.	.	.	
etc.								

FIG. 8. Derivation of flexible patterns from multiple alignment and secondary structure assignment for seven globin sequences, showing a partial pattern. The secondary structure column gives the position of α helices A and B. Alignment is based on tertiary structure comparisons. Flexible pattern elements are shown by vertical bars. Patterns 2–8 have progressively fewer elements until only the most highly conserved positions remain (see Fig. 10).

Scan number	Source of query	Method (Gap penalty)	Additional Structural Information?	Globins before first nonglobin	Globins remaining in top 500 scores	Globins not in top 500 scores
1	Single sequence (HAHU)	FASTP	No	297	7	41
2		NW(16)	No	306	8	31
3		NW–SS(16)	Yes	311	9	25
4	7 Globins (3D structure alignment)	BS(16)	No	309	19	17
5		BS–SS(16)	Yes	318	12	15
6		FP	Yes	345	0	0
7	Single sequence (HAHU)	FP	Yes	337	7	1
8	Two sequences (HAHU, GGICE3)	FP	Yes	344	1	0
9	7 globins (automatic multiple alignment)	FP	No	327	18	0

FIG. 9. Comparison of alignment procedures by database scanning with queries derived from globin sequences.

sensitive than the NW single sequence method (scan 2). Similarly, the BS–SS procedure (scan 5) further improved on the results obtained by the NW–SS method (scan 3) by identifying 318 globins before the first non-globin (cf. 311), with only 15 globin sequences not among the top 500 scores (cf. 25). The most startling improvement in performance, however, was obtained by the flexible pattern method (FP, scan 6) which gave perfect selectivity for globins with no nonglobin sequences scoring higher than the 345 whole globins in the database.

The successful scan 6 used a pattern that consisted of 107 pattern elements and 5 flexible gaps (Pattern 1, Fig. 8). The elements consisted of all aligned positions that had no gaps in any of the sequences and were also within secondary structural regions, while the scoring scheme took mean values from the MDM over all seven aligned sequences. Scan 7 used the same pattern elements as scan 6, but the scores were derived from only the human α-hemoglobin sequence (HAHU). This scan also performed significantly better than the multiple alignment method (scan 5), with only eight sequences not scoring higher than a nonglobin. It confirms that the bulk of the improvement from using the FP method comes from discarding the variable regions of the protein sequence, rather than from the use of multiple sequences.

The small deficiency in scan 7 is virtually eliminated by including one further sequence when describing the pattern elements. Scan 8 illustrates the result of this scan using a pattern derived from HAHU and GGICE3; only the bacterial hemoglobin fails to score higher than a nonglobin.

Pattern 1 has 107 elements or 79% of the shortest sequence. In order to investigate whether this high percentage of the alignment was actually required, a series of seven patterns (see Fig. 8) with successively fewer elements was derived and tested against the database. The result of scanning each pattern is summarized in Fig. 10. As expected, the overall trend in sensitivity and selectivity is downward as fewer elements are included. However, even pattern 6, which contains only 28 elements (21% of the shortest sequence), performs better than the full multiple alignment BS–SS method (335 globins before first nonglobin, cf. 318 for scan 5). Values in parentheses are for patterns in which the flexible gaps are unconstrained. The poorer performance of these patterns demonstrates the importance of defining flexible gaps to model the observed variation in sequence length within a protein family.

Derivation of Flexible Pattern When No Three-Dimensional Structure Is Known

In general, a sequence family may be known but with no details of three-dimensional structure available to guide the alignment or derivation of a pattern. Can an effective pattern be derived from just the sequences? To answer this question the seven globins used for scans 4–6 were multiply aligned by the single order algorithm described above; pairwise scores clustered at 7.9 S.D., suggesting confidence in the alignment. All positions at which gaps occurred were discarded, and, of the remaining positions, only those that had conservation values above 0.4 were maintained. Finally, gaps were made flexible between elements where insertions and

Pattern Number	Conservation Number Cutoff	Number of pattern elements	Percentage of shortest sequence	Globins before first nonglobin (total in PIR = 346)	Globins remaining in top 500 scores	Globins not in top 500 scores
1	0.0	107	79	345 (343)	0 (2)	0 (0)
2	0.2	87	64	345 (343)	0 (0)	0 (2)
3	0.3	61	45	343 (341)	2 (2)	0 (2)
4	0.4	46	34	344 (329)	1 (13)	0 (3)
5	0.5	38	28	343 (318)	1 (22)	1 (5)
6	0.6	28	21	335 (306)	9 (19)	1 (20)
7	0.7	15	11	295 (0)	33 (281)	18 (64)
8	0.8	8	6	1 (0)	298 (281)	46 (67)

FIG. 10. Result of scans using patterns with progressively fewer elements (see Fig. 8) derived from the seven-globin alignment[38] at increasing conservation value cutoffs.

deletions had been included by the automatic alignment algorithm, but were kept to fixed lengths where no insertions/deletions were observed. The resulting pattern consisted of 39 elements and, when scanned against the PIR database (scan 9), scored all globins in the top 500, with 327 globins giving scores above nonglobins. Thus, flexible patterns can be derived purely from sequence information and show a useful improvement in sensitivity and selectivity over the multiple alignment method (scan 4, 17 globins not in top 500 scores, only 309 globins before first nonglobin), or conventional single sequence methods.

Implementation and Availability of Programs

The techniques described in this chapter are all implemented in the AMPS package (alignment of multiple protein sequences), which provides the functions described together with additional features for multiple sequence manipulation and analysis within an easy-to-use environment. The package is available for a nominal fee to academic users. It is implemented on a VAX/VMS or Sun 3 with fp68881 coprocessor, and the current program limits for multiple alignment are 250 sequences of up to 1200 amino acids in length.

Acknowledgments

I would like to thank Dr. Mike Sternberg for support and advice. I would also like to thank Drs. Marketa Zvelebil, Chris Rawlings, and Prof. Tom Blundell. This work was supported by the Science and Engineering Research Council (UK) and The Imperial Cancer Research Fund.

[26] Genomic Divergence through Gene Rearrangement

By DAVID SANKOFF, ROBERT CEDERGREN, AND YVON ABEL

Introduction

Measures of similarities and distances among nucleotide or amino acid sequences have been used to find related regions in long sequences, to test for homology, to assess phylogenetic and functional relationships, and to estimate divergence time between pairs of evolutionarily related sequences. This last use especially invokes a model, possibly implicit and including a random component, of sequence change through nucleotide (or amino acid) replacement and insertion or deletion of single nucleotides or small blocks of contiguous nucleotides. Modeling genetic events by these mechanisms, however, cannot reflect the more macroscopic processes of evolutionary divergence of organisms, such as duplication, inversion, and transposition (shuffling) of parts of the genome. Although little comprehensive data are available as yet to study evolution at the level of entire genomes, the megasequencing projects now being set up will be producing genomic sequences in the near future.

In this chapter we discuss simple probabilistic models for genome shuffling introduced by Sankoff and Goldstein[1] and apply them to the assessment of relationships among a number of bacterial genomes. Lacking complete nucleotide sequences at this level, we assess our methodology on genetic map data. See Nadeau and Taylor[2] and Sakharov and Valeev[3] for comparable approaches.

---

[1] D. Sankoff and M. Goldstein, *Bull. Math. Biol.* **51,** 117 (1988).
[2] J. H. Nadeau and B. A. Taylor, *Proc. Natl. Acad. Sci. U.S.A.* **81,** 814 (1984).
[3] E. A. Sakharov and A. K. Valeev, *Dokl. Akad. USSR* **301,** 1213 (1988).

Shuffling Models

In the simplest model of random genome shuffling, we assume that the genome consists of n fragments, linearly disposed. These fragments may consist of genes, entire operons, or other larger or smaller regions of the genome. In the absence of evidence to the contrary, we assume that each of these fragments has the same probability per unit time of being transposed elsewhere in the genome, even for genomes of very different sizes. To ensure this we postulate that shuffling events occur at regular time intervals, inversely proportional to n (i.e., the rate of shuffling events is proportional to n), and each event sees one fragment chosen at random, moved to some other randomly chosen point on the genome between two other fragments, or to one end, and inserted there. The same process can also be used to model shuffling of a circular genome, though in this case we need not worry about the possibility of moving to the end of the genome.

More general models, which we do not discuss here, would favor relatively short-range fragment migration according to some probability distribution over distance along the genome. Such models would allow incorporation of empirically obtained parameters on rates of transposition, the distribution of transposition hot spots, and the tendency of contiguous fragments to be transposed as a unit.

Measures of Divergence

For models of sequence divergence through replacement, insertion, and deletion, the usual way of measuring sequence similarity or difference is to write one sequence above the other as in Fig. 1 and to draw a series of trace lines connecting pairs of terms, one in each sequence, such that no two lines cross and such that an optimality criterion is satisfied. Both elements of a pair of terms connected in the trace are inferred to originate in the same term in the ancestral sequence. The optimality criterion basically sums the similarity or difference scores of each pair of terms connected by the trace lines, plus a score for each unconnected term, i.e., insertion or deletion.

AUUACAGGUUCGUC

UUAGGAGGCGAC

FIG. 1. Trace between two sequences implying two replacement mutations (dotted lines) and four insertions and deletions (unconnected terms).

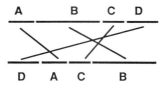

FIG. 2. Trace between two genomes consisting of four fragments. There are four intersections in this trace.

In models where fragment migration is the mechanism of evolutionary divergence, there will be exactly n trace lines connecting two genomes of common ancestry, and these will necessarily cross, as in Fig. 2, unless no net evolution has occurred. In fact, the more fragments have moved, the more intersections there tend to be. Thus, we can count the number of such intersections and use this as an indicator of the extent of divergence. Note that, in contrast to gene-level models involving replacement, insertion, and deletion, in transposition models we assume that we know which fragments in one genome are related to which ones in the other, so that there is no necessity of finding the optimal trace. There is no "alignment problem"; we know the true trace.

For circular genomes, the trace is constructed by drawing two concentric circles, and then connecting corresponding fragments by trace lines proceeding in a clockwise or counterclockwise direction within the ring between the circles, as in Fig. 3. Note that, in contrast to linear genomes, even though we may know which fragments correspond to each other in the two circular genomes, there are many ways of constructing the trace because of the possibility of choosing the clockwise or counterclockwise direction for each connecting line. As we shall see, this leads to a problem

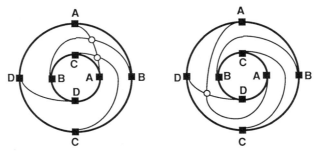

FIG. 3. Two ways of constructing a trace for the same pair of circular genomes. Filled squares represent midpoints of fragments A, B, C, and D. Open circles indicate intersections.

of optimal alignment, though not of the same sort as in the gene-level models.

Where few transposition events differentiate between two related genomes, most of the fragments will be in the same linear (or circular) order in the two sequences, but as the number of migrations increase it will become increasingly difficult to find a long subset of fragments that are in the same order in the two. Thus, another way of assessing the similarity between two genomes is to find the largest subset of the fragments which are in the same order in both genomes. The size of this subset, in comparison to n, is an indicator of the relationship between the two genomes. This indicator does not depend on how the trace is constructed.

Initial Behavior and Limit Theory

Sankoff and Goldstein investigated some of the properties of the number of intersections in a genome consisting of n fragments, under the random shuffling model.[1] In the linear model, in each of the first few shuffling operations, the number of intersections can be expected to increase by somewhat less than $3n/8$ on the average. This may be seen by considering first a fragment at one end of the genome. When it is moved randomly, this will give rise to between 0 (if it is put back into its original position) and $n - 1$ intersections (if it is moved to the opposite end of the genome). The average will be $(n - 1)/2$. For a fragment at the center of the genome, its movement will result in between 0 and $(n - 1)/2$ intersections (assuming n is odd), for an average of $(n - 1)/4$. Thus, the movement of a randomly chosen fragment will result in a number of intersections midway between the two extreme cases, namely $3(n - 1)/8$. Because the rate of shuffling events has also been assumed linear with n, we may expect the number of intersections to increase at an initial rate proportional to n^2. After many shuffles (of the order of $n \log n$), it can be proved that the number of intersections will approach $n(n - 1)/4$.

In Fig. 3, it can be seen that the choice of clockwise versus counterclockwise for a trace line can affect the number of intersections, and that the choice of the shortest route (i.e., $< 180°$) does not minimize the number of intersections. Thus, the two A fragments give rise to two intersections when connected by the shortest route and only one when connected by the more circuitous route. Nevertheless, it seems reasonable to make the convention that all trace lines travel no more than 180° (this is always possible). In this case, the initial increase per shuffle should be of the order of $3n/16$, using the same reasoning as in the linear case, so that the initial rate per unit time should also be quadratic in n. The asymptotic expectation can be shown to be of the order of $n^2/6$. Note, however, that this latter

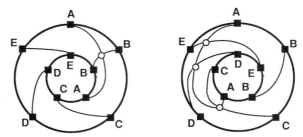

Fig. 4. Changing number of intersections as one genome is rotated with respect to the other.

value is calculated for the trace of two thoroughly shuffled circular genomes randomly rotated with respect to each other. However, by slowly rotating one genome while holding the other fixed, and examining how this changes the trace (as some clockwise lines get too long, i.e., travel more than 180°, and are thus replaced by shorter counterclockwise ones, and/or vice versa), we inevitably find some traces with a lot more intersections and some with less, as in Fig. 4.

Which rotation should we use to calculate the number of intersections? Perhaps the most natural choice is the one which minimizes the number of intersections in the trace since, in preasymptotic behavior, the minimizing rotation will presumably be close to the original homologous alignment of the two genomes. The minimizing value in the completely randomized case would be hard to predict analytically, but simulations are quite feasible, as discussed below.

As for the largest ordered fragment subset, the initial few shuffles should each reduce its size by about 1, in both the circular and linear models. The limiting value for the expected size of this subset is $2\sqrt{n}$ in the linear model[4] and is very close to the same value in the case of random circular genomes.

Data

There are of course few data sets at present representing entire genomes. On the nucleic acid level, these are confined to viruses and some mitochondria and chloroplasts. It is thus necessary to make use of other kinds of data in developing and testing our methodology. In this study, we identify the genome with a linear (or, rather, circular) set of genetic

[4] D. Sankoff and S. Mainville, *in* "Time Warps, String Edits and Macromolecules" (D. Sankoff and J. B. Kruskal, eds.), pp 363–365. Addison-Wesley, Reading, MA, 1983.

markers of bacteria:[5] *Escherichia coli, Salmonella typhimurium, Bacillus subtilis, Caulobacter crescentus,* and *Pseudomonas aeruginosa.*

The shuffling of the relative positions of the markers in one organism with respect to the corresponding markers in the other reflects the genomic history of transposition. There are a number of problems associated with the use of these kinds of data. The most serious has to do with the comparability of the different databases. There is little difficulty in identifying corresponding markers on the maps of *E. coli* and *Salmonella,* but whether similarly labeled markers in other genomes represent homologous genes is not always clear.[6] Likewise, two homologous markers may be labeled in slightly, or even completely, different ways in two different genomes. Thus, we systematically examined the functional descriptors of the genes in order to construct a normalized labeling across all five organisms. Identical labels for clearly unrelated genes in two organisms, caused by orthographic coincidence, were altered to avoid artifactual correspondences. Conversely, homologous genes differently labeled in two experimental traditions were relabeled to reflect this homology. Markers missing from either the map or the list of descriptions were discarded, and orthographic inconsistencies were regularized.

Functionality, of course, especially as inferred from the brief descriptions accompanying the genetic maps, is not always a reliable guide to homology, but it seems clear that, at least in a statistical sense, the labels in our normalized data base[7] constitute a better reflection of marker identities and differences across organisms than the uncorrected maps.

The fact that there may not be many comparable markers in two genomes as inferred from the four-letter marker labels contained in the genetic map data base led us to experiment as well with the first three letters only, given the possibility that slightly different labeling conventions might obscure genuine homologies, either because these are unknown or because they cannot be inferred from the functional descriptors in the original data. Thus, we repeat all calculations once requiring the first three letters of the marker labels to be identical in order to determine "homology" and once requiring all four symbols to be identical.

Independent of labeling considerations, maps of different genomes have different sets of markers, if only due to more extensive research on one organism compared to another. This, however, presents no difficulty within the framework of our model. By discarding, in each pairwise com-

[5] S. J. O'Brien, ed., "Genetic Maps." Cold Spring Harbor Laboratory, Cold Spring Harbor, NY, 1987.

[6] K. E. Sanderson and J. R. Roth, *Microbiol. Rev.* **52**, 485 (1988).

[7] R. Cedergren, Y. Abel, and D. Sankoff, unpublished work.

parison of genomes, all markers not present in either one or the other, we arrive at a value of n which is perfectly appropriate for that particular comparison. Our model has been constructed, especially with regard to the rates of occurrence of transposition events, so that whatever we can infer from the n comparable markers does not depend on what has happened to the markers which have been discarded. This is actually an advantage of using genetic maps in this type of study rather than physical maps or complete sequences. One of the problems in applying shuffling models to sequence-level genomic data is the difficulty of identifying the fragments. It is not usually obvious where to place the boundaries around a region of the sequence which can potentially be transposed, unless it is already located in different places within the genome in a number of related organisms. This becomes feasible only with large bodies of comparative data. The genetic markers, however, are generally independent, discrete entities, though in some cases the location of one marker may be tied functionally to the position of another.

Another problem has to do with gene duplication, represented in this data set by repeated labels in the same genome. Gene duplication should not be widespread in bacterial genomes, however, so that most apparent duplications in our data are probably due to similarity in three-letter marker labels of nonhomologous genes within the same operon. Nevertheless, in this exercise, an ad hoc solution to the duplicate label problem is to pick one marker at random out of each set of repetitions independently in each organism prior to counting intersections or finding largest ordered subsets. This may be repeated a few times to produce an average score.

Computational Methods

Not all the genomes we are studying are known to be circular; nevertheless, for the purposes of testing our methods, we treat them all in the same way. We compare the five genomes two at a time. As a first step we list each of the genetic markers in each genome. We retain only those which occur in both, according to either a three-symbol or a four-symbol criterion for matching their labels. If there are duplicate labels in one genome, one is chosen at random to correspond with that label in the other genome. If there are duplicate labels in both genomes, as many of such random correspondences are set up as possible. This protocol gives us a set of n markers. The analysis to follow is repeated for 10 such sets, created by different random choices among the duplicate labels, and the scores of the following calculations are averaged over the 10.

As mentioned above, given two circular genomes derived from the same ancestor by a random shuffling process, it is not necessarily clear how they were aligned (rotationally) at the moment of their divergence. Thus, we repeat the following analysis using every possible rotation of one genome with respect to the other.

A trace line is defined for each corresponding pair of markers in the two genomes. Because of the circular configuration, each trace line can be drawn clockwise or counterclockwise; we always choose the shorter path ($<180°$, as in Fig. 4). A routine then enumerates the number of intersecting trace lines in computing time quadratic with n. The search for the largest subset of markers with the same order in the two genomes is carried out by a dynamic programming algorithm similar to that used for matching nucleotide or amino acid sequences. It also requires quadratic time. The rotation with the smallest average intersection score is inferred to reflect the original alignment of the two genomes. In further studies, another approach might be to use the largest ordered subset of markers as a basis for the alignment, since this subset is likely to contain just those markers which have not undergone shuffling.

Simulations

Figure 5 portrays the difference between $n^2/6$ (the limiting approximation for the expected number of intersections under random rotation), the simulated expected value (based on 1000 samples), and the simulated minimum rotation expected value (based on 1000 samples for n less than 60, on 100 samples for n less than 150, and 10 samples for larger n). Both curves are approximately linear, confirming that, as a proportion of $n^2/6$, both discrepancies tend toward zero. In addition, the standard deviation of the simulated number of intersections also tends toward zero as a function of $n^2/6$.

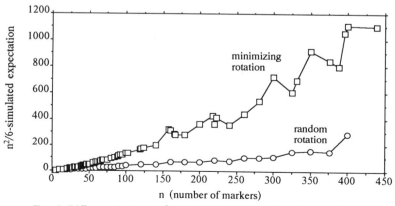

FIG. 5. Difference between $n^2/6$ and simulated number of intersections.

Comparisons

Figure 6 shows the number of intersections in an optimally rotated alignment of each pair of genomes, divided by the simulated expected value under the null hypothesis of completely randomly shuffled genomes, as a function of n, number of markers in common to each pair (based on the first three symbols of the label). The error bars represent the standard deviation of the predicted number of intersections as estimated in the simulations, plus the standard deviation of the observed number of intersections as estimated from the 10 sets of markers with randomly matched duplicate labels, all divided by the simulated expectation.

The relationship between *E. coli* and *Salmonella* can be seen clearly in the figure with the number of intersections being only 20% of that expected under complete randomization. This is true despite an inversion of a large segment of the genome that is evident when comparing one organism with the other, probably contributing a large portion of the intersections.

The other comparisons, except some of those involving *B. subtilis,* also show fewer intersections (i.e., a greater relationship) than randomly shuffled genomes, though only those involving *Caulobacter* are convincing.

Figure 7 compares the four-symbol intersection rates with the three-symbol ones of Fig. 6.

In all cases, again with the exception of comparisons involving *B. subtilis,* the increased confidence in marker homology when all four letters of the corresponding labels must be identical is reflected in a much greater level of relationship as detected through a decrease in the normalized rate of intersection. It should be noted that the error ranges for the four-symbol comparisons are distinctly less than those for three-symbol comparisons

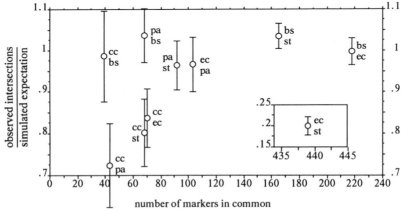

FIG. 6. Pairwise comparisons of genomes, ec, *Escherichia coli;* st, *Salmonella typhimurium;* bs, *Bacillus subtilis;* cc, *Caulobacter crescentus;* pa, *Pseudomonas aeruginosa.*

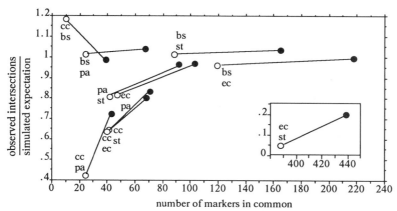

FIG. 7. Normalized intersection rates using three-symbol (filled dots) and four-symbol (open dots) correspondences.

for similar values of n, since there is little or no variation due to random correspondence of duplicate labels.

Both the three-symbol and four-symbol analyses permit us to make the same phylogenetic inferences. *E. coli* and *Salmonella* are closely related, *Caulobacter* and *Pseudomonas* are fairly closely related, and the first pair are more closely related to the second pair than either is to *B. subtilis.* In addition, the results suggest that *Pseudomonas* has been evolving at a faster rate than *Caulobacter.* None of these inferences is contrary to what is known about bacterial phylogeny.

The size of the largest marker subset with a common order in the two genomes, divided by $2\sqrt{n}$, gives the identical phylogenetic picture as does the normalized rate of intersection.

Discussion

The genetic maps currently available contain too few markers in common to enable us to test our methods on organisms which are evolutionarily very divergent. The closely related (around 115 million years) *E. coli* and *Salmonella* genomes have many of the same markers mapped, and the relationships among these and *Caulobacter* and *Pseudomonas* can be assessed, but between all of these and *B. subtilis,* representing a time depth of about 800 million years, our measures show no detectable difference from a completely shuffled model. It does seem important, however, to develop this type of methodology in anticipation of the quantity of experimental results which are expected in the next few years.

Further mathematics and/or simulations are needed to understand the behavior of our measures based on optimal rotations, instead of under the (usually false) hypothesis that we know the original alignment of the two genomes.

Normalized rate of intersection and largest common ordered subset measures of divergence through rearrangement should be compared for sensitivity and accuracy to indices based on the least number of rearrangements necessary to convert one genome into another.[8]

Much work is needed to collate the labels for markers used in distantly related genomes.

Empirical and theoretical research on fragment transposition distances is needed to produce better models of this phenomenon. Generalizing the term "synteny" used in chromosomal genetics to encompass the notion of proximate fragments in one genome tending to be close together as well on another would give an appropriate label to this field of study.

Acknowledgments

This work was supported by individual operating grants and to David Sankoff and Robert Cedergren, as well as an infrastructure grant and a CRAY computer time allotment, from the National Sciences and Engineering Research Council of Canada. David Sankoff and Robert Cedergren are Fellows in the Evolutionary Biology Program of the Canadian Institute for Advanced Research.

[8] D. Sankoff, *Bul. Int. Stat. Inst.* **53(3),** 461 (1989).

[27] Multiple Sequence Comparison

By DAVID J. BACON and WAYNE F. ANDERSON

Introduction

One reason for performing amino acid sequence comparisons is to discover structural and/or functional similarities among proteins. Because structural similarity may be present in proteins that do not exhibit a strong sequence similarity (e.g., see Matthews *et al.*[1]), one would like to be able to recognize the structural resemblance even when the sequences are very different.

This chapter addresses the problem of finding weak similarities or distant relationships among proteins for which only the sequences are known. Comparing just two sequences at a time by current methods does

[1] B. W. Matthews, M. G. Grutter, W. F. Anderson, and S. J. Remington, *Nature (London)* **290,** 334 (1981).

not allow for a sufficiently sensitive test of similarity. Apparent pairwise similarities frequently fail the statistical tests for significance. Simultaneous intercomparison of several sequences, on the other hand, often yields a significantly nonrandom signal that in turn provides a statistical basis for assertions about structure and/or function.

Fixed-Length Sequence Comparison

The sequence matching method described here has its roots in the procedures of Fitch[2] and Cantor and Jukes[3] for comparing two protein sequences: every run of k consecutive residues from one sequence is compared with every such run in the other sequence. The score for each run comparison is a sum of k residue similarity scores as looked up from a 20×20 table (4×4 in the case of polynucleotides). Typically k is chosen to be between 15 and 30 residues. The objective is to find the runs of length k which match best according to the similarity scores in the table. The fact that k is a constant for any given experiment is what gives this procedure the name fixed-length sequence comparison. It is also possible to have a variable-length method in which the search is made for runs of many different lengths[4] or to allow for insertions or deletions in the sequences.[5]

Multiple fixed-length sequence comparison (MSC) is conceptually no different from pairwise fixed-length comparison, except that the scoring function is generalized and there are many more ways to form combinations of runs. Given m runs, one run from each sequence, we define the mutual matching score to be the sum of the $m(m - 1)/2$ pairwise run comparison scores. This reduces to the usual definition when $m = 2$. The combinations of runs that should theoretically be considered in the search for the best one are described by the following recursive formula: for every run from one sequence chosen from a set of m, put this run with every combination of runs from the remaining $m - 1$ sequences. This is just the usual set of pairwise combinations when $m = 2$.

The number of combinations of runs grows rather explosively as m increases. For $m = 2$ sequences, having lengths L_1 and L_2, there are $(L_1 - k + 1)(L_2 - k + 1)$ combinations, or a little fewer than n^2 for sequences of equal length n. For $m = 3$, there are nearly n^3 such combinations and, in general, on the order of n^m. If $n = 319$, $k = 20$, and $m = 5$, this comes to some 2.4×10^{12} combinations. It is clearly not practicable to evaluate them all.

---

[2] W. M. Fitch, *J. Mol. Biol.* **16,** 9 (1966).
[3] C. R. Cantor and T. H. Jukes, *Proc. Natl. Acad. Sci. U.S.A.* **56,** 177 (1966).
[4] P. Argos, *J. Mol. Biol.* **193,** 385 (1987).
[5] S. B. Needleman and C. D. Wunsch, *J. Mol. Biol.* **48,** 443 (1970).

It is also not necessary to evaluate all combinations. In this chapter, we present a "heuristic" search method which, though not absolutely guaranteed to find the best matches, almost always does manage to find them.[6] Moreover, it does so in a number of search steps proportional to only nm^2, a function that grows much more slowly with respect to the number of sequences, m, than does the n^m required for exhaustive evaluation of all combinations.

The main use of the heuristic method is in discovering regions of similarity and evaluating them statistically. It does not attempt to construct full sequence alignments in the Needleman and Wunsch[5] sense and, indeed, is generally quite unnecessary in the cases of strong sequence similarity for which full alignments are likely to be meaningful. It has great value, however, in uncovering distant relationships among special regions in proteins even when evolution has eroded the homology of the remaining portions of the polypeptide chains to the point of virtual unrecognizability.

Sensitivity of Sequence Comparison

There are many 20×20 tables of amino acid similarity scores in common use. One example is the mutation data (MD) matrix;[7] another is a structural–functional table which reflects properties such as size, presence of polar or hydrophobic groups, and general preferences for forming different kinds of secondary structure.[6]

There is growing evidence that, even if there were such a thing as an optimal scoring table by some definition, pairwise sequence comparison would still be too weak on its own to discern many real protein relationships unambiguously. This problem is a statistical one. A run in one sequence may appear to be related to a run in another sequence, but not to an extent that gives one confidence in the relationship being more than that which would be expected by chance in a population of about n^2 possible run pairs. Yet in many of these cases it does later turn out that there is a structural relationship.

Multiple sequence comparison offers a way out of this dilemma. It occurs quite often that there is some third sequence, apparently but not convincingly related to each of the first two, for which the mutual intercomparison of all three sequences gives a highly significant score. One also finds that the addition of an unrelated sequence reduces the significance, reflecting a highly desirable behavior of the statistics of multiple sequence

[6] D. J. Bacon and W. F. Anderson, *J. Mol. Biol.* **191,** 153 (1986).
[7] B. C. Orcutt, M. O. Dayhoff, D. A. George, and W. C. Barker, "User's Guide for the Alignment Score Program of the Protein Identification Resource (PIR)," PIR Report ALI-1284. National Biomedical Research Foundation, Washington, D.C. 1984.

comparison: the signal-to-noise ratio rises sharply when a postulated mutual relationship is real and drops when it is not.

Heuristic Multiple Sequence Comparison Method

The objective, formally, is to find mutually similar runs of length k among sequences $S_1, S_2, \ldots, S_m$ that are of lengths $L_1, L_2, \ldots, L_m$, respectively. The mutual similarity of the runs starting at positions p_1, $p_2, \ldots, p_m$, respectively, is defined by the scoring function

$$\sum_{i=1}^{m} \sum_{j=i+1}^{m} \sum_{q=0}^{k-1} T[S_i(p_i + q), S_j(p_j + q)]$$

where T is the 20×20 table of residue similarities. An exhaustive combinatorial algorithm to find the highest matching scores would evaluate this formula for all possible combinations of the starting positions and would note which combinations yield the highest scores.

In practice, it is necessary to limit the combinatorial "explosion" that occurs as the number of sequences increases. The approach is to try to guess where the best scores will be in the space mapped out by the run combinations. The guessing game works as follows. In any region of mutual similarity, consisting of a set of m runs, each constituent subset of sequences should also have a reasonable level of mutual similarity. This is true, though to a progressively weaker extent, right down to subsets consisting of 2 runs. So the procedure starts by examining a pair of sequences, and performing the exhaustive search algorithm on them. This is not unreasonably time consuming for only two sequences. The run pairs that yield the top matching scores are saved in a "heap," denoted H_2. The number of run combinations, M, saved in a heap is typically chosen to be about 1000. Then a third sequence, S_3, is selected, and the scoring function is evaluated using run triplets made up from pairs in H_2 combined with every possible run in S_3. Of the $M(L_3 - k + 1)$ scores that this step produces, only the triplets associated with the highest M scores are saved in a new heap, H_3. This process continues, with each new heap H_i being created from heap H_{i-1} and sequences S_i, until all the sequences are used up. The final heap H_m should include the region of highest mutual similarity of length k among all the sequences.

Clearly, the success of this procedure depends critically on the strength of the matching in the particular subsets used to build up the final heap, and it is thus dependent on the order in which sequences are selected. It turns out that with an adequate heap size, the algorithm is remarkably robust and even works quite well with random sequences.[6] Moreover, the success rate of the algorithm improves as the strength of the run matching increases, because the better the match is, the more likely it is to contain submatches that are strong enough to be in the first two or three heaps. This can be defeated by making the heaps too small, but a constant heap size of 1000 proves to be adequate in all but the most pathological cases.

Because the heap size M and the run length k are constants for any given application of the heuristic search procedure, and each new heap H_i is constructed by comparing each of approximately n runs in a new sequence S_i against all the $M(i-1)$ runs in the previous heap H_{i-1}, the total number of run comparisons leading to H_m is proportional to nm^2, a function that is only quadratic in the number of sequences. The factor n (average sequence length) tends to be constant to within an order of magnitude in practice. The execution time of the heuristic procedure is much better than that of exhaustive search even for three sequences; the savings are more dramatic for larger groups, where an experiment on five sequences may take minutes instead of months.

Statistics

The highest score arising from the intercomparison of sequences is statistically significant only if it is higher than what chance alone would be expected to produce. Even this does not guarantee that there is anything interesting about the region of mutual similarity, but it does suggest strongly the possibility of a common polypeptide fold.

The major probability model used in the MSC program is based on a "generating functions" model developed for two sequences by McLachlan.[8] It yields, for any given score s, the frequency with which a score of s or higher is expected to occur by chance for a set of sequences of lengths L_1, $L_2, \ldots, L_m$. Also accounted for in this probability model are the run length k, the particular 20×20 table of nonnegative integer residue similarity scores in use, and information on the relative abundances of the residues. The population being nonrandomly sampled consists of all the regions that the exhaustive combinatorial search algorithm would have examined, so it is the size of this population that is multiplied by the probabilities to get the expected frequencies. If the observed number of occurrences of a given score of s or higher is much more than the expected number (including the case where the top score occurs once and is expected to occur less than once), then the score is considered significant. As a rule, we take the cumulative frequency of a score to be significant if it is at least 100 times that which the model predicts to occur by chance.

It is important not to rely on such crude measures of score quality as "number of standard deviations above the mean" for multiple compari-

---

[8] A. D. McLachlan, *J. Mol. Biol.* **61**, 409 (1971).

sons. Not only does the search procedure sample the population nonrandomly, but the score frequency distribution of the population itself is not normal when more than two sequences are intercompared. This is because when a region A of one random sequence happens to compare well with a region B in another random sequence and also well with a region C in a third random sequence, the chances of a higher than average score between B and C are distinctly greater than 50%, even though all three sequences are random. Yet all three pairs A–B, A–C, and B–C contribute to the score for the mutual comparison. Even though the three sequences are independent, their pairwise comparison scores are not. The distribution of the three-way scores for random sequences is higher in the tail regions than a normal distribution with the same mean and standard deviation would be. If a normal distribution were used as a reference instead of the "correct" one given by the MSC probability model, there is the serious danger of overinterpreting high scores.

Operation of MSC Program

The FORTRAN program that implements the sequence intercomparison algorithm requires, in addition to the sequences, a few parameters that control its operation. The run length, k, should be set to around 20 to give the most sensitive sequence comparison. Smaller numbers tend not to produce scores that stand out as well above the background noise of essentially random run comparisons. Larger numbers suffer from the problem that gaps, which might improve local alignments, are not detected by the algorithm. Another MSC parameter is the heap size. It has already been suggested that 1000 is usually an adequate number. Some economy of computer time can be realized with the use of smaller heaps, but pushing this too far can impair the ability of the program to find the best regions of mutual similarity. The only hazard associated with raising the heap size beyond 1000 is that the program will take longer to run and require a little more memory.

Whenever more than five sequences or several long sequences are to be intercompared, it is advisable to run the program twice, with the sequences supplied in reverse order the second time, and check to make sure the same top matches are found in both cases. This is to allow for the possibility that the region of best mutual matching happens to be particularly weak in the first two or three sequences supplied to the program. In extreme circumstances, a few more permutations of the sequence order may be tried in order to produce a clear result. This should be necessary only if there are many sequences with little similarity.

There are some program parameters that control details of the probability models to be used by MSC in estimating significance. The user may select the usual probability model based on generating functions, a model which causes the program to intercompare scrambled sequences a number of times to obtain reference frequency data directly, or both models at once. In addition, these may be used with either the residue counts of the input sequences or natural residue abundances as read from a library file, or both. Thus, up to four ways of estimating significance may be invoked simultaneously. It is worth noting here that the generating functions model in conjunction with the residue counts of the input sequences is rather expensive in terms of computer time when there are more than five sequences to compare. This is the most conservative of the four possible models in that it is the least likely to overestimate significance, but, because of the cost in computer time, the library values are more practicable when more than five sequences are intercompared. Bacon and Anderson[6] have illustrated the behavior of the four probability models for all subsets of two or more sequences from a set of five unrelated sequences. All the models are conservative enough that overinterpretation of high scores is unlikely.

Finally, there is a parameter to tell the program how many regions of mutual similarity to print from the final heap. These are sorted starting with the region having the highest score.

In addition to the above control parameters, MSC requires a library file that specifies the 20×20 table of residue similarities, the natural residue abundances, and the basic generating function coefficients corresponding to these abundances. The default library that comes with the program has amino acid abundances derived from the NEWAT database[9] and a table of amino acid similarities based on structural and functional considerations.[6] There is an auxiliary program for generating a new library when the similarity table or abundances are to be changed, to facilitate experimentation with different amino acid distributions and comparison regimes.

A suitable computer configuration for running MSC is a mainframe, minicomputer, or workstation. A small personal computer can even be used to good effect, because the memory requirements of the program are quite modest. Most of the memory is used by the heaps, and at most two of these are actually retained by the program at any given time. However, the program can take a few hours to perform a large sequence intercomparison on a slow computer.

To simplify the use of MSC, there is an interactive program to help set up input. This program allows one to supply sequences in virtually any format and ensures that all control parameters are specified.

[9] R. F. Doolittle, *Science* **214**, 149 (1981).

Examples

A control experiment with five ribonucleases demonstrated that the significance levels reported by MSC are an accurate indication of the presence or absence of protein relationships.[6] Two pairs in the set of five proteins had strong enough similarity to be detected clearly with pairwise comparison, yet when any larger subset of the five were intercompared, the significance dropped.

In a set of five mutually related DNA-binding proteins, there was a very strong tendency for the significance of mutual intercomparisons to rise as proteins were added to the set, even though the similarity of some of the pairs in the set was well below the level normally considered significant. The overall matched region turned out to be the well-known helix–turn–helix DNA-binding motif in all the proteins.

The application of MSC to a set of enzymes employing the coenzyme flavin adenine dinucleotide (FAD) yielded a clear similarity when three or more sequences were compared despite uncertain pairwise similarities.[6,10] Because two of these sequences were from enzymes of known three-dimensional structure, it was possible to associate the similar region with the FAD-binding site.

Intercomparison of the sequences of the dimethyl sulfoxide reductase A subunit, biotin sulfoxide reductase, and formate dehydrogenase revealed that these three enzymes, all of which utilize molydopterin cofactors, contained four regions of amino acid sequence similarity.[11] In this case, overlapping fixed length runs were extracted from the heap to extend the regions of similarity beyond the initial runs of 20 residues. The three-dimensional structures of these enzymes are not known, so it is not yet possible to determine the functions of the four regions.

Discussion

The MSC program is very useful in finding regions of similarity among sequences and determining whether the similarity is significant, but in some ways it might appear not to go far enough. For example, it does not account for insertions or deletions of residues in the sequences, and its run length is a fixed parameter for any given experiment. These restrictions are deliberate, and the chief reasons for making them are statistical. If a range of run lengths were automatically sampled, not only would the execution time and memory requirements increase, but significance could be ob-

[10] S. T. Cole, K. Eiglmeier, S. Ahmed, N. Honors, M. L. Elmes, W. F. Anderson, and J. H. Weiner, *J. Bacteriol.* **170,** 2448 (1988).
[11] P. T. Bilous, S. T. Cole, W. F. Anderson, and J. H. Weiner, *Mol. Microbiol.* **2,** 785 (1988).

scured unnecessarily by the larger reference population that would have to be considered. Allowing "gaps" would make this problem much worse.

A very real concern with MSC is that the ratio of observed to expected high score frequencies is not really the same thing as statistical significance. What is needed is an accurate model of the variability in the high scores for random sequence intercomparisons. This would allow significance to be more properly quoted as the probability of the highest score occurring by chance. The problem is not that the frequency ratios cannot be trusted, only that they are not an ideal measure of significance.

Although it is unclear that full sequence alignments are useful when similarity is weak, full alignments can occasionally be meaningful if the aligned regions are of roughly equal length. There are many possible approaches to this task. Needleman and Wunsch[5] mention a multiple sequence generalization of their algorithm which appears to imply an implementation requiring computation time proportional to n^m for m sequences of length n and a constant gap penalty or to n^{2m-1} if the gap penalty is an arbitrary function. We have been experimenting with an implementation of this kind for the purpose of finding best paths between regions of similarity, keeping the computing time down by imposing an upper limit on the extent to which gaps can change the register of the sequences. Murata[12] uses a constant gap penalty to constrain the computation time to something proportional to n^3 for three sequences. Gotoh[13] describes fairly general gapping conditions that allow alignments to be made in time proportional to n^2 for two sequences. The profile analysis method of Gribskov and Eisenberg[14] may be a particularly effective way of dealing rationally with large numbers of sequences.

Sequence alignments based on sequence alone, of course, usually do not reflect the best way of aligning the three-dimensional structures unless the homology is well above the levels MSC is typically used to detect.[15] On the other hand, a very common reason for aligning protein sequences is to deduce information on the three-dimensional structure of one protein based on the structure of a related protein. The variable gap penalty method of Chothia and Lesk[16] may offer a better approximation to a structurally correct alignment than the unmodified Needleman–Wunsch[5] procedure, although the tests we have made with this so far do not indicate much improvement. Certainly, the general idea of using as much structural information as possible in order to improve alignments has great merit.

[12] M. Murata, this volume, [22].
[13] J. Gotoh, *J. Mol. Biol.* **162,** 705 (1982).
[14] M. Gribskov, R. Lüthy, and D. Eisenberg, this volume, [9].
[15] R. J. Read, G. D. Brayer, L. Jurášek, and M. N. G. James, *Biochemistry* **23,** 6570 (1984).
[16] A. M. Lesk, M. Levitt, and C. Chothia, *Protein Eng.* **1,** 77 (1986).

Conclusion

Multiple sequence comparison is most useful when sequence similarity is weak. The technique outlined here does not attempt to produce an overall sequence alignment, but it is easy to use, finds results reasonably quickly if they are to be found at all, extends sequence comparison to far more than two or three sequences, and provides a statistical basis for assertions about regions of protein similarity.

Acknowledgments

This work was supported by the Medical Research Council of Canada through a grant to the MRC Group on Protein Structure and Function, Department of Biochemistry, University of Alberta.

[28] Simultaneous Comparison of Several Sequences

By Mauno Vihinen

Introduction

Numerous methods have been developed to compare and align two or more nucleic acid or protein sequences. The difference between sequence comparison and alignment is that the former indicates all similarities between the sequences whereas the latter method aligns the matching bases or residues. Sequence alignments are valuable for sequence divergence studies and for computer modeling based on homologous counterparts. The comparisons give overall sequence similarity regardless of alignment. Dot-plot figures, the graphic presentation of sequence comparison, show conserved regions as well as the alignment, which can be seen around the main diagonal.

The number of known sequences has increased drastically, and often several sequences having the same function as the sequence under study can be found in databases. This has lead to the need for methods analyzing simultaneously several sequences. Most of these techniques are for aligning several sequences (see other chapters in this volume), but some are for simultaneous comparison of several sequences.[1,2] In particular, a new

[1] M. Vihinen, *Comput. Appl. Biosci.* **4**, 89 (1988).
[2] G. Krishnan, K. K. Rajinder, and P. Jagadeeswaran, *Nucleic Acids Res.* **14**, 543 (1986).

method to study sequence similarities by comparing one sequence to several others was developed.[1] In this approach pairwise comparisons of aligned sequences are superimposed to search conserved regions of the query sequence. The method can be used to compare DNA, RNA, or protein sequences.

The three-dimensional structures of related proteins are known to be more conserved than their primary sequences. Also, the secondary structures comprising the tertiary structure are conserved. To be able to use these data the algorithm was modified so that it can simultaneously compare predicted secondary structural features of several sequences and thus extend comparisons into a wholly new dimension.

Methods

General Description

The main idea of the algorithm is to compare one sequence X with several other sequences. The classic window/stringency method is used in pairwise comparisons between sequence X and each of the sequences $Y_1 \ldots Y_n$. The observation arrays from the pairwise comparisons are superimposed, and those points scoring equal or higher than some predetermined stringency value are recorded (Fig. 1). The same idea is used to compare secondary structural predictions, where numerical values are compared instead of characters.

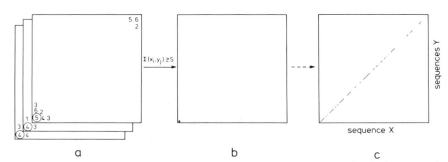

FIG. 1. Multiple sequence comparison scheme. Observation arrays of pairwise comparisons are superimposed, and values of cells (x_i, y_j) in each array are summed (a). If the total number of matches for a certain cell is equal to or greater than the stringency S, the point is accepted and drawn on the dot-plot (b). This is repeated for each cell of the observation arrays to get the relationships between the sequence X on the x axis and the sequences Y on y axis (c).

Sequence Alignment

Before the sequence comparison can be performed all sequences Y have to be aligned with a sequence X, since the conserved residues can have different spacing even in closely related sequences. There are many methods from which to choose. All the insertions and deletions causing gaps in either of the aligned sequences are taken into account in the sequence Y, because gaps could occur in different places in the sequence X when different Y sequences are compared to it. Gaps appearing in the sequences Y are filled with dots; those in the sequence X lead to deletion of the insertion from the sequence Y. In this way conserved characters have the same spacing and distance in the sequences Y. Dots used to adjust the conserved characters are ignored unless one occurs in the middle of the window, where they are not allowed.

The goodness of the alignments can be tested by doing the usual comparison of the two sequences. If the conserved residues are in the main diagonal, the alignment can be used. If the conserved regions wander around the main diagonal, the parameters of alignment should be adjusted. To avoid excessive gaps, however, the gap weight and the length weight should not be too small. As a rule several alignments should be tested with each sequence Y, and the one with highest weight values giving a reasonable alignment should be chosen. Doolittle gives guidelines for estimating the number of gaps allowed in related sequences.[3]

Simultaneous Comparison of Several Sequences

The aligned sequences Y are compared pairwise with the sequence X, the query. The value of the window determines how many successive characters from each sequence are compared at a time. A certain number of matches, the stringency, must be reached for a point to be recorded in the middle of the window. The window is slid in steps of 1 through both sequences so that every character in sequence X will be compared to every character in sequence Y. This is repeated for each pair.

Comparison tables that take into account relatedness of amino acid residues give much better results for protein sequence comparisons than a unitary matrix, which is based only on identities. The residues can mutate to others with different frequencies. The most often used comparison table, the Dayhoff mutation matrix,[4] is based on observed amino acid replacements between similar proteins from closely related organisms.

---

[3] R. F. Doolittle, *Science* **214**, 149 (1981).
[4] M. O. Dayhoff, R. M. Schwartz, and B. C. Orcutt, *in* "Atlas of Protein Sequence and Structure" (M. O. Dayhoff, ed), Vol. 5, Suppl. 3, p. 345. National Biomedical Research Foundation, Washington, D.C., 1978.

The observation arrays containing results of pairwise comparisons are superimposed in the following step. The values of each cell (x_i, y_j) of all arrays are summed, and only those points having a score equal to or higher than a second stringency are recorded (Fig. 1). The value of the second stringency can be either the sum of the sequences Y having scores equal to or higher than the first stringency or the total number of matches of the pairwise comparisons. Both implementations can be used without problem, although they can give different results, especially when the first stringency is zero and the total number of matches is used. It seems preferable to use the first stringency even when the second stringency is used to calculate the total number of matches, because highly homologous sequences can cause a bias.

Actually, the technique using observation arrays containing the results of all pairwise comparisons was not implemented as described here. Each cell (x_i, y_j) is calculated simultaneously for all pairwise comparisons, because in this way storage of the observation arrays in memory is avoided. The end result is the same as if the described method, which is easier to understand, were used.

The choice of the window and stringency parameters has great effect on the results. A too wide window and low stringency give an enormous number of points appearing by chance (noise), which hides the conserved regions. Too high stringency is also disadvantageous, because important signals are lost. Use of this algorithm, as well as any comparison method, requires testing of several parameter sets. The value of the second stringency should probably be low in initial analyses to see whether similarities occur, and be higher when the window and the first stringency have been adjusted.

Comparison of Secondary Structure Predictions

Secondary structure predictions have to be done prior to a comparison. Any of the numerous predictions for α, β, and turn structures, hydropathy, flexibility,[5] acrophilicity,[6] surface probabilities,[7] and other features can be used. All these methods use as the input primary amino acid sequences, for which numerical values are calculated residue by residue. Some predictions are made for a fixed number of consecutive residues, whereas the prediction window can vary greatly for others, such as hydropathy. The size of the window used in a prediction should be carefully chosen, because

---

[5] P. A. Karplus and G. E. Schulz, *Naturwissenschaften* **72**, 212 (1985).

[6] T. P. Hopp, *in* "Synthetic Peptides in Biology and Medicine" (K. Alitalo, P. Partanen, and A. Vaheri, eds.), p. 3. Elsevier, Amsterdam, 1985.

[7] J. Janin, S. Wodak, M. Levitt, and M. Maigret, *J. Mol. Biol.* **125**, 357 (1978).

too large a number of residues can hide important peaks by reducing the magnitude of alterations and by smoothing sharp peaks.

The values obtained from structural predictions are compared by using the window/stringency method. The alignment of the sequences have to be taken into account. The same alignment should be used both in sequences as well as in secondary structure comparisons.

Because different sequences having similar secondary structural features can have slightly different predicted values, an extra value, a limit, is introduced to measure similarity of the sequences. The limit facilitates determination of the degree of similarities as comparison tables used in sequence comparisons. The absolute difference between the values for residues in sequence X and sequence Y must be smaller than or equal to the value of the limit to be accepted as a match.

Since all important secondary structural features have a length of at least a few residues, the window is used to increase the signal-to-noise ratio. Similar values can be found all around the sequence, and the noise could prevent finding the conserved regions. Use of this second window is analogous to the use of a window in multiple sequence comparison. Noise can be reduced by concentrating on a certain bandwidth around the main diagonal. However, some important data can be missed.

The observation arrays containing the data for matches in the compared secondary structure predictions are superimposed as described for multiple sequence comparisons. The method is very sensitive for the parameters used. The value of the limit must be changed for each prediction, because the amplitude of the values varies. The smaller the difference between the highest and lowest peaks in prediction, the smaller the value of the limit should be. The window should usually be smaller than in sequence comparisons.

Implementation

The programs are written in FORTRAN to function with the University of Wisconsin software package environment.[8] However, the software package is not required to run the programs. The programs called MULTICOMP, for multiple sequence comparisons, and SSCOMP, for secondary structure comparisons, as well as the program to calculate the secondary structural predictions are interactive. Alignments were done with the Needleman–Wunsch algorithm,[9] although other alignments can be used.

---

[8] J. Devereux, P. Haeberli, and O. Smithies, *Nucleic Acids Res.* **12**, 387 (1984).
[9] S. B. Needleman and C. D. Wunsch, *J. Mol. Biol.* **48**, 443 (1970).

The programs were run under the VAX/VMS operating system. They are available from the author for a nominal charge.

Results of comparisons are shown in the form of a dot-plot picture, which has on the x axis the sequence X and on the y axis all the sequences $Y_1 \ldots Y_n$. This kind of picture shows the conserved regions of sequence X in relation to all the other sequences. The dot-plot illustration should not be used to look for conserved sites in any of the sequences Y, since the alignments have changed the spacing in sequences Y by introducing gaps and deleting stretches of sequences. If conserved sites of any of the sequences Y are needed, a new comparison is required as the interesting sequence Y in the x axis. Another presentation can be a line drawing or a list of conserved regions in sequence X, although these do not show the spacing of conserved regions in relation to the main diagonal.

Use of Multiple Sequence Comparisons to Study Saccharifying and Liquefying α-Amylases

α-Amylases have been divided into two categories, saccharifying and liquefying enzymes, according to the extent of hydrolysis of starch.[10] Taka-amylase A of *Aspergillus oryzae* is saccharifying whereas the *Bacillus stearothermophilus* enzyme is liquefying. Amylases are known to have some sequence similarity.[11-13] As a test of applicability of the multiple sequence analysis, the method was used to find whether liquefying α-amylases could be distinguished from saccharifying ones on the basis of appearance of similarities in different regions when compared to other starch-hydrolyzing enzymes.

The sequences for Taka-amylase A[14] and *B. stearothermophilus* α-amylase[15] were compared with each other (Fig. 2a). The main diagonal indicates significant homology, as previously reported. The similarities in the hydropathy predictions (Fig. 2b) appear in the same regions as sequence similarities. The empty space between about residues 100 and 150 in the y axis is due to an insert in the *B. stearothermophilus* enzyme.

Both α-amylase sequences were compared to several amylolytic en-

[10] J. Fukumoto, *J. Ferment. Technol.* **41,** 427 (1963).

[11] J. C. Rogers, *Biochem, Biophys. Res. Commun.* **128,** 470 (1985).

[12] R. M. Mackay, S. Baird, M. J. Dove, J. A. Erratt, M. Gines, F. Moranelli, A. Nasim, G. E. Willick, M. Yaguchi, and V. L. Seligy, *Biosystems* **18,** 279 (1985).

[13] B. Svensson, *FEBS Lett.* **230,** 72 (1988).

[14] H. Toda, K. Kondo, and K. Narita, *Proc. Jpn. Acad.* **58B,** 208 (1982).

[15] I. Suominen, M. Karp, J. Lautamo, J. Knowles, and P. Mäntsälä, *in* "Extracellular Enzymes of Microorganisms" (J. Chaloupka and V. Krumphanzl, eds.), p. 129. Plenum, New York, 1987.

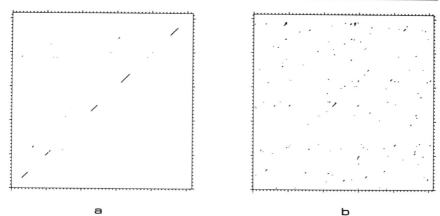

FIG. 2. Comparison of (a) sequences and (b) hydropathy profiles of *B. stearothermophilus* and *A. oryzae* α-amylases. The sequence of the liquefying *B. stearothermophilus* enzyme is on the *x* axis and that of the saccharifying Taka-amylase A on the *y* axis. The comparison window was 20, the first stringency 0, and the second stringency 12 for sequence comparison. The hydropathy profiles were calculated with a window of seven residues by the method of Hopp and Woods.[22] The values of the parameters were 6 for the window, 0.15 for the limit, 0 for the first stringency, and 5 for the second stringency.

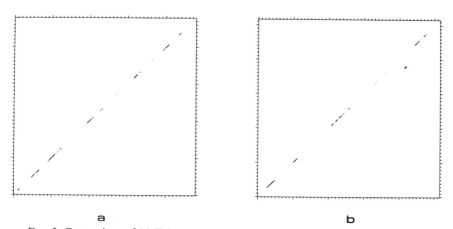

FIG. 3. Comparison of (a) Taka-amylase A and (b) *B. stearothermophilus* α-amylase to glucoamylase, isoamylase, β-amylase, pullulanase, and cgtase sequences. The window was 20, the first stringency 8, and the second stringency 43.

zymes: *Aspergillus niger* glucoamylase I,[16] *Pseudomonas amyloderamosa* isoamylase,[17] *Klebsiella aerogenes* pullulanase (α-dextrin endo-1,6-α-glucosidase),[18] *Bacillus circulans* β-amylase,[19] and *Bacillus macerans* cyclodextrin glucosyltransferase (cgtase)[20] (Fig. 3). The conserved regions shared by *B. stearothermophilus* α-amylase and Taka-amylase A are 33–49, 101–112, 220–236, 444–456 and 47–67, 113–126, 194–210, 434–438, respectively, whereas the liquefying enzyme has conserved regions also at 26–30, 208–214, and 469–473 and the saccharifying enzyme at 11–15, 91–109, 227–234, 316–325, and 392–407.

Two of the common regions of similarities between *B. stearothermophilus* and Taka-amylase A were also reported by Svensson.[13] The residues thought to be involved in the catalytic site and in substrate binding of Taka-amylase A[21] are conserved as well as the corresponding sites in the *B. stearothermophilus* α-amylase. Similarities were also found in the amino- and carboxy-terminal parts of both enzymes. The *B. stearothermophilus* and *A. oryzae* α-amylases were used as representatives of their classes. The sequences for liquefying α-amylases in particular are very similar.[1] Knowledge about sequence similarities and dissimilarities in saccharifying and liquefying α-amylases is useful for modeling the three-dimensional structures of α-amylases and for studying and modifying properties of the enzymes by site-directed mutagenesis. These kinds of data can be used in study of phylogenetic relationships, too.

The comparison of secondary structural features of these sequences did not show conserved regions. This is understandable, since the accuracy of the secondary structure predictions varies from about 50 to 70%, and the sequences were only distantly related. The comparisons of secondary structure cannot be any better than the predictions used. However, the secondary structural features can be valuable in cases of more conserved sequences (Fig. 2b).

Comparison of Hydropathy Scales

The comparison of secondary structural features can be extended to study predictive methods. Several methods and scales are available, espe-

[16] E. Boel, I. Hjort, B. Svensson, F. Norris, K. E. Norris, and N. P. Fiil, *EMBO J.* **3**, 1097 (1984).
[17] A. Amemura, R. Chakraborty, M. Fujita, T. Noumi, and M. Futai, *J. Biol. Chem.* **263**, 9271 (1988).
[18] N. Katsuragi, N. Takizawa, and Y. Murooka, *J. Bacteriol.* **169**, 2301 (1987).
[19] K. W. Siggens, *Mol. Microbiol.* **1**, 86 (1987).
[20] T. Takano, M. Fukuda, M. Monma, S. Kobayashi, K. Kainuma, and K. Yamane, *J. Bacteriol.* **166**, 1118 (1986).
[21] Y. Matsuura, M. Kusunoki, W. Harada, and M. Kakudo, *J. Biochem.* **95**, 697 (1984).

FIG. 4. Comparison of hydropathy profiles of *B. stearothermophilus* α-amylase calculated by values of Hopp and Woods (*x* axis) and Kyte and Doolittle (*y* axis). The window was 6, the limit 0.4, the first stringency 0, and the second stringency 6.

cially for predictions of α-helix, β-sheet, and turn and for hydropathy calculations. Methods to predict secondary structural features can be evaluated and their accuracies estimated by comparison to known three-dimensional structures. Here, two hydropathy scales, that of Hopp and Woods[22] and that of Kyte and Doolittle,[23] are compared for the *B. stearothermophilus* α-amylase in Fig. 4. The two methods predict many of the peaks and valleys similarly. The scales were normalized to have the same average of absolute values according to Cornette *et al.*[24]

Comparison of structural predictions can be used to study different methods and scales and even to search for relationships between different features of structure. The ends of α and β-structures are often hydrophilic,[25] and these kinds of data can be easily obtained with the method.

[22] T. P. Hopp and K. Woods, *Proc. Natl. Acad. Sci. U.S.A.* **78**, 3824 (1981).

[23] J. Kyte and R. F. Doolittle, *J. Mol. Biol.* **157**, 105 (1982).

[24] J. L. Cornette, K. B. Cease, H. Margalit, J. L. Spouge, J. A. Berzofsky, and C. DeLisi, *J. Mol. Biol.* **195**, 659 (1987).

[25] T. P. Hopp, *in* "Proteins: Structure and Function" (J. J. L'Italien, ed.), p. 437. Plenum, New York and London, 1987.

Conclusion

Here, several protein sequences were compared simultaneously with a multiple sequence comparison method to show conserved regions of one sequence. However, the method is applicable for comparing DNA and RNA sequences, too. Pairwise comparisons are superimposed; thus, enormous computer power is not required. This approach is very fruitful because a multiple sequence comparison gives more data than individual pairwise analyses. As a test, sequences for liquefying and saccharifying α-amylases were shown to share some common regions with other amylolytic enzymes, but each shared different conserved regions with others.

Several secondary structural features can be predicted, and their use to compare protein sequences can point to important features possibly invisible by sequence comparisons. However, some caution is necessary in the use of secondary structure predictions, because predictive methods do not always give correct results. In connection with sequence analysis, structure predictions can be valuable in searching for functional sites, e.g., for protein engineering studies. The relatedness of two hydropathy prediction scales was analyzed with the method. Such applications extended to the analysis of different predictive methods may have numerous applications in the study of the relationships between different structural features. The method can even be used to analyze three-dimensional structures, either refined or modeled.

Acknowledgments

Antti Euranto and Petri Luostarinen are thanked for implementation of the algorithm. This work was supported by a grant from Neste Oy Foundation.

[29] Hierarchical Method to Align Large Numbers of Biological Sequences

By WILLIAM R. TAYLOR

Introduction

The rapidly increasing determinations of biological sequences have increased the corresponding need for a fast and effective multiple sequence alignment computer program for their analysis. The contents of this volume indicate that this need has not been neglected by those who write programs. The number of sequence alignment programs that have appeared in the last few years, in particular those directed toward aligning more than two sequences, seem to have kept pace (at least in variety) with the growing sequence data banks.

In this chapter I describe the computer program that resulted in response to my own need to align more than two protein sequences. However, with the exception of similar contemporary developments, I have made no attempt to review the rapidly changing multiple sequence alignment field.

Background

Several years ago, if the need arose to align more than two sequences I used the ALIGN program[1] to align pairs of sequences and then combined these pair alignments together in a text editor to produce a multiple alignment. If the sequences were not very similar (e.g., <50% identity) then some care was necessary to choose the best alignments. For this I followed this simple heuristic: take the best alignment, say A with B (written AB), as a core and then add on the next best alignment that contains either A or B, say, BC, giving the multiple alignment ABC. This can be repeated using the edge sequences in the alignment until all the sequences are aligned (see Fig. 1). However, as editing several large sequences together in this way can prove to be rather tedious I wrote a simple program, based on the results of the ALIGN program, to do this. Strangely, repeating what I had been doing in the text editor with a computer program, although conceptually straightforward, proved to be surprisingly difficult to code.

Chained Pairwise Alignment

Recursive Alignment Linking

The program that linked together alignments[2] was written in FORTRAN, and the routine that performed the merging required more than 100 lines of code. (The source of the complexity lay in placing insertions in regions that already contain inserts.) A more aesthetically acceptable solution to this problem was found when the program was recoded in the

---

[1] B. C. Orcutt, M. O. Dayhoff, D. A. George, and W. C. Barker, "User's Guide for the Alignment Score Program of the Protein Identification Resource (PIR)," *PIR Report ALI-1284,* National Biomedical Research Foundation, Washington, D.C., 1984.
[2] W. R. Taylor, *CABIOS* 3, 81 (1987).

a

	ISRBT	ISCHT	ISLAT	ISBYT	ISBSTF
ISRBT		4.40 —1→	4.29	2.80	2.28
ISCHT	4.40	3	4.13	2.76	2.40
ISLAT	4.29	4.13 ↓2		2.78	2.41
ISBYT	2.80	2.76	2.78		2.00
ISBSTF	2.28	2.40	2.41	2.00	

b

```
ISBYT         - Triosephosphate isomerase (EC 5.3.1.1) - Baker's yeast
       2.78
ISLAT         - Triosephosphate isomerase (EC 5.3.1.1) - Coelacanth
       4.29
ISRBT         - Triosephosphate isomerase (EC 5.3.1.1) - Rabbit
       4.40
ISCHT         - Triosephosphate isomerase (EC 5.3.1.1) - Chicken
       2.40
ISBSTF        - Triosephosphate isomerase (EC 5.3.1.1) - Bacillus
                stearothermophilus
```

FIG. 1. (a) Relatedness matrix obtained from the alignment of five sequences of triose-phosphate isomerase. The best alignment (ISRBT with ISCHT) is indicated by a thick box. From this pair, other overlapping sequences are found (indicated by finer boxes), producing the ordered list of sequences in (b).

computer language C. This treated the combined alignments as a linked tree of pointers which was then recursively processed by one simple routine. The tree has its root at the beginning of the first sequence (A) in the alignment. This residue was considered to be linked to the second residue in the same sequence and any residue against which it had been aligned. The aligned residue in, say, sequence B in the alignment AB is linked only to itself in the following alignment in the chain of pair alignments, say, in sequence B in the alignment BC. This second occurrence of the residue in B is, like the initial residue, linked to the following residue (in B) and any residue with which it aligns (in C). However, the residue in B is not necessarily the first in the sequence so it must also have a link to its preceding residue. These links define a tree spanning the whole alignment that has at most three branches from any node (see Fig. 2). The branches are referred to as LEFT and RIGHT (along a sequence) and DOWN when jumping from one pair alignment to the next. The order or processing the branches is LEFT, DOWN, then RIGHT, and the same procedure processes each branch. The residues are entered in the final alignment only after each branch has hit a dead end since at this stage the total number of insertions required between the last and current positions is known.

a

```
Sequence 1 = ALGORITHM
Sequence 2 = ALIGNING
Sequence 3 = ELEMENTARY
sequence 4 = ALIGNMENTS
```

b

```
A  L  *  G  O  R  I  T  H  M
:  :  :  :  :     :  :  :
A  L  I  G  N  *  I  N  G

|  |  |  |  |     \  \  \  \

A  L  I  G  N  *  *  *  I  N  G
:  :  :  :  :  :  :  :  :  :  :
E  L  *  *  E  M  E  N  T  A  R  Y

|  |  |  |  |  |  |  |  |  |  |

E  L  *  *  E  M  E  N  T  A  R  Y
:  :  :  :  :  :  :  :  :  :
A  L  I  G  N  M  E  N  T  S
```

FIG. 2. (a) Four sequences to be aligned by the algorithm described in the text. (b) The three best alignments resulting from the pairwise alignment of the sequences in (a) are linked together. These links, including those to the right and left along the first sequence in each pair, form a tree shown in (c). (c) The tree defined in (b) is processed by the following recursive routine:

```
enter (aln, col, row, pos)
{
    col = bump_check (row, col);
    if (clear(LEFT)) enter(aln, col − 1, row, pos − 1);
    if (down) col = enter(aln + 1, col, row + 1, down);
    if (clear(RIGHT)) enter(aln, col + 1, row, pos + 1);
    alignment[row][col] = residue[a];
    return col;
}
```

The routine (shown only in skeleton C code) calls itself with three different sets of parameters depending on the direction of branching. The parameters are as follows: aln, a pointer to an ordered list of pairwise alignments (AB, BC, CD); col, the column position in the output array (alignment); row, the row position in the same array; and pos, the position in the alignment pointed to by aln. The variable down is a pointer from one pair alignment to another and has the value of the new position in the following alignment. The routine clear simply checks end conditions when moving right and left along a sequence, while bump_check keeps check on the maximum current column position and increments this if there is danger of overwriting existing entries in the alignment array. Because of this possible change, the value of col must be returned. The growing multiple alignment is filled up only on the return path from a series of calls since only at this point is the number of required columns known. (d) The resulting alignment.

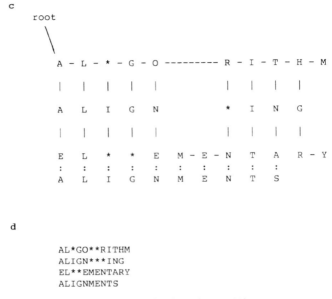

c

```
root
    \
    A - L - * - G - O --------- R - I - T - H - M
    |   |   |   |   |           |   |   |   |
    A   L   I   G   N           *   I   N   G
    |   |   |   |   |           |   |   |   |
    E   L   *   *   E   M - E - N   T   A   R - Y
    :   :   :   :   :   :   :   :   :   :
    A   L   I   G   N   M   E   N   T   S
```

d

```
AL*GO**RITHM
ALIGN***ING
EL**EMENTARY
ALIGNMENTS
```

FIG. 2c and d. See legend on p. 459.

Comparison to True Multiple Alignment Method

The problem of multiple sequence alignment had meanwhile been approached in a more rigorous way through the generalization of the basic Needleman and Wunsch algorithm[3] to higher dimensions, each corresponding to a sequence (e.g., Refs. 4, 5, and this volume). These "true" multiple alignment programs can produce a mathematically defined "best" alignment since they simultaneously consider all the sequences, whereas the concatenation of pairwise alignments considers only two at a time.

Unfortunately, the "true" multiple alignment programs require considerable computer resources (both time and memory) to solve problems of a useful size. When aligning protein sequences, this expenditure is not necessarily justified since there is no clear definition of how two protein sequences should, in principle, be related. This theoretical weakness arises because the amino acids in protein sequences realize their full functional

[3] S. B. Needleman and C. D. Wunsch, *J. Mol. Biol.* **48,** 444 (1970).

[4] M. Murata, J. S. Richardson, and I. L. Sussman, *Proc. Natl. Acad. Sci. U.S.A.* **82,** 3073 (1985).

[5] M. S. Johnson and R. F. Doolittle, *J. Mol. Evol.* **23,** 267 (1986).

significance only in the three-dimensional structure of the protein. Since this structure is generally unknown, the constraints on the mutability of each residue cannot be fully known. Consequently, residues in different sequences cannot be compared without an inherent degree of uncertainty. The work of Dayhoff and Barker[6] provides a guide to the likelihood that two residues in different sequences are equivalent; however, compared to this rough measure, the subtleties in preference discriminated by the Needleman–Wunsch algorithm[3] are very fine. In practical terms, there- fore, it is not justified to spend a large amount of (computer) time seeking a mathematically best solution to the multiple sequence alignment problem while the underlying model of amino acid relatedness remains ill defined.

As an example, I repeated the alignment of three small copper-binding proteins that had been aligned by the method of Murata *et al.*[4] The alignment produced by linking pairwise alignments[2] contained all the essential features of the solution arrived at by the more rigorous method and arrived at effectively the same solution 30 times faster (on a slower computer).

Incorporating Cluster Algorithm

The method of Taylor[2] linked together pairwise alignments to produce a multiple alignment by selecting the best aligned pair of sequences from a family and adding further alignments onto this core, in order of their relatedness, until all the sequences had been incorporated. For a single homogeneously related family of sequences this method is sufficient. If the family contains two distinct subfamilies, however, then the alignments of these subfamilies will be linked through a poorly aligned pair. A simple change was made to the earlier algorithm to avoid this situation.

Beginning again with the best aligned pair of sequences within the group (i.e., those with the highest score), further alignments are considered in order of decreasing score. If the alignments overlap (e.g., *AB* and *BC* form *ABC*, as above) then the chain of linked alignments is extended.[2] If, however, the alignment cannot be added to the end, then, rather than be rejected, it begins a new chain (unless either of the pair of sequences is already buried in a chain). Subsequent alignments are compared to all free ends for potential sites of addition, and where an alignment can join two ends, the chains are fused into one (i.e., the alignment *HI* joins chains *IJK* and *EGFH* forming *EGFHIJK*) Run to its conclusion (when the align- ments of all pairs of sequences have been considered) this algorithm will

---

[6] M. O. Dayhoff and W. C. Barker, "Atlas of Protein Sequence and Structure," Suppl. 3. National Biomedical Research Foundation, Washington, D.C., 1978.

generally produce an ordered list of sequences very similar to that produced by the previous algorithm. The new method, however, has the advantage that a cutoff can be introduced for the score value, below which alignments will not be allowed to join the chains of sequences. This implies that the subfamilies need not necessarily be linked and that isolated sequences may remain excluded (see Fig. 3).

An interesting variation of the above algorithm is to allow a chain to be linked to itself. That is, given the alignment *ABC*, if the next best alignment is *AC* then the list *ABC* can be considered to be a circle, closed to further additions. This condition results in smaller groups, but these are

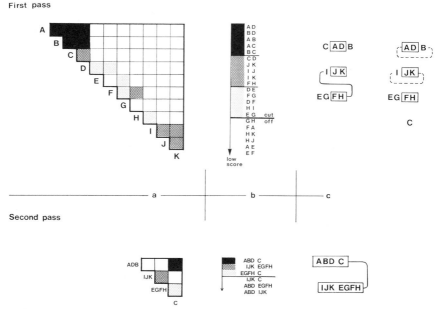

FIG. 3. The basic algorithm illustrated for a two-pass run (two cycles) on 11 sequences. The sequences are designated *A* to *K*, and the matrix (a) shows the relatedness of each pair of sequences. The relatedness (or score) for a pair of sequences is indicated graphically, with solid being strongly related through shades of gray to blank representing weak or no similarity. The scores for each pair are then sorted (b), and those above a given cutoff (i.e., black or gray) are used to define clusters of sequences (c). Two different clusters are shown. In both, the initial pair of sequences in the cluster is boxed, with the others having been added to this pair. In the first example three groups were initially defined but the lower two fused into one. The second example shows the effect of the circle parameter (see text). The dashed lines indicate groups that are closed to addition. This configuration is followed through a second pass of the algorithm in which the alignment of each group behaves as a single (consensus) sequence (see text for details). This second pass quickly unifies the sequences in one group.

then more likely to be closely related to each other than to nonmembers of the group.

Consensus Sequence Definition

In a parallel development, I had been using consensus patterns (templates) to align remotely related sequences.[7] These patterns were matched to a new sequence using either functional classes of amino acids or a score based on the frequency of amino acid occurrence at a given position in a sequence alignment. Similar ideas were developed by Grioskov et al.[8] and Barton and Sternberg[9] using the Needleman–Wunsch algorithm[3] (see elsewhere in this volume). These methods defined a consensus profile from a family of proteins against which a new sequence was matched using the frequency of amino acid occurrence as a score.

Each sequence alignment defined by the above cluster algorithm can be reduced to a consensus sequence. This approach has the attractive feature that by reducing each subalignment to a single consensus sequence, the cluster algorithm can be reapplied to both consensus and single sequences to produce further groupings until nothing more will align. The score between two positions in a pair of alignments was defined as follows:

$$s = \sum_{i=1}^{N_A} \sum_{j=1}^{N_B} M_{XY}$$

where N_A is the number of sequences in alignment A, N_B the number of sequences in alignment B, and M_{XY} the measure of relatedness between residues X and Y (A_i and B_j for positions i and j in alignments A and B, respectively).

The overall score (S) for the alignment of the two alignments is given by the sum of scores (s) on the best path through the score matrix (see Fig. 4) and normalized by both the number of sequences in each alignment and the length of the shorter alignment, i.e.,·

$$S = s/[N_A N_B (L_A, L_B)]$$

where L_A and L_B are the lengths of alignments (or sequences) A and B, respectively.

[7] W. R. Taylor, J. Mol. Biol. **188,** 233 (1986).
[8] M. Gribskov, A. D. McLachlan, and D. Eisenberg, Proc. Natl. Acad. Sci. U.S.A. **84,** 4355 (1987).
[9] J. G. Barton and M. J. E. Sternberg. J. Mol. Biol. **198,** 327 (1987).

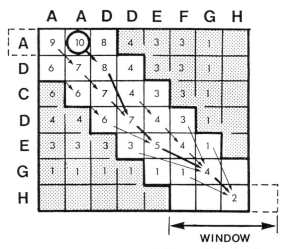

FIG. 4. The effect of the window parameter is illustrated for the alignment of two short sequences (AADDEFGH with ADCDEGH). Using a simple scoring scheme (identity scores 2 and a gap costs 1), the resulting values accumulated in the cells of the score matrix are shown with some trace-back routes (arrows). The window parameter with a value of 4 instructs the program to ignore the shaded cells.

Implementation

Needleman–Wunsch Algorithm Recoded

The algorithm described above has been implemented as a computer program called MULTAL.[10] As before, all alignments, whether between previous alignments, single sequences, or both, are determined by a standard, Needleman and Wunsch-like algorithm.[3] In the simpler method of Taylor[2] the routines from the program ALIGN[1] were used. In this later version these were completely rewritten to improve performance, resulting in roughly an order of magnitude increase in the speed of the basic algorithm.

A window was also defined about the diagonal of the score matrix to ignore unreasonably large insertions (see Fig. 4). This is a standard technique, variations of which are described in Kruskal and Sankoff.[11] The window is centered on the diagonal if the sequences are of equal length;

[10] W. R. Taylor, *J. Mol. Evol.* **28,** 161 (1988).

[11] J. B. Kruskal and D. Sankoff, "Times Warps, String Edits and Macromolecules" D. Sankoff and J. B. Kruskal, eds.) Chap. 10. Addison-Wesley, Reading, Massachusetts, 1983.

otherwise, it is offset in proportion to the difference in their lengths, and if this exceeds the window size, the comparison is not calculated.

MULTAL was written in C, and internal sequence storage was controlled using the dynamic memory allocation features of this language. Thus, there is effectively no limit on the number of sequences that can be considered, providing total storage requirements remain within the virtual memory limit of the system. The current version of the program interacts directly with the University of Wisconsin (UWGCG) sequence analysis package.

Alignment Parameters

Sequence alignment is sensitive to the choice of gap penalty and the form of the relatedness matrix, and it is often desirable to vary these in response to the similarity of the sequences considered. As the effective overall similarity of the sequences (and clusters) can vary through the cycles of the above algorithm, the program allows both the gap penalty and the relatedness matrix to be altered between cycles.

Variation of the relatedness matrix is restricted to an initial choice of two matrices, say, the identity matrix (ID) and the Dayhoff and Barker matrix[6] of amino acid relatedness (MD). A parameter (matwt) was defined that instructs the program to apply a weight to each matrix when calculating the distance between two amino acids. For example, if matwt = 2 then the score is calculated as 0.8ID + 0.2MD. (The ID matrix scores 10 for a match so as to be numerically more equivalent to MD.) Typically, a run might begin using full weight on the first matrix (matwt = 0) when aligning closely related sequences and, on subsequent cycles, gradually shift emphasis onto the second matrix by increasing matwt. In addition, it is common practice to allow a constant to be added to the matrix. In the implementation described here a constant can be added to each matrix at the start of a run but not varied between cycles (6 was added to MD in the results discussed below).

To allow complete freedom in the variation of these parameters, the program reads a complete set afresh between cycles. A typical parameter specification file is shown in Table I.

Interactive Alignment

Because automatic alignment programs are not guaranteed to produce a result that conforms with a user's expectations, the program can also be run interactively, with the ability to examine the intermediate results and vary the parameters accordingly if these are not satisfactory. As alignments

TABLE II
PARAMETER SPECIFICATION FILE[a]

Parameter					Sequences plus families	Families
matwt	gapen	span	window	cutoff		
					5727	0
0	20	1	5	990	5653	52
0	20	2	5	950	5370	222
0	25	4	10	900	5152	317
0	25	8	15	850	4840	438
0	30	16	20	800	4354	545
0	35	32	25	750	4105	606
0	40	64	30	700	3796	648
0	45	128	35	650	3599	679
0	50	256	40	600	3245	729
0	50	512	45	550	3032	758
0	50	1024	50	500	2613	750
0	50	2048	50	500	2366	724
0	50	4096	50	500	2319	714
0	50	8192	50	500	2318	714

[a] The condensation of the NBRF data bank into families is tabulated for progressive cycles of the program MULTAL. The gap penalty (gapen) was slowly increased over the cycles, and there was a progressive shift from the identity (ID) matrix toward an average of this with the Dayhoff and Barker[6] relatedness matrix (MD) controlled by the matwt parameter. The parameters window, span, and cutoff control the degree of clustering, with the latter setting a limit to the score between sequences considered for alignment. See Fig. 7 for plot.

can be evaluated only after they have been calculated, the program retains the state of the alignments in previous cycles to be returned to if desired.

This interactive facility was encoded by calling the alignment program recursively from itself, rather than successively in a loop. This trivial change results in the automatic storage of the previous states of the program, which can then be returned to by exiting the current state.

Strategies for Large Numbers of Sequences

A fast multiple alignment program was developed to enable interactive use as described above, which is especially useful when aligning remotely related sequences. However, a fast program also allows greater numbers of sequences to be considered if one is still prepared to wait as long as before.

Local Clustering

It is often found that the sequences to be aligned are encountered already ordered (perhaps taxonomically classified) with the more similar sequences adjacent. In this situation it is unnecessary to compare all pairs of sequences in the first cycle. The program, therefore, allows only pairs to be compared that lie within a preset separation of each other, referred to below as the span (Fig. 5). It might be expected that this could allow the alignment of closely related but remotely placed sequences to be missed. However, if this occurs in the early cycles, on subsequent cycles the span can be increased and, combined with the contraction caused by the clustering of sequences, the parameters can be chosen so that in the final cycles any isolated sequence is compared to all others, either as an individual or as a component in a consensus.

The window size can also be altered between cycles of the program, as can the cutoff level on the score. Together these features allow similar sequences (also of similar length) to be initially aligned into subfamilies, then on subsequent cycles the similarity cutoff can be lowered (to allow the

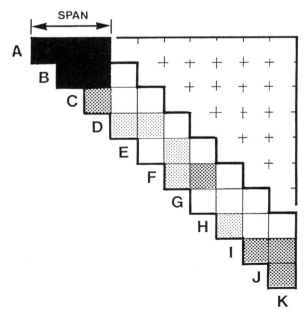

Fig. 5. Using the matrix described in Fig. 3, the effect of the span parameter is illustrated (for the value of span = 3).

subfamilies to align) while simultaneously widening the window. Thus, although there is now more to calculate with each comparison, there are fewer comparisons to perform.

Using all these features together allows large numbers of sequences to be initially scanned and partially condensed into closely related families. For subsequent cycles a parameter profile can be chosen to fit the size of the group of sequences and their rate of condensation.

Rough Presorting of Sequence Order

If the sequences are encountered in a random order, then the above strategy will still produce a complete alignment, only the rate of condensation will be much slower. A general approach to this problem is to use a rough but fast measure of sequence relatedness for the first pass over the sequences. To compare each sequence with every other would typically involve many millions of comparisons, so any form of exact sequence alignment is impractical. However, it is possible to precalculate the location of all sequence fragments of a fixed length for a large number of sequences. A rough measure of sequence relatedness can thus be relatively quickly calculated as the number of common peptides. The resulting list of scores for all pairwise comparisons between sequences could be used directly by the program MULTAL; however, this would have the undesirable effect of forcing an alignment order based on a rough measure of relatedness. Instead, the scores were sorted and the cluster algorithm applied without a cutoff to provide a complete ordered list of sequences.

A separate program called SEQSORT implements this method using the tripeptide lookup table provided with the PIR database. A peptide length of 4 was used (i.e., 16,000 peptides from AAAA to YYYY), which greatly reduces the number of matches (relative to a length of 3) and thus saves time when incrementing the count of hits common to each pair of sequences (which is the rate-limiting step).

Clustering Tests

Cytochrome c Superfamily Alignment

The effect of variation in the degree and rate of sequence clustering was investigated using the cytochrome c family. This extensive family (> 100 sequences) consists of several subfamilies, the interrelationships of which have been analyzed by Schwartz and Dayhoff[12] (see also Dayhoff and

---

[12] R. M. Schwartz and M. O. Dayhoff, *Science* **199**, 395 (1978).

Barker[6] and Fig. 6a). The program MULTAL was applied to this family using different alignment strategies and the results compared.

Two different strategies were tried. The first allowed no families to align that scored less than 500 using the ID matrix (for a pair of sequences this is ~50% identity). Using the same cutoff, the second approach gradually shifts weight from the ID matrix to the MD matrix as described above. As this allows higher overall scores to be obtained, the gap penalty (gapen) was gradually increased throughout this run in compensation. The approach using the "softer" scoring scheme created the largest families. These agreed well with the Schwartz and Dayhoff[12] classification by defining a large group of sequences consisting of the eukaryotic mitochondrial and photosynthetic bacterial sequences (80 sequences including cytochromes c, c-550, and most of the c_2 sequences). The other sequences fall into two main groups; the c_6 sequences from cyanobacteria and chloroplasts (13 sequences) and the aerobic bacterial sequences (c-551 and some c_2, 8 sequences). The remainder includes a small family of two c-555 sequences, remotely related to a c_5 sequence. These minor families were eventually joined into one large family (see Fig. 6b).

Using the exact matching scheme (ID matrix), the families were more fragmented (e.g., the photosynthetic bacterial and eukaryotic mitochondrial sequences remained distinct). Although the major subfamilies were clearly defined, 14 sequences remained unclassified (Fig. 6c) compared to 4 in the previous run.

Both the above runs began with the order of sequences as found in the PIR database, which is already almost optimally ordered. As a more exacting test, the alignment program was provided with the sequences in the order determined by the program SEQSORT and rerun using the "softer" scoring scheme described above. The order imposed by SEQSORT reflected all the major subfamilies, but these were arranged in different positions relative to the order in the PIR database. The results of the alignment starting from this order again identified the major subfamilies and, surprisingly, arrived at this grouping by a route that is largely isostructural with the alignment trees described above (Fig. 6d). Some differences were found, however, the most extreme example being the early alignment of CCMST and CCRF2V, proteins which are separated by 55 sequences in the PIR order. This large displacement was a result of the two sequences being adjacent in the SEQSORT-derived order. Despite this misplacement the resulting alignment was correct.

NBRF Data Bank Alignment

As a test of speed and capacity, the program was applied to the current NBRF (PIR) protein sequence database (Release 16.0). To avoid polypro-

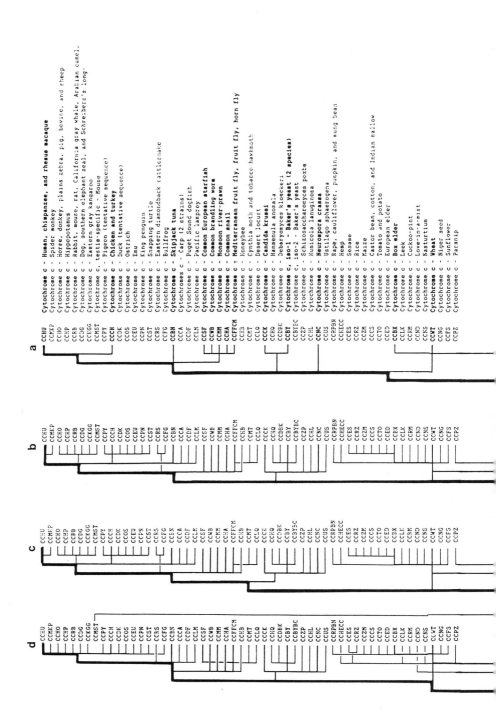

			Cytochrome c - **Human, chimpanzee, and rhesus macaque**
		CCHU	Cytochrome c - Spider monkey
		CCMKP	Cytochrome c - Horse, donkey, plains zebra, pig, bovine, and sheep
		CCHO	Cytochrome c - Hippopotamus
		CCRP	Cytochrome c - Rabbit, mouse, rat, California gray whale, Arabian camel,
		CCRB	Cytochrome c - Dog, southern elephant seal, and Schreibers's long-
		CCDG	Cytochrome c, testis-specific - Mouse
		CCKGG	Cytochrome c - Eastern gray kangaroo
		CCMST	Cytochrome c - Pigeon (tentative sequence)
		CCPY	Cytochrome c - **Chicken and turkey**
		CCCH	Cytochrome c - Duck (tentative sequence)
		CCDK	Cytochrome c - Ostrich
		CCOS	Cytochrome c - Emu
		CCEU	Cytochrome c - King penguin
		CCPN	Cytochrome c - Snapping turtle
		CCST	Cytochrome c - Eastern diamondback rattlesnake
		CCRS	Cytochrome c - Bullfrog
		CCFG	Cytochrome c - **Skipjack tuna**
		CCBN	Cytochrome c - Carp (2 strains)
		CCCA	Cytochrome c - Puget Sound dogfish
		CCDF	Cytochrome c - Pacific lamprey
		CCLM	Cytochrome c - **Common European starfish**
		CCSF	Cytochrome c - **Common brandling worm**
		CCWB	Cytochrome c - **Monsoon river-prawn**
		CCMM	Cytochrome c - **Common snail**
		CCHA	Cytochrome c - **Mediterranean fruit fly, fruit fly, horn fly**
		CCFFCM	Cytochrome c - honeybee
		CCB3	Cytochrome c - Desert locust
		CCMT	Cytochrome c - **Candida krusei**
		CCLQ	Cytochrome c - Hansenula anomala
		CCCK	Cytochrome c, iso-1 - **Baker's yeast (2 species)**
		CCDBK	Cytochrome c, iso-2 - Baker's yeast
		CCBY	Cytochrome c - Debaryomyces kloeckeri
		CCBYEC	Cytochrome c - Schizosaccharomyces pombe
		CC2P	Cytochrome c - Humicola lanuginosa
		CCHL	Cytochrome c - **Neurospora crassa**
		CCNC	Cytochrome c - Ustilago sphaerogena
		CCUS	Cytochrome c - **Rape, cauliflower, pumpkin, and mung bean**
		CCRPBN	Cytochrome c - Hemp
		CCHECC	Cytochrome c - Sesame
		CCES	Cytochrome c - Rice
		CC2M	Cytochrome c - Maize
		CCS	Cytochrome c - Castor bean, cotton, and Indian mallow
		CCTO	Cytochrome c - Tomato and potato
		CCED	Cytochrome c - European elder
		CCBX	Cytochrome c - **Box elder**
		CCLK	Cytochrome c - Leek
		CCRM	Cytochrome c - Cuckoo-pint
		CCND	Cytochrome c - Love-in-a-mist
		CCNS	Cytochrome c - Nasturtium
		CCVT	Cytochrome c - **Wheat**
		CCNG	Cytochrome c - Niger seed
		CCFS	Cytochrome c - Sunflower
		CCPZ	Cytochrome c - Parsnip

a b c d

470

CCFA — Cytochrome c — Buckwheat
CCSP — Cytochrome c — Spinach
CCGK — **Cytochrome c — Ginkgo biloba**
CCEI — Cytochrome c — Enteromorpha intestinalis
CCEG — **Cytochrome c — Euglena gracilis**
CCCRCO — **Cytochrome c — Crithidia oncopelti**
CCCRCF — **Cytochrome c — Crithidia fasciculata**
CCTE — **Cytochrome c — Tetrahymena pyriformis**
CCRF2G — Cytochrome c2 — Rhodopseudomonas globiformis
CCAG2 — Cytochrome c2 — Agrobacterium tumefaciens (strain II Chrys)
CCRD2 — **Cytochrome c2 — Rhodomicrobium vannielii**
CCRF2V — **Cytochrome c2 — Rhodopseudomonas viridis**
CCRF2A — **Cytochrome c2 — Rhodopseudomonas acidophila**
CCNA5A — Cytochrome c550 — Nitrobacter agilis
CCQF2F — **Cytochrome c2, iso-1 — Rhodospirillum fulvum**
CCQF2M — **Cytochrome c2, iso-1 — Rhodospirillum molischianum**
CCQFM2 — **Cytochrome c2, iso-2 — Rhodospirillum molischianum**
CCQFF2 — **Cytochrome c2, iso-2 — Rhodospirillum fulvum**
CCRF2P — **Cytochrome c2 — Rhodopseudomonas palustris (strain 2.1.6)**
CCRF7P — **Cytochrome c2 — Rhodopseudomonas palustris (strain 2.1.37)**
CCQF2R — **Cytochrome c2 — Rhodospirillum rubrum**
CCQF2P — **Cytochrome c2 — Rhodospirillum photometricum**
CCRF2C — **Cytochrome c2 — Rhodopseudomonas capsulata (2 strains)**
CCRF2S — **Cytochrome c2 — Rhodopseudomonas sphaeroides**
CCPC50 — Cytochrome c550 — Paracoccus denitrificans
CCRFG2 — **Cytochrome c2 — Rhodopseudomonas gelatinosa**
CCQF2T — **Cytochrome c2 — Rhodospirillum tenue**
CCPS5A — **Cytochrome c551 — Pseudomonas aeruginosa**
CCPS5F — **Cytochrome c551 — Pseudomonas fluorescens biotype C**
CCPS5S — **Cytochrome c551 — Pseudomonas stutzeri**
CCPS5M — **Cytochrome c551 — Pseudomonas mendocina**
CCPS5D — **Cytochrome c551 — Pseudomonas denitrificans**
CCAV5 — Cytochrome c551 — Azotobacter vinelandii O
CCPC54 — Cytochrome c554 — Paracoccus sp.
CCPC54A — Cytochrome c4 — Pseudomonas aeruginosa
CCML6 — Cytochrome c6 — Monochrysis lutheri (soluble f, c553)
CCPF6 — **Cytochrome c6 — Petalonia fascia (soluble f, c553)**
CCAU6 — **Cytochrome c6 — Alaria esculenta (soluble f, c553)**
CC3F6 — Cytochrome c6 — Bumilleriopsis filiformis (soluble f, c553)
CCPR6 — **Cytochrome c6 — Porphyra tenera (soluble f, c553)**
CCIA6 — **Cytochrome c6 — Microcystis aeruginosa (soluble f, c553)**
CCA16 — Cytochrome c6 — Anabaena variabilis (soluble f, c553)
CCYC6L — Cytochrome c6 — Synechococcus lividus (soluble f, c553)
CCYC6 — Cytochrome c6 — Synechococcus sp. (soluble f, c553)
CCFZ6 — Cytochrome c6 — Aphanizomenon flos-aquae (soluble f, c553)
CCPB6 — Cytochrome c6 — Plectonema boryanum (soluble f, c553)
CCSG6 — **Cytochrome c6 — Spirulina maxima (soluble f, c553)**
CCEG6 — **Cytochrome c6 — Euglena gracilis (soluble f, c553)**
CCTW5T — Cytochrome c552 — Thermus thermophilus
CCDV5M — Cytochrome c553 — Desulfovibrio vulgaris (strain Miyazaki)
CCPS5VM — Cytochrome c5 — Pseudomonas mendocina
CCER51 — Cytochrome c551 — Ectothiorhodospira halophila
CCCF55 — **Cytochrome c555 — Chlorobium limicola f.sp. thiosulfatophilum**
CCPH55 — **Cytochrome c555 — Prosthecochloris aestuarii**
CCBo1 — Cytochrome c1, heme protein — Bovine
CCBo11 — Cytochrome c1, nonheme c1 protein — Bovine

FIG. 6. See legend on p. 472.

teins and large repetitive sequences, a maximum sequence length of 500 residues was set. Similarly, to avoid short peptides, a minimum length of 50 residues was set, leaving 5727 sequences. It might be noted at this point that because the program controls its own storage requirements, no changes were required to read in this volume of data.

The first two passes over the data base proceeded cautiously, aligning only closely adjacent sequences within 90% identity and of lengths within 5 residues. In subsequent cycles, the score cutoff was linearly reduced to 500 while the window size was similarly increased until a maximum insertion size of 50 residues was allowed. The range of adjacent sequences compared (span) was doubled in each cycle until this exceeded the size of the data bank. In the final cycles, the span was larger than the size of the condensed data bank; thus, all sequences had been compared to each other, either as individuals or as members of a family. The progressive condensation of the data bank is plotted in Fig. 7 and tabulated, with details of the parameter for each cycle, in Table I. In the final cycles the condensation of the data base had converged, with little reduction being made in the number of distinct entries (single plus consensus sequences). The data base condensed to 2318 distinct entries, of which 714 contained more than one sequence. This degree of compression is similar to a less complete analysis on an older version of the same data bank.[10]

[13] W. C. Barker, L. T. Hunt, B. C. Orcutt, D. G. George, L. S. Yeh, H. R. Chen, M. C. Blomquist, G. C. Johnson, E. I. Seibel-Ross, M. K. Hong, and R. S. Ledley, "Protein Identification Resource," Version 4.3. National Biomedical Research Foundation, Washington, D.C., 1984.

FIG. 6. Progressive clustering of the cytochrome *c* superfamily by the program MULTAL using different parameters. The ill sequences are taken from the PIR data bank (Version 4.3)[12] in the order in which they occur there and are designated by a code followed by the protein name and their source. (a) For comparison the phylogenetic tree derived by Schwartz and Dayhoff[13] is shown in a simplified form (adpated from Dayhoff and Barker[6]). Those sequences known in 1978 are shown in bold-face type. In the following trees (b–d) each level of the tree represents the state of clustering after a cycle of the program. For example, in the run described in (b), CCHO, CCHP, CCRB, CCDG, CCKGG, and CCMST were aligned in the first cycle of the program. The order of the sequences within such groups (and hence their position in the alignment) is determined by the algorithm described in the text. For clarity, however, the order as found in the PIR data bank is retained. To facilitate comparison with (a), elements that correspond to that tree are emphasized. (b) Clusters defined using the parameters specified in the MD matrix (using the "softer" scoring scheme). Careful examination shows that the trees are remarkably similar. (c) Clusters defined using the parameters specified in the ID matrix (scored by amino acid identity). The tree is largely isostructural with (b) but less complete. (d) As (b) but using the sequence order generated by the program SEQSORT (see text). The tree is largely isostructural with (b) barring some localized rearrangement. One unexpectedly long connection is made between CCMST and CCRF2V; for clarity, it is drawn behind the list of sequence codes.

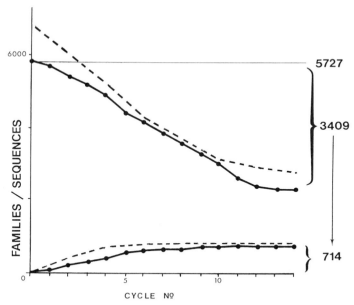

FIG. 7. Plot of the condensation of the NBRF data bank. The 5727 starting sequences are condensed into consensus sequences (families), the number of which rises to 714, while the total distinct entries (single plus consensus sequences) falls to 2318, leaving only 1604 unattached sequences (see Table I for full details). The dashed line plots the course of a similar run[10] using an older (smaller) version of the data bank but on a doubled scale.

The above run took 10 days (234 CPU-hr) on a SUN-4 computer and performed roughly 13 million alignments. Such a large run gives a good estimate of the average speed of the basic pairwise alignment algorithm on typical protein sequences (between 50 and 500 residues) and gives a value of $\frac{1}{15}$ sec to align a pair of sequences (or consensus sequences). As this step is rate determining for large runs, this value can be used to calculate the total time for any run once the number of pairwise alignments has been estimated.

Summary

The method presented here is intended as a compromise between finding a good overall alignment and the time taken to do so. Many multiple alignment algorithms spend an excessively large amount of effort trying to find the best global alignment. This time is often ill spent because the results of the standard dynamic programming alignment algorithm are dominated by the choice of gap penalty and the form of the score matrix, both of which have a poor theoretical foundation. Nonetheless, it is important that savings in time do not compromise the quality of the alignment. By using the consensus sequence approach, this danger is largely avoided as the conserved features of the sequences are quickly identified and preserved through further cycles.

In the alignment of existing alignments, which is one of the more novel aspects of the method, each alignment was treated as an averaged consensus sequence with gaps making no contribution. This gives rise to the advantageous property that gaps will have a greater propensity to be inserted where there are already gaps and is equivalent to a local change in the gap penalty. This type of behavior represents a transition away from the homogeneous scoring schemes used in aligning two sequences toward a scoring scheme that depends on position in the sequence. The alignment of consensus sequences thus forms a bridge between simple pair alignment and the alignment of discrete patterns in which sequence features and allowed gap locations are exaggerated.

To complete this transition the program described above has been integrated into the earlier pattern matching (template) program.[7] Such templates can reliably locate sequence similarities that are too weak or scattered to be found by the more standard alignment methods[14] and should therefore produce a further condensation of the sequence data bank. Only by continually extending our knowledge of the relationships between sequences to increasingly distant similarities can we hope to avoid being overwhelmed by the increasing amount of data.

[14] L. H. Pearl and W. R. Taylor, *Nature (London)* **329**, 351 (1987).

[30] Significance of Protein Sequence Similarities

By JOHN F. COLLINS and ANDREW F. W. COULSON

Introduction

Similarities between pairs of protein sequences are expressed as alignments of subsequences such as:

```
      * *  *          * *  *    * *      * *   *        *          *     *     *
  2 -NILEDPVPVKRHSDAVFTD    NYTGLRKQMAVKKYLN S ILNGK-  42
137 -NIVEE  LR  RRHADGSFSDEMNYV  L  DS LATRDF INWLLQTK- 175
```

An offset between the sequences and a set of insertions/deletions has been found which aligns pairs of residues at several positions in a short region. Beyond the ends of the regions shown, the frequency of matches is much lower. The score of the similarity is the sum of scores for each pairwise alignment of residues, reduced by a penalty for each gap. A simple scheme might score 10 for each match, 1 for each mismatch, and −10 for each unpaired residue in a postulated insertion/deletion.

Such similarities are potentially important because they may indicate some functional, structural, or evolutionary relationship between the proteins. Biological significance of this kind does not necessarily imply a strong statistical significance. Similarities as strong as the example occur "by chance" when typical proteins are compared exhaustively with the presently known database of protein sequences. However, these particular sequences[1] represent hormonal peptides which are known to display similar activities. Furthermore, the pair can be connected by a chain of similarities each of which is individually so strong that there can be no doubt about the evolutionary relationship between successive members of the chain, and therefore between the sequences at the two ends.

A low statistical significance does not imply that a similarity is not biologically important; on the other hand, a similarity which is very improbable does imply that sequences are related, even if this relationship is not yet understood. The main value of assessment of the statistical significance of a similarity is therefore to provide additional evidence that a biological relationship exists. In addition, if a large number of subsequence similarities is found as the result of a database search, it is prudent to examine first those which are least likely to have occurred by chance.

It is sometimes implied that the improbability of an alignment has an absolute and fixed value. However, the frequency with which alignments with a given level of similarity will be found in comparisons of unrelated sequences depends on many aspects of how the search has been carried out. Different algorithms allow for different patterns of insertions and deletions; clearly, more insertions will allow more matches, and therefore reduce their significance. The scheme used to score similarity also has an effect. In the above example, three of the first five positions match; if L/V and D/E are counted as (partial) matches rather than as mismatches, the score might rise to four or five out of five, a much more significant result. The larger the database which has been searched, the more probable it is that a given level of similarity will occur by chance.

All these factors complicate any attempt to calculate the improbability

---

[1] The top sequence is from a human vasoactive intestinal peptide precursor, the bottom from bovine glucagon.

of an alignment *a priori,* and, in particular, the dependence on the search algorithm leads to severe mathematical difficulties. A further factor which makes accurate calculation in practice impossible is the nature of protein sequences themselves. Protein sequences are not random; the requirement that a protein have a compact, functional, and foldable structure imposes constraints on the sequence which affect the probability that a given subsequence similarity might occur between unrelated proteins. Some of these patterns are known and can be accounted for in calculation (e.g., the nonuniform overall frequency distribution of amino acid residues). Others, though well known, are hard to allow for (e.g., the balance of hydrophobic to hydrophilic residues in proteins of different size). Still others, of course, remain to be discovered.

It might be thought that all but the simplest of these factors are too subtle to have much effect on the estimate of statistical significance; the requirement, after all, is only for a method to give the correct order of magnitude. High precision is rarely important. However, the only way to be sure that the cumulative effect of many small effects is not large is to take full account of all of them.

The operational question we would like to have answered is, "Given that a certain subsequence alignment has been found in comparing a query sequence against a database of sequences, what is the chance that an alignment of the same degree of similarity (or better) would have been found if the query sequence had been compared with a database just like the database which was used (in terms of amino acid composition and other sequence patterns arising from physicochemical constraints) except that it contained no sequences which are significantly related to the query sequence?" "Significantly" here implies biological significance, which arises when two sequences are more similar than is required by the appropriate constraints to produce folded functional molecules. In this chapter, "by chance" (in quotes) means "by chance, but subject to these constraints."

It is clear that this question cannot be answered satisfactorily in an analytical way. The semiempirical method we describe here has the virtues that it answers exactly the question posed; that the method is robust, reliable, and easily understood; that it takes account of all nonrandomness of sequence patterns; that it requires only a trivial extension of computing effort beyond that necessary for the original search; and that it can be applied to virtually any general search algorithm.

The method was devised in association with the development of programs[2] for protein sequence database searching which employ an exhaus-

---

[2] A. F. W. Coulson, J. F. Collins and A. Lyall, *Comput. J.* **30,** 420 (1987).

tive algorithm related to that of Smith and Waterman.[3] This method is known to be mathematically complete,[4] and for this reason it is generally regarded as a satisfactory solution to the problem of searching sequence databases for sequence relationships. In effect, the use of this algorithm reduces the problem of finding sequence similarities to the problem of defining biological similarity in terms of the substitutabilities of amino acids and the probabilities of insertions/deletions. On machines of conventional architecture, programs incorporating this algorithm generally run so slowly that complete searches even of the current database are impracticable.

The most efficient implementations of the method use SIMD[5] machines such as the AMT Distributed Array Processor (DAP) 510 (7×10^6 matrix entries per second) and the Thinking Machines Corporation CM-2 32K (2.5×10^7 matrix entries per second[6]).

The practical significance of this performance is that complete searches of the current protein sequence database using the most exhaustive algorithm known can be performed interactively, and we have made many hundred such searches. The methods for assessment of significance of matches will be described solely for the exhaustive algorithm, but we should stress that the same methods can be adapted to virtually any search method.

Outline of Method

The aim of database searching is to pick out the cases of significant similarity from a larger background of unrelated sequences. By definition, therefore, most of the sequences in a database are not significantly related to the query sequence. There are some families of proteins of which many sequences are known, for example, globins, cytochromes c, calcium-binding proteins, and peptide hormones related to glucagon. However, none of these families constitutes more than 4% of the known database. A search with a protein belonging to one of these families may therefore produce up to about 400 significant alignments. Provided a search method generates substantially more than this number of similarities, most of these will have originated "by chance." Current versions of the DAP search program collect statistics (the numbers of subsequence alignments achieving a given

[3] T. F. Smith and M. S. Waterman, *J. Mol. Biol.* **141,** 645 (1981).
[4] P. H. Sellers, *J. Algorithms* **1,** 359 (1980).
[5] SIMD refers to a class of parallel computers in which a single stream of instructions is applied to multiple streams of data.
[6] E. Lander and P. Mesirov, *Proc. Int. Conf. Parallel Processing, 1988* (in press) (1989).

score) for about 16,000 local similarities. The lower scoring 15,600 of these results are taken as representing the distribution of similarity scores which occur "by chance." The statistical significance of the 50–100 top scoring alignments is assessed by asking whether they plausibly belong to the same distribution as the "by chance" results; if any clearly do not, then these results are statistically significant.

It is simplest to make this assessment if the "by chance" distribution can be linearized, and a plausible argument suggests a simple way in which this can be done. Most search algorithms start by comparing a single pair of residues; if this pairing has a positive score, an attempt is made to extend the similarity, with or without insertions/deletions, and this process continues while the score continues to grow. The number of local similarities achieving a given score (in a particular search) will therefore depend on the number having achieved a slightly smaller score, and the probability that each of these will be extended a little further (rather than stop).

Under most circumstances, it seems likely that in the "by chance," distribution, this probability of extension is constant. It is possible that patterns of secondary structure will sometimes invalidate this assumption. For example, some "by chance" similarities will arise because both the query sequence and the database sequence happen to contain an α helix with the same pattern of hydrophobic and hydrophilic residues. Helices are observed to have a nonuniform length distribution in globular proteins, and the probability of extending such a similarity will therefore also be nonuniform with respect to length.

Such effects are most likely to be of significant size only for very short similarities. In any case, we have now accumulated a great deal of empirical evidence that they are of only trivial effect in real database searches. If the probability of extending a "by chance" similarity is a constant, then the logarithm of the number of similarities achieving a given score should be a linear function of the score itself. Figure 1 shows that this is very closely true and demonstrates that this linearization can be used to determine by visual inspection whether particular similarities belong to the "by chance" distribution.

It is more convenient to make this assessment numerically, and this can be done by least-squares fitting a straight line to the linearized "by chance" distribution. Extrapolation of this line allows the statistical significance of an alignment to be assessed in two ways. The intuitively simpler of these is to calculate the expected frequency of a similarity achieving a given score. A significant result is one for which the expected frequency is significantly less than 1; for example, a value of 0.01 implies that such a result would be likely to occur only "by chance" in a search of a database 100 times larger than that actually used. Alternatively, the significance of a given similarity

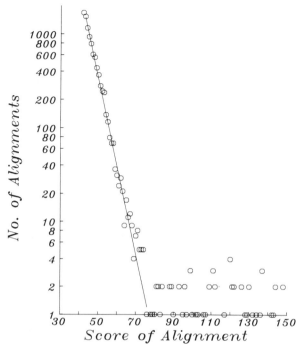

FIG. 1. Number of reported similarities (logarithmic scale) achieving a given score when all members of a large protein sequence database were compared with the sequence of bovine glucagon. The straight line was least-squares fitted to the lower part of the distribution as described in the text.

can be specified in terms of the number of standard deviations (of the fitted points) that the candidate value lies from the fitted straight line. This latter method has the virtue that it will not attribute high significance to a similarity if the linearization of the "by chance" distribution fails for some reason. We have never observed such a case in practice.

The method depends only on collecting enough results to have an adequate representation of the "by chance" distribution. The parameters of the distribution depend on the frequency distribution of residues in the query sequence. If this is nonuniform within the sequence, then the improbability of a given match will also vary within the sequence. However, provided enough results are collected covering every region of the query sequence, the same method can be applied to give a "local" expected frequency, which takes account of compositional inhomogeneities.

Procedures

Calculation of Overall Expected Frequency of an Alignment

Step 1: Collect Data. During the search, maintain an array of M variables with the similarity score values from the M highest scoring alignments found so far. In current versions of the DAP program searching a database containing 2.8×10^6 amino acid residues, M is equal to 16,384; similar results were obtained in earlier versions using smaller databases with $M = 4096$. When lower scores have to be dropped from this array, take care to remove every member of the current lowest value class.

Step 2: Block into Histogram. When the search is complete, use the final data to generate values for the integer array S, where $S(q)$ is the number of alignments in the set whose score lies between $q - \delta$ and $q + \delta$. If the pairwise amino acid scores are scaled to be integers with absolute values less than 100, $\delta = 0.5$ is generally satisfactory.

Step 3: Choose Upper and Lower Bounds for the Fit, and Check for Zero Values. The lowest value of q is usually used as the lower bound. In cases where the collected data include the peak of the distribution (e.g., where a very small database is used) the lower bound must be shifted well above the peak value. The upper bound is set to a value which ensures that no biologically significant results are included. In our case, the top 500 results are excluded. This figure was chosen on the basis that the current database contains no more than 500 entries from any closely related family of proteins. After this step, there should be at least nine classes between the upper and lower bounds, and no values of S should be zero. If either of these conditions is not met, adjust the histogram blocking or the upper and lower bounds.

Step 4: Transform and Least-Squares Fit. Fit the values of log $S(q)$ as a linear function of the values of q. Press et al.[7] (p. 508) give explicit FORTRAN code in the subroutine FIT, which returns values of the slope and intercept of the best fitting straight line, together with χ^2 for the fit and values for the expected standard deviation in slope and intercept (σ_A and σ_B). The statistical weights of all points are assumed equal.

Step 5: Report Expectation Frequency. For each alignment whose statistical significance is required, calculate the expected number of alignments with this score as antilog ($A + B \times$ score), where A is the intercept and B the slope of the regression line. For alignments expected less than once, calculate the number of standard deviations the point lies above the line as log (expected number of occurrences)/σ_A.

---

[7] W. S. Press, B. P. Flannery, S. A. Teukolsky, and W. T. Vetterling, "Numerical Recipes: The Art of Scientific Computing." C.U.P., Cambridge, England, 1986.

Calculation of Local Expected Frequency

Step 1: Collect Data. During the search, maintain an array which contains the start and finish position of each alignment within the query sequence. In considering the local significance of an individual alignment, generate a secondary array *M* of collected alignments which have any residues in common with the target.

Steps 2–5: Perform Steps 2–5 with these data as above, provided at least 80 results remain.

Alternative Approaches

The basic method of assessing statistical significance is always the same: generate the appropriate "by chance" frequency distribution of alignments and test whether the candidate alignment plausibly belongs to it. The method advocated here is to generate the "by chance" distribution in the same operation as the main search. Two other methods have been used to try to discover the form of this distribution.

One of these methods is to repeat the search or alignment algorithm many times, using a different randomly shuffled version of the query sequence each time. The objection in principle to this method is that a randomly shuffled protein sequence is not "like" a real protein sequence. In the case of a database search algorithm, it must always be better to devote any extra computing effort to collecting more data from the real "by chance" distribution than to collecting factitious data from random sequences of characters.

Sometimes the question arises of the significance of an alignment which is not the result of a search (e.g., two individual sequences might be compared because they are known to have related functions). If the similarity is represented by a subsequence alignment, its significance should be assessed with respect to the distribution of other subsequence alignments in the same comparison. If this set is not large enough, a search (using the same algorithm and scoring scheme) of the sequence database or a representative sample will give a plot like Fig. 1 from which the significance of any alignment, however discovered, can be assessed.

Attempts have also been made to estimate the form of the "by chance" distribution *a priori*. This requires specific consideration of the search/alignment algorithm and, in principle, also of the statistical properties of the sequences of real proteins. In practice, only the nonuniform frequency distribution of real proteins is allowed for. Table I shows amino acid frequency tables calculated for large databases of protein sequences. If the observed frequencies of amino acid residues A and B are f_A and f_B, respec-

TABLE I

COMPARISON OF AMINO ACID COMPOSITION IN
PROTEIN SEQUENCE DATABASE[a] AND IN
ALIGNMENTS USED TO PROVIDE THE INITIAL
DAYHOFF SUBSTITUTION DATA[b]

Amino Acid	Symbol	PIR 17	Dayhoff
Ala	A	0.0759	0.0869
Cys	C	0.0197	0.0330
Asp	D	0.0519	0.0469
Glu	E	0.0616	0.0499
Phe	F	0.0398	0.0399
Gly	G	0.0727	0.0889
His	H	0.0233	0.0340
Ile	I	0.0525	0.0370
Lys	K	0.0587	0.0809
Met	M	0.0917	0.0849
Asn	N	0.0437	0.0400
Pro	P	0.0520	0.0509
Gln	Q	0.0408	0.0380
Arg	R	0.0519	0.0409
Ser	S	0.0710	0.0699
Thr	T	0.0589	0.0579
Val	V	0.0648	0.0649
Trp	W	0.0138	0.0100
Tyr	Y	0.0326	0.0300

[a] D. G. George, W. C. Barker, and L. T. Hunt. *Nucleic Acids Res.* **14,** 11 (1986).
[b] M. O. Dayhoff, R. M. Schwarz, and B. C. Orcutt, *in* "Atlas of Protein Sequence and Structure" (M. O. Dayhoff, ed.), Vol. 5, Suppl. 3, p. 345. National Biomedical Research Foundation, Washington, D.C., 1978.

tively, the expected frequency with which A will align with B in random comparisons is $f_A f_B$.

Three types of search algorithms have been analyzed statistically. The simplest is the comparison of every subsequence of a fixed "window" length from the query sequence with every such subsequence from one or more reference sequences.[8] Each window comparison is scored by summing the appropriate score for each pair of aligned amino acid residues.

[8] A. D. MaLachlan, *J. Mol. Biol.* **61,** 409 (1971).

The results of this type of comparison are often displayed graphically by diagonal or dot-plot programs. An efficient algorithm is given below for calculating the appropriate probabilities in this case; a longer explanation is given elsewhere.[9]

Algorithm

Define W as the subsequence, "window," length of comparison. Define $fD(k)$ and $fQ(k)$, $0 < k < 21$, as the frequencies of amino acids of type k in the database and query sequence, respectively. Define $S(i,j)$ (assumed integral) as the score for alignment of amino acid residues of types i and j. Define L and M as the minimum and maximum values of S.

Stage 1: Compute Values for the "Transition Probability" Array, P(m), $L - 1 < m < M + 1$

```
Initialize all values of P(m) to 0.
For n1: = 1 to 20:
  For n2: = 1 to 20:
    P(S(n1,n2)): = P(S(n1,n2)) + fD(n1).fQ(n2)
  end
end
```

Stage 2: Compute Probability of Each Score at Each Position.

```
Define an array T(l,m), 0 < l < W + 1, W.L − 1 < m < W.M + 1
Initialize all values of T(l,m) to 0.
Initalize the iteration:
For n1: = L to M:
  T(1,n1): = P(n1)
end
Interation:
For n1: = 1 to W-1:
  For n2: = n1.L to n1.M:
    For n3: = L to m:
      T(n1 + 1, n2 + n3): = T(n1 + 1, n2 + n3) + T(n1,n2).P(n3)
    end
  end
end
```

---

[9] J. F. Collins and A. F. W. Coulson, *in* "Nucleic Acid and Protein Sequence Analysis: A Practical Approach" (M. J. Bishop and C. J. Rawlings, eds.), p. 323. IRL Press, Oxford, England, 1987.

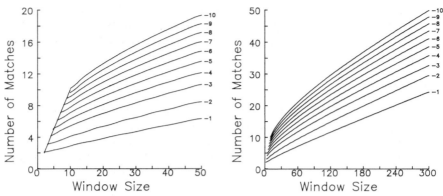

FIG. 2. Contour diagrams for the log probability of matching at least (Number of Matches) when two randomly chosen protein subsequences of length (Window Size) are aligned.

Stage 3: Compute Probabilities of Score-or-Better at Each Position.

For n1: = 1 to W:
    For n2: = n1.M − 1 to n1.L:
        T(n1,n2): = T(n1,n2) + T(n1,n2 + 1)
    end
  end

    At the end of Stage 2, $T(l,m)$ contains the value of the probability of achieving score m in one trial with a window of length l; at the end of Stage 3, $T(l,m)$ contains the value of the probability of achieving a score of at least m in one trial with a window of length l.

    Figure 2 shows contour diagrams for the values of $T(l,m)$ for $W = 100$ and with the "unitary" scoring matrix, $S(i,j) = 1$ if $i = j$; $S(i,j) = 0$ if $i \neq j$. This kind of diagram has an important practical application in designing dot plots to detect weak similarity. It is usually desirable in these cases to examine a range of window lengths, and the probability diagram can save trial and error by suggesting the threshold value that will give a suitable number of dots on the plot. For example, Fig. 3 shows a 5-out-of-10 dot plot for a self-comparison. Figure 2 suggests that 14-out-of-100 will be comparably sparse, and Fig. 4 confirms this and shows the presence of a repeat in the sequence which was not detected at the shorter window length.

    Methods for estimating the statistical significance of similarities found by the Korn–Queen algorithm were outlined in the original publication[10]

---

[10] J. L. Korn, C. L. Queen, and M. N. Wegman, *Proc. Natl. Acad. Sci. U.S.A.* **74**, 4401 (1977).

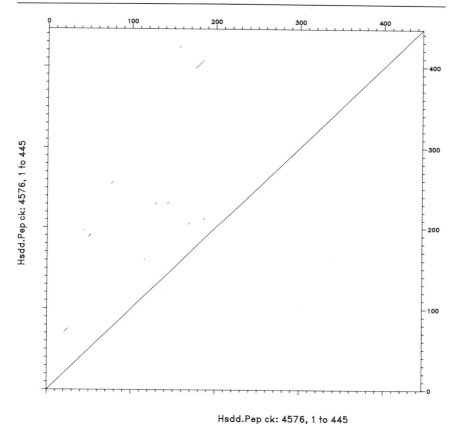

Hsdd.Pep ck: 4576, 1 to 445

FIG. 3. Dot-plot diagram for the self-comparison of the sequence of a type I restriction enzyme specificity subunit. A dot on the diagram indicates that at least 5 out of 10 adjacent residues match at that position.

and elaborated by Brutlag when the algorithm was ported to the Intelli-Genetics software. The analysis is complex, and since the algorithm is now recognized to be of limited value, the calculation is not described further.

The third search method that has been analyzed theoretically is the so-called best local similarity algorithm of Smith and Waterman.[3] This is similar to our exhaustive search algorithm, except that only a single local similarity is collected for each pairwise comparison. The appropriate distribution is therefore related to the distribution of, for example, the longest run of heads per set of n coin tosses, in a large number of such sets. The theoretical analysis is difficult, and it has been applied only to the case where matches, mismatches, and each inserted/deleted residue make the

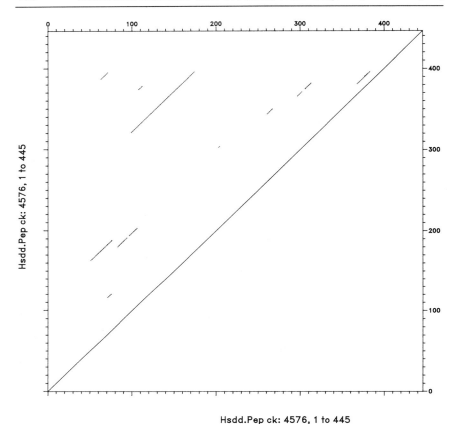

Hsdd.Pep ck: 4576, 1 to 445

Fig. 4. As in Fig. 3, except each dot indicates that at least 14 out of 100 adjacent residues match at that position.

same contribution to the overall score. In fact, published results refer only to a single set of such scores, appropriate to nucleic acid comparisons. Smith *et al.*[11] proposed a theoretical distribution under these restricted conditions which provides that the expected maximum score in a pairwise comparison should grow as $\log(nm)$ for random sequences of lengths n and m. They obtained experimental data that confirmed the general nature of the distribution and provided estimates of its parameters. These allow the expected value and variance to be calculated for pairwise nucleic acid comparisons. The significance of the maximum observed similarity can then be assessed in terms of the expected values.

[11] T. F. Smith, M. S. Waterman, and C. Burks, *Nucleic Acids Res.* **13**, 645 (1985).

This analysis was taken further by Waterman *et al.*,[12] who showed that there are regions in the plane defined by the ratios of the mismatch-to-match and gap-to-match scores in which the maximum expected score grows as n^2 rather than as $\log n$ for the comparison of sequences of length n.

[12] M. S. Waterman, L. Gordon, and R. Arratia, *Proc. Natl. Acad. Sci. U.S.A.* **84**, 1239 (1987).

[31] Fast Alignment of DNA and Protein Sequences

By GAD M. LANDAU, UZI VISHKIN, and RUTH NUSSINOV

Introduction

The library of sequenced DNA already totals about 20 million nucleotides and is likely to increase dramatically over the next decade. Searching for some "key subsequences" with presumably special coding or regulatory function is a basic problem often arising in the analysis of the biological significance of these data. Similar problems arise in the study of evolution, where we need to assess the degree of similarity between sections of DNA often belonging to different species or genes. The numerous possible matchings of long sequences generate, for these data, a problem of computational difficulty. In particular, a very large number of elementary operations, namely, checking if two bases are identical, counting mismatches, and comparing numbers of mismatches, is required. New efficient, packaged strategies that employ new algorithms are needed in order to achieve the desired goals. Our algorithms suggest a more efficient organization of the matching task and the involved bookkeeping of the quality of partial matchings. In this way they minimize the total number of elementary operations required. Compared with a naive, simplistic approach, factors of hundreds in efficiency can in principle be gained.

We describe below, in general outline and omitting detailed proofs, two such algorithms. Some extra details, proofs, and further references to the extensive literature can be found in our original paper on this subject[1] on which the present work is based. We describe the algorithms in DNA context, but these algorithms are obviously applicable to proteins as well. Both algorithms can be viewed as text analysis algorithms, with the text being the complete nucleotide library, a fraction thereof, or some specific virus or gene.

[1] G. M. Landau, U. Vishkin, and R. Nussinov, *CABIOS* **4**, 19 (1988).

In order to account for mutations, the first algorithm (A) allows imperfect matchings of the standard subsequence (pattern), which is m nucleotides long, with the text, of total length n. Up to k differences are allowed. In the second algorithm (B) we compare two texts of length n_1 and n_2 and search for matchings of any subsequence in one text with the other.

Differences between the pattern and the text can be due to (1) substitution, (2) deletion, and (3) insertion mutations. Thus, we indicate an imperfect match between the pattern TGATACT ($m = 7$) and a subsection of the text GTCAAGCTC ($n = 9$) that starts at the second location and ends at the 8th location of the text. For example, in the alignment $\frac{\text{TGATA\ CT}}{\text{TCA\ AGCT}}$, there are $k = 3$ differences arising from a substitution of a C for a G, deletion of T, and insertion of a G.

The number of basic operations, and correspondingly the total computation time T, increases with the size of the parameters specifying the problem. For Algorithm A these are n, the total size of DNA searched; m, describing the size of the patterns investigated; and k, indicating roughly the allowed evolutionary distance. For very large values of n, m, and k, only those algorithms for which the computation time T increases most slowly could be feasible. For Algorithm A, our analysis proved that T is proportional to nk, namely, $T \leq$ constant $\times nk$, the constant depending on the computer being used. We denote this by $T = O(nk)$. The important point is that, unlike previous algorithms, the time bound T is independent of m, the size of the pattern matched. For $k \ll m$, this constitutes a major improvement. The linear increase of T with the text(s) length(s) is clearly unavoidable. Finding a weaker than linear dependence on the total number of allowed mutations k seems to be difficult (or perhaps even impossible). Thus, Algorithm A, presented first by Landau and Vishkin[2,3] (building on, and improving results)[4-8] may, in fact, be the best obtainable.

For Algorithm B the computation time is $T = O(n_1 n_2 k)$. Algorithm B is particularly simple and easy to program. The problem solved by Algorithm B lies at the heart of numerous homology searches, since its solution can be used as a database for retrieving many interesting properties regarding relations between the pattern and the text. In particular, the accumulated information allows (1) recombining nonconsecutive optimally aligned subsequences and (2) extracting most frequently repeating sequence

[2] G. M. Landau and U. Vishkin, *Proc. ACM Symp. Theory Comput., 18th* 220 (1986).
[3] G. M. Landau and U. Vishkin, *J. Algorithms* **10**, 157 (1989).
[4] G. M. Landau and U. Vishkin, *Proc. IEEE Symp. Found. Comput. Sci., 26th* 126 (1985).
[5] G. M. Landau and U. Vishkin, *Theor. Comput. Sci.* **43**, 239 (1986).
[6] G. M. Landau and U. Vishkin, *J. of Comput. Syst. Sci.* **37**, 63 (1988).
[7] G. M. Landau, U. Vishkin, and R. Nussinov, *Nucleic Acids Res.* **14**, 31 (1986).
[8] G. M. Landau, U. Vishkin, and R. Nussinov, *J. Theor. Biol.* **126**, 483 (1987).

motifs. This can be done for those motifs (from the pattern) which repeat (in the text) with l differences for any $l \le k$.

From the standpoint of molecular biology, the advantages of such an algorithm are clear. Suppose a newly sequenced gene is compared with a presumably functionally or evolutionarily related one. If only searches for overall optimal alignments are instituted, the result can be disappointing since some badly matched sections imply an overall low homology score. Our algorithm, on the other hand, has the property that badly matched segments will be overlooked and only the functionally better ones noted (point 1 above). Of special interest is point 2, that is, the information which is computed by our algorithm can be used to locate repeating sequence motifs. Thus, regulatory repeats in upstream regions, such as in the yeast amino acid biosynthetic pathway,[9] will be noted.

Algorithm B is reminiscent of the algorithm of Sankoff,[10] where a related problem was considered. The algorithm developed by Goad and Kanehisa[11] also addresses the same issues as our Algorithm B. However, the mathematical problem solved by them is not identical to the one presented here. Whereas their procedure looks for acceptable "average" homology score between two aligned subsequences, we give the alignments between any two subsequences with $0, 1, 2, \ldots, k$ differences. Thus, their algorithm is not fully comparable to ours.

In this chapter, we deal only with the problem of matching biological sequences for regions where the number of differences is relatively small. In practice, it is also often required to consider possibilities where sizable gaps between matching regions occur. Formally, costs are assigned to various gaps among matched subsequences, and, unlike our problems, these costs grow considerably slower than the length of a gap. Such extended problems are discussed in many papers. We refer the reader to the survey article by Waterman[12] that gives an excellent review of the literature until 1984. More recent work on this topic includes Lipman and Pearson[13] and Altschul and Erickson.[14-16]

Algorithm A is described in the next section. Algorithm B is given in the third section. Considerations regarding retrieval of further information from the database provided by Algorithm B are also given.

[9] G. Lucchini, A. G. Hinnebusch, C. Chen, and G. R. Fink, *J. Mol. Cell Biol.* **4**, 1326 (1984).
[10] D. Sankoff, *Proc. Natl. Acad. Sci. U.S.A.* **69**, 4 (1972).
[11] W. B. Goad and M. I. Kanehisa, *Nucleic Acids Res.* **10**, 247 (1982).
[12] M. S. Waterman, *Bull. Math. Biol.* **46**, 473 (1984).
[13] D. J. Lipman and W. R. Pearson, *Science* **227**, 1435 (1985).
[14] S. F. Altschul and B. W. Erickson, *Bull. Math. Biol.* **48**, 603 (1986).
[15] S. F. Altschul and B. W. Erickson, *Bull. Math. Biol.* **48**, 617 (1986).
[16] S. F. Altschul and B. W. Erickson, *Bull Math. Biol.* **48**, 633 (1986).

Algorithm A

Let $b_1 b_2 \ldots b_n$ be the base sequence of the text and $r_1 r_2 \ldots r_m$ that of the pattern. The task addressed by the algorithm is to find all matches of the pattern in the text with at most k differences. Any such match will be viewed as a successful match (within the prescribed standards).

To set the stage for presenting Algorithm A, we describe first a simple dynamic programming algorithm which achieves this goal in $O(mn)$ steps. (It was given independently in nine different papers; a list of these papers can be found in Ref. 17.) This is next improved by a method based on that of Ukkonen,[18] and finally Algorithm A is described.

O(mn) Dynamic Programming Algorithm

We can consider instead of the complete task of matching the entire pattern, $r_1 r_2 \ldots r_m$, with the text, $b_1 b_2 \ldots b_n$, a series of subtasks. The latter are matchings of increasing (in length) contiguous substrings of the pattern with increasing contiguous substrings of the text.

Let us construct an $(m + 1) \times (n + 1)$ matrix D of "distances." Its elements $D_{i,l}$ are entered at the intersection of the ith row, labeled by r_i, the ith base in the pattern, and the lth column, labeled by b_l, the lth base in the text. For convenience, add a zero row and a zero column, corresponding to an empty text or pattern. The $D_{i,l}$ values are the minimal number of differences in matchings of the subsection $r_1 r_2 \ldots r_i$ and any contiguous substring of the text ending at b_l. Evidently, any element in the last row satisfying $D_{m,l} \leq k$ indicates that a "successful" match of the complete pattern and a subsection of the text ending at b_l has been obtained.

In Fig. 1 we illustrate the D matrix for a concrete example of an $n = 7$ text, GGGTCTA, and $m = 4$ pattern, GTTC, while the number of allowed differences is $k = 2$. We repeatedly refer to this example as we introduce more sophisticated algorithms.

From the defintion of $D_{i,l}$ it is evident that all $D_{0,l} = 0$, since we can always match an empty pattern by moving it all the way to the left in front of the text. Also, $D_{i,0} = i$ since i deletions are required to obtain a match. Given this initialization of the 0th row and 0th column, we proceed to fill up the rest of the matrix by computing $D_{i,l}$ from $D_{i-1,l}$, $D_{i,l-1}$, and $D_{i-1,l-1}$, the elements above, to the left of, and diagonally preceding $D_{i,l}$. Given the "distances," i.e., number of differences in the best matches of

[17] D. Sankoff and J. B. Kruskal, eds., "Time Warps, String Edits, and Macromolecules: The Theory and Practice of Sequence Comparison." Addison-Wesley, Reading, Massachusetts, 1983.
[18] E. Ukkonen, *Proc. Int. Conf. Found. Comput. Theor., Lect. Notes Comput. Sci.* **158,** 487 (1983).

		G	G	G	T	C	T	A
	0	0	0	0	0	0	0	0
G	1	0	0	0	1	1	1	1
T	2	1	1	1	0	1	1	2
T	3	2	2	2	1	1	1	2
C	4	3	3	3	2	1	2	2

FIG. 1. In our example the text is GGGTCTA (length $n = 7$) and the pattern is GTTC (length $m = 4$). We add an auxiliary row and an auxiliary column to our matrix to get an $(n + 1) \times (m + 1)$ matrix with entries $D_{i,l}$. These entries give the minimal number of differences between the pattern portion, $r_1 r_2 \ldots r_i$ and any contiguous substring of the text ending at b_l. Suppose we allow at most $k = 2$ differences. There are occurrences of the pattern in the text with $\leq k$ differences ending at locations b_4, b_5, b_6, and b_7.

$r_1 r_2 \ldots r_{i-1}$ and a subsection of the text ending at b_l, $r_1 r_2 \ldots r_i$ and a subsection of the text ending at b_{l-1}, and $r_1 r_2 \ldots r_{i-1}$ and a subsection of the text ending at b_{l-1}, respectively, we need to assess the minimal number of differences in the best match of $r_1 r_2 \ldots r_i$ and a subsection of the text ending at b_l. We do this as follows: if $r_i = b_l$, we can match r_i with b_l and adjoin it to the preexisting match $r_1 r_2 \ldots r_{i-1}$ and a subsection of the text ending at b_{l-1} (so that $D_{i,l} = D_{i-1,l-1}$). If $r_i \neq b_l$, we can choose the best partial match of $r_1 r_2 \ldots r_{i-1}$ and a subsection of the text ending at b_{l-1} and settle for a substitution of r_i and b_l (so that $D_{i,l} = D_{i-1,l-1} + 1$). We could, however, refrain from matching r_i and b_l and use the best available match of $r_1 r_2 \ldots r_i$ and a subsection of the text ending at b_{l-1} or of $r_1 r_2 \ldots r_{i-1}$ and a subsection of the text ending at b_l, in which case we will have an extra insertion or deletion and $D_{i,l} = D_{i,l-1} + 1$ or $D_{i-1,l} + 1$. We choose among these alternatives the one yielding the minimal $D_{i,l}$.

This is summarized in the following program:

Initialization   for all l, $0 \leq l \leq n$, $D_{0,l} := 0$
            for all i, $1 \leq i \leq m$, $D_{i,0} := i$
for $i := 1$ to m do
   for $l := 1$ to n do
      $D_{i,l} := \min(D_{i-1,l} + 1, D_{i,l-1} + 1, D_{i-1,l-1}$ if $r_i = b_l$ or $D_{i-1,l-1} + 1$
      otherwise)

where $D_{i,l}$ is the minimum of three numbers. These three numbers are obtained from the predecessors of $D_{i,l}$ on its column, row, and diagonal, respectively. Evidently it requires $O(mn)$ basic steps to fill in the $(m + 1) \times (n + 1)$ elements $D_{i,l}$.

Improved Algorithm[18]

Given a partial match of $r_1 r_2 \ldots r_i$ and a subsection of the text ending at b_l, it is advantageous to try matching additional equal length stretches of pattern and text. This corresponds to proceeding in the $D_{i,l}$ matrix along diagonal d, defined by $l - i = d$. This observation motivates the improved algorithm, in which we abstract from $D_{i,l}$ a new set of quantities $L_{d,e}$ and directly compute the latter. Obviously, the distances along any diagonal are nondecreasing (either $D_{i,l} = D_{i-1,l-1}$ or $D_{i,l} = D_{i-1,l-1} + 1$). Therefore, there will be some maximal row number i for which $D_{i,i+d}$ equals some integer e. We label this row number $L_{d,e}$. Thus, e is the (minimum) number of differences between $r_1 r_2 \ldots r_{L_{d,e}}$ (of the pattern) and a subsection of the text ending at $b_{L_{d,e}+d}$. Also, we must have $r_{L_{d,e}+1} \neq b_{L_{d,e}+d+1}$, since otherwise we could extend the match without increasing the number of differences beyond e.

For our k differences problem we need to consider only $L_{d,e}$ values for which $e \leq k$. Having any such $L_{d,e} = m$ indicates a successful match of the whole pattern and the portion of the text ending at b_{m+d}. Returning again to our concrete example, we indicate in the legend to Fig. 2 the value of $L_{d,e}$ for the third ($d = 3$) diagonal as inferred from elements $D_{i,i+d}$ in Fig. 2.

Next we come to the evaluation of $L_{d,e}$. We will constantly attempt at each stage to maximize the row number for any allowed number of differences and along any diagonal d that we follow. We assume that the $L_{x,y}$ values were already computed for all diagonals and for $x < e$. We want to find the value of $L_{d,e} = i$, where i is some row number. From the discussion of the simple algorithm, $D_{i,l}$ could have been assigned its value e if one of the following conditions is satisfied: (1) $D_{i-1,l-1}$, the element preceding $D_{i,l}$ on the diagonal d, is also e and $r_i = b_l$ (the match proceeds and we move along the diagonal) or (2)(a)$D_{i-1,l-1}$ is $e - 1$ and $r_i \neq b_l$, (b)$D_{i,l-1}$ (the

		G	G	G	T	C	T	A
				0				
G					1			
T							1	
T								1
C								2

FIG. 2. Example of $L_{d,e}$ values. The $D_{i,l}$ values from Fig. 1 along the third, $d = 3$, diagonal are reproduced. We deduce that the corresponding $L_{3,e}$ values are $L_{3,0} = 0, L_{3,1} = 3$, and $L_{3,2} = 4$.

element preceding $D_{i,l}$ on the ith row, which is on the $d-1$ diagonal) is $e-1$, or (c)$D_{i-1,l}$ (the element preceding $D_{i,l}$ on the lth column, which is on the $d+1$ diagonal) is $e-1$.

Let us trace backward our computation toward the first row of the matrix. We can start from $D_{i,l}$ and follow its predecessors on diagonal d by option 1 above until the first time option 2 above occurs. The following program "inverts" this description in order to compute the $L_{d,e}$ values. $L_{d,e-1}$, $L_{d-1,e-1}$, and $L_{d+1,e-1}$ are used to initalize the variable row (corresponding to options a, b, and c in 2 above), which is then increased by one at a time until it hits the correct value of $L_{d,e}$ (corresponding to option 1 above).

As we scan the matrix along successive diagonals and gradually increase the allowed e (for $e \le k$), we may reach the last row of the matrix. This means that we can declare a successful match in the following iterative program:

1. Initialization   for all d, $0 \le d \le n+1$, $L_{d,-1} := -1$
   for all d, $-(k+1) \le d \le -1$, do
   $L_{d,|d|-1} := |d|-1$
   $L_{d,|d|-2} := |d|-2$
2. for $e := 0$ to k do
   for $d := -e$ to n do
   3. $row := \max [(L_{d,e-1}+1), (L_{d-1,e-1}), (L_{d+1,e-1}+1)]$
   4. while $r_{row+1} = b_{row+1+d}$ do
      $row := row+1$
   5. $L_{d,e} := row$
   6. if $L_{d,e} = m$ then
      print *THERE IS AN OCCURRENCE ENDING AT b_{d+m}*

We skip the detailed proof of the correctness of the improved algorithm,[18] as well as the fairly obvious initialization. The present algorithm evaluates $L_{d,e}$ for $n+k+1$ diagonals, and for each diagonal the variable row can get (at most) $m+1$ different values. Therefore, it still requires $O(mn)$ steps.

New Algorithm A

We wish to accelerate the above modified algorithm to operate in $O(nk)$ computation time. We do so by modifing the innermost loop (line 4) of the preceding program and jumping over to $L_{d,e}$ in $O(1)$ steps (rather than at least m/k, on the average, for the first mismatch in a continuous run when there are altogether k differences in a pattern of length m).

To this end, however, we need to first reconstruct the information in the pattern and text in a more efficient form. This is done by using a suffix tree, an extremely powerful and versatile tool that enables retrieval of information regarding similarities between contiguous substrings of the same string. Let us then detour briefly to define a suffix tree for a general sequence $c_1, c_2, \ldots, c_l$. (For convenience, we assume c_l to be a "special base" differing from all the above bases.) Consider the l subsequences $C_i = c_i, \ldots, c_l$ of the sequence C. Each such C_i is called a "suffix" of C. Let C_i and C_j be two suffixes. Assume that $c_i, \ldots, c_{i+f}$ is the longest common "prefix" of C_i and C_j. That is, $c_i, \ldots c_{i+f}$ is the longest contiguous substring starting at the leftmost end of C_i (and C_j) so that $c_i, \ldots, c_{i+f} = c_j, \ldots, c_{j+f}$ and $c_{i+f+1} \neq c_{j+f+1}$. The merit of the suffix tree "data structure" can be summarized as follows: the suffix tree of C enables us to find very fast the longest common prefix for each pair of suffixes.

Now that we understand the functionality of suffix trees, we can proceed to describing how they look (see also Fig. 3). The suffix tree has a root, (an unpredictable number of) internal nodes, and l "leaves" (i.e., termination nodes). Each of the l suffixes defines exactly one leaf. It remains to define the internal nodes of the tree. Recall our two suffixes C_i and C_j. The longest common subsequence $c_i, \ldots, c_{i+f}$ defines an internal node (node which is not a leaf) or a branching point of the tree. We note that the contiguous sequence of $f + 1$ letters that match exactly the subsequence $c_1, \ldots, c_{i+f}$ may occur more than once in C. Still, all these subsequences together will define only one internal node. So far we have not said anything about the edges of the tree. Each edge connects a pair of nodes. The rationale behind putting the edges of the tree is as follows. There will

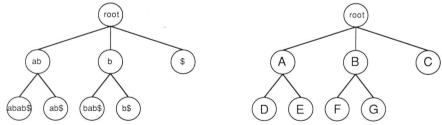

FIG. 3. Suffix tree for a concrete example, the string $abab\$$, where the special symbol $\$$ is different than a and b. The five suffixes of decreasing length with fixed ending $\$$ are denoted as follows: $abab\$ = D$, $bab\$ = F$, $ab\$ = E$, $b\$ = G$, and $\$ = C$. These suffixes form the extreme tips or tree "leaves." The nodes $ab = A$ and $b = B$ correspond, according to the rule of tree formation, to $LCA(D,E)$ and $LCA(F,G)$. The LENGTH and START values of the various elements are indicated as follows: $START(A) = 1$, $LENGTH(A) = 2$, $START(B) = 4$, $LENGTH(B) = 1$, $START(C) = 5$, $LENGTH(C) = 1$, $START(D) = 1$, $LENGTH(D) = 5$, $START(E) = 3$, $LENGTH(E) = 3$, $START(F) = 2$, $LENGTH(F) = 4$, $START(G) = 4$, $LENGTH(G) = 2$.

be a path in the tree from the root to the leaf defined by C_i and another path from the root to the leaf of C_j. This path must pass through the (internal) node defined by $c_i, \ldots, c_{i+f}$. This completes the definition of the suffix tree. An alternative (more formal) definition can be found in Ref. 1.

Example. The suffix tree is best illustrated by a concrete example, the sequence *abab* (see also Fig. 3). We add to it the symbol $ (it is crucial that this symbol is different than both *a* and *b* to form *abab$*. It contains then five suffixes, *abab$*, *bab$*, *ab$*, *b$*, and $, which form the leaves of the suffix tree. For instance, the suffixes *abab$* and *ab$* can track back to a common prefix *ab*. This indeed will be their lowest common ancestor in the suffix tree. Another instance will be the suffixes *bab$* and *b$* whose lowest common ancestor in the tree is defined by *b*. Finally, *ab*, *b*, and $ are joined to the root.

On construction of the suffix tree, we require that for each node v of the tree a contiguous subsequence $c_i, \ldots, c_{i+f}$ which defines it will be stored as follows: START$(v) := i$ and LENGTH $(v) := f + 1$. Below, we show how the suffix tree of the above example will be stored.

Example (Continued). We denote the five suffixes that form the leaves of the tree: $D = abab$$, $F = bab$$, $E = ab$$, $G = b$$, and $C = $$. We denote the internal nodes $A = ab$ and $B = b$. Figure 3 gives the full details of how the tree is stored using the START and LENGTH arrays.

To set the stage for the improvement, two elements are required. First, the suffix tree of the composite sequence $b_1, \ldots, b_n \, ¢ \, r_1, \ldots, r_m $$, consisting of the text $b_1, \ldots, b_n$, the pattern $r_1, \ldots, r_m$, and two extra fictitious bases $¢$ and $, different from all bases in both text and pattern. This computation of the suffix tree can be achieved via the standard algorithm of Weiner[19] or McCreight[20] in $O(n)$ time and $O(n)$ space requirements. Second, a problem often faced in analyzing the suffix trees is finding the lowest common ancestor of two suffixes (or, pictorially, finding the branching point furthest removed from the root that still leads to both leaves). Remarkably, after some initial overall organization of the suffix tree, queries for the lowest common ancestor can be answered in $O(1)$ time.

We now return to the crucial step (line 4) of the previous algorithm. For a diagonal d, the situation following line 3 is that we matched (with e differences) $r_1, \ldots, r_{row}$ of the pattern with some substring of the text that ends at b_{row+d}. We want to find the largest q for which $r_{row+1}, \ldots, r_{row+q}$ equals $b_{row+d+1}, \ldots, b_{row+d+q}$. We can directly increase row to

[19] P. Weiner, *Proc. 14th IEEE Symp. Switching Automata Theory,* 1 (1973).
[20] E. M. McCreight, *J. Assoc. Comput. Mach.* **23,** 262 (1976).

$row + q$, rather than wasting on this a potentially large number of computations in line 4.

Let $LCA_{row,d}$ be the lowest common ancestor (LCA) of the leaves of the suffixes $b_{row+d+1}, \ldots$ and $r_{row+1}, \ldots$ in the suffix tree. The desired q is simply LENGTH ($LCA_{row,d}$). Thus, the problem of finding q is reduced to finding $LCA_{row,d}$. We use either the algorithm of Harel and Tarjan[21] or the later simpler algorithm of Schieber and Vishkin[22] for the purpose of computing LCAs in the suffix tree whenever we need to find such a q throughout the algorithm.

In a query of the LCA, the above algorithms require $O(1)$ time [after a reprocessing of the suffix tree which takes $O(n)$ time]. For each of the $n + k + 1$ diagonals we evaluate in Algorithm A a total of $k + 1$ $L_{d,e}$ values. Therefore, we have $O(nk)$ LCA queries. It will take $O(nk)$ time to process then, yielding the claimed $O(nk)$ time algorithm.

To illustrate how this procedure works, we refer again to our example with GGGTCTA as the text and GTTC as the pattern (Fig. 4). Specifically, let us explain how we compute $L_{3,1}$. For this, we use $L_{2,0}, L_{3,0}$, and $L_{4,0}$. Specifically, $L_{2,0} = 2$, $L_{3,0} = 0$, and $L_{4,0} = 0$. Our algorithm (line 3) initializes row to $\max(L_{2,0}, L_{3,0} + 1, L_{4,0} + 1) = 2$. This is reflected in the box in which "Initially $row = 2$" is written. From the suffix tree we get that $q = 1$ (since $r_3 = b_6 = T$ and $r_4 \neq b_7$). Therefore, $L_{3,1} := 3$.

Algorithm B

Algorithm B looks for the longest alignment with l differences, $l \leq k$, of any consecutive subsequence in one sequence with a consecutive subsequence of another sequence. In order to have uniform notations, we will call the first sequence the pattern and assume the total base number $n_1 = m$. The second sequence will be the text of length $n_2 = n$.

The basic construct here is a three-dimensional, $m \times n \times k + 1$, array $D_{[i=1 \ldots m; \ j=1 \ldots n; \ l=0 \ldots k]}$. In this array the pattern $R = r_1, \ldots r_m$ is stretched along the y axis, the sequence $B = b_1, \ldots, b_n$ along the x axis, and the number of differences $l = 0 \ldots k$ is stretched along the z axis. In each call we store a pair of numbers, $D_{i,j,l} = (f,g)$. $i + f$ and $j + g$ mark the termination points of the "longest" alignment (as defined in the next paragraph) with l differences, starting at i and j, in the pattern and text, respectively. If $r_i \neq b_j$ and $l = 0$, there is no such alignment, and we therefore assign $D_{i,j,0} = (f,g) = (-1,-1)$.

Given the two locations r_i and b_j; the "longest alignment starting there

[21] D. Harel and R. E. Tarjan, *SIAM J. Comput.* **13**, 338 (1984).
[22] B. Schieber and U. Vishkin, *SIAM J. Comput.* **17**, 1253 (1988).

		G	G	G	T	C	T	A
				() $(L_{3,0})$	() $(L_{4,0})$			
G								
T				() $(L_{2,0})$	(Initially $row = 2$)			
T							1 $(L_{3,1})$	
C								

FIG. 4. Illustration of how to find the elements $L_{d,e}$. As before, e is some allowed number of differences, and $L_{d,e}$ equals i if i is the maximal row number on diagonal d such that $D_{i,i+d} = e$. For instance, the entry $L_{2,0}$ reflects the fact that the immediate next bases C and T do not match. We explain how to compute $L_{3,1}$. We use $L_{2,0}$ (=2), $L_{3,0}$ (=0), and $L_{4,0}$ (=0). Our algorithm (line 3) initializes row to max($L_{2,0}$, $L_{3,0} + 1$, $L_{4,0} + 1$) = 2 (see the box in which "Initially $row = 2$" is written). Next, we turn to the suffix tree. We need to find the LCA of the suffix of the pattern r_3, r_4 (=TC) and the suffix of the text b_6, b_7 (=TA). From the suffix tree we get that the length of the matching prefix between these suffixes is $q = 1$ (since $r_3 = b_6 = $ T and $r_4 \neq b_7$). Therefore, $L_{3,1} := 3$.

(with $l \leq k$ differences) can be naturally defined as that extending furthest to the right in the text (i.e., with g maximal). The following algorithm searches for such alignments. However, other definitions of longest alignments can be easily incorporated by some modification.

To visualize the following procedure, imagine slicing the three-dimensional array into horizontal layers. Each layer with a fixed height l is then displayed as a separate $D_{[i,j]}$ matrix. For every matrix we compute the entries $D_{i,j} = (f,g)$ along diagonals $i + j = d$ (orthogonal to those used earlier), proceeding along any diagonal d from bottom up (with decreasing i). We start with the furthest diagonal $d = n + m$ and conclude with $d = 2$. This is illustrated in Fig. 5 for our standard example, $m = 4$, $n = 7$. At each location (i,j) of the diagonal, we compute all the values $D_{i,j,0}, \ldots, D_{i,j,k}$ climbing "up" along the vertical column in the three-dimensional array, directly above i,j.

Let us now explain how any $D_{i,j,l}$ (e.g., cell X in Fig. 6) is computed. Any $D_{i,j,l}$ consists of two entries f and g denoted by Xf and Xg for all X. This computation is done using information in neighboring cells (Y, W, Z) in Fig. 6, which, according to the above ordering has already been computed.

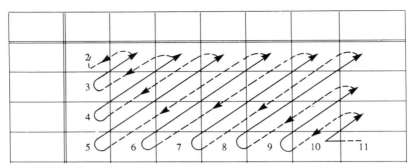

Fig. 5. Illustration of the order of computation. Each continuous line represents a diagonal. Each diagonal has a number. The broken arrows guide us to start at diagonal 11 and end at diagonal 2. The arrow on each continuous line indicates the order of computation on a diagonal.

Recall that, at this point, we are searching for the longest alignment of subsequences of pattern and text, starting with r_i and b_j, respectively, and with l differences. The following alternatives exhaust all possibilities:

1. The b_j is not matched with r_i and is therefore an "insertion" into the text.
2. The r_i is not matched with b_j. It is "deleted" from the pattern.
3. $r_i = b_j$ and we have a correct match of these two bases.
4. $r_i \neq b_j$ and we still match the two bases, reviewing the situation as a substitution (of b_j for r_i).

In Case 1 consider $D_{i,j+1,l-1}$ at the Y cell. It represents an alignment with $l - 1$ differences starting at r_i (in the pattern) and b_{j+1} (in the text). By declaring b_j to be an insertion we extend this alignment one more base to the left in the text. This yields a tentative alignment with l differences starting at i and j. The g value in cell X (termed Xg, i.e., the length of the

				X	Y			
				W	Z			

Fig. 6. Location of the $X,W,Z,$ and Y cells utilized in Algorithm B. To compute the values of the X cell, we utilize the Y cell if b_j is an "insert." The W cell is used when r_i is deleted from the text. For a match or a substitution, we utilize the Z cell.

present alignment), will then be one unit bigger than in Y. In Case 2 consider $D_{i+1,j,l-1}$ at the W cell, representing the longest alignment with $l-1$ differences, starting with r_{i+1} and b_j. If we declare that r_i was deleted in the text, we generate an alignment of the subsection $b_j, \ldots, b_{j+Wg}$ with l differences, so that $Xg = Wg$. In Case 3 we can start from $D_{i+1,j+1,l}$, in cell Z, indicating the l differences match starting at r_{i+1}, b_{j+1} and extend it by one base so that $Xg = Zg + 1$. Finally, in Case 4 $D_{i+1,j+1,l-1}$, also at the cell Z, can serve as a candidate for fixing $D_{i,j,l}$, and again $Xg = Zg + 1$. Since we are looking for the longest alignment starting at r_i, b_j, we will choose the alternative, yielding the largest Xg (and take over the corresponding Xf as well). If there are several candidates of maximum g, we further choose the one having the maximal f. The following is a concise, more formal representation of Algorithm B.

Algorithm for Computing the Matrix $D_{[1, \ldots, m; 1, \ldots, n; 0, \ldots, k]}$

for $d := n + m$ downto 2 do
   if $d > m$(*see Remark below*)
     then $i := m; j := d - m$
     else $i := d - 1; j := 1$
   While $i > 0$ and $j < n + 1$ do
     for $l := 0$ to k do
       $W := D_{i+1,j,l-1}; Y := D_{i,j+1,l-1}$
       $Z := (D_{i+1,j+1,l}$ if $r_i = b_j$ or $D_{i+1,j+1,l-1}$ otherwise)
       $Xg := \max(Wg, Zg + 1, Yg + 1)$
       $Xf := \max[(Wf + 1$ if $Wg = Xg$ and -1 otherwise$)$, $(Yf$ if $Yg + 1 = Xg$ and -1 otherwise$)$, $(Zf + 1$ if $Zg + 1 = Xg$ and -1 otherwise$)]$
       $D_{i,j,l} := X$
       $i := i - 1; j := j + 1$

Remark. Diagonals $m + 1, \ldots, m + n$ start in row m, diagonals $2, \ldots, m$ start in column 1. Observe that we handle both cases.

We refer one more time to our GTTC = pattern and GGGTCTA = text example and take the maximal number of differences allowed to be 2. In Fig. 7 we give the values (f,g) of each cell in the matrix D. In the $l = 0$ case (exact match) and $b_j \neq r_j$, $D_{i,j,0}$ gets the values $(-1, -1)$.

The algorithm performs $O(nmk)$ iterations. Each takes constant time. Therefore, the algorithm requires $O(nmk)$ time. This running time is the best possible since merely filling in the nmk cells of matrix D requires nmk time.

a

	G	G	G	T	C	T	A
G	0,0	0,0	1,1	-1,-1	-1,-1	-1,-1	-1,-1
T	-1,-1	-1,-1	-1,-1	0,0	-1,-1	0,0	-1,-1
T	-1,-1	-1,-1	-1,-1	1,1	-1,-1	0,0	-1,-1
C	-1,-1	-1,-1	-1,-1	-1,-1	0,0	-1,-1	-1,-1

b

	G	G	G	T	C	T	A
G	1,1	3,3	2,3	1,0	1,1	1,0	0,0
T	0,0	0,0	2,2	1,2	1,1	1,1	0,0
T	0,0	0,0	1,2	1,2	0,1	1,1	0,0
C	0,0	0,0	0,0	0,1	0,1	0,0	0,0

c

	G	G	G	T	C	T	A
G	3,4	3,4	3,4	2,2	2,2	1,1	0,0
T	1,1	2,3	2,3	2,3	2,2	2,1	1,0
T	1,1	1,3	1,3	1,3	1,2	1,1	1,0
C	0,1	0,1	0,2	0,2	0,2	0,1	0,0

FIG. 7. Three consecutive levels of the $D_{[i=1 \ldots m; j=1 \ldots n; l=0 \ldots k]}$ array: (a) is the first layer ($l = 0$), (b) is $l = 1$, and (c) is $l = 2$. The l indicates the maximal number of differences allowed. At each square $D_{i,j,l}$ there is a pair (f,g). The pair indicates that the longest alignment with l differences starting with bases r_i of the pattern and b_j of the text ends at r_{i+f} (in the pattern) and b_{j+g} (in the text).

Finally, we note how we extract the information mentioned in points 1 and 2 in the Introduction. (1) Optimally matched sequences may be nonconsecutive on B. Thus, in order to obtain a good alignment it is essential to locate and skip over badly matched subsequences. This can be done by using the information already stored in the three-dimensional array D. Suppose we have a good alignment ending at b_{j-1} and r_{i-1} and we look for a good alignment for the subsequence $r_i, \ldots, r_m$ and parts of $b_j, \ldots, b_n$. We scan row i in D (with user-defined l) and find some good alignments of $r_i, \ldots$. Suppose that such an alignment starts at b_{j+v}. We skip from diagonal $i + j$ to diagonal $i + j + v$ on row i (see Fig. 8). We may

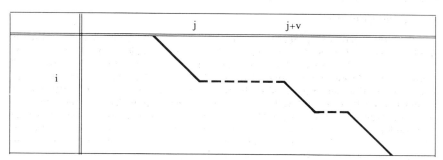

FIG. 8. Illustration of point 1. The solid diagonal lines represent good matches, and the horizontal broken lines "skip" to the next good match.

repeat this skip more than once. (2) Let i, $1 \leq i \leq m$, be a location in the pattern R and $l, 1 \leq l \leq k$, be a number of differences. $D_{i,j,l}$, $1 \leq j \leq n$, provides the longest match between r_i, r_{i+1}, . . . and b_j, b_{j+1}, . . . with at most l differences. Checking all these entries $(D_{i;1}, \ldots, {}_{n;l})$ reveals whether a prefix of "substantial length" of the string r_i, r_{i+1}, . . . repeats often in the text B.

Conclusions

The search for optimal local or global alignments of different DNA and/or protein sequences is a major research endeavor in molecular biology. Since unexpected homologies are often found, an extensive purely syntactical approach, making no *a priori* assumptions, is required. The efficiency of the algorithm used is often, along with powerful computers, a precondition for the attainability of our desired goals.

The existing arsenal of algorithms utilized in molecular biology is quite extensive, and often some modification of an existing packaged program suffices. It is most useful, however, to appeal, in interdisciplinary efforts, to novel, more advanced approaches to this general problem developed by computer scientists. The algorithms presented in this work are fast and address practical problems. Still, a major unresolved difficulty is that of applying various weights to differences. Specifically, a substitution may have to carry a different weight than an insertion or deletion. Another example is that several consecutive differences between DNA or protein fragments may have to be weighed differently than taking the sum of each of the differences separately. The hope is that the use of computer programs resulting from our algorithms as building blocks in some general enough software package (possibly interactive) will enable computation of such more general problems.

Acknowledgments

Uzi Vishkin has been supported by National Science Foundation Grants NSF-CCR-8615337, NSF-CCR-8906949, the Office of Naval Research under contract N00014-85-K-0046, the Applied Mathematical Sciences subprogram of the Office of Energy Research, U.S. Department of Energy, under Contract DE-AC02-76ER03077 at the Department of Computer Science, Courant Institute of Mathematical Sciences, New York University, and the Foundation for Research in Electronics, Computers and Communication, administered by the Israeli Academy of Sciences and Humanities at Tel Aviv University. Gad M. Landau has been supported by National Science Foundation Grant NSF-CCR-8908286, and by New York State Science & Technology Foundation, Center for Advanced Technology in Telecommunications, Polytechnic University, Brooklyn, NY.

Section VI

Estimating Sequence Divergence

[32] Statistical Geometry on Sequence Space

By MANFRED EIGEN and RUTHILD WINKLER-OSWATITSCH

Introduction

Classic methods of comparative sequence analysis are based on distances that are obtained by counting, in two correctly aligned sequences, the number of positions occupied by different symbols. This concept of distance was introduced in information theory.[1] It is straightforward if only binary symbols are involved that substitute one another with uniform probability. The concept has been extended to sequences made up of more than two classes of symbols, even if these change according to different substitution rates,[2] yielding a kind of "information distance." Severe limitations with respect to nucleic acid and protein sequences, however, result from the fact that tolerance and hence fixation of a mutation depends on the location of the position in the sequence. Establishment of a distance requires the correct alignment of both sequences compared. In the presence of insertions and/or deletions, alignment requires consensus criteria that can be met only by larger sets of sequences. Although the methods described in this chapter are based on proper alignment, and although the results obtained may contribute to improve the realization of optimal alignment, we do not deal explicitly with this problem here, but rather refer to those articles which specifically focus on the subject of alignment. We hence start from the assumption that proper alignment is possible.

Alignment as such represents a two-dimensional matrix and thus invites horizontal and vertical inspection. Distance is calculated by horizontal summing of differences between two sequences. Positional nonuniformities of mutation or fixation manifest themselves in vertical deviations from consensus occupation. In this chapter we describe methods of comparative sequence analysis that combine horizontal and vertical criteria. They are used to construct geometries that are more complex, but at the same time also more informative than simple distance dendrograms. We start by introducing the concept of sequence space,[3] a high-dimensional space that is most appropriate for representing sequence relations.

[1] R. W. Hamming, *Bell Syst. Tech. J.* **29,** 147 (1950).
[2] M. Kimura, *J. Mol. Evol.* **16,** 111 (1980).
[3] M. Eigen, R. Winkler-Oswatitsch and A. W. M. Dress, *Proc. Natl. Acad. Sci. U.S.A.* **85,** 5913 (1988).

METHODS IN ENZYMOLOGY, VOL. 183

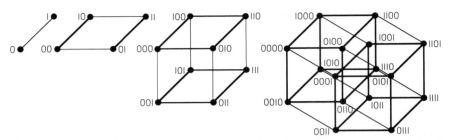

FIG. 1. Iterative buildup of sequence space, starting with one position. Each additional position requires a doubling of the former diagram and connection of corresponding points in both diagrams (which represent nearest neighbors). The final hypercube of dimension v contains as subspaces $\binom{v}{k} 2^{v-k}$ hypercubes of dimension k. (Reprinted by permission of the authors.)

Sequence Space

The concept of sequence space was first used in coding theory.[1] Applications to proteins and nucleic acids were suggested[4,5] and expressed in quantitative terms.[3,6,7] In order to introduce the concept, let us start with binary sequences.

Two conditions have to be met: (1) any of the 2^v possible sequences of length v should be represented by a point, and (2) the points are to be ordered in such a way that all kinship neighborhoods are correctly represented. As Fig. 1 shows, by iterative buildup this leads inevitably to a point space represented by a hypercube of dimension v. The metric of this space is the so-called Hamming metric,[1] where distances are counted in units like street blocks, the particular length of which remains indefinite. All 2^v points in this sequence space are located at the surface of the hypercube whose $v2^{v-1}$ edges connect nearest neighbors (i.e., sequences that differ by one point mutation). The main features of this space that differentiates it from spaces more amenable to our comprehension are its enormous volume encompassing 2^v discrete points, its tremendous connectivity, providing through mutational jumps direct access from any reference sequence to $\sum_{\kappa=1}^{k} \binom{v}{\kappa}$ mutant sequences of distance $\leq k$, and the shortness of detour-free

[4] J. Maynard-Smith, *Nature (London)* **225**, 563 (1970).

[5] I. Rechenberg, "Evolutionsstrategie." Frommann-Holzboog, Stuttgart-Bad Canstatt, West Germany, 1973.

[6] M. Eigen and P. Schuster, *Naturwissenschaften* **64**, 541 (1977).

[7] M. Eigen, J. McCaskill, and P. Schuster, *J. Phys. Chem.* **92**, 6881 (1988).

paths between any two points in the hypercube, never exceeding a distance that equals the dimension v.

The transition probabilities among any two points in this space depend on both the fitness values to be assigned to the corresponding sequences (related to their replication parameters and lifetimes) and the error matrix. This quadratic matrix comprises 2^{2v} terms and is symmetric for uniform error rates, a fact that becomes obvious[8] if terms are ordered according to the iteration scheme shown in Fig. 1. Such details are important with respect to evolutionary theory. We mention them in order to demonstrate that fixation rates for mutations must depend heavily on the structural and functional properties of the particular positions. The fact that uniform substitution rates in certain cases are observed at all may be due to the fact that they refer to neutral mutations.

The binary sequence space model immediately applies to nucleic acids if only transversions (i.e., changes between purines R and pyrimidines Y) are taken into consideration. Four sequences in this space are always related to one another through boxlike structures, of which examples will be shown below. The binary model also provides the basis for considering more than two digit classes.

Nucleic acids are composed of four different monomeric units, in their usual notation: A, U/T, G, and C. A sequence of length v then has $4^v = 2^{2v}$ possible alternatives requiring $2v$ binary decisions or a sequence space of dimension $2v$. Identification of a sequence requires two successive decisions: (1) assignment of the base class R or Y to each position, requiring a v-dimensional hypercube, and (2) assignment of the specific nucleotide (R = A or G, Y = U/T or C), requiring for each point in the first hypercube a subspace that again is a hypercube of dimension v. If four sequences are compared in this space they may involve triangle or tetrahedral structures in their representative geometries (see examples below).

The model may be further generalized by adding successive binary steps. In this way it would be possible to include the consideration of insertions or deletions. One would have to start from a maximum number of positions and decide in the first step whether a position is occupied or not, then follow the procedure described above.

Geometrical representations of sequence space diagrams become less lucid if more than four symbol classes which are not equivalent in their substitution behavior are involved. Formally, however, the concept applies, as does the distance concept, to proteins as well, i.e., to sequences with 20 or more classes of symbols.

[8] D. S. Rumschitzki, *J. Math. Biol.* **24,** 667 (1987).

Distance Statistics

The Meaning of Distance

The vagueness of the term distance, simply defined as the number of positions in (supposedly correctly) aligned sequences that are occupied by different symbols, is obvious from the above considerations. Not only does it mean jumping to and from high-dimensional subspaces that on the whole would differ (if a unit length in metric terms were definable), but even such a definition for a given subspace does not make any sense because of the context-dependent substitution probabilities at individual positions. The true distance between any two points in this space depends on historical details of the interplay of mutation and selection. Dendrograms that refer to distances usually do not reflect these uncertainties correctly, even if the correlation between segment length and time is taken as having a statistical nature (involving large fluctuations).

Distance correlations, e.g., as expressed by dendrograms, look even more problematic if viewed in high-dimensional sequence space. What one generally should expect is a highly interwoven network, which, if projected on a plane, would look like the (schematic) diagram in Fig. 2a. What one would expect to find is some basic topology that is characteristic for the divergence of the sequences under consideration. For instance, one would like to know whether divergence occurred simultaneously and independently for each individual sequence, as is represented, in the ideal case, by a bundlelike diagram (Fig. 2b), or whether divergence reflects consecutive

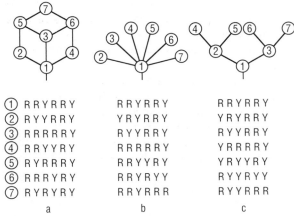

		a			b		c
①	RRYRRY			RRYRRY			RRYRRY
②	RYYRRY			YRYRRY			YRYRRY
③	RRRRRY			RYYRRY			RYYRRY
④	RRYYRY			RRRRRY			YRRRRY
⑤	RYRRRY			RRYYRY			YRYYRY
⑥	RRRYRY			RRYRYY			RYYRYY
⑦	RYRYRY			RRYRRR			RYYRRR

FIG. 2. Examples of a net-, b bundle-, and c treelike divergences. (Reprinted by permission of the authors).

evolution as reflected in phylogeny and represented (ideally) by treelike diagrams (Fig. 2c). Furthermore, there are all sorts of transitions between bundles and trees, as well as all sorts of nonsymmetrical trees.

Physically, the reasons for uncertainties in assigning correct topologies are the following: (1) Substitution probabilities at different positions are nonuniform. (2) Parallel and reverse mutations in individual pairs introduce loops into dendrograms. (3) The number of (known) pair distances, i.e., $n(n-1)/2$ in a set of n sequences, generally does not match the number of unknowns, e.g., $2n-3$ segments of a dendrogram.

Horizontal and Vertical Distances

Even if substitution probabilities are extremely nonuniform, statistics of distances usually show homogeneous distributions. In Fig. 3 such a distribution is seen for a set of tRNAs from yeast. The histogram refers to the $n(n-1)/2$ pair distances d_{ik}. Instead of pair distances, we may also introduce distances of individual to consensus sequences d_{ck}. The consensus sequence records for every position the symbol that appears most frequently. The distribution of the distances d_{ck} has a very similar shape, except that the average is shifted to lower values. The divergence of individual tRNAs should be essentially of a parallel (simultaneous) nature and represented by a bundle (as schematized in Fig. 2b). In the case of an ideal bundle (Fig. 2b) the average pair distance $(\bar{d}_{ik})$ should be twice as large as the average consensus distance $(\bar{d}_{ck})$. However, owing to the presence of

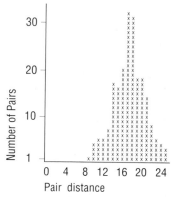

FIG. 3. Distribution of the frequency of pair distances d_{ik} for yeast tRNA sequences. Such a distribution is typical for all species families of tRNAs. (Reprinted by permission of the authors).

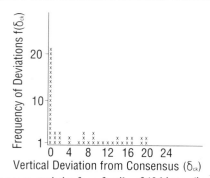

FIG. 4. Column divergence statistics for a family of 40 binary (i.e., R, Y) tRNA sequences of *Halobacterium volcanii*. Again, as in Fig. 3, the distribution is typical for all species families of tRNAs that have been studied thus far. The distribution $f(\delta)$ refers to the present state of divergence δ. It does not resemble a Gaussian form, as would be expected for uniform substitution probabilities. The peak at abscissa value zero indicates a large fraction of constant positions. The 30 variable positions in these sequences show all degrees of variance up to "complete randomization." (Reprinted by permission of the authors.)

parallel and reverse mutations (see below), such a relation generally does not hold.[9] For both $\bar{d}_{ik}$ and $\bar{d}_{ck}$ the shape of the distribution does not reflect positional nonuniformities of substitution probability. The histogram looks very similar to one that would be obtained for uniform substitution rates. This is also true for treelike (i.e., consecutive) divergence, although the histogram as such would look less bell shaped (favoring larger over smaller distances in the distribution).

How do we know that substitution probabilities in this case are extremely nonuniform? We only have to look at the alignment in a different way. Instead of summing differences (i.e., between an individual and the consensus sequence) horizontally, we inspect the alignment vertically for each position. Figure 4 shows the distribution $f(\delta_{ck})$ of vertical deviations δ_{ck} from the consensus sequence for a set of sequences ($\nu = 52$ reference positions, where base pairs are counted as one position) analogous to that shown in Fig. 3. The vertical distance δ_{ck} counts at any position k of the aligned sequences the number of symbols that deviate from the consensus symbol. For binary sequences complete randomization is equivalent to $\bar{\delta}_{ck} = 0.5n$. Histograms of this sort for various tRNA species families reveal that 21 ± 2 positions appear with a zero probability of change ($\delta_{ck} = 0$), while the rest of the positions show any degree of variability between zero and $0.5n$.

Use of the unweighted vertical distance δ, of course, is meaningful only

[9] M. Eigen and R. Winkler-Oswatitsch, *Naturwissenschaften* **68**, 217 (1981).

if either no more than two symbol classes (e.g., R, Y) are involved or, in the case of more than two symbol classes, substitution probabilities of all symbol classes are uniform. Otherwise the vertical distances have to be modified, which may be effected in a similar way as in information theory through the Shannon entropy (which correspondingly modifies the bit number in case of nonuniform frequencies of symbol usage). Both δ and d, in fact, are information distances.[10]

Obviously horizontal distances d reflect some average behavior, if substitution probabilities are nonuniform, while vertical distances δ specify the degree of nonuniformity. If substitution probabilities were uniform, both histograms for bundlelike divergence should yield bell-shaped distributions with defined averages. Distance statistics with nonuniform substitution probabilities may yield misleading results if parallel and reverse mutations come into play.

Effect of Parallel and Reverse Mutations

With increasing divergence, the chance increases that a position which has mutated once will undergo another mutation. For binary sequences this will always mean a reversal of the first mutation and hence will reduce the distance. If more than two symbols are involved such a substitution may leave the distance unchanged. Similarly, a mutation may occur at a position at which other sequences have already undergone substitutional changes, and hence is called a "parallel" mutation. With respect to the initial sequence such a change always increases the distance, whereas with respect to the "parallel" mutant it decreases the distance (or, with more than two symbols, leaves it unchanged).

Parallel and reverse mutations are responsible if distances do not match ideal dendrograms (see below). They also cause the distances not to increase linearly with time, or better, with the total number of mutational events that occurred. Let us call this number "mutation distance" Δ, which for uniform substitution rates, indeed, would be a measure of time. In the literature the mutation distance is often counted per 100 residues and called PAMs (percent accepted mutations = number of accepted point mutations per hundred residues). If substitution probabilities were uniform at all positions, one could readily estimate expectation values for parallel and reverse mutations and thereby calculate Δ from distances d. In Fig. 5 we show for $n = 30$ sequences (diverging from a common precursor) the expected functional dependence of $\bar{d}_{ik}$ and $\bar{d}_{ok}$, the distance between initial (precursor) and individual sequences, on Δ. The curves refer to a (bundle-

[10] C. L. Manske and D. J. Chapman, *J. Mol. Evol.* **26**, 226 (1987).

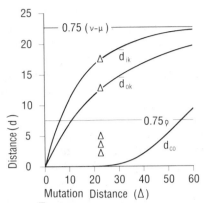

FIG. 5. Average pair distance $\bar{d}_{ik}$, average distance of initial from individual sequences $\bar{d}_{ok}$, and distance between initial and consensus sequence d_{co} as functions of mutation distance Δ. Curves represent computer simulations for $n = 30$ sequences and $v = 30$ positions. (For analytical expressions, see Refs. 3 and 9). Experimental values of tRNA families match these curves only if they are reduced to independently variable positions. The three lower triangles give the spread of experimental distances of three master sequences (deviations from average), indicating that d_{co} has reached some saturation value ρ referring to about 10 hyperdiverged positions and thereby contradicting the model of positionally uniform divergence. [tRNA comprises $v = 52$ reference positions if the anticodon is omitted and base pairs are counted as one position. By subtracting the $\mu = 21$ constant positions we arrive at $(v = \mu) = 31$ independently variable positions.] (Reprinted by permission of the authors.)

like) parallel divergence assuming uniform substitution probability. We may also, under the same assumptions, calculate how well the consensus sequence represents the initial sequence [i.e., how the distance d_{co} between consensus and initial sequence depends on Δ and thereby how well d_{ok} represents the (measurable) values of $\bar{d}_{ck}$].

It is surprising how well precursor sequences can be reconstituted from consensus sequences at even fairly high degrees of randomization (i.e., for large relative contributions of parallel and reverse mutations). Unfortunately, those relations break down if substitution probabilities are not uniform. First, $\bar{d}_{ik}$ and $\bar{d}_{ck}$ do not level off at $0.5v$ as would be expected for binary sequences. Given two symbol classes only, 50% deviation means random appearance of R or Y. In fact, it was to be expected from the histogram in Fig. 4, which revealed $\mu = 20$ invariable positions, that the curves should level off at $\bar{d} = 0.5$ $(v - \mu)$ rather than at $d = 0.5v$. Second, the consensus sequence need not be identical with the initial sequence $(d_{co} = 0)$, as is suggested by Fig. 5. There may be ρ hypervariable positions that have already randomized completely $(d_{co} = 0.5\rho)$. In this case one could not infer a precursor sequence simply by comparing the magnitudes of $\bar{d}_{ik}$ and $\bar{d}_{ck}$. In the example of tRNA sequences, the consensus sequences of different species differ appreciably, while for an expectation value of d_{co} of zero they all should be identical. If substitution rates were uniform, a value of 3/2 for $\bar{d}_{ik}/\bar{d}_{ck}$ would require d_{co} to be zero (see Fig. 5).[9]

The above example refers to bundlelike divergence, as would be expected for parallel and independent evolution. For consecutive branching, vertical analysis would have to be carried out differently. Nevertheless, positional nonuniformities would perturb the correct topology of branching in a similar way as they do in bundlelike divergences. Fortunately, phylogenetic branching seems to involve essentially neutral mutations (cf. molecular clock) that might be accepted in a quite uniform manner. This may explain why tree constructions based on distances have been so successful. Statistical geometry may complement the classic methods in testing the significance of dendrogram construction. In fact, statistical geometry is directly applicable to the metric space of distances, as shown below.

Statistical Geometry in Distance Space

Two correctly aligned sequences A and B define a distance AB. If we add a third sequence C we could formally always construct a tripodal dendrogram in which the three unknowns a, b, and c are matched as to correctly represent the three known distances AB, AC, and AD (Fig. 6). However, the dendrogram thus obtained may be meaningless in that the suggested precursor (i.e., a sequence to be attributed to the branching point

2 aligned sequences **A,B** define a distance :

A •————————• B

3 aligned sequences **A,B,C** define a tripod :

A •————┬————• B
            |
            C

A 4th sequence **D** may fit the tripod in three alternative ways :

None of the resulting dendrograms need to fit the Hamming metric :

FIG. 6. Construction of dendrograms. (Reprinted by permission of the authors.)

or node) may not exist. The truth, i.e., whether or not the construction was correct, comes with a fourth (related) sequence D that, according to Fig. 6, may be matched with the tripodal diagram in three possible ways. A difficulty arising at the same time is the fact that we have six (known) distances (AB, AC, AD, BC, BD, and CD), but only five (unknown) segments for an exactly matching dendrogram. Mathematically, one can prove that in a set of $n > 4$ sequences, an ideal dendrogram exists for the total set if all $\binom{n}{4}$ quartets (i.e., four sequence combinations) are exact dendrograms.[11] Because of parallel and reverse mutations this is almost never the case, and therefore tree constructions, where $n(n - 1)/2$ pair distances are used to match $3n - 2$ dendrogram segments, are an optimization problem, i.e., a problem in which the best approximation, given certain constraints, is searched for.

There is a direct way to look at the quality of dendrogram matching. If for any quartet combination six distances are known, why not represent the quartet by a diagram that possesses six segments? How this is to be effected is shown in Fig. 7. The procedure starts from the six distances, given in numerical form, combines them into distance sums, and constructs a diagram consisting of six segments which follow as the solutions of the system of six equations. The following cases may be distinguished: (1) All three distance sums are equal: $x = 0$, $y = 0$. (2) Two distance sums are equal and larger than the third: $x = 0$, $y =$ finite. (3) Two distance sums are equal and smaller than the third: $x = y =$ finite. (4) All three distance sums differ: $x =$ finite $\neq y =$ finite. The resulting geometries can then be interpreted in the following way: if both x and y are zero, the system represents an ideal bundle; if only one of these parameters is zero the system represents an ideal tree. By definition, the smaller of the two segments is called x and the larger one y, so that x is a measure of "deviation from ideal treelikeness," and both x and y together are a measure of "deviation from bundlelikeness."

The method becomes a "statistical" one by looking at all $\binom{n}{4}$ quartets of the set of n sequences and by constructing their average geometries. While for single diagrams the method is not very reliable, the statistical averages obtained for all $\binom{n}{4}$ quartets are self-consistent. In particular, one can prove that a distribution is ideally bundlelike if all individual x and y values are zero, while it is ideally treelike if all x values are zero and y values remain finite. Hence, the averages $\bar{x}$ and $\bar{y}$ are measures of departure from ideal topologies. As an example, the average distance space dendrogram of 19 sequences of the envelope gene of human immunodeficiency virus (HIV)

[11] J. M. S. Simões-Periera, *J. Comb. Theory* **6**, 303 (1969).

4 Sequences : **A,B,C,D,**

6 Distances : **AB,AC,AD,BC,BD,CD,**

3 Distance Sums : Small **AB+CD = a+b+c+d+2x**

Medium **AC+BD = a+b+c+d+2y**

Large **AD+BC = a+b+c+d+2x+2y**

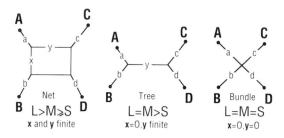

Distance Classes

Deviation from Treelikeness : x = (L-M) /2

Branching Distance : y = (L- S) /2

Mean Branch Length : (a+b+c+d) /4 = (S+M-L) /4

FIG. 7. Analysis of four-sequence combinations in distance space. (Reprinted by permission of the authors.)

is represented in Fig. 8.[12,13] Such a diagram looks almost like an ideal dendrogram. However, Fig. 9 demonstrates with a bundle divergence how insensitively the parameter $\bar{x}$ reacts to randomization.

While it is true that a finite x value for an individual quartet is indicative of parallel or reverse mutations, the converse of this statement does not necessarily hold for geometries in distance space. Consider the four aligned binary sequences containing the following nucleotide triplets

$$A: ----RRR----$$
$$B: ----RYY----$$
$$C: ----YRY----$$
$$D: ----YYR----$$

which resulted from noncongruent parallel mutations at all three positions. The six distances are of magnitude 2 so that, according to the criteria in

[12] K. Nieselt, personal communication (Dissertation, Göttingen, 1989/1990).
[13] GenBank Release 58.0 (12/88), Los Alamos National Laboratory, Los Alamos, New Mexico.

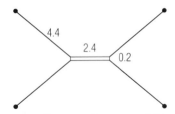

FIG. 8. Average distance space diagram of quartets of 19 sequences of the *env* gene of HIV[13] (1918 positions were chosen, for which no insertions or deletions occur). The average relative pair distance $\bar{d}_{ik}/v$ is 10.6%. (Reprinted by permission of the authors.)

Fig. 7, both x and y are zero, corresponding to an ideal bundle diagram. The reason for this misinterpretation is obvious: there is no sequence that would correspond to the central node having a distance 1 to all four sequences A, B, C, and D. Each of the other four possible sequences

$$E: \text{-----YYY-----}$$
$$F: \text{-----YRR-----}$$
$$G: \text{-----RYR-----}$$
$$H: \text{-----RRY-----}$$

has a distance 1 to only three of the sequences A, B, C, or D but a distance 3 to the residual one. The presence of any of the four sequences E, F, G, and H in a set of $n > 4$ sequences comprising A, B, C, and D would immediately reveal this fact. Hence, there are pitfalls in correlating mere distances that, owing to their cumulative nature, are not easily realized,

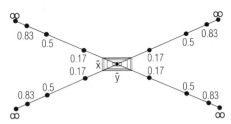

FIG. 9. Simulation of average distance diagrams as functions of Δ/v. Five instances are depicted starting with $\Delta/v = 0$ and extending to $\Delta/v = \infty$. The Δ/v values are inserted along the protrusions. The $\bar{x}$ and $\bar{y}$ values define a rectangle the size of which increases from zero (for $\Delta/v = 0$) to the outer frame for $\Delta/v = \infty$ (each frame corresponding to one of the points marked along the protrusions). The protrusions (cf. marked points) represent the averages $\frac{1}{4}(a + b + c + d)$. The diagrams demonstrate how small $\bar{x}$ and $\bar{y}$ are in comparison to the average protrusions, and hence how insensitively randomization is recorded in distance space diagrams. (Reprinted by permission of the authors.)

except by checking positions individually, which leads us directly to sequence space analysis.

Statistical Geometry in Sequence Space

Binary (RY) Sequence Space

As far as analysis of quartet combinations of sequences is concerned, geometries can readily be derived that represent true correlations in sequence space. We proceed successively from binary (R, Y) to uniform and finally to nonuniform quaternary sequence space, distinguishing the four nucleotides (A, U, G, C) and possible differences in substitution rates for transversions $(R \rightleftharpoons Y)$ and transitions $(Y_1 \rightleftharpoons Y_2; R_1 \rightleftharpoons R_2)$. The main difference between distance and sequence space analysis lies in the fact that the latter utilizes positional information and hence represents, so to speak, a combination of horizontal and vertical analysis.

Figure 10 shows how to proceed. Instead of summing all differences between any two sequences horizontally, we inspect first the four aligned sequences vertically and specify eight position classes, one class as "four of a kind" (where all four sequences show identical symbols), four classes as "three of a kind" (with either A, B, C, or D deviating from the rest), and three classes as "two pairs" (with pairwise congruences, A and B versus C and D, A and C versus B and D, A and D versus B and C). The representative geometry is now a box with four protrusions $(\alpha, \beta, \gamma,$ and $\delta)$. If statistical

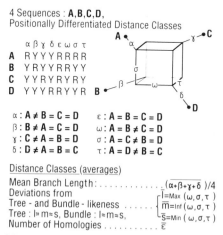

FIG. 10. Analysis of four-sequence combinations in RY sequence space. (Reprinted by permission of the authors.)

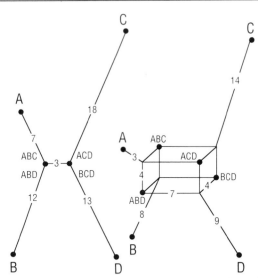

FIG. 11. Comparison of (left) distance and (right) sequence space analysis of four individual 5SrRNAs[14] in RY notation.[14] [Data from G. E. Fox, K. R. Luehrsen, and C. R. Woese, *Zentralbl. Bakteriol., Parasitenkd. Infektionskrankh. Abt. 1* **3**, 330 (1982); V. A. Erdmann, J. Wolters, E. Huysmans, A. Vandenberghe, and R. de Wachter, *Nucleic Acids Res.* **12**, 3133 (1984): *A, Bacillus pasteurii; B, Halobacterium salinarium; C, Anacystis nidulans; D, Methanococcus vannielli.*] The apparently ideal distance dendrogram is fictitious. The range of uncertainty of nodes is revealed more reliably in the sequence space diagram. (Reprinted by permission of the authors.)

averages are taken for all $\binom{n}{4}$ quartet combinations, they refer to average protrusions, i.e., to $\frac{1}{4}\{\alpha + \beta + \gamma + \delta\}$, while the average box dimensions may be identified according to their sizes: $\bar{l}$ (largest), $\bar{m}$ (medium), and $\bar{s}$ (smallest). The topology of divergence again is deduced from the diagrams (rather than presupposed). A tree will show one large box dimension and a bundle three box dimensions of similar magnitude. The degree of randomization can be inferred from the relative sizes of the three (or two smaller) box dimensions as compared to the average length of the protrusions. If they are of similar magnitude, or if the (two smaller) box dimensions exceed the protrusions in length, the particular topology is highly randomized (netlike) and can no longer be unequivocally assigned.

Figure 11 demonstrates with an example (four ribosomal 5 S RNA sequences)[14] the advantage of sequence over distance space analysis. The

[14] M. Eigen, B. Lindemann, R. Winkler-Oswatitsch, and C. H. Clarke, *Proc. Natl. Acad. Sci. U.S.A.* **82**, 2437 (1985)

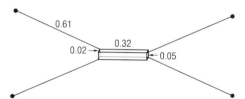

FIG. 12. Average RY sequence space diagram of quartets of sequences of the *gag* gene of HIV (1485 positions were chosen, for which no insertions or deletions occur).[12,13] (Reprinted by permission of the authors.)

distance space diagram suggests an ideal tree, while the sequence space diagram indicates the true uncertainties (as caused by parallel and reverse mutations) in assigning the nodes of divergence. In fact, it suggests that no safe assignment is possible for this early divergence of two archaebacteria and two eubacteria (one of which is a cyanobacterium). The assignment obtained here by matching mere distances is very likely to be wrong. Three more examples of average RY sequence space diagrams are shown in Figs. 12–14.[12,15]

If we want to obtain a sensitive distinction of residual tree from bundle divergences, we could use the following method. In the alignment of the total set of sequences all symbols at a given position are vertically mixed. Note that no symbols are to be exchanged horizontally, nor are symbols to be added to or taken away from columns, so that the consensus symbol for every position remains exactly the same as before mixing. The vertical mixing will randomize any residual tree correlation, so that finally it can no longer be distinguished from a bundle correlation. One plots the quantity $\bar{l} + \bar{s} - 2\bar{m}$ as a function of the number of changes introduced by vertical mixing. This quantity, if any treelikeness were present, should decay toward a value close to zero, where it then fluctuates randomly.

Quarternary (AUGC) Sequence Space

The extension of the procedure to quaternary sequences is demonstrated in Fig. 15. In addition to the three categories of positional combination mentioned above, i.e., "four of a kind," "three of a kind," "two pairs," we have to add two more: "one pair" (with six possible realizations, i.e., $A = B \neq C \neq D$, $A = C \neq B \neq D$, $A = D \neq B \neq C$, $B = C \neq A \neq D$, $B = D \neq A \neq C$, and $C = D \neq A \neq B$) and "no pair" ($A \neq B \neq C \neq D$). These five "poker combinations" comprise all possible vertical specifications in

[15] M. Eigen, B. Lindemann, M. Tietze, R. Winkler-Oswatitsch, A. W. M. Dress, and A. von Haeseler, *Science* **244**, 673 (1989).

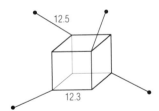

FIG. 13. Average diagram of $\binom{28}{4}$ = 20475 quartet combinations of the 10 highly diverged positions in the species family of 28 tRNAs of *Bacillus subtilis* (binary R, Y sequences). The divergence is almost completely randomized, as indicated by the large average box dimensions $\bar{l}$, $\bar{m}$, and $\bar{s}$ which show similar lengths and therefore are given as uniform averages. The (relative) average box length (12.3%) is about as large as the (relative) average protrusion length (12.5%). The average (relative) Hamming distance between two sequences is $2 \times 12.5 + 2 \times 12.3 = 49.6\%$, as to be expected for randomized binary sequences. (Reprinted by permission of the authors.)

quartets of sequences made up of (at least) four symbol classes. A representative geometry for a highly randomized bundlelike divergence is shown in Fig. 15. The regular triangular faces again result from averaging over all six contributions to "one pair." Each side of a triangle comprises 5/6 of this average of the six-pair combinations plus the contribution from the "no pair" combination. (Note that it is not always possible to draw individual triangles because the sum of two sides may be smaller than the third side. This is just a matter of geometrical representation which does not influence the numerical evaluation of data.) Four representative average AUGC sequence space diagrams are presented in Figs. 16–18.[12,15,16]

Transversions versus Transitions

We are now in a position to proceed to further refinement, namely, counting transversions and transitions separately. In order to gain some understanding of the physical procedure, let us go back to the concept as it was introduced in an earlier section.

We recall that statistical geometry in binary (RY) sequence space yields the extent of divergence that is exclusively due to transversions. We used the logical basis of the binary concept in order to obtain a model of quaternary (AUGC) sequence space by assigning to each point of the hypercube of dimension v, representing RY space, a subspace that itself is a hypercube of dimension v. Following the evolutionary change of a sequence now means jumping back and forth between different hypercubes.

[16] P. Palese, *in* "Evolutionary Processes and Theory" (S. Karlin and E. Nevo, eds.), p. 53. Academic Press, New York, 1986.

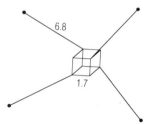

FIG. 14. Average diagram of $\binom{24}{4} = 10626$ quartet combinations of the 21 moderately diverged positions in 24 sequences of an ancestor family referring to an early node of divergence (node II, i.e., the tripod of eubacteria, archaebacteria, and eukaryotes). The diagram represents an almost ideal bundle. The average lengths of protrusions are much larger than the average box dimensions $\bar{l}$, $\bar{m}$, and $\bar{s}$, which themselves turn out to be of similar size, so that a cube with edge length $(\bar{l} + \bar{m} + \bar{s})/3$ was constructed. The average (relative) Hamming distance between two sequences is $2 \times 6.8 + 2 \times 1.7 = 17\%$.[15] (Reprinted by permission of the authors.)

4 Sequences : **A,B,C,D,**
Positionally Differentiated Distance Categories

		Examples	Degeneracies
0	Four of a Kind	**A = B = C = D**	4 x 1
1	Three of a Kind	**A = B = C ≠ D**	12 x 4
2	Two Pairs	**A = B ≠ C = D**	12 x 3
3	One Pair	**A = B ≠ C ≠ D**	12 x 6
4	No Pair	**A ≠ B ≠ C ≠ D**	24 x 1

Category **0** refers to constant positions .
Category **1** refers to conservative positions with small variation .
Category **2** refers to positions of medium variability .
Category **3** and **4** refer to positions of large variability .
One large dimension in category **2** indicates treelikeness .

FIG. 15. Analysis of four-sequence combinations in AUGC sequence space. (Reprinted by permission of the authors.)

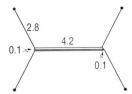

FIG. 16. Average quartet divergence (since 1933) of 16 sequences of the 890 nucleotides comprising the *NS* gene of influenza A virus (dimensions are not drawn in true proportions). The two smaller box dimensions are not visible in this representation (4% of triangle dimension).[16] (Reprinted by permission of the authors.)

We ask if there is also a space that is exclusively characteristic of transitions in the same way as the first hypercube is exclusively representative of transverions. If we can assign such a space, we would be able to compare directly divergences based on transversions with those based on transitions.

For this purpose we define a subspace of the RY hypercube that includes only those coordinates which refer to homologous positions of a given quartet (i.e., "four of a kind": RRRR or YYYY). In this subspace the four sequences compared are represented by the same point. We then construct the "transitional" subspaces for this point. If the total number of homologous quartet positions ("four of a kind") in the aligned RY sequences is d_0, then the hypercube representing the "transitional" subspaces is of dimension d_0. We call the second subspace "transitional" because any change of positions in this subspace is entirely of transitional nature. The positions in this subspace (all referring to either RRRR or YYYY) discriminate *transitional* changes that include "four of a kind" (like UUUU),

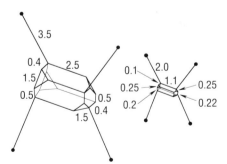

FIG. 17. Average AUGC sequence space diagram of quartets of (left) 19 sequences of the *env* gene (cf. Fig. 6) and (right) 11 sequences of the *gag* gene (cf. Fig. 9) of HIV.[12,13] (Reprinted by permission of the authors.)

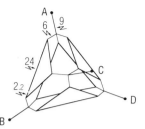

FIG. 18. Average quartet divergence of 40 tRNA sequences of *Halobacterium volcanii* in AUGC space.[15] (Reprinted by permission of the authors.)

"three of a kind" (like UUUC), and "two pairs" (such as UUCC). Looking at relative divergence in this (d_0-dimensional) space and comparing them with relative divergences in the (ν-dimensional) RY hypercube yields directly the ratio of individual rates for transition and transversion.

Now we have to construct the combined space, using Fig. 19 to specify the various categories (distance segments). There are now eight categories that can be found in quartets of aligned sequences if transitions and transversions are discriminated in assigning the five "poker" combinations. In order to facilitate some intuitive comprehension, we quote for each of the eight categories one example of symbol combination (Fig. 20). Note that the first of the two subscripts in d_{qb} always refers to the specification in quaternary sequence space (0 = "four of a kind," 1 = "three of a kind," 2 = "two pairs," 3 = "one pair," and 4 = "no pair"), while the second index denotes one of the three possible combinations in binary RY space.

RY AUGC	Four of a kind	Three of a kind	Two pairs
Four of a kind	d_{00}	/	/
Three of a kind	d_{10}	d_{11}	/
Two pairs	d_{20}	/	d_{22}
One pair	/	d_{31}	d_{32}
No pair	/	/	d_{42}

FIG. 19. Distance segments for discrete analysis of transversions and transitions in AUGC sequence space. The five categories defined in Fig. 15 are ordered according to the specifications given in Fig. 10. Eight combinations d_{qb} are possible, where the first index refers to quaternary (AUGC), the second to binary (RY) specification. (Reprinted by permission of the authors.)

d_{q0}	d_{00}	d_{10}	d_{20}		d_{q1}	d_{11}	d_{31}		d_{q2}	d_{22}	d_{32}	d_{42}
R	A	G	G		Y	U	C		Y	U	U	U
R	A	A	G		R	A	G		Y	U	U	C
R	A	A	A		R	A	A		R	A	A	A
R	A	A	A		R	A	A		R	A	G	G

FIG. 20. Examples of representation of the distance segments d_{qb}. (Reprinted by permission of the authors.)

Data Evaluation

It is obvious that data processing must be done by a computer. A set of n sequences comprises $\binom{n}{4}$ quartet combinations that have to be looked up for their individual nature. With $n = 50$ sequences, for instance, the number of quartets amounts to 230 300. Moreover, the analysis has to start from properly aligned sequences. Specific computer programs that have been adapted to particular sets of sequences (tRNAs, 5 S rRNAs, viral genes up to lengths of several thousand) will be available in the near future and may also be worked out for general usage with any kind of sequence.[12]

Average distance segments then may be used to characterize the stage of divergence by comparing them with model curves obtained by simulating various (e.g., bundlelike or more or less symmetric treelike) divergence topologies. All segments are to be normalized by dividing them by the number of reference positions (v) and expressing them in percentages. The fact that for each of the sequences one can obtain a set of segments that refer to the same stage of divergence (Δ/v) allows for internal calibration. Some of the segments are more sensitive to small, others to large degrees of divergence. Averaging, of course, is possible only if individual distance segments are combined into classes. This is done in the following, separately for the four statistical geometries introduced above.

Distance Space

The quantities to be obtained for a quartet of sequences characterized by six distances (cf. Fig. 7) are four protrusions a, b, c, and d and two box dimensions x and y. Averaging over all $\binom{n}{4}$ combinations yields the following parameters: $\frac{1}{4}(a + b + c + d)$, $\bar{x}$, and $\bar{y}$, where by definition x comprises the smaller and y the larger rectangle sites. In addition, one may record the two average distances $\bar{d}_{ik}$ and $\bar{d}_{ck}$.

We have simulated a bundlelike (parallel) divergence (Fig. 21, top) and a symmetric tree-like divergence (Fig. 21, bottom) and recorded the five mentioned parameters. As is seen, the curves in the top diagram fit the experimental data obtained for 30 tRNA sequences from *Escherichia coli*,

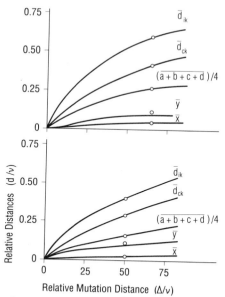

FIG. 21. Segments of distance space diagrams as functions of relative mutation distance Δ/v. Top: Simulation of bundlelike divergence. Bottom: Simulation of symmetric treelike divergence. The examples of experimental data in the top diagram refer to species families of tRNA (*E. coli, H. volcanii*, and yeast), those in the bottom diagram to phylogenies of tRNAs (mti and Phe).[15] (Reprinted by permission of the authors.)

which after early differentiation should have had a parallel divergence of over 3 billion years. On the other hand, the curves would not fit a treelike phylogeny, whereas simulations of a treelike divergence in the bottom diagram do fit those data [here for the phylogeny of the tRNA (methionine initiator [mti]) and tRNA (phenylanine [Phe])]. The $\bar{y}$ parameter, however, indicates that the tree segments are less symmetric than assumed in the simulated model. Parameters obtained for distance space generally change quite monotonously with Δ/v (or time) and become very insensitive at large degrees of divergence. Since $\bar{x}$ reaches a relatively small saturation value (cf. Fig. 9), it is always possible to construct dendrograms which do not sensitively reflect the degree of randomization.

Binary Sequence Space

The segment parameters that can be deduced from Fig. 10 are the four protrusions α, β, γ, δ and the three box dimensions in addition to the number d_0 of positions which for a particular quartet of sequences are homologous. In the statistical averages we combine them into $\bar{d}_0$, the average number of "four of a kind," $\bar{d}_1 = \overline{\alpha + \beta + \gamma + \delta}$, the average number of "three of a kind," and $\bar{d}_2 = \overline{\sigma + \tau + \omega}$, the average number of "two pairs." For the latter parameter we could also record independently the largest, medium, and smallest box dimension if we want to differentiate between bundle- and treelike topologies of divergence.

Figure 22 shows a simulation of a bundlelike (parallel) divergence. The set of curves is fitted by the obviously parallel divergence of tRNA species families. In fact, these curves are matched by 15 tRNA families analyzed, for which two classes of positions, i.e., moderately and highly diverged, have been distinguished and separately evaluated.[15] The value for Δ/ν obtained congruently for all species families at moderately diverged positions was 0.23.

Changes of $\bar{d}_0$, $\bar{d}_1$, and $\bar{d}_2$ with Δ/ν for RY sequence space are still fairly monotonous. However, $\bar{d}_2$ (or the two smaller box parameters contributing to $\bar{d}_2$) reaches saturation more slowly and yields larger values than $\bar{x}$ so that randomization can be detected more sensitively than in distance space, as was already suggested by Fig. 11.

Quaternary Sequence Space

As an example, we simulate again a bundlelike divergence for the five distance categories: d_0 = "four of a kind," d_1 = "three of a kind," d_2 = "two pairs," d_3 = "one pair," and d_4 = "no pair." The d values are related to the characteristic average geometry shown in Fig. 15, which combines a cube and a regular tetrahedron, in the following way: the average protru-

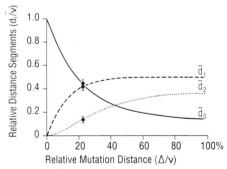

FIG. 22. The three lines describe a computer simulation of the three average distance segments of geometries of quartet combinations of RY sequences (statistical averages), assuming parallel divergence with uniform substitution rates. The experimental values (dots) shown are for the three average segments $\bar{d}_0$, $\bar{d}_1$, $\bar{d}_2$ of present tRNA species families and refer to moderately diverged positions. All experimental segments $\bar{l}$, $\bar{m}$, and $\bar{s}$ that contribute to $\bar{d}_2$ were almost of the same magnitude, identifying the divergence as bundlelike with negligible residual treelikeness.[15] (Reprinted by permission of the authors.)

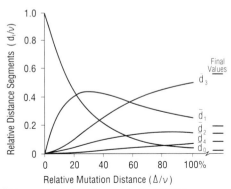

Fɪɢ. 23. The five distance categories of quartets in AUGC sequence space as functions of relative mutation distance Δ/v (%) (computer simulation for $n = 30$ sequences with $v = 30$ positions and uniform probabilities of substitution).[3] (Reprinted by permission of the authors.)

sion lengths are $\bar{d}_1/4$; the average box segment lengths are $\bar{d}_2/3$; the average edge length of the tetrahedron is $\bar{d}_4 + 5\bar{d}_3/6$. The parameters $\bar{d}_1$, $\bar{d}_2$, $\bar{d}_3$, and $\bar{d}_4$ with $\bar{d}_0 = v - \{\bar{d}_1 + \bar{d}_2 + \bar{d}_3 + \bar{d}_4\}$ result directly from evaluation of experimental data. Figure 23 shows the curves obtained from analytical calculations[17] that are identical with averages obtained by computer simulation of the diffusion model (parallel divergence).

The curves shown in Fig. 23 reveal a sophisticated differentiation of distances into segments that respond much more sensitively to different regimes of differentiation than plain distances do. The segment $\bar{d}_1$ passes through a maximum at about 30% divergence, even causing the quantity $\bar{d}_1/\bar{d}_2$ to change its sign relative to one at a high level of divergence ($> 50\%$). This increase of diversification is even more obvious if transitions and transversions are counted separately as suggested in an earlier section.

Figure 24 shows the simulation of bundlelike (parallel) divergence for the eight distance segments specified in Fig. 19. The experimental data for tRNA species families clearly show larger values for transitional than for transversional substitution rates.[15] Hence, there is no consistent fit with the curves in Fig. 23 which suppose uniform rates for all types of substitution. The best fit with the curves in Fig. 24 was obtained if transition is assumed to occur (2.5 ± 0.3) times faster than (individual) transversion. Since there are two types of transversions versus one type of transition, the relative mutation distances for the experimental data appear at values that are $(2 + 2.5)/2 = 2.25$ times larger than those found for exclusively transver-

[17] A. W. M. Dress, *Adv. Math.* **53**, 321 (1984).

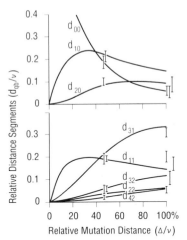

FIG. 24. The eight lines describe a computer simulation of the eight average distance segments of 30 sequences, each comprising 30 positions in AUGC space, assuming different rates of transition and transversion. The distance segments are defined in the text. The ratio of individual rates of transition to transversion in the simulation is 2.5. Experimental values match a common Δ/v value of 0.48 for all distance segments of moderately variable positions. Distance segments for highly diverged positions show large fluctuations around values that correspond to complete randomization. The bars indicate the maximal spread of data. They are placed so as to yield an optimal compromise for all eight distance segments, which is reached far outside the frame at the right-hand side of the picture. (Reprinted by permission of the authors.)

sional changes ($\Delta/v = 0.48$ versus 0.23). The curves in Fig. 24, in particular those for d_{10} and d_{11} (which pass through a maximum), react most sensitively to the assumed ratio of transitional and transversional changes, which accordingly can be determined quite accurately.

Statistical Geometry and Tree Construction

The assignment of early nodes in the construction of phylogenetic trees often involves difficulties arising from the relatively large contribution of parallel and reverse mutations. Distances compensate those effects and therefore are not suitable to recover the correct nodal sequences. Statistical geometry in sequence space reduces the uncertainties, and also shows which part is irreducible due to the inherently stochastic nature of data. The following procedure is suggested. The data first are arranged according to known kinships. They are assigned to (four) groups (say, A, B, C and D), which may be the kingdoms or taxa that are most separated. Each of these four groups should include several members. One now constructs sequence space geometries by combining quartets that always contain one member

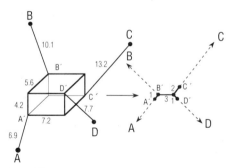

Fig. 25. Left: Average RY sequence space diagram for quartets of tRNA sequences referring to the anticodon GUG (histidine). *A*, Eubacteria; *B*, mitochondria; *C*, archaebacteria; *D*, eukaryotes. Right: Superposition of 256 corner point sequences *A'*, *B'*, *C'*, and *D'* yields consensus sequences that form the diagram. (Reprinted by permission of the authors.)

of each group (i.e., *ABCD*). If each group includes k members, one thus obtains k^4 geometries. As an example, we show in the left-hand side of Fig. 25 the average geometry obtained for the $4^4 = 256$ quartet combinations of each four eubacterial, archaebacterial, eukaryotic (cytoplasmic), and mitochondrial tRNA sequences (histidine adaptor with anticodon GUG). The large box dimensions indicate the average true uncertainty of single node assignments. We then reconstruct for each of the 256 quartets the sequences that refer to the corner points *A'*, *B'*, *C'*, and *D'* and superimpose these sequences. Their consensus sequences then are taken to construct the diagram in the right-hand side of Fig. 25. In the present case we obtain an ideal dendrogram, though in general the diagram obtained may be a rectangle or again a box, however, with greatly reduced sizes. The precision in reconstructing early nodes depends on how many members, congruently for all four boxes, one has available. The procedure may be iterated for nodes within the groups *A*, *B*, *C*, and *D* if a sufficient number of sequences can be assigned to subgroups in which one of the four constituents is always given by the corresponding consensus corner point sequence *A'*, *B'*, *C'*, and *D'*, that has been determined in the preceding step. Nodes in trees may then be represented by boxes or other geometries to demonstrate their irreducible uncertainties.

Conclusions

The method introduced in this chapter is relatively new so that applications are still sparse. We have described the models in qualitative terms, particularly for those who are more interested in possible applications than in the mathematical foundation. The mathematical basis, in particular the connections between evolutionary dynamics on sequence space and tensor products of representative spaces, is given in more detail in papers by A. Dress and co-workers.[17-20] The methods of statistical geometry are not just alternatives to existing methods of comparative sequence analysis; rather, they represent a new approach and are meant to complement rather than to compete with classic procedures of phylogenetic analysis. Their aim is mainly to check the predicative power of data.

Distances as such, by their stochastic and cumulative nature, have inherent and irreducible uncertainties. The advantage of the sequence space method is that it is of a statistical nature, referring to large sets of data, and that it allows reliable assessment of averages and higher moments, thereby providing an independent check on metric uniformity and topology assignment even at high degrees of randomization. Statistical geometry may be generalized further to include correlations among more than four sequences and to account for more than four symbols (e.g., for the 20 symbols of proteins). The method has been successfully applied to tRNAs and RNA viruses to yield clues about the early evolution of the genetic code[15] as well about the quasi-species[21] nature of viruses.[12] It may as well prove useful for a study of other gene families. As data are accumulating at an ever-increasing rate, it is important to have reliable methods that make large-scale comparisons possible and uncover kinship relationships still hidden in our genes. What might result in the end is a broad overview of the evolution of life manifested in all sequences in a congruent way and as such recovered by statistically sound methods.

Acknowledgments

We wish to thank Andreas Dress for cooperation on mathematical aspects of statistical geometry, Katja Nieselt and Björn Lindemann for providing the computer evaluation of sequence data, and John McCaskill for reading the manuscript.

[18] A. W. M. Dress, in "Die Bedeutung der von Berlin ausgehenden Mathematik in Vergangenheit und Gegenwart" (H. Begehr, ed.), in press. Kolloquium-Verlag, Berlin, 1989.
[19] A. W. M. Dress and D. S. Rumschitzki, *Acta Appl. Math.* **11**, 103 (1988).
[20] H. J. Bandelt and A. W. M. Dress, *Adv. Appl. Math.* **7**, 309 (1986).
[21] M. Eigen, J. McCaskill, and P. Schuster, *Adv. Chem. Phys.* **75**, 149 (1989).

[33] Statistical Methods for Estimating Sequence Divergence

By TAKASHI GOJOBORI, ETSUKO N. MORIYAMA, and MOTOO KIMURA

General Principle

Methods for estimating the number of nucleotide base substitutions are crucial for studies of molecular evolution. Knowledge of the number of base substitutions is particularly important for computing the evolutionary rate and constructing phylogenetic trees at the DNA level.

From the perspective of population genetics, nucleotide and amino acid substitutions can both be treated as stochastic processes.[1] Methods for estimating the number of DNA base substitutions, however, are different from those for estimating the number of amino acid substitutions. Because only four kinds of nucleotide bases (usually denoted A, T, C, and G) exist, multiple and superimposed nucleotide substitutions at the same site may occur undetected, especially when sequence divergence is great. For example, when two comparable DNA sequences have different bases, say A and G, at the corresponding (i.e., homologous) site, the changes of A → T → G for one sequence and no change for the other may have occurred at the site (Fig. 1). The observed number of nucleotide differences between the two DNA sequences is thus frequently different from the total number of nucleotide substitutions that have actually occurred during their divergence. Statistical methods for estimating the number of nucleotide substitutions are therefore required for comparative studies of DNA sequences. In this chapter, we first describe various methods for estimating the number of nucleotide substitutions and then discuss the advantages and disadvantages of these methods.

Methods for Estimating Total Number of Nucleotide Substitutions

One-Parameter Method

In order to derive an estimation formula for the number of nucleotide substitutions, we must assume a specific model for the pattern of base substitutions among the four kinds of nucleotide bases. We start with the

---

[1] M. Kimura, "The Neutral Theory of Molecular Evolution." Cambridge University Press, Cambridge, England, 1983.

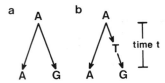

FIG. 1. Examples of nucleotide base changes in the course of evolution, where the ancestral base at a given site is assumed to be A. In these examples, two descendant DNA sequences compared t years later have nucleotides A and G at this site. In example a, one nucleotide substitution (A → G) has occurred, while in b two nucleotide substitutions (A → T and T → G) have occurred.

one-parameter model, in which the rate of substitution is assumed to be equal between any pair of nucleotides.[2,3] Thus, the process of substitution is described by a single parameter α, which stands for the rate of substitution of one particular base (say, A) for another (say, T), and this is assumed to be constant over evolutionary time (Table I). The estimation formula based on this model may be derived as follows.

Consider two homologous DNA sequences which diverged from a common ancestor t years ago. We denote by $I(t)$ the probability that two nucleotide bases at corresponding (homologous) sites at time t are identical to each other. At the two corresponding sites, two bases are either identical to or different from each other. Thus, their probabilities are, respectively, $I(t)$ and $1 - I(t)$. Let us derive a recurrence formula for $I(t)$.

We first consider the probability that two homologous nucleotide sites at time t are identical to each other and are also identical at time $t + 1$. This consists of two mutually exclusive events: One represents the case in which both nucleotide bases change into two other identical bases, and the other represents the case in which both remain unchanged. The probability of the former event is $3\alpha^2$ and that of the latter is $(1 - 3\alpha)^2$. Therefore, the probability that the two nucleotide bases remain identical at the site at time $t + 1$ when they are identical at time t is given by $[(1 - 3\alpha)^2 + 3\alpha^2]I(t)$.

Next, we consider the probability that bases at two corresponding sites at time t are different from each other, but that they become identical at time $t + 1$. This also consists of two mutually exclusive events. The first is the case in which a change occurs at one of the two corresponding sites but the other site remains unchanged. This probability is $2\alpha(1 - 3\alpha)$. The second is the case in which both nucleotide bases change into two other identical bases simultaneously. This probability is $2\alpha^2$. Therefore, the

[2] T. H. Jukes and C. R. Cantor, *in* "Mammalian Protein Metabolism III" (H. N. Munro, ed.), p. 21, Academic Press, New York, 1969.
[3] M. Kimura and T. Ohta, *J. Mol. Evol.* **2,** 87 (1972).

TABLE I
PATTERN OF NUCLEOTIDE SUBSTITUTION[a]

Original nucleotide	Substituted nucleotide			
	A	T	C	G
One-parameter model				
A	—	α	α	α
T	α	—	α	α
C	α	α	—	α
G	α	α	α	—
Two-parameter model				
A	—	β	β	α
T	β	—	α	β
C	β	α	—	β
G	α	β	β	—
Three-parameter model				
A	—	β	γ	α
T	β	—	α	γ
C	γ	α	—	β
G	α	γ	β	—
Four-parameter model				
A	—	γ	$\theta\alpha$	α
T	γ	—	α	$\theta\alpha$
C	$\theta\beta$	β	—	γ
G	β	$\theta\beta$	γ	—
Six-parameter model				
A	—	α_1	α	α
T	β_1	—	α	α
C	β	β	—	α_2
G	β	β	β_2	—

[a] Each element represents the rate of substitution between a given pair of nucleotides per site per year.

probability that the two nucleotide sites become identical at time $t + 1$ when they are different from each other at time t is $[2\alpha(1 - 3\alpha) + 2\alpha^2][1 - I(t)]$. Thus, we obtain

$$I(t + 1) = [(1 - 3\alpha)^2 + 3\alpha^2]I(t) + [2\alpha(1 - 3\alpha) + 2\alpha^2][1 - I(t)] \quad (1)$$

The solution of this equation which satisfies the initial condition $I(0) = 1$ is

$$I(t) = [1 + 3(1 - 8\alpha + 16\alpha^2)^t]/4 \quad (2)$$

Since α is generally very small (e.g., $\alpha = 10^{-9}$), we may neglect terms of α^2 so that we get

$$I(t) = [1 + 3(1 - 8\alpha)^t]/4 \tag{3}$$

This may also be expressed as

$$I(t) = 1 - \tfrac{3}{4}(1 - e^{-8\alpha t}) \tag{4}$$

The same result can also be obtained by solving a differential equation derived by substituting $dI(t)/dt$ for $I(t + 1) - I(t)$ in Eq. (1).[4]

The mean number of nucleotide substitutions accumulated per site at time t is given by $2 \times 3\alpha t$, where the factor 2 comes from the fact that the divergence of two sequences always involves two evolutionary lines each having the time length t (Fig. 1). Denoting the evolutionary distance in terms of accumulated changes (i.e., $2 \times 3\alpha t$) by K, we have

$$K = -\tfrac{3}{4} \ln(1 - \tfrac{4}{3} F_D) \tag{5}$$

where $F_D = 1 - I(t)$. In Eq. (5), F_D stands for the fraction of different sites between the two DNA sequences compared. Equation (5), first presented by Jukes and Cantor[2] (see also Kimura and Ohta[3]), is not only simple but also very useful because it gives a reasonable estimate when sequence divergence is small (note that $K \approx F_D$ when $F_D \ll 1$). Moreover, the standard error (σ_K) of this estimate (K) is given by

$$\sigma_K = \frac{\sqrt{\tfrac{1}{n} F_D(1 - F_D)}}{1 - \tfrac{4}{3} F_D} \tag{6}$$

where n is the total number of sites compared.[3]

Two-Parameter Method

The molecular structures of nucleotide bases A and G (purines) are similar, and those of C and T (pyrimidines) are also similar. Evolutionary base substitutions between purines (A $\rightarrow$ G or G $\rightarrow$ A) and between pyrimidines (C $\rightarrow$ T or T $\rightarrow$ C) may be called transition-type substitutions, while those between purines and pyrimidines are called transversion-type substitutions. It has been shown that for many nuclear genes the rate of transition-type changes is much higher than that of transversion-type ones.[5] It has also been reported that more than 90% of nucleotide substitutions in mitochondrial DNAs of primates such as humans are of the

[4] M. Nei, "Molecular Population Genetics and Evolution." North-Holland, Amsterdam, 1975.

[5] T. Gojobori, W.-H. Li, and D. Grant, *J. Mol. Evol.* **18**, 360 (1982).

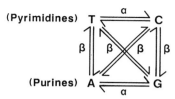

FIG. 2. Two-parameter model of base substitutions. The rates of transition and transversion substitutions are represented by α and β, respectively.

transition type.[6,7] To deal with such a situation, Kimura[8] developed a two-parameter method for estimating the number of nucleotide substitutions as follows.

As shown in Fig. 2 (also see Table I), α and β represent the rates of transition and transversion substitutions (per site per year), respectively. Let P and Q be, respectively, the fractions of nucleotide sites showing transition and transversion differences between the two sequences compared [e.g., P is the fraction of sites showing AG (or GA) and CT (or TC) nucleotide pairs when the two DNA sequences are compared]. A set of differential equations for P and Q at time t are

$$\frac{dP(t)}{dt} = 2\alpha - 4(\alpha + \beta)P(t) - 2(\alpha - \beta)Q(t) \tag{7}$$

$$\frac{dQ(t)}{dt} = 4\beta - 8\beta Q(t) \tag{8}$$

For details concerning the derivation of Eqs. (7) and (8), see Kimura.[8] Under the initial conditions $P(0) = Q(0) = 0$, the relevant solutions are

$$P(t) = \tfrac{1}{4} - \tfrac{1}{2}e^{-4(\alpha+\beta)t} + \tfrac{1}{4}e^{-8\beta t} \tag{9}$$

$$Q(t) = \tfrac{1}{2} - \tfrac{1}{2}e^{-8\beta t} \tag{10}$$

The rate (k) of nucleotide substitution, which is the number of nucleotide substitutions per site per year, is given by $k = \alpha + 2\beta$. This is based on the following consideration. Suppose that a given site is occupied by a particular base, say, A. It changes to one of the other three bases (T, C, or G). In this case, either A changes to G at the rate α or A changes to T or C at the rate of 2β. Thus, we have $k = \alpha + 2\beta$ per year.

Therefore, the total number (K) of nucleotide substitutions per site between the two sequences which diverged from the common ancestor t

[6] W. M. Brown, E. M. Prager, A. Wang, and A. C. Wilson, *J. Mol. Evol.* **18**, 225 (1982).
[7] C. F. Aquadro and B. D. Greenberg, *Genetics* **103**, 287 (1983).
[8] M. Kimura, *J. Mol. Evol.* **16**, 111 (1980).

years ago is given by $K = 2kt = 2\alpha t + 4\beta t$. From Eqs. (9) and (10), we have

$$8\beta t = -\ln[1 - 2Q(t)] \tag{11}$$

$$4\alpha t = -\ln[1 - 2P(t) - Q(t)] + \tfrac{1}{2}\ln[1 - 2Q(t)] \tag{12}$$

Thus,

$$K = -\tfrac{1}{2}\ln[(1 - 2P - Q)\sqrt{1 - 2Q}] \tag{13}$$

In Eq. (13) the letter t in $P(t)$ and $Q(t)$ is dropped for simplicity.

From Eqs. (9) and (10), we note that when $\alpha = \beta$, we have $P = (1/2)Q$. Substituting $P = (1/2)Q$ into Eq. (13) and noting that $P + Q$ is equal to the fraction of sites having different bases between two DNA sequences (i.e., F_D), Eq. (13) reduces to Eq. (5), the equation of the one-parameter model.

The standard error of the estimate K obtained by using Eq. (13) is given by

$$\sigma_K = \frac{1}{\sqrt{n}}\sqrt{(a^2 P + b^2 Q) - (aP + bQ)^2} \tag{14}$$

where $a = 1/(1 - 2P - Q)$, $b = [1/(1 - 2P - Q) + 1/(1 - 2Q)]/2$, and n is the total number of sites compared.[8]

Three-Parameter Method

Kimura[9] also studied a method assuming three parameters in order to estimate the number of nucleotide substitutions. He derived the estimation formula in the same way as for the two-parameter method. He originally called this model "the 3ST model" (three substitution-type model) (see Table I).

We denote by $\underline{P(t)}$ the fraction of sites having TC or AG nucleotide pairs at time t, by $\overline{Q(t)}$ the fraction of sites having TA or CG nucleotide pairs, and by $R(t)$ the fraction of sites having TG or CA nucleotide pairs. Then, $P(t)$, $\overline{Q(t)}$, and $R(t)$ may be expressed as follows:

$$P(t) = [1 - e^{-4(\alpha+\beta)t} - e^{-4(\alpha+\gamma)t} + e^{-4(\beta+\gamma)t}]/4 \tag{15}$$

$$\overline{Q}(t) = [1 - e^{-4(\alpha+\beta)t} + e^{-4(\alpha+\gamma)t} - e^{-4(\beta+\gamma)t}]/4 \tag{16}$$

$$R(t) = [1 + e^{-4(\alpha+\beta)t} - e^{-4(\alpha+\gamma)t} - e^{-4(\beta+\gamma)t}]/4 \tag{17}$$

As shown in Table I, the distance or the total number of nucleotide substitutions per site is given by $K = 2(\alpha + \beta + \gamma)t$. From Eqs. (15)–(17), we have

$$4(\alpha + \beta)t = -\ln\{1 - 2[P(t) + \overline{Q}(t)]\} \tag{18}$$

[9] M. Kimura, *Proc. Natl. Acad. Sci. U.S.A.* **78**, 454 (1981).

$$4(\alpha + \gamma)t = -\ln\{1 - 2[P(t) + R(t)]\} \tag{19}$$

$$4(\beta + \gamma)t = -\ln\{1 - 2[\overline{Q}(t) + R(t)]\} \tag{20}$$

Thus, the distance (K) may be estimated by

$$K = -\tfrac{1}{4}\ln[(1 - 2P - 2\overline{Q})(1 - 2P - 2R)(1 - 2\overline{Q} - 2R)] \tag{21}$$

where the argument t is omitted from $P(t)$, $\overline{Q}(t)$, and $R(t)$ to simplify the presentation.

Equation (21) is the formula for the three-parameter model. Note that the two-parameter model is a special case in which $\gamma = \beta$, and therefore Q of the two-parameter method is given by $Q = 2\overline{Q} = 2R$. Under these conditions, it is clear that Eq. (21) reduces to Eq. (13).

The standard error of K estimated by Eq. (21) is given by

$$\sigma_K = \frac{1}{\sqrt{n}} \sqrt{(a^2 P + b^2\overline{Q} + c^2 R) - (aP + b\overline{Q} + cR)^2} \tag{22}$$

where $a = (C_{12} + C_{13})/2$, $b = (C_{12} + C_{23})/2$, and $c = (C_{13} + C_{23})/2$, in which $C_{12} = 1/(1 - 2P - 2\overline{Q})$, $C_{13} = 1/(1 - 2P - 2R)$, and $C_{23} = 1/(1 - 2\overline{Q} - 2R)$.[9]

Four-Parameter Method

Takahata and Kimura[10] proposed a method for estimating the number of substitutions based on a four-parameter model. The model is shown in Table I. The total number (K) of nucleotide substitutions may be estimated by

$$K = -\frac{1}{4}\ln\left\{\frac{(S_{13} - \overline{Q}_1)(S_{24} - \overline{Q}_2) - [(P - R)/2]^2}{\omega(1 - \omega)}\right.$$
$$\left. \times \left[1 - \frac{P + R}{2\omega(1 - \omega)}\right]^{8\omega(1-\omega)-1}\right\} \tag{23}$$

where ω stands for the fraction of A + T in the two DNA sequences compared (i.e., A + T content). In this formula, S_{13} represents the fraction of sites having AA or TT nucleotide pairs, S_{24} represents that of CC or GG pairs, and $\overline{Q}_1$ and $\overline{Q}_2$, respectively, denote the proportions of sites having AT and GC pairs. Finally, P stands for the fraction of sites having CT or AG nucleotide pairs, and R stands for that of GT or AC pairs. Because the formula for the standard error of K, as estimated by Eq. (23), is complicated, a computer is required for actual use.

A different type of four-parameter method was proposed by Tajima

[10] N. Takahata and M. Kimura, *Genetics* **98**, 641 (1981).

and Nei[11] to derive a formula similar to Eq. (5), assuming a so-called equal input model. In this model, the rate of substitution to the ith nucleotide is assumed to be the same, regardless of the original bases, i.e., $\lambda_{Ai} = \lambda_{Ti} = \lambda_{Ci} = \lambda_{Gi} = \alpha_i$, excluding the cases which correspond to λ_{ii}, where λ_{ij} (i, $j = $ A, T, C, and G) is the element of the matrix representing the pattern of base substitution as shown in Table I. Although this model appears to be rather artificial, it leads to better estimates than the one-parameter model.

Six-Parameter Method

The six-parameter model is based on the model originally proposed by Kimura,[9] who called it "the 2FC model" (two frequency class model). Its exact formulation was given by Gojobori et al.,[12] who showed that the number of nucleotide substitutions per site between two sequences compared is given by

$$K = -pq \ln\left(\frac{B_1}{pq}\right) - \frac{2q_A q_T}{p} \ln\left[\frac{p}{3q_A q_T}\left(F_{12} - B_1 - \frac{3E_{12}}{B_1}\right)\right]$$
$$- \frac{2q_C q_G}{q} \ln\left[\frac{q}{3q_C q_G}\left(F_{34} - B_1 + \frac{3E_{34}}{B_1}\right)\right] \qquad (24)$$

In Eq. (24), q_A, q_T, q_C, and q_G stand, respectively, for the contents of A, T, C, and G in the DNA sequences compared. Also, we use the following notations: $p = q_A + q_T$, $q = q_C + q_G$, $B_1 = pq - (x_{AC} + x_{AG} + x_{TC} + x_{TG})$, $E_{12} = (q_A q - x_{AC} - x_{AG})(q_T q - x_{TC} - x_{TG})$, $E_{34} = (q_C p - x_{AC} - x_{TC})(q_G p - x_{AG} - x_{TG})$, $F_{12} = x_{AA} + x_{TT} - x_{AT} - p^2 + 3q_A q_T$, and $F_{34} = x_{CC} + x_{GG} - x_{CG} - q^2 + 3q_C q_G$, where x_{ii} represents the fraction of sites having the same base pairs i, and $2x_{ij}$ ($i \neq j$) represents the fraction of sites having different base pairs i and j ($i, j = $ A, T, C, and G).

In the special case in which all six parameters are equal to each other ($\alpha = \alpha_1 = \alpha_2 = \beta = \beta_1 = \beta_2$), Eq. (24) reduces to Eq. (5) of the one-parameter method. In fact, in this case, we can show that $q_i = \frac{1}{4}$, $x_{ii} = x_{AA}$, $x_{ij} = x_{AT}$ ($i \neq j$), $4x_{AA} + 12x_{AT} = 1$, $p = q = \frac{1}{2}$, $B_1 = \frac{1}{4} - 4x_{AT}$, $E_{12} = E_{34} = (\frac{1}{8} - 2x_{AT})^2$, and $F_{12} = F_{34} = 2x_{AA} - x_{AT} - \frac{1}{16}$. By substitution, Eq. (24) reduces to Eq. (5). The formula for the standard error of K in this case is complicated, so a computer program is required for actual use.

Taking into account the fact that the base content varies from one DNA sequence to another, Hasegawa et al.[13] modified Kimura's two-parameter model. They considered not only transition and transversion substitutions separately but also difference of contents of the four nucleo-

[11] F. Tajima and M. Nei, Mol. Biol. Evol. **1**, 269 (1984).
[12] T. Gojobori, K. Ishii, and M. Nei, J. Mol. Evol. **18**, 414 (1982).
[13] M. Hasegawa, H. Kishino, and T. Yano, J. Mol. Evol. **22**, 160 (1985).

TABLE II
Formulas for Estimating Number of Nucleotide Substitutions[a]

Method	Estimating formula
One-parameter	$K = -\frac{3}{4} \ln(1 - \frac{4}{3}F_D)$
Two-parameter	$K = -\frac{1}{2} \ln[(1 - 2P - Q)\sqrt{(1 - 2Q)}]$
Three-parameter	$K = -\frac{1}{4} \ln[1 - 2P - 2\overline{Q})(1 - 2P - 2R)(1 - 2\overline{Q} - 2R)]$
Four-parameter	$K = -\frac{1}{4} \ln\left\{ \dfrac{(S_{13} - \overline{Q}_1)(S_{24} - \overline{Q}_2) - [(P - R)/2]^2}{\omega(1 - \omega)} \left[1 - \dfrac{P + R}{2\omega(1 - \omega)} \right]^{8\omega(1-\omega)-1} \right\}$
Six-parameter	$K = -pq \ln\left(\dfrac{B_1}{pq}\right) - \dfrac{2q_A q_T}{p} \ln\left[\dfrac{p}{3q_A q_T}\left(F_{12} - B_1 + \dfrac{3E_{12}}{B_1}\right) \right]$
	$\qquad - \dfrac{2q_C q_G}{q} \ln\left[\dfrac{q}{3q_C q_G}\left(F_{34} - B_1 + \dfrac{3E_{34}}{B_1}\right) \right]$

[a] Definitions of each parameter are given in Table III.

tide bases. Thus, their method may be regarded as a six-parameter method. Since their method is quite intricate mathematically, we recommend that interested readers consult Kishino and Hasegawa ([34], this volume).

Analyzing the nucleotide substitution patterns of mammalian mitochondrial genes, Lanave et al.[14] developed a method for estimating the evolutionary distance between gene sequences compared. In their method, all parameters representing the pattern of base substitution (see Table I) are estimated from actual data. Therefore, their method may be called the 12-parameter method. Readers who are interested in their method are invited to consult Saccone et al. ([35], this volume).

Application

Various formulas for estimating the divergence of DNA sequences in terms of the number of base substitutions per site (i.e., K) are summarized in Table II. The definitions of parameters appearing in these formulas are listed in Table III. The application of these methods to actual data is demonstrated below.

For this purpose, let us compare DNA sequences of the hemoglobin β chain gene between mouse and rabbit (data from Konkel et al.[15] and Hardison et al.[16]). In Fig. 3, the two sequences are aligned codon by codon with respect to the coding region of this gene. We estimate the distance in terms of the number of nucleotide substitutions at the third codon position

[14] C. Lanave, G. Preparata, C. Saccone, and G. Serio, *J. Mol. Evol.* **20**, 86 (1984).
[15] D. A. Konkel, J. V. Maizel, Jr., and P. Leder, *Cell* **18**, 865 (1979).
[16] R. C. Hardison, E. T. Butler III, E. Lacy, T. Maniatis, N. Rosenthal, and A. Efstratiadis, *Cell* **18**, 1285 (1979).

TABLE III

OBSERVED VALUES OF VARIOUS QUANTITIES REQUIRED FOR ESTIMATING NUMBER OF NUCLEOTIDE SUBSTITUTIONS FROM SEQUENCE COMPARISON SHOWN IN FIG. 3

Type of base pair and number of sites[a]

Method	Identical				Transitional		Transversional				K
	T→T, 26	A→A, 1	C→C, 33	G→G, 40	T→C, 20	A→G, 9	T→A, 4	C→G, 5	T→G, 6	A→C, 2	
One-parameter					←———————————— $F_D = 0.315$ ————————————→						0.409
Two-parameter					←$P = 0.199$→		←————————— $Q = 0.116$ —————————→				0.426
Three-parameter					←$P = 0.199$→		←$\bar{Q} = 0.061$→		←$R = 0.055$→		0.426
Four-parameter	←$S_{13} = 0.185$→		←$S_{24} = 0.500$→		←$P = 0.199$→		$\bar{Q}_1 =$ 0.027	$\bar{Q}_2 =$ 0.034	←$R = 0.055$→		0.463[b]
Six-parameter	$X_{TT} =$ 0.178	$X_{AA} =$ 0.007	$X_{CC} =$ 0.226	$X_{GG} =$ 0.274	$2X_{TC} =$ 0.137	$2X_{AG} =$ 0.062	$2X_{TA} =$ 0.027	$2X_{CG} =$ 0.034	$2X_{TG} =$ 0.041	$2X_{AC} =$ 0.014	0.564[c]

[a] The total number of sites considered is 146.

[b] $\omega = q_A + q_T = 0.34$.

[c] $q_A = 0.059$, $q_T = 0.281$, $q_C = 0.318$, $q_G = 0.342$, $p = 0.34$, $q = 0.66$, $q_A = 0.342$, $q_G = 0.318$, $q_C = 0.281$, $q_T = 0.059$, $B_1 = 0.0974$, $E_{12} = 9.1 \times 10^{-5}$, $F_{12} = 0.1056$, $F_{34} = 0.3737$.

```
                 1                                      10                                       20
mouse   ATG GTG CAC CTG ACT GAT GCT GAG AAG GCT GCT GTC TCT TGC CTG TGG GGA AAG GTG AAC TCC
             *        * * ** **           *       *       * **              *          * ***
rabbit  ATG GTG CAT CTG TCC AGT GAG GAG AAG TCT GCG GTC ACT GCC CTG TGG GGC AAG GTG AAT GTG

                                                30                                       40
mouse   GAT GAA GTT GGT GGT GAG GCC CTG GGC AGG CTG CTG GTT GTC TAC CCT TGG ACC CAG CGG
             *                                                        *               *
rabbit  GAA GAA GTT GGT GGT GAG GCC CTG GGC AGG CTG CTG GTT GTC TAC CCA TGG ACC CAG AGG

                                                50                                       60
mouse   TAC TTT GAT AGC TTT GGA GAC CTA TCC TCT GCC TCT GCT ATC ATG GGT AAT GCC AAA GTG
             *   *   * **   *         *       * **       * *       ***     * *    *
rabbit  TTC TTC GAG TCC TTT GGG GAC CTG TCC TCT GCA AAT GCT GTT ATG AAC AAT CCT AAG GTG

                                                70                                       80
mouse   AAG GCC CAT GGC AAG AAG GTG ATA ACT GCC TTT AAC GAT GGC CTG AAT CAC TTG GAC AGC
             *                       * * *       *  **   *   *       *       *       *
rabbit  AAG GCT CAT GGC AAG AAG GTG CTG GCT GCC TTC AGT GAG GGT CTG AGT CAC CTG GAC AAC

                                                90                                       100
mouse   CTC AAG GGC ACC TTT GCC AGC CTC AGT GAG CTC CAC TGT GAC AAG CTG CAT GTG GAT CCT
             *                 * * **   *         * *   *                           *
rabbit  CTC AAA GGC ACC TTT GCT AAG CTG AGT GAA CTG CAC TGT GAC AAG CTG CAC GTG GAT CCT

                                                110                                      120
mouse   GAG AAC TTC AGG CTC CTG GGC AAT ATG ATC GTG ATT GTG CTG GGC CAC CAC CTT GGC AAG
                                         * *  * *   *            ***   *   * *        *
rabbit  GAG AAC TTC AGG CTC CTG GGC AAC GTG CTG GTT ATT GTG CTG TCT CAT CAT TTT GGC AAA

                                                130                                      140
mouse   GAT TTC ACC CCC GCT GCA CAG GCT GCC TTC CAG AAG GTG GTG GCT GGA GTG GCC ACT GCC
             *       *   * ***  **              **                    *       *
rabbit  GAA TTC ACT CCT CAG GCG CAG GCT GCC TAT CAG AAG GTG GTG GCT GGT GTG GCC AAT GCC

                 146
mouse   TTG GCT CAC AAG TAC CAC TAA
             *            *        *
rabbit  CTG GCT CAC AAA TAC CAC TGA
```

FIG. 3. Comparison of nucleotide sequences of hemoglobin β chain genes of mouse and rabbit. The two sequences are aligned with each other codon by codon. The coding region used in the present analysis is bracketed, and asterisks indicate the sites where the two sequences have different bases.

between these two sequences. The initiation and termination codons are excluded from the comparison, because the former is usually invariant and changes of the latter are quite restrictive. Thus, the total number of sites compared is 146. In Fig. 3, the sites where two sequences differ are indicated by asterisks. Table III lists observed values of various quantities required for each method.

One-Parameter Method. Because nucleotide bases are different at 46 of 146 sites compared at the third codon positions, the fraction of different sites is $F_D = \frac{46}{146} = 0.315$. Putting this value into Eq. (5), we obtain

$$K = -\tfrac{3}{4} \ln(1 - \tfrac{4}{3} \times 0.315) = 0.409$$

namely, the estimated number of substitutions per site is 0.409. Note that the estimated value (K) is much larger than the observed value ($F_D =$

0.315). In general, the more the value of F_D increases, the more the difference between K and F_D increases. This is the reason why statistical methods for estimating the number of nucleotide substitutions are required, particularly when the divergence between the DNA sequences compared is large.

Note that the estimation formula breaks down when the observed value F_D happens to exceed $\frac{3}{4}$, because the argument of the logarithmic function in Eq. (5) becomes negative. This is due to the fact that the expected maximum value of F_D is $\frac{3}{4}$ at the equilibrium state of random substitutions. Similar problems exist more or less in other methods, too, because the estimation formulas similarly contain logarithmic functions which come from correction of multiple changes in the Poisson process.

Two-Parameter Method. In the β hemoglobin example, 29 of 46 different sites show transition-type differences whereas the remaining 17 show transversion-type differences. Thus, we have $P = \frac{29}{146} = 0.199$ and $Q = \frac{17}{146} = 0.116$. Plugging these values into Eq. (13), we obtain $K = 0.426$.

Three-Parameter Method. Of 17 sites having transversion-type differences, 9 have TA or CG nucleotide pairs and the remaining 8 have TG or AC pairs. This means that $\overline{Q} = \frac{9}{146} = 0.061$, $R = 0.055$, and $P = 0.199$. Substituting these values into Eq. (21), we get $K = 0.426$, which happens to be the same value as that obtained from the two-parameter method.

Four-Parameter Method. As shown in Table III, $\overline{Q}$ of the three-parameter method is separated into two parameters, $\overline{Q}_1$ and $\overline{Q}_2$. The number of sites showing TA nucleotide pairs is 4, and the number of sites showing CG pairs is 5. Thus, we have $\overline{Q}_1 = \frac{4}{146} = 0.027$ and $\overline{Q}_2 = \frac{5}{146} = 0.034$. The number of sites having TT or AA nucleotide pairs is 27 and that having CC or GG pairs is 73. This means $S_{13} = \frac{27}{146} = 0.185$ and $S_{24} = \frac{73}{146} = 0.5$. We have already obtained $P = 0.199$ and $R = 0.055$ (see the three-parameter method). Also, we find that the A + T content of two DNA sequences is $\omega = 0.34$ (see footnote b in Table III). Therefore, we obtain $K = 0.463$ from Eq. (23).

Six-Parameter Method. The six-parameter method requires the numbers of all types of nucleotide pairs, as shown in Table III. The frequencies are $x_{TT} = \frac{26}{146} = 0.178$, $x_{AA} = \frac{1}{146} = 0.007$, . . ., $2x_{TG} = \frac{6}{146} = 0.041$, and $2x_{AC} = \frac{2}{146} = 0.014$. Then, the contents of various bases (q_A, q_T, q_C, and q_G) of the two sequences are

$$q_A = x_{AA} + x_{AT} + x_{AC} + x_{AG} = 0.059$$

$$q_T = x_{AT} + x_{TT} + x_{TC} + x_{TG} = 0.281$$

$$q_C = x_{AC} + x_{TC} + x_{CC} + x_{CG} = 0.318$$

$$q_G = x_{AG} + x_{TG} + x_{CG} + x_{GG} = 0.342$$

Also, we compute the A + T content $p = q_A + q_T = 0.34$, and the C + G content $q = q_C + q_G = 0.66$. In addition, we have $B_1 = pq - (x_{AC} + x_{AG} + x_{TC} + x_{TG}) = 0.0974$, $E_{12} = (q_A q - x_{AC} - x_{AG})(q_T q - x_{TC} - x_{TG}) = 9.1 \times 10^{-5}$, $E_{34} = (q_C p - x_{AC} - x_{TC})(q_G p - x_{AG} - x_{TG}) = 0.0021$, $F_{12} = x_{AA} + x_{TT} - x_{AT} - p^2 + 3q_A q_T = 0.1056$, and $F_{34} = x_{CC} + x_{GG} - x_{CG} - q^2 + 3q_C q_G = 0.3737$. Putting all these values into Eq. (24), we obtain $K = 0.564$.

The rate of nucleotide substitution per site per year (k) may be computed by $k = K/(2t)$, where K represents the number of nucleotide substitutions that have occurred between the two DNA sequences which diverged t years ago. In this example, the estimated rate of nucleotide substitution turns out to be $k = 0.564/(2 \times 80 \times 10^6) = 3.53 \times 10^{-9}$/site/year, where we assumed $t = 80 \times 10^6$ as the divergence time of mice and rabbits from their common ancestor.

Advantages and Disadvantages of Various Methods

As shown by the above treatments of the example given in Table III, the estimated value of K (0.409) by the one-parameter method is much smaller than that of the six-parameter method ($K = 0.564$). This is due to the fact that the rate of nucleotide substitution is much different between different base pairs. In fact, as shown in Table III, the number of sites showing base pair AC is only 2 while the number showing base pair TC is 20. When the numbers of nucleotide differences are quite large between pairs of nucleotides, the one-parameter method tends to give underestimates. However, studies based on computer simulations show that when the value of K is much less than 1.0, the difference between the estimated and true values is not very large.[10,12]

As mentioned earlier, it is known that for the mitochondrial genomes of mammals, particularly of primates, more than 90% of nucleotide substitutions are transition-type while less than 10% are transversion-type changes. In such cases, the two-parameter method may be suitable, and, as shown in Table III, the three-parameter method will lead to results not much different from those of the two-parameter method.

When the number of differences varies depending on the types of base pairs, and furthermore when the value of K is expected to become more than 1.0, the four- and six-parameter methods are more suitable than other methods. In particular, the estimate obtained by the six-parameter method is often close to the true value even though the value of K becomes much larger. The formula for the six-parameter method, however, frequently tends to become inapplicable owing to sampling and stochastic errors unless the DNA sequences compared are sufficiently long.

Methods for Estimating Numbers of Synonymous and Nonsynonymous Substitutions

When we compare two DNA sequences containing a protein-coding region, it is of particular interest to distinguish the rate of silent or, more precisely, synonymous substitutions (not causing amino acid changes) from those that lead to amino acid alterations (nonsynonymous substitutions). If the number of nucleotide differences between two DNA sequences is very small, the number of synonymous and nonsynonymous substitutions can be obtained simply by counting synonymous and nonsynonymous nucleotide differences. However, when two or three nucleotide differences exist between corresponding codons of the two sequences, the distinction between synonymous and nonsynonymous substitutions must be inferred using appropriate statistical methods.

Figure 4 illustrates an example of an underlying or hidden process of change in the sequences compared. When there is only one nucleotide difference between two codons compared, we can immediately decide whether the difference is synonymous or nonsynonymous. When we observe two nucleotide differences between the two codons being compared,

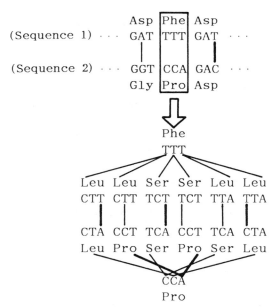

FIG. 4. Example illustrating synonymous and nonsynonymous nucleotide differences when homologous codons are compared. Thick and thin lines, respectively, indicate synonymous and nonsynonymous nucleotide differences. For details, see text.

there are two possible pathways that lead to the observed differences. On the other hand, when there are three nucleotide differences between the two corresponding codons, six underlying pathways are possible between the two codons (see Fig. 4). As an example, consider the comparison between the codon TTT in Sequence 1 and the codon CCA in Sequence 2 shown in Fig. 4. There exist six different possible pathways to change from one to the other. The numbers of synonymous and nonsynonymous changes in the codons can be inferred from these possible evolutionary pathways. The thick lines represent synonymous changes that do not cause amino acid changes, while the thin lines represent nonsynonymous (i.e., amino acid altering) nucleotide changes. For the purpose of estimating the substitution numbers, we have two different types of methods available: the "unweighted pathway method" and the "weighted pathway method." The main difference between these two types of methods is that in the former an equal weight is given to two or more evolutionary pathways, whereas in the latter a greater weight is given to an evolutionary pathway involving synonymous substitutions than to one involving nonsynonymous substitutions.

In this section, we describe briefly the essence of several methods proposed for estimating the numbers of synonymous and nonsynonymous substitutions. Readers interested in the detailed procedures and whether computer programs are available should consult the original papers cited.

Unweighted Pathway Methods

Method of Perler et al. Perler *et al.*[17] developed an unweighted pathway method for estimating synonymous substitutions. In their method, nucleotide sites are separated into groups of synonymous and nonsynonymous sites, and each group consists of three different categories. These categories are based on the numbers (1, 2, and 3) of possible nucleotide changes from the nucleotide present at a site. Nucleotide differences are also classified into synonymous and nonsynonymous differences, each with three categories, making comparison quite complicated.

Moreover, the method tends to give serious underestimates of synonymous substitutions. Using their own method, Perler *et al.*[17] claimed that, in the evolution of preproinsulin and globin genes, the synonymous changes accumulate about 7 times more rapidly than nonsynonymous changes, but the accumulation saturates in 85–100 million years. However, Kimura[1,9] pointed out that such a "saturation" phenomenon of synonymous substitutions is a statistical artifact, and then the computer simulation of Gojo-

---

[17] F. Perler, A. Efstratiadis, P. Lomedico, W. Gilbert, R. Kolodner, and J. Dodgson, *Cell* **20**, 555 (1980).

bori[18] showed that the method of Perler *et al.* gives serious underestimates at the synonymous sites, particularly when $K > 0.6$.

Method of Nei and Gojobori. The method of Nei and Gojobori[19] (N–G method) is essentially an unweighted version of the method of Miyata and Yasunaga.[20] The N–G method is much simpler than that of Perler *et al.* but gives a better estimate.

In the N–G method, the proportion of synonymous changes at the ith position of a codon is denoted by f_i ($i = 1$, 2, and 3). The number of synonymous sites for this codon is given by $s = \Sigma_{i=1}^{3} f_i$ and that of nonsynonymous sites by $n = 3 - s$. For example, consider codon TTA (Leu). For this codon, we have $f_1 = \frac{1}{3}, f_2 = 0$, and $f_3 = \frac{1}{3}$, as shown by changing bases at the first, second, and third positions. Thus, $s = \frac{2}{3}$ and $n = \frac{7}{3}$. For a DNA sequence consisting of r codons, the total numbers of synonymous and nonsynonymous sites are, therefore, given by $S = \Sigma_{j=1}^{r} s_j$ and $N = 3r - S$, where s_j is the value of s for the jth codon.

We now explain how to compute the numbers of synonymous and nonsynonymous differences between two nucleotide sequences. Let us denote the numbers of synonymous and nonsynonymous differences per codon by s_d and n_d, respectively. For example, if the codons compared are GAT (Asp) and GAC (Asp), $s_d = 1$ and $n_d = 0$ [see Fig. 4 (top, right)]. The comparison of TTT (Phe) and CCA (Pro) [Fig. 4 (top, middle)] is more complicated. There are six pathways by which to reach one codon from the other. Of the six, five contain one synonymous and two nonsynonymous nucleotide differences in each pathway, whereas for the remaining pathway, three nonsynonymous nucleotide differences are involved. Taking the average over these pathways (giving equal weight to each), we have $s_d = (1 \times 5 + 0 \times 1)/6 = \frac{5}{6}$ and $n_d = (2 \times 5 + 3 \times 1)/6 = \frac{13}{6}$ (see Fig. 4). Similarly, for the case of two nucleotide differences, s_d and n_d can be computed.

The total numbers of synonymous and nonsynonymous nucleotide differences can be obtained by summing these values over all codons. That is, $S_d = \Sigma_{j=1}^{r} s_{dj}$ and $N_d = \Sigma_{j=1}^{r} n_{dj}$, where s_{dj} and n_{dj} are the numbers of synonymous and nonsynonymous differences for the jth codon and r is the number of codons compared. Note that $S_d + N_d$ is equal to the total number of nucleotide differences between the two DNA sequences compared.

Thus, we estimate the fractions of synonymous (F_{SD}) and nonsynonymous (F_{ND}) differences by $F_{SD} = S_d/S$ and $F_{ND} = N_d/N$, where S and N are the average numbers of synonymous and nonsynonymous sites in the

---

[18] T. Gojobori, *Genetics* **105**, 1011 (1983).
[19] M. Nei and T. Gojobori, *Mol. Biol. Evol.* **3**, 418 (1986).
[20] T. Miyata and T. Yasunaga, *J. Mol. Evol.* **16**, 23 (1980).

two sequences compared. Finally, in order to estimate the numbers of synonymous (K_S) and nonsynonymous (K_N) substitutions per site, we use the one-parameter method and replace F_D by F_{SD} and F_{ND}, respectively, in Eq. (5). This is done to correct for multiple substitutions.

Weighted Pathway Methods

Method of Miyata and Yasunaga. In molecular evolution, pathways containing more synonymous nucleotide differences tend to occur with higher probability than those having fewer synonymous nucleotide differences. Moreover, with respect to nonsynonymous (amino acid altering) nucleotide differences, we note that the replacement of similar amino acids tends to occur more frequently than that of dissimilar amino acids. Taking such observations into account, Miyata and Yasunaga[20] developed a weighted pathway method (M–Y method) for estimating the numbers of synonymous and nonsynonymous substitutions. In the M–Y method, the weight given for each nucleotide difference depends on the biochemical similarity of the amino acid replacement involved. The estimated numbers of synonymous and nonsynonymous sites may also be obtained by suitably weighting different pathways. Then, using the estimated fractions of synonymous and nonsynonymous changes, we can compute the numbers of evolutionary substitutions by the formula for the one-parameter method.

Although the M–Y method appears to give better estimates of synonymous and nonsynonymous substitutions than the N–G method, computer simulations have shown that the two methods give very similar estimates and that the estimates obtained by the former method are not substantially better than those obtained by the latter. The reason for this is that the genetic code seems to have such a property that the weights for different pathways should be quite similar.[19,21]

Method of Li et al. Li et al.[22] developed another weighted pathway method. They used a different weighting scheme based on the expected and observed frequencies of nucleotide substitutions. They also considered nondegenerate, 2-fold degenerate, and 4-fold degenerate sites separately. However, their method again leads to essentially the same estimates of K_s and K_N as obtained by the N–G method.

Application, Advantages, and Disadvantages of Each Method. Although we can rely on hand calculations to obtain estimates by methods such as that of Perler *et al.* and the N–G method, in general, computer programs are required or preferable when we use the above methods. In the following, we show, using the sequence comparison given in Fig. 3, the estimated

[21] M. Nei, "Molecular Evolutionary Genetics." Columbia University Press, New York, 1987.
[22] W.-H. Li, C.-I Wu, and C.-C. Luo, *Mol. Biol. Evol.* **2**, 150 (1985).

numbers of synonymous and nonsynonymous substitutions as obtained by various methods.

If we use the N–G unweighted pathway method, the estimated total numbers of synonymous and nonsynonymous sites are $S = 104.3$ and $N = 333.7$, and the total numbers of synonymous and nonsynonymous nucleotide differences are $S_d = 41.8$ and $N_d = 42.2$, respectively. Thus, the fractions of synonymous and nonsynonymous differences are given by $F_{SD} = S_d/S = 0.401$ and $F_{ND} = N_d/N = 0.126$, respectively. Substituting these values into Eq. (5), we get $K_S = 0.574$ for synonymous substitutions and $K_N = 0.138$ for nonsynonymous substitutions.

If we use the M–Y weighted pathway method, we obtain $S = 104.4$ and $N = 333.6$, and also $S_d = 42.3$ and $N_d = 41.7$. Thus, the proportions of synonymous and nonsynonymous differences are given by $F_{SD} = S_d/S = 0.405$ and $F_{ND} = N_d/N = 0.125$, respectively. Putting these values into Eq. (5), we have $K_S = 0.582$ for synonymous substitution and $K_N = 0.137$ for nonsynonymous substitution.

Through comparison of the sets of K values obtained by the two methods (N–G and M–Y methods), it is clear that they give very similar results. Therefore, we recommend the N–G method rather than the M–Y method, because the former is much simpler to use than the latter.

The method of Perler et al. is not recommended since it tends to give serious underestimates, particularly for the number of synonymous substitutions. For example, when the nucleotide sequence of the Drosophila pupal cuticle protein gene is compared with that of the Drosophila larval cuticle protein gene (data from Henikoff et al.[23] and Snyder et al.[24]), the method of Perler et al. gives $K_S = 1.227$ whereas the N–G method gives $K_S = 1.518$, the difference in K_S thus being quite significant. The method of Li et al. may be useful when one wants to know the numbers of synonymous substitutions at 4- and 2-fold degeneracy sites separately.

Based on consideration of the set of synonymous codons and the variance of the estimated frequencies of the different synonymous codons, Lewontin[25] proposed a method for estimating the number of synonymous substitutions per codon. This method, however, is quite impractical because it is very sensitive to small differences in codon usage of the DNA sequences compared.

---

[23] S. Henikoff, M. A. Keene, J. S. Sloan, J. Bleskan, R. Hards, and D. Patterson, *Proc. Natl. Acad. Sci. U.S.A.* **83**, 720 (1986).

[24] M. Snyder, M. Hunkapiller, D. Yuen, D. Silvert, J. Fristrome, and N. Davidson, *Cell* **29**, 1027 (1982).

[25] R. C. Lewontin, *Mol. Biol. Evol.* **6**, 15 (1989).

Discussion

When the distance or the number of evolutionary base substitutions that separate the two DNA sequences is rather small, various methods as explained above (the one-, two-, and three-parameter, etc., methods) give similar results, with similar estimates of the K values. In such cases, more elaborate methods do not necessarily have much advantage over a simpler one.

As the real distance increases, however, and particularly when the real value of K is larger than unity, exact estimation of the actual number of base substitutions becomes very difficult. This is because, as the number of multiple substitutions increases, it becomes increasingly difficult to estimate superimposed or hidden substitutions. It is in such a situation that the real merit of the more exact method becomes apparent. Actually, already at the range of intermediate K values, the advantage of the six-parameter method over the one-parameter method becomes evident as revealed by the examples in Table III, in which one-parameter method gives $K = 0.409$ whereas six-parameter method gives $K = 0.564$.

One important but unsolved problem inherent in the estimation of K, when K is large, is how to treat "inapplicable cases." All the methods so far developed treat DNA sequences as if they were infinite in length. In other words, they disregard the sampling effect which arises from the fact that all actual sequences are finite in length. This causes trouble when the two sequences compared have a large distance. In such a case, the probability that the observed number of base differences happens to exceed the theoretical "saturation level" increases, and this makes the estimation formula inapplicable. For example, in the one-parameter method, if the observed fraction of difference (F_D) happens to exceed $\frac{3}{4}$, the argument of the logarithmic term becomes negative, and Eq. (5) can no longer be applied. Such inapplicable cases increase as the true K value increases, and discarding inapplicable cases leads to serious underestimates, as pointed out by Kimura.[9]

There has been some invalid criticism that the estimation methods constructed under the above models would not apply to actual data because the real patterns of nucleotide substitutions may be different from the model. Similarly, Lewontin[25] claims that the above methods cannot be applied to codons in the protein-coding region, because amino acid changes are generally much constrained and synonymous and nonsynonymous changes belong to different classes. We cannot accept such an extreme view. Furthermore, the method of Lewontin[25] itself contains some problems, and, even for estimating the number of synonymous changes, this method cannot be recommended. Needless to say, it is important to know the real pattern of nucleotide substitutions, and, taking this into account, to develop better methods for estimating the number of nucleotide substitutions. In fact, some authors have attempted to estimate the pattern of nucleotide substitutions from actual data.[12,22] We should also add that some of the widely used methods, particularly the method of Perler *et al.*,[17] lead to serious underestimates and therefore cannot be recommended.

In applying various estimation methods, we note that as long as we know the statistical behavior and level of the estimated value of K, as well as the application range of the formulas, the methods reviewed above are sufficiently useful for our purposes. However, considering the great importance of the estimation of sequence divergence for studies of molecular evolution, further efforts are needed to develop better statistical methods. Also, in order to assess the validity of various statistical methods already available, extensive simulation experiments will have to be conducted. In other words, we should perform a systematic investigation using the Monte Carlo method and assuming diverse patterns of evolutionary mutant substitutions with known parameters and with known outcomes of multiple substitutions to see how well these statistical methods are able to estimate the true values. In fact, much work remains to be done in this field.

[34] Converting Distance to Time: Application to Human Evolution

By HIROHISA KISHINO and MASAMI HASEGAWA

Introduction

DNA sequence data provide us with a good source of information on the evolutionary history of organisms. Since nucleotide substitution in evolution is best regarded as stochastic, statistical methods based on probabilistic models are required for data analysis. Among these, the maximum likelihood method has a firm basis in probability theory. It requires an explicit probabilistic model of nucleotide substitution to calculate the likelihood function and helps to make the assumptions clear.

In molecular phylogenetics, we want to solve two problems: (1) inference of branching order, and (2) estimation of branching dates. This chapter is concerned mainly with the latter problem, but the two are interrelated. The estimation of branching dates has been based on an approximate constancy of the molecular evolutionary rate that is called the

METHODS IN ENZYMOLOGY, VOL. 183

molecular clock.[1,2] However, it has recently been suggested that the rate sometimes differs among taxonomic units.[3] A relative rate test is applicable to examine whether the rate differs among lineages.[1,4] The test, however, depends on the branching order. To avoid circularity we should first infer the branching order without assuming constancy of the rate. The maximum likelihood method developed by Felsenstein[5] is suitable for this purpose.[6]

After establishing the branching order, we estimate the branching dates by the method presented below. This is an extension of our earlier method based on the assumption of rate constancy,[4,7] and it has been described briefly.[8] The rate variation does not itself invalidate the molecular clock analysis, provided the variation can be incorporated into the model as is done in this chapter.

Statistical Model

Let us consider s homologous nucleotide sequences that consist of n nucleotide sites. Our data set can be represented in tabular form as follows:

$$
\begin{aligned}
&X_{11}, X_{12}, X_{13}, \ldots, X_{1q}, \ldots, X_{1n}\\
&X_{21}, X_{22}, X_{23}, \ldots, X_{2q}, \ldots, X_{2n}\\
&\qquad\qquad\qquad\qquad\qquad\qquad\qquad\\
&X_{p1}, X_{p2}, X_{p3}, \ldots, X_{pq}, \ldots, X_{pn}\\
&\qquad\qquad\qquad\qquad\qquad\qquad\qquad\\
&X_{s1}, X_{s2}, X_{s3}, \ldots, X_{sq}, \ldots, X_{sn}
\end{aligned}
\tag{1}
$$

where X_{pq} is T, C, A, or G and denotes the state of the qth site of the pth species. The tabular form of Eq. (1) can be rewritten as

$$
\mathbf{X}_1, \mathbf{X}_2, \mathbf{X}_3, \ldots, \mathbf{X}_q, \ldots, \mathbf{X}_n
\tag{2}
$$

[1] A. C. Wilson, S. S. Carlson, and T. J. White, *Annu. Rev. Biochem.* **46,** 573 (1977).
[2] M. Kimura, "The Neutral Theory of Molecular Evolution." Cambridge University Press, Cambridge, England, 1983.
[3] R. J. Britten, *Science* **231,** 1393 (1986).
[4] M. Hasegawa, H. Kishino, and T. Yano, *J. Mol. Evol.* **26,** 132 (1987).
[5] J. Felsenstein, *J. Mol. Evol.* **17,** 368 (1981).
[6] H. Kishino and M. Hasegawa, *J. Mol. Evol.* **29,** 170 (1989).
[7] M. Hasegawa, H. Kishino, and T. Yano, *J. Mol. Evol.* **22,** 160 (1985).
[8] M. Hasegawa, H. Kishino, and T. Yano, *J. Human Evol.* (in press) (1989).

where $\mathbf{X}_q$ represents the qth column of Eq. (1). There are 4^s possible outcomes for each site. Basic assumptions are that each site changes homogeneously and independently of others; that is, the process of nucleotide substitution follows an independently identical distribution (i.i.d.), and the process is in equilibrium. Let $\mathbf{X} = (X_1, \ldots, X_s)^T$ (superscript T denotes a transposed vector) be a representative variable of a site. From the data, we obtain information on the following probabilities:

$$\Pr(X_1 = i_1, \ldots, X_s = i_s) = P_{i_1, \ldots, i_s} \qquad (i_1, \ldots, i_s = \text{T, C, A, G})$$

We specify a model for the process of nucleotide substitution and represent these probabilities in terms of parameters such as the branching dates and substitution rates, estimating the parameters by the maximum likelihood method.

We denote by $n_{i_1, \ldots, i_s}$ the number of sites that have a value of $(i_1, \ldots, i_s)$. This follows a multinomial (4^s-nomial) distribution,

$$\text{Multi}(n; P_{i_1, \ldots, i_s}, i_1, \ldots, i_s = \text{T, C, A, G})$$

and represents the most detailed information about the data under the i.i.d. assumption. The average and the variance–covariance of these statistics are given by

$$\mathrm{E}\{n_{i_1, \ldots, i_s}\} = nP_{i_1, \ldots, i_s} \tag{3}$$

$$\mathrm{Cov}\{n_{i_1, \ldots, i_s}, n_{i'_1, \ldots, i'_s}\}$$
$$= n(\delta_{i_1, \ldots, i_s; i'_1, \ldots, i'_s} P_{i_1, \ldots, i_s} - P_{i_1, \ldots, i_s} P_{i'_1, \ldots, i'_s}) \tag{4}$$

where $\delta_{i_1, \ldots, i_s; i'_1, \ldots, i'_s}$ equals 1 when $i_1 = i'_1, \ldots, i_s = i'_s$ (variance), and 0 otherwise (covariance).

The number of states increases explosively as s increases. Therefore, we reduce the data to differences and compare the differences with a probability distribution to which they conform.

Markov Model

We assume that each site evolves according to a stationary Markov process in which a nucleotide i (T, C, A, or G) is replaced by another nucleotide j in an infinitesimally short time interval, dt, with a probability of $P_{ij}(dt)$, as follows:

$$P_{ij}(dt) = \begin{cases} \alpha \pi_j dt & \text{(for transition)} \\ \beta \pi_j dt & \text{(for transversion)} \end{cases} \tag{5}$$

where π_j is the composition of nucleotide j, and α and β are parameters that determine transition and transversion rate, respectively.

Defining the substitution rate matrix R by

$$R = \begin{array}{c} \\ \text{T} \\ \text{C} \\ \text{A} \\ \text{G} \end{array} \begin{array}{cccc} \text{T} & \text{C} & \text{A} & \text{G} \\ \left(\begin{array}{cccc} -\alpha\pi_C - \beta\pi_R & \alpha\pi_C & \beta\pi_A & \beta\pi_G \\ \alpha\pi_T & -\alpha\pi_T - \beta\pi_R & \beta\pi_A & \beta\pi_G \\ \beta\pi_T & \beta\pi_C & -\alpha\pi_G - \beta\pi_Y & \alpha\pi_G \\ \beta\pi_T & \beta\pi_C & \alpha\pi_A & -\alpha\pi_A - \beta\pi_Y \end{array} \right) \end{array} \quad (6)$$

where $\pi_Y = \pi_T + \pi_C$ and $\pi_R = \pi_A + \pi_G$, the function $P(t)$ for an arbitrary time interval t is given by

$$P(t) = e^{tR} \quad (7)$$

Denoting the eigenvalues and eigenvectors of R by λ_i and $\mathbf{u}_i (i = 1, 2, 3, 4)$, and defining $U = (\mathbf{u}_1, \mathbf{u}_2, \mathbf{u}_3, \mathbf{u}_4)$, we can write the right-hand side of Eq. (7) as

$$e^{tR} = U \begin{pmatrix} e^{t\lambda_1} & & & \\ & e^{t\lambda_2} & & \\ & & e^{t\lambda_3} & \\ & & & e^{t\lambda_4} \end{pmatrix} U^{-1} \quad (8)$$

The eigenvalues and eigenvectors are given by

$$\begin{array}{cc} \lambda_1 = 0 & \lambda_2 = -\beta \\ \lambda_3 = -(\pi_Y\beta + \pi_R\alpha) & \lambda_4 = -(\pi_Y\alpha + \pi_R\beta) \end{array} \quad (9)$$

$$\mathbf{u}_1 = \begin{pmatrix} 1 \\ 1 \\ 1 \\ 1 \end{pmatrix} \quad \mathbf{u}_2 = \begin{pmatrix} 1/\pi_Y \\ 1/\pi_Y \\ -1/\pi_R \\ -1/\pi_R \end{pmatrix}$$

$$\mathbf{u}_3 = \begin{pmatrix} 0 \\ 0 \\ \pi_G/\pi_R \\ -\pi_A/\pi_R \end{pmatrix} \quad \mathbf{u}_4 = \begin{pmatrix} \pi_C/\pi_Y \\ -\pi_T/\pi_Y \\ 0 \\ 0 \end{pmatrix} \quad (10)$$

Now, the numbers of transition differences S_{ij} and transversion differences V_{ij} between the ith and jth sequences are defined as follows:

$$S_{ij} = \underbrace{\sum_{l_1 l_2}}_{\text{transition}} n_{\ldots l_1 \ldots l_2 \ldots}^{\quad\; \hat{i} \qquad \hat{j}} \quad (11)$$

$$V_{ij} = \underbrace{\sum_{l_1 l_2}}_{\text{transversion}} n_{\ldots l_1 \ldots l_2 \ldots}^{\quad\; \hat{i} \qquad \hat{j}}$$

where $n_{\ldots \underset{i}{T} \ldots \underset{j}{C} \ldots}$ indicates the number of sites that have T in the ith sequence and C in the jth sequence irrespective of the other sequences.

Let us consider a pair of species i and j which separated t million years (Myr) ago. When the transition and transversion rates remain unchanged during evolution, from time reversibility and the Chapman–Kolmogorov equation, the expected number of transition differences between i and j is given by

$$\overline{S}_{ij} = \underbrace{\sum_{l_1 l_2}}_{\text{transition}} n P_{\ldots \underset{i}{l_1} \ldots \underset{j}{l_2} \ldots}$$

$$= n \underbrace{\sum_{l_1 l_2}}_{\text{transition}} \sum_h \pi_h P_{hl_1}(t) P_{hl_2}(t) \tag{12}$$

$$= 2n\{(\pi_T\pi_C + \pi_A\pi_G) + (\pi_T\pi_C\pi_R/\pi_Y + \pi_A\pi_G\pi_Y/\pi_R)\, e^{-2\beta t} \\ - (\pi_T\pi_C/\pi_Y)\, e^{2t(\alpha\pi_Y + \beta\pi_R)} - (\pi_A\pi_G/\pi_R)\, e^{2t(\alpha\pi_R + \beta\pi_Y)}\}$$

Similarly, the expected number of transversion differences is given by

$$\overline{V}_{ij} = 2n\pi_Y\pi_R\{1 - e^{-2\beta t}\} \tag{13}$$

Furthermore, variances of differences and covariances between differences are calculated similarly from Eq. (4).[7]

Likelihood Function

The parameters are estimated by the maximum likelihood function. From the central limit theorem, the random vector $\mathbf{D} = (V_{12}, \ldots, V_{1s}, V_{23}, \ldots, V_{2s}, \ldots, V_{(s-1)s}, S_{12}, \ldots, S_{1s}, S_{23}, \ldots, S_{2s}, \ldots, S_{(s-1)s})^T$ follows asymptotically a multivariate normal distribution: $N(\overline{\mathbf{D}}, \Omega)$, where $\overline{\mathbf{D}}$ and Ω are the mean vector and the variance–covariance matrix of $\mathbf{D}$. Denoting the unknown parameters by $\theta = (t_1, t_2, t_3, \ldots, \alpha, \beta)^T$, the likelihood function is given by

$$L(\theta|\mathbf{D}) = \frac{1}{(2\pi)^{s(s-1)}\sqrt{\det \Omega(\theta)}} \exp\left\{-\frac{1}{2}[\mathbf{D} - \overline{\mathbf{D}}(\theta)]^T\Omega(\theta)^{-1}[\mathbf{D} - \overline{\mathbf{D}}(\theta)]\right\} \tag{14}$$

Here, we should note that Ω involves the unknown parameter θ. Therefore, we adopt the following updating algorithm for estimating θ:

1. Input the initial value θ_0. Suppose now that we are at the nth step of the updating algorithm of estimation of θ, and $\hat{\theta}_n$ is available. Then,
2. Calculate $\hat{\Omega}_n = \Omega(\hat{\theta}_n)$.

3. Generalized least-squares estimation of θ: using $\hat{\Omega}_n = \Omega(\hat{\theta}_n)$, we minimize

$$R_{n+1} = [\mathbf{D} - \overline{\mathbf{D}}(\theta)]^T \hat{\Omega}_n^{-1} [\mathbf{D} - \overline{\mathbf{D}}(\theta)]$$

by the Newton method. The estimate of θ thus obtained is denoted by $\hat{\theta}_{n+1}$.
4. Iterate the Steps 2 and 3 above until convergence is achieved.

Defining the sum of squares of residuals (S.S.R.) by

$$R(\mathbf{D},\theta) = [\mathbf{D} - \overline{\mathbf{D}}(\theta)]^T \Omega^{-1} [\mathbf{D} - \overline{\mathbf{D}}(\theta)] \tag{15}$$

the variance of the estimates $\mathrm{Var}(\hat{\theta})$ is given by

$$\widehat{\mathrm{Var}}(\hat{\theta}) = B^{-1}(\hat{\theta})A(\hat{\theta})\Omega(\hat{\theta})A^T(\hat{\theta})B^{-1}(\hat{\theta}) \tag{16}$$

where

$$A(\theta) = \frac{\partial^2}{\partial\theta\,\partial\mathbf{D}^T} R(\theta|\mathbf{D})$$

$$B(\theta) = \frac{\partial^2}{\partial\theta\,\partial\theta^T} R(\theta|\mathbf{D})$$

Variable Rate

When the evolutionary rate differs among different lineages, the parameters α and β can be different among lineages, but $\overline{V}_{ij}$ and $\overline{S}_{ij}$ are given by formulas similar to Eqs. (12) and (13). Consider the case shown in Fig. 1 as an example, where i and j diverged t Myr ago and evolution proceeded with the transition and transversion rates of α_1 and β_1 at first along the two lineages, but along the line leading to i, the rates changed to α_2 and β_2 at t_1 Myr ago and again changed to α_3 and β_3 at t_2 Myr ago.

We denote by $P^{(k)}(t)$ the transition probability on the branch to which the parameters α_k and β_k are assigned. The eigenvalues, $\lambda^{(k)}$'s, and eigenvectors, $\mathbf{u}^{(k)}$'s, of the matrix of substitution rate $R^{(k)}$ defined by $P^{(k)}(dt) = I + R^{(k)}\,dt$ are represented by the same formulas as Eqs. (9) and (10), where α and β are replaced by α_k and β_k. $P^{(k)}(t)$ is represented as

$$P^{(k)}(t) = e^{tR(k)} = U \begin{pmatrix} e^{t\lambda_1^{(k)}} & & & \\ & e^{t\lambda_2^{(k)}} & & \\ & & e^{t\lambda_3^{(k)}} & \\ & & & e^{t\lambda_4^{(k)}} \end{pmatrix} U^{-1} \equiv UT^{(k)}U^{-1} \tag{17}$$

What is important here is that the eigenvectors do not depend on the parameters α_k and β_k and are common for all branches.

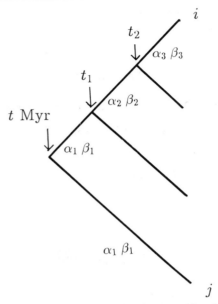

FIG. 1. Example used in the discussion of estimating branching dates when evolutionary rates vary among different branches. [From Hasegawa *et al.*, *J. Human Evol.* (in press) (1989).]

Therefore, we get, for example, the following reduction:

$$P^{(2)}(t_1 - t_2)P^{(3)}(t_2) = UT^{(2)}U^{-1}UT^{(3)}U^{-1}$$

$$= U \begin{pmatrix} 1 & & & \\ & e^{(t_1-t_2)\lambda_2^{(2)}+t_2\lambda_2^{(3)}} & & \\ & & e^{(t_1-t_2)\lambda_3^{(2)}+t_2\lambda_3^{(3)}} & \\ & & & e^{(t_1-t_2)\lambda_4^{(2)}+t_2\lambda_4^{(3)}} \end{pmatrix} U^{-1}$$

$$= U \begin{pmatrix} 1 & & & \\ & \exp\{-(t_1 - t_2)\beta_2 \\ & \quad - t_2\beta_3\} & & \\ & & \exp\{-(t_1 - t_2)(\pi_Y\beta_2 + \pi_R\alpha_2) \\ & & \quad - t_2(\pi_Y\beta_3 + \pi_R\alpha_3)\} & \\ & & & \exp\{-(t_1 - t_2)(\pi_Y\alpha_2 + \pi_R\beta_2) \\ & & & \quad - t_2(\pi_Y\alpha_3 + \pi_R\beta_3)\} \end{pmatrix} U^{-1}$$

(18)

Thus, it is enough to replace $2\alpha t$ and $2\beta t$ in Eqs. (12) and (13) by

$(2t - t_1)\alpha_1 + (t_1 - t_2)\alpha_2 + t_2\alpha_3$ and $(2t - t_1)\beta_1 + (t_1 - t_2)\beta_2 + t_2\beta_3$, respectively. The covariances can be derived similarly.

Consideration of Ω

Up to now, we have assumed the independence and homogeneity among sites. However, this assumption might not be strictly valid. The composition of nucleotides may not be homogeneous along the sequence even in pseudogenes. Furthermore, there might be a correlation between sites. Even if the assumption of homogeneity and independence is violated, our estimates of branching dates and the average substitution rates may still be good approximations. Nevertheless, the standard errors of the estimates should be modified, because the variance–covariance matrix Ω is calculated under the assumption of a multinomial distribution for the frequencies of the states. When the neighboring sites in the genes have positive correlations, the variance would be larger than that obtained under the assumption of independence. Conversely, if they are negatively correlated, it is smaller.

It is quite difficult, and perhaps is not fruitful at present, to incorporate the many factors affecting the correction of Ω into the model. Here, we take a simpler approach. When the fitting of the model to the data is unsatisfactory, by introducing an additional parameter σ we modify the variance–covariance matrix as $\sigma^2\Omega$, preserving the correlation structure of Ω obtained under the assumption of i.i.d. The parameter σ is a correction factor. Under this modification, the estimates themselves do not change, but the standard errors are modified by a factor of $\hat{\sigma}$. Now that the theoretical basis of Ω is weakened, we should use a common Ω in comparing between models, rather than estimating it for each model.

Model Selection: Akaike Information Criterion

Different tree topologies constitute different models. Further, when we take into account a variable rate of substitution, several stages of heterogeneity among different branches can be considered. An extreme case is every branch having its own parameters α_k and β_k, and the other extreme is rate constancy throughout the tree. If we introduce many parameters, we get a flexible model, but the precision of the estimates worsens. On the other hand, models that are too simple give biased estimates. Minimizing both the bias and the variance is a trade-off requirement. It is important to adopt a criterion of model selection based on a firm theoretical background, the results of which are persuasive for almost everyone.

A likelihood ratio test has been widely used for model selection. It provides comparison between two or among a set of nested models. Suppose we have two nested models 1 and 2 ($\supset$ 1), whose likelihood functions are $L_1(\theta_1)$ and $L_2(\theta_2)$, respectively. Under the truth of the restricted model 1, the difference of estimated log likelihoods between the models,

$$2\{\log L_2(\hat{\theta}_2) - \log L_1(\hat{\theta}_1)\} \qquad (19)$$

where $\hat{\theta}_1$ and $\hat{\theta}_2$ are maximum likelihood estimates, follows a χ^2 distribution with the degrees of freedom being the difference of numbers of free parameters between the two models. However, the models to be compared are often nonnested as is the case in comparing different bifurcating trees.

Akaike derived a criterion of model selection through the study of time series problems.[9,10] This criterion is based on the information theory and can be applied even in comparing nonnested models. He obtained the concept of AIC (an information criterion, now called the Akaike information criterion),

$$\text{AIC} = -2 \times (\text{estimated log likelihood of the model})$$
$$+ 2 \times (\text{number of free parameters of the model}) \qquad (20)$$

as an asymptotically unbiased estimator of the expected Kullback–Leibler information quantity[11] of the true distribution to the distribution estimated by the maximum likelihood method under the model. A model which minimizes the AIC is considered to be the most appropriate model. The minimum AIC estimate is a natural extension of the classic maximum likelihood estimate. It becomes possible to compare the adequacy of the approximation of different types of models even with different numbers of parameters.

In our analysis, the log likelihood is estimated by

$$\log L(\hat{\theta}) = -\tfrac{1}{2} \log \det \Omega(\hat{\theta}) - \tfrac{1}{2}[\mathbf{D} - \overline{\mathbf{D}}(\hat{\theta})]^T \Omega(\hat{\theta})^{-1}[\mathbf{D} - \overline{\mathbf{D}}(\hat{\theta})] \qquad (21)$$

when we assume that $\sigma^2 = 1$. If we take $\sigma^2 \Omega$ as the variance–covariance matrix of $\mathbf{D}$, σ^2 is estimated by

$$\hat{\sigma}^2 = \frac{1}{s(s-1)}[\mathbf{D} - \overline{\mathbf{D}}(\hat{\theta})]^T \Omega(\hat{\theta})^{-1}[\mathbf{D} - \overline{\mathbf{D}}(\hat{\theta})] \qquad (22)$$

Then, the log likelihood is estimated by

$$\log L(\hat{\theta}) = -\tfrac{1}{2} \log \det \Omega(\hat{\theta}) - \tfrac{1}{2} \log \hat{\sigma}^2 - \frac{s(s-1)}{2} \qquad (23)$$

[9] H. Akaike, *IEEE Trans. Autom. Control* **AC-19**, 716 (1974).
[10] Y. Sakamoto, M. Ishiguro, and G. Kitagawa, "Akaike Information Criterion Statistics." Reidel, Dordrecht, 1986.
[11] S. Kullback and R. A. Leibler, *Ann. Math. Stat.* **22**, 79 (1951).

Data Analysis: Application to Human Evolution

One of the most interesting applications of molecular phylogenetics has been on human evolution. We shall apply the method described above to the $\psi\eta$-globin gene sequenced by Koop et al.,[12] Miyamoto et al.,[13,14] and Fitch et al.,[15] and to mitochondrial DNA (mtDNA) sequenced by Brown et al.[16] Since insertion/deletion events are not taken into account in our Markov model, sites for which insertion/deletion events took place in any species under consideration are excluded. To calibrate a molecular clock, the branching of orangutans from the African apes–human clade, taken as a reference, is assumed to be 13 Myr ago.[8,17]

$\psi\eta$-Globin Gene

Figure 2 gives the maximum likelihood tree of the $\psi\eta$-globin genes from human, chimpanzee, gorilla, orangutan, rhesus monkey, spider monkey, and owl monkey estimated by DNAML in Felsenstein's program package PHYLIP (Version 3.1), where no constraint on the evolutionary rate is imposed. The number of homologous sites provided for the analysis is 2036. The location of the root is arbitrary by this method. The length of each branch is proportional to the estimated number of nucleotide substitutions. This figure is useful in constructing models to be included in model selection. Since the lengths of branches in hominoids (humans and apes) are shorter than those in Old World monkeys (rhesus monkeys), Model 2, where slowdown in hominoids is allowed (by assigning to the hominoid lines a set of parameters, α_2 and β_2, different from α_1 and β_1 assigned to other lines), should be compared with Model 1 of rate constancy (α_1 and β_1 are assigned to all lines). Furthermore, shorter length of the human branch than those of apes suggests further slowdown in humans (Model 3). Within New World monkeys, the shorter length of the spider monkey branch compared to that of owl monkeys suggests another model (Model 4), where slowdown also took place in spider monkeys relative to owl monkeys. In Table I, these models are compared by the AIC.

Model 2 has a lower AIC value than Model 1, consistent with earlier observations that the evolutionary rate of nuclear DNA has decreased in

[12] B. F. Koop, M. Goodman, P. Xu, K. Chan, and J. L. Slightom, *Nature (London)* **319**, 234 (1986).
[13] M. M. Miyamoto, J. L. Slightom, and M. Goodman, *Science* **238**, 369 (1987).
[14] M. M. Miyamoto, B. F. Koop, J. L. Slightom, and M. Goodman, *Proc. Natl. Acad. Sci. U.S.A.* **85**, 7626 (1988).
[15] D. H. A. Fitch, C. Mainone, J. L. Slightom, and M. Goodman, *Genomics* **3**, 237 (1988).
[16] W. M. Brown, E. M. Prager, A. Wang, and A. C. Wilson, *J. Mol. Evol.* **18**, 225 (1982).
[17] P. Andrews, *Cold Spring Harbor Symp. Quant. Biol.* **52**, 419 (1986).

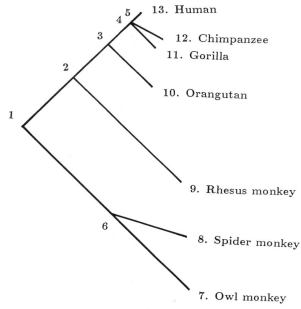

FIG. 2. Maximum likelihood tree of the $\psi\eta$-globin gene (2036 nucleotides).

hominoids relative to Old World monkeys.[3,4] Model 3, where the rate has further decreased in humans compared to apes, is better than Model 2.

Models based on the trifurcation of branchings among humans, chimpanzees, and gorillas can be compared with models of bifurcation, the former designated by primes. Model 3' has a lower AIC value than Model 3, indicating that the trichotomy cannot be resolved by these data. Model 4' has lower AIC than Model 3', suggesting that the rate has varied within New World monkeys as well as in hominoids. Instead of slowdown in spider monkeys, Model 5', where acceleration has occurred in owl monkeys relative to spider monkeys, is examined. Although not shown in Table I, it gives an AIC value of 128.97, higher than 127.80 of Model 4'.

From the analysis of Model 4', it turns out that $\hat{\beta}_1 = 1.195 \times 10^{-3}$/Myr (transversion in owl and rhesus monkeys) and $\hat{\beta}_4 = 1.091 \times 10^{-3}$/Myr (transversion in spider monkeys) are nearly the same, but $\hat{\alpha}_1 = 6.160 \times 10^{-3}$/Myr (transition in owl and rhesus monkeys) and $\hat{\alpha}_4 = 3.517 \times 10^{-3}$/Myr (transition in spider monkeys) are distinguishable. Model 4B', where the number of parameters is reduced by assigning α_4 and β_1 (not β_4) to the spider monkey lineage, has an AIC value lower by 1.9 than Model 4'. As

TABLE I

BRANCHING DATES AND EVOLUTIONARY RATES ESTIMATED FROM ψη-GLOBIN GENE SEQUENCES OF 2036 NUCLEOTIDES[a]

Parameter	1	2	3	3'	4'	4B'
	Constant rate in NWM, OWM, and hominoids	Constant within hominoids	Slowdown in hominoids		Slowdown in humans	Slowdown in spider monkey
	Bifurcation among man, chimp, and gorilla	Same rate between NWM and OWM		Trifurcation among man, chimp, and gorilla		
Rates in branches						
2-3, 3-10, 3-4 } 4-11, 4-5, 5-12]	α_1, β_1	α_2, β_2	α_2, β_2	α_2, β_2	α_2, β_2	α_2, β_2
5-13 (human)	α_1, β_1	α_2, β_2	α_3, β_3	α_3, β_3	α_3, β_3	α_3, β_3
1-2, 2-9 (OWM) } 1-6, 6-7 (NWM)]	α_1, β_1	α_1, β_1	α_1, β_1	α_1, β_1	α_1, β_1	α_1, β_1
6-8 (spider monkey)	α_1, β_1	α_1, β_1	α_1, β_1	α_1, β_1	α_4, β_4	α_4, β_1
Branching dates (Myr)						
$\hat{t}_1$ (NWM/OWM)	41.56 ± 4.35	31.80 ± 4.63	30.86 ± 4.28	30.86 ± 4.28	32.05 ± 4.51	31.91 ± 4.47
$\hat{t}_2$ (spider/owl monkey)	18.90 ± 2.67	13.23 ± 2.55	12.84 ± 2.41	12.84 ± 2.41	15.29 ± 3.05	15.03 ± 2.92
$\hat{t}_3$ (rhesus monkey)	25.62 ± 2.42	23.31 ± 2.70	22.55 ± 2.42	22.55 ± 2.42	22.50 ± 2.42	22.50 ± 2.42
$\hat{t}_4$ (orangutan)	13	13	13	13	13	13
$\hat{t}_5$ (gorilla)	5.40 ± 0.75	5.65 ± 0.83	6.96 ± 0.93 }	6.95 ± 0.93	6.95 ± 0.93	6.95 ± 0.93
$\hat{t}_6$ (chimpanzee)	5.28 ± 0.75	5.52 ± 0.83	6.92 ± 0.95 }			
Rates (10^{-3}/Myr)						
$\hat{\alpha}_1$	4.124 ± 0.448	5.791 ± 0.983	5.989 ± 0.979	5.989 ± 0.979	6.160 ± 1.005	6.189 ± 1.004
$\hat{\alpha}_2$	—	3.961 ± 0.520	4.272 ± 0.555	4.271 ± 0.555	4.259 ± 0.556	4.259 ± 0.556
$\hat{\alpha}_3$	—	—	2.151 ± 0.856	2.148 ± 0.851	2.150 ± 0.850	2.149 ± 0.850
$\hat{\alpha}_4$	—	—	—	—	3.517 ± 1.153	3.583 ± 1.161

(Continued)

TABLE I

BRANCHING DATES AND EVOLUTIONARY RATES ESTIMATED FROM $\psi\eta$-GLOBIN GENE SEQUENCES OF 2036 NUCLEOTIDES[a]

(Continued)

Parameter	1	2	3	3'	4'	4B'
			Slowdown in hominoids			
	Constant rate in NWM, OWM, and hominoids	Constant within hominoids	Same rate between NWM and OWM		Slowdown in humans	Slowdown in spider monkey
	Bifurcation among man, chimp, and gorilla			Trifurcation among man, chimp, and gorilla	Slowdown in man, chimp, and gorilla	
$\hat{\beta}_1$	0.824 ± 0.097	1.220 ± 0.229	1.255 ± 0.229	1.255 ± 0.229	1.195 ± 0.227	1.182 ± 0.223
$\hat{\beta}_2$	—	0.707 ± 0.129	0.813 ± 0.142	0.813 ± 0.141	0.815 ± 0.142	0.815 ± 0.142
$\hat{\beta}_3$	—	—	0.275 ± 0.203	0.274 ± 0.202	0.275 ± 0.202	0.274 ± 0.202
$\hat{\beta}_4$	—	—	—	—	1.091 ± 0.351	—
$\hat{v}_1$	1.423 ± 0.142	2.029 ± 0.336	2.095 ± 0.332	2.095 ± 0.332	2.108 ± 0.336	2.108 ± 0.335
$\hat{v}_2$	—	1.325 ± 0.147	1.453 ± 0.156	1.454 ± 0.156	1.452 ± 0.156	1.452 ± 0.156
$\hat{v}_3$	—	—	0.665 ± 0.239	0.664 ± 0.237	0.664 ± 0.237	0.664 ± 0.237
$\hat{v}_4$	—	—	—	—	1.407 ± 0.387	1.468 ± 0.351
S.S.R.	42.79	35.89	28.65	28.68	24.14	24.24
df	35	33	31	32	30	31
P	0.17	0.33	0.59	0.63	0.76	0.79
AIC	136.45	133.55	130.31	128.34	127.80	125.90[b]

[a] The tree topology and numbering of nodes and operational taxonomic units (OTUs) are shown in Fig. 2. Ω of Model 3 was used in all cases. Standard errors are given. In addition to $\hat{\alpha}$ and $\hat{\beta}$, the average rate of nucleotide substitution per site (transition plus transversion), $\hat{v} = 2[(\pi_T\pi_C + \pi_A\pi_G)\hat{\alpha} + \pi_Y\pi_R\hat{\beta}]$, is given. NWM, New World monkeys; OWM, Old World monkeys; df, degrees of freedom of a χ^2 distribution of S.S.R.

[b] Minimum AIC.

far as we have examined various models, including others not mentioned here, Model 4B′ has the minimum AIC.

In this analysis, Ω of Model 3 was used for all models. When the Ω value of another model is used, the absolute value of the AIC changes, but relative AIC differences between models, which is relevant to model selection, are little affected. Under the given model, S.S.R. follows a χ^2 distribution with the degree of freedom equal to $s(s - 1)$ minus the number of free parameters. The P value of Model 4B′ is 0.79, and the fitting of the model to the data is satisfactory.

Excluding owl monkeys, we have data for 6166 nucleotide sites of the $\psi\eta$-globin gene from the remaining species, and in Table II several models are compared for these data. Consistent with the previous analysis (Table I), Model 3′ gives the minimum AIC. Given the lack of owl monkey data, the rate variation in spider monkeys cannot be evaluated. In this case, the P value is only 0.0004, and even the best model is not a good representation of the actual process of evolution. In this situation, we introduce σ as described in the preceding section. The introduction of the new parameter constitutes another model, and its AIC is denoted by AIC(2), which is compared with AIC(1) defined without the introduction of σ (denoted simply by AIC in Table I). Since AIC(2) is always lower than AIC(1) for these data, standard errors of the estimates are multiplied by $\hat{\sigma}$.

Mitochondrial DNA

The mtDNA data analyzed here are from human, chimpanzee, gorilla, orangutan, gibbon, bovine, and mouse and consist of two classes of sites, class 1 (third codon position of protein-coding genes; 232 sites) and class 2 (the remaining sites in protein-coding genes and sites in tRNA genes; 664 sites).[7] Of the 664 sites of class 2, 267 sites are regarded as variable[18] and are provided for the analysis. Figure 3 gives the maximum likelihood tree of the mtDNA estimated by DNAML in PHYLIP, where branch lengths are proportional to the estimated number of substitutions in the class 2 sites.

Several models are compared in Table III for the mtDNA data. Although the transversion rate of mtDNA is significantly lower in the bovine lineage than in the murid and primate lineages,[19] this variation is not taken into account in this analysis, because it affects little the estimates of branching dates in hominoids, when the orangutan divergence is taken as a reference. Therefore, the estimates of the branching dates of mouse and

---

[18] M. Hasegawa and H. Kishino, *Evolution* **43**, 672 (1989).
[19] M. Hasegawa and H. Kishino, *Jpn. J. Genet.* **64**, 243 (1989).

TABLE II

BRANCHING DATES AND EVOLUTIONARY RATES ESTIMATED FROM $\psi\eta$-GLOBIN GENE SEQUENCES OF 6166 NUCLEOTIDES[a]

Parameter	Constant rate		Constant within hominoids		Slowdown in hominoids / Slowdown in human	
Model	1	1'	2	2'	3	3'
	Bifurcation	Trifurcation	Bifurcation	Trifurcation	Bifurcation	Trifurcation
Rates in branches						
1–8 (NWM)						
1–2, 2–9 (OWM)	α_1, β_1	α_1, β_1	α_1, β_1	α_1, β_1	α_1, β_1	α_1, β_1
2–3, 3–10, 3–4						
4–11, 4–5, 5–12	α_1, β_1	α_1, β_1	α_2, β_2	α_2, β_2	α_2, β_2	α_2, β_2
5–13 (human)	α_1, β_1	α_1, β_1	α_2, β_2	α_2, β_2	α_3, β_3	α_3, β_3
Branching dates (Myr)						
$\hat{t}_1$ (NWM/OWM)	43.21 ± 4.50	43.35 ± 4.56	32.89 ± 4.14	32.93 ± 4.19	32.20 ± 3.60	32.20 ± 3.62
$\hat{t}_2$ (rhesus monkey)	26.10 ± 2.40	26.16 ± 2.43	23.74 ± 2.32	23.76 ± 2.35	23.23 ± 1.98	23.23 ± 2.00

t_3 (orangutan)	13	13	13	13	13	13
t_4 (gorilla)	6.49 ± 0.79 }	6.40 ± 0.79	6.74 ± 0.76 }	6.66 ± 0.76	7.49 ± 0.76 }	7.45 ± 0.76
t_5 (chimpanzee)	6.27 ± 0.80 }		6.53 ± 0.77 }		7.34 ± 0.78 }	
Rates (10^{-3}/Myr)						
$\hat{\alpha}_1$	3.614 ± 0.379	3.605 ± 0.382	5.175 ± 0.773	5.168 ± 0.780	5.293 ± 0.707	5.291 ± 0.711
$\hat{\alpha}_2$	—	—	3.405 ± 0.374	3.403 ± 0.377	3.484 ± 0.364	3.488 ± 0.365
$\hat{\alpha}_3$	—	—	—	—	2.647 ± 0.705	2.620 ± 0.696
$\hat{\beta}_1$	0.761 ± 0.089	0.756 ± 0.089	1.096 ± 0.173	1.095 ± 0.174	1.117 ± 0.158	1.117 ± 0.159
$\hat{\beta}_2$	—	—	0.684 ± 0.102	0.676 ± 0.102	0.778 ± 0.100	0.774 ± 0.101
$\hat{\beta}_3$	—	—	—	—	0.327 ± 0.160	0.323 ± 0.158
$\hat{v}_1$	1.236 ± 0.118	1.231 ± 0.119	1.772 ± 0.250	1.770 ± 0.253	1.810 ± 0.228	1.810 ± 0.229
$\hat{v}_2$	—	—	1.147 ± 0.107	1.143 ± 0.108	1.213 ± 0.103	1.212 ± 0.104
$\hat{v}_3$	—	—	—	—	0.790 ± 0.192	0.781 ± 0.189
S.S.R.	77.18	78.44	59.16	60.17	49.30	49.90
df	24	25	22	23	20	21
P	<0.00001	<0.00001	~0.00003	~0.00004	~0.0003	~0.0004
$\hat{\sigma}$	1.604	1.617	1.404	1.416	1.282	1.290
AIC(1)	180.15	179.41	166.12	165.14	160.34	158.87
AIC(2)	**163.32**	**161.80**	**159.34**	**157.85**	**157.94**	**156.23**[b]

[a] The tree topology and numbering of nodes and OTUs are shown in Fig. 2. Trifurcation and bifurcation refer to branching among humans, chimpanzees, and gorillas. Ω of Model 3 was used in all cases. The smaller of the AIC(1) and AIC(2) values is indicated by boldface. Standard errors are given.

[b] Minimum AIC.

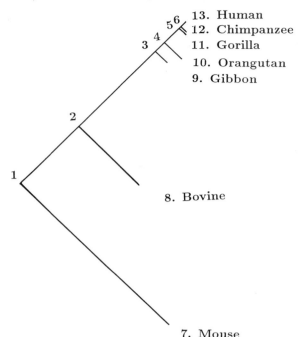

FIG. 3. Maximum likelihood tree of mtDNA (class 2 sites).

bovine are not reliable in this analysis. The fitting of the model to these data is generally satisfactory, and we need not introduce σ in this case.

In Table III, the constant rate model of bifurcating branchings of gorillas and later of chimpanzees from humans (Model 1) is compared with that of trifurcation among the trio (Model 1'). Contrary to the case of the $\psi\eta$-globin genes, bifurcation is preferred to trifurcation, with the AIC difference of 2.42. Thus, under the assumption of rate constancy among the trio, a closer relationship of chimpanzees of humans rather than to gorillas is suggested.

The branches in the great apes–human clade (hominoids excluding gibbons) are longer than that of gibbons in Fig. 3, and therefore several models allowing different rates between these taxa are examined. Model 2, where transition and transversion rates differ only in class 2 sites, gives nearly the same AIC value as Model 1. In Model 2, however, it turns out that estimates of transversion rates are nearly the same between the two taxa. Model 2B, where only the transition rate of class 2 sites differs, gives AIC values lower by 2.0 than Model 2, and it is the minimum AIC model.

TABLE III

BRANCHING DATES AND EVOLUTIONARY RATES ESTIMATED FROM MITOCHONDRIAL DNA SEQUENCES[a]

	Model			
	Constant rate		Rate change (class 2 sites) in great apes–human clade	
Parameter	1 Bifurcation	1' Trifurcation	2 Bifurcation	2B Bifurcation
Rates in branches				
Class 1 sites				
all branches	$\alpha^{(1)}, \beta^{(1)}$		$\alpha^{(1)}, \beta^{(1)}$	$\alpha^{(1)}, \beta^{(1)}$
Class 2 sites				
1–7, 1–2, 2–8				
2–3, 3–9 (gibbon)	$\alpha_1^{(2)}, \beta_1^{(2)}$	$\alpha_1^{(2)}, \beta_1^{(2)}$	$\alpha_1^{(2)}, \beta_1^{(2)}$	$\alpha_1^{(2)}, \beta_1^{(2)}$
3–4, 4–10, 4–5, 5–11				
5–6, 6–12, 6–13				
(great apes and human)	$\alpha_1^{(2)}, \beta_1^{(2)}$	$\alpha_1^{(2)}, \beta_1^{(2)}$	$\alpha_2^{(2)}, \beta_2^{(2)}$	$\alpha_2^{(2)}, \beta_1^{(2)}$
Branching dates (Myr)				
$\hat{t}_1$ (mouse)	100.22 ± 18.20	100.52 ± 18.90	113.38 ± 24.53	113.85 ± 22.27
$\hat{t}_2$ (bovine)	62.44 ± 10.90	64.15 ± 11.33	70.39 ± 14.85	70.67 ± 13.34
$\hat{t}_3$ (gibbon)	16.22 ± 1.94	16.36 ± 1.98	18.61 ± 2.60	18.64 ± 2.54
t_4 (orangutan)	13	13	13	13
$\hat{t}_5$ (gorilla)	4.64 ± 0.84	} 4.10 ± 0.76	4.52 ± 0.85	4.51 ± 0.85
t_6 (chimpanzee)	3.51 ± 0.73	}	3.44 ± 0.74	3.44 ± 0.74
Rates (10^{-3}/Myr)				
Class 1 sites				
$\hat{\alpha}^{(1)}$	305.43 ± 105.31	311.60 ± 107.50	314.23 ± 108.47	314.28 ± 108.42

(*Continued*)

TABLE III
BRANCHING DATES AND EVOLUTIONARY RATES ESTIMATED FROM MITOCHONDRIAL DNA SEQUENCES[a]
(Continued)

	Model			
	Constant rate		Rate change (class 2 sites) in great apes–human clade	
	1	1'	2	2B
Parameter	Bifurcation	Trifurcation	Bifurcation	Bifurcation
$\hat{\beta}^{(1)}$	7.47 ± 1.23	7.23 ± 1.20	6.84 ± 1.33	6.82 ± 1.25
$\hat{v}$	56.10 ± 18.37	57.04 ± 18.75	57.31 ± 18.90	57.31 ± 18.90
Class 2 sites (variable sites only)				
$\hat{\alpha}_1^{(2)}$	57.68 ± 8.44	57.29 ± 8.42	28.35 ± 9.94	28.23 ± 9.48
$\hat{\alpha}_2^{(2)}$	—	—	63.56 ± 9.47	63.56 ± 9.47
$\hat{\beta}_1^{(2)}$	5.36 ± 0.87	5.22 ± 0.86	4.89 ± 1.15	4.86 ± 0.90
$\hat{\beta}_2^{(2)}$	—	—	4.83 ± 1.24	—
$\hat{v}_1^{(2)}$	16.05 ± 2.13	16.89 ± 2.12	8.97 ± 2.66	8.93 ± 2.45
$\hat{v}_2^{(2)}$	—	—	17.17 ± 2.41	17.19 ± 2.38
S.S.R.	79.74	84.18	75.48	75.48
df	79	80	77	78
P	0.46	0.36	0.53	0.56
AIC	320.05	322.47	319.78	317.78[b]

[a] The tree topology and the numbering of nodes and OTUs are shown in Fig. 3. Trifurcation and bifurcation refer to branching among humans, chimpanzees, and gorillas. Ω of Model 1 was used in all cases. Standard errors are given.
[b] Minimum AIC.

This suggests that the transition rate in the class 2 sites of the great apes – human clade is higher than that of gibbons. Although it may also be the case in class 1 sites, this allowance of rate variation does not improve the model as measured by the AIC, probably because, in class 1 sites, the transition differences among species under consideration have nearly reached the saturation level,[7] and does not contain sufficient information on the probable rate variation between the two taxa.

Discussion

In our method sequence data are reduced to the form of transition and transversion differences. Although this reduction causes a loss of information contained in $n_{i_1, \ldots, i_k}$, it has several merits. Figure 4 shows a relationship between S/n and V/n for the class 2 sites of mtDNA based on Model 1. This graphical representation helps us to grasp intuitively what factors are important to fit the model to the data and to improve the model as described in our earlier work.[7] Another merit is that, when the multino-

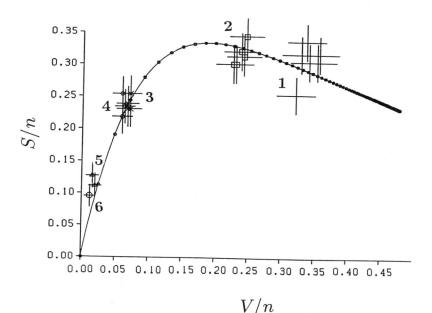

FIG. 4. Relationship between S/n and V/n for the class 2 sites of mtDNA. The numbering corresponds to that of nodes in Fig. 3. Vertical and horizontal lines indicate standard errors of S_{ij}/n and V_{ij}/n, respectively. The interval between neighboring small circles along the curve is 5 Myr.

mial model gives an unsatisfactory explanation of the data, the correction of variance–covariance structure is easily implemented in the generalized least-squares estimation procedure.

The branching dates of chimpanzees and gorillas from humans estimated from mtDNA data seem to be incongruent to those from the $\psi\eta$-globin gene. A gene tree sometimes differs from a species tree for several reasons.[4,8,20] Further analysis of DNA sequence data from many independent gene loci must be performed in the future to clarify the phylogenetic position of humans. We think that the method presented in this chapter would be useful for this purpose. Our Markov model may be too simple to describe the actual pattern of nucleotide substitution in evolution. However, the advantage of our method is that it is based on a clearly described model. Therefore, we can incorporate complexity into our model and construct alternative models as we gain more knowledge about the actual process of nucleotide substitution in the future. In addition, the AIC provides a basis for model selection among these alternatives.

Acknowledgments

We thank Dr. T. Yano for helpful discussions. We are grateful to Mr. J. Adachi for help with computer programs. This work was supported by grants from the Ministry of Education, Science, and Culture of Japan.

[20] M. Nei, "Molecular Evolutionary Genetics." Columbia University Press, New York, 1987.

[35] Influence of Base Composition on Quantitative Estimates of Gene Evolution

By Cecilia Saccone, Cecilia Lanave, Graziano Pesole, and Giuliano Preparata

Introduction

The measure of the genetic distance between organisms is one of the most challenging and difficult issues in molecular evolution. The concept of the molecular clock at the protein level proposed in 1962 by Zuckerkandl and Pauling has received wide support from experimental data obtained in the last 25 years with the sequencing of proteins and nucleic

METHODS IN ENZYMOLOGY, VOL. 183

acids. However, many problems, such as whether there is a universal clock and how it ticks in different lineages, still remain controversial. In this scenario, the construction of simple models of molecular evolution appears as a necessary and scientifically appropriate first step in any methodological approach.

A few years ago we proposed a simple stochastic model of molecular evolution, the stationary Markov model, and we demonstrated that it is at work in a large variety of types of evolutionary dynamics operating at the gene level. In this chapter we present the theoretical basis of the model, its mathematical formulation, and a few experimental applications.

Theoretical Basis

It is our opinion that any model of molecular evolution having a chance of being realistic must be stochastic. This does not mean that molecular evolution cannot obey deterministic laws, but, as our experience in statistical physics has taught us time and again, the only realistic approach for the description of intricate phenomena which involve myriads of interactions (as, e.g., in the physical analysis of gases) is a statistical one. The necessity to follow a stochastic approach in the study of gene evolution is also dictated by purely biological reasons. When the genetic distance among organisms is measured at the level of nucleic acids instead of proteins, one is faced with a number of problems due to the wide fluctuation in nucleotide composition among different organisms and to well-established properties of the genetic code, such as its triplet structure and degeneracy. An important implication of the above aspects is that the rate of evolution in mRNA-coding genes by far exceeds (particularly for silent codon positions) that of the corresponding proteins. As a consequence, any purely divergent model of molecular evolution becomes untenable when referred to nucleic acids. Deterministic models like the distance matrix and the maximum parsimony approach belong to this class (for review, see Ref. 1), for they are based on divergent evolutionary changes only and therefore cannot be considered realistic; such models are destined to fail in any quantitative analysis of evolutionary processes at the molecular level.

This leaves us with the stochastical approach as the only appropriate means to describe molecular evolution at the nucleotide level. Numerous attempts to formulate a stochastic model for nucleic acid evolution have appeared in the literature.[1,2] However, the majority of models suffer from two severe drawbacks: *a priori* assumptions as to nucleotide substitution

---

[1] A. E. Friday, *Oxford Surv. Evol. Biol.* **4**, 61 (1987).
[2] Special Issue, "Molecular Evolutionary Clock," *J. Mol. Evol.* **26** (1987).

probabilities, which are often inconsistent with the data to which the methods are applied, and lack of an accurate statistical analysis. The model we propose is free from these limitations.

Our basic starting hypothesis is that in a nucleotide sequence at a given time T, any particular nucleotide is the product of a stochastic process whose properties depend on the particular position in the sequence occupied by the nucleotide under consideration. The probability of mutation and the subsequent mutation fixation of that nucleotide do not depend on the history of the nucleotide but only on the environmental conditions and on the biological constraints prevailing at the time. Thus, the accumulation of successful mutations along a nucleotide sequence becomes the result of a complex of physicobiological causes, which can be mathematically described by a 4×4 matrix $R_{ij}(T)$, the rate matrix, which fully describes at time T the propensity of a nucleotide $j(i, j = \text{A, C, G, T})$ to be substituted by nucleotide i.

A general stochastic process, however, is still a very complicated description depending on, in addition to the particular nucleotide position, the time T of observation and the phylogenetic lineage one is considering, something that makes it quite intractable from the mathematical standpoint. To become an effective means to measure quantitatively gene evolution having a clocklike behavior, a stochastic model should fulfill the following conditions: (1) the rate matrix R_{ij} should depend only on the particular site and not on the phylogenetic lineage and on the time T, and (2) the nucleotide frequencies q_i ($i = \text{A, C, G, T}$) under comparison at equivalent (first, second, or third) codon positions should be constant both in the extant species considered and in their ancestors, a situation that we call "stationary." A stochastic process that meets these requirements is called a "stationary Markov process" and has a simple mathematic description (see below).

In spite of these restrictive conditions, which make the probability of finding in nature something that ticks as a "Markov clock" extremely low, we have shown that Markov clocks are a rather general feature of molecular evolution. We have also found evidence that not all molecular evolution proceeds with clocklike accuracy. The occurrence of processes affecting the regular tick of molecular clocks can be revealed by a loss of stationarity, which leads to "base drift." There seems to be a well-defined relation between the velocity of a given evolutionary process and the time span in which it behaves as a Markov clock in the sense that the faster the process the shorter the time in which it is Markovian. This latter concept, which we call the base-drift hypothesis,[3] suggests that in order to analyze

[3] G. Preparata and C. Saccone, *J. Mol. Evol.* **26**, 7 (1987).

quantitatively molecular evolution over long times (i.e., when comparing distant species) one should focus on processes having slow dynamics. It must be recalled, however, that besides the inherent fluctuations of stochastic evolution leading to a drift in the base composition behaving according to our base-drift hypothesis, there may be other, selective reasons for base drift, like molecular drive,[4] directional mutation pressure,[5] and genome compartmentalization.[6]

Finally, it is necessary to mention that we define "evolutionary dynamics" as the process of evolution of a set of nucleotide sites, which should be reasonably subjected to similar constraints and thus to similar evolutionary mechanisms. In other words, instead of considering each nucleotide in the sequence as representing a distinct dynamic process, in order to make statistically meaningful statements we group a number of sites together so as to obtain a kind of "average" dynamics. In protein-coding genes, for example, all the nucleotide sites at the first codon position are grouped together and their collective pattern and tempo of substitution is analyzed. Similarly, the collective behavior of nucleotide sites at the second and third, silent codon positions defines other evolutionary dynamics, all distinct from each other. Our model applies to each dynamic separately (e.g., to the silent codon position in a given set of homologous gene sequences). Different nucleotide populations q_i and different rate matrices R_{ik} correspond to different dynamics.

Stochastic Model[7]

Let us consider two codon sequences of length L from homologous genes of two different species originating from a common ancestor. We call the ancestor A and the two species under study B and C, and let us suppose that B and C are at a time distance T from A. We denote the four nucleotide bases A, C, G, and T as 1, 2, 3, and 4, respectively, and concentrate our attention on the equivalent positions (i.e., first, second, or third silent codon position) of each homologous gene of A, B, and C.

The problem we want to solve is to determine the rate of evolutionarily effective substitutions from the rate of divergence between different coding sequences at the silent codon position. To achieve this, we construct a Markov chain in which the four bases form a finite set of possible states.

[4] G. A. Dover, *J. Mol. Evol.* **26**, 47 (1987).
[5] N. Sueoka, *Proc. Natl. Acad. Sci. U.S.A.* **85**, 2653 (1988).
[6] G. Bernardi, B. Olofsson, J. Filipski, M. Zerial, J. Salinas, G. Cuny, M. Meunier-Rovital, and F. Rodier, *Science* **228**, 95 (1985).
[7] C. Lanave, G. Preparata, C. Saccone, and G. Serio, *J. Mol. Evol.* **20**, 86 (1984).

For each member of the sequences we introduce the matrix $P_{ik}(T)$, which represents the probability that base k of A has undergone mutation to base i of B or C. The argument T is the evolutionary divergence time, defined above. By definition we have

$$\sum_{i=1}^{4} P_{ik}(T) = 1 \qquad (1)$$

for each $k = 1, \ldots, 4$. If we denote by $q_i(T)_{B,C}$, $i = 1, 2, 3, 4$, the frequencies of nucleotide i in each of the sequences B and C, and by $q_i(0)$ the same quantities for the ancestor A, we obviously have

$$q_i(T)_B = q_i(T)_C = \sum_k P_{ik}(T) q_k(0) \qquad (2)$$

The hypothesis of stationarity that, according to our discussion in the Introduction, turns the Markov process into a clocklike phenomenon, implies that we must have

$$q_i(0) = q_i(T)_B = q_i(T)_C \qquad (3)$$

Any failure of the data to satisfy Eq. (3) will thus lead to the inapplicability of the analysis.

Let us now indicate by R_{ik} the time-independent rate matrix which characterizes the stationary Markov process through the following Kolmogorov differential equations:

$$\frac{d}{dT} P_{ik}(T) = \sum_{r=1}^{4} P_{ir}(T) R_{rk} \qquad (4)$$

with the initial conditions

$$P_{ik}(0) = \delta_{ik} \qquad (5)$$

From Eqs. (4) and (5) we get:

$$R_{rk} = \frac{d}{dT} P_{rk}(T)|_{T=0} \qquad (6)$$

In matrix notation the solution of Eq. (4), with the boundary conditions of Eq. (5), can be written as

$$P(T) = \exp(TR) \qquad (7)$$

To carry out the analysis, it is necessary to determine explicitly the individual transition probabilities $P_{ik}(T)$ obeying the differential equations [Eq. (4)]. These functions depend on the eigenvalues and eigenvectors of the rate matrix R in the manner we now describe. In the case under study R possesses four eigenvalues λ_α ($\alpha = 0, 1, 2, 3$), one of which is equal to zero and the others real and nonpositive. Thus, we may write

$$R_{ik} = \sum_{\alpha=0}^{3} \lambda_\alpha u_i^{(\alpha)} v_k^{(\alpha)} \tag{8}$$

where $u_i^{(\alpha)}$ and $v_k^{(\alpha)}$ are the left and right eigenvectors, respectively, and satisfy the orthogonality relation:

$$\sum_i u_i^{(\alpha)} v_i^{(\beta)} = \delta_{\alpha\beta} \tag{9}$$

It is easy to see that for $\lambda = 0$ the eigenvectors can be determined explicitly and are given by

$$v_i^{(0)} \equiv (1, 1, 1, 1)$$
$$u_i^{(0)} \equiv (q_1, q_2, q_3, q_4) \tag{10}$$

For $\alpha = r = 1, 2, 3$, we introduce the vectors $w_i^{(r)}$, according to the definitions

$$v_i^{(r)} = \frac{1}{(q_i)^{1/2}} w_i^{(r)}$$
$$u_i^{(r)} = (q_i)^{1/2} w_i^{(r)} \tag{11}$$

It is easily verified that the orthogonality conditions in Eq. (9) are satisfied if

$$\sum_{i=1}^{4} (q_i)^{1/2} w_i^{(r)} = 0 \tag{12}$$

$$\sum_{i=1}^{4} w_i^{(r)} w_i^{(s)} = \delta_{rs} \tag{13}$$

By making use of Eqs. (10)–(13), we can rewrite Eq. (8) as

$$R_{ik} = \sum_{r=1}^{3} \lambda_r \left(\frac{q_i}{q_k}\right)^{1/2} w_i^{(r)} w_k^{(r)} \tag{14}$$

and through Eq. (7) we can give the matrix $P_{ik}(T)$ the representation

$$P_{ik}(T) = u_i^{(0)} v_i^{(0)} + \sum_{r=1}^{3} \exp(\lambda_r T) \left(\frac{q_i}{q_k}\right)^{1/2} w_i^{(r)} w_k^{(r)} \tag{15}$$

Let us now consider the two sequences B and C evolved from the common ancestor A after time T. The probability that at a given site one finds nucleotide i in the sequence of B and nucleotide j in C is given by

$$S_{ij}(T) = \sum_{k=1}^{4} P_{ik}(T) P_{jk}(T) q_k \tag{16}$$

where $S_{ij}(T)$ is obviously symmetric. On substituting Eq. (15) in Eq. (16) we obtain

$$S_{ij}(T) = q_i q_j + (q_i q_j)^{1/2} \sum_{r=1}^{3} \exp(2T\lambda_r) \mathbf{w}_i^{(r)} \mathbf{w}_j^{(r)} \tag{17}$$

This expression suggests the introduction of a new matrix

$$M_{ij}(T) = \frac{S_{ij}(T)}{(q_i q_j)^{1/2}} = (q_i q_j)^{1/2} + \sum_{r=1}^{3} \exp(2T\lambda_r) \mathbf{w}_i^{(r)} \mathbf{w}_j^{(r)} \tag{18}$$

Equation (18) relates the experimental data N_{ij} (see next section) to the evolution time T and to the structure of the rate matrix R through eigenvalues λ_r and the vectors $\mathbf{w}_i^{(r)}$. Finally, defining the average substitution rate per site for unit time

$$v_s = \sum_{\substack{i,k \\ i \neq k}} R_{ik} q_k \tag{19}$$

we get easily

$$v_s = -\sum_{i,r} q_i \mathbf{w}_i^{(r)} \mathbf{w}_i^{(r)} \lambda_r \tag{20}$$

Data Handling

The procedure to be followed to apply the method is as follows:

1. Align all the sequences under consideration.
2. Select one of the possible couples of sequences S1 and S2 having the common ancestral sequence S, with divergence time T.
3. Isolate L (allegedly) equivalent positions in each sequence (e.g., in a protein-coding sequence the first, second, or third silent codon position).
4. Compute the substitution matrix N_{ij} for the L positions by counting the number of time that nucleotide i in sequence S1 has become j in sequence S2.
5. Check the stationarity by computing $q_i(S1)$ and $q_i(S2)$, where q_i is the normalized frequency of the ith nucleotide ($i = $ A, C, G, T, $\Sigma_i q_i = 1$). Assuming a multinomial distribution, we form the χ^2 distribution per degree of freedom:

$$\chi^2(S1,S2) = \tfrac{1}{3} \sum_{i=1}^{4} \left\{ \frac{[q_i(S1) - \bar{q}_i]}{[\Delta q_i(S1)]^2} + \frac{[q_i(S2) - \bar{q}_i]}{[\Delta q_i(S2)]^2} \right\} \tag{21}$$

with $\bar{q}_i = [q_i(S1) + q_i(S2)]/2$ and $\Delta q_i = [q_i(1 - q_i)/L]^{1/2}$
We say the two sequences S1 and S2 satisfy the stationarity condi-

tion if $\chi^2 \leq 1.5$. Failure to meet this condition leads to the inapplicability of our analysis.

6. In our application it has always turned out that, once the stationarity condition is obeyed, the N_{ij} values are, within the statistical fluctuation arising from the finiteness of the length of sequences, symmetrical. Form the symmetrized matrix S_{ij} [Eq. (16)]:

$$S_{ij} = \frac{N_{ij} + N_{ji}}{2L} \qquad (22)$$

7. Compute from S_{ij} the matrix M_{ij} [Eq. (18)].
8. Diagonalize M_{ij} and obtain the eigenvalues Λ_r ($\Lambda_r = \exp(2T\lambda_r)$), $r = 1, 2, 3$) and the eigenvectors $w_i^{(r)}$.
9. In the analysis of p species we must get, within the statistical fluctuation,

$$\lambda_r = \frac{1}{2T_a} \log \Lambda_r^{(a)} = \frac{1}{2T_b} \log \Lambda_r^{(b)} = \cdots = \frac{1}{2T_p} \log \Lambda_r^{(p)} \quad (23)$$

where T_a and T_b are the divergence times of the ath and bth pairs, respectively. Fixing any single T_k value, from the most trustworthy set of paleontological data, calculate all other divergence times as

$$\left(\frac{T_m}{T_k}\right)_r = \frac{\log \Lambda_r^{(m)}}{\log \Lambda_r^{(k)}} \qquad m(\neq k) = 1, \ldots, p \qquad (24)$$

To obtain a more precise determination of the ratios T_m/T_k combine the three different determinations $(T_m/T_k)_r$ and their standard deviations σ_r, for each eigenvalue, with weights p_r, which are taken inversely proportional to σ_r.

10. Calculate, given the divergence time T, the rate matrix R_{ij} [Eq. (14)].
11. Compute the average substitution rate v_s per site per unit of time [Eq. (19)].
12. Calculate the statistical fluctuation related to our stochastic process by applying different simulation procedures generating matrices N_{ij} on the basis of the observed values. We give a schematic description for the Monte Carlo simulation method:
    a. Iterate this procedure for the column index $j = 1, 4$.
    b. Extract randomly a row value $i = k$ ($k = 1, 4$) and compute

$$L_j = \sum_{i=1}^{4} N_{ij} \qquad (25)$$

    c. Generate a set of three elements N_{ij} ($i \neq k$) according to a

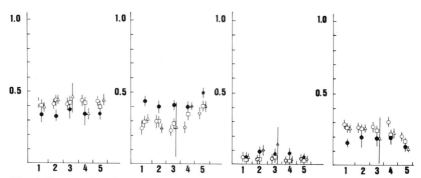

FIG. 1. Frequencies of the four bases (left to right: A, C, G, T) at the silent positions in the genes CoI (1), CoII (2), ATP6 (3), CoIII (4), and Cyt b (5) for rat ($\square$), mouse ($\bigcirc$), cow ($\triangle$), and human ($\bullet$). Error bars give the standard deviation. (Adapted from Lanave *et al.*[7])

Poisson law with parameter $\langle N_{ij} \rangle$ equal to the experimentally observed value.

d. Calculate N_{kj} from $L_j - \Sigma_{i \neq k} N_{ij}$.

e. Check if the element N_{kj} is negative. In this case the Monte Carlo event is to be rejected.

This procedure is repeated a large number of times, to obtain, for instance, 1000 simulated N_{ij}. On this sample of matrices N_{ij}, one performs the appropriate operations, thus obtaining averages and standard deviations.

Application

We have applied the method to various mitochondrial and nuclear genes.[7-9] We report here as examples the results obtained by analyzing two data sets.

Mitochondrial Genes

The third silent codon positions of five mitochondrial mammalian genes, coding the cytochrome oxidase subunits (CoI, CoII, CoIII), the 25-kDa ATPase subunit (ATP6), and the cytochrome b (Cyt b) subunit of four mammalian mitochondrial genomes [mouse (M), rat (R), cow (B), human (H)], were analyzed. In Fig. 1 we report the frequencies q_i of the four bases A, C, G, and T for the genes CoI, CoII, ATP6, CoIII, and Cyt b

[8] C. Lanave, G. Preparata, and C. Saccone, *J. Mol. Evol.* **21**, 346 (1985).

[9] C. Lanave, S. Tommasi, G. Preparata, and C. Saccone, *BioSystems* **19**, 273 (1986).

of rat, mouse, cow, and human. We observe that, whereas for the triplet rat, mouse, cow, the stationarity condition is verified and the q_i values, within statistical fluctuations, coincide, independently, of the genes considered, for humans the q_i values are again gene independent but the stationarity condition is never verified and the q_i values differ from those of the other three species. The observed gene independence of the q_i values suggests the usefulness of considering a "supergene" obtained by linking the genes one after the other. In this way we can substantially reduce the statistical errors affecting our analysis. In Table I we report the q_i values we obtained by analyzing the supergene in the four mammalian species. From now on, we consider only the supergenes.

In Table II we report, together with the statistical errors, the eigenvalues Λ_r for the comparisons of each of the possible pairs of species. The divergence times for the pairs rat–cow (T_{RB}) and mouse–cow (T_{MB}) are determined as follows. We report the determinations of the ratios T_{MB}/T_{RM} (Fig. 2A) and T_{RB}/T_{RM} (Fig. 2B) afforded by each of the three eigenvalues Λ_r, and by combining the different determinations $(T'/T)_r$, and their errors we get

$$\frac{T_{RB}}{T_{RM}} = 1.45 \pm 0.19$$

$$\frac{T_{MB}}{T_{RM}} = 1.35 \pm 0.16$$

which, keeping T_{RM} fixed at 35 million years (MY), yield

$$T_{RB} = 50.7 \pm 6.6 \quad \text{MY}$$

$$T_{MB} = 47.2 \pm 5.6 \quad \text{MY}$$

which lead to the phylogenetic tree shown in Fig. 3.

TABLE I

BASE FREQUENCIES AND STANDARD DEVIATIONS AT SILENT
POSITIONS IN "SUPERGENE" FOR FOUR MAMMALIAN SPECIES[a]

	Species			
Base	Rat	Mouse	Cow	Human
A	0.41 ± 0.02	0.43 ± 0.03	0.41 ± 0.03	0.34 ± 0.02
C	0.33 ± 0.02	0.27 ± 0.01	0.32 ± 0.03	0.43 ± 0.02
G	0.03 ± 0.01	0.03 ± 0.01	0.06 ± 0.03	0.06 ± 0.01
T	0.22 ± 0.01	0.26 ± 0.01	0.21 ± 0.03	0.17 ± 0.02

[a] Adapted from Lanave et al.[7]

TABLE II
EIGENVALUES Λ_r $(\Lambda_1 \leq \Lambda_2 \leq \Lambda_3)$ FROM COMPARISONS
OF "SUPERGENE" FOR FOUR MAMMALIAN SPECIES[a]

Pair	Λ_1	Λ_2	Λ_3
Rat–mouse	0.06 ± 0.04	0.15 ± 0.05	0.65 ± 0.03
Rat–cow	0.01 ± 0.04	0.10 ± 0.04	0.50 ± 0.04
Mouse–cow	0.03 ± 0.04	0.12 ± 0.04	0.51 ± 0.03

[a] Adapted from Lanave et al.[7]

The high degree of consistency between the data and the predictions of the stochastic model allows us to calculate, given one divergence time T, the rate matrix R_{ij} and the statistical error affecting it. Introducing T_{RB} and T_{MB}, we obtain the rate matrices for the rat–cow and the mouse–cow comparisons (Table III). It is worth noting that, owing to the low content of G in the supergene ($q_G = 0.04 \pm 0.02$), the statistical errors on the substitutions from G are necessarily quite large. However, from the data

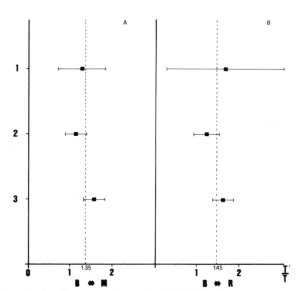

FIG. 2. Values for the divergence time ratios T'/T for (A) cow–mouse, and (B) cow–rat, rat–mouse obtained using the three different eigenvalues (1, 2, 3) taken in increasing order. The average value for R and M is equal to 1.40. (Adapted from Lanave et al.[7])

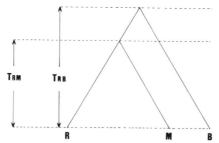

FIG. 3. Phylogenetic tree for rat, mouse, and cow. T_{RM} is the divergence time between rat and mouse, while T_{RB} denotes the divergence time between rodents and cow. (Adapted from Lanave *et al.*[7])

reported in Table III it emerges quite clearly that the rates of the transitions globally exceed those of transversions by at least one order of magnitude. Finally, we determine the average substitution rate v_s of the silent positions of the mitochondrial genes of mammalian to be

$$v_s = 1.4 \times 10^{-8} \qquad \text{substitution/site/year}$$

Nuclear Genes

The 2000-base pair $\psi\eta$-globin gene region of human, common chimpanzee, gorilla, orangutan, rhesus monkey, and owl monkey was analyzed.

TABLE III
RATE MATRICES R_{ik} OBTAINED FOR COMPARISONS RAT–COW (TOP)
AND MOUSE–COW (BOTTOM)[a,b]

	A	C	G	T
A	-5.39 ± 1.41	2.73 ± 1.63	16.31 ± 11.84	3.24 ± 2.62
	-5.79 ± 1.42	3.02 ± 1.33	19.34 ± 12.19	3.39 ± 1.66
C	2.10 ± 1.25	-11.17 ± 4.58	1.88 ± 12.05	12.35 ± 6.76
	2.09 ± 0.92	-12.47 ± 3.28	0.66 ± 8.45	12.05 ± 4.36
G	1.64 ± 1.22	0.22 ± 1.61	-20.46 ± 11.72	0.41 ± 2.41
	1.88 ± 1.23	0.10 ± 1.24	-23.08 ± 12.03	0.53 ± 1.71
T	1.52 ± 1.20	7.94 ± 4.24	1.89 ± 11.72	-16.34 ± 7.52
	1.83 ± 0.90	9.34 ± 3.30	3.07 ± 9.07	-15.95 ± 4.69

[a] Adapted from Lanave *et al.*[7]
[b] All values should be multiplied by 10^{-9}.

	A	C	G	T
A	549	7	40	6
C	6	340	3	39
G	44	7	424	11
T	10	28	13	513

Sites compared = 2040

Substitutions = 214

$\chi^2 = 0.069$

$\text{prob}(\chi^2) = 0.98$

Stationarity =VERIFIED

Human   : $q_A=0.295$ $q_C=0.190$ $q_G=0.238$ $q_T=0.276$
Owl monkey: $q_A=0.299$ $q_C=0.187$ $q_G=0.235$ $q_T=0.279$

FIG. 4. N_{ij} substitution matrix from the comparison between human and owl monkey sequences, also showing the stationarity check and the nucleotide frequencies for each species.

The sequence alignment used was that of Koop et al.[10] For this data set, owing to the pseudogene character of these sequences, we have made no distinction among the three codon positions, which should undergo the same evolutionary process.

Figure 4 shows, as an explanatory example, the N_{ij} matrix and the stationary check (expressed as χ^2 value) constructed from the comparison between human and owl monkey. The χ^2 values reported in Table IV show that the nucleotide sequences of this data set are highly stationary, always being less than 0.25. The phylogenetic tree (Fig. 5) can be reconstructed by following the procedure described in this previous paragraph, fixing to 14.5 MY the divergence time of orangutans with respect to the other primates. An average value of $1.44 \pm 0.31 \times 10^{-9}$ substitution/site/year was found for the rate of the nucleotide substitution process.

Conclusion

The purpose of this chapter has been to illustrate a method for determining some relevant parameters defining stochastic dynamics, which is assumed to operate in a large number of gene evolutionary processes. After presenting the theoretical basis of our approach, we pointed out that in order to have a general stochastic process amenable to mathematical analysis, it is absolutely necessary that in the course of the evolution the base compositions not become altered. This "stationarity condition" filters away a number of possible gene comparisons, but this seems to be the price that one must pay in order to obtain believable numbers. A few significant examples have been presented to make the case that such a price is a reasonable one. We have also emphasized the importance of a correct

[10] B. F. Koop, M. Goodman, P. Xu, K. Chan, and J. L. Slightom, Nature (London) 319, 234 (1985).

TABLE IV
χ^2 VALUES FOR CHECKING THE STATIONARITY CONDITION OF
PRIMATE PSEUDOGENES

	Common chimp	Gorilla	Orangutan	Rhesus monkey	Owl monkey
Human	0.025	0.0808	0.0493	0.0618	0.0649
Common chimp		0.0208	0.1390	0.1206	0.0916
Gorilla			0.2307	0.1702	0.1652
Orangutan				0.0948	0.1815
Rhesus monkey					0.0863

statistical analysis, for otherwise the numbers fluctuate too much, in a noncontrollable way. The restrictiveness of the "stationarity condition" clearly indicates that the full potential of the method described will only be exploited when the information on gene sequences becomes much more extensive.

Acknowledgments

This work was supported by grants from Ministers Pubblica Istituzione, Italy, and Progetto Finalizzato Biotecnologie e Biostrumentazione, CNR, Italy.

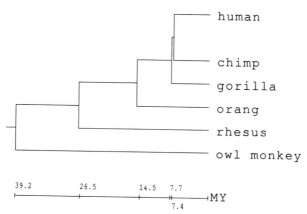

FIG. 5. Phylogenetic tree for the 2000-base pair $\psi\eta$-globin data set calibrated by assuming a divergence time of orangutans from African apes and humans of 14.5 million years (MY).

[36] Maximum Likelihood Methods

By Naruya Saitou

General Principle

Application of the maximum likelihood (ML) method to the problem of phylogenetic tree reconstruction was first studied for the case of gene frequency data.[1] Later, an ML algorithm for constructing unrooted phylogenetic trees from nucleotide sequence data was developed by Felsenstein.[2] Recently, Saitou[3] proposed a stepwise tree-searching algorithm for the ML method. This is similar to that of the neighbor-joining method,[4] in which distance matrices are used.

Let us first explain the general principle of the ML method for nucleotide sequence data. Consider the tree in Fig. 1. Let us ignore the root R and consider the tree as an unrooted tree. We first restrict our attention to a specific nucleotide site, and assume that nucleotide N_i was observed at sequence (node) i ($i =$ A, B, C, D, or E). On the other hand, nucleotide N_j at node j ($j =$ X, Y, or Z) is unknown, and it can be one of four nucleotides. Thus, the likelihood (L) for this site becomes

$$L = \sum_{N_Y} \left[g_Y P_{YC} \left(\sum_{N_X} P_{YX} P_{XA} P_{XB} \right) \left(\sum_{N_Z} P_{YZ} P_{ZD} P_{ZE} \right) \right] \tag{1}$$

where g_Y is the probability that node Y has nucleotide N_Y, $P_{ij} \equiv P(N_i, N_j, d_{ij})$ is the probability of observing nucleotides N_i and N_j in sequences i and j, respectively, and d_{ij} is the expected number of nucleotide substitutions between these two sequences. Summation is for four possible nucleotides, because N_X, N_Y, and N_Z are variables. To obtain $P(N_i, N_j, d_{ij})$, we must specify the pattern of nucleotide substitution. If we use the random substitution model (the one-parameter model),[5]

$$P(N_i, N_j, d_{ij}) = \tfrac{1}{4} + (\tfrac{3}{4}) \exp(-4d_{ij}/3) \qquad \text{(if } N_i = N_j) \tag{2a}$$

$$P(N_i, N_j, d_{ij}) = \tfrac{1}{4} - (\tfrac{1}{4}) \exp(-4d_{ij}/3) \qquad \text{(if } N_i \neq N_j) \tag{2b}$$

[1] L. L. Cavalli-Sforza and A. W. F. Edwards, *Am. J. Hum. Genet.* **19,** 233 (1967).

[2] J. Felsenstein, *J. Mol. Evol.* **17,** 368 (1981).

[3] N. Saitou, *J. Mol. Evol.* **27,** 261 (1988).

[4] N. Saitou and M. Nei, *Mol. Biol. Evol.* **4,** 406 (1987).

[5] T. H. Jukes and C. R. Cantor, *in* "Mammalian Protein Metabolism" (H. N. Munro, ed.), p. 21. Academic Press, New York, 1969.

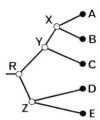

FIG. 1. Phylogenetic tree for five nucleotide sequences.

If we assume the two-parameter model,[6] in which transitions and transversions can occur at different rates, the expressions that correspond to Eqs. (2a) and (2b) are somewhat complicated [see Eqs. (9a)–(9c)]. It is also true when other kinds of data such as amino acid sequences[7] or restriction site data[8] are considered. However, the essential nature of the likelihood function remains the same.

In any case, the likelihood for each nucleotide site defined by Eq. (1) is then multiplied for all sites and computed for many combinations of branch lengths for a given tree topology, and the combination that shows the highest likelihood is chosen as the ML solution. It has been proved that there exists a single ML point in the possible parameter range for a given tree topology when the one-parameter model of nucleotide substitution was assumed.[9]

For the case of a rooted tree, it is necessary to assume a constant rate of nucleotide substitution. For example, the following constraints should be invoked for the tree in Fig. 1: $d_{XA} = d_{XB}$, $d_{ZD} = d_{ZE}$, $d_{YC} = d_{YX} + d_{XA}$, and $d_{RY} + d_{YC} = d_{RZ} + d_{ZD}$. Because of this, the number of parameters to be estimated for n sequences is $n - 1$ for rooted trees, compared to $2n - 3$ parameters (branch lengths) for unrooted trees.

Each nucleotide site is considered separately in Felsenstein's method.[2] When each site is assumed to evolve at the same evolutionary rate, however, a more essential unit of comparison for the ML method is the "nucleotide configuration." A nucleotide configuration is the distribution pattern of nucleotides for a given set of sequences (see Tables I and II for examples). When the one-parameter model is assumed, the possible number of configurations for n sequences is $(4^{n-1} + 3 \times 2^{n-1} + 2)/6$.[10] There

[6] M. Kimura, *J. Mol. Evol.* **16**, 111 (1980).
[7] R. L. Kashyap and S. Subas, *J. Theor. Biol.* **47**, 75 (1974).
[8] P. E. Smouse and W.-H. Li, *Evolution* **41**, 1162 (1987).
[9] K. Fukami and Y. Tateno, *J. Mol. Evol.* **28**, 460 (1989).
[10] N. Saitou and M. Nei, *J. Mol. Evol.* **24**, 189 (1986).

TABLE I
NUCLEOTIDE CONFIGURATIONS FOR
THREE SEQUENCES

Configuration[a]			Observed no.	
No.	A	B	C	Case a	Case b
1	i	i	i	789	40
2	i	i	j	98	20
3	i	j	i	59	15
4	j	i	i	50	5
5	i	j	k	4	20
				1,000	100

[a] A, B, and C are different sequences, and i, j, and k are nucleotides that are different from each other.

TABLE II
NUCLEOTIDE CONFIGURATIONS FOR
FOUR SEQUENCES

Configuration[a]				
No.	A	B	C	D	Observed no.
1	i	i	i	i	196
2	i	i	i	j	67
3	i	i	j	i	40
4	i	j	i	i	77
5	j	i	i	i	22
6	i	i	j	j	12
7	i	j	i	j	5
8	i	j	j	i	8
9	i	i	j	k	11
10	i	j	i	k	21
11	j	i	i	k	10
12	j	k	i	i	12
13	j	i	k	i	3
14	i	j	k	i	15
15	i	j	k	l	1
					500

[a] A, B, C, and D are sequences of Fig. 4, and i, j, k, and l are different nucleotides.

are 5, 15, and 51 configurations for three, four, and five sequences, respectively.

If we consider the nucleotide configuration, the general formula for the likelihood becomes

$$L = \prod_{i=1}^{c} U_i^{m_i} \times \frac{m!}{m_1! m_2! \ldots m_c!} \tag{3}$$

where c is the number of possible nucleotide configurations, U_1 the probability of obtaining the ith configuration, m_i the observed number of the ith configuration, and m the sum of m_i values, or the total number of nucleotides compared. Because the number of configurations depends on the number of sequences compared, we consider the cases of two, three, four, and five sequences separately. The case of unrooted trees is first considered.

Two Sequences

When we have only two sequences, there are only one unrooted and one rooted trees. Thus the ML solution for the rooted tree becomes identical to that for the unrooted tree.

Let us consider the one-parameter model of nucleotide substitution. There are only two configurations for this case: two nucleotides are the same (configuration 1) or different each other (configuration 2). Therefore, Eq. (3) becomes

$$L = U_1^{m_1} \times U_2^{m_2} \times \frac{(m_1 + m_2)!}{m_1! m_2!} \tag{4}$$

where $U_1 = 1/4 + (3/4) \exp(-4d_{ij}/3)$ [see Eq. (2a)], and $U_2 = 3/4 - (3/4) \exp(-4d_{ij}/3)$ [see Eq. (2b)]. First, we take the logarithm,

$$\log L = m_1 \log[(1 + 3x)/4] + m_2 \log[(3 - 3x)/4] + \text{constant} \tag{5}$$

where $x = \exp(-4d_{ij}/3)$. To obtain the maximum likelihood estimate of d_{ij}, we differentiate $\log L$ with respect to x, and equate it to zero. This gives

$$\frac{-4m_1 x}{1 + 3x} + \frac{4m_2 x}{3 - 3x} = 0 \tag{6}$$

Solving Eq. (6) (note that $x \neq 0$ while d_{ij} is finite),

$$\hat{x} = 1 - 4m_2/3(m_1 + m_2) \tag{7}$$

Hence,

$$d_{ij} = -(3/4) \log(1 - 4\pi_{ij}/3) \tag{8}$$

where $\pi_{ij} = m_2/(m_1 + m_2)$. Equation (8) is identical to the famous formula of Jukes and Cantor.[5]

When the two-parameter model[6] is assumed, we have to distinguish transitional and transversional differences. Thus, the number of configurations becomes three. Probabilities U_i $(i = 1, 2, 3)$ are given by

$$U_1 = 1/4 + p^2/4 + pq/2 \qquad \text{(no difference)} \qquad \text{(9a)}$$
$$U_2 = 1/4 + p^2/4 - pq/2 \qquad \text{(transitional difference)} \qquad \text{(9b)}$$
$$U_3 = 1/2 - p^2/2 \qquad \text{(transversional difference)} \qquad \text{(9c)}$$

where $p = \exp(-2\beta)$ and $q = \exp(-2\alpha)$. α and β are the expected numbers of transitions and transversions, respectively. A likelihood function can be obtained using these U_i values, and we equate the differentiations of the log likelihood with p and q to be zero, as in the case of the one-parameter model. We solve the two equations,

$$\hat{p} = (1 - 2m_3/m)^{1/2} \qquad \text{(10a)}$$
$$\hat{q} = [1 - (2m_2 + m_3)/m]/\hat{p} \qquad \text{(10b)}$$

where m_i is the observed number of the ith configuration and $m = m_1 + m_2 + m_3$. Thus, the total number of nucleotide substitutions $d \equiv \alpha + 2\beta$ is estimated by

$$\hat{d} = -(1/2) \log(\hat{p}^2 \hat{q})$$
$$= -(1/2) \log[(1 - 2P - Q)(1 - 2Q)^{1/2}] \qquad \text{(11)}$$

where $P = m_2/m$ and $Q = m_3/m$. Equation (11) is equivalent to Kimura's estimation formula of evolutionary distance.[6]

It is hypothesized that a similar correspondence between the ML estimate and the distance formula above can be applied to other types of nucleotide substitution models (see this volume, [33] for other models). If this conjecture is true, the ML method is identical with the distance method for the case of two sequences.

Three Sequences

Unrooted Trees

There is only one unrooted tree for three sequences (Fig. 2a). Let us assume the one-parameter model of nucleotide substitution. There are five nucleotide configurations (Table I). Probabilities U_i for a given set of three distances $(d_{XA}, d_{XB}, \text{ and } d_{XC})$ are given by

$$U_1 = 4h(i,i,i), \qquad \text{(12a)}$$
$$U_2 = 12h(i,i,j) \qquad (i \neq j) \qquad \text{(12b)}$$

$$U_3 = 12h(i,j,i) \qquad (i \neq j) \tag{12c}$$

$$U_4 = 12h(j,i,i) \qquad (i \neq j) \tag{12d}$$

$$U_5 = 24h(i,j,k) \qquad (i \neq j \neq k) \tag{12e}$$

where i, j, and k are any one of the four nucleotides and

$$h(i,j,k) = \sum_X g_x P_{xi}(d_{XA}) P_{xj}(d_{XB}) P_{xk}(d_{XC}) \tag{13}$$

where g_x is the probability of observing nucleotide x at node X and $P_{xi}(d_{XA}) = P(x,i,d_{XA})$ [see Eqs. (2a) and (2b)]. By substituting these U_i values into Eq. (3), the likelihood function is obtained.

An analytical solution for the ML value is not easily found, but the ML value can be obtained numerically by changing three distances (d_{XA}, d_{XB}, and d_{XC}). In this case, however, the ML estimates of these distances are no longer the same as those obtained by the distance method,[11] in which case the branch length estimates are given by:

$$\hat{d}_{AX} = (d_{AB} + d_{AC} - d_{BC})/2 \tag{14a}$$

$$\hat{d}_{BX} = (d_{AB} + d_{BC} - d_{AC})/2 \tag{14b}$$

$$\hat{d}_{CX} = (d_{AC} + d_{BC} - d_{AB})/2 \tag{14c}$$

where d_{AB}, d_{AC}, and d_{BC} are given using Eq. (8), which is the ML estimate of each distance value. In applying Eq. (8) the proportion of nucleotide differences (π_{ij}) between sequences i and j is estimated by

$$\pi_{AB} = (m_3 + m_4 + m_5)/m \tag{15a}$$

$$\pi_{AC} = (m_2 + m_4 + m_5)/m \tag{15b}$$

$$\pi_{BC} = (m_2 + m_3 + m_5)/m \tag{15c}$$

It is clear from Eqs. (15a)–(15c) that different sets of m_i values ($1 \leq i \leq 5$) can give the same set of π_{AB}, π_{AC}, and π_{BC} values. Thus, the estimates of d_{XA}, d_{XB}, and d_{XC} by the distance method are not the ML estimates. However, it has been shown that the ML estimates of branch lengths are quite similar to those obtained by the distance method.[12]

Rooted Trees

There are three rooted trees for three sequences, and we have two parameters (d_1 and d_2) to be estimated (see Fig. 2b). Let us designate the tree in Fig. 2b as "tree 1," in which sequences A and B are closer to each other than to C. As for the other two possible trees, sequences A and C are closer in tree 2, and B and C are closer in tree 3.

[11] W. M. Fitch and E. Margoliash, *Science* **155**, 279 (1967).
[12] N. Saitou, Ph.D. dissertation, University of Texas at Houston, 1986.

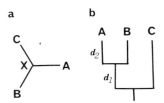

FIG. 2. (a) Unrooted tree and (b) rooted tree for three sequences.

Assuming the one-parameter model, expressions for U_i for tree 1 become

$$U_1 = (A + Ba^2)/16 \qquad (16a)$$
$$U_2 = (3A - Ba^2)/16 \qquad (16b)$$
$$U_3 = (C + Da^2)/16 \qquad (16c)$$
$$U_4 = U_3 \qquad (16d)$$
$$U_5 = 2(C - Da^2)/16 \qquad (16e)$$

where $A = 1 + 3b^2$, $B = 6(1 + b)b^2$, $C = 3(1 - b^2$, $D = 6(1 - b)b^2$, and $a = \exp(-4d_1/3)$ and $b = \exp(-4d_2/3)$.[3] [Equation (16e) corresponds to Eq. (4e) of Saitou,[3] which was incorrect.] From these probabilities, one can evaluate the likelihood, $L(1)$, of tree 1 under specific d_1 and d_2 values. Computation of $L(2)$ and $L(3)$ for trees 2 and 3, respectively, is done in a similar manner.

Saitou[3] derived the conditions for obtaining tree 1 as the ML estimate, or for obtaining the relationship $L(1) > L(2)$ and $L(1) > L(3)$, i.e., $m_2 > m_3$ and $m_2 > m_4$, where m_i is the observed number of the ith nucleotide configuration (see Table I). When this condition is satisfied, d_{AB} becomes the smallest among d_{AB}, d_{AC}, and d_{BC}. Thus, tree 1 is chosen if we use UPGMA (unweighted pair-group method),[13] in which the pair of sequences with the smallest distance is first clustered. This means that the topology of the UPGMA tree is always identical with that of the maximum likelihood tree, though the estimates of branch lengths (d_1 and d_2) may be different to some extent.

Figure 3 shows two examples of likelihood surfaces for artificial data (Cases a and b of Table I).[3] In Case a, the above condition ($m_2 > m_3$ and $m_2 > m_4$) is satisfied and tree 1 has the ML value at $d_1 = 0.0259$, whereas the trifurcating tree ($d_1 = 0$) gives the ML value for the other two trees (Fig.

[13] R. Sokal and P. H. A. Sneath, "Principles of Numerical Taxonomy." Freeman, San Francisco, California, 1963.

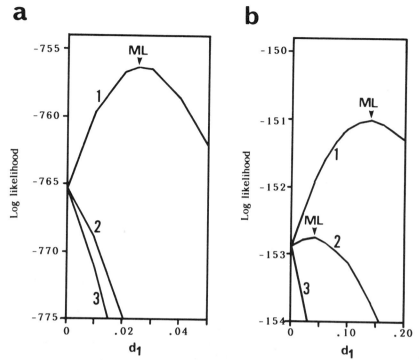

FIG. 3. Two examples of likelihood surfaces for three possible trees. (a) and (b) correspond to Cases a and b in Table I. See text for details.

3a). The likelihood surfaces represent the ML values for given d_1 values (an ML value was computed by varying d_2 for a given d_1 value).

The above condition is also satisfied in Case b, and tree 1 gives the highest ML value. However, tree 2 also gives an ML value higher than that for the trifurcating tree (Fig. 3b). In this case the following relationship is satisfied:

$$d_{AC} < (d_{AB} + d_{BC})/2 \qquad (17)$$

Thus, if sequences A and C are first clustered (tree 2), a positive estimate of d_1 is obtained. This situation is different from that for Case a, in which a negative estimate of d_1 is obtained if tree 2 is considered. It seems that a positive estimate of d_1 corresponds to the ML value higher than that for the trifurcating tree.[3] For tree 3, however, d_1 becomes negative, and the log-likelihood value decreases as the value of d_1 increases (Fig. 3b).

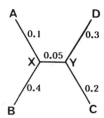

FIG. 4. Unrooted tree for four sequences. Numbers are lengths for each branch.

Four Sequences

Unrooted Trees

There are three unrooted trees for four sequences, and we have to consider 15 nucleotide configurations (Table II). The likelihood function (not shown) can be obtained in a manner similar to the case of three sequences.

Let us consider a numerical example. This is a result based on a simulated data set (see Table II), assuming the tree of Fig. 4.[3] The tree in which sequences A and B are clustered (i.e., the true tree as in Fig. 4) had the highest ML value (log likelihood is -1004.2) with $\hat{d}_{XY} = 0.026$. The estimates of other branch lengths were $\hat{d}_{XA} = 0.110$, $\hat{d}_{XB} = 0.361$, $\hat{d}_{YC} = 0.188$, and $\hat{d}_{YD} = 0.303$. On the other hand, the same tree was obtained when we applied the neighbor-joining method for a distance matrix. Branch length estimates were $\hat{d}_{XY} = 0.033$, $\hat{d}_{XA} = 0.101$, $\hat{d}_{XB} = 0.366$, $\hat{d}_{XC} = 0.179$, and $\hat{d}_{XD} = 0.316$, which were close to the estimates obtained by the ML method. The corresponding log-likelihood value for these estimates was -1004.5, slightly lower than that of the ML estimate.

The maximum likelihood value for the other two unrooted trees was obtained for the case with no interior branch (a quadrifurcating tree). This is analogous to the trifurcation of the rooted tree for three sequences. However, other trees may have ML values with positive estimates of the interior branch.

Rooted Trees

The number of possible rooted trees for four sequences is 15, and 3 of them (trees 3a, 3b, and 3c) are shown in Fig. 5. U_i of Eq. (3) is computed in a manner similar to the case of three sequences, and the likelihood function can be obtained using these U_i values.[3]

Although the ML method and UPGMA may no longer give the same tree on every occasion, it has been shown that the tree estimated by the ML

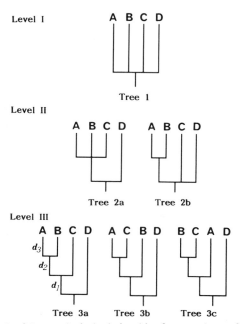

FIG. 5. Three levels of the topological relationships for rooted trees for four sequences.

method is similar to that obtained by UPGMA.[3] In this connection, it should be noted that UPGMA gives least-squares estimates of branch lengths for the tree obtained.[14] That is, UPGMA minimizes the quantity, $\Sigma(d_{ij} - 2\lambda t_{ij})^2$, where λ is the rate of nucleotide substitution and t_{ij} is the time since divergence between sequences i and j.

Let us explain Saitou's algorithm[3] using Fig. 5. This algorithm is reminiscent of the neighbor-joining method. First the ML value for tree 1 (Level I) is computed. There is no information on the clustering of sequences with tree 1, whereas there are trifurcating and bifurcating points in trees at the next step (Level II). The ML values for all 10 trees at Level II are computed and are compared with each other. We choose the tree with the highest ML value among the 10. Suppose it is tree 2a. We then compute the ML values for trees 3a, 3b, and 3c, which are produced when the trifurcating point of tree 2a is resolved. In any case, the tree with the highest ML value is chosen as the final tree. In this way, we may be able to find the tree with the highest ML value. When the number of sequences is

---

[14] R. Chakraborty, *Can. J. Genet. Cytol.* **19,** 217 (1977).

large, this algorithm requires much less computational time compared with tree-by-tree examination.

The two trees (2a and 2b) at Level II are related to tree 3a at Level III, and they may be used as null trees to test the significance of d_1 and d_2 of tree 3a. If d_1 is not significantly greater than zero, we assume that tree 2a is the correct one and proceed to test the significance of d_2 with tree 1 as the null tree. In general, there are $n - 1$ levels for trees with n sequences, and we can successively test the significance of all interior branches of a tree.

Five Sequences

There are 15 unrooted trees and 105 rooted trees for five sequences, and the number of possible nucleotide configurations under the one-parameter model of nucleotide substitution is now 51. Construction of the likelihood function is straightforward as before. We consider only unrooted trees.

Algorithm for Finding ML Tree

Let us explain Saitou's algorithm[3] for the ML method for unrooted trees, with a numerical example of five nucleotide sequences of higher primates (human, common chimpanzee, pygmy chimpanzee, gorilla, and orangutan).[15] As in the case of rooted trees, we first compute the ML estimate for the Level I tree of Fig. 6. The initial value (d_{xi}) for the length of the branch between node X and sequence i may be computed from the distance matrix, applying the neighbor-joining method as follows:[3]

$$d_{xi} = \sum_{j=1}^{5} d_{ij} - \frac{3}{4} \sum_{j<k} d_{jk} \qquad (18)$$

The maximum log-likelihood value for the Level I tree was -638 for the example data. Ten different trees (Level II trees) are then considered, and the ML value for each tree is computed. These trees have one trifurcation and one bifurcation. The clustering of common and pygmy chimpanzees gave the highest log-likelihood value of -591. From each tree at Level II, three trees are produced if the trifurcation is resolved (Level III). The ML estimates of six branch lengths for the case of Level II are used as the initial values for the computation of the ML estimation for a tree at Level III (see Fig. 6). In this case, the seven-dimensional likelihood surface is numerically examined for three possible trees. The log-likelihood values were -586 for tree 1 (chimpanzee and gorilla clustered), -584 for tree 2

[15] J. Hixson and W. M. Brown, *Mol. Biol. Evol.* **3**, 1 (1986).

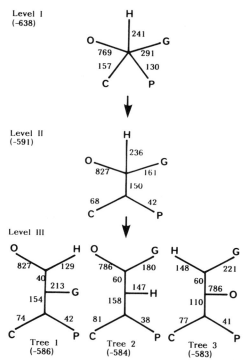

FIG. 6. Three steps to obtain the maximum likelihood tree for the data of Hixson and Brown.[16] H, Human; C, common chimpanzee; P, pygmy chimpanzee; G, gorilla; O, orangutan. Estimated branch lengths for each tree should be divided by 10,000. Figures in parentheses are the maximum log-likelihood values.

(human and chimpanzee clustered), and −583 for tree 3 (human and gorilla clustered). Thus, tree 3 was found to be the ML tree under this algorithm. This new algorithm can be extended to any number of sequences, as in the case for rooted trees.

Comparison with Other Methods

When we applied other tree-making methods to the same data set, the choice of the tree depended on the method used.[3] UPGMA chose tree 1; the neighbor-joining method found tree 3 as the best tree, and most of the branch lengths estimated by this method were quite similar to those of the ML method. On the other hand, the distance Wagner method[16] chose tree 2 when the proportion of different nucleotides was used as distance.

[16] J. Farris, *Am. Nat.* **106**, 645 (1972).

TABLE III
RESULTS OF FOUR METHODS FOR PRIMATE SEQUENCE DATA[a]

Method	Tree 1	Tree 2	Tree 3	Tree 4
Fitch–Margoliash	0	+0.60	+0.47	+25.33
Minimum evolution	+1.10	+0.35	0	+6.59
Maximum parsimony	+1	0	0	+8
Maximum likelihood	−3.97	−2.98	0	−33.86

[a] Values for the Fitch–Margoliash and minimum evolution methods are percent standard deviation and sum of branch lengths × 1000, respectively. Values for the maximum parsimony method are the required number of nucleotide substitutions, and those for the maximum likelihood method are the log likelihood. Values of the best tree are set to zero, and the other values represent differences from the best tree. From Saitou and Imanishi.[17]

Phylogenetic trees are constructed step by step for methods such as UPGMA, the neighbor-joining method, and the distance Wagner method, and finally a single tree is obtained. This type of method can be called a "stepwise" clustering method.[17] In contrast, all possible trees (or a limited number of plausible trees) are compared under a certain criterion for Fitch–Margoliash method,[11] the maximum parsimony method,[18] the minimum evolution method,[17] and the standard ML method.[2] Results of these methods applied for the same data set as above are presented in Table III.[17] Trees 1–3 in Table III correspond to those of Fig. 6, whereas pygmy chimpanzee and human as well as gorilla and orangutan are clustered in tree 4.

The program DNAML of PHYLIP Version 3.1, developed by J. Felsenstein, was used for the ML method. The transition/transversion ratio was set to be 5.0, and observed frequencies of nucleotides were used. Trees 1–4 were examined using the "user tree" option. Because a different model of nucleotide substitution was used in this case, the log-likelihood values for trees 1–3 are different from those obtained using Saitou's algorithm (Fig. 6). However, the order of the maximum log-likelihood values among the three trees is the same as that obtained above, and tree 3 was chosen. The ML value for tree 4 was considerably lower than those for the other three trees. This is apparently because the highly probable clustering of common and pygmy chimpanzees was not realized in this tree.

As for the other methods, trees 1 and 3 were chosen by the method of

[17] N. Saitou and T. Imanishi, *Mol. Biol. Evol.* **6**, 514 (1989).
[18] W. M. Fitch, *Am. Natur.* **111**, 223 (1977).

Fitch–Margoliash and the minimum evolution method, respectively. Interestingly, the rank of these trees in terms of the minimum evolution method was the same as that of the ML method (see Table III). Finally, the maximum parsimony method found both trees 2 and 3 as equally parsimonious. These inconsistent results indicate that the nucleotide sequence data used are not sufficient to determine the branching order among human, chimpanzee, and gorilla. This seems to be consistent with the simulation result of Saitou and Nei.[10]

Discussion

The original formulation of the ML method[1] included the probability of tree topology, which is given by assuming a Yule process. Felsenstein[2] took a different approach, in which the ML value for each tree is computed and the tree with the highest ML value is chosen. However, the likelihood function to be used varies from tree to tree, so the ML values for different trees are conditional and cannot be compared in the standard way.[3,19] It has been claimed[20] that this problem can be avoided by applying information theory. Yet there still remains a paradox.

Let us consider Fig. 3b. In this case both trees 1 and 2 had higher ML values than that for the trifurcating tree where d_1 is necessarily zero. Because this trifurcating tree is a submodel of either tree 1 or 2, we may use the likelihood ratio test for the significance of these two trees. As Felsenstein[21] indicated, however, it is possible that both trees 1 and 2 are significantly better than the trifurcating tree. Then which tree should we choose? This paradox seems to apply in trees 1–3 of Fig. 6, since the log-likelihood values for all these trees are considerably larger than that of Level II tree. Therefore, we should be cautious in applying the ML method for the problem of phylogenetic tree estimation.

Nevertheless, the ML method may still be useful for practical purposes. For example, it has been shown by computer simulations[3,22] that the ML method can find the correct tree with an appreciable proportion even when the maximum parsimony method is positively misleading under certain conditions for four sequences.[23] Saitou and Imanishi[17] compared the relative efficiency of the five tree-making methods (maximum parsimony, maximum likelihood, Fitch–Margoliash, minimum evolution, and neigh-

[19] M. Nei, "Molecular Evolutionary Genetics." Columbia Univ. Press, New York, 1987.
[20] T. Kishino and M. Hasegawa, *J. Mol. Evol.* **29**, 170 (1989).
[21] J. Felsenstein, *in* "Statistical Analysis of DNA Sequence Data" (B. S. Weir, ed.), p. 133. Dekker, New York, 1983.
[22] M. Hasegawa and T. Yano, *Bull. Biomed. Soc. Jpn.* **5**, 1 (1984).
[23] J. Felsenstein, *Syst. Zool.* **27**, 401 (1978).

bor-joining methods). They used a computer simulation under the model tree for six sequences, and showed that the neighbor-joining method, the minimum evolution method, and the ML method performed better than Fitch–Margoliash and maximum parsimony methods.

From a practical point of view, however, a distance matrix method, such as the neighbor-joining method, seems to be the first choice for determining tree topology. The maximum likelihood method is time consuming, and should better be used after a certain number of prospective trees are chosen by some distance methods. For a review of distance matrix methods for phylogenetic tree construction, readers may refer to Nei[19] and Saitou.[24]

[24] N. Saitou, in "Handbook of Statistics, Volume 8: Statistical Methods for Biological and Medical Sciences" (C. R. Rao and R. Chakraborty, eds.). North-Holland, New York, in press.

Section VII

Phylogenetic Trees

[37] Maximum Parsimony Approach to Construction of Evolutionary Trees from Aligned Homologous Sequences

By John Czelusniak, Morris Goodman, Nancy D. Moncrief, and Suzanne M. Kehoe

Introduction

One of the most effective principles used for constructing evolutionary trees from protein and DNA sequence data is that of minimum evolution[1,2] of maximum parsimony,[3-5] also appropriately called maximum homology.[6] A maximum parsimony tree accounts for the evolutionary descent or related sequences by the fewest possible genic changes. Such a tree maximizes the genetic likenesses associated with common ancestry while minimizing the incidence of convergent mutations. Since common ancestry rather than convergent evolution is the most probable explanation for extensive interspecies matches of either nucleotide sequences or encoded amino acid sequences, algorithms that search for evolutionary trees by the maximum parsimony or maximum homology principle provide a way to find from sequence data the best supported genealogical hypotheses.

The correct evolutionary tree is a rooted tree that graphically depicts the cladistic relationships which exist among the operational taxonomic units (OTUs), i.e., in our case contemporary sequences. As defined in Moore *et al.*[5] a tree consists of nodes connected by lines, and a rooted tree starts from the root, which is the node ancestral to all other nodes. Each ancestral node gives rise to two descendant nodes. Terminal or exterior nodes have no further descendants; these nodes correspond to OTUs. All nonroot, nonexterior nodes are interior nodes; they correspond to ancestral sequences such as may be inferred from the contemporary sequences by the maximum parsimony principle. The connecting line between two adjacent nodes is a link in the tree. The number of nonmatching sites

[1] A. F. W. Edwards and L. L. Cavalli-Sforza, *Ann. Hum. Genet.* **27**, 104 (1963).

[2] E. Zuckerkandl, *Protides Biol. Fluids* **12**, 102 (1964).

[3] J. S. Farris, *Syst. Zool.* **19**, 83 (1970).

[4] W. M. Fitch, *Syst. Zool.* **20**, 406 (1971).

[5] G. W. Moore, J. Barnabas, and M. Goodman, *J. Theor. Biol.* **38**, 459 (1973).

[6] G. W. Moore, *in* "Molecular Anthropology" (M. Goodman and R. E. Tashian, eds.), p. 117. Plenum, New York, 1976.

between the aligned sequences of the two adjacent nodes is the length of that link. The number of these sequence changes counted over all links constitutes the length of the tree.

Calculation of tree length is simplified by removing the root from the tree. Such an unrooted tree or network still retains the interior nodes and the exterior nodes (the OTUs). A network of N exterior nodes has $N - 2$ interior nodes and $2N - 3$ links. Thus a network of N OTUs can be converted to $2N - 3$ dendrograms, since a dendrogram can be created by putting the root on any one of the network's links. The maximum parsimony procedure can reconstruct ancestral sequences for each interior node of a tree but cannot determine which interior node or which pair of adjacent interior nodes is closest to the root. However, after the search for the maximum parsimony network is terminated, one can use the full weight of evidence on the cladistic relationships of the OTUs to assign a root to the whole maximum parsimony tree or, at least, to some of its branches.

The problem of finding the maximum parsimony tree can be broken down into two parts. The first part proved to be easy and was solved by Fitch[4] for homologous nucleotide sequences. The algorithm requires as input data both the OTUs, which are contemporary homologous nucleotide sequences already aligned against one another (see Section V of this volume for alignment algorithms), and the instructions for a tree or dendrogram specifying any one of the possible dichotomous branching orders for the OTUs; the algorithm then constructs maximum parsimony ancestral sequences at the interior nodes of the tree and calculates its minimum NR (nucleotide replacement) length. Proofs that Fitch's algorithm does calculate the minimum NR length of any given bifurcating tree for a set of aligned nucleotide sequences were provided independently by Hartigan,[7] Sankoff,[8] and Moore et al.[5] Moreover, Moore et al.[5] presented a maximum parsimony algorithm which calculated mRNA sequence ancestors from contemporary amino acid sequences by generalizing the Fitch procedure for nucleotides to codons. The algorithm of Moore et al.,[5] which we use in our computer programs for constructing maximum parsimony trees from amino acid sequence data, accurately accounts for the degeneracy of the genetic code and is guaranteed by mathematical proof to give correct solutions.

As just reviewed, the Fitch algorithm for nucleotide sequences[4] and the Moore et al. extension of this algorithm to amino acid sequences[5] have

[7] J. A. Hartigan, *Biometrics* **29**, 53 (1973).
[8] D. Sankoff, *Publ. Cent. Rech. Math. Tech. Rep.* **262.** University of Montreal, Montreal, 1973.

solved the first and easy part of the maximum parsimony problem, which is to calculate the minimum NR length of any given tree among the unrooted trees that the OTUs can possibly form. The second part of the maximum parsimony problem is the hard part. It is to find among all these possible unrooted trees that tree or minute subset of trees at the lowest NR length. Proving that one has found the tree or trees of lowest NR length is not possible in actual practice for data sets with more than 10 OTUs because the only known way to find the lowest NR length among different trees is to determine in the total set of trees the NR length of each tree. Such a brute force approach, even with supercomputers, cannot work when impossibly large numbers of unrooted trees have to be examined. A 10-OTU network itself has 2,027,025 possible alternative branching arrangements while a 20-OTU network has over 10^{20} possible alternative arrangements. Since there is no proven procedure for restricting to manageable proportions the number of trees to be examined in the search for the tree or trees of lowest length when there are astronomical numbers of alternatives, we employ an heuristic strategy in which an initial tree is iteratively improved (i.e., converted to lower length trees) by branch swapping.[9] We stop the search for the maximum parsimony tree when further branch swapping after extensive trials with different starting dendrograms fails to find any new trees of lower NR length.

Because the lowest lengths found may only represent a local minimum and not the global minimum, the effectiveness of the maximum parsimony approach for inferring phylogenetic relationships among the OTUs in large data sets has been questioned. Rather than carry out time-consuming open-ended searches for maximum parsimony trees, some of our colleagues working with large data sets containing many OTUs use pairwise distances among the OTUs and an arbitrary clustering procedure such as the unweighted pair group method[10] to construct trees from the distance data. They have presented these trees constructed from distance data as though they were correct evolutionary trees or the best dendrogram constructions under the circumstances. On the basis of our experiences in searching for maximum parsimony trees with large data sets of many sequences, we adamantly disagree with those colleagues who use only distance matrices for dendrogram constructions. We present results which illustrate that trees with the most lowered NR lengths found by our branch-swapping procedures, using the discrete character-state information in the aligned sequences, are much more congruent with the overall

[9] M. Goodman, J. Czelusniak, G. W. Moore, A. E. Romero-Herrera, and G. Matsuda, *Syst. Zool.* **28**, 32 (1979).
[10] R. R. Sokal and C. D. Michener, *Univ. Kan., Sci. Bull.* **38**, 1409 (1958).

phylogenetic evidence on the species that the sequences represent than are the trees constructed from distance matrices by the unweighted pair-group method.[10] The trees constructed by the unweighted pair-group method and by certain other clustering algorithms applied to distance matrices, however, do contain considerable phylogenetic structure and thus provide good starting dendrograms in the search for lower NR length trees. In fact with the data set that we used in the present study, one of the clustering algorithms — the neighbor-joining method of Saitou and Nei[11] — produced a tree that was relatively close to the most parsimonious trees that we found.

Sequences, Computer Programs, and Heuristic Strategy

The data set used for the present study consists of hemoglobin amino acid sequences from 116 vertebrate species, each species represented by the amino acid sequence of α and β subunits of the principal hemoglobin synthesized in adult life. The tandemly combined α and β sequence yielded an alignment of 289 amino acid positions, 143 positions for aligned α sequences and 146 positions for aligned β sequences. Comprising the 116 vertebrates or OTUs are 1 shark, 1 lungfish, 2 teleosts, 1 toad, 3 crocodilians, 19 birds, and 89 mammals. Of the 89 mammals, 85 are eutherians (placentals) belonging to 14 orders, 2 are metatherians (marsupials), and 2 are prototherians (monotremes). (For a nearly up to date compilation of the hemoglobin sequences from these vertebrates see Kleinschmidt and Sgouros[12] and also Czelusniak et al.[13])

Using our computer program TPMMD, we constructed a minimum mutation distance matrix[14,15] from pairwise comparisons of the 116 OTUs. We then produced from this distance matrix an unweighted pair-group tree[10] by our computer program UPGMA (unweighted pair-group method by arithmetic averaging), a distance Wagner tree by our computer program FTE (Farris tree estimation) after the algorithm described by Farris,[16] and a neighbor-joining tree[11] by our computer program NJ. These three trees produced by clustering algorithms from the distance matrix served as starting points in the search by our computer programs MPAFEP and

[11] N. Saitou and M. Nei, *Mol. Biol. Evol.* **4**, 406 (1987).
[12] T. K. Kleinschmidt and J. G. Sgouros, *Biol. Chem. Hoppe-Seyler* **368**, 579 (1987).
[13] J. Czelusniak, M. Goodman, B. F. Koop, D. A. Tagle, J. Shoshani, G. Braunitzer, T. K. Kleinschmidt, W. W. de Jong, and G. Matsuda, *in* "Current Mammalogy" (H. H. Genoways, ed.). Plenum, New York, in press.
[14] T. H. Jukes, *Adv. Biol. Med. Phys.* **9**, 1 (1963).
[15] W. M. Fitch and E. Margoliash, *Science* **155**, 279 (1967).
[16] J. S. Farris, *Am. Nat.* **106**, 645 (1972).

MPALMX for the lowest NR length trees. MPAFEP and MPALMX both use the maximum parsimony algorithm[5,6,9] that calculates from amino acid sequences the lowest NR length for each tree examined.

MPAFEP examines all nearest neighbor single-step changes (NNSSCs) in the network topology, i.e., it examines for each two adjacent interior nodes the three swaps for the four branches originating from the two nodes. For the N exterior nodes of the network, there are $2(N - 3)$ alternative trees in each round of swaps. The tree having the lowest NR score initiates the next round of swaps. This iterative lowering of NR score stops when two consecutive rounds of NNSSCs fail to find a tree of lower NR count. At this time, since only a local minimum may have been found, the investigator should use new input trees with more extensive branch changes than the nearest neighbor ones to search for even shorter trees. Our experience shows that a reliable way to find shorter length trees is to use input trees with branch changes that agree with strongly supported hypotheses on the genealogical relationships of the OTUs. An option that may be employed with program MPAFEP in such an heuristic search is to designate subtrees within which no branch-swaps are permitted; MPAFEP then computes the set of codons corresponding to the maximum parsimony solution for the ancestor of each designated subtree and treats each of these subtree ancestors as a terminal taxon.

The program MPALMX examines all possible trees for eight terminal taxa. A terminal taxon can either be an exterior node (one of the OTUs) or a designated subtree. MPALMX computes the NR score of each of the 10,395 possible trees that eight terminal taxa can form and then lists the trees in order of their distribution from lowest to highest NR score. In our heuristic strategy, we use the shortest length tree found in the search using program MPAFEP as the starting tree for the further search, now using program MPALMX, for yet lower length trees. This starting tree is converted to a variety of rooted trees, each divided into a different set of eight terminal taxa or fixed subtrees. Moving the starting tree's root to different links, which we do with our computer program MROOT, facilitates examining the different sets of eight terminal taxa.

For example, among the 116 vertebrate OTUs represented by α and β hemoglobin sequences, there are 7 bats. In one of the MPALMX runs we placed the root on the link separating bats from all other OTUs; we then specified the interior node for this outgroup of bats (the 110 nonbat OTUs) as a terminal taxon or fixed subtree, and the 7 other terminal taxa were the exterior nodes for the bat OTUs. This MPALMX run allowed us to find the most parsimonious arrangement for the 7 bats on the assumption that Chiroptera, the order of bats, is monophyletic and that the chiropteran branch was correctly positioned in the network with respect to other eutherian orders. In another MPALMX run, the root was moved to the link separating seven of the eutherian orders (Primates, Lagomorpha, Scandentia, Rodentia, Chiroptera, Insectivora, and Carnivora) from the other vertebrate OTUs, and the eight fixed subtrees were the nodes for these seven orders and their outgroup.

If such MPALMX runs find a network with length shorter than that of the starting network, it is advisable to submit this new starting tree with the full number of terminal taxa (i.e., without fixed subtrees) to an MPAFEP run and also to further MPALMX runs after again dividing it into different sets of eight terminal taxa. Usually the MPALMX runs achieve very little shortening of tree length over that found after extensive MPAFEP runs. However, the MPALMX runs are still very useful because they provide information on the relative strengths of the putative cladistic groups in the parsimonious trees. By listing the trees according to increasing NR score, the MPALMX results reveal for each putative cladistic group, allowed to break apart in MPALMX runs, the minimum number of NRs needed above the lowest NR score which finds a tree that breaks up that group.

Once having found an optimal evolutionary tree (either one at lowest NR length or a near-parsimonious tree that interests us because of its congruence with the overall phylogenetic evidence), we can construct ancestral codons and link lengths by our computer program TPAB, which determines these sequences and lengths by the maximum parsimony method.[5] Ambiguities in parsimony assignments of codons, i.e., different sets of ancestral codons each giving the same NR score, are resolved by choosing the set that minimizes the sum of the distances on the tree for every pair of exterior nodes.[17] The distance between a pair of exterior nodes is the sum of the lengths of the links connecting the two exterior nodes. After obtaining the TPAB results, we can use our computer program TAVA to correct the lengths of the links for hidden superimposed mutations. TAVA propagates mutational information from pairs of nodes more populated by intervening links to those less populated.[18,19]

Programs TPMMD, UPGMA, FTE, NJ, MPAFEP, MPALMX, MROOT, TPAB, and TAVA are written in FORTRAN and run on the Unix system with 386 computers (copies of these programs are available from J.C.).

[17] M. Goodman, G. W. Moore, J. Barnabas, and G. Matsuda, *J. Mol. Evol.* **3,** 1 (1974).
[18] G. W. Moore, *J. Theor. Biol.* **66,** 95 (1977).
[19] M. L. Baba, L. L. Darga, M. Goodman, and J. Czelusniak, *J. Mol. Evol.* **17,** 197 (1981).

Results

We had completed an heuristic maximum parsimony search on the data set of 116 OTUs and written a draft of this chapter before implementing our NJ computer program. Thus, we first describe the results of this heuristic search that was started with UPGMA and FTE trees and then turn to the additional results obtained very recently with the starting NJ tree.

Figure 1 shows the UPGMA tree constructed from the minimum mutation distance matrix for the 116 OTUs. Considerable phylogenetic structure is apparent in this tree. For example, among taxa generally recognized as monophyletic groups and so depicted in the tree are Amniota (crocodilians, birds, and mammals), Crocodilia, Aves (birds), Galliformes (chicken, pheasant), Falconiformes (eagle, condor), Passiformes (starling, sparrow), Mammalia (monotremes, marsupials, and eutherians), Monotremata, Marsupialia, Eutheria (the eutherians or placental mammals), Proboscidea (elephants), Perissodactyla (equines, tapir, and rhinoceros), Cetacea (whales), Pecora (bovines, sheep, goat, and moose), Anthropoidea (simian primates), Hominoidea, Ceboidea (New World monkeys), and a number of other lower ranking taxa.

There are also, however, many incongruences with well-established phylogenetic relationships. Some examples of these incongruencies in the tree are the following: (1) the teleost (carp and goldfish) branch rather than the amphibian branch is next to Amniota; (2) alligator is closest to crocodile when it should be closest to caiman; (3) ostrich fails to group closest to rhea; (4) guinea pig is three links away from other rodents; (5) within the main rodent branch, rat *(Rattus)* should be next to mouse *(Mus)* but is farthest away, and the Muroidea is further scrambled by the mixing in of a Sciuromorpha (squirrels) branch; (6) hedgehog fails to join other lipotyphlous insectivores, being five links away from the branch of musk shrew and mole; (7) Carnivora fails to be monophyletic, a felid branch being seven links away from other carnivores, and among the nonfelid carnivores, Procyonidae fails to be monophyletic, racoon failing to group closest to coatimundi; (8) Artiodactyla fails to be monophyletic, camelids being four links away from other artiodactyls; (9) the Indian false vampire bat fails to group monophyletically with other bats; (10) lemur is four links away from other primates; and (11) within the main group of primates, Lorisiformes (galago and lorises) and Cercopithecoidea (Old World monkeys) fail to be monophyletic groups.

The UPGMA tree had a length of 3034 NRs. The program MPAFEP brought this tree after 34 iterations of NNSCCs to a score of 2956 NRs

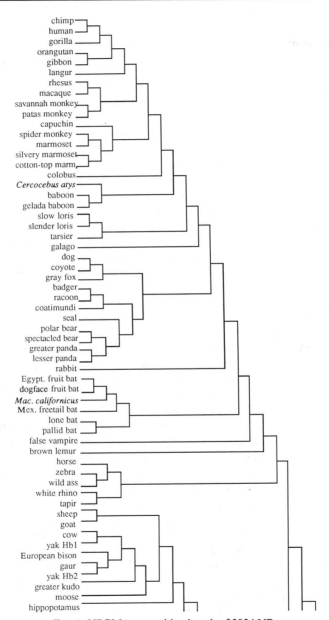

FIG. 1. UPGMA tree, with a length of 3034 NRs.

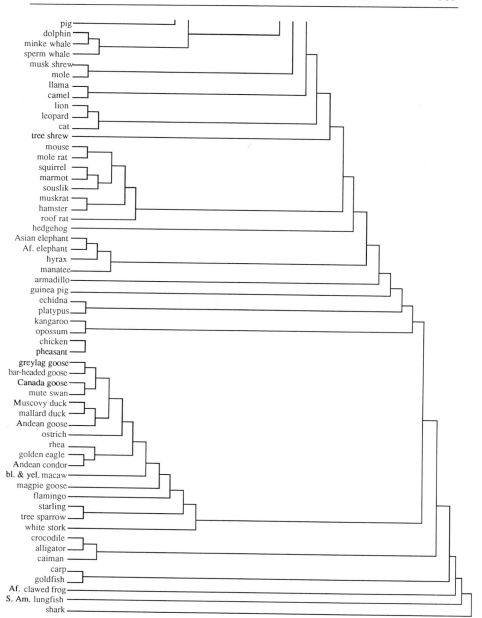

FIG. 1. *Continued.*

which was in a local valley that terminated the iterations. As judged by congruencies with well-established phylogenetic relationships, there were at least seven improvements in the output 2956 NR tree over the input 3034 NR tree: (1) the amphibian branch moved next to the amniote branch, yielding a monophyletic Tetrapoda; (2) alligator grouped closest to caiman; (3) crocodilians joined Aves, resulting in a monophyletic Archosauria; (4) among rodents, *Rattus* moved next to *Mus;* (5) the Indian false vampire bat joined other bats, resulting in a monophyletic Chiroptera; (6) among carnivores, racoon moved next to coatimundi; and (7) among primates, Lorisiformes became monophyletic (galago joining the slow and slender loris).

We obtained similar results with the input distance Wagner or FTE tree. Its NR length was 2986, and it showed many incongruencies with established phylogenetic relationships. For example, within Mammalia Artiodactyla, Lipotyphola (the lipotypholous insectivores), Chiroptera, Rodentia, Muroidea, Lorisiformes, and Ceboidea all failed to be monophyletic groups. Program MPAFEP produced from this input FTE tree an output tree, at score 2934 and in a local valley entered after 26 iterations, which showed far fewer phylogenetic incongruencies but not so few as the lowest NR length trees found after continuing the search with input trees that were more congruent with prevailing views on phylogeny. Using both MPAFEP and MPALMX, we were able to shorten the tree length to NR score 2916.

The trees at the lowest length of all trees examined, as illustrated by the 2916 NR tree shown in Fig. 2, depict a branching pattern for the OTUs that overall corresponds closely to prevailing views on the phylogenetic relationships of the species represented by hemoglobin sequences. Teleostei, Tetrapoda, Amniota, Crocodilia, Aves, Mammalia, Monotremata, Marsupialia, and Eutheria (taxa which divide Vertebrata into major clades) all appear as monophyletic groups. Within Crocodilia, alligator and caiman group together into family Alligatoridae which then joins Crocodylidae (crocodile). The 19 avian OTUs belong to 7 orders, 6 of which are represented by more than single OTUs. The maximum parsimony results (Fig. 2) depict 5 of these 6 orders as monophyletic groups. They are Passiformes (starling, sparrow), Galliformes (chicken, pheasant), Ansiformes (geese, ducks, swan), Palaeognathiformes (ostrich, rhea), and Falconiformes (eagle, condor). The sixth order is Ciconciformes, with two of its suborders represented by hemoglobin sequences: Ciconiae (stork) and Phoenicopteri (flamingo). These two suborders fail to group together; instead, Phoenicopteri joins Ansiformes, which is incongruent with the traditional arrangement of birds but not with some views on flamingo and ansiform phylogeny.

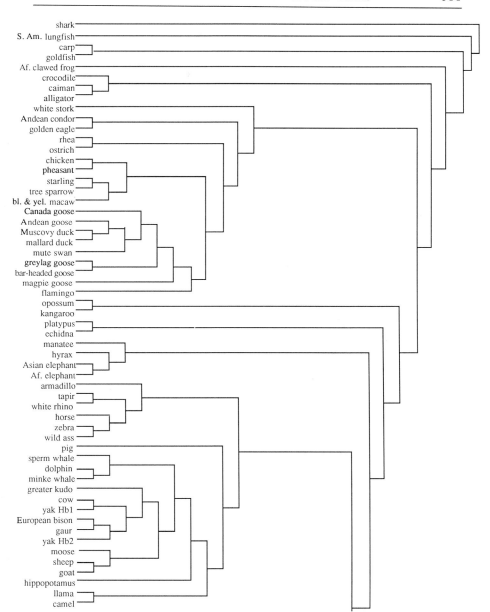

FIG. 2. Representative maximum parsimony tree, with a length of 2916 NRs.

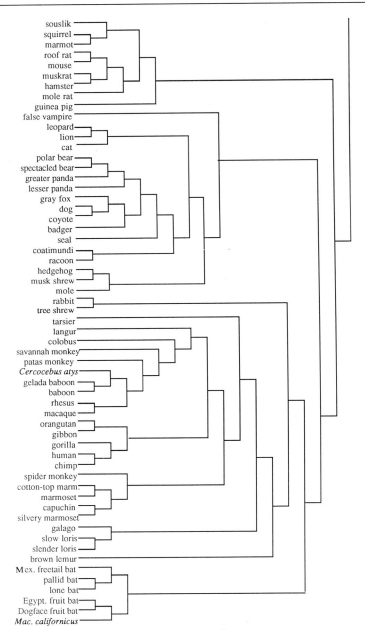

FIG. 2. *Continued.*

The 86 eutherian OTUs belong to 14 orders, 9 of which are represented by more than single OTUs. The maximum parsimony results (Fig. 2) depict 7 of these 9 orders to be monophyletic groups. They are Proboscidea (African and Asian elephants), Perissodactyla, Cetacea, Rodentia, Insectivora, Carnivora, and Primates. Moreover, the branchings below the ordinal level are in most respects concordant with prevailing views on the phylogeny of these groups. For example, Perissodactyla divides into Ceratomorpha (tapir, rhinoceros) and Hippomorpha (horse, zebra, ass); Rodentia divides into Caviomorpha (guinea pig), Scuiromorpha, and Muroidea (mole rat, the cricetids hamster and muskrat, and the murids *Mus* and *Rattus*); Carnivora divides into Feloidea (cat, leopard, lion) and Arctoidea (bears, pandas, procyonids, seal, badger, canids). Within Primates, suborder Anthropoidea divides into infraorders Platyrrhini (ceboids or New World monkeys) and Catarrhini, and Catarrhini divides into superfamilies Cercopithecoidea (Old World monkeys) and Hominoidea.

With respect to the branching pattern of arctoids it is worth noting that in the UPGMA tree (Fig. 1) greater and lesser pandas join each other. However, in the maximum parsimony tree (Fig. 2) the greater panda joins the ursid or bear branch, which agrees with prevailing views[20] on the ursid affinities of the giant panda. In an earlier study before the spectacled bear sequence was added to our data set of hemoglobin sequences, the maximum parsimony tree as well as the UPGMA tree joined greater and lesser pandas together first, in disagreement with prevailing views on ursid and panda phylogeny. Increasing the phylogenetic density of sequence data has apparently resulted in a more accurate maximum parsimony reconstruction.

These results clearly demonstrate that the lowest NR length trees found by our heuristic maximum parsimony search procedures are much closer to the correct evolutionary tree than the trees produced from the distance matrix by such arbitrary clustering algorithms as UPGMA and FTE. However, our results also indicate that certain trees near but not at the lowest NR length are even closer to the correct evolutionary tree. For example, in the 2916 NR tree (Fig. 2) the Indian false vampire bat joined the insectivore–carnivore branch rather than the main bat branch. Having one monophyletic bat group by joining the main bat branch and the Indian false vampire bat together either next to the insectivore–carnivore branch or next to the primate–tree shrew–rabbit branch adds 2 NRs to the tree length. Program MPALMX found the first 2916 NR tree using a 2918 NR tree in which all bats were grouped together but not as a fixed subtree, i.e., the Indian false vampire bat was free to leave this grouping.

[20] E. Mayr, *Nature (London)* **323,** 769 (1986).

Several other branching arrangements at NR length 2916 that are probably phylogenetically incorrect are as follows: (1) the avian branch is closer to the mammalian branch than to the crocodilian branch; (2) the eutherian branch is closer to the monotreme branch than to the marsupial branch; and (3) the artiodactyl pecoran branch is closer to the cetacean branch than to camelid, hippopotamus, and pig branches. In each case, the more commonly agreed on phylogenetic arrangement adds 1 to 2 NRs to the tree length. In fact, a tree at 2921 NR length depicted a monophyletic Archosauria (Aves and Crocodilia joined together), a monophyletic Theria (Eutheria and Marsupialia joined together), and a monophyletic Artiodactyla (a pig–camelid–hippo branch joined to the pecoran branch) and otherwise either exactly followed or was as congruent with prevailing phylogenetic views as the 2916 NR tree in Fig. 2.

On recently reformating the neighbor-joined algorithm[11] to give output dendrograms of the type that the programs MPAFEP and MPALMX can use as input dendrograms with our data sets of aligned amino acid sequences, we constructed an NJ tree from the minimum mutation distance matrix for the 116 OTUs. This NJ tree had a length of 2950 NRs. Not only was its length shorter than the UPGMA and FTE trees, but it was also much more congruent with prevailing views on the phylogenetic relationships of the OTUs. For example, except for Ciconciformes, Ansiformes, Artiodactyla, and Lipotyphla, all other bird and mammalian orders were depicted as monophyletic groups by the NJ tree. Program MPAFEP, starting with this NJ tree, produced output trees after the twenty-second and twenty-third iterations that were at the same low length, 2916 NR, as the tree in Fig. 2 but in a different valley. The 2916 trees found from iterating the NJ tree, while very similar to the previously found 2916 trees, are not quite so good phylogenetically. The Australian magpie goose did not join other ansiformes, hedgehog did not join other lipotyphlous insectivores, cricetids did not join murids, galago did not join other lorisiforms, and capuchin monkey did not join other ceboids. It is worth noting that the lowest NR trees found for myoglobin sequences[21] strongly group galago with other lorisiforms and capuchin monkey with other ceboids.

Conclusion

We feel we have demonstrated (by example but not by rigorous proof) that our heuristic maximum parsimony search procedures can approach

[21] M. Goodman, A. E. Romero-Herrera, H. Dene, J. Czelusniak, and R. E. Tashian, *in* "Macromolecular Sequences in Systematic and Evolutionary Biology" (M. Goodman, ed.), p. 115. Plenum, New York, 1982.

the correct evolutionary tree. The particular data set providing the example presented in this chapter contained not enough aligned orthologous sequence positions and perhaps still too sparse a number of OTUs to ensure that the maximum parsimony search would converge on a single lowest NR length tree that was the correct evolutionary tree. Our experiences, however, indicate that when these conditions are met the maximum parsimony trees found will be extremely close to the correct evolutionary trees.

Acknowledgments

This work was supported by National Science Foundation Grant NSF 86-07202 and National Institutes of Health Grant HL 33940. We thank Drs. Gerhard Braunitzer, Traute Kleinschmidt, and Genji Matsuda for providing us with preprints and copies of some of the hemoglobin sequences before publication.

[38] Phylogeny Determination Using Dynamically Weighted Parsimony Method

By PATRICK L. WILLIAMS and WALTER M. FITCH

Introduction

Workers in the field frequently do not, for drawing phylogenetic inferences, wish to use regions of sequences that they find difficult to align, and thus they discard them. This is an extreme form of character weighting in which all characters (positions) are weighted either zero or one. Workers also occasionally wish to weight those nucleotide substitutions that are transversions more heavily than those that are transitions because transversions may be much less common than transitions. Both kinds of weighting are subject to the charge of investigator bias in the absence of some procedure for assigning weights and computing minimal-length values (scores) for the various possible trees. This chapter presents various methods for assigning both kinds of weights plus a method for evaluating trees given the weights. These were described earlier.[1]

Definitions

A *character* is a nucleotide position, and a *character state* is the particular nucleotide in that position. For example, the third character of a coding

---

[1] P. L. Williams and W. M. Fitch, *in* "Hierarchy of Life" (H. Jörnvall, ed.), p. 453. Nobel Symposium. Elsevier, Amsterdam, The Netherlands, 1989.

sequence is the third position and its state is G (because coding sequences start with the methionine codon, AUG). Each character of a sequence is represented by a four-bit nibble in which A = 1000, C = 0100, G = 0010, and T = U = 0001. This permits ambiguity to be easily represented. For example, purine = R = (A and/or G) = 1010, pyrimidine = Y = (C and/or U) = 0101, etc. This is especially useful because AND and OR operations can be performed on the decimal numbers themselves which range from 1 to 15 = N.

There are two kinds of weighting. The first is *character weighting* in which the weight or importance to be given to any position may differ from that given to other positions. These weights are stored in a *character weight vector*, sometimes referred to hereafter simply as the *C vector*. The method of assigning weights is described shortly.

The second kind of weighting is *character state transformation weighting* in which the weight or value assigned for one character state transformation, say, A → C, may differ from that of another, say, A → T. These values are stored in the *character state transformation matrix* or, more simply, the *T matrix*.

Given that one wants to find the best tree, it would be desirable to examine the space, Q, of all possible trees, a space whose size increases rapidly as the number of sequences, r, increases, namely, $Q = \Pi(2i - 1)$ for all i from 1 to $r - 2$. This rapidly becomes computationally impossible ($Q = 135,135$ unrooted trees for only nine sequences). Thus, one must search selectively. This program conducts its search by starting with a tree provided by the investigator (called the *seed tree*) and doing branch swapping. Specifically, it does *neighbor branch swapping* in which it looks at the seed tree and all possible interchanges of two subtrees that attach to the opposite ends of a single branch (i.e., two subgraphs that are incident to adjacent vertices). Whenever it finds a better tree, that tree is recorded along with any others that are equally good. All these equally good best trees then become the basis of further neighbor branch swapping until no better trees are found. This constitutes one *pass*. Best trees are those with lowest scores.

In all these discussions, a *tree* is a graph consisting of points *(nodes)* joined by lines *(branches)*. The graph is *connected*, meaning that one can get from any point to any other point by traversing one or more contiguous branches. It is a tree rather than a network because there is only one possible path between any two nodes. A node is said to be *exterior* if only one branch connects to it and *interior* otherwise. The tips of a tree represent the sequences analyzed. In an *unrooted tree*, all interior nodes have at least three branches incident to them (if all interior nodes have exactly three incident branches, the tree is said to be *strictly bifurcating*). The interior nodes represent ancestors, but the ancestral relation among interior nodes is unstated in an unrooted tree. A *rooted tree* is obtained by adding an extra node, called the root, onto one of the branches of the tree to give an interior node with only two incident branches. The root is usually intended to represent the ancestor of all other portions of the tree and gives temporal directionality to the tree.

Dynamic weighting means altering the values of the C vector and/or the T matrix as a consequence of the results of the previous pass. With these new sets of weights, one makes another pass repeating the search for the best tree. The user has the option of starting subsequent passes using either the seed tree or the best tree of the previous pass. The procedure is repeated until the same best tree has been obtained on two successive passes or 20 passes have been completed.

The *score, S,* equals $\Sigma c_i t_{jk}$ where the cost for a transformation (substitution) from nucleotide j to nucleotide k is t_{jk} and the character weight given for the ith character is c_i, i.e., c_i is an element of the C vector and t_{jk} is an element of the T matrix. The sum is over all positions and over all transformations required, each weighted by the character weight of the position in which it occurs.

Method

Data

What follows assumes that one has a set of nucleotide sequences that have been prealigned so that the ith position of each sequence is presumed to be homologous across all sequences.

As presently constituted, all gapped positions are changed to N (=any nucleotide) so that gaps have no part in the algorithm. While the program could be rewritten to treat each gapped nucleotide as a fifth element, this would not be sound practice for it would then treat all gaps of length one as no more important than a nucleotide substitution and treat long gaps as if they were many independent *indels* (insertions and/or deletions). For the investigator who wishes to include indel events, this can better be done by adding additional nucleotides at the end of the sequences, say, A and T (for Attendant and Truant as the nucleotides are present or not), with each gap represented by a different added position in the sequence. A gap can be assigned additional weight by representing it with more than one (identical) position in the sequence. Overlapping gaps must be treated carefully as in Fitch and Yasunobu.[2]

---

[2] W. M. Fitch and T. K. Yasunobu, *J. Mol. Evol.* **5**, 124 (1975).

Initial Weighting

One must assign values for the C vector and the T matrix before doing the first pass. The C vector is the simplest. As of the time of writing, the only possible starting values are *uniform weights,* i.e., $c_i = 1$ for all i. It remains to be investigated whether other starting values would be useful. One possibility is to weight the positions inversely proportional to the number of different nucleotides that position possesses when all sequences are examined.

The T matrix can be initialized in any one of three ways. The first is *uniform weighting,* i.e., $t_{jk} = 1$ for all j not equal to k.

The second way depends on counting the number of positions at which nucleotides j and k both occur (exist, hence *existential weighting*) and assigning weights inversely proportional to the number of positions containing both j and k. The rationale is that the more positions having both a C and U, the more often one would expect there to have been C $\leftrightarrow$ U interchanges. Moreover, the more often one would expect to see them compared to some other interchange, say, A $\leftrightarrow$ U, the less relative weight it should have.

The third way depends upon counting the number of different ways one might get nucleotides j and k at a position, summed over all positions. Thus, a position with 5 C's and 4 U's contributes $5 \times 4 = 20$ toward the CU count. This therefore is called *combinational weighting.* As for existential weighting, the weights are inversely proportional to the count.

For all T matrices, since there must be no cost for a nucleotide not changing, the values along the major diagonal are set equal to zero, i.e., $t_{jj} = 0$ for all j. For all matrices, initial or otherwise, the off-diagonal values are normalized to an average of one. For the C vector, the values are also normalized to an average of one. Thus, scores may be compared among trees that use the same aligned sequences even though the C vector and T matrix differ. These values are all multiplied by 100 to permit integer rather than floating point arithmetic.

Dynamic Weighting

The best tree from the first (previous) pass is then examined, and two counts are made. One is the number of times, c_i', each position has had a nucleotide substitution, the other is the number of times, over all positions, that each nucleotide was substituted by another, t_{jk}'.

The C vector may then be changed at the user's option, in one of two ways. In *linear weighting,* $c_i = 1/c_i'$. This is the method of Farris.[3] In *quadratic weighting,* $c_i = 1/(c_i')^2$. Positions that don't change are ignored.

[3] J. S. Farris, *Syst. Zool.* **18**, 374 (1969).

The vector is normalized as above. Quadratic weighting is obviously more drastic. If there were five positions with 1, 2, 3, 4, and 5 changes, their normalized linear weights would be 2.19, 1.09, 0.73, 0.55, and 0.44 whereas their quadratic weights would be 3.41, 0.85, 0.38, 0.21, and 0.14. In the realistic case, there will be many more positions with one change than with five changes, and the weights will change accordingly even if the range is still one to five.

The T matrix may also be changed but only linearly. Thus, $t_{jk} = 1/t'_{jk}$, but there remains a choice. The preceding T matrix is almost certainly asymmetric. One may wish $t_{jk} = t_{kj}$ (e.g., C $\rightarrow$ U = U $\rightarrow$ C). In that case, we set $t_{jk} = t_{kj} = 2/(t'_{jk} + t'_{kj})$, and the T matrix becomes symmetric.

Parsimony

Once the T matrix contains numbers other than zero and one, the usual parsimony method[4] is no longer usable. The procedure we developed proved to be identical to that previously developed by Sankoff and Cedergren.[5] The idea is simple. Consider a single nucleotide position. An ancestor has two immediate descendants, and the score for that ancestor is the lowest score for any of the four nucleotides at that ancestor. These four scores are obtained using the scores already known for its immediate descendants. Let h and i be the immediate descendants of ancestor j, let $S(g,k)$ be the score for nucleotide g (1, 2, 3, 4 = A, C, G, U, respectively). If we want to know the score for nucleotide C, $S(2j)$, in the ancestor, we need to know what it costs to get there from h (and i too, but we will ignore this for the moment). $S(2j)$ could come from any of the four nucleotides of h, and we wish to choose the nucleotide(s) that give the lowest score. Figure 1 shows an example. $S(1h) = 20$ and the T matrix value for changing A $\rightarrow$ C is 3.

Thus, h would contribute 23 to the score of $2j$ if no other nucleotide in h could contribute less. For the example given in Fig. 1, no other nucleotide from h can give a lower score and so 23 is the contribution of h to the C score in j. Of course, the lowest possible contribution from i must be added as well, and the same procedure must be applied to the other three nucleotides that must be scored for j. Stated more mathematically,

$$S(kj) = \min[S(mh) + T(mk)] + \min[S(ni) + T(nk)] \qquad (1)$$

where k, m, and n are nucleotides in j, h, and i, respectively, $S( )$ is a score,

[4] W. M. Fitch, *Syst. Zool.* **20**, 406 (1971).
[5] D. Sankoff and R. J. Cedergren, in "Time Warps, String Edits and Macromolecules: The Theory and Practice of Sequence Comparison" (D. Sankoff and J. B. Kruskal, eds.), p. 253. Addison-Wesley, Reading, Massachusetts, 1983.

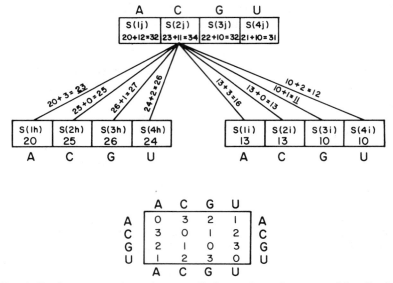

FIG. 1. Parsimony procedure when not all changes have the same weight. Consider a region of a tree consisting of an ancestor, *j*, and its two immediate descendants, *h* and *i*, and the associated *h* and *i* scores, *S*, for the four letters shown. The problem is to calculate the scores for *j* using the matrix of arbitrary substitution values shown at the bottom. The example shown is for the estimate of $S(2,j)$ which is the sum of two minima, each of which is from four sums of two terms representing the score of a descendant nucleotide and the substitution value from C (2 denotes C) to that descendant nucleotide. See text for further details.

$T(\ )$ is a T matrix value, and the minima are to be taken over all four possible values of *m* and *n* for each of the four possible values of *k*.

The procedure begins at the tips and works upward, as one can only perform the operation on ancestors whose scores for its immediate descendants are known. The scores for the tips, the known sequenced nucleotides, are zero for any nucleotide that could be there and very large for any nucleotide that could not be there. "Very large" in the previous sentence means larger than any value in the T matrix. This assures that all descents end on a permissible nucleotide at the tip.

The process just described is for a single nucleotide position. It must be repeated for each nucleotide position and the scores summed over all positions. If *j* in Fig. 1 were the root of the tree, then the minimum score among its four scores would become the score (length) of the tree.

Caveats

The dynamic weighting procedures described above can produce a T matrix that is not metric, that is, does not obey the triangle inequality. For example, if $T(AC) = 2$, $T(CU) = 1$, and $T(AU) = 4$, it becomes possible for one to get from A to U at a lower cost by going via C than by going directly to U from A. This is particularly acute at the root where the interposition of the extra node converting an unrooted tree to a rooted tree gives one an especially good opportunity to put in an intermediate. We have therefore introduced a "numbing" procedure such that Eq. (1) is modified to require that, in obtaining the minima, either m or n must equal k; in other words, the score for a given nucleotide in an ancestor must be obtained using that nucleotide in at least one of its two immediate descendants. This numbing may be invoked at the user's option at the root alone or at all ancestral nodes.

A second problem involves asymmetric T matrices. In normal parsimony procedures, all nucleotide changes have the same cost, namely, one. Thus, the length of the tree is independent of the location of the root. If, however, $T(AC)$ does not equal $T(CA)$, then the length of the tree (score) may be altered by shifting the root so as to change whether the nucleotide substitution is $A \rightarrow C$ or $C \rightarrow A$. It is therefore recommended that an asymmetric T matrix not be used when the location of the root is not well circumscribed.

A third problem is cycling. This appears to be very rare. The program is designed to detect cycling with a periodicity of eight or fewer passes or, in the worst case, to stop after 20 passes if convergence has not occurred. Convergence has never required more than 6 passes. In the rare cases where the program went 20 passes, there were two trees, each of which induced weights that made the other tree better given those weights. Even in those cases, the two trees that cycled were obtained within 6 passes.

Parameters

One may use either a batch mode or an interactive mode. The parameters are the same, but they are introduced one at a time in a query–answer mode in the latter case.

NP and LS are the number of sequences and the length of those sequences, respectively.

KPTFL controls how much printout one receives. Default is zero which gives less ouput.

IALT is set to 1 if branch swapping is desired, to 0 if branch swapping is to be suppressed. This permits looking at just one tree of your choice if it is so desired. Default is zero.

IDYNWT is set to 0 if no dynamic weighting is wanted. If dynamic weighting is wanted, it is set to 1 or 2 as one specifies, on succeeding passes, to use the best tree of the previous pass or the seed tree to start the branch-swapping process. Default is one.

ISYM is 0 or 1 as the T matrix is to be asymmetric or symmetric, respectively. Default is one.

IWT determines whether the initial T matrix is to be read in $(-4,-3)$, uniform $(-2,-1)$, existential (0), or combinatorial (1). If the value is -3 or -1, the initial matrix is used for all passes. For all other cases, dynamic weighting is used for all passes after the first. Default is zero.

ICHW is the indicator for character weighting and is 0 if weighting is to be uniform on all passes, 1 if it is to be linear after the first pass, and 2 if it is to be quadratic after the first pass. Default is one.

KODE is set to 1 if one wishes to indicate the sequences are exons so that separate T matrices are used for each coding position. If, to speed up the analysis, the sequences have been shortened by removing all unvaried positions, the coding positions of each nucleotide must be provided. These are read in as $X.Y$ where X is the codon number and Y the position number (e.g., 37.2 means position 2 of the thirty-seventh codon).

IROOTS refers to the numbing procedure. A 0 causes no numbing, a 1 invokes numbing at the root only, and a 2 invokes it at all ancestral nodes. Default is one.

Example Result

We used the program on 14 superoxide dismutase amino acid sequences from 6 mammals, swordfish, fruit fly, 3 plants, 2 fungi, and a bacterium. The sequences were back translated into ambiguous codons. The results are shown in Table I. In every case, the seed tree was that which seemed most biologically reasonable. In no case was the "best" tree in accord with biological opinion (Fig. 2). In all cases the mammals were badly related among themselves, and the single monocotyledenous plant was the sister group of one rather than of both dicots (while the monophyletic nature of the dicots is not established, no one seems to doubt that spinach and cabbage should be sister groups relative to corn). Thus, the superoxide dismutases are a particularly noisy set of data as reflected in the preceeding results and in the fact that, of 21 different combinations of parameters, we obtained 12 different "best" trees. Moreover, all but the two sets of results shown with asterisks are even more unbiological in that, with the bacterial sequence as the outgroup, the root thus suggested divided the eukaryotes into two groups frequently making the fruit fly the sister

TABLE I
DYNAMIC WEIGHTING FOR 14 SUPEROXIDE DISMUTASE SEQUENCES[a]

| Transformation | Character | | | Initialization |
	U	L	Q	
U	1-631	1-443	1-347	u
Ls	2-278	3-186	4-142	u
	7-278	5-188	6-144	e
	7-273	3-186	8-149	c
La	2-231	9-167	10-124*	u
	2-231	11-163	6-129	e
	2-231	9-167	12-133*	c

[a] Sequences are nucleotide sequences back-translated from amino acid sequences by a program (AATRANS) designed to choose ambiguous codons for arginine, leucine, and serine for each position that will minimize the difference from the codons of other amino acids known to be at that position. The sequences were from *Homo sapiens* (man), *Rattus norvegicus* (rat), *Sus scrofa* (pig), *Bos taurus* (cow), *Ovis aries* (sheep), *Equus caballus* (horse), *Xiphias gladius* (swordfish), *Drosophila melanogaster* (fruit fly), *Brassica oleracea* (white cabbage), *Spinacia oleracea* (spinach), *Zea mays* (corn), *Neurospora crassa, Saccharomyces cerevisiae* (baker's yeast), and *Photobacterium leiognathi.* A total of 21 combinations of weighting procedures were used. The positions could be weighted uniformly for all passes or inverse linearly or inverse quadratically after the first pass, as designated by the columns labeled U, L, and Q, respectively. The transformation matrix could be initialized uniformly, existentially, of combinatorially as designated by the rows labeled u, e, and c at the right. Successive passes could have the T matrix be uniform or inverse linearly, symmetric or asymmetric as designated by the row labels U, Ls, and La at the left. The numbers in the table are of the form tree-score. Tree 1 is the lower tree of Fig. 2. Tree 12 the upper tree.

group of all other eukaryotes or, even more frequently, clustering them with the plants and fungi making vertebrates the sister group of all other animals. On two occasions swordfish went with fruit fly into the plant – fungi group, making mammals the sister group of all other eukaryotes. Only the asterisked analyses represent cases where the result is biologically reasonable, with plants and fungi the sister group to animals, fruit fly the sister group of vertebrates, and swordfish the sister group of mammals.

The upper left corner value in Table I is for the usual parsimony procedure. The asterisked results arise from the heavier forms of weighting (inverse quadratic for positions and inverse linear asymmetric for transformations). Weighting only the characters or only the transformations tended not to make a great difference. Analyses of other data tend to provide similar indications that weighting is beneficial.

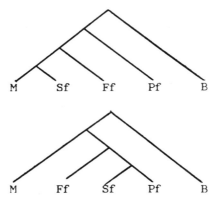

FIG. 2. Two possible relationships of the superoxide dismutases. B, Bacteria; M, mammals; Ff, fruit fly; Pf, plants plus fungi; Sf, swordfish. The bottom tree is the tree obtained by standard parsimony (C vector and T matrix all ones). The top tree is what biologists believe to be correct and is the result for trees 10 and 12 in Table I (asterisked). Trees 10 and 12 differ in how they relate the mammals; however, neither of these relationships is likely to be correct.

Discussion

The basic principle is that rare events are the most reliable for ascertaining relationships. If vertebrae were only invented once, then all animals possessing them are a *monophyletic group,* that is, every vertebrate is more closely related to every other vertebrate than to any invertebrate or other "lower" organism. Ideally, we would like to analyze that set of characters that had changed only once for they would give the correct tree by almost every method. Since we do not know which characters those are, this procedure tries a successive weighting approach that uses a reasonable approximation of the correct tree as a way of estimating how many times a character has changed (or a state transformed) to filter out some of the noise from characters that appear to have changed often. This gets a better tree from which a better refinement can be made and the process repeated until convergence is achieved.

The theory is fine, but does it make any practical difference, and are the results more likely to reflect the true phylogeny? The example provided was chosen specifically because of its extremity. It thus illustrates that the tree obtained may depend crucially upon weighting. For very good data, the kind where most methods get the same answer, weighting does not change the answer. Very good means that the number of differences between two sequences closely approximates the amount of change, the number of nucleotide substitutions since their common ancestor. Thus, weighting has an effect in precisely those instances where noise is apparent in the data. It does make a practical difference.

Whether the result is more likely correct than when unweighted parsimony is used has not been proved. It ought to be true in principle, but the improvement might be marginal; we only have the impression that, with noisy data, the weighted trees are more often more similar to current biological opinion. Certainly the example shown illustrates this since no biologist would place the fruit fly as the sister taxon of all eukaryotes or of plants plus fungi.

Inspection of Table I reveals that the scores decrease as one goes from left to right and from U to Ls to La. This does not necessarily mean that such weighting gives a better result. One would see such an effect even if the best tree were always the same; this is true despite the normalization that keeps the average weight equal to one. Nevertheless, where the trees do change, we have the distinct impression that the more dramatic weightings give better trees in that, even if they are not identical, they are closer to what was expected *a priori*. Simulations are in progress to test the generality of this perception.

One effect of the scoring procedure is that one has fewer trees of the same length. Thus, five equally parsimonious trees (i.e., having identical scores) under uniform weighting might have five rather different scores with this method. Thus, the method is discriminating in that it will choose among trees that are otherwise undifferentiated. This is as it should be and is advantageous, but one must be cautious in that some differences may not be statistically significant.

The method employed here uses only strictly bifurcating trees. That means every ancestral score is the result of examining its two immediate descendant's score vector and the T matrix. There is nothing, however, to prevent Eq. (1) from having additional terms, as for example, if the ancestor had more than two immediate ancestors. We have not been motivated to do this because such cases are merely special cases of bifurcating trees with a branch that has length zero, meaning that the sequences at either end of the branch are the same. This will be discovered in the process of using the present method.

Program Availability

This program, WTSUBS, is available along with two others. The first program, AUTSUBS, examines a suite of parameter options. Thus, the results in Table I were obtained in only two runs, i.e., there was no need to run the WTSUBS program 21 times. The second program, ALLTOPS, will, for a fixed set of parameters, run all possible unrooted trees for up to eight taxa and give a distribution of their scores. All three programs are written in FORTRAN and have been compiled on IBM-compatible machines under the Ryan/McFarland compiler and on the Sun using the f77

Unix compiler. They can no doubt be made to run on other systems, but the commands for bit extraction will probably need to be changed (we pack 8 nucleotides per 32-bit word). These programs will be provided in their FORTRAN form and, if desired, in compiled form along with sample data and output. One 1.2-Mbyte floppy disk should be sent to the author (W. F.) for each of the three programs wanted.

[39] Unified Approach to Alignment and Phylogenies

By Jotun Hein

Introduction

Two important considerations in the analysis of molecular sequences are alignment and phylogeny reconstruction. In both cases the history of the molecules must be reconstructed. The principle of parsimony states that the solution which minimizes the amount of evolution used to explain the extant sequences (sequences from the past cannot be obtained) approximates the real history well. This principle has dominated previous methods used to solve these problems.

An alignment of two short DNA sequences is shown below:

$$\text{TTAGTCC - AT}$$
$$\text{T - - GTGCGAT}$$

Each sequence is padded with gap signs so similar regions are matched. To define the most parsimonious alignment, the weight assigned to basic evolutionary events, substitutions, and insertion/deletions, must be specified. By choosing these weights to reflect actual occurrences, the usefulness and reliability of the results can be greatly enhanced. The weight, g_k, for an insertion/deletion of length k frequently takes the form $a + bk$ (a and b are nonnegative parameters), since this allows the use of a faster algorithm. When a is strictly positive, then runs of gap signs are interpreted as one insertion/deletion; if a is zero, then a stretch of gap signs can be interpreted as composed of several insertions/deletions. Since the parameter a is usually positive, this alignment would postulate an insertion/deletion of length 2 and another of length 1. Only one substitution is postulated, a C for a G or vice versa. Weights of substitutions have been developed from empirical tables of mutation frequencies.[1,2] This is done efficiently by various dynamic programming algorithms developed since the late 1960s. The typical running time for these algorithms are $O(l^2)$ or $O(ld)$, where l is the length of the sequences and d the distance between the two sequences.

Five sequences related by an unrooted tree are shown in Fig. 1. Typically, the procedure to find the phylogeny would presume an alignment of all sequences obtained manually. Each column would be treated independently and then be explained by a series of substitutions. Different branching configurations or tree topologies would be checked, and the one allowing for shortest total branch length would be chosen.

Conventionally, the alignment problem involves two sequences and must consider both substitutions and insertions/deletions. the phylogeny problem involves more sequences but usually requires that the insertions/deletions be taken care of beforehand. The accomplishment of the method presented here is to solve both problems simultaneously.

The phylogenetic alignment problem is computationally very difficult. It is harder than two special cases of it: First, the traditional parsimony phylogeny problem is known to be NP-complete,[3] i.e., no algorithm is likely to exist that can guarantee to find the most parsimonious solution in polynomial time. Second, it is harder than the phylogenetic alignment problem, when the phylogeny has been given and only insertions/deletions of length 1 occur, where the method to solve it needs computing time proportional to $2^n l^n n$.[4] If longer gaps are allowed, the situation becomes even worse.[5] The method presented here is therefore a fast approximation algorithm.

Method

The method has two aspects: finding the phylogeny and aligning the sequences according to the phylogeny. The central approximation and novelty consists of decomposing the many-sequence alignment problem to a series of pairwise problems of aligning sequences, reconstructing ancestral sequences, and then aligning the ancestral sequences. At each such cycle, the set of most parsimonious ancestral sequences is investigated.

[1] M. O. Dayhoff, "Atlas of Protein Sequences and Structure," Vol. 5, Suppl. 3. National Biomedical Research Foundation, Washington, D.C., 1978.
[2] R. F. Doolittle, "Of ORFs and URFs." University Science Books, Mill Valley, California, 1986.
[3] L. R. Foulds and R. L. Graham, *Adv. Appl. Math.* **3**, 43 (1982).
[4] D. Sankoff, *SIAM J. Appl. Math.* **78**, 35 (1975).
[5] J. J. Hein, *J. Theor. Biol.* (submitted for publication).

```
s1          TGTGGTA
s2          TGAGCAA
s3          TCTGAAG
s4          CCGTAAG
s5          CCCTTTG
```

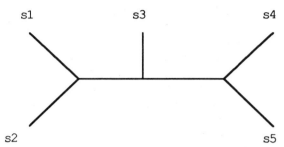

Fig. 1. Five short sequences and one possible relationship (of 15 possibilities) that happens to allow the most parsimonious history of the sequences. Columns one and two indicate that the tree topology must have s4 and s5 as brothers, column four indicates that s1 and s2 are brothers, and then the tree is fully determined. Column six would fit better on a tree topology where s1 and s5 are brothers. This, however, cannot outweigh the first three mentioned sites. There exist algorithms that will evaluate the weight for a column on a tree topology. Using such algorithms, different tree topologies can be tested and the best topology retained.

This is accomplished by a generalization of the traditional sequence comparison algorithm that allows comparison of graphs with edges labeled with sets of nucleotides/amino acids. In the present context the new algorithm is used to align alignments or even groups of alignments and thus braid alignments of a few sequences into alignments of more sequences. This method allows for a more thorough search for likely ancestral sequences by not having to choose between equally good alignments arbitrarily. The algorithm that achieves this can best be described as a generalization of a combination of a string comparison algorithm of Gotoh[6] with a graph comparison algorithm of Sankoff.[7]

The method is described in a simple situation, where only insertions/deletions of length 1 occur and all events are weighted 1.

---

[6] O. Gotoh, *J. Mol. Biol.* **162**, 705 (1981).

[7] D. Sankoff, *in* "Time Warps, String Edits and Macromolecules" (D. Sankoff and J. Kruskal, eds.). Addison-Wesley, Reading, Massachusetts, 1983.

New Algorithm

Let $s1$ and $s2$ be two sequences of length $l1$ and $l2$, respectively. The substring consisting of the first i elements of sk ($k = 1, 2$) is denoted sk_i, and the ith element in $sk[i]$ refers to the ith element of sk. D_{ij} refers to the distance between $s1_i$ and $s2_j$.

Sellers[8] and Sankoff[9] gave an algorithm that allowed calculation of the distance between $s1$ and $s2$ and the corresponding alignment. From Eq. (1) the distance is determined as

$$D_{i,j} = \min\{D_{i-1,j-1} + d(s1[i], s2[j]), D_{i,j+1} + 1, D_{i-1,j} + 1\} \qquad (1)$$

subject to the initial condition $D_{0,0} = 0$, and all $D_{i,j}$ with negative arguments are defined to be infinity.

Aligning two sequences corresponds to finding the shortest path in a graph with nodes indicated by the grid $\{0, \ldots, l1\} \times \{0, \ldots, l2\}$. Each node (i,j) had three predecessors $(i-1,j-1)$, $(i,j-1)$, and $(i,j-1)$. The two last edges are weighted 1, since they correspond to insertions/deletions. The first edge is weighted as $d(s1[i], s2[j])$.

The use of Eq. (1) is illustrated for two short nucleotide sequences (Fig. 2), in which $D_{i,j}$ values are calculated. Backtracking from $D_{l1,l2}$ and connecting entries on the path will define a backtracking graph. By choosing a path from $(l1,l2)$ to $(0,0)$ a minimal alignment is chosen. Then, by choosing at each position in the alignment (or path) where the two sequences have different signs, a set of potential ancestral sequences is generated according to the alignment. The set of potential ancestral sequences can be defined by a slight modification of the backtracking graph. This graph, G, will be directed, noncyclic, and connected. The nodes of G will be the nodes (i,j) that were visited in the backtracking. The edges will correspond to one position in an alignment; $(i,j) - (i-1,j-1)$ will correspond to a match and will be associated with the nucleotide set that is the union of the two nucleotides that are being matched. Positions corresponding to insertions/deletions $(i,j) - (i,j-1)$ [or $(i-1,j)$] will be associated with the nucleotide of the nondeleted sequence and a gap sign. By traversing G from $(l1,l2)$ to $(0,0)$ and choosing an element at each edge (and jumping over gap signs), a set of sequences, $S(G)$, is generated. $S(G)$ happens to be identical to the following set of sequences:

$$S = \{s: D(s1,s) + D(s,s2) = D(s1,s2)\} \qquad (2)$$

In other words, these are the sequences that can be postulated as ancestral to $s1$ and $s2$ without necessitating any extra evolution. Sequences in S are

[8] P. Sellers, *J. Comb. Theor.* **16**, 253 (1974).
[9] D. Sankoff, *Proc. Natl. Acad. Sci. U.S.A.* **69**, 4 (1972).

s1 = TCA   s2 = TG

Minimal alignments:

i)   TCA   ii)   TCA
       TG-   T-G

Ancestral Sequences:

i) TC,   TGA   ii)   TA,   TCG

Sequence Graph Generating Ancestral Sequences:

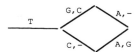

Fɪɢ. 2. Illustration of how the simplest two-sequence alignment problem is solved by dynamic programming in a graph. Two short strings are given, and the distance between them is calculated by applying Eq. (1) and registering all distances between subsequences in an integer matrix of dimensions $\{0, \ldots, l1\} \times \{0, \ldots, l2\}$. The alignments corresponding to the minimal distance can be obtained by backtracking in this matrix from $(l1,l2)$ to $(0,0)$. The two obtained alignments are shown. Given an alignment, a set of potential ancestral sequences is obtained by choosing between two states at each position in the alignment where the two sequences differ. These ancestral sequences, excluding the original sequences themselves, are shown for the two alignments. The sequence graph corresponding to the backtracking is shown at the bottom of the figure.

said to be between $s1$ and $s2$,[10] in analogy with points on an interval which obeys the same equation relative to the interval ends.

The traditional dynamic programming algorithm can be generalized to compare sequence graphs and to find sequence subgraphs that are close to each other according to some sequence metric. Simultaneously, the subgraphs are aligned. This permits a rational representation of potential ancestral sequences and selection among these sets of sequences, because

[10] W. H. E. Day, *Bull. Math. Biol.* **46,** 327 (1984).

more information is available in the form of closely related sequences. These graphs are used to represent potential ancestor sequences at each node in the phylogeny. The following generalization of an algorithm of Sankoff[7] allows this. Let $G1$ and $G2$ be two sequence graphs. The distance between two such graphs is defined as the distance between the two closest nucleotide sequences in $S(G1)$ and $S(G2)$, i.e.,

$$d(G1,G2) = \min\{[D(s1,s2)]: s1 \in S(G1), s2 \in S(G2)\} \qquad (3)$$

Let $G1_i$ and $G2_j$ be the subgraphs consisting only of nodes that can be reached from node i and node j, respectively (including these nodes), and only edges involving these nodes. Let $D(G1_i,G2_j)$ be the distance between these two subgraphs. The presence of gap signs leads to a nonstandard definition of a predecessor: Node i' is a predecessor to node i if there is an edge pointing from i and i' or if it is a predecessor to a node that has an edge with a gap sign pointing to it from node i. Between each node, i, and a predecessor, i', to it there will be exactly one edge with a set of nucleotides associated, and this is denoted $n(i, j)$. Let $P(i,1)$ and $P(j,2)$ be the predecessors to i and j in $G1$ and $G2$, respectively. The distances between subgraphs will now obey an equation similar to Eq. (1):

$$D(G1_i,G2_j) = \min\{D(G1_{i'},G2_{j'}) + \text{dist}[n(i,i'),n(j,j')], \qquad (4)$$
$$D(G1_i,G2_{j'}) + 1,$$
$$D(G1_{i'},G2_j) + 1\}$$

The minimum is taken over i' in $P(1,i)$ and j' in $P(2,j)$. The function $\text{dist}(x,y)$ will now be the distance function on sets of nucleotides defined by the two closest members in the two sets according to $d(x,y)$. The initial condition is $d(G1_0,G2_0) = 0$. Again, successive applications of Eq. (4) allow calculation of $d(G1,G2)$. Backtracking picks out the pairs of sequences in $S(G1)$ and $S(G2)$ closest to each other and simultaneously aligns them. If $G1$ and $G2$ represent ordinary sequences, then Eq. (4) reduces to Eq. (1).

The method is illustrated in Fig. 3 for the sequence graph from Fig. 2 and a new sequence. To see the reasoning behind Eq. (4), focus on how $D_{4,3}$ is calculated, when all other entries in the matrix are known. The last position in the alignment of the closest sequence in $S(G2)$, say, s', and $s3$ [$S(G1)$] must either be a match between s' and $s3$, a deletion in $s3$, or a deletion in s'. If it is a match the nucleotide in $s3$ must be G, since $s3$ is a traditional sequence. The last nucleotide in s' could be an A from edge $(2,4)$ or an A or G from $(3,4)$. It could also be a G or C from edge $(1,2)$ if the gap sign at edge $(2,4)$ had been chosen. In total, there are five ways to choose the last nucleotide in s'. By going through these possibilities, it is easily verified that the most parsimonious solution for making a match is to choose the G from edge (3,4), which gives $D_{4,3} = D_{3,2} + d(\text{G,G}) = 1 + 0$. Completely analogous reasoning applies to the cases of deletions in $s3$ or s'. The new sequence, $s3$, in effect selects in the set of equally good alignments of $s1$ and $s2$ and also which ancestor sequences would demand the least extra evolution.

This generalization of the string comparison algorithm becomes extremely useful because it allows the integration of alignments of different sequences. The method actually used has some facets not described above. First, it allows for longer insertions/deletions that are weighted by $g_k = a + bk$. This is done by applying the reasoning of Gotoh[6] to the above graph comparison algorithm. One consequence is that a special type of arrow must be introduced to take care of insertions/deletions, since adding a minus sign to an edge is no longer sufficient. The reason is that adding a series of minus signs to consecutive edges would allow them to be chosen independently, while the insertion/deletion must either be chosen or not in its full length. Second, the method of finding sequence graphs from backtracking is generalized to graph comparison and not just string comparison as described above. Details of this method are reported in Hein.[11]

Most implementations of sequence comparison algorithms use an array of dimensions $\{0, \ldots, l1\} \times \{0, \ldots, l2\}$ to store the $D_{i,j}$ values. As noted by Ukkonen[12] calculations can be restricted to a band around the diagonal $i = j$ (Fig. 4a), which reduces the memory requirements correspondingly. If only the distance between two sequences is needed, only two horizontal rows of the band are needed, which allows the calculation of the distance between two very long sequences, with only little memory. However, if the alignment of the sequences is needed, the introduction of this trick will transform the $O(l2)$ into an $O(l3)$ alignment. When the penultimate position in the alignment is to be calculated the rows corresponding to $j = l2 - 1$ and $j = l2 - 2$ are needed, but the values in the last row have been "forgotten" and must be recalculated starting from row $j = 0$ and up to row $j = l2 - 2$. These repeated recalculations are very time consuming.

It is possible to modify the memory storage and still have an $O(l^2)$ algorithm for much longer sequences. The point is to avoid recalculating all the rows. This can be done by storing the values of evenly spaced rows up through the band (Fig. 4b). Assume the rows are 100 bp long and the computer allows storage for 200 of them. Using 100 rows and spacing them 100 apart and then using the remaining 100 rows to recalculate rows between the saved rows would allow the comparison of two sequences

[11] J. J. Hein, *Mol. Biol. Evol.* In Press .
[12] E. Ukkonen, *Inf. Control* **64**, 100 (1985).

a

Two sequence graphs:

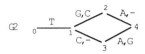

b

Dynamic programming:

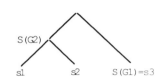

c

Minimal alignment of weight 1:

TCG
ACG

d

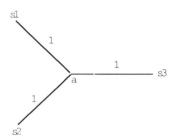

e   Proposed history of s1, s2 and s3

s1 TCA
a   TCG
s2 T-G
s3 ACG

FIG. 3. Sequence graph comparison algorithm. In (a) two small sequence graphs are given. The length of a sequence graph is the number of nodes minus one. (b) Analogously to ordinary sequence comparison the distance between subgraphs are tabulated in a $\{0, \ldots, l1\} \times \{0, \ldots, l2\}$ integer matrix. The nodes belonging to a minimal path are underlined. Minimal edges are not underlined since, in contrast to simple sequence comparison, they can jump over entries. (c) Sequence pair and their alignment. (d) Overall use of the graph comparison algorithm. Two sequence graphs are to be compared. $S(G2)$ represents $s1$, $s2$, and four sequences between $s1$ and $s2$. $S(G1)$ represents only one sequence, $s3$. The effect of introducing $s3$ is to select one alignment of $s1$ and $s2$ as more likely (the second alignment in Fig. 2) and also to select the most likely ancestral sequence compatible with that alignment. The use of this algorithm makes it possible to reconstruct a complete history of the sequences (e). Since the comparison of $G2$ and $s3$ chose one alignment of $s1$ and $s2$ and also their immediate common ancestor, a, and $s3$ was simultaneously aligned to a, all three sequences have been aligned, and the one ancestral sequence in the tree relating them has been predicted.

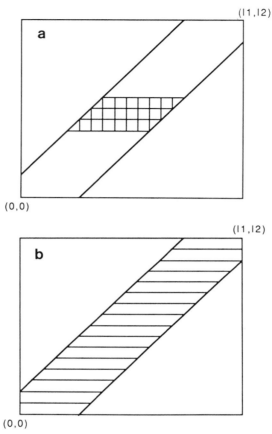

FIG. 4. The complete matrix is not necessary in the dynamic programming algorithm, but only a band around the diagonal up through the matrix (a). Sometimes this band has an area too large for a computer to allocate, in which case it is then possible to formulate the algorithm so it uses only two rows in the band simultaneously. The algorithm becomes much slower, however, since rows must be recalculated when performing the backtracking. The two rows will have to start in the bottom of the matrix; each time an unknown row has to be recalculated. If it is possible to allocate a series of rows, but still not enough to cover the complete band, these rows can be used rationally to store the values of rows uniformly spaced in the matrix, so fewer recalculations need to be performed.

10 kb long instead of 200 bp, a very significant jump indeed. Some recalculations would have to be made; they would roughly double the calculation time, which is still much faster than a cubic algorithm. The memory requirements for this strategy grows proportional to $l^{3/2}$.

The implementation of this method is a bit more complex for sequence

Topology: ((1,2)4,3)5

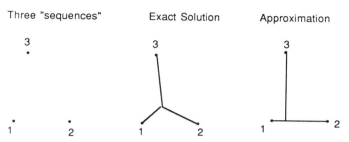

Fig. 5. Nature of the approximation for three sequences represented by points in the plane. First, three unconnected points (sequences) are shown. Then an auxiliary (ancestral sequence) point is introduced so the total length of the resulting tree with the correct topology is minimized. The last tree illustrates the nature of the approximation: all auxiliary points must lie on lines between previously defined points or sets of points.

graphs than for sequences. The complexity of the sequence graph concept prevents the use of a very space-efficient algorithm of Hirschberg[13] and Myers and Miller,[14] the memory requirements of which grow only linearly as a function of sequence length.

The nature of the approximation implicit in the graph comparison algorithm is shown in Fig. 5. It is seen that the best overall solution is not necessarily what seems best with only a subset of the data. For closely related sequences, the approximation probably coincides with the truly minimal solution. The algorithm is most likely better in the space of sequences than in the Euclidean plane.

The following example demonstrates that the algorithm is indeed approximate and does not always give a minimal solution to the phylogenetic alignment problem. Let $s1$ = TATCG, $s2$ = TCGAG, and $s3$ = TCG. The minimal alignment of $s1$ and $s2$ has weight 3 and is

TATCG

TCGAG

The smallest distance from $s3$ to an ancestor according to this alignment is 2, giving a complete history of weight 5. If the suboptimal alignment

TATCG - -

- - TCGAG

[13] D. S. Hirschberg, *Commun. ACM* **18**, 341 (1975).
[14] E. Myers and W. Miller, *CABIOS* **4**, 19 (1988).

of weight 4 was chosen, s3 would belong to one of the ancestors (distance 0), yielding a history of weight 4. This counterexample also demonstrates that suboptimal alignments within e of the optimal can have ancestral sequences that are more distant than e from any optimal ancestral sequence.

Complete Method for Phylogenetic Alignment

The overall strategy is to use pairwise distances, which can be calculated by traditional pairwise alignment methods, then to make an initial tree, to align the sequences following this tree, and finally to switch to the parsimony criteria for the goodness of the tree. The order of calculations is as follows:

1. The most informative distances for the tree construction process are calculated. When the method is applied to larger sequence numbers the complete distance matrix is not calculated, since this is wasteful. Thus, the input to the distance tree method is a sparse matrix. The reason for not using a specific distance is simple: Assume a large tree for many globins has been constructed and now an additional sequence, say, orangutan β-globin, must be added to the tree. A single comparison with a leghemoglobin will reveal that is not a plant globin, thus making comparisons with the other plant globins unnecessary. A different comparison will reveal that it is not a myoglobin, etc. Each such comparison will reveal many distances to be uninformative. This is not important when fewer than 20 sequences are analyzed, but it gives a major speedup when more than 100 sequences are analyzed.

2. A distance tree is constructed for the sequences by adding sequences one by one to a growing tree. There exists a variety of tree construction methods. The method used is new and described in detail in Hein.[11,15]

3. Rearrangements are performed on the obtained tree to improve the overall fit of the tree to the distance data. There exists a variety of possible rearrangements. The simplest is nearest neighbor interchange as first described by Robinson.[16] Each internal edge in tree is visited several times, and each time the three possible configurations of subtrees around the edge are tested, with the best configuration being kept. Here the criterion for best is how well the tree matches the distance data. Felsenstein[17] allows an

---

[15] J. J. Hein, *Bull. Math. Biol.,* in press (1989).

[16] D. F. Robinson, *J. Comb. Theor.* **11,** 105 (1971).

[17] J. Felsenstein, "PHYLIP." Department of Genetics, University of Washington, Seattle, Washington, 1988.

[18] R. Cedergren, M. W. Gray, Y. Abel, and D. Sankoff, *J. Mol. Evol.* **28,** 98 (1988).

option of moving subtrees much further than to a neighbor edge as in nearest neighbor interchange. Cedergren *et al.*[18] has generalized nearest neighbor interchange to include more than the four subtrees flanking one edge. Both these operations allow a more thorough search of the tree topologies, but they slow down the tree construction considerably.

4. The resulting tree is used to guide the alignment algorithm such that a parsimony tree is obtained that has the same topology as the distance tree. It is at this point that the graph comparison algorithm is used. Thus, a complete history is obtained and the ancestral sequences are determined. The new tree might have different branch lengths than the previous distance tree; it will also have an arbitrary root, but it will have the same topology.

5. The criterion for goodness of the history of the sequences is now parsimony, i.e., the fewer events the better. Again, rearrangements are performed on the tree to improve it in an effort to make it more parsimonious.

6. One history/phylogenetic alignment among a set of equally parsimonious possibilities is chosen. Statistics of interest, such as tables of substitutions and insertions/deletions, and the exact dimensions of the phylogeny are calculated.

Program

The program implementing this method consists of about 4000 lines written in standard C language. The code is distributed on 16 modules. It is highly portable, having been transported between different computers a number of times without change. The program accepts a series of user-specified parameters: (1) Are the sequences DNA or protein? (2) A mutation distance matrix: The default for proteins is derived from Doolittle's similarity measure.[2] For DNA, transversions are weighted 5 and transitions 2. (3) Two parameters determining the insertion/deletion weight, g_k, for an insertion/deletion of length k of the form $a + bk$. A typical g_k value is $11 + 3k$. (4) Input modifiers: Are the sequences in GenBank or PIR format? (5) Output modifiers: Are ancestral sequences of interest? Are the different tables of events of interest? (6) A user-specified tree can be used, and this accelerates the program considerably. It allows the user to investigate the mutational weight of different evolutionary hypotheses.

The results of the program are now illustrated on two sets of sequences: 5 S RNAs and globins (proteins). This is not an optimal analysis of these molecules, both of which belong to families with 3–600 sequenced members. Because parsimony is more reliable if branch lengths are short, it would be rational to use a larger number of sequences than the number used. When that many sequences are known, it is also possible to get a detailed picture of the pattern of mutability within the molecule, something that can be profitably used in guiding the alignment. However, the groups of sequences actually used can fit on one page and illustrates the potential of the method nicely.

5 S RNA

Ten 5 S RNAs were taken from GenBank Version 57. The gap penalty used was $g_k = 11 + 3k$, and the output was as extensive as possible. It took 75 sec on a VAX 11/750 to analyze these sequences. The phylogeny is shown in Fig. 6; it is noncontroversial. The data set is too small to be able to determine unambiguously which pair among fungi, animals, and plants is related closest. The statistics of the sequences and their proposed history

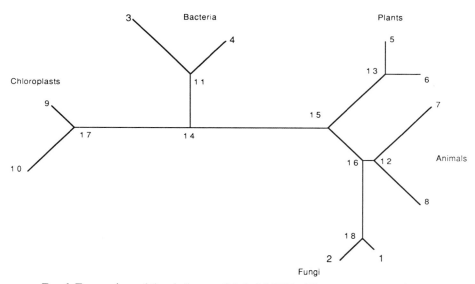

FIG. 6. Ten species and the phylogeny of their 5 S RNAs. The sequences come from fungi: *Auricularia auricula-judae* (1), *Auricularia edulis* (2); bacteria: *Bacillus brevis* (3), *Bacillus firmus* (4); plants: *Equisitum arvense* (5), *Cycad revoluta* (6); animals: *C. elegans* (7), *Gallus gallus* (chicken) (8); chloroplasts: *Jungermannia subulata* (9) and *Dryopteris acuminata* (10). The total branch length of the phylogeny is 843. The only point of contention is that the relationship between fungi, animals, and plants is still subject to debate. Animals and fungi have predominantly showed up as the most closely related, but the sequences are too short to settle the issue. Parsimony trees are likely to underestimate the lengths of long branches. The most probable, biologically and from this analysis, is on the edge between 14 and 15, separating prokaryotes from eukaryotes.

is shown in Fig. 7. The weight of this history is 843, and animals and fungi are the closest kingdoms among the eukaryotes. If the user-tree option is used, it is seen that the weight of the history of the sequences will be 847 if animals and plants are placed as brothers, and also 847 if plants and fungi are placed as closest kingdoms. More details about the evolution of these sequences can be obtained from the alignments shown in Fig. 8.

Globins

Six α-globins, five β-globins, five myoglobins, and four leghemoglobins (including the bacterial globin from *Vitreoscilla*) were chosen for analysis. The same gap penalty was used as for the RNA examples, but in this case the output consisted only of the alignment (Fig. 9) of the extant sequences (no ancestral sequences) and the phylogeny (Fig. 10).

The positions of the kingdoms are completely as should be expected. The leghemoglobins constitute an outgroup. The α- and β-globins are the most recently created groups. The only point of controversy would be the relationship of mammals, birds, and reptiles. In the α- and β-globins the reptiles and birds are sister taxa, while in the myoglobins the mammals and birds are sister taxa. Taxonomy favors the first: birds and reptiles as sister taxa. The weight of the obtained history is 2410.

Remarks

There are things the program cannot do. (1) The homology search the program performs is global, not local. If the sequences differ in length, it should be due to evolutionary changes, not that flanking regions have been unequally sequenced. This restriction can sometimes be circumvented by running the program twice: first to define which regions are common to all sequences and second to run the program only on these regions. (2) It accepts only standard nucleotides and amino acids and simply ignores characters in the sequences that are not standard. (3) The program is not in itself a homology tester. The sequences to be compared should be homologous, and there must be a history to be reconstructed.

Discussion

In summary a method has been described that simultaneously finds the phylogeny and aligns any number of homologous sequences. It introduces new algorithms that permit a more thorough search for optimal alignments. The program is fast and requires less memory than ordinary pairwise alignment programs.

The alignment of multiple sequences is a problem of central impor-

	1	2	3	4	5	6	7	8	9	10
1		90	42	50	56	54	70	68	43	46
2	36		43	50	60	59	69	70	47	47
3	282	266		75	55	52	52	48	58	57
4	269	265	108		52	48	54	56	60	55
5	205	185	247	249		87	59	66	44	46
6	206	182	257	258	52		64	64	43	45
7	135	145	232	238	193	184		72	49	46
8	141	143	266	236	170	158	116		45	46
9	273	255	218	209	279	282	281	276		85
10	268	251	236	228	272	273	279	262	84	

Substitutions:

	A	C	G
C	25		
G	63	40	
T	22	61	13

Insertion-deletions

Length	1	2
Number	13	2

```
a:8
b:3
ds:6
number of sequences:10
total number of pairwise comparisons: 54
total length of alignment: 133
number of conserved sites: 20
total weight of history: 843
```

FIG. 7. At top is a matrix giving the distances (lower triangle) and percent homology (upper triangle) between all pairs of sequences. Visual inspection immediately affirms that sequences from the same kingdom have small distances between them and have a high percent homology. The parameters a and b are those from the gap penalty function. ds is the width of the band within which the dynamic programming is performed; the smaller this number, the faster the program will run; however, it must be sufficiently large to not exclude the best alignment. Then, the genetic events occurring in this history is tabulated. Since the tree is unrooted it is not possible to determine the direction of events. The notable feature is the high occurrence of transitions (A $\leftrightarrow$ G and C $\leftrightarrow$ T) relative to transversions. The insertions/deletions are all very short. The total number of comparisons is $54 = (10 \times 9)/2$ (distance tree construction) $+ (10 - 1)$ (number of ancestral sequences reconstructed, including one arbitrary root). The first term, $n(n - 1)/2$, is unnecessarily large when n is large.

```
10  tatt-ctggtgtcccaggcgtagaggaaccacaccgatccatctcgaacttggtggtgaaactctg  65
17  t--t-ctggtgtcccaggcgtagaggaaccacaccaatccatcccgaacttggtggtgaaactctg  63
 9  t--t-ctggtgtctcaggcgtggaggaaccacaccaatccatcccgaacttggtggtgaaactcta  63
14  t----ctggtggccatggcgtagaggaaacaccccatcccataccgaactcggcagttaagctctg  62
 3  t----ctggtgatgatggcggaggggacacacccgttcccataccgaacacggccgttaagccctc  62
11  t----ctggtggcgatggcgaagaggacacacccgttcccataccgaacacggcagttaagctctc  62
 4  t----ctggtggcgatagcgagaaggtcacacccgttcccataccgaacacggaagttaagcttct  61
15  g----cctgcggccatagcaccgtgaaagcaccccatcccat-ccgaactcggcagttaagcacgg  61
 8  g----cctacggccatcccaccctggtaacgcccgatctcgt-ctgatctcggaagctaagcaggg  61
12  g----cctacggccataccccctgaaagcaccccatcccgt-ccgatctgggaagttaagcaggg  61
 7  g----cttacggaccatatcacgttgaatgcacgccatcccgt-ccgatctggcaagttaagcaacg  61
16  g----cctacggccatagcaccctgaaagcaccccatcccgt-ccgatctgggaagttaagcaggg  61
 1  a----tccacggccataggactctgaaagcactgcatcccgt-ccgatctgcaaagttaaccagag  61
18  a----tccacggccataggactctgaaagcaccgcatcccgt-ccgatctgcgaagttaaacagag  61
 2  a----tccacggccataggactgtgaaagcaccgcatcccgt-ctgatctgcgcagttaaacacag  61
 5  g---tggtgcggtcataccagcgctaatgcaccgcatcccat-cagaactccgcagttaagcgcgc  62
13  g----ggtgcggtcataccagcgttaatgcaccggatcccat-cagaactccgcagttaagcgcgc  61
 6  g----ggtgcgatcataccagcgttaatgcaccggatcccat-cagaactccgcagttaagcgcgc  61
          *                   *  *          * *   ** *        *  ** *
```

```
10  ccgcggt--aaccaatact-cg-gg-gggggccct-gcggaaaaatagctcgatgccagga--ta  122
17  ctgcggt--ga-cgatact-tg-gg-gggagcccg-atggaaaaatagctcgatgccagga--t-  118
 9  ttgcggt--ga-cgatactgta-gg-ggaagcccg-atggaaaaatagctcgatgccagga--t-  119
14  ctgcgcc--ga-tggtact-tg-gg-gggagcccg-ctgggaaaataggacgctgccag-a--t-  116
 3  cagcgcc--aa-tggtact-tgctc-cgcagggag-ccgggagagtaggacgtcgccag-g--c-  117
11  cagcgcc--ga-tggtact-tg-gg-ggcagtccg-ctgggagagtaggacgctgccag-g--c-  116
 4  cagcgcc--ga-tggtagt-ta-gg-ggctgtccc-ctgtgagagtaggacgctgccag-g--c-  116
15  ttgcgcccaga-tagtact-tg-ggtgggagaccgcctgggaaacctggatgctgcaag-c--t-  119
 8  tcgggcctggt-tagtact-tg-gatgggagacctcctgggaataccgggtgctgtagg-ct-t-  120
12  ttgagcccagt-tagtact-tg-gatgggagaccgcctgggaatcctgggtgctgtagg-c--t-  119
 7  ttgagtccagt-tagtact-tg-gatcggagacggcctgggaatcctggatgttgtaag-c--t-  119
16  ttgcgcccagt-tagtact-tg-ggtgggagaccgcctgggaatcctgggtgctgtagg-c--t-  119
 1  taccgcccagt-tagtacc-ac-ggtggggggaccacgcgggaatcctgggtgctgt-gg-t--t-  118
18  taccgcccagt-tagtacc-ac-ggtgggggaccacatgggaatcctgggtgctgt-gg-t--t-  118
 2  tgccgcctagt-tagtacc-at-ggtgggggaccacatgggaatcctgggtgctgt-gg-t--t-  118
 5  ttgggccagaa-cagtact-gg-gatgggtgacctcccggggaagtcctggtgccgcacc-c--c-  120
13  ttgggccagcc-tagtact-ag-gatgggtgacctcctgggaagtcctgatgctgcacc-c--t-  119
 6  ttgggttggag-tagtact-ag-gatgggtgacctcctgggaagtcctaatattgcacc-c-tt-  120
        *          **        *  *        *  *              *
```

FIG. 8. Phylogenetic alignment of the ten 5 S RNA sequences. Numbers 1 to 10 represent the extant sequences (see Fig. 6), while 11 to 18 represent reconstructed ancestor sequences.

tance to any molecular biologist who has acquired a sequence with known homologous variants. It could be a member of a multigene family or a sequence that has been determined for the first time in a given species, or both. There are alternative approaches to aligning many sequences than the phylogenetic approach adopted here, but a phylogenetic method has several advantages. First and most obvious, it solves both individual problems, but also better than if they were treated separately. The phylogeny reconstruction gains from incorporating alignment into the method. Previously, an alignment had to be presented to the phylogeny program, which meant that the position of insertions/deletions had already been

```
18  g--v-ltdvqvalvkssfeefnanipknthrfftlvleiapgakdlfsflkgssevpqnnpdlqahagkvfkltveaaiqlevn
17  g----ftekqealvnsssqlfkqnpsnusvlfytiilqkaptakamfsflkdsagvv-dspklgahaekvfgmvrdsavqlrat
19  vnkv-fteeqealvvkawavmkknsaelglqflki-feiapsaknlfsylkdspvpleqnpklkphattfvmttesavqlrkag
20  ---m-ldqqtiniikatvpvlkehgvtitttfyknlfakhpevrplfdmgtq--esleqpkalamtvlaaaqnienlpailpav

 7  ---vhltpe----eksavtalwgkvnvd--evggealgrllvvypwtqrffesfgdlstpdavmgnpkvkahgkkvlgafsdgl
 8  ---vhltse----ekncittiwskvqvd--qtggealgrmlvvypwttrffgsfgdlsspgavmnsskvqahgakvltsfgeav
 9  ---vhwtae----ekqlitglwgkvnva--ecgaealarllivypwtqrffasfgnlssptailgnpmvrahgkkvltsfgdav
10  ---asfdph----ekqligdlwhkvdva--hcggealsrmlivypwkrryfenfgdisnaqaimhnekvqahgkkvlasfgeav
11  ---gs----------dlvsgfwgkvdag--kiggealarllvvyqwtqryfttfgnlgsadaichnakvlahgekvlaaegegl

 2  ---v-lsaa----dknnvkgawekvgthageygaealermflsfpttktyfphf------dlthgsaqnkahgqkvgaaltkav
 1  ---v-lspa----dktnvkaawgkvgahageygaealermflsfptthtyfphf------dlshgsaqnkghgkkvadaltnav
 3  ---v-lsaa----dkghvkaiwgkvgghageyaaeglertfhsfptktyfphf------dlshgsaqiqahgkkiadalgqav
 4  ---v-lsgt----dktnvkgifskisshaeeygaetlermfitypqtktyfphf------dlhhgsaqikahgkkvanalieav
 5  ---v-lssd----dkcnvkavwskvaghleeygaealermfcaypqtkiyfphf------dlshgsaqirahgkkvfaalheav
 6  ---s-lsas----ekaavlsivgkigsqgsalgsealtrlflsfpqtktyfphf------dltpgsadlnthggkiinalagaa

16  ---g-lsdd----ewhhvlgiwakvepdlsahgqeviirlfqvhpetqerfakfknlktidelrsseevkkhgttvltalgril
15  ---g-lsdq----ewqqvltiwgkveadiaghghevlmrlfhdhpetldrfdkfkglktepdmkgsedlkkhgqtvltalgaql
14  ---g-lsdg----ewqlvlnawgkveadipghgqevlirlfkghpetlekfdkfkhlksedemkasedlkkhgatvltalgnil
12  ---g-lsdg----ewqlvlnvwgkveadipghgqevlirlfkghpetlekfdkfkhlksedemkasedlkkhgatvltalggil
13  ---g-lsdg----ewqlvlnvwgkveadvaghgqdilirlfkghpetlekfdkfkhlkteadmkasedlkkhgdtvltalgail
```

```
18  gqvas-datlkslgsvhvsk-vvdahfppvvkeailktikevvgd-kwseelntawtiaydelaiiikke-----mk-da---a
17  gevvl-dgk---dgsihiqk-gvldphfvvvkeallktikeasgd-kwseelsaawevaydglataik-----------a---a
19  kvtvk-esdlkrigaihfkt-gvvnehfevtrfalletikeavpe-mwspemknawgvaydqkvaaikfe-----mkpss---t
20  ----------kkiavkhcqa-gvaaahypivgqellgaikevlgd-aatddildawgkayqviadvfiqveadlyaq-av---e

 7  ahldnlkgtfatlselhcdklhvdpenfrllgnvlvcnlahhfgk-eftppvqaayqkvvagvanalahk-----yh-------
 8  khlddlkgtyaklselhcdklhvdpenfkmlg-iiviclaehfgk-dftpecvaw--klvvagvahalhk-----yh-------
 9  knldniknt fsqlselhcdklhvdpenfrllgdiliivlaahfsk-dftpecqaawqklvrvvahalark-----yh-------
10  chldgirahfanlsklhceklhvdpenfkllgdiiiivlaahypk-dfglechaayqklvrqvaaalaae-----yh-------
11  khpenlkahyaklseyhsnklhvdpanfrllgnvfitvlarhfqh-eftpelqhaleahfcavgdalaka-----yh-------

 2  ghlddlpnalsdlsdlhahklrvdpvnfkllshcllvtlsrhlpeqeftpavhasldkffsnvstvltsk-----yr-------
 1  ahvddmpnalsalsdlhahklrvdpvnfkllshcllvtlaahlpa-eftpavnasldkflasvstvltsk-----yr-------
 3  ehiddlpgtlslsdlhahklrvdpvnfkllshcllvtfaahlgd-aftpevhasldkflaavstvltsk-----yr-------
 4  nhiddisgalsklsdlhaqklrvdpvnfkllgqcflvvvaihhps-altpevhasldkflcavgavltak-----yr-------
 5  nhiddlpgalcrlselhahslrvdpvnfkflaqcvlvvvakhhpg-sltpevhasldkflcavsvvltsk-----yr-------
 6  nhlddlagnlsslsdlhaynlrvdpgnfpllahiiqvvlathfpg-dftaevqaawdkflalvsavltsk-----yr-------

16  klknnhepelkplaeshatkhhipvkyleficeiivkviaekhps-dfgadsqaamrkalelfrndmask-----yk-efgfqg
15  kkkghheadlkqlaqthatkhhipvkylefiseviikviaekhaa-dfgadsqaamkkalelfrddmask-----yk-efgfqg
14  kkkgnheaelkplaqshatkhhisvqflefiseaiiqviqskhpg-dfggdaqaamgkalelfrndmaak-----yk-elgfqg
12  kkkgnheaelkplaqshatkhhipvkylefiseciiqvqskhpg-dfgadaqgamnkalelfrkdmasn-----yk-elgfqg
13  kkkghhdaelkplaqshatkhhipikylefiseaiihvlhsrhpa-qfgadaqgamnkalelfrkdiaak-----yk-elgfhg
```

FIG. 9. Marginal alignment of the globins without ancestral sequences; different types of globins are separated by blank lines. See the legend to Fig. 10 for identification of species 1–20.

fixed, even though they are as important a class of events as substitutions/mutations.

A phylogenetic alignment mimics the result of an infinitely efficient manual aligner very well: first by comparing closely related sequences, and then progressing to more distantly related sequences. Nonphylogenetic methods assume a symmetric relationship between all sequences, which implicitly bases the alignment on the special phylogeny that has one inter-

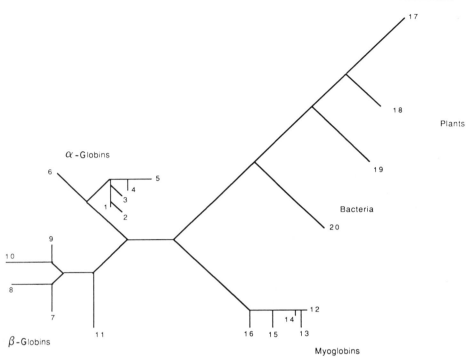

FIG. 10. The phylogeny of the 20 globins has four major clusters: α-, β-, myo-, and leghemoglobin. The sequences are from the following species: α-Globins: human (1), Abyssinian hyrax (2), Eastern Gray kangaroo (3), ostrich (4), Nile crocodile (5), bullfrog (6); β-globins: human (7), North American opossum (8), chicken (9), Nile crocodile (10), edible frog (11); myoglobins: human (12), saddleback dolphin (13), North American opossum (14), chicken (15), map turtle (16); leghemoglobins: broad bean I (17), yellow lupine I (18), *Parasponia andersonii* (19), and bacterial globin from *Vitreoscilla* (20). The most likely position of the root is on the branch leading to leghemoglobins (including *Vitreoscilla*) or possibly on the branch leading to *Vitreoscilla*.

nal node and all extant sequences connected to this node. Second, homologous sequences do have a history, and that history is of interest, whatever the motivation for aligning the sequences. When nonphylogenetic methods are used, the results are very frequently interpreted as statements about evolution anyway. Last, compared to manual alignment, an algorithm is time saving, highly practical, and permits computer simulations to evaluate the significance of obtained results.

The program presented is very useful and reliable in many everyday applications. Since the program can analyze very large sets of sequences, an auxiliary program, CONDENSE.C, was developed that can write out marginal phylogenies or phylogenies that have compressed a set of leaves into one leaf in order not to lose the overview of a very large phylogeny. A modification to the program facilitating comparison of more distantly related groups of internally similar sequences has also been developed.

The central graph comparison algorithm could be extended in two useful ways to make the approximation even better. First, it would be possible to compare more than two graphs simultaneously. This would be very straightforward and completely analogous to the generalization of two-sequence comparison to three-sequence comparison.[5] Second, near-optimum methods[19] could be used to investigate not only sequences between two sequence sets, but also sequences almost between two sequence sets.

The limits to the reliability of the method are set by being a parsimony method. The solution to the parsimony problem is an almost complete history of the sequences. The question is, "How reliable is this history?" The history of the sequences has two connected facets: the alignment and the phylogeny of the sequences. The usefulness of alignments extends far beyond the point where the actual assignment of insertions/deletions and mutations is to believed. It might still match similar regions in an informative way and can also be used to test whether homologies have arisen by chance or not. It is known that traditional parsimony methods can be led astray if rates of mutations are highly uneven in different lineages.[20,21] Since the method reduces to the traditional parsimony method, if insertions/deletions do not occur, it suffers from this shortcoming as well.

A serious charge against parsimony methods is their nonstatistical nature: There is no inherent information about the reliability of the obtained results. Felsenstein[22] has used the method of bootstrapping to counter this. Central to this method is the notion that the input is a series of independently observed vectors of character states (columns in the alignment) that can be sampled. It is not possible to use bootstrapping in any combined phylogeny–alignment method. Each complete sequence is now the character state. However, statistical methods must be given sequences in a prealigned state at present, making multiple alignment methods indispensable.

In conclusion, the method and program presented here should be very useful to researchers in molecular biology and evolution, especially because of its user friendliness and modest running time and memory requirements.

[19] T. H. Byers and M. S. Waterman, *Math. Biosci.* **77**, 179 (1985).
[20] J. Felsenstein, *Syst. Zool.* **27**, 401 (1978).
[21] J. A. Lake, *J. Mol. Evol.* **26**, 59 (1987).
[22] J. Felsenstein, *Evolution* **39**, 783 (1985).

Acknowledgments

The author is grateful for the encouragement and advice received from Dr. Russell Doolittle, Dr. Charles Langley, and Dr. Michael Waterman.

[40] Statistical Tests of Molecular Phylogenies

By WEN-HSIUNG LI and MANOLO GOUY

Introduction

Although phylogenetic reconstruction has long been recognized as a problem in statistical inference,[1] few methods have been developed for evaluating the confidence level of estimated phylogenies. This deficiency has occurred largely because of the complexity of the problem. This problem has become important because the rapid accumulation of molecular data has generated much interest in phylogenetic studies.[2,3] In fact, it is no longer acceptable to throw sequences through any available tree-making method and to publish the results without some evaluation of the reliability of the inferred phylogeny.[4]

The purpose of this chapter is to provide a summary of methods for testing the significance of inferred phylogenies. We classify analytic methods into the character state approach and the distance matrix approach. We also review the bootstrap approach, which resamples the data to infer the variability of the estimate obtained by a tree-making method.[5]

Character State Approach

In a pioneering work, Cavender[6] studied the confidence limits that can be placed on a phylogeny inferred from a four-species data set using parsimony methods. The results were given in terms of the total number of characters. Felsenstein[7] studied the same problem under the assumption of

[1] A. W. F. Edwards and L. L. Cavalli-Sforza, *in* "Phenetic and Phylogenetic Classification" (V. H. Heywood and J. McNeill, eds.), p. 67 (Syst. Assoc. Publ. No. 6). Systematics Association, London, 1964.
[2] M. Nei, "Molecular Evolutionary Genetics." Columbia Univ. Press, New York, 1987.
[3] J. Felsenstein, *Annu. Rev. Genet.* **22,** 521 (1988).
[4] D. Penny, *Nature (London)* **331,** 111 (1988).
[5] J. Felsenstein, *Evolution* **39,** 783 (1985).
[6] J. A. Cavender, *Math. Biosci.* **40,** 271 (1978).
[7] J. Felsenstein, *Syst. Zool.* **34,** 152 (1985).

METHODS IN ENZYMOLOGY, VOL. 183

a constant rate of evolution (an evolutionary clock) and showed that the confidence limits are narrower than those obtained by Cavender. He considered inferences made by using parsimony methods, and the results were given in terms of the number of "phylogenetically informative" characters (see definition below). More recently, Lake[8] proposed the evolutionary parsimony method. These two methods are reviewed below.

Maximum Parsimony

Consider the problem of inferring the branching order of three species with an outgroup (species 4). The three possible rooted trees are shown in Fig. 1a–c. In terms of parsimony a nucleotide site is informative (useful) for choosing among the three trees only if it is in the same state in two of the four species and in another state in the other two species. For example, if the site has the configuration AAGG among the four species, then it supports the first tree (Fig. 1a) because this tree requires (at least) only one nucleotide substitution to explain the configuration, whereas the other trees (Fig. 1b,c) each require two substitutions. As another example, if the configuration is AAGC, then it is not informative because each of the three trees requires two substitutions. Under the assumption of rate constancy, an informative site has a higher probability (say p) of supporting the true tree and a lower probability (say q) of supporting each of the two other trees. The probability that among n random informative sites the number (S) of sites supporting a particular incorrect tree is m or larger is given by

$$\text{Prob}(S \geq m) = \sum_{i=m}^{n} \frac{n!}{i!(n-i)!} q^i (p+q)^{n-i} \qquad (1)$$

which is obtained by expanding $[q + (p + q)]^n$. Since $p + 2q = 1$ and since $q \leq p$, Eq. (1) assumes the maximum value when $p = q = \frac{1}{3}$, i.e., if the three species represent a true trichotomy (Fig. 1d).

Now suppose in a sequence data set with n informative sites the best supported tree is favored by m sites. Is this tree the true tree? Let us assume that this tree is incorrect and has by chance been supported by so many sites. The probability for a particular incorrect tree to be supported by m or more sites is given by Eq. (1). Therefore, the probability for one of the two incorrect trees to be supported by m or more sites is

$$P = 2 \, \text{Prob}(S \geq m) \qquad (2)$$

If P is smaller than α, then the observed tree is significant at the level of α. In practice q is unknown, and it is necessary to use the least favorable

[8] J. A. Lake, Mol. Biol. Evol. 4, 167 (1987).

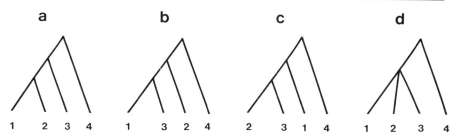

FIG. 1. a, b, and c represent the three possible rooted, bifurcating trees for three species with one outgroup, and d represents the case of trichotomy.

value, $\frac{1}{3}$. The above theory was developed by Felsenstein.[7] He has also developed a statistic for testing whether the best supported topology is significantly better than the next best topology. The computation of this statistic is more complicated, but the powers of the two tests are about the same.[7] This is not surprising because if the best supported tree is significant, it is likely to be significantly better than the next best tree (and also the worst tree), and vice versa.

Equation (2) assumes that a true trichotomy is impossible so that only two of the three trees can be incorrect. In practice, however, the most difficult cases to resolve are those that are close to a true trichotomy; a good example is the branching order of humans, chimpanzees, and gorillas. Therefore, a more rigorous analysis is to take trichotomy as the null hypothesis. The probability for rejecting this hypothesis is

$$P = 3 \operatorname{Prob}(S \geq m) \tag{3}$$

Under this hypothesis $q = \frac{1}{3}$. We suggest that an inferred phylogeny should be taken with caution if the probability obtained from Eq. (2) is less than 0.05 whereas that obtained from Eq. (3) is larger than 0.05.

Although the above results were derived under the assumption that one of the four species is known to be an outgroup, they apply equally well in the absence of this knowledge if one considers unrooted trees. Furthermore, the results should hold approximately even under nonconstant rates of evolution if the sequences are fairly closely related so that parallel and back mutations are rare.

Table I shows the minimum m value required for Eq. (1) to be smaller than or equal to a given probability, assuming that n informative sites are observed and that $q = p = \frac{1}{3}$. It can be used to determine the significance level. For example, if $n = 10$ and $m = 8$, then the significance level is approximately $2 \times 0.5\% = 1\%$ or $3 \times 0.5\% = 1.5\%$, depending on whether Eq. (2) or Eq. (3) is used. If n is greater than 100, Eq. (1) can be easily

TABLE I

Minimum m Value Required for Prob($S \geq m$) in Eq. (1) to Be Smaller than or
Equal to δ^a

n	δ (%)					n	δ (%)				
	0.05	0.30	0.50	1.00	2.50		0.05	0.30	0.50	1.00	2.50
10	9	9	8	8	7	34	22	20	20	19	18
11	10	9	9	8	8	36	23	21	20	20	19
12	11	10	9	9	8	38	24	22	21	21	19
13	11	10	10	9	9	40	24	23	22	21	20
14	12	11	10	10	9	43	26	24	24	23	22
15	12	11	11	10	10	47	28	26	25	24	23
16	13	12	11	11	10	50	29	27	26	26	24
17	13	12	12	11	11	53	30	28	28	27	26
18	14	13	12	12	11	57	32	30	29	28	27
19	14	13	13	12	11	60	33	31	31	30	28
20	15	14	13	13	12	63	35	33	32	31	29
21	15	14	14	13	12	67	36	34	34	33	31
22	16	15	14	14	13	70	38	35	35	34	32
23	16	15	15	14	13	73	39	37	36	35	33
24	17	16	15	15	14	77	41	38	38	36	35
25	17	16	16	15	14	80	42	40	39	38	36
26	18	16	16	15	14	83	43	41	40	39	37
27	18	17	17	16	15	87	45	42	42	40	39
28	19	17	17	16	15	90	46	44	43	42	40
29	19	18	17	17	16	93	47	45	44	43	41
30	20	18	18	17	16	97	49	46	46	44	43
32	21	19	19	18	17	100	50	48	47	46	44

[a] It is assumed that $q = p = \frac{1}{3}$. n, Number of informative sites observed.

computed by assuming that m/n follows a normal distribution with mean
equal to $\frac{1}{3}$ and variance equal to $\frac{1}{3} \times \frac{2}{3}/n$.

Evolutionary Parsimony

The evolutionary parsimony method also applies to only four lines of
descent. Since it assumes that all nucleotides are equivalent and makes a
distinction only between transitional and transversional substitutions, the
representation of sequence data can be simplified as follows.[8] Label the
four DNA sequences under study by 1, 2, 3, and 4. We work along the
sequences from the 5′ to the 3′ end and, at each position, recode the
nucleotides in all four sequences in terms of their relationships to the
nucleotide in sequence 1 at that position. All nucleotides that are the same
as the one in sequence 1 are represented by "1," those related by a

transitional mutation are represented by "2," and the two possible trans-versions are represented by "3" and "4" in the order we come across them, if at all. With this notation, any configuration of four nucleotides can be represented by 1 of 36 four-dimensional vectors. For example, the two sets of nucleotides ATCG and GCCA are recoded as vectors 1342 and 1332, respectively.

Let us assume that the tree root is known so that the three possible alternative trees can be represented by Fig. 1a–c, though the theory applies equally well without this knowledge. For a set of nucleotide data, let N_{ijkl} be the total score of the configuration $ijkl$ over the length of the sequences. Lake[8] defined the following three quantities:

$$X = N_{1133} + N_{1234} - N_{1233} - N_{1134} \qquad (4)$$

$$Y = N_{1313} + N_{1324} - N_{1323} - N_{1314} \qquad (5)$$

$$Z = N_{1331} + N_{1342} - N_{1332} - N_{1341} \qquad (6)$$

where X is the score associated with the tree in Fig. 1a, Y with the tree in Fig. 1b, and Z with the tree in Fig. 1c. The reason for subtracting the last two terms in each score is that this would make the total score for an erroneous tree statistically 0. Lake[8] called, for the tree in Fig. 1a, $A = N_{1133} + N_{1234}$ the parsimony-like term and $B = N_{1233} + N_{1134}$ the back-ground term, and proposed $(A - B)^2/(A + B)$ as a χ^2 with one degree of freedom for testing whether the X score is significantly greater than 0; if the number of sites is small, one should test the equality of A and B by the binomial test.[9] He suggested that a tree is "proven" if its associated χ^2 value is significant whereas the other two χ^2 values are not. We note, however, since three independent χ^2 values are tested, the significant level is not α, but is approximately $1 - (1 - \alpha)^3 \approx 3\alpha$. Lake[8] has extended the theory to the case of multiple species in each lineage, but the mathematical treat-ment does not appear to be rigorous.

The method of Lake may be more robust against unequal rates of evolution than the maximum parsimony method (see later). However, it utilizes a relatively small part (only transversional differences) of the se-quence information and is likely to be less efficient than the maximum parsimony method.

Distance Matrix Approach

Instead of testing the significance of an inferred phylogeny, it is simpler to test the significance of estimated internodal distances. As explained

[9] R. Holmquist, M. M. Miyamoto, and M. Goodman, *Mol. Biol. Evol.* 5, 217 (1988).

below, in the case of four taxa, significance of the internodal distance can be taken as significance of the inferred phylogeny. When the number of taxa under study is more than four, the two problems are no longer equivalent, and the requirement of all internodal distances being significantly greater than 0 seems to be a too stringent test for the significance of the inferred branching order. A simple way to test the significance of internodal distances is to study their variances.

Variances of Internodal Distances

Mueller and Ayala[10] proposed computing these variances by the jack-knife method, while Nei *et al.*[11] derived analytic formulas for the case of a UPGMA tree, i.e., a tree estimated by the unweighted pair-group method.[12] The UPGMA method assumes a constant rate of evolution, but there is now strong evidence that this assumption is often violated.[13] It is therefore desirable to consider an approach that does not make this assumption. Li[14] proposed a two-step approach. The first step is to infer the branching order. One can use the transformed distance method,[15-17] the neighbor-joining method,[18] or any other method that does not assume rate constancy and that has been shown to be effective for obtaining the correct tree. The second step is to estimate the branch lengths by the least-squares method.[19,20] The variances of internodal distances are then obtained from the equations derived from the least-squares method. The variances for the cases of four and five taxa are given below.

Four Taxa. Suppose that the inferred tree topology is as shown in Fig. 2a; the root of the tree can be determined if one of the four taxa is an outgroup. The branch lengths estimated by the least-squares method are

$$a = \tfrac{1}{2}d_{12} + \tfrac{1}{4}(d_{13} - d_{23} + d_{14} - d_{24}) \tag{7}$$

$$b = d_{12} - a \tag{8}$$

[10] L. D. Mueller and F. J. Ayala, *Genet. Res.* **40**, 127 (1982).
[11] M. Nei, J. C. Stephens, and N. Saitou, *Mol. Biol. Evol.* **2**, 66 (1985).
[12] P. H. A. Sneath and R. R. Sokal, "Numerical Taxonomy." Freeman, San Francisco, California, 1973.
[13] W.-H. Li, M. Tanimura, and P. M. Sharp, *J. Mol. Evol.* **25**, 330 (1987).
[14] W.-H. Li, *Mol. Biol. Evol.* **6**, 424 (1989).
[15] J. S. Farris, *in* "Major Patterns in Vertebrate Evolution" (M. K. Hecht, P. C. Goody and B. M. Hecht, eds.), p. 823. Plenum, New York, 1977.
[16] L. C. Klotz, N. Komar, R. L. Blanken, and R. M. Mitchell, *Proc. Natl. Acad. Sci. U.S.A.* **76**, 4516 (1979).
[17] W.-H. Li, *Proc. Natl. Acad. Sci. U.S.A.* **78**, 1085 (1981).
[18] N. Saitou and M. Nei, *Mol. Biol. Evol.* **4**, 406 (1987).
[19] L. L. Cavalli-Sforza and A. W. F. Edwards, *Am. J. Hum. Genet.* **19**, 233 (1967).
[20] R. Chakraborty, *Can. J. Genet. Cytol.* **19**, 217 (1977).

a   b

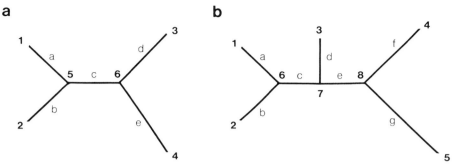

FIG. 2. Model trees used in the derivation of the mean and variance of the branch lengths.

$$c = \tfrac{1}{4}(d_{13} + d_{23} + d_{14} + d_{24}) - \tfrac{1}{2}(d_{12} + d_{34}) \qquad (9)$$

$$d = \tfrac{1}{2}d_{34} + \tfrac{1}{4}(d_{13} + d_{23} - d_{14} - d_{24}) \qquad (10)$$

$$e = d_{34} - d \qquad (11)$$

and the variance of c is

$$\begin{aligned} V(c) = \tfrac{1}{16}[&V(d_{13}) + V(d_{23}) + V(d_{14}) + V(d_{24}) \\ &+ 2V(d_{16}) + 2V(d_{26}) + 4V(d_{56}) + 2V(d_{53}) + 2V(d_{54})] \\ &- \tfrac{1}{8}[V(d_{15}) + V(d_{25}) + V(d_{63}) + V(d_{64})] \\ &+ \tfrac{1}{4}V(d_{12}) + \tfrac{1}{4}V(d_{34}) \end{aligned} \qquad (12)$$

where $V(d_{ij})$ denotes the variance of the estimate of the distance between sequences i and j.

For protein sequence data the mean and variance of the number (d_{ij}) of amino acid replacements per site between sequences i and j can be estimated by

$$d_{ij} = -w \ln(1 - p/w) \qquad (13)$$

$$V(d_{ij}) = p(1 - p)/[L(1 - p/w)^2] \qquad (14)$$

where $w = \tfrac{19}{20}$, p is the proportion of different amino acids between the two sequences, and L is the number of residue sites compared. For a pair of extant sequences, Eqs. (13) and (14) are readily applicable. However, the sequences at nodes 5 and 6 do not exist, and thus variances such as $V(d_{16})$ and $V(d_{56})$ cannot be estimated directly from actual data. However, they can be estimated as follows,[11] using $V(d_{16})$ as an example. From Eq. (13),

$$p = w(1 - e^{-d_{ij}/w}) \qquad (15)$$

Since $d_{16} = a + c$, $p = w[1 - e^{-(a+c)/w}]$; $a + c$ can be obtained from Eqs. (7) and (8). Putting p into Eq. (14) one readily obtains $V(d_{16})$.

Next, consider nucleotide sequence data. Under the assumption of random substitution among the four types of nucleotides, i.e., the one-parameter model, the mean and variance of the number of substitutions per nucleotide site between sequences i and j are also given by Eqs. (13) and (14) except that now $w = \frac{3}{4}$, p is the proportion of different nucleotides between the two sequences, and L is the number of nucleotide sites compared.[21,22] Under the two-parameter model,[23] the formulas corresponding to Eqs. (13) and (14) are

$$d_{ij} = A + B \tag{16}$$

$$V(d_{ij}) = [x^2 P + z^2 Q - (xP + zQ)^2]/L \tag{17}$$

where P and Q are, respectively, the proportions of transitional and transversional differences between sequences i and j, $A = \frac{1}{2} \ln(x) - \frac{1}{4} \ln(y)$ is the number of transitional substitutions per site, $B = \frac{1}{4} \ln(y)$ is the number of transversional substitutions per site, $x = 1/(1 - 2P - Q)$, $y = 1/(1 - 2Q)$, and $z = (x + y)/2$. Note that, unlike Eq. (14), Eq. (17) involves two parameters, P and Q. The formulas corresponding to Eq. (15) are given by

$$Q = \frac{1}{2}(1 - e^{-2B}) \tag{18}$$

$$P = \frac{1}{2}[1 - Q - e^{-(2A+B)}] \tag{19}$$

Five Taxa. Suppose that the inferred branching order is as shown in Fig. 2b. The branch lengths are then given by

$$a = \frac{1}{2}d_{12} + \frac{1}{6}[d_{13} - d_{23} + d_{14} - d_{24} + d_{15} - d_{25}] \tag{20}$$

$$b = d_{12} - a \tag{21}$$

$$c = \frac{1}{4}(d_{13} + d_{23}) + \frac{1}{8}(d_{14} + d_{24} + d_{15} + d_{25}) - \frac{1}{4}(d_{34} + d_{35}) - \frac{1}{2}d_{12} \tag{22}$$

$$d = \frac{1}{2}(d_{13} + d_{23}) - \frac{1}{2}d_{12} - c \tag{23}$$

$$e = \frac{1}{4}(d_{34} + d_{35} - d_{13} - d_{23}) + \frac{1}{8}(d_{14} + d_{24} + d_{15} + d_{25}) - \frac{1}{2}d_{45} \tag{24}$$

$$f = d_{34} - d - e \tag{25}$$

$$g = d_{45} - f \tag{26}$$

If two of the five taxa, say, taxa 4 and 5, are known to be outgroups, then

[21] T. H. Jukes and C. R. Cantor, *in* "Mammalian Protein Metabolism" (H. N. Munro, ed.), p. 21. Academic Press, New York, 1969.
[22] M. Kimura and T. Ohta, *J. Mol. Evol.* **2**, 87 (1972).
[23] M. Kimura, *J. Mol. Evol.* **16**, 111 (1980).

one needs to obtain only the variance of c:

$$V(c) = \tfrac{1}{16}[V(d_{13}) + V(d_{23}) + 2V(d_{17}) + 2V(d_{27}) + 4V(d_{67}) + 2V(d_{63})]$$
$$+ \tfrac{1}{64}[V(d_{14}) + V(d_{24}) + V(d_{15}) + V(d_{25}) + 2V(d_{18})$$
$$+ 2V(d_{28}) + 4V(d_{68}) + 2V(d_{64}) + 2V(d_{65})]$$
$$- \tfrac{1}{2}[V(d_{16}) + V(d_{26}) + V(d_{73})]$$
$$- \tfrac{1}{8}[2V(d_{78}) + V(d_{74}) + V(d_{75})]$$
$$+ \tfrac{1}{16}[V(d_{34}) + V(d_{35}) + 2V(d_{38})] + \tfrac{1}{4}V(d_{12}) \tag{27}$$

If only one or no outgroup is known, then one needs also to obtain the variance of e:

$$V(e) = \tfrac{1}{16}[V(d_{34}) + V(d_{35}) + V(d_{13}) + V(d_{23})]$$
$$+ \tfrac{1}{64}[V(d_{14}) + V(d_{24}) + V(d_{15}) + V(d_{25})] + \tfrac{1}{4}V(d_{45})$$
$$- \tfrac{1}{8}[V(d_{17}) + V(d_{27}) - V(d_{63}) - V(d_{38}) - V(d_{74}) - V(d_{75})]$$
$$- \tfrac{1}{4}[V(d_{67}) + 2V(d_{73}) - V(d_{78}) + 2V(d_{84}) + 2V(d_{85})]$$
$$+ \tfrac{1}{32}[V(d_{18}) + V(d_{28}) + 2V(d_{68}) + V(d_{64}) + V(d_{65})] \tag{28}$$

Computer programs for the above formulas are available on request by sending a floppy disk to the authors.

Test of Significance of Inferred Phylogeny

In the case of three taxa with one or two outgroups the above results can be used to test the significance of an inferred phylogeny. Since in this case there is only one internal branch, i.e., branch c, testing the significance of the internal branch is equivalent to testing the significance of the inferred phylogeny. More explicitly, the null hypothesis is that the true phylogeny is a trichotomy, i.e., the three taxa diverged at the same time. This hypothesis is the same as the hypothesis of $c = 0$. Therefore, if the estimated c is significantly greater than 0, the null hypothesis of trichotomy is rejected and the inferred branching order can be taken as statistically significant. If one considers unrooted trees, the same argument applies to the case of four taxa with no outgroup because in this case there is also only one internal branch (Fig. 2a).

The above formulas for the mean and variance of c were derived under the assumption that the inferred branching order of the three taxa was [(1,2)3], which means that lineage 3 branched off earlier than did lineages 1 and 2. If, instead, the inferred branching pattern is [(1,3)2], then the subscripts 2 and 3 in the above formulas should be exchanged, and if the inferred branching pattern is [(2,3)1], subscripts 1 and 3 should be exchanged. Under the null hypothesis of trichotomy, the three branching patterns [(1,2)3], [(1,3)2], and [(2,3)1] occur with equal probability. However, in each data set only one pattern can occur, and only one c can be positive and tested for significant deviation from 0. Regardless of which pattern occurs, the probability for c to assume a particular (nonnegative) value is the same. If the distribution of c is the same as the distribution of $|x|$, where x is a standard normal random variate, then the standard statistical test based on the standard normal distribution can be applied. In particular, the estimated c value is significant at the 5% level if the ratio of mean to standard error is at least 2, and is significant at the 1% level if the ratio is at least 2.60. Obviously, the case of four taxa with no outgroup can be treated in the same manner, if one considers unrooted trees. A computer simulation[14] for the case of three taxa with one outgroup shows that the significance level defined by the above criteria is generally applicable, though the distribution of c is not strictly normal.

When the number of taxa under study is more than four, the situation becomes complicated. For example, in the case of five taxa there are two internal branches (Fig. 2b), and the probability for (only) one of them to become by chance significantly greater than 0 at the level of $\alpha = 5\%$ is $2\alpha = 10\%$. Thus, in this case one cannot reject the null hypothesis that all the internal branches have zero length, i.e., all the taxa diverged at the same time point and form a "star" phylogeny; of course, this null hypothesis can be rejected if $\alpha \leq 2.5\%$. On the other hand, the probability for both internal branches to be by chance significant at the level of $\alpha = 5\%$ is only approximately $\alpha^2 = 0.0025$ (it is approximate because the two internodal distances are not estimated independently). Hence, the requirement of all internal branches being significant seems to be a too stringent test for the significance of the inferred topology. However, one cannot draw a conclusion about the significance of an inferred tree topology as long as one or more of the internodal distances are nonsignificant; of course, the uncertainty can be restricted to a subset of taxa.

Numerical Examples

The following examples assume that the rate of nucleotide substitution is constant over time and that the observed number of substitutions between each pair of sequences is equal to the expected value.

Table II shows the case of three species with an outgroup (taxon 4) (Fig. 2a). In Table II, α denotes the proportion of transitional changes; $\alpha = \frac{1}{3}$ if substitutions occur randomly. The standard error (S.E.), which is the square root of $V(c)$, is larger for $\alpha = \frac{2}{3}$ than for $\alpha = \frac{1}{3}$. Since transitional

TABLE II
STANDARD ERROR OF THE ESTIMATE OF LENGTH
OF BRANCH c IN FIG. 2a[a]

c	α	S.E.	c/S.E.	L'
0.010	$\frac{1}{3}$	0.0046	2.17	850
	$\frac{2}{3}$	0.0050	2.00	1,000
0.005	$\frac{1}{3}$	0.0039	1.28	2,500
	$\frac{2}{3}$	0.0043	1.16	3,000
0.001	$\frac{1}{3}$	0.0032	0.31	41,000
	$\frac{2}{3}$	0.0037	0.27	54,000

[a] The branch lengths in Fig. 2a are $a = b = 0.05$, $d = 0.05 + c$, $e = 0.15 - c$. α, Proportion of transitional substitutions. S.E. was computed under the assumption that the number of nucleotide sites (L) studied is 1000. L' is the number of nucleotide sites required for the ratio c/S.E. to be 2 or larger (i.e., to be ~5% significant). From W.-H. Li, *Mol. Biol. Evol.* **6,** 424 (1989).

changes generally occur more often than transversional changes,[24,25] the two-parameter model is more realistic than the one-parameter model; the former is applicable to all α values whereas the latter only to $\alpha = \frac{1}{3}$.

The ratio c/S.E. can be used to test whether c is significantly greater than 0. A ratio of 2 can be taken as significant as the 5% level. All the values in Table II were obtained for $L = 1000$. When $c = 0.01$, the ratio is 2 or larger if $\alpha \leq \frac{2}{3}$. Thus, this case requires only a small amount of sequence data to resolve the branching order of the three species. When c is 0.005, then the ratio is considerably smaller than 2; e.g., the ratio is 1.28 for $\alpha = \frac{1}{3}$. Equations (14) and (17) imply that $V(c)$ is inversely proportional to L. Therefore, for the ratio to increase from 1.28 to 2 the L value should increase from 1000 to $L' = 1000 \times (2/1.28)^2 = 2500$, approximately. The other L' values in Table II were obtained in the same manner. If c is 0.001, then the number of nucleotide sites to be studied is rather large, greater than 50,000. Saitou and Nei[26] have considered this problem from a different angle. They studied the probability of obtaining the correct topology as

[24] W. M. Brown, E. M. Prager, A. Wang, and A. C. Wilson, *J. Mol. Evol.* **18,** 225, (1982).
[25] W.-H. Li, C.-I. Wu, and C.-C. Luo, *J. Mol. Evol.* **21,** 58 (1984).
[26] N. Saitou and M. Nei, *J. Mol. Evol.* **24,** 189 (1986).

a function of the number of nucleotides studied under various tree-making methods.

Table III shows the amount of reduction in $V(c)$ when a second outgroup (taxon 5) is added (Fig. 2b); the $V(c)$ value for the case of one outgroup is denoted by $V_1(c)$ and that for the case of two outgroups by $V_2(c)$. The reduction increases as $V_1(c)$ becomes larger. Since $V(c)$ is inversely proportional to L, a reduction in $V(c)$ can also be achieved by increasing L. Is it more advantageous to increase L or to add a second outgroup? The total number of nucleotides sequenced is $4L$ for the case of one outgroup and $5L$ for the case of two outgroups, the latter being 1.25 times the former. Therefore, if the same total number of nucleotides is to be sequenced, it is less advantageous to add a second outgroup than to increase L, if $V_1(c)/V_2(c)$ is smaller than 1.25, whereas the reverse is true if the ratio is larger than 1.25. In Table III the ratio is smaller than or equal to 1.25 for the first six cases and is larger than 1.25 for the last six cases. Since the ratio tends to increase with $V_1(c)$, in general it is more advantageous to increase L if $V_1(c)$ is relatively small but more advantageous to add a second outgroup if $V_1(c)$ is relatively large. In all the cases in Table III the distances from sequences 4 and 5 to the other three are the same, i.e., $g = f$ in Fig. 2b, so that the fifth sequence is as good a reference as the fourth one. If the fifth is more distantly related to the other three than the fourth

TABLE III
VARIANCE $[V_2(c)]$ OF THE ESTIMATE OF LENGTH OF BRANCH c IN FIG. 2b[a]

$a = b$	d	$f = g$	e	c	α	$V_1(c)$ ($\times 10^{-4}$)	$V_2(c)$ ($\times 10^{-4}$)	$V_1(c)/V_2(c)$
0.02	0.025	0.05	0.025	0.005	$\frac{1}{3}$	0.072	0.066	1.09
					$\frac{2}{3}$	0.079	0.071	1.11
			0.029	0.001	$\frac{1}{3}$	0.028	0.023	1.22
					$\frac{2}{3}$	0.033	0.027	1.22
0.05	0.055	0.10	0.045	0.005	$\frac{1}{3}$	0.153	0.126	1.21
					$\frac{2}{3}$	0.188	0.150	1.25
			0.049	0.001	$\frac{1}{3}$	0.104	0.079	1.32
					$\frac{2}{3}$	0.135	0.100	1.35
0.10	0.110	0.15	0.040	0.010	$\frac{1}{3}$	0.445	0.351	1.27
					$\frac{2}{3}$	0.577	0.444	1.30
			0.045	0.005	$\frac{1}{3}$	0.376	0.287	1.31
					$\frac{2}{3}$	0.501	0.374	1.34

[a] α, Proportion of transitional substitutions. $V_1(c)$ was obtained under the assumption that the second outgroup (taxon 5 in Fig. 2b) is not available. From W.-H. Li, *Mol. Biol. Evol.* **6**, 424 (1989).

sequence is, then the reduction in $V(c)$ is expected to be smaller than those shown in Table III. Further, the effect will also be reduced if sequences 4 and 5 are closely related to each other.

Bootstrap Method

Like the jackknife, the bootstrap is a method of resampling the data under study to infer the variability of the estimate.[27] This method was introduced into phylogenetic studies by Felsenstein.[5] It is not a tree-making method but a statistical method that can be used to evaluate the confidence level of a phylogenetic estimate obtained from a data set by a tree-making method. Suppose that L nucleotides from each of the species under study are obtained and that a certain tree-making method, which can be either a cladistic or a distance matrix method, is used to analyze the data. We randomly resample L sites from the data table with replacement and then reconstruct a tree. This process is repeated a large number (r) of times. We then search for all subsets of species that occur on 95% or more of the r estimated phylogenies. Each of these subsets is considered to be supported in the sense that its alternatives are rejected. A computer program is available from J. Felsenstein.

When the number of species involved is more than five, the bootstrap seems to be the best method now available, for it is rather difficult to treat such a case analytically. However, there are difficulties in the interpretation of results obtained by this approach.[5] First, the confidence statements are not joint confidence statements: for two subsets each supported at the 95% level, we might have lower than 90% confidence in the statement that they are both present in the true tree. Second, there is the "multiple tests" problem: if there are 20 subsets, then on average one should show significance at the 5% level purely by chance. One way to overcome these difficulties is to use a high level of confidence, say, 99% or even higher. In the literature, the resampling process is usually repeated only 100 times. This is too low; at least several hundred times should be conducted, particularly when many species are involved. Of course, this can be very time consuming, especially if a computation-intense method such as the maximum parsimony method is used. In the case of four species, the above-mentioned difficulties do not occur because each tree contains only two subsets, so that when one of them is determined, so is the other. However, as just noted, the advantage of the bootstrap method is its ability to deal with more than four species.

[27] B. Efron and G. Gong, *Am. Stat.* **37**, 36 (1983).

Concluding Remarks

Problem of Statistical Inconsistency

When the degree of sequence divergence is small or moderate (say, 50% or lower), the maximum parsimony method, the transformed distance method, and the neighbor-joining method are likely to be statistically consistent in the sense that the probability for each of these methods to give the correct tree increases with the amount of data. However, when the degree of divergence is high and when the assumption of rate constancy is violated, these methods may not be consistent,[28,29] and thus a highly significant tree may in fact be an erroneous one. (This problem can also occur even with equal rates of evolution.[30]) The evolutionary parsimony method was intended to overcome the effects of unequal rates of evolution. However, this method can also become statistically inconsistent if the transitional or transversional rates are unequal. Indeed, in an application of this method to the small subunit rRNA sequences from human, *Drosophila*, rice, and *Physarum*, we obtained a χ^2 of 4.8 for the tree with human and rice as a sister group, the χ^2 values for the other two trees being 0.6 and 0.3. The favored tree is obviously erroneous because humans are much more closely related to *Drosophila* than to rice. The bootstrap approach does not correct the statistical inconsistency problem that is inherent in the tree-making method used.[5] Thus, regardless of which approach is used to test the significance of a phylogenetic estimate, one should be aware of the possibility of statistical inconsistency, particularly when highly divergent sequences are used.

Heterogeneous Data

Phylogenetic studies often use sequence data from different DNA regions. If all the regions studied have similar rates of nucleotide substitution, then all the data can be combined together into a single set. If considerable variation in rates exists, however, regions with different rates should be treated separately, particularly when the distance matrix approach is used because the internodal distances are dependent on the rate of evolution. The question then arises as to how to test the significance when the results from different data sets are combined. A simple test

---

[28] J. Felsenstein, *Syst. Zool.* **27**, 401 (1978).
[29] W.-H. Li, K. H. Wolfe, J. Sourdis, and P. M. Sharp, *Cold Spring Harbor Symp. Quant. Biol.* **52**, 847 (1987).
[30] D. Penny, M. D. Hendy, and I. M. Henderson, *Cold Spring Harbor Symp. Quant. Biol.* **52**, 857 (1987).

procedure is the inverse χ^2 method.[31] Suppose that there are k different data sets. Let p_i be the significance level (probability) estimated from the ith data set. If the null hypothesis is true, then $-2 \ln (p_i)$ has a χ^2 distribution with 2 degrees of freedom and

$$P = -2 \sum_{i=1}^{k} \ln(p_i)$$

has a χ^2 distribution with $2k$ degrees of freedom.

Acknowledgments

This study was supported by National Institutes of Health Grant GM30998.

[31] R. A. Fisher, "Statistical Methods for Research Workers," 4th Ed. Oliver and Boyd, London, 1932.

[41] Nearest Neighbor Procedure for Relating Progressively Aligned Amino Acid Sequences

By RUSSELL F. DOOLITTLE and DA-FEI FENG

Introduction

Generally speaking, there are two quite different approaches to constructing phylogenies from sequence data: the one matrix based, and the other column-by-column character analysis. The latter procedure is a kind of parsimony approach. Matrix methods, on the other hand, strive for arrangements that most readily accommodate pairwise distances.

Both methods have strengths and weaknesses. For example, it has been noted that the tendency for overall optimization inherent in the matrix approach systematically foreshortens the ancient (deeper) branches.[1] Nonetheless, matrix methods are usually more sensitive than common ancestor schemes, and, in the case of protein sequences, are much more widely used. Although less popular, column-by-column methods have been used with protein sequences in the past, particularly in some early

[1] M. O. Dayhoff, C. M. Park, and P. J. McLaughlin, *in* "Atlas of Protein Sequence and Structure" (M. O. Dayhoff, ed.), Vol. 5, pp. 7–16. National Biomedical Research Foundation, Silver Spring, Maryland, 1972.

TABLE I
TWO WAYS OF FINDING EVOLUTIONARY RELATIONSHIPS WITH SEQUENCES

Distance matrix method
    Align the sequences properly
    Compute the pairwise distances
    Set up a difference matrix
    Find topology by clustering the most similar (least different) sequences
    Calculate branch lengths by finding the best fit of pairwise distances to topology
Character analysis
    Align the sequences properly
    Examine each column of residues four taxa at a time for evidence of common ancestry
    Calculate the most likely nearest neighbors
    Reconstruct phylogeny on the basis of nearest neighbors
    Calculate branch lengths from four-taxon parsimony

studies by Dayhoff and colleagues.[2] A summary of the different operations involved in matrix-derived and character-based phylogenies is presented in Table I.

In a typical matrix method, the pairwise distances are determined for a set of aligned sequences and the values assembled in a matrix. The distance values must then be accommodated to the inner and outer branch segments that constitute the tree. Of the vast number of possible trees, the best ones are thought to be those with the smallest total branch lengths or the smallest deviation of the pairwise distances obtained from the final tree compared with the original input data. Ordinarily, only the most likely trees are constructed and compared. In the method of Fitch and Margoliash,[3] a good starting topology is found by choosing the smallest pairwise distance in the matrix, combining and averaging the contributions of its two members, and then compiling a new matrix smaller by one. The procedure is repeated iteratively until the entire matrix is reduced to the last member. This "good tree" can be modified and many slightly differing versions examined.

In the progressive alignment procedure,[4,5] alternate topologies are actually being explored as the alignment is being made, and we have made the point that the tree that emerges ought to be among the best from the start. Nonetheless, negative branch lengths occur occasionally, indicating an uncomfortably arranged tree. Moreover, sometimes the relative positions of member sequences may change significantly merely because some dis-

[2] M. O. Dayhoff and R. V. Eck, "Atlas of Protein Sequence and Structure," pp. 33–41. National Biomedical Research Foundation, Silver Spring, Maryland, 1968.

[3] W. M. Fitch and E. Margoliash, *Science* **155,** 279 (1967).

[4] D.-F. Feng and R. F. Doolittle, *J. Mol. Evol.* **25,** 351 (1987).

[5] D.-F. Feng and R. F. Doolittle, this volume, [23].

tantly related sequence is or is not included in the starting set. Finally, we have witnessed some situations where different gene products from the same set of organisms have yielded slightly different topologies. For all these reasons, we sought an independent check for the matrix method after progressive alignment. Early studies suggested that the common ancestor method actually gave better topologies when closely related sequences were involved,[1] and it was with that need in mind that we resurrected a simple parsimony scheme. In this regard, we have developed a nearest neighbor method that we call PAPA, for parsimony after progressive alignment.

Methods

The PAPA approach to finding a phylogeny by a parsimony, or common ancestor, procedure begins with the same progressive alignment employed with a matrix method.[4,5] Beyond that, we have more or less followed a conventional four-taxon scheme throughout,[6] the usual three arrangements for four taxa being denoted, in line with recent schemes used for nucleic acids,[7] as the "X," "Y," and "Z" trees (Fig. 1).

With regard to the character states themselves, we have not attempted to use any kind of weighted scale for similar or frequently substituted amino acids; rather, we have limited the scheme to the presence or absence of identical residues. There are only 15 situations for the four-taxon arrangement under these circumstances,[6] and for ease of computation these were assigned capital letter designations (Table II). The frequencies of these occurrences are easily tallied by computer and can be manipulated in various ways. In deciding which arrangement is most likely for a set of four sequences, we used the expressions

$$\text{``}X\text{''} = 1.5\text{E} + \text{H} + \text{K}$$
$$\text{``}Y\text{''} = 1.5\text{F} + \text{I} + \text{L}$$
$$\text{``}Z\text{''} = 1.5\text{G} + \text{J} + \text{M}$$

Thus, the numbers of times each of the arrangements listed in Table II occurs in a given four-taxon set is tallied, and the tree (X, Y, or Z) with the highest score after the summation is taken as the best. The rationale for E, F, and G counting 50% more than the H–M arrangements is explained below.

After the tally is complete and the assignment of X, Y, or Z trees made for each set of four sequences, the nearest neighbors for each of the trees are identified, and the number of times each pair of sequences is so found

[6] J. A. Cavender, *Math. Biosci.* **40,** 271 (1978).
[7] J. A. Lake, *Mol. Biol. Evol.* **4,** 167 (1987).

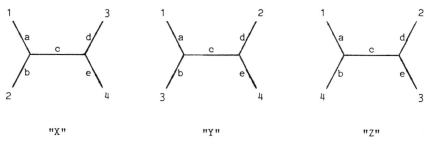

"X"   "Y"   "Z"

FIG. 1. The three possible (unrooted) trees for four taxa, which in the cases under discussion here are sequences 1–4. The five distances, a–e, can be estimated by considering the mutational probability factors shown in Table IV.

is entered in a matrix. As has been noted by Fitch,[8] the number of nearest neighbor pairs is equal to $2 \times (n!)/[(n-4)! \times 4!]$, and the maximum number of times two sequences can find themselves nearest neighbors is $(n-3)(n-2)/2$ (Table III). Thus, if there are 10 sequences, there are 420 nearest neighbor pairs, and the most any pair will be considered together in the same set of four is 28. A phylogeny derived from the nearest neighbor matrix is easily constructed merely by following the agglomerative scheme of Fitch and Margoliash,[3] in this case averaging the highest numbers and reducing the matrix iteratively.

Once the topology is determined by neighborliness, it is a relatively simple matter to find all the branch lengths of the tree from a consideration of distances observed in the various four-taxon trees.[8] To facilitate the latter process, we have assigned a set of the most parsimonious mutational parameters to each of the 15 columnar situations (Table IV). These depend on the type of arrangement and the particular tree (X, Y, or Z) under consideration. The rationale for the values is straightforward; the E, F, and G situations have a value of 1 set to the inner (c) branch for X, Y, and Z, respectively, because each of these situations is explained most parsimoniously by a single mutation on the inner segment. In contrast, the other two situations in each of these cases (i.e., F or G for tree X; E or G for Y; E or F for Z) require a minimum of two changes, the probabilities of which are spread equally over the four outer branches. Accordingly, their values are set at 0.5 ($\frac{2}{4}$ in Table IV). The H–M situations require a minimum of two changes even for the best tree. The two changes could have occurred in any of three segments (including the inner segment c), so values of 0.67 ($\frac{2}{3}$ in Table IV) are assigned. This latter point explains why E, F, or G count 50% more than H–M in choosing the most likely tree.

The A–D situations (Table II) are most parsimoniously explained by a

[8] W. M. Fitch, *J. Mol. Evol.* **18**, 30 (1981).

TABLE II
FOUR-TAXON ANALYSIS AFTER PROGRESSIVE ALIGNMENT[a]

Sequence	O	A	B	C	D	E	F	G	H	I	J	K	L	M	P
1	x	y	x	x	x	x	x	x	x	x	x	y	y	y	x
2	x	x	y	x	x	x	y	y	x	y	y	z	x	x	y
3	x	x	x	y	x	y	x	y	y	x	z	x	z	x	z
4	x	x	x	x	y	y	y	x	z	z	x	x	x	z	w

$$X = 1.5E + H + K$$
$$Y = 1.5F + I + L$$
$$Z = 1.5G + J + M$$

[a] The 15 possible arrangements of "character-presence" in a column of four aligned sequences are shown, where x, y, z, and w are different amino acid residues. Thus, in the O column, all four residues are identical, and in the P column all four are different. The most informative columns are E–G (two pairs of identities). Columns H–M have two identical residues and two others. A tally of the 15 types in an alignment of four sequences allows a judgment to be made as to which tree is most likely (X, Y, or Z in Fig. 1). Columns with gaps are not included in the tally.

TABLE III
SOME INTERESTING NUMBERS ASSOCIATED WITH n SEQUENCES

n	Possible unrooted trees[a]	Sets of four taxa[b]	Nearest neighbors[c]	Most possible[d]	Pairwise comparisons[e]	Segments in tree[f]
4	3	1	2	1	6	5
5	15	5	10	3	10	7
6	105	15	30	6	15	9
7	945	35	70	10	21	11
8	10,395	70	140	15	28	13
9	135,135	126	252	21	36	15
10	2.0×10^6	210	420	28	45	17
11	3.4×10^7	330	660	36	55	19
12	6.6×10^8	495	990	45	66	21
13	1.4×10^9	715	1430	55	78	23
14	3.2×10^{11}	1001	2002	66	91	25
15	7.9×10^{12}	1365	2730	78	105	27

[a] The number of unrooted trees is equal to $[1 \times 3 \times 5 \cdots (2n - 5)]$.

[b] The number of four-taxon arrangements is equal to $n!/[(n - 4)! \times 4!]$.

[c] Two sets of nearest neighbors are provided by each four-taxon tree.

[d] The maximum number of times two taxa may be nearest neighbors is $(n - 3)(n - 2)/2$.

[e] The number of pairwise comparisons is equal to $n(n - 1)/2$.

[f] The total number of segments in any unrooted tree is equal to $2n - 3$.

TABLE IV

MUTATIONAL PROBABILITY FACTORS USED FOR CALCULATING
BRANCH LENGTHS IN X, Y, OR Z TREES[a]

Tree	O	A	B	C	D	E	F	G	H	I	J	K	L	M	P
Xa	–	1/1	–	–	–	–	2/4	2/4	–	–	–	2/3	1/1	1/1	3/5
Xb	–	–	1/1	–	–	–	2/4	2/4	–	1/1	1/1	2/3	–	–	3/5
Xc	–	–	–	–	–	1/1	–	–	2/3	–	–	2/3	–	–	3/5
Xd	–	–	–	1/1	–	–	2/4	2/4	2/3	–	1/1	–	1/1	–	3/5
Xe	–	–	–	–	1/1	–	2/4	2/4	2/3	1/1	–	–	–	1/1	3/5
Ya	–	1/1	–	–	–	2/4	–	2/4	–	–	–	1/1	2/3	1/1	3/5
Yb	–	–	–	1/1	–	2/4	–	2/4	1/1	–	1/1	–	2/3	–	3/5
Yc	–	–	–	–	–	–	1/1	–	–	2/3	–	–	2/3	–	3/5
Yd	–	–	1/1	–	–	2/4	–	2/4	–	2/3	1/1	1/1	–	–	3/5
Ye	–	–	–	–	1/1	2/4	–	2/4	1/1	2/3	–	–	–	1/1	3/5
Za	–	1/1	–	–	–	2/4	2/4	–	–	–	–	1/1	1/1	2/3	3/5
Zb	–	–	–	–	1/1	2/4	2/4	–	1/1	1/1	–	–	–	2/3	3/5
Zc	–	–	–	–	–	–	–	1/1	–	–	2/3	–	–	2/3	3/5
Zd	–	–	1/1	–	–	2/4	2/4	–	–	1/1	2/3	1/1	–	–	3/5
Ze	–	–	–	1/1	–	2/4	2/4	–	1/1	–	2/3	–	1/1	–	3/5

[a] As an example, if a column position in a four-taxon alignment has an E arrangement (Fig. 1), then the X tree is most parsimoniously achieved by a single mutation on the inner segment, so the c value is equal to 1.0. In that same tree, an F arrangement requires at least two mutations, and those two could have occurred on any of the four outer branches. Accordingly, $2/4 = 0.5$ for the segments a, b, d, and e.

single change, and in each case it is fixed to the appropriate outer segment. The P column, in which all four residues are different, requires a minimum of three changes, the probabilities of which are spread over all five segments, the resulting value being 0.6 ($\frac{3}{5}$ in Table IV). Note that although A–D and P are not informative with regard to the topology, they are necessary for calculating proper branch lengths.

After the A–P situations have been tallied, they are multiplied by the appropriate mutation probability parameter for each tree (X, Y, and Z), and the products are summed and then divided by the total number of residue sets employed (residue columns containing gaps are not used). The most probable tree in each case is the one with the longest inner segment, which is another way of saying that $X = 1.5E + H + K$, $Y = 1.5F + I + L$ and $Z = 1.5G + J + M$. A listing of the five segments for each four-taxon tree is saved and used for the construction of the overall tree. In this regard, because the topology is already known from the nearest-neighbor analysis, segments can be assigned directly, beginning with nearest neighbors and then adding nested interior segments progressively.

Using the PAPA Program

The starting file for the PAPA program contains all the sequences to be considered, consecutively, but with neutral elements (X's) in all places where gaps were inserted by the progressive alignment. If the program TREE was used to obtain the alignment,[5] the appropriate file is the one designated "wf1" (write-file 1). The format is what we call the "Old Atlas" format and is patterned on what was used in early versions of the *Atlas of Protein Sequence and Structure.*[2] A simple program called FORMAT takes any other format and converts it to this style.

The output from the various PAPA programs can be either detailed or abbreviated. In the detailed version, PAPA1, a complete tabulation of the various counts for each of the four-taxon sets is provided, as well as a summary matrix of nearest-neighbor tallies and a branching order. In the abbreviated form, PAPA2, only the summary matrix and branching order are printed. This latter version is used especially when the number of sequences is large and the number of four-taxon units too cumbersome for printing (Table III). PAPA1 and PAPA2 use only the simple equalities in Table II and, as such, deal only with topologies. A third version of the program, PAPA3, utilizes the mutational probability parameters listed in Table IV and provides all the information for constructing a phylogenetic tree, including branch lengths.

Currently, the programs are set to handle as many as 50 aligned protein sequences up to 500 residues in length. It should be remembered that, although the computing requirements for the PAPA programs are modest, the number of four-taxon trees grows much more rapidly with an increased number of sequences than do pairwise comparisons (Table III).

Comparison of Matrix and Character Approaches

The PAPA programs have been applied to a wide variety of phylogenies that had previously been determined by a matrix method after progressive alignment. For the most part, the agreement has been excellent. For example, a set of 31 reverse transcriptase sequences gave exactly the same topology by PAPA1 as had been found by the matrix procedure.[9]

An informative illustration of the kind of agreement that can be expected is afforded by a comparison of phylogenies for the same seven globin sequences that were used as an example in our chapter on progressive alignment.[5] Thus, the input file contained seven progressively aligned globins with neutral elements filling in for any gaps. The tally of the frequency of the 15 arrangements A–O as determined by PAPA1 is shown

[9] R. F. Doolittle, D.-F. Feng, M. S. Johnson, and M. A. McClure, *Quart. Rev. Biol.* **64**, 1 (1989).

TABLE V
OUTPUT FROM PAPA1: TALLY OF ARRANGEMENTS[a]

Set	O	A	B	C	D	E	F	G	H	I	J	K	L	M	P	Q	X	Y	Z
1,2,3,4	20	2	3	8	26	6	1	2	35	3	3	2	2	7	9	39	46	7	13
1,2 3 5	24	3	4	5	23	11	1	2	32	2	2	4	0	6	10	39	53	4	11
1,2,3,6	24	3	2	4	28	8	2	1	40	3	4	0	2	7	11	29	52	8	13
1,2,3,7	19	3	1	5	32	10	1	2	37	5	3	0	2	6	12	30	52	9	12
1,2,4,5	22	2	5	6	6	21	1	0	42	2	3	9	1	3	10	35	83	5	6
1,2,4,6	10	3	2	17	18	8	3	0	45	3	5	1	1	2	15	35	58	9	7
1,2,4,7	12	3	1	12	16	8	1	1	49	6	4	2	2	1	14	36	63	10	7
1,2,5,6	12	4	2	15	17	11	1	1	43	6	4	0	3	0	16	33	60	11	6
1,2,5,7	13	1	3	11	16	8	0	0	49	6	4	1	5	3	13	35	62	11	7
1,2,6,7	14	2	1	11	14	10	0	1	57	6	5	2	4	4	12	25	74	10	11
1,3,4,5	21	7	6	6	2	9	3	0	15	4	3	15	8	3	26	40	44	17	6
1,3,4,6	11	4	1	14	12	2	3	2	14	9	6	6	3	4	38	39	23	17	13
1,3,4,7	11	3	2	9	12	3	2	0	17	9	7	7	7	7	32	40	29	19	14
1,3,5,6	13	6	1	12	15	3	2	3	11	6	5	3	2	9	38	39	19	11	19
1,3,5,7	13	4	2	7	15	3	1	1	15	6	6	2	7	13	33	40	22	15	21
1,3,6,7	13	3	2	7	13	3	0	2	22	7	6	5	10	6	39	30	32	17	15
1,4,5,6	11	4	3	1	16	2	0	10	6	6	9	8	8	18	30	36	17	14	42
1,4,5,7	11	4	4	2	16	1	0	5	6	5	8	5	9	22	33	37	13	14	38
1,4,6,7	7	3	8	6	5	3	1	1	15	13	8	7	9	8	38	36	27	24	18
1,5,6,7	8	3	7	8	6	2	2	1	14	11	7	8	5	10	41	35	25	19	19
2,3,4,5	21	7	3	5	1	10	2	1	20	6	2	17	10	3	20	40	52	19	7
2,3,4,6	12	3	1	14	10	2	2	2	19	8	5	6	4	6	35	39	28	15	14
2,3,4,7	12	2	3	10	10	2	1	0	22	7	4	7	7	9	32	40	32	16	13
2,3,5,6	14	5	1	12	13	2	1	2	17	4	5	4	3	12	34	39	24	9	20
2,3,5,7	13	4	1	9	14	4	1	1	17	4	5	2	5	14	34	40	25	11	21
2,3,6,7	14	2	2	8	13	2	0	2	25	6	4	6	9	6	39	30	34	15	13
2,4,5,6	11	4	4	2	13	1	0	8	6	4	9	8	7	23	32	36	16	11	44
2,4,5,7	12	3	2	3	12	1	0	5	5	6	8	7	8	26	33	37	14	14	42
2,4,6,7	9	1	7	6	4	3	2	1	11	12	7	8	8	9	44	36	24	23	18

TABLE V *(continued)*

Set	O	A	B	C	D	E	F	G	H	I	J	K	L	M	P	Q	X	Y	Z
2,5,6,7	9	2	7	5	6	2	2	1	11	10	8	8	8	10	44	35	22	21	20
3,4,5,6	13	2	6	2	15	1	1	6	2	10	9	4	6	23	30	38	8	18	<u>41</u>
3,4,5,7	12	3	5	2	16	0	2	6	3	10	8	4	7	21	30	39	7	20	<u>38</u>
3,4,6,7	8	2	9	6	7	2	1	3	10	11	7	5	9	4	46	38	18	<u>22</u>	16
3,5,6,7	10	0	7	7	9	3	3	1	17	7	8	6	4	6	42	38	28	16	16
4,5,6,7	7	4	3	8	8	8	1	1	28	5	7	7	5	8	33	35	<u>47</u>	12	17

[a] Seven prealigned globin sequences were subjected to PAPA analysis and the choices of best tree made for each set of four (1, human hemoglobin γ chain; 2, human hemoglobin β chain; 3, human hemoglobin α chain; 4, hagfish hemoglobin; 5, lamprey hemoglobin; 6, human myoglobin; 7, gastropod myoglobin). Q is a count of columns containing gaps. The seven "non-X" trees are underlined.

in Table V, and the nearest-neighbor frequencies in Table VI. The branching order (Fig. 2) is the same as was found for the second (reconsidered) tree determined by the matrix method.[5] The branch lengths determined by PAPA3 were significantly different, however, several inner segments being longer than were found by the matrix method (Fig. 2).

Comment

Recent innovations in the analysis of nucleic acid sequences by character-based methods[7] prompted us to reexamine the column-by-column analysis of protein sequences. In this regard, although we have unabashedly

TABLE VI
NEAREST-NEIGHBOR COUNT FOR SEVEN GLOBIN SEQUENCES[a]

	HBGH	HBBH	HAHU	HEHA	HBRL	MYOH	MYCR
HBGH	—	10	6	1	1	1	1
HBBH		—	6	1	1	1	1
HAHU			—	1	2	3	2
HEHA				—	10	3	4
HBRL					—	3	3
MYOH						—	9
MYCR							—

[a] HBGH, Human hemoglobin γ chain; HBBH, human hemoglobin β chain; HAHU, human hemoglobin α chain; HEHA, hagfish hemoglobin; HBRL, lamprey hemoglobin; MYOH, human myoglobin; MYCR, gastropod myoglobin.

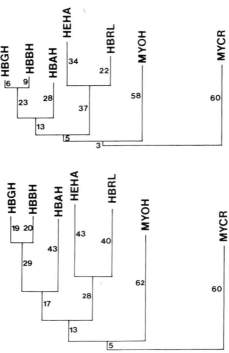

FIG. 2. Comparison of matrix-derived tree (top) with PAPA tree (bottom) for seven globins. In both cases, a progressive alignment was performed first.[5] HBGH, Human hemoglobin γ chain; HBBH, human hemoglobin β chain; HAHU, human hemoglobin α chain; HEHA, hagfish hemoglobin; HBRL, lamprey hemoglobin; MYOH, human myoglobin; MYCR, gastropod myoglobin.

adopted some of the notation used by Lake,[7] it must be emphasized that the procedure described here is in no way linked to the principle feature of Lake's method, which is concerned with operator invariants and their application to rate problems.

There are some distinct differences in dealing with amino acids, of course, mostly centering around the fact that there are 20 amino acids but only four bases. Thus, in the case of nucleic acids it is a relatively simple matter to distinguish between transitions and transversions; weighting all the different categories of amino acid interconversions, however, would be considerably more cumbersome. Accordingly, we have limited our character state analysis of amino acids to identities and nonidentities (presence or absence) and have not attempted to include weighted scales.

Ordinarily, parsimony schemes for nucleic acid data limit themselves to those four-taxon situations in which "two pairs" of identical bases are arrayed, as in arrangements E–G in Table II. The number of occurrences of such two-pair sets is low in most protein alignments of interest, however, and one is forced to consider the one-pair sets (H–M in Table II) as well in order to obtain sufficient information for reliable trees. That the procedure works without further weighting is well illustrated by our direct comparisons with the matrix method.

Fitch[10] has taken the position that, although parsimony schemes can be made to work with amino acid sequences, they are wasteful of information if they do not take into account the genetic code. In fact, we have found[11] that consideration of the genetic code alone is not useful in amino acid alignment schemes. This is not so much because many common amino acid replacements require more than a single base change, although a significant number do, but rather because fully one-half of all single-base change amino acid interconversions involve dissimilar amino acids and are relatively uncommon. In any event, we have found that the mere tallying of amino acids provides phylogenies that are in remarkable agreement with matrix methods, providing the starting alignment is proper.

The question arises, When should one use the PAPA procedure and when the matrix method? In fact, we currently perform both, since, if one has an alignment, the PAPA programs require very little computer power. If the topologies that emerge from the two approaches differ, we promptly plug the PAPA branching order into the matrix data to see if a "better" tree, as determined by the criterion of a lower percent standard deviation, is obtained. If the two topologies that arise are the same by these two very independent procedures, we regard the tree, in the vernacular of the day, as robust.

Availability

All the programs described in this chapter are written in the C programming language and have been used, for the most part, on VAX computers running the Unix operating system. They are readily adapted for the VMS operating system, however. We also have executable versions of the PAPA programs that run on the IBM-PC (or equivalents). The latter, however, do not perform the initial progressive alignment. Any alignment may be used, of course, including manually concocted ones, so long as neutral elements are inserted wherever gaps occur. The programs are available simply by writing to the authors.

[10] W. M. Fitch, *Am. Nat.* **111**, 223 (1977).
[11] D.-F. Feng, M. S. Johnson, and R. F. Doolittle, *J. Mol. Evol.* **21**, 112 (1985).

[42] Phylogenetic Relationships from Three-Dimensional Protein Structures

By Mark S. Johnson, Andrej Šali, and Tom L. Blundell

Introduction

As evolution progresses, accumulated changes in DNA and RNA lead to differences in amino acid sequences and corresponding alterations in the tertiary structure of proteins. Although these changes take place most often on the protein surface exposed to the solvent, mutations can be accepted within the inaccessible hydrophobic interior formed from packed secondary structure elements. Thus, relative translations and rotations of α helices and β strands do occur while their general spatial relationships remain highly conserved.[1,2] As a result, when comparisons are made among more distantly related structures, fewer topologically equivalent positions are found and lead to greater root mean square (RMS) deviations.[3,4]

Consequently, one should be able to chart the evolution of proteins from a comparison of their structures. Indeed, as protein structures are generally more conserved in evolution than are amino acid sequences,[5] they can be used to infer relationships among proteins where an alignment of their sequences is not statistically significant.[6] The first phylogenetic tree derived from structural information was based on the number of topologically equivalent positions in several dinucleotide- and mononucleotide-binding proteins[7] (for a review, see Matthews and Rossmann[8]). Following this work, Johnson *et al.*[6] compared six families of homologous structures and sequences (immunoglobulins, *c*-type cytochromes, globins, serine proteinases, eyelens γ-crystallins, and nucleotide-binding domains) and showed that trees based on sequence and structure are generally congruent.

Pattern matching and comparison methods for proteins can be employed to derive both the equivalences between residues and the overall

---

[1] A. M. Lesk and C. Chothia, *J. Mol. Biol.* **136**, 225 (1980).

[2] A. M. Lesk and C. Chothia, *J. Mol. Biol.* **160**, 325 (1982).

[3] C. Chothia and A. M. Lesk, *EMBO J.* **5**, 823 (1986).

[4] T. J. P. Hubbard and T. L. Blundell, *Protein Eng.* **1**, 159 (1987).

[5] M. Bajaj and T. L. Blundell, *Annu. Rev. Biophys. Bioeng.* **13**, 453 (1984).

[6] M. S. Johnson, M. J. Sutcliffe, and T. L. Blundell, *J. Mol. Evol.,* in press (1990).

[7] W. Eventoff and M. G. Rossmann, *CRC Crit. Rev. Biochem.* **3**, 111 (1975).

[8] B. W. Matthews and M. G. Rossmann, this series, Vol. 115, p. 397.

similarity scores which can then be used to construct trees. It is important to recognize what information about the proteins is included in the comparison. This information generally focuses on only a few aspects, and there are two major groups of comparison methods: those using sequences of the amino acid residues and those employing three-dimensional structures of proteins.

The most familiar sequence comparison methods are the dynamic programming procedures based on the algorithm of Needleman and Wunsch,[9] which has recently been exploited to obtain multiple sequence alignments.[10-13] These methods usually consider the mutation rates of amino acid residues[14,15] to derive optimal comparison scores and corresponding alignments. The utility of many amino acid properties for the alignment of amino acid sequences was systematically explored by Argos[16] using his own optimization algorithm. Physical properties of amino acid residues are also considered in the pattern matching technique of Taylor,[17] which can align several protein sequences simultaneously. The hierarchical aspects of sequence and secondary structure organization were implemented in the program ARIADNE[18] for the matching of a given sequence pattern to a protein sequence and in the approach developed by Rawlings et al.[19] for reasoning about protein topology.

Alternatively, methods which compare tertiary structures of proteins are dominated by the rigid-body least-squares superposition of the α-carbon (C_α) positions (see Matthews and Rossmann[8] for a review). However, Rao and Rossmann[20] also included the main chain direction in their pairwise comparison procedure, which enabled the alignment of more divergent protein structures. The rigid-body approach was recently extended by Sutcliffe et al.[21] for the simultaneous comparison of several

[9] S. B. Needleman and C. D. Wunsch, *J. Mol. Biol.* **48**, 443 (1970).

[10] D.-F. Feng and R. F. Doolittle, *J. Mol. Evol.* **25**, 351 (1987).

[11] D.-F. Feng and R. F. Doolittle, this volume, [23].

[12] G. Barton and M. J. E. Sternberg, *J. Mol. Biol.* **198**, 327 (1987).

[13] G. J. Barton, this volume, [25].

[14] R. M. Schwartz and M. O. Dayhoff, *in* "Atlas of Protein Sequence and Structure" (M. O. Dayhoff, ed.), Vol. 5, Suppl. 3, p. 353. National Biomedical Research Foundation, Washington, D.C., 1978.

[15] D.-F. Feng, M. S. Johnson, and R. F. Doolittle *J. Mol. Evol.* **21**, 112 (1985).

[16] P. Argos, *J. Mol. Biol.* **193**, 385 (1987).

[17] W. R. Taylor, *J. Mol. Biol.* **188**, 233 (1986).

[18] R. H. Lathrop, T. A. Webster, and T. F. Smith, *Commun. ACM* **30**, 909 (1988).

[19] C. J. Rawlings, W. R. Taylor, J. Nyakairu, J. Fox, and M. J. E. Sternberg, *J. Mol. Graph.* **3**, 151 (1985).

[20] S. T. Rao and M. G. Rossmann, *J. Mol. Biol.* **76**, 241 (1973).

[21] M. J. Sutcliffe, I. Haneef, D. Carney, and T. L. Blundell, *Prot. Engineer.* **1**, 377 (1987).

protein structures. Sippl[22] has used information from intramolecular C_α distance matrices to compare protein structures, while Murthy[23] applied dynamic programming to compare the secondary structure organization of proteins, taking into account the absolute angles and distances between the idealized secondary structure segments in approximately superposed proteins. Similarly, Richards and Kundrot[24] took into account the internal relationships between the elements of secondary structure to search for a given pattern in a protein structure database. Sheridan *et al.*[25] considered the residue secondary structure type to define the weights for each residue pair, which were later used in a dynamic programming procedure to find the alignment of two proteins. Recently, Barton and Sternberg[26] applied a dynamic programming procedure and an intermolecular distance matrix for roughly superposed loops to obtain the alignment of hypervariable regions.

In this chapter, we first describe a method for the multiple rigid-body superposition of structures and show the usefulness of the pairwise rigid-body superposition in determining relationships among homologous protein structures.[6] We continue with a description of a more flexible alignment procedure that compares a number of structural properties and relationships through simulated annealing and dynamic programming algorithms.[27] This latter technique escapes the limitations imposed by rigid-body alignments: it can taken into account deformations and translocations of secondary structure elements such as those illustrated in Fig. 1 between the two sets of aspartic proteinase domains and in Fig. 2 for the cytochromes *c*. From both approaches, the rigid-body and multifeature methods for the comparison of structures, phylogenetic trees are derived for the proteins listed in Table I. In general, these trees are isomorphous to those that may be obtained from the alignment of the corresponding amino acid sequences.[28]

[22] M. J. Sippl, *J. Mol. Biol.* **156**, 359 (1982).
[23] M. R. N. Murthy, *FEBS Lett.* **168**, 97 (1984).
[24] F. M. Richards and C. E. Kundrot, *Proteins* **3**, 71 (1988).
[25] R. P. Sheridan, J. S. Dixon, and R. Venkataraghavan, *Int. J. Peptide Protein Res.* **25**, 132 (1986).
[26] G. Barton and M. J. E. Sternberg, *J. Mol. Graph.* **6**, 190 (1988).
[27] A. Šali and T. L. Blundell, *J. Mol. Biol.*, in press (1990).
[28] The programs described here were written in the C and FORTRAN programming languages on a MICROVAX II computer (Digital Equipment Corporation) running the VMS operating system, version 4.5 or higher. These programs will be available from the authors in the near future.

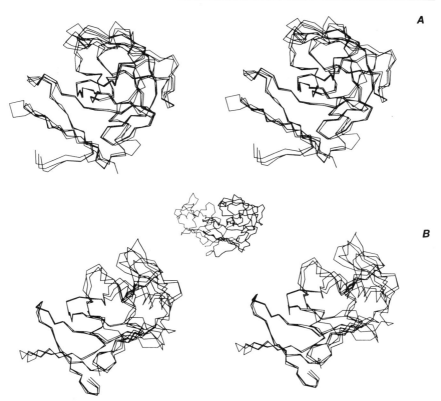

FIG. 1. Stereo views for the rigid-body superposed aspartic proteinase domains: (A) amino-terminal halves and (B) carboxy-terminal halves of endothiapepsin, penicillopepsin, and rhizopuspepsin (Table I). The amino- and carboxy-terminal domains are themselves a result of a duplication event and pack with C_2 symmetry. To show this, the C_α backbone for the entire two-domain structure of endothiapepsin (amino-terminal domain: thick line; carboxy-terminal: thin line) is inserted in the center; the C_2 rotation axis is located at the domain interface, approximately vertical and in the plane of the paper. Endothiapepsin, penicillopepsin, and rhizopuspepsin were split into domains at residues 174–175, 174–175, and 178–179 (crystallographic numbering), respectively.

Protein Structure Comparison by Rigid-Body Superposition

Topological Equivalence

The optimal superposition of two sets of coordinates is a common problem where the goal is to obtain the "best" fit of an object A to an object B over some set of coordinates said to be topologically equivalent for

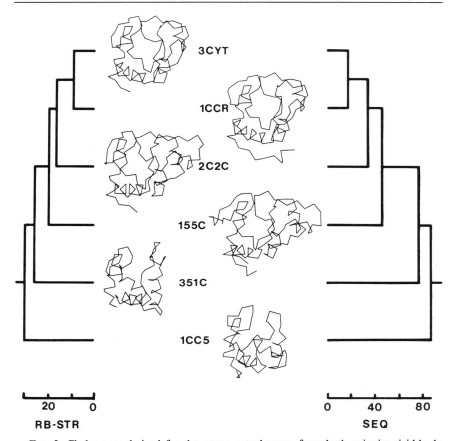

FIG. 2. Cladograms derived for the c-type cytochromes from both pairwise rigid-body comparisons (RB-STR) and multiple alignment of sequences (SEQ). C_α backbones are depicted for each of the structures after they have been fit with the multiple rigid-body superposition procedure; the structures have been translated only within the plane of the page. See Table I for identification of structures and sequences. The C_α coordinates belonging to the "unknown" (UNK) residues, as listed in the Brookhaven file for 155C, were excluded from the comparisons.

the two objects.[7,8,20,29,30] For proteins, these topologically equivalent positions can be defined as those C_α atoms in the two superposed structures that lie within a specified distance of each other, provided that these equivalences are colinear and hence obey the "no-knot" constraint.

[29] D. R. Ferro and J. Hermans, *Acta Crystallogr.* **A33,** 345 (1977).
[30] A. D. McLachlan, *Acta Crystallogr.* **A34,** 871 (1982).

TABLE I
PROTEIN STRUCTURES ALIGNED IN THIS STUDY

Brookhaven code[a]	Description	Resolution (Å)
Aspartic proteinases		
4APE	Endothiapepsin	2.1
2APP	Penicillopepsin	1.8
2APR	Rhizopuspepsin	1.8
Cytochromes c		
3CYT	Albacore tuna heart ferricytochrome c (oxidized)	1.8
1CCR	Rice embryo ferricytochrome c	1.5
2C2C	*Rhodospirillum rubrum* ferricytochrome c_2	2.0
155C	*Paracoccus denitrificans* cytochrome c-550	2.5
351C	*Pseudomonas aeruginosa* ferricytochrome c-551	1.6
1CC5	*Azotobacter vinelandii* ferricytochrome c_5	2.5
Globins		
2HHB	Human deoxyhemoglobin α and β chains	1.7
2HCO	Human carbonmonoxyhemoglobin α and β chains	2.7
1HHO	Human oxyhemoglobin α and β chains	2.1
1HBS	Human sickle cell hemoglobin	3.0
1FDH	Human deoxyhemoglobin α and γ fetal chains	2.5
2DHB	Horse deoxyhemoglobin α and β chains	2.8
1HDS	Deer sickle cell hemoglobin	2.0
2LHB	Sea lamprey hemoglobin V (cyano/met)	2.0
2MBN	Sperm whale metmyoglobin	2.0
3MBN	Sperm whale deoxymyoglobin	2.0
1ECD	*Chironomous thummi thummi* erythrocruorin	1.4
1LH1	*Lupinus luteus* leghemoglobin	2.0

[a] F. C. Bernstein, T. F. Koetzle, G. J. B. Williams, E. F. Meyer, M. D. Brice, J. R. Rodgers, O. Kennard, T. Shimanouchi, and M. Tasumi, *J. Mol. Biol.* **112,** 535 (1977).

Typically for proteins, one does not have *a priori* knowledge as to the extent of the topological equivalence, but only an idea of positions that are likely to be equivalent and which may be useful as a starting point for a comparison. This initial set of equivalences can be determined from an examination of atomic coordinates on a graphics device, from residues highly conserved in a sequence alignment, or from positions that are known to be crucial to the structural integrity, catalysis, or ligand binding of a protein. For the initial fit of two structures, three or more C_α positions must be specified to orient the structures. After the initial superposition of the structures, topological equivalences are redetermined, and the structures are again superposed based on this newly determined set of equivalences. The procedure iterates until stability in both the topological equivalences and the RMS distance over the equivalent positions is obtained.

The Residual

The superposition of the structures hinges on the minimization of a function, δ (the residual), of the general form

$$\delta = \sum_{i=1}^{a} w_i (\vec{X}_i - \mathcal{R}\vec{Y}_i)^2 \tag{1}$$

where w_i is the weight for the ith pair of equivalenced positions $\vec{X}_i$ and $\vec{Y}_i$ of the two structures; $\mathcal{R}$ is the 3 by 3 rotation matrix that superposes structure B onto structure A. The use of this function presupposes that the coordinates for the two structures have been translated so that the centers of gravity for the two sets of equivalenced positions are located at the origin of the coordinate system. For the rigid-body comparison of structures in this chapter we have used a modification[6] of a program MNYFIT[21] that was designed for the multiple alignment of structures by superposition.

Iterative Weighted Superposition

In the iterative weighted superposition procedure, one of the structures is chosen initially as the first approximation to the average of all structures (the framework), and each of the other structures is then fitted to it pairwise with the rapid procedure of McLachlan.[30] A new framework can then be calculated for b molecules from

$$\vec{F}_i^k = \frac{\displaystyle\sum_{j=1}^{b} w_{ij}^{k-1} \vec{Z}_{ij}}{\displaystyle\sum_{j=1}^{b} w_{ij}^{k-1}} \tag{2}$$

where $\vec{F}_i^k$ are the coordinates of point i on the framework at iteration k and $\vec{Z}_{ij} = \mathcal{R}_j \vec{Y}_{ij}$ are the coordinates of atom i from molecule j fitted to the previous framework F^{k-1}. The weight factors

$$w_{ij}^{k-1} = \frac{1}{\alpha_j^2 + \sigma_i d_{ij}} \tag{3}$$

include α_j as an estimate of the error in the coordinates of molecule j, σ_i, the standard deviation of the distance from the framework for the ith set of topologically equivalent positions a, and d_{ij}, the distance between atom i of molecule j to point i on F^{k-1}.

The residual is then calculated as

$$\delta^k = \frac{\sum\limits_{j=1}^{b}\sum\limits_{i=1}^{a} w_{ij}^{k-1}(\vec{F}_i^k - \vec{Z}_{ij})^2}{\sum\limits_{j=1}^{b}\sum\limits_{i=1}^{a} w_{ij}^{k-1}} \tag{4}$$

If the RMS distance between F^k and F^{k-1} is less than 10^{-5} Å and the difference between δ^k and δ^{k-1} is less than 10^{-5} Å^2, then a minimum has been obtained and the equivalences can be updated.

These topologically equivalent atoms are determined from the optimal path through a matrix of Euclidean distances between all main-chain C_α positions from the proteins. To trace this path, a dynamic programming technique is used.[9,31] If the equivalences have not changed from the previous iteration, the superposition is complete. Otherwise, the molecules are fitted to F^k pairwise, a new framework determined, the residual calculated, and this process repeated until convergence is attained.

Distance Metric from Rigid-Body Superposition

From the pairwise rigid-body comparisons, two pieces of information can be obtained: the topological equivalences and the RMS distance over these equivalent C_α positions (Table II). The number of topologically equivalent positions is converted to a pairwise fractional topological equivalence (*PFTE*) by dividing by the length of the smaller structure. The RMS is a distance measure and is converted to a similarity score, the *SRMS*, calculated as $1 - \text{RMS (Å)}/3.5$ (Å); a 3.5-Å cutoff is used in the definition of topological equivalence. The distance metric, D, employed in this study involves a weighted contribution [Eq. (5)] of these two parameters:[6]

$$D = -100 \ln(w_1 PFTE + w_2 SRMS) \tag{5}$$

where the weights, w_1 and w_2, are calculated from

$$w_1 = [(1 - PFTE) + (1 - SRMS)]/2 \tag{6}$$

$$w_2 = (PFTE + SRMS)/2 \tag{7}$$

with

$$w_1 + w_2 = 1 \tag{8}$$

For more closely related sequences, our experience has shown that the fraction of topologically equivalent positions may not differentiate be-

[31] M. L. Fredman, *Bull. Math. Biol.* **46**, 553 (1984).

TABLE II
FEATURES USED IN COMPARISON OF PROTEIN STRUCTURES[a]

Rigid-body structural comparisons:
Number of topologically equivalent positions
RMS distance over the topological equivalences

Multifeature structural comparisons:

Residues	Segments
Properties	
Identity	Secondary structure type
Physical properties	Amphipathicity
Local conformation	Improper dihedral angle
Distance from gravity center	Distance from gravity center
Side-chain orientation	Orientation relative to gravity center
Main-chain orientation	Side-chain accessibility
Side-chain accessibility	Main-chain accessibility
Main-chain accessibility	Position in space
Position in space	Global orientation
Global direction in space	
Main-chain dihedral angles	
Relations	
Hydrogen bond	
Distances to one or more nearest neighbors	Distances to one or more nearest neighbors
Disulfide bond	Relative orientation of two or more segments
Ionic bond	
Hydrophobic cluster	

[a] Structural features that are considered by the rigid-body and the multifeature approach to the comparison of protein structures and the determination of phylogenetic relationships are given. For the multifeature approach, various features are represented by rows and different levels of protein organization by columns. Only residue and secondary structure levels are shown here. The term property is used for all protein features that imply comparison of only one element from each protein. Conversely, the term relationship is used for a feature that implies comparison of at least two elements from each protein.

tween the structures.[4,6] Conversely, the RMS distance does provide a good measure of the difference between structures where the relationship is close. As a result, the weights, w_1 and w_2, are used to inversely modulate the contribution of the *PFTE* and the *SRMS* to the distance score [Eqs. (6)–(8)].[6] A matrix containing all pairwise distances can then be used directly by clustering or tree-generating techniques to display the relationships derived from the rigid-body structural comparisons.

Multifeature Comparison of Proteins

Although proteins within homologous families have the same tertiary folding, the elements of secondary structure undergo deformations, relative translations, and rotations to optimize packing of side chains and to adapt to evolutionary pressure. Thus, for two proteins with 30% sequence identity, the topologically equivalent residues defined by rigid-body superposition have a root mean square difference of approximately 1.5 Å and may comprise as few as one-third of the total number of residues.[4,6] This may not provide a sufficient basis for structural comparisons and emphasizes the requirement for a more flexible procedure for defining topological equivalence, one that can take into account relative movements and distortions of the secondary structure elements.

Structural Aspects of Proteins: Properties, Relationships, and Hierarchy

To achieve this flexibility, we include in the comparison method a number of protein features from several levels of the protein structure hierarchy (Table II). The protein is treated as a sequence of elements where each element is associated with a series of properties and may be engaged in a number of relationships with other elements. Additionally, these elements may exist at any level of the hierarchy of protein structure: residue, secondary structure, supersecondary structure, motif, domain, or globular protomer. For example, at the residue level, properties such as the local conformation and relationships like hydrogen bonds can be included. At the level of secondary structure, properties like segment solvent accessibility and relationships such as the relative spatial orientation of two segments can be incorporated.

Alignments from Dynamic Programming and Simulated Annealing

The comparison method is based on the dynamic programming technique generally used for sequence alignments.[9] In this method, one starts with the calculation of an N by M weight matrix $\mathcal{W}$ where N and M are the numbers of residues in the two compared proteins. This matrix is calculated in such a way that every element W_{ij} is proportional to the sum of the differences between various features of the residues i and j:

$$W_{ij} = \sum_l \left(\sum_p \rho^{pl} w_{ij}^{pl} + \sum_r \rho^{rl} w_{ij}^{rl} \right) \tag{9}$$

The contributions w_{ij} are the differences between individual features of the residues i and j, and factors ρ determine their relative weights. Superscript l runs over all levels of protein structure, superscript p stands for properties, and superscript r for relationships. When features at the secondary structure level are considered, the residues inherit the weights from the secondary structure segments involved, the corresponding α helices or β strands. It is trivial to define w_{ij}^{pl} for properties p; for example, the seventh property at the first level of structure, $w_{ij}^{7,1}$, describes the difference in the residue main-chain solvent accessibilities and is simply an absolute difference in the fractional main-chain accessibilities for the residues i and j from the first and second protein, respectively. In addition to properties, specific relationships such as hydrogen bonding interactions, which tend to be conserved in protein folds, can also be used in our comparison method. However, a relationship by its very nature affects more than one element in a sequence, and this precludes the simple procedure for the inclusion of this information into the residue-by-residue weight matrix $\mathcal{W}$.

To incorporate the information about relationships into the derivation of the final equivalences, we first use simulated annealing optimization[32] to obtain pairwise alignments based on relationships alone.[27] The underlying goal in the implementation of simulated annealing optimization is to maximize the number of equivalent relationships and minimize violations of a "no-knot" constraint. Since simulated annealing does not necessarily produce a global optimum, the optimization for every pair of structures is repeated several times. The fractional numbers of matching of residues i and j from proteins A and B, which can be obtained from several relationship alignments of proteins A and B in a straightforward way, are used to define the relationship weights w_{ij}^{rl}. These weights can be introduced directly into the residue by residue weight matrix $\mathcal{W}$ [Eq. (9)].

The dynamic programming algorithm then uses the matrix $\mathcal{W}$ to derive the most parsimonious alignment of the two structures. The overall distance score, which reflects the dissimilarity in selected features of the two proteins, is also obtained. A detailed description of these algorithms and their implementation in the program COMPARER may be found elsewhere.[27]

Multiple Structural Alignments

In the above description of the multifeature alignment method, we have assumed that the three-dimensional structures would be compared in a pairwise manner. However, such pairwise comparisons of several proteins may not be self-consistent, in the same way as pairwise sequence-based alignments may not be self-consistent. For this reason, we proceed by simultaneously aligning all structures. In COMPARER, we have adopted a strategy that employs a combination of the approaches by Feng and Doolittle[10,11] and Barton and Sternberg.[12,13] The procedure is divided

[32] S. Kirkpatrick, C. D. Gelatt, and M. P. Vecchi, *Science* **220**, 671, (1983).

into two parts. The first part is the construction of a dendrogram relating the homologous proteins, either *ad hoc* or from distance scores from pairwise comparisons.[10,11] The second part involves the gradual addition of new proteins, as imposed by the tree topology, into a growing multiple alignment. The most similar proteins and groups of proteins are structurally aligned first, and the gaps that are introduced do not change in later stages.[10,11] The weight matrix for the dynamic alignment of the two groups of previously aligned proteins is defined on the basis of the pairwise weight matrices relating the proteins from the two groups.

Distance Metric from Multifeature Comparison

With COMPARER, a pairwise distance score for each protein pair is obtained from the corresponding pairwise alignment implied by the multiple alignment. First, a sum of the weights W_{ij} that relate the residues equivalent in the pairwise comparison is found. This sum is then normalized via division by the number of equivalent residues in the pairwise comparison to give the intermediate score e. The final pairwise distance, E, that is used in the clustering procedure is then defined as

$$E = -100 \ln(1 - e/D_c) \tag{10}$$

D_c is a constant equal to the random value of the distance score e and is obtained for each protein pair by increasing the average of the weight matrix elements W_{ij} by three standard deviations of these elements divided by the square root of the number of equivalent residues. In addition to the distance score E, which does not incorporate information about gaps in the alignment, we considered a similar score that does incorporate gap penalties. Tree topologies for the two distance measures were the same in all cases.

Trees reflecting the evolution of different aspects of the proteins can be obtained by calculating the pairwise score E from the weights W_{ij} that were derived from different combinations of protein features. Thus, evolutionarily variable sequence features such as residue identity can be used for classification of similar proteins, and more conserved structural features, like hydrogen bonding, can be used for more divergent structures. Conversely, the clustering can also be used to infer the variability of a given protein feature in evolution.

Methods

The alignment of sequence data was produced by the "historical" multiple alignment procedure of Feng and Doolittle;[10,11] trees were derived from the distance metric of Feng *et al.*[15] Structures and sequences were obtained from the structure files of the Brookhaven Protein Data Bank[33] with the following exceptions: the sequence of cytochrome c-550 from *Paracoccus denitrificans* is as revised by Ambler *et al.*;[34] the globin sequences from *Parasponia*[35] and *Vitreoscilla*,[36] for which there are no structures, were obtained from the NEWAT[37] sequence data bank.

Tree topologies and branch lengths were determined from the distance matrices using the program KITSCH from the phylogeny inference package (PHYLIP) of Felsenstein.[38] This procedure is a modification of the original Fitch–Margoliash[39] method and accounts for unequal rates of change among the proteins by adjusting distances so that the total branch lengths from the root of the tree to the tips of each of the leaves are equidistant. In addition, numerous topologies are explored by swapping branches locally. The "best" tree is defined as the one that minimizes the sum of the squared differences between the equivalent distances from the tree and the input matrix, where each squared difference is also normalized by the corresponding squared distance from the input matrix.

Phylogenetic Trees from Structural Comparisons

In this chapter, we concentrate on the phylogenies that can be inferred from distance scores obtained from the pairwise rigid-body superposition and the alignment of structures based on many features. The results stemming from these analyses of three-dimensional protein structures are then compared with those obtained from an alignment of the amino acid sequences. Three homologous families serve as examples; these include the amino- and carboxy-terminal domains of the aspartic proteinases, eukaryotic and microbial c-type cytochromes, and globins.

Aspartic Proteinases

Structures of three fungal aspartic proteinases (Table I) have been solved to high resolution: endothiapepsin (4APE), penicillopepsin (2APP),

[33] F. C. Bernstein, T. F. Koetzle, G. J. B. Williams, E. F. Meyer, M. D. Brice, J. R. Rodgers, O. Kennard, T. Shimanouchi, and M. Tasumi, *J. Mol. Biol.* **112**, 535 (1977).

[34] R. P. Ambler, T. E. Meyer, M. D. Kamen, S. A. Schichman, and L. Sawyer, *J. Mol. Biol.* **147**, 351 (1981).

[35] A. A. Kortt, J. E. Burns, M. J. Trinick, and C. A. Appleby, *FEBS Lett.* **180**, 55 (1985).

[36] S. Wakabayashi, H. Matsubara, and D. A. Webster, *Nature (London)* **322**, 481 (1986).

[37] R. F. Doolittle, *Science* **214**, 149 (1981).

[38] J. Felsenstein, *Evolution* **39**, 783 (1985).

[39] W. M. Fitch and E. Margoliash, *Science* **15**, 279 (1967).

and rhizopuspepsin (2APR). The sequence identity between any pair of these three proteins is roughly 40%. Additionally, Tang et al.[40] have shown that amino-terminal and carboxy-terminal domains, each comprising about one-half of the molecule, are related by C_2 symmetry.

The multifeature alignment of amino- and carboxy-terminal domains (Fig. 3a) gives equivalences that are generally identical to those obtained from careful inspection of the structures on a graphics terminal (Fig. 1). This is so even though fragments of secondary structure have undergone translations, rotations, distortions, and numerous changes in sequence. This alignment stands in sharp contrast to that obtained from either the rigid-body multiple structure superposition or from the multiple sequence alignment (Fig. 3). The multiple structure rigid-body technique locates only 43 topologically equivalent positions among the six domains (asterisked positions in Fig. 3a), each domain consisting of approximately 150 residues. These positions are in complete agreement with the COMPARER alignment (Fig. 3a). Pairwise rigid-body superposition of structures was used to derive the distances for tree construction and led to between 67 and 77 topologically equivalent positions between the two sets of domains. A comparison of the sequences of the amino-terminal domains with the carboxy-terminal domains of the aspartyl proteinases aligns the conserved active-site amino acid triad Asp-Thr-Gly that is present in both domains. Outside of the neighborhood surrounding the catalytic region, only one other section is aligned similarly to that obtained from the multifeature comparison method (Fig. 3).

The trees (Fig. 4) derived from the structural comparisons, either the rigid-body or the multifeature procedure, are congruent: the amino-terminal domains clearly branch apart from the carboxy-terminal domains. For the cluster of either domain, the shorter distance is between endothiapepsin and penicillopepsin (Fig. 4) and is consistent with an alignment of the three full-length sequences. By sequence, the branch order within each domain's cluster is well determined. However, the sequence similarity between the domains is not statistically significant at the level of 3σ; this illustrates the power of the structural comparison method where unequivocal relationships are found. The numerous structural features that are common to both sets of domains, most notably the hydrogen bonding patterns, are consistent with the notion that the domains result from gene duplication.[40]

[40] J. Tang, M. N. G. James, I. N. Hsu, J. A. Jenkins, and T. L. Blundell, *Nature (London)* **271**, 618 (1978).

(a) F-STR

```
          *****              ****              ***********  ******
4APE-N  ---STGSATTTPIDSLDDAYITPVQ-IGT-----PAQTLNLDFDTGSSDLWVFSSETTASEVDGQTIYTPSK
2APP-N  --AASGVATNTPTA-NDEEYITPVT-IG-------GTTLNLNFDTGSADLWVFSTELPASQQSGHSVYNPSA
2APR-N  --AGVGTVPMTDYG-NDIEYYGQVT-IGT-----PGKKFNLDFDTGSSDLWIASTLCT-NCGSGQTKYDPNQ

4APE-C  YTGSITYTAVSTKQ---GFWEWTSTGYAVGSGTFKSTSIDGIADTGTTLLYLPATVVSA---------YWAQ
2APP-C  YTGSLTYTGVDNSQ---GFWSFNVDSYTAGSQ-SGDG-FSGIADTGTTLLLLDDSVVSQ---------YYSQ
2APR-C  FKGSLTTVPIDNSR---GWWGITVDRATVGTSTVAS-SFDGILDTGTTLILPNNIAAS---------VARA


                                 ****  *
4APE-N  STTAKLLSGATWSISYGDGSSSSGD----VYTDTVSVGGLTVTGQ-----------------AVESAKKVS
2APP-N  --TGKELSGYTWSISYGDGSSASGN----VFTDSVTVGGVTAHGQ-----------------AVQAAQQIS
2APR-N  SSTYQAD-GRTWSISYGDGSSASGI----LAKDNVNLGGLLIKGQ-----------------TIELAKREA

4APE-C  VSGAKSSSSV--------GGYVFPCSA-TLPSFTFGVGSARIVIPGDYIDFGPISTGSSSCFGGIQSSA---
4APE-C  VSGAQQDSNA--------GGYVFDCST-NLPDFSVSISGYTATVPGSLINYGPSGD-GSTCLGGIQSNS---
4APE-C  Y-GASDNGD---------GTYTISCDTSAFKPLVFSINGASFQVSPDSLVFEEF---QGQCIAGFGYG----


               ****                                 ****              ****
4APE-N  SSFTEDSTIDGLLGLAFSTLNTVSPTQQKTFFDNAKAS--LDSPVFTADLGY---HAPGTYNFGFIDTTA
2APC-N  AQFQQDTNNDGLLGLAFSSINTVQPQSQTTFFDTVKSS--LAQPLFAVALKH---QQPGVYDFGFIDSSK
2APR-N  ASFASG-PNDGLLGLGFDTITTVRG--VKTPMDNLISQGLISRPIFGVYLGKAKNGGGGEYIFGGYDSTK

4APE-C  ------GIGINIFGD-------------VALKAA---------FVVFNGA-----TTPTLGFASK----
2APP-C  ------GIGFSIFGD-------------IFLKSQ---------YVVFDSD-----G-PQLGFAPQA---
2APR-C  ------NWGFAIIGD-------------TFLKNN---------YVVFNQG-----V-PEVQIAPVA--E
```

(b) SEQ

```
4APE-N  -STGSATTTPIDSLD-------DAYITPVQIGT-P-AQTLNLDFDTGSSDL----------WVFSSETTAS
2APP-N  AASGVATNTPTAN-D-------EEYITPVTIG----GTTLNLNFDTGSADL----------WVFSTELPAS
2APR-N  AGVGTVPMTDYGN-D-------IEYYGQVTIGT-P-GKKFNLDFDTGSSDL----------WI-ASTLCTN

4APE-C  -YTGSITYTAVSTKQGFWEWTSTGY--AVGSGTFK-STSIDGIADTGTTLLYLPATVVSAYWAQVSGAKSS
2APP-C  -YTGSLTYTGVDNSQGFWSFNVDSYTAGSQSG-----DGFSGIADTGTTLLLLDDSVVSQYYSQVSGAQQD
2APR-C  -FKGSLTTVPIDNSRGWW----GITVDRATVGTSTVASSFDGILDTGTTLLILPNNIAASV-ARAYGASDN


4APE-N  EVDGQTIYT-PSKSTTAKLLSGATWSISYG-----DGSS---SSGDVYTD--TVSVGGLTVTGQAVESAKK
2APP-N  QQSGHSVYN-P--SATGKELSGYTWSISYG-----DGSS---ASGNVFTD--SVTVGGVTAHGQAVQAAQQ
2APR-N  CGSGQTKYD-PNQSSTYQA DGRTWSISYG-----DGSS---ASGILAKD--NVNLGGLLIKGQTIELAKR

4APE-C  SSVGG--YVFPC-SAT-LP------SFTFG-----VGSARIVIPGD-YIDFGPISTGSSSCFGGIQSSAGI
2APP-C  SNAGG--YVFDC-S-T-N-LPDFSVSIS-GYTATVPGSL--INYGP-SGD------G-STCLGGIQSNSGI
2APR-C  GD-GT--YTI---SCDTSAFKPLVFSI--------NGASFQVSPDSLVFEEFQ--G-QCIAG----F-GY


4APE-N  VSSSFTEDSTIDGLLGLAFSTLNTVSPTQQKTFFDNAKASLDSPVFTADL---GYHAPGTYNFGFIDTTA
2APP-N  ISAQFQQDTNNDGLLGLAFSSINTVQPQSQTTFFDTVKSSLAQPLFAVAL---KHQQPGVYDFGFIDSSK
2APR-N  EAASFASGPN-DGLLGLGFDTITTVRGVKTPMDNLISQGLISRPIFGVYLGKAKNGGGGEYIFGGYDSTK

4APE-C  GINIFG-----DVALKAAF----VVFNGA------------TTP----TL--------G---FASK--
2APP-C  GFSIFG-----DIFLKSQY----VVFD-S----------G-DGP----QL--------G---FAPQA-
2APR-C  GNWGFAIIG--DTFLKNNY----VVFN-Q----------G-GVP--------------G---EVQIAPVAE
```

FIG. 3. Alignments of the aspartic proteinase domains from (a) multifeature (F-STR) and (b) multiple sequence (SEQ) comparisons; amino- and carboxy-terminal domains are labeled N and C, respectively (see Table I for identification of structures and sequences). The multifeature alignment was derived from the following properties and relationships: physical properties of amino acid residues (0.05), residue main-chain accessibilities (0.20), hydrogen bonding pattern (0.80), residue identities (0.10), Φ (0.05) and Ψ (0.10) dihedral angles, absolute C_α positions in space (0.15), and absolute main-chain direction in space (0.20). Numbers in parentheses represent the relative weights of the corresponding features. Asterisks indicate those positions that are considered topologically equivalent after multiple rigid-body superposition.

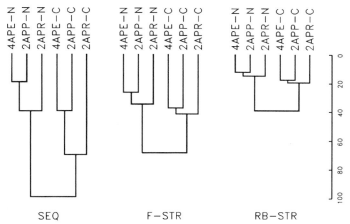

FIG. 4. Dendrograms derived for the domains of the aspartic proteinases from multiple sequence (SEQ), pairwise rigid-body (RB-STR), and multifeature (F-STR) alignments. For the multifeature tree, the same properties and relations were used as for the alignment in Fig. 3. See Table I for the identification of structures and sequences. The amino- and carboxy-terminal domains are labeled N and C, respectively.

Cytochromes c

Six c-type cytochromes that have either eukaryotic (bony fish and plant) or microbial origins are listed in Table I. These proteins cover a wide range of relationships; the percent identity obtained from the sequence alignment ranges between 62% for a comparison of albacore cytochrome c (3CYT) with rice embryo cytochrome c (1CCR), and 15% for a comparison of the rice cytochrome and the cytochrome c_5 from *Azotobacter vine-landii* (1CC5).[6] Trees constructed from pairwise rigid-body fits, the overall tree based on the COMPARER approach, and that from the multiple alignment of the corresponding sequences are in complete agreement with regard to topology (Figs. 2 and 5). In each case, the eukaryotic mitochon-drial cytochromes c segregate from the microbial c_2, c-550, c-551, and c_5 cytochromes. The dendrograms agree with sequence-based trees described by others.[41]

A multiple structure rigid-body alignment identifies only 10 topologi-cally equivalent residues out of the roughly 100 residues in each structure. The fitted structures, translated within the plane of the page, are shown in Fig. 2. It is clear from the view shown in Fig. 2 that there is more in

[41] R. M. Schwartz and M. O. Dayhoff, *in* "Atlas of Protein Sequence and Structure" (M. O. Dayhoff, ed.), Vol. 5, Suppl. 3, p. 29. National Biomedical Research Foundation, Washing-ton, D.C., 1978.

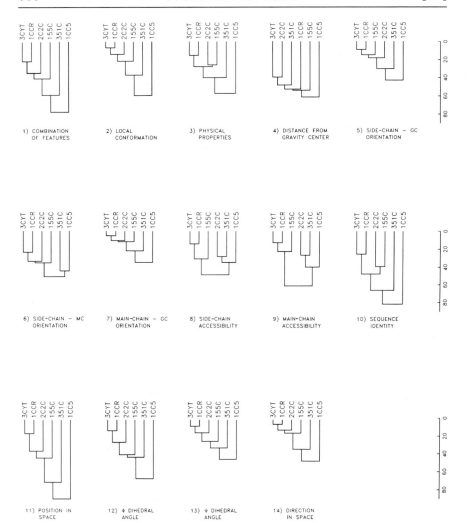

FIG. 5. Cladograms constructed from a multifeature alignment of the c-type cytochromes. Trees 2–14 were derived from the pairwise distances E [Eq. (10)] that were calculated using the individual features only. On the other hand, tree 1 was constructed from the same weight matrix elements W_{ij} that were used to derive the multiple alignment (data not shown): features included were physical properties of amino acid residues (0.20), distance of the C_α from the molecular gravity center (0.20), residue identities (0.10), absolute distance in space (0.30), and absolute main-chain directions in space (0.20). GC, Molecular gravity center; MC, main chain.

common among the structures than the small number of equivalences would suggest. The multifeature comparison method provides a technique for establishing a complete alignment of the structures (results not shown).

Trees reflecting various aspects of c-type cytochromes can be derived from their multiple alignment (Fig. 5). Trees 2 through 14 were obtained from the pairwise distances E that were calculated by considering only one feature in the derivation of weight matrix elements W_{ij} [Eq. (9)]. Tree 1 was constructed from the distances E obtained from the combination of features used to align the structures (Fig. 5). Of 14 trees, 7 have the same topology as the trees based on the multiple sequence and pairwise rigid-body comparison (Fig. 2); these include the trees derived from the local conformation of the main chain, orientation of the main chain relative to the molecular gravity center, absolute position of C_α atoms, main-chain direction in approximately superposed structures, Φ and Ψ dihedral angles, and the tree constructed from a combination of distances calculated from the features used to obtain the multiple alignment.

Trees that have a topology different from the most frequent one include the two trees derived from sequence information: the first of these two trees is based on the five physical characteristics of amino acid residues (such as hydrophobicity) that were found useful in construction of sequence alignments by Argos,[16] and the second tree is derived from a consideration of residue identities only. It may be noted that the clustering in these two trees is the same and corresponds to the subjective impression obtained from a consideration of the shape of the cytochrome structure in Fig. 2. The two trees reflecting similarities in the main-chain and side-chain accessibilities are also congruent with each other, but they are different from the two topologies mentioned above. In contrast, the unique and self-inconsistent topologies of the two trees that involve the orientation of side chains relative to the main chain and relative to the molecular gravity center imply that the orientation of the side chains is not a useful indicator for establishing relationships between divergent protein structures.

The three tree topologies (Fig. 5) based on the combination of features, sequence criteria, and solvent accessibility, demonstrate that evolutionary pressure does not act on all aspects of protein structure in the same way; thus, different criteria may be better for different purposes. For example, trees constructed from rigid-body superpositions are suitable in the selection of structures for determination of a framework[6] in homology-based protein modeling,[21] sequence-based trees are convenient for the description of evolutionary relationships among relatively similar proteins, while trees based both on pairwise rigid-body and multifeature comparison may be better for the analysis of more divergent structures.

Globins

The 19 globin structures listed in Table I were compared by the two structural procedures. Included within these comparisons are a number of globin structures that are identical in sequence (2HHB, 2HCO, and 1HHO; 2MBN and 3MBN) but differ in ligand binding [deoxy-, carbon-monoxy-, and oxyhemoglobins (human); deoxy- and metmyoglobins (sperm whale) as well as globins that are very distantly related to the vertebrate globins [the erythrocruorin from the larva of *Chironomus thummi thummi* (1ECD) and the leghemoglobin from the leguminous plant *Lupinus luteus*]. From the sequence alignments,[6] the vertebrate sequences are between 15 and 24% identical to the insect erythrocruorin and plant leghemoglobin. In contrast to the alignment of the aspartic proteinase domains (Fig. 3), the alignment of the globins using the multifeature approach is very similar to the multiple alignment constructed from the amino acid sequences (Fig. 6). On the other hand, the multistructure rigid-body comparison leads to only 15 topologically equivalent positions.

The sequence-derived tree (Fig. 7) agrees with trees based on sequence.[10,42,43] Overall, the trees constructed from either set of structural data are also in agreement (Fig. 7). The α chains cluster apart from the β/γ cluster; the sea lamprey hemoglobin clusters at about the same position as the human myoglobin, and each of these structures and sequences clusters apart from the erythrocruorin and leghemoglobins as one would expect on the basis of classic organismal evolution. COMPARER produces a tree that coincides completely with the sequence tree, but differentiation among the structures in cases where the sequences are identical also takes place. This also occurs with the pairwise rigid-body procedure, where all deoxy chains cluster together (2HHB, 1FDH, 1HBS, and 2DHB) and those subunits that have oxygen or carbon monoxide bound also cluster together; this occurs for both α and β chains and presumably reflects the conformational changes that occur on ligand binding. For all three trees, the deer sickle cell subunits are outliers to the α chains and the β/γ chains.

Concluding Remarks

About 30 new protein structures are determined each year and added to the nearly 400 structures that have already been deposited in the Brookhaven Protein Data Bank.[33] While this is far from the number of known

[42] M. Goodman, G. W. Moore, and G. Masuda, *Nature (London)* **253,** 603 (1975).
[43] L. T. Hunt, S. Hurst-Calderone, and M. O. Dayhoff, *in* "Atlas of Protein Sequence and Structure" (M. O. Dayhoff, ed.) Vol. 5, Suppl. 3, p. 229. National Biomedical Research Foundation, Washington, D.C., 1978.

(a) F-STR

```
2HHB (A)    ---------VLSPADKTNVKAAWGKVGAHAGEYGAEALERMFLSFPTTKTYFPHF-DLSH-----GSAQVKGHGKKVADALTNAVA
2DHB (A)    ---------VLSAADKTNVKAAWSKVGGHAGEYGAEALERMFLGFPTTKTYFPHF-DLSH-----GSAQVKAHGKKVGDALTLAVG
1HDS (A)    ---------VLSAANKSNVKAAWGKVGGNAPAYGAQALQRMFLSFPTTKTYFPHF-DLSH-----GSAQQKAHGQKVANALTKAQG
2HHB (B)    --------VHLTPEEKSAVTALWGKV--NVDEVGGEALGRLLVVYPWTQRFFESFGDLSTPDAVMGNPKVKAHGKKVLGAFSDGLA
2DHB (B)    --------VQLSGEEKAAVLALWDKV--NEEEVGGEALGRLLVVYPWTQRFFDSFGDLSNPGAVMGNPKVKAHGKKVLHSFGEGVH
1FDH (B)    --------GHFTEEDKATITSLWGKV--NVEDAGGETLGRLLVVYPWTQRFFDSFGNLSSASAIMGNPKVKAHGKKVLTSLGDAIK
1HDS (B)    --------MLTAEEKAAVTGFWGKV--DVDVVGAQALGRLLVVYPWTQRFFQHFGNLSSAGAVMNNPKVKAHGKRVLDAFTQGLK
2LHB        PIVDTGSVAPLSAAEKTKIRSAWAPVYSTYETSGVDILVKFFTSTPAAQEFFPKFKGLTTADELKKSADVRWHAERIINAVDDAVA
2MBN        ---------VLSEGEWQLVLHVWAKVEADVAGHGQDILIRLFKSHPETLEKFDRFKHLKTEAEMKASEDLKKHGVTVLTALGAIILK
1ECD        ---------LSADQISTVQASFDKVKGD----PVGILYAVFKADPSIMAKFTQFAG-KDLESIKGTAPFETHANRIVGFFSKIIG
1LH1        --------GALTESQAALVKSSWEEFNANIPKHTHRFFILVLEIAPAAKDLFSFLKGTSEVP--QNNPELQAHAGKVFKLVYEAAI
```

```
2HHB (A)    HVDD---M--PNALSALSDLHAHKLRVDPVNFKLLSHCLLVTLAAHLPAEFTPAVHASLDKFLASVSTVLTSKYR--------
2DHB (A)    HLDD---L--PGALSNLSDLHAHKLRVDPVNFKLLSHCLLSTLAVHLPNDFTPAVHASLDKFLSSVSTVLTSKYR--------
1HDS (A)    HLND---L--PGTLSNLSNLHAHKLRVNPVNFKLLSHSLLVTLASHLPTNFTPAVHANLNKFLANDSTVLTSKYR--------
2HHB (B)    HLDN---L--KGTFATLSELHCDKLHVDPENFRLLGNVLVCVLAHHFGKEFTPPVQAAYQKVVAGVANALAHKYH--------
2DHB (B)    HLDN---L--KGTFAALSELHCDKLHVDPENFRLLGNVLVVVLARHFGKDFTPELQASYQKVVAGVANALAHKYH--------
1FDH (B)    HLDD---L--KGTFAQLSELHCDKLHVDPENFKLLGNVLVTVLAIHFGKEFTPEVQASWQKMVTAVASALSSRYH--------
1HDS (B)    HLDD---L--KGAFAQLSGLHCNKLHVNPQNFRLLGNVLALVVARNFGGQFTPNVQALFQKVVAGVANALAHKYH--------
2LHB        SMDDTEKM--SMKLRNLSGKHAKSFQVDPEYFKVLAAVIADTVAAG--------DAGFEKLMSMICILLRSAY--------
2MBN        KKGH---H--EAELKPLAQSHATKHKIPIKYLEFISEAIIHVLHSRHPGDFGADAQGAMNKALELFRKDIAAKYKELGYQG---
1ECD        ELPN---I--EADVNTFVASH-KPRGVTHDQLNNFRAGFVSYMKAHTD--FA-GAEAAWGATLDTFFGMIFSK--------M--
1LH1        QLEVTGVVVTDATLKNLGSVHV-SKGVADAHFPVVKEAILKTIKEVVGAKWSEELNSAWTIAYDELAIVIKKEMDD------AA
```

(b) SEQ

```
2HHB (A)    ---------VLSPADKTNVKAAWGKVGAHAGEYGAEALERMFLSFPTTKTYFPHF-DLSH-----GSAQVKGHGKKVADALTNAVA
2DHB (A)    ---------VLSAADKTNVKAAWSKVGGHAGEYGAEALERMFLGFPTTKTYFPHF-DLSH-----GSAQVKAHGKKVGDALTLAVG
1HDS (A)    ---------VLSAANKSNVKAAWGKVGGNAPAYGAQALQRMFLSFPTTKTYFPHF-DLSH-----GSAQQKAHGQKVANALTKAQG
2HHB (B)    --------VHLTPEEKSAVTALWGKV--NVDEVGGEALGRLLVVYPWTQRFFESFGDLSTPDAVMGNPKVKAHGKKVLGAFSDGLA
2DHB (B)    --------VQLSGEEKAAVLALWDKV--NEEEVGGEALGRLLVVYPWTQRFFDSFGDLSNPGAVMGNPKVKAHGKKVLHSFGEGVH
1FDH (B)    --------GHFTEEDKATITSLWGKV--NVEDAGGETLGRLLVVYPWTQRFFDSFGNLSSASAIMGNPKVKAHGKKVLTSLGDAIK
1HDS (B)    --------MLTAEEKAAVTGFWGKV--DVDVVGAQALGRLLVVYPWTQRFFQHFGNLSSAGAVMNNPKVKAHGKRVLDAFTQGLK
2LHB        PIVDTGSVAPLSAAEKTKIRSAWAPVYSTYETSGVDILVKFFTSTPAAQEFFPKFKGLTTADELKKSADVRWHAERIINAVDDAVA
2MBN        ---------VLSEGEWQLVLHVWAKVEADVAGHGQDILIRLFKSHPETLEKFDRFKHLKTEAEMKASEDLKKHGVTVLTALGAIILK
1ECD        ---------LSADQISTVQASFDKV--KGDPVG---ILYAVFKADPSIMAKFTQFAG-KDLESIKGTAPFETHANRIVGFFSKIIG
1LH1        --------GALTESQAALVKSSWEEFNANIPKHTHRFFILVLEIAPAAKDLFSFLKGTSEVP--QNNPELQAHAGKVFKLVYEAAI
PARA        ---------MLDQQTINIIKATVPVLKEHGVTITTTFYKNLFAKHPEVRPLF----DMGRQE---SLEQPKALAMTVLAA-AQNIE
VITR        ------VNKVFTEEQEALVVKAWAVMKKNSAELGLQFL-KIFEIAPSAKNLFSYLKDSPVP----LEQNPLKPHATTFVMTTESAVQ
```

```
2HHB (A)    HVDD---M--PNALSALSDLHAHKLRVDPVNFKLLSHCLLVTLAAHLPAEFTPAVHASLDKFLASVSTVL----TSKYR------
2DHB (A)    HLDD---L--PGALSNLSDLHAHKLRVDPV.NFKLLSHCLLSTLAVHLPNDFTPAVHASLDKFLSSVSTVL----TSKYR------
1HDS (A)    HLND---L--PGTLSNLSNLHAHKLRVNPVNFKLLSHSLLVTLASHLPTNFTPAVHANLNKFLANDSTVL----TSKYR------
2HHB (B)    HLDN---L--KGTFATLSELHCDKLHVDPENFRLLGNVLVCVLAHHFGKEFTPPVQAAYQKVVAGVANAL----AHKYH------
2DHB (B)    HLDN---L--KGTFAALSELHCDKLHVDPENFRLLGNVLVVVLARHFGKDFTPELQASYQKVVAGVANAL----AHKYH------
1FDH (B)    HLDD---L--KGTFAQLSELHCDKLHVDPENFKLLGNVLVTVLAIHFGKEFTPEVQASWQKMVTAVASAL----SSRYH------
1HDS (B)    HLDD---L--KGAFAQLSGLHCNKLHVNPQNFRLLGNVLALVVARNFGGQFTPNVQALFQKVVAGVANAL----AHKYH------
2LHB        SMDDTEKM--SMKLRNLSGKHAKSFQVDPEYFKVLAAVI---------ADTVAAGDAGFEKLMSMICILL----RSAY-------
2MBN        KKGH---H--EAELKPLAQSHATKHKIPIKYLEFISEAIIHVLHSRHPGDFGADAQGAMNKALELFRKDI----AAKYKELGYQG
1ECD        ELPN---I--EADVNTFVASHKPR-GVTHDQLNNFRAGFVSYMKAH--TDF-AGAEAAWGATLDTFFGMI----FSKM-------
1LH1        QLEVTGVVVTDATLKNLGSVHV-SKGVADAHFPVVKEAILKTIKEVVGAKWSEELNSAWTIAYDELAIVI----KKEMDDAA---
PARA        NLPA---I--LPAVKKIAVKHCQA-GVAAAHYPIVGQELLGAIKEVLGDAATDDILDAWGKAYGVIADVFIQVEADLYAQAVE--
VITR        LRKAGKVTVKESDLKRIGAIH-FKTGVVNEHFEVTRFALLETIKEAVPEMWSPEMKNAWGVAYDQLVAAIKFEMKPSST------
```

FIG. 6. (a) Multifeature alignment (F-STR) and (b) sequence-based alignment (SEQ) for the globins of Table I. The following properties were used in the multifeature procedure: physical properties of amino acid residues (0.05), residue main-chain accessibilities (0.20), residue identities (0.10), and absolute main-chain directions in space (0.20). A, α chains; B, β and γ chains.

primary sequences, the structural data contain considerably more information per protein than do the sequence data. This information may be used to resolve evolutionary relationships among proteins. The following statements summarize the major conclusions of this chapter.

(1) Phylogenetic trees that are generally in agreement with trees derived from amino acid sequences can be constructed from structural information.

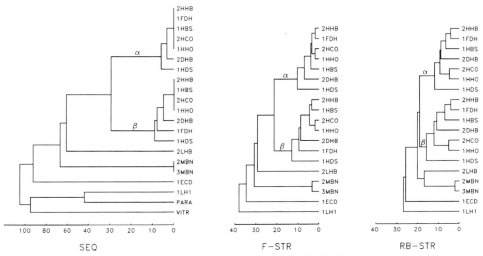

FIG. 7. Trees constructed for 19 globin structures (Table I) from pairwise rigid-body (RB-STR), multifeature (F-STR), and multiple sequence (SEQ) alignments. The tree for the multifeature approach was calculated from the same combination of properties that was used to derive the alignment in Fig. 6. α, α chains; β, β and γ chains.

(2) Trees that could not be obtained from the comparison of sequences alone may be established from structural comparisons: Evolutionary trees for divergent proteins, where the sequence relationships are not statistically significant, can be derived from structures because tertiary structure is more conserved in evolution than sequence. For example, even though the comparison of the amino- and carboxy-terminal domains of the aspartic proteinases reveals many common structural features, the sequence similarity is not statistically significant. Structural procedures can distinguish between different crystal structures for the same sequence. For example, perturbations of hemoglobin structures induced by ligand binding are seen in the tree constructed from rigid-body superpositions, where all deoxy chains cluster apart from those with bound ligands.

Acknowledgments

We thank our colleague John P. Overington for valuable discussions. We also thank Michael J. Sutcliffe. We thank the American Cancer Society (M.S.J.) for a fellowship and the ORS Awards Scheme and the Research Council of Slovenia for support (A.Š.).

Author Index

Numbers in parentheses are footnote reference numbers and indicate that an author's work is referred to although the name is not cited in the text.

Subject Index

A